A SOCIAL
AND ECONOMIC
ATLAS
OF
INDIA

DELHI
OXFORD UNIVERSITY PRESS
OXFORD NEW YORK
1987

Oxford University Press, Walton Street, Oxford OX2 6DP

New York Toronto
Delhi Bombay Calcutta Madras Karachi
Petaling Jaya Singapore Hong Kong Tokyo
Nairobi Dar es Salaam
Melbourne Auckland

and associates in
Beirut Berlin Ibadan Nicosia

SBN 19 562041 0

ACKNOWLEDGEMENTS

The following should be noted for all maps in this Atlas:

© *Government of India Copyright 1986*

*Based upon Survey of India map with the permission of the Surveyor General of India.
The responsibility for the correctness of internal details rests with the publisher.*

*The territorial waters of India extend into the sea to a distance of
twelve nautical miles measured from the appropriate base line.*

The administrative headquarters of Chandigarh, Haryana and Punjab are at Chandigarh.

*The boundary of Meghalaya shown on this map is as interpreted from the North-Eastern Areas
(Reorganization) Act, 1971, but has yet to be verified.*

Ref
G
2281
.G1
T8
1987

Designed, compiled, cartographed and printed by
TT. Maps & Publications Private Ltd.
328, G.S.T. Road, Chromepet, Madras 600044
and published by S.K. Mookerjee, Oxford University Press
YMCA Library Building, Jai Singh Road
New Delhi 110001, India.

PREFACE

This *Atlas* is dedicated to the man who inspired it — T.T. Krishnamachari. Between 1947 and 1965 he was a member of the committee that drafted the Indian Constitution, Minister for Industries, Commerce Minister, and twice India's Finance Minister.

It was in the late sixties, several years after he had retired to his house in suburban Madras, that I met TTK. During his visits to TT. MAPS he would often remark that India lacked a proper data base. 'Data in India are too scattered,' he would say. It is often a struggle to retrieve them, they are seldom presented within a comprehensible format, and never in a visually attractive manner that enables one to take in the whole picture at a glance. What is needed, he felt, is to present planners and administrators, industrialists and marketing men, students and scholars, engineers and scientists, social workers, advertising executives and politicians with a more complete picture of India.

TT. MAPS, founded in 1965, was then in its infancy. To oblige TTK would have been a Herculean task at that time. Twenty years later it has still been a major effort, extending over two years, but I am happy that we have at last been able to fulfil TTK's wish. I hope all those he listed as in need of such a data base will find in this *Atlas* many of the answers they seek.

S. MUTHIAH

CONTENTS

INTRODUCTION

The Atlas

This *Atlas* provides an up-to-date picture of India's habitat, society and economy through maps, charts and tables supplemented by a brief but comprehensive text. It also looks at India's development potential in the light of the Seventh Five Year Plan and examines what the economic and social scene is likely to be in India in the nineties.

The *Atlas* shows in considerable detail the land and the people of India, the available resources, the infrastructure developed to utilize these resources, and the two equations that inevitably follow — the impact of the development of resources on the national economy and on India's economic status in the world. In analysing this impact, the *Atlas* also provides information on threats to development by over-use and over-exploitation of the country's resources.

All this information is in nine sections, each with a specific objective.

THE LAND (6 maps and 9 charts) shows the physical features and the administrative framework for the assessment of regional development and differences.

THE PEOPLE (41 maps and 54 charts) offers a broad but detailed picture of Indian society and relates development with the population.

THE CLIMATE (10 maps and 5 charts) provides information on the factors generally beyond human control which must be considered when selecting developmental projects for implementation.

THE NATURAL RESOURCES (23 maps and 26 charts) assesses the potential of land, water, forest and mineral resources and offers information on the extent to which exploitation is possible.

THE INFRASTRUCTURE (40 maps and 55 charts) examines the facilities developed and being developed for optimum utilization of human and natural resources.

THE PRODUCE (78 maps and 132 charts) examines what man and nature, resources and infrastructure have been able to produce for India's people and assesses how far production meets the country's requirements.

THE TOURIST VISTA (4 maps and 3 charts) identifies the potential of yet another aspect of development, an area now considered the fastest-growing industry in the world.

THE NATIONAL ECONOMY (23 maps and 25 charts) evaluates the present level of development and points to the potential for the future.

THE INTERNATIONAL EQUATION (21 maps and 61 charts) demonstrates the country's continued dependence on the world outside but indicates areas where this dependence is likely to be reduced.

The nine sections are presented with the help of authentic, qualitative and quantitative information from a variety of sources and interpreted by skilled cartography.

The maps and charts in the *Atlas* examine different degrees of development at the national, state and district levels. All the maps and charts have a basic simplicity of design but they have been made as informative as possible. Page x contains explanatory notes on how to read some of the more complex data presented in the maps. The following points should also be noted when studying the information provided in this *Atlas*.

All the maps of India in the *Atlas* are based on a common projection and have a similar layout and cartographic presentation. However, each map is designed in a manner which permits the most effective presentation of the data. The maps used are on the Lambert azimuthal equal area projection and are in four scales: the full page map, 1:15,000,000 (1 cm = 150 kms); the half page map, 1:21,000,000 (1 cm = 210 kms); the quarter page map, 1:30,000,000 (1 cm = 300 kms) and the inset map, 1:35,000,000 (1 cm = 350 kms). The world maps are on Mercator projection and in two scales: the full-page map is 1:120,000,000 (1 cm = 1200 kms) and the one-third page map is 1:205,000,000 (1 cm = 2050 kms). All the maps have been approved by the Survey of India and the SOI-approved latitudes and longitudes appear only where they are necessary — on the physical, political, administration and parliamentary constituencies maps on pages 3, 5, 7 and 13. All the maps are accompanied by legends and notes.

Dates such as 1984-85 refer to the Indian financial year — the 12-month period from 1 April to 31 March. Dates such as 1984 and 1985 refer to a calendar year — the twelve months from 1 January to 31 December.

Decimals 0.5 and above are rounded off upwards, decimals below 0.5 are rounded off downwards. It should be noted that figures are in most cases approximate measures. The degree of accuracy thus varies according to the state and the item being measured. For example, even though Census data can be regarded as reasonably accurate, considerable variations are known to exist in the data. Moreover, most figures for the later years are provisional and subject to revision even when provided officially by the states/departments concerned. Similarly, rates of growth are affected by the choice of starting and finishing points. They are higher when starting from a relatively low point. The natural course of events is for a percentage growth rate to be high in the early stages of a new development, but to slow down as the total size increases.

States have been classified as major and minor throughout the *Atlas* on the basis of population. The major states (15) are: Andhra Pradesh, Assam, Bihar, Gujarat, Haryana, Karnataka, Kerala, Madhya Pradesh, Maharashtra, Orissa, Punjab, Rajasthan, Tamil Nadu, Uttar Pradesh and West Bengal. The minor states (7) are: Himachal Pradesh, Jammu & Kashmir, Manipur, Meghalaya, Nagaland, Sikkim and Tripura.

In 1986, the Indian Parliament passed Acts making the Union Territories of Arunachal Pradesh and Mizoram into states. These came into effect in February 1987. Goa became a state in May 1987. But since the data for this *Atlas* were compiled prior to these announcements, these three states have been treated as Union Territories in the *Atlas*.

The district boundaries on the maps are according to information available in 1985. Information superimposed on each district is based on the 1981 configuration of the districts. Several 1985 districts were, in 1981, parts of bigger districts. At the time of the Census in 1981 there were 412 districts. In 1985 there were 439 districts, of which 437 are marked on the map. (District boundaries for two districts — one in Assam and one in Arunachal Pradesh — were not available at the time of going to press.) The transparent overlay provided with the *Atlas* shows district boundaries and makes easier identification of the developmental process wherever district information is given.

All population figures are based on the last official Census (1981), and extrapolations are based on trends. Most of the data on which the maps are based pertain to the 1980s, but in a few cases earlier data have been used as more recent information is not available. The data used for compiling the maps were the latest available at the time the maps were drawn. However, more recent statistical information has been included in the text and the tables. The differences which result, between text and maps, are few and of minor significance.

Population figures used in the *Atlas* for post-1981 periods are estimates based on average growth, and the per capita figures for such periods are based upon estimated population. All estimated and projected data are accompanied by supportive notes. Most of the 1989-90 projections are based on official estimates for the Seventh Plan period. Money value figures for 1989-90 in these projections are at 1984-85 prices.

As the 1981 Census could not be held in Assam because of disturbed conditions in the state, the compilers have estimated the 1981 population figures for Assam on the basis of average growth. The 1981 Census also did not provide figures for certain areas of Jammu and Kashmir and hence these have not been included.

The explanatory text facing each map describes the map, supplements the information on it and brings some of the data up to date. These notes are further supported by detailed statistical tables with the latest information available at the time of going to press. Abbreviations used in this *Atlas* are listed below:

STATES			UNION TERRITORIES			m. t.	=	million tonnes
A.P.	=	Andhra Pradesh	A. & N.	=	Andaman & Nicobar	mm	=	millimetre
Ass.	=	Assam	Ar. P.	=	Arunachal Pradesh	cm	=	centimetre
Bih.	=	Bihar	Cha.	=	Chandigarh	cms	=	centimetres
Guj.	=	Gujarat	D. & N.H.	=	Dadra & Nagar Haveli	°C	=	°Celsius
Har.	=	Haryana	Del.	=	Delhi	m.cu.m.	=	million cubic metres
H.P.	=	Himachal Pradesh	G.D. & D.	=	Goa, Daman & Diu	m.ha.m.	=	million hectare metre
J. & K.	=	Jammu & Kashmir	Lak.	=	Lakshadweep	MW	=	Megawatt
Kar.	=	Karnataka	Miz.	=	Mizoram	KW	=	Kilowatt
Ker.	=	Kerala	Pon.	=	Pondicherry	kwh	=	kilowatt hour
M.P.	=	Madhya Pradesh	**UNITS**			**OTHERS**		
Mah.	=	Maharashtra	m.	=	million	GNP	=	Gross National Product
Man.	=	Manipur	th.	=	thousand	GDP	=	Gross Domestic Product
Meg.	=	Meghalaya	km	=	kilometre	SDP	=	State Domestic Product
Nag.	=	Nagaland	kms	=	kilometres	N.A.	=	Not Available
Ori.	=	Orissa	sq km	=	square kilometre	Neg.	=	Negligible
Pun.	=	Punjab	sq kms	=	square kilometres	Incl.	=	Including
Raj.	=	Rajasthan	kg	=	kilogram	avg.	=	average
Sik.	=	Sikkim	kgs	=	kilograms	Max.	=	Maximum
T.N.	=	Tamil Nadu	g	=	gram	Min.	=	Minimum
Tri.	=	Tripura	ha	=	hectare	App.	=	Approximate
U.P.	=	Uttar Pradesh	Rs	=	Rupees	No.	=	Number
W.B.	=	West Bengal	t.	=	tonne/tonnes	Regd.	=	Registered
UTs.	=	Union Territories	th. t.	=	thousand tonnes	Pop.	=	Population

The Basis

The composite picture of India that this *Atlas* reveals is of a developing country that is complex, dynamic, yet sluggish in parts. The basis for this picture is the maps on the facing page, which show the levels of economic development throughout India. These maps are a synthesis of all the information found in this *Atlas*; conversely, the information provided in these maps forms the basis for the analyses in the pages that follow.

The main map on the opposite page indicates the varying degrees of development at the district level. Certain districts have index values of more than 250 and are quite strong on both rural and urban base indicators. All urban districts demonstrate very high levels of development. Comparable levels of economic development occur in Faridkot, Ludhiana and Patiala (Punjab), Surendranagar (Gujarat) and Kodagu (Karnataka). The next level of economic development is found in Jalandhar and Bhatinda (Punjab), Sirsa (Haryana), Vadodara (Gujarat), Pune (Maharashtra), Chikmagalur and Bangalore (Karnataka), Coimbatore (Tamil Nadu) and Ernakulam (Kerala). These districts are noted for agricultural, industrial and service sector development. Districts which are predominantly agricultural (deltaic as well as interior districts) also show greater levels of development.

An average of this districtwise information, taken with a view to getting an idea of the state-level economic development, indicates a range from 1075 for the union territory of Delhi down to 10 for the state of Manipur. Eight states are higher than the all-India average of 100, while thirteen states and five union territories are lower. The better-developed parts of the country are Goa, Gujarat, Haryana, Karnataka, Kerala, Maharashtra, Punjab, Tamil Nadu and West Bengal. The union territories of Chandigarh and Pondicherry also fall into this category. Union territories and states which are below average and underdeveloped are Andhra Pradesh, the Andaman & Nicobar Islands, Arunachal Pradesh, Assam, Bihar, Dadra & Nagar Haveli, Himachal Pradesh, Lakshadweep, Madhya Pradesh, Manipur, Meghalaya, Mizoram, Nagaland, Orissa, Rajasthan, Sikkim, Tripura and Uttar Pradesh.

The greater part of the interior peninsula, Rajasthan, Jammu & Kashmir, the plains districts of Madhya Pradesh, Uttar Pradesh, Bihar and West Bengal and most of the north-eastern hill districts have levels of economic development below the all-India average of 100. The districts of the Gangetic plain and the north-eastern hills have indices of 76-100 while Ladakh is entirely underdeveloped. The districts that are at the bottom of the scale are essentially those that are drought-prone.

The pattern is clear: the overpopulated states in the Indo-Gangetic Plain and the tribal-dominated areas, such as the north-eastern states and Orissa, are, generally speaking, the less developed states. The states with lesser population problems and fewer tribal areas as well as the urban union territories are better developed.

It should be noted that the maps on the facing page are based on aggregate indices of economic development arrived at by using district data on a variety of aspects of development. The indices are a very rough proxy of key indicators for the GNP of districts.

Acknowledgements

OFFICIAL SOURCES

Census of India
Department of Agriculture
  & Cooperation
Department of Economic Analysis,
  Reserve Bank of India
Department of Family Welfare
Department of Statistics (Statistical
  Abstracts), Ministry of Planning
Department of Tourism
Meteorological Department
Ministry of Fisheries
Ministry of Information and
  Broadcasting
Ministry of Industry and Company
  Affairs

National Atlas and Thematic Mapping
  Organisation
National Remote Sensing Agency
Planning Commission
Survey of India
World Bank

NON-OFFICIAL SOURCES

Association of Indian Engineering
  Industry
Centre for Monitoring Indian Economy
Department of Economics and Statistics,
  Tata Services Limited
Economic Times
Fertiliser Association of India
Times of India Directory

Information has occasionally been taken from other sources as well. In such cases, the sources are acknowledged in the respective texts.

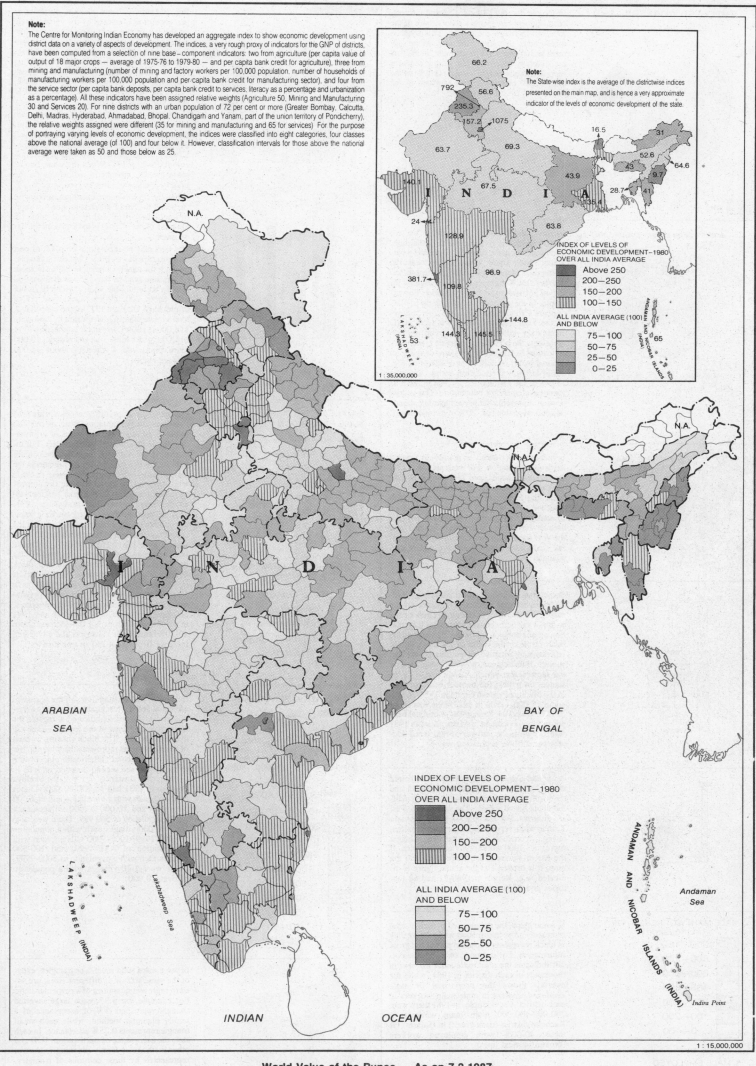

Note:
The Centre for Monitoring Indian Economy has developed an aggregate index to show economic development using district data on a variety of aspects of development. The indices, a very rough proxy of indicators for the GNP of districts, have been computed from a selection of nine base – component indicators: two from agriculture (per capita value of output of 18 major crops — average of 1975-76 to 1979-80 — and per capita bank credit for agriculture), three from mining and manufacturing (number of mining and factory workers per 100,000 population, number of households of manufacturing workers per 100,000 population and per capita bank credit for manufacturing sector), and four from the service sector (per capita bank deposits, per capita bank credit to services, literacy as a percentage and urbanization as a percentage). All these indicators have been assigned relative weights (Agriculture 50, Mining and Manufacturing 30 and Services 20). For nine districts with an urban population of 72 per cent or more (Greater Bombay, Calcutta, Delhi, Madras, Hyderabad, Ahmadabad, Bhopal, Chandigarh and Yanam, part of the union territory of Pondicherry), the relative weights assigned were different (35 for mining and manufacturing and 65 for services). For the purpose of portraying varying levels of economic development, the indices were classified into eight categories, four classes above the national average (of 100) and four below it. However, classification intervals for those above the national average were taken as 50 and those below as 25.

Note:
The State-wise index is the average of the districtwise indices presented on the main map, and is hence a very approximate indicator of the levels of economic development of the state.

INDEX OF LEVELS OF
ECONOMIC DEVELOPMENT–1980
OVER ALL INDIA AVERAGE

- Above 250
- 200—250
- 150—200
- 100—150

ALL INDIA AVERAGE (100)
AND BELOW

- 75—100
- 50—75
- 25—50
- 0—25

1 : 35,000,000

ARABIAN SEA

BAY OF BENGAL

INDEX OF LEVELS OF
ECONOMIC DEVELOPMENT—1980
OVER ALL INDIA AVERAGE

- Above 250
- 200—250
- 150—200
- 100—150

ALL INDIA AVERAGE (100)
AND BELOW

- 75—100
- 50—75
- 25—50
- 0—25

ANDAMAN Sea

INDIAN OCEAN

1 : 15,000,000

World Value of the Rupee — As on 7.2.1987

Country	Currency	Unit	Appx. value of the Rupee	Country	Currency	Unit	Appx. value of the Rupee	Country	Currency	Unit	Appx. value of the Rupee
Australia	Dollar	1	8.79	Italy	Lira	100	0.93	Singapore	Dollar	1	5.99
Bangladesh	Taka	100	43.30	Japan	Yen	100	8.58	Sri Lanka	Rupee	1	0.45
Brazil[1]	Cruzado	1	0.88	Kenya	Shilling	1	0.80	Sweden	Kroner	1	2.01
Canada	Dollar	1	9.89	Kuwait	Dinar	1	44.62	Switzerland	Franc	1	8.50
Egypt	Pound	1	9.52	Malaysia	Dollar	1	5.21	U.K.	Pound	1	19.96
France	Franc	1	2.15	The Netherlands	Guilder	1	6.35	U.S.A.	Dollar	1	13.19
Hong Kong	Dollar	1	1.69	Nigeria[1]	Naira	1	12.34	U.S.S.R.	Rouble	1	9.45
Iran	Rial	100	17.16	Pakistan	Rupee	1	0.75	W. Germany	Deutschmark	1	7.16

Note: [1] March 1986.

EXPLANATORY NOTES

EMPLOYED WORKERS—1981
MALE

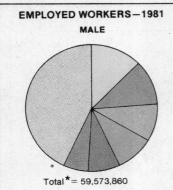

Total* = 59,573,860

▨ Maharashtra 12.9%

In the pie diagrams (the segmented circles), each segment represents a state's share of the total (given below the circle). This share is measured by the ratio of the *angle* of the segment to 360 degrees. For example, in the first pie diagram on p. 55, the yellow segment represents Maharashtra's share (12.9%) of the 59,573,860 employed male workers in the country in 1981.

ADULT & ELEMENTARY EDUCATION

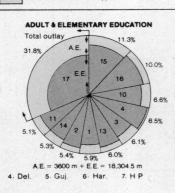

A.E. = 3600 m + E.E. = 18,304.5 m
4. Del. 5. Guj. 6. Har. 7. H P

In some pie diagrams, information on two related subjects is given on the same diagram. Each segment in the diagram represents a particular state's share. This share is measured by the ratio of the *angle* of the segment to 360 degrees. Each segment is further divided into two portions measured by *radius*. For example, in the second pie diagram on p. 101, segment 15 represents UP's share (percentage) of the *total* national outlay on adult AND elementary education. The green portion represents the percentage spent on elementary education in UP out of its share of the total national outlay on both types of education (*i.e.* out of UP's total education outlay, which is 11.3% of the national Rs. 21,904 m, 80% is spent on elementary education). The yellow portion represents the percentage UP spends on adult education (*i.e.* 20% of its total share).

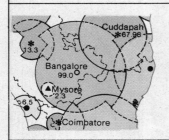

Sometimes an overlay can provide several items of information. For example, on p.125, the coloured shapes give information (by colour) about population covered by radio stations; the furthest distance covered in each direction — E, W, N, S — in km (the distance of each cardinal point from the centre); the area covered in sq km (figure indicated); and the year the station was commissioned (symbol used for the central point, *e.g.* circle, star, triangle, etc.).

→ 1982-83
→ 1975-76
25,000
50,000 telephones

6
Doctors
12 Beds

Sometimes additional information is provided by overlaying circles or squares on the maps. The diameter of the outer circle or the width of the outer square represents information on one subject, while the diameter of the inner circle or the width of the inner square represents information on a different but allied subject. Both data are, however, measured on the same scale, which is indicated. For instance, on p.121, the number of telephones in the Madhya Pradesh Circle in 1975-76 was about 25,000, while in 1982-83 it was about 50,000. On p.103, the number of hospital beds per million population in Maharashtra in 1981 was 12, while the number of registered doctors per million population was 6.

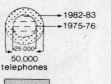

2228
68

A bar and a square may be joined. The height of the bar provides data on one subject and the width of the square provides data on a different but linked subject. The scales, however, are different. For example, on p.179, a square 1 cm in width represents 20 factories manufacturing fish products, while a bar 1 cm in height represents 100 workers in these factories. If the bar or square is too large to fit into the map, it is broken and the actual figure is indicated, *e.g.* Kerala in 1980-81 had 68 factories and 2228 workers.

MALE MAIN WORKERS—1981
(in thousand)

▨ 1000—1600
▨ 800—1000
▨ 600—800
▨ 400—600
▨ 200—400
▨ 100—200
▨ 0—100

In some maps, the colour used in each district represents gross information while the overlay in black represents a specific part of the gross information. For instance, on p. 55, the colour indicates the *total male main workers* in thousands in each district in 1981, while the overlay shows the *proportion of male employed workers* in male main workers in each district. For example, in 1981 there were 200,000-400,000 male main workers in Kachchh (dark green) in Gujarat. Of these, 40-60% were employed workers (broken vertical lines).

% MALE EMPLOYED WORKERS* IN TOTAL MALE WORKERS — 1981

▨ 80—100
▨ 60—80
▨ 40—60
▨ 30—40
▨ 20—30
▨ 10—20
▨ Below 10

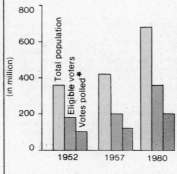

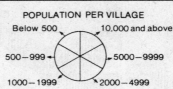

Bar diagrams indicate absolute numbers or percentages as measured by the vertical scale on the left or the right. The horizontal line at the base of the bars indicates the distribution, *e.g.* by year, or state, or product, etc. When there are two or more bars indicating two or more subjects, each subject is given in a different colour and the subject is marked on the bar diagram itself or indicated in the accompanying legend.

In graphs where information relates to two subjects, and where the scales are different, the scale on the left refers to one graph and the scale on the right refers to the other graph. Usually the graphs are depicted in different colours, *e.g.* p.191. When there are two or more graphs, alphabetic indications on the graphs or colour coding is used, *e.g.* p.217.

When the figures exceed the highest point on the scale, the bar or graph is broken just short of the highest point and the actual value (in the units of the Y axis) is indicated on the bar or graph.

The bars may be separate as on p.13, or part of a broader bar as on p.191. For example, on p.13, the total population in 1980 was about 675 million (yellow bar), of which about 360 million were eligible voters (green bar), but only about 200 million votes were polled (violet bar). On page 191, on the other hand, total vehicle sales in Maharashtra in 1984 were 158,000 (yellow), of which 2-wheelers were 115,000 (green left), 3-wheelers about 10,000 (dark pink) and 4-wheelers about 34,000 (orange right).

% MALE LITERATES IN MALE POPULATION —1981

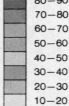

▨ 80—90
▨ 70—80
▨ 60—70
▨ 50—60
▨ 40—50
▨ 30—40
▨ 20—30
▨ 10—20

INCREASING OR DECREASING MALE LITERACY (1971 — 1981)

▦ 3 Step increase
▦ 2 Step increase
▦ 1 Step increase
  No change
▦ 1 Step decrease

In some of the maps, different colours are used to indicate district status and these are overlaid in black. In this way, two types of information are given about the same district. For instance, in the first map on p. 49, the colours indicate the percentage of male literates in the male population in a particular district in 1981, while the overlay shows whether the number of male literates in that district has increased, decreased or stayed the same since 1971. This increase or decrease is measured in steps, the number of steps being related to what the colour scheme would have been in 1971 and what it is in 1981. These steps — the move from one colour to the next or beyond — may be measured using the key for the black overlay. For example, on p. 49, 13 districts in Maharashtra (coloured dark pink) have 50-60% male literates in their male population. There was, according to the overlays, no change in status between 1971 and 1981 in six of these districts, a 1 step increase (from 40-50%) in six other districts and a two step increase (from 30-40%) in one district.

POPULATION PER VILLAGE

Below 500   10,000 and above
500 — 999   5000 — 9999
1000 — 1999   2000 — 4999

NUMBER OF VILLAGES

▦ 40,000—60,000
▦ 20,000—40,000
▦ 10,000—20,000
▦ 5000—10,000
▦ 1000—5000
▦ 500—1000
▦ 100—500
▦ 50—100
▦ 10—50
▦ 1—10

In some of the pie diagrams, all the segments are equal and each segment represents a particular classification (indicated alongside the respective segments in the legend — see top diagram on left). The black shading in each of these segments represents the total number of units of a second classification (according to the key — see second diagram on left) in a state. For instance, on p. 19, Madhya Pradesh in 1981 had 40,000-60,000 villages (overlay) each with a population less than 500 (segment) and 10,000-20,000 villages each with a population of 500-999. There were also 5000-10,000 villages each with a population of 1000-1999, 1000-5000 villages each with a population of 2000-4999, and 100-500 villages each with a population of 5000-9999. There were 1-10 villages with a population of over 10,000.

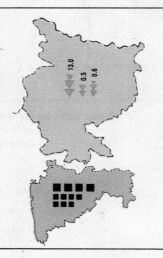

In some maps solid squares or triangles, either of the same size or of different sizes, are used to represent quantities of a particular item. For example, on p.37, each large inverted triangle represents 5% of acceptance of a family planning method, while each small triangle represents 0.25% acceptance. In each state the acceptance of sterilization, IUD and other methods of family planning is shown, represented by three columns of triangles. Thus, in Bihar, 13% of family planning acceptors were sterilized, 0.5% used the IUD, and 0.6% took other methods.

In the first map on p.161, each large square represents 1 million tonnes and each small square represents 100,000 tonnes of jowar produced. Thus, the production of jowar in Maharashtra in 1982-83 was 4 million plus 700,000 (4.7 million) tonnes.

The Land

This section comprises 6 maps and 9 charts. It presents a picture of how India is administered and provides a framework for assessing regional development and disparities.

Note:

The maps of India in this atlas are on the Lambert azimuthal equal area projection and are in four scales: the full page map, 1:15,000,000 (1 cm=150 kms); the half page map, 1:21,000,000 (1 cm=210 kms); the quarter page map, 1:30,000,000 (1 cm=300 kms) and the inset map, 1:35,000,000 (1 cm=350 kms). All maps have been approved by the Survey of India and the SOI approved latitudes and longitudes appear only in this section.

PHYSICAL

India lies to the north of the equator, between 8° 4′ and 37° 6′ North and 68° 7′ and 97° 25′ East. It is bounded in the south by the Indian Ocean, in the west by the Arabian Sea, in the east by the Bay of Bengal, in the north-east, north and a part of the north-west by the Himalayan ranges, and in the rest of the north-west by the Great Indian Desert.

There are three broad tectonic divisions in India but the forms of the landscape are infinite. The Great Himalaya and the associated young fold-mountains; the Indo-Gangetic Plain (broadly comprising the states of Rajasthan, Uttar Pradesh, West Bengal and parts of Punjab, Haryana, Madhya Pradesh and Bihar); the ancient plateau of Peninsular India including the eastern and western coastal plains (all the states south of 23° N latitude)—these are the most striking of India's physical divisions.

The geologically recent Himalaya display a great variety of relief. The Indo-Gangetic Plain forms a long stretch of monotonous aggradational surface, the result of continuous deposits from the hills and mountains on both sides of the plain. The Peninsula displays an open, more ancient topography, though it has its own more recent landforms created during the great Tertiary orogeny.

Physiographically, India can be divided into seven regions: the Northern Mountains, constituting the Himalayan ranges; the great Indo-Gangetic Plain; the Central Highlands (including the Vindhya and Aravalli ranges); the Peninsular Plateau in the states of Maharashtra, Andhra Pradesh, Karnataka and Tamil Nadu; the East Coast; the West Coast; and the Islands.

The Himalaya, stretching over 22 degrees of longitude and extending 2400 kms between the Indus and Brahmaputra rivers, are conventionally divided into five parallel belts: the Shiwalik range or the Lesser Himalaya at 1800-3000 m; the zone of spurs with a deeply dissected planation surface at 4570 m; the Great Himalaya with many peaks over 6000 m; the Outer Himalaya at 3660-4300 m; and the edge of the Tibetan Plateau reaching up to 5790 m. The Himalaya ('the abode of snow') has some of the highest peaks in the world: Everest in Nepal (8848 m), Mount Godwin-Austen or K² (8611 m), Kanchenjunga in Nepal (8598 m), Makalu in Nepal (8481 m), Dhaulagiri in Nepal (8172 m) and Nanga Parbat (8126 m).

The Great Himalaya form a vast arc, which curves down southwards; deep gorges and channels are formed in the range by the headwaters of the Ganga and other snow-fed rivers. Throughout the Himalaya, the southern slopes are steeper than the northern ones.

While Nanga Parbat lies at its western extremity, at the eastern end of the Great Himalaya is the peak of Namcha Barroa. Yet further east, the range swings to a north-south axis under the influence of the Shillong Plateau and the old Yunnan block. The Assam-Burma ranges seem to correspond in a general way with the north-eastern wedge of the Peninsular block and the resistance of the Yunnan block. This general alignment is attributable to the continental drift involving two land masses, Gondwana and Laurasia. The Himalaya, a young fold mountain range, were formed by the collision of the two land masses and the alignment is the result of local drift-related forces.

The Indo-Gangetic Plain, a highly productive agricultural area, is densely populated. The Plain is a great crescent of alluvium that stretches from the delta of the Indus to that of the Ganga. The sediments of the Plain are some 4500 m deep, deposited in a great rift that has sunk under the weight of alluvium, but the filling is of very unequal depth. The Plain is remarkably homogeneous topographically, the only relief being the floodplain bluffs and other related features of river erosion. The only topographical changes on this vast aggradational surface are those associated with the numerous changes in the course of the rivers. The Plain, at a height of 150-300 m, comprises the western half of the upper Gangetic Plain (Uttar Pradesh), Punjab and Haryana, the Thar Desert (Rajasthan) and the upland river plains of the Peninsula.

Peninsular India's northern boundary is an irregular line running from Kachchh, along the western flank of the Aravalli range to near Delhi, and from there roughly parallel to the Yamuna and the Ganga as far as the Hazaribag Plateau and the Ganga Delta. The Gangetic alluvium penetrates south of this line as well.

The average elevation of the plateau that dominates most of Peninsular India ranges between 300 and 1800 m. It is possible to distinguish between elevation — based subdivisions of 300-600 m, 600-900 m, and 900-1800 m. A low elevation plateau (300-600 m) occupies the greater part of the Deccan, Central India and the Chota Nagpur Plateau, and extends into Tamil Nadu, Karnataka and Andhra Pradesh. A plateau zone between 600 and 900 m (the Nilgiris) lies in the southern part of the Deccan, where the Eastern Ghats approach and meet the Western Ghats. There are hill ranges in the Nilgiris going up to 1800 m with a peak of 2637 m at Doda Betta, near Udagamandalam. The residual blocks of the Aravalli and the Satpura also belong to this subdivision. A zone ranging in height from 900 to 1800 m comprises hills and plateaux and is to be found in the Sahyadri range (near Ajanta) and in Orissa. A similar elevation is found in the north-eastern hills.

That the Ghats owe their origin to the subsidence of a land mass to the west appears to be supported by the absence of a simple eastward tilting of the whole block. The wide and almost senile valleys of the east-flowing rivers are, on the whole, graded almost to their heads, and form a striking contrast with the youthful valleys of the west-flowing streams. The latter fall 600 m or more in 80 kms, to base level. The anomalous direction of the west-flowing Narmada and Tapi rivers can only be explained by the fact that they occupy two rifts formed by sag-faulting. The major Peninsular rivers, such as the Godavari, Krishna and Kaveri (Cauvery), are remarkably graded and marked by interruptions in profile where they cut through the Eastern Ghats.

The river systems of India may be broadly classified into two groups: the Himalayan and the Deccan. The main rivers of the Himalayan group, which are snow-fed, are the Indus, Ganga and Brahmaputra. The Indus has five major tributaries. The Ganga and its numerous tributaries occupy one-fourth of the total geographical area of the country. The Brahmaputra flows about 2580 kms from western Tibet into the Bay of Bengal, but only 725 kms of its course lies in India. Within India the Brahmaputra has a basin of 240,000 sq kms (total 580,000 sq kms). The Deccan system includes the west-flowing Narmada, Tapi and Periyar and the east-flowing Mahanadi, Godavari, Krishna, Penneru and Kaveri.

Note: m = metres.

Source: *National Atlas of India, Vol. II*, NATMO.

Table 1

INDIA — PHYSIOGRAPHIC DIVISIONS

Major Divisions	Sub-divisions	Local regions
The Northern Mountains	Western Himalaya	N. Kashmir Himalaya; S. Kashmir Himalaya; Punjab Himalaya; Kumaon Himalaya (north-west corner of U.P.)
	Central Himalaya	Nepal Himalaya East; Nepal Himalaya West
	Eastern Himalaya	Western — including Bhutan, Sikkim & Darjiling; Eastern — including Assam, Arunachal P.
	North-eastern Himalaya	Purvachal (north-east region); Meghalaya; Assam Valley
The Indo-Gangetic Plain	Western Plain	Marusthali (west Rajasthan); Rajasthan Bagar
	Northern Plain	Punjab Plain; Ganga-Yamuna Doab (west U.P.); Rohilkhand Plain (north-west U.P.); Avadh Plain (east U.P.)
	Eastern Plain	N. Bihar Plain; S. Bihar Plain; Bengal Basin (West Bengal)
The Central Highlands	North Central Highlands	Aravalli Range (east Rajasthan); East Rajasthan Uplands; Madhya Bharat Pathar (parts of Rajasthan, north-west M.P.); Bundelkhand Uplands (south-west U.P. and north-west M.P.)
	South Central Highlands	Malwa plateau (north M.P.); Vindhya Scarplands, Vindhya Range & Narmada Valley (M.P.)
The Peninsular Plateau	Western Hills	North Sahyadri (Western Ghats); Central Sahyadri (west of Belgaum to Kodagu district of Karnataka); South Sahyadri (Anai Mudi in Tamil Nadu)
	North Deccan	Satpura Range (Boundary between M.P. & Maharashtra); Maikala Plateau; Maikala Range (Balaghat & Mandla dts., M.P.); Maharashtra Plateau
	South Deccan	Telangana Plateau (north and west A.P.); Karnataka Plateau
	Eastern Plateau	Baghelkhand Plateau (Shahdol dt., M.P.); Chota Nagpur Plateau (S. Bihar); Garhjat Hills (Sundargarh, Sambalpur dts., Orissa); Mahanadi Basin (south-east dts. of M.P. & Balangir dt. Orissa); Dandakaranya (Bastar dt., M.P.)
	Eastern Hills	North-eastern Ghats (central A.P. & Orissa); South-eastern Ghats (Coimbatore dt., Tamil Nadu and Chittoor, Cuddapah dts., A.P.); Tamil Nadu Uplands; Nilgiri (Doda Betta)
The East Coast	The East Coastal Plain	Utkal Plain (Mahanadi delta); Andhra Plain (Godavari & Krishna delta); Tamil Nadu Plain (Kaveri delta)
	The Eastern Continental Shelf	
The West Coast	The West Coastal Plain	Kachchh Peninsula; Kathiawar Peninsula (west Gujarat); Gujarat Plain; Konkan Coast (Coastal Maharashtra); Karnataka Coast; Kerala Plain
	The Western Continental Shelf	
Islands	Bay Islands	Andaman Islands; Nicobar Islands
	Arabian Sea Islands	Lakshadweep & Amindivi Islands; Minicoy Islands

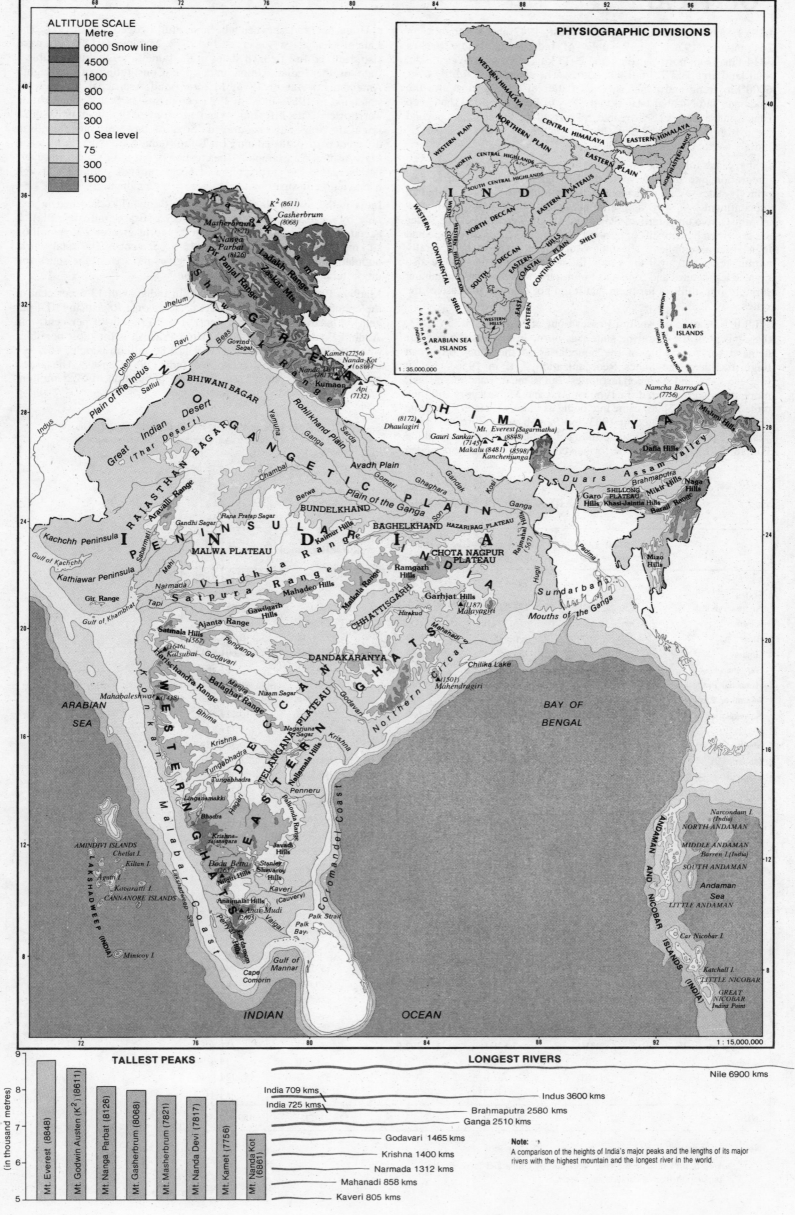

POLITICAL

India became independent on 15 August, 1947 and a Republic on 26 January, 1950. The Republic of India's mainland stretches 3214 kms from north to south and 2933 kms from west to east. The total land area is 3.29 million sq kms. The coastline is a little over 6000 kms long and it has a land frontier of 15,200 kms. But the Andaman and Nicobar Islands in the Bay of Bengal and Lakshadweep Islands in the Arabian Sea are integral parts of the Indian Union beyond these frontiers. Indira Point (Pygmalion Point) in the Nicobars is India's southernmost point.

In the years following Indian independence from British rule, the integration into the newly constituted Indian Union of the former princely states — 560 principalities constituting two-fifths of the Indian subcontinent — was a preliminary to the formation of new administrative units termed States and Union Territories. The states are governed by councils of ministers (cabinets) responsible to elected legislative assemblies. The union territories were to be centrally administered but, over the years, the larger among them were also granted legislatures and councils of ministers, with the Centre retaining some responsibilities for them that it did not enjoy in the case of the states.

With the creation of Andhra Pradesh out of Madras State in 1953, the concept of the linguistic state was born. In response to popular demand for redrawing of the boundaries of the states on broad linguistic lines, the States Reorganization Act, in 1956, created fourteen states and six union territories. But as the demands continued, more states were created. In 1960 Gujarat and Maharashtra (till then united as Bombay State) came into being as linguistic states. The state of Nagaland was created in 1963 and Punjab was bifurcated, after years of agitation, into Hindi-speaking Haryana and Punjabi-speaking Punjab. The hill areas of the former composite Punjab were added to Himachal Pradesh when it became a full-fledged state in 1971. The state of Meghalaya was created in 1972. Manipur and Tripura were also declared full-fledged states at the same time. Sikkim became a state of the Indian Union under the Constitution (Thirty-eighth Amendment) Act of 1975; Mizoram and Arunachal Pradesh got statehood in 1987. India in 1987 comprises 24 States and 7 Union Territories. The bifurcation and creation of political units on a linguistic basis has encouraged 'Sons of the Soil' movements in different parts of the country (in Tamil Nadu, Karnataka, Punjab and Maharashtra, for example). The creation of linguistic states, however, has increased the participation of people in state politics, although it has led to fissiparous tendencies of regionalism.

India is the seventh largest country in the world (exceeded in size only by the USSR, Canada, China, USA, Brazil and Australia). It accounts for 2.5 per cent of the total world area of 133.9 million sq kms. Its population (1984) is 15.5 per cent of the total world population of 4837 million. Table 2 gives the area of the states and union territories and their population.

Madhya Pradesh is the largest state in India, with 13.5 per cent of the country's total geographic area. Next come Rajasthan (10.4 per cent), Maharashtra (9.4 per cent), Uttar Pradesh (9 per cent) and Andhra Pradesh (8.4 per cent). Uttar Pradesh is the most populous of all the states of the Union with 110.86 million people, 16.2 per cent of the 685 million population of India (1981 Census). The largest state in India, Madhya Pradesh, has a population of 52.2 million, merely 7.6 per cent of the total.

Sources: 1. *Primary Census Abstract, Part II B (i)*, Census of India 1981.
2. *India 1985*, Ministry of Information and Broadcasting.

Table 2   **AREA & POPULATION — 1981**

	Total Area (in sq km)	Rural Area (in sq. km)	Urban Area (in sq. km)	Rural Population	Urban Population
States					
Andhra Pradesh	275,068	271,022.0	4,045.9	41,062,097	12,487,576
Assam	78,438	78,391.0	435.0	17,849,657[1]	2,047,186[1]
Bihar	173,877	170,678.5	3,198.5	61,195,744	8,718,990
Gujarat	196,024	191,259.4	4,764.6	23,484,146	10,601,653
Haryana	44,212	43,448.2	763.8	10,095,231	2,827,387
Himachal Pradesh	55,673	55,460.6	212.4	3,954,847	325,971
Jammu & Kashmir[2]	222,236	221,648.8	587.2	4,726,986	1,260,403
Karnataka	191,791	188,108.2	3,682.8	26,406,108	10,729,606
Kerala	38,863	37,075.0	1,788.0	20,682,405	4,771,275
Madhya Pradesh	443,446	438,567.7	4,878.3	41,592,385	10,586,459
Maharashtra	307,690	301,802.2	5,887.8	40,790,577	21,993,594
Manipur	22,327	22,175.5	151.5	1,045,493	375,460
Meghalaya	22,429	22,344.2	84.8	1,094,486	241,333
Nagaland	16,579	16,470.2	108.8	654,696	120,234
Orissa	155,707	153,418.9	2,288.1	23,259,984	3,110,287
Punjab	50,362	49,162.6	1,199.4	12,141,158	4,647,757
Rajasthan	342,239	337,741.7	4,497.3	27,051,354	7,210,508
Sikkim	7,096	N.A.	N.A.	265,301	51,084
Tamil Nadu	130,058	124,197.3	5,860.7	32,456,202	15,951,875
Tripura	10,486	10,431.6	54.4	1,827,490	225,568
Uttar Pradesh	294,411	289,850.6	4,560.0	90,962,898	19,899,115
West Bengal	88,752	86,106.0	2,646.0	40,133,926	14,446,721
Union Territories					
Andaman & Nicobar	8,249	8,234.9	14.1	139,107	49,634
Arunachal Pradesh	83,743	N.A.	N.A.	590,411	41,428
Chandigarh	114	45.7	68.3	28,769	422,841
Dadra & Nagar Havveli	491	484.3	6.7	96,762	6,914
Delhi	1,483	891.1	591.9	452,206	5,768,200
Goa, Daman & Diu	3,814	3,621.1	192.9	734,922	351,808
Lakshadweep	32	21.4	10.6	21,620	18,629
Mizoram	21,081	20,762.0	319.0	371,943	121,814
Pondicherry	492	392.0	100.0	288,424	316,047
INDIA	3,287,263	3,143,240.9	53,183.1	525,457,335[1]	159,727,357[1]

Note: [1] Estimated

[2] Area figures include the areas occupied by China and Pakistan, but population figures exclude such areas.

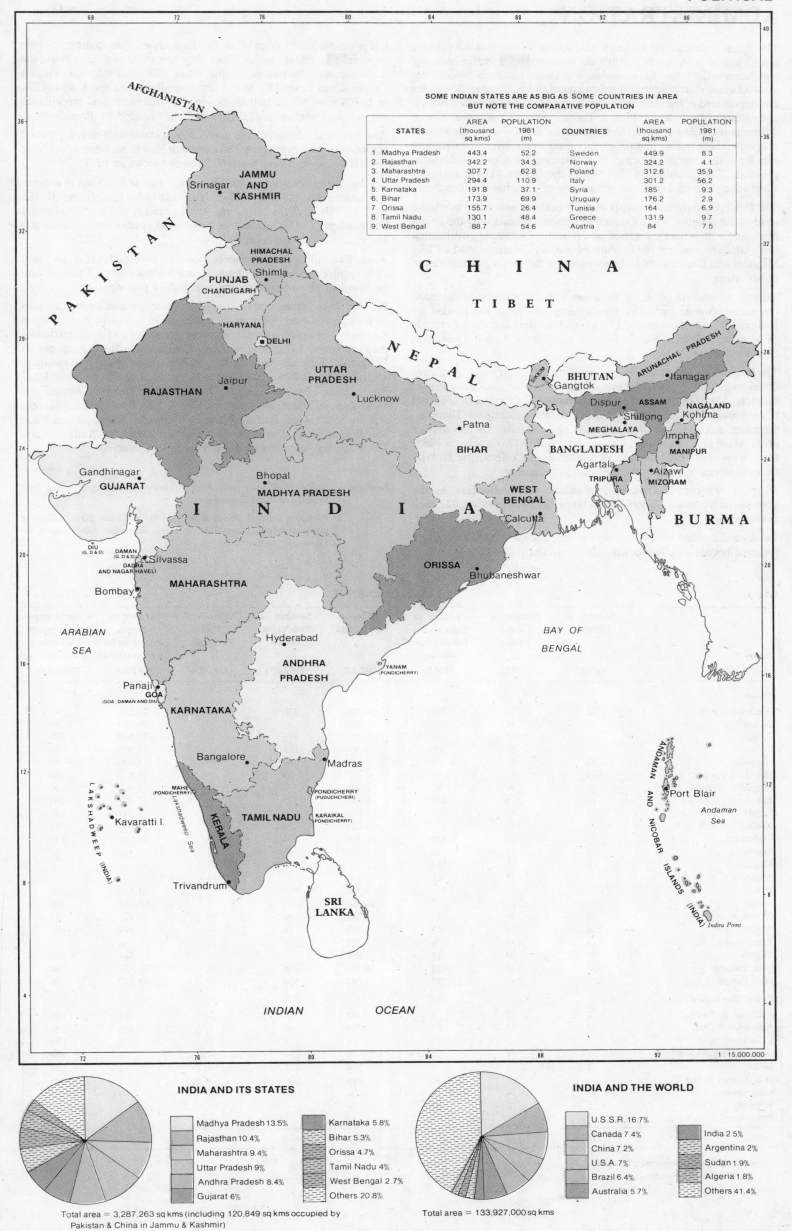

SOME INDIAN STATES ARE AS BIG AS SOME COUNTRIES IN AREA
BUT NOTE THE COMPARATIVE POPULATION

STATES	AREA (thousand sq kms)	POPULATION 1981 (m)	COUNTRIES	AREA (thousand sq kms)	POPULATION 1981 (m)
1. Madhya Pradesh	443.4	52.2	Sweden	449.9	8.3
2. Rajasthan	342.2	34.3	Norway	324.2	4.1
3. Maharashtra	307.7	62.8	Poland	312.6	35.9
4. Uttar Pradesh	294.4	110.9	Italy	301.2	56.2
5. Karnataka	191.8	37.1	Syria	185	9.3
6. Bihar	173.9	69.9	Uruguay	176.2	2.9
7. Orissa	155.7	26.4	Tunisia	164	6.9
8. Tamil Nadu	130.1	48.4	Greece	131.9	9.7
9. West Bengal	88.7	54.6	Austria	84	7.5

1 : 15,000,000

INDIA AND ITS STATES

Madhya Pradesh 13.5%	Karnataka 5.8%
Rajasthan 10.4%	Bihar 5.3%
Maharashtra 9.4%	Orissa 4.7%
Uttar Pradesh 9%	Tamil Nadu 4%
Andhra Pradesh 8.4%	West Bengal 2.7%
Gujarat 6%	Others 20.8%

Total area = 3,287,263 sq kms (including 120,849 sq kms occupied by Pakistan & China in Jammu & Kashmir)

INDIA AND THE WORLD

U.S.S.R. 16.7%	India 2.5%
Canada 7.4%	Argentina 2%
China 7.2%	Sudan 1.9%
U.S.A. 7%	Algeria 1.8%
Brazil 6.4%	Others 41.4%
Australia 5.7%	

Total area = 133,927,000 sq kms

ADMINISTRATION

The 7th Schedule of the Indian Constitution, which was adopted by the Constituent Assembly in 1950, allocates 97 specific areas of legislation to the Union Government. The Union List includes defence, armed forces, atomic energy, foreign affairs and citizenship. The states are empowered to legislate on 66 items, including police, public health, agriculture, forests and fisheries. There is also a Concurrent List of 47 items including criminal law, marriage and divorce, transfer of property, trade unions and newspapers, on which either government can legislate though, in case of a dispute, the Centre's decision prevails. The Constitution also allows the Centre to rule against a state's decision, on a state subject, in case of a dispute between them.

The Union Government consists of the President, who is the executive head of the country, and a Council of Ministers, headed by the Prime Minister, with whom the actual power rests. Each state government has a similar structure; the executive head is a Governor, and a Chief Minister and Council of Ministers carry on the actual administration in his name.

There is no uniform pattern in the organizational structure of the union territories. Some, including Pondicherry and Goa, are headed by Lieutenant Governors assisted by Chief Ministers and their Councils of Ministers; others, like Delhi, are administered by a Lieutenant Governor alone.

The large populations and vast areas of these primary administrative divisions require subdivision into smaller administrative units termed districts. Many of the districts in the country are too big in area and have populations too large for efficient administration. There is, therefore, a continuing process of breaking up the districts into smaller ones. At the time of the 1981 Census there were 412 districts. In 1985, there were 439. Each district is headed by a Collector or District Commissioner appointed by the state government.

Districts are further divided into taluks with tahsildars as the administrative heads. In 1981 there were 3342 taluks in the 412 then existing districts.

Taluks are divided into yet smaller units which are administered by regional bodies locally elected. These elected bodies are known as Corporations, Municipalities or Panchayats, depending on the economic levels of the localities they are to administer. Thus, the administrative bodies of major cities (73 in 1985) are elected Corporations headed by Mayors; the smaller cities and towns (1274 in 1980) have Municipalities headed by Chairmen; and the villages are run by Panchayats (217,319 in 1985) headed by Presidents.

Another local administrative division concerned with developmental activities is the Community Development Block, each block covering several villages. There were 5272 such blocks in 1982.

In many states 'Panchayati Raj' in the rural areas functions as a three-tier system operating at village (*gram*), block (*samiti*) and district (*zilla*) levels. But the states are free to modify the system. Jammu and Kashmir and Kerala follow a one-level system while Orissa and Haryana have a two-level system.

Areas like military cantonments, ports or 'Notified Areas' in every state require special types of administrative machinery. These too are designed to allow some measure of public participation.

Besides these administrative divisions, there are also Zonal Councils at a national level, each with a high-level advisory body, which discusses matters of common interest to the states and union territories they represent. There are five Zonal Councils: the North Zone covers Haryana, Himachal Pradesh, Jammu & Kashmir, Punjab, Rajasthan, Chandigarh and Delhi; the Central Zone, Madhya Pradesh and Uttar Pradesh; the East Zone, Bihar, Orissa, Sikkim and West Bengal; the West Zone, Gujarat, Maharashtra, Goa and Dadra and Nagar Haveli; and the South Zone, Andhra Pradesh, Karnataka, Kerala, Tamil Nadu and Pondicherry. The north-east and the islands of Andaman and Lakshadweep are not represented in these Councils.

Sources: 1. *Statistical Abstract India 1984*. Ministry of Planning. Quoting from Ministry of Rural Development.

2. *Profiles of Districts, Parts I & II, July 1985*, and *Basic Statistics Relating to the Indian Economy, Vol.2: States, September 1986*, Centre for Monitoring Indian Economy.

Table 3

LOCAL BODIES

States	Number of Districts (1985)	Number of Taluks/ Tahsils (1981)	Number of Community Development Blocks (1982)	Number of Panchayats (1982-83)	Number of Panchayat Samitis (1982-83)	Number of Zilla Parishads (1982-83)	Number of villages covered under Panchayati Raj (1982-83)	% Rural population covered by Gram Panchayats (1982-83)
States								
Andhra Pradesh	23	318	324	19,550	330	22	27,221	90
Assam	17	26	130	714	—	20	20,799	96
Bihar	38	595	587	11,367	588	33	77,848	100
Gujarat	19	183	250	12,965	182	19	18,697	100
Haryana	12	41	83	5,541	96	—	6,690	99
Himachal Pradesh	12	74	69	2,357	69	12	16,916	100
Jammu & Kashmir	14	50	92	1,469	—	—	6,900	100
Karnataka	19	176	268	8,402	175	19[1]	27,028	100
Kerala	14	62	144	1,002	—	—	1,362	100
Madhya Pradesh	45	195	457	16,229	459	16	75,544	98
Maharashtra	30	236	426	24,504	296	29	38,791	100
Manipur	8	27	26	107	6	—	512	57
Meghalaya	5	34	24	—	—	—	—	—
Nagaland	7	77	21	—	—	—	—	—
Orissa	13	114	314	4,390	314	—	51,574	100
Punjab	12	45	117	10,950	—	—	12,188	100
Rajasthan	27	199	232	7,292	236	26	37,124	100
Sikkim	4	4	10	215	—	—	404	100
Tamil Nadu	18	159	374	13,075	378	24[1]	—	—
Tripura	3	10	17	689	—	—	844	100
Uttar Pradesh	57	248	876	74,102	896	56	112,624	100
West Bengal	16	307	335	3,242	324	15	38,047	98
Union Territories								
Andaman & Nicobar	2	7	5	42	—	—	183	100
Arunachal Pradesh	10	97	43	704	48	9	3,096	100
Chandigarh	1	1	1	21	1	1	22	100
Dadra & Nagar Haveli	1	1	2	10	—	—	72	100
Delhi	1	1	5	204	5	—	258	100
Goa, Daman & Diu	3	13	12	196	—	—	462	100
Lakshadweep	1	4	4	—	—	—	—	—
Mizoram	3	22	20	—	—	—	—	—
Pondicherry	4	15	4	—	11[2]	—	334	—
INDIA	439	3,342	5,272	217,339	4,414	301	575,540	92

Note: [1] District Development Councils.
[2] Commune Panchayats

6

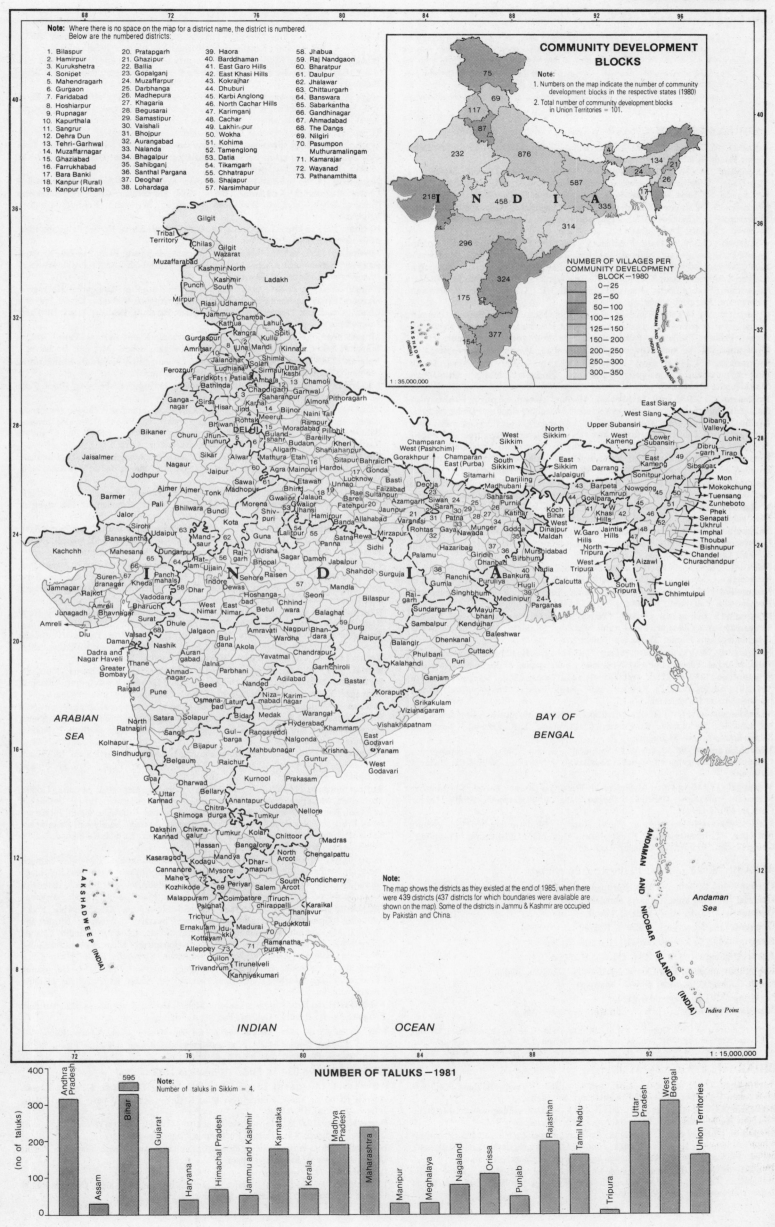

LIST OF TALUK NAMES

Note: 1. The districts listed here are according to the 1981 Census. In states where districts have been divided into smaller units or combined into larger ones, or the names of districts have been changed, the 1985 status is given at the end of the relevant state's list of taluks.

2. The name of the district headquarters is given beside the name of the district wherever it is different from the district name.

3. Some districts, like the Dangs in Gujarat and Madras in Tamil Nadu, are themselves taluks and hence do not have taluks listed.

ANDHRA PRADESH (275,068 sq kms, 23 districts, 318 taluks)

Adilabad (16,128 sq kms): Lakshettipet; Nirmal; Sirpur; Mudhol; Chinnur; Asifabad; Utnur; Boath; Khanapur; Wankadi.

Anantapur (19,130 sq kms): Hindupur; Gooty; Tadpatri; Kadiri; Madakasira; Nallamada; Dharmavaram; Singanamala; Uravakonda; Rayadurg; Kambadur; Penukonda; Sri Satya Sai; Chnnekothapalle; Kalyandurg; Kanekallu.

Chittoor (15,152 sq kms): Chandragiri; Puttur; Punganuru; Madanapalle; Palmaner; Bangarupalem; Chinagottigallu; Vayalpad; Kuppam; Tamballapalle; Srikalahasti; Satyavedu; Nagari; Thottambadu.

Cuddapah (15,359 sq kms): Proddatur; Rayachoti; Pulivendla; Badvel; Koduru; Rajampet; Kamalapuram; Sidhout; Lakkireddipalle; Jammalamadugu; Muddanuru.

East Godavari (10,807 sq kms, *Kakinada*): Rajahmundry; Amalapuram; Pithapuram; Alamur; Prattipadu; Tuni; Korukonda; Mummidivgram; Kottapeta; Ruzole; Tallarevu; Rayavaram; Rangampeta; Ramachandrapuram; P. Gannavaram; Peddapuram; Yellavaram; Rampachodavaram.

Guntur (11,391 sq kms): Tenali; Chilakalurupet; Narasaraopet; Bapatla; Ponnuru; Mangalagiri; Repalle; Piduguralla; Sattenapalle; Macherla; Pallapatla; Rajupalem; Emani; Talluru; Tadikonda; Palnad; Vinukonda; Prathipadu; Amruthalur; Ipuru.

Hyderabad (217 sq kms): Golconda; Secunderabad; Mushirabad; Charminar.

Karimnagar (11,823 sq kms): Peddapalli; Metpalli; Jagtial; Hazurabad; Sirsilla; Bheemadevarapalle; Vemalwada; Sultanabad; Husnabad; Mallial; Gangadhara; Mahadeopur; Manthani.

Khammam (16,029 sq kms): Kothagudem; Madhira; Thirumalayapalem; Bhadrachalam; Sathupalli; Sudimalle; Yellandu; Aswaraopet; Mangoor; Bhoorgampadu; Nugur.

Krishna (8727 sq kms, *Machilipatnam*): Vijayawada; Nuzvid; Gudivada; Divi; Nandigama; Kaikalur; Mylavaram; Movva; Gannavaram; Vuyyuru; Vissanapeta; Tiruvuru; Kanchikacherla; Jaggayyapet; Pamarru; Bantumilli; Mandavalli.

Kurnool (17,658 sq kms): Adoni; Nandyal; Dhone; Allagadda; Pattikonda; Atmakur; Emmiganuru; Nandikotkur; Koilkuntla; Alur; Kodumuru; Banganapalle.

Mahbubnagar (18,432 sq kms): Kalwakurti; Wanparti; Nagarkarnul; Kondangal; Gadwal; Makhtal; Kolhapur; Shadnagar; Atmakur; Alampur; Achampet; Jadcherla.

Medak (9699 sq kms, *Sangareddi*): Andol; Zahirabad; Gajwel; Narsapur; Siddipet; Ramayampet; Dubak; Narayankher; Sadasivpet; Medak.

Nalgonda (14,240 sq kms): Suriapet; Kodad; Miriyalguda; Devarkonda; Bhongir; Huzurnagar; Nidamanur; Ramannapeta; Thungathurthi; Nakrekal; Yadagirigutta; Motkur; Chandur; Nampalli.

Nellore (13,076 sq kms): Kavali; Gudur; Batchireddipalem; Kovur; Atmakur; Indukurpet; Venkatagiri; Udayagiri; Nayudupeta; Sullurpet; Vinjamur; Vakadu; Podalakur; Rapur.

Nizamabad (7956 sq kms): Bodhan; Armur; Bheemgal; Banswada; Kamareddi; Domakonda; Yallareddi; Madnur.

Prakasam (17,626 sq kms, *Ongole*): Chirala; Kandukur; Addanki; Kondapi; Parchur; Santhamaguluru; Giddalur; Maddipadu; Markapur; Kanigiri; Darsi; Bestavaripeta; Erragondapalem; Podile; Tarlupadu; Pamur.

Rangareddi (7493 sq kms, *Hyderabad*): Vallabhnagar; Hyathnagar; Rajendranagar; Pargi; Medchal; Tandur; Chevella; Vikarabad; Ibrahimpatan; Marpalli; Maheshwaram.

Srikakulam (5837 sq kms): Palkohda; Razam; Pathapatnam; Sompeta; Ponduru; Tekkali; Amadalavalasa; Kotabommali; Ranasthalam; Ichchapuram; Palasa; Narasannapeta; Hiramandalam.

Vishakhapatnam (11,161 sq kms): Anakapalle; Paderu; Narsipatnam; Bhimunipatnam; Elamanchili; Nakkapalle; Kotauratla; Chodavaram; K Kothapadu; Chintapalle; Sabbavaram; Araku; Madugula.

Vizianagaram (6539 sq kms): Parvatipuram; Bobbili; Salur; Gajapatinagaram; Chipurupalle; Bhogapuram; Srungavarapukota; Viyyampeta; Nellimarla; Badangi; Kurupam.

Warangal (12,846 sq kms): Mahbubabad; Wanangal (Rural); Parkal; Vardannapet; Maripeda; Narsampet; Cheriyal; Ghanpur; Jangaon; Kodakondia; Mulug; Gudur; Chityal; Eturnagaram.

West Godavari (7742 sq kms, *Eluru*): Tanuku; Bhimavaram; Kovvur; Poduru; Narsapur; Tadepallegudem; Penumantra; Akividu; Chintalapudi; Polavaram; Ganapavaram; Bhimadole; Gopalapuram.

ASSAM (78,523 sq kms, 10 districts, 26 taluks)

Cachar (6962 sq kms, *Silchar*): Karimganj; Hailakandi.

Darrang (8775 sq kms, *Tezpur*): Mangaldai.

Dibrugarh (7000 sq kms): Sadiya; Tinsukia.

Goalpara (10,359 sq kms, *Dhubri*): Kokrajhar; Goalpara.

Kamrup (9863 sq kms, *Guwahati*): Barpeta; Nalbari.

Karbi Anglong (10,332 sq kms, *Diphu*): Hamram.

Lakhimpur (6000 sq kms, *North Lakhimpur*): Dhemaji; Jonai.

North Cachar Hills (4890 sq kms, *Haflong*).

Nowgong (5561 sq kms): Marigaon.

Sibsagar (8989 sq kms, *Jorhat*): Sibsagar; Golaghat; Majuli.

Note: 17 districts in 1985. Goalpara was divided into Dhuburi, Kokrajhar, Goalpara; Darrang into Sonitpur, Darrang; Sibsagar into Jorhat, Sibsagar; Cachar into Karimganj, Silchar, Pragjyotishpur, a new district for which boundaries are not available, was also formed.

BIHAR (173,877 sq kms, 31 districts, 595 taluks)

Aurangabad (3305 sq kms): Jaina; Amba; Sillod; Kannad; Paithan; Vaijapur; Bhokardan; Gangapur; Jafferabad; Khuldabad; Soegaon.

Begusarai (1918 sq kms): Barauni; Bakhri; Balia; Bachwara; Bhagwanpur; Matihani; Sahebpur Kamal; Teghra; Cheria Bariarpur; Khudabandpur.

Bhagalpur (5589 sq kms): Jagdishpur; Colgong; Pirpainti; Naugachhia; Amarpur; Bihpur; Sultanganj; Gopalpur; Banka; Sabour; Dhuraiya; Shambhuganj; Shahkund; Belhar; Rajaun; Katuria; Bausi; Sonhaula; Barahat; Chanan; Nathnagar.

Bhojpur (4098 sq kms, *Ara*): Jagdishpur; Dumraon; Buxar; Rajpur; Piro; Sahar; Brahampur; Barhara; Shahpur; Simri; Nawanagar; Tarari; Koliwar; Sandesh; Bihia; Itarhi; Udawantnagar; Charpokhari.

Darbhanga (2279 sq kms): Benipur; Maniagachhi; Ghanshyampur; Haiyaghat; Biraul; Singhwara; Baheri; Jale; Keotiranway; Kusheshwar Asthan; Bahadurpur.

Dhanbad (2996 sq kms): Chas; Jharia-cum-Jorapokhar-cum-Sindri; Nirsa-cum-Chirwanda; Baghmara-cum-Katras; Chandan Kioari; Gobindpur; Topchanchi; Tundi; Baliapur.

Gaya (6545 sq kms): Arwal; Kurtha; Ghosi; Wazirganj; Tikari; Makhdumpur; Karpi; Atri; Fatehpur; Jehanabad; Konoh; Sherghati; Belaganj; Bodh Gaya; Paraiya; Imamganj; Amas; Mohanpur; Barachatti; Khizarsarai; Kako; Garpa; Manpur; Dumaria.

Giridih (6892 sq kms): Bermo; Bagodar; Gumia; Jamua; Dhanwar; Dumri; Nawadih; Deori; Gande; Birni; Peterbar; Bengabad; Gawan; Pirtanr; Jaridih; Kasmar; Tisri.

Gopalganj (2033 sq kms): Barauli; Kuchaikote; Baikunthpur; Uchkagaon; Manjha; Hathua; Katiya; Bhorey; Bijaipur.

Hazaribag (11,165 sq kms): Patratu; Ramgarh; Kodarma; Mandu; Chauparan; Itkhori; Chatra; Bishungarh; Katkamsari; Barhi; Gola; Hunterganj; Churchoo; Barakatha; Simaria; Ichak; Jainagar; Barkakana; Pratappur; Markacho; Keredari; Tandwa; Satgawan.

Katihar (3057 sq kms): Bakhri; Kadwa; Barsoi; Azamnagar; Korha; Falka; Pranpur; Manihari; Amdabad; Balrampur.

Madhubani (3501 sq kms): Madhepur; Phulparas; Benipatti; Bisfi; Pandaul; Khajauli; Rajnagar; Babubarhi; Laukaha; Laukahi; Jaynagar; Harlakhi; Andhratharhi; Basopatti; Jhanjharpur; Ladania; Madhwapur.

Munger (7908 sq kms): Lukeesarai; Jamalpur; Khagaria; Surajgarha; Kharagpur; Sikandra; Gogri; Lachhimipur; Jhajha; Alauli; Shaikhpura; Barbigha; Khaira; Parbatta; Chakai; Chautham; Tarapur; Halsi; Jamui; Sono; Barahiya; Beldaur; Ariari; Dharhara; Sangrampur.

Muzzaffarpur (3172 sq kms): Musahri; Kurhani; Kanti; Baruraj; Paro; Saraiya; Minapur; Sakra; Aurai; Gaighatti; Katra; Sahebganj; Dholi (Moraul); Bochacha.

Nalanda (2367 sq kms, *Bihar Sharif*): Rajgir; Chandi; Hilsa; Asthawan; Ekangarsarai; Islampur; Harnaut; Noorsaria; Rahvi; Giriak; Sarmera.

Nawada (2494 sq kms): Waris Aliganj; Sirdala; Akbarpur; Gobindpur; Hisua; Pakri Barawan; Kauakol; Rajauli; Narhat.

Palamu (12,749 sq kms, *Daltenganj*): Hussainabad; Bishrampur; Chainpur (Bashandi); Chhatarpur; Garhwa; Patan; Bhaunathpur; Untari; Majhiaon; Meral (Piprakalan); Balumath; Panki; Ranka; Latehar; Leslieganj; Manatu; Barwadih; Chandwa; Hariharganj; Mahuadanr; Manika; Ghuki; Bhandaria; Garu.

Paschim Champaran (5228 sq kms, *Bettiah*): Narkatiaganj; Bagaha; Sindhaw; Majwalia; Chanpatia; Ramnagar; Lauriya; Jogapatti; Nautan; Bairia; Gawnaha; Sikta; Mainatanr; Thakrahan; Madhubani.

Patna (3202 sq kms): Patna (Rural); Danapur-cum-Khagaul; Fatwah; Barh; Mokama; Bikram; Phulwari; Paliganj; Bihta; Masaurhi; Maner; Naubatpur; Bakhtiyarpur; Dhanarua; Pandarak; Punpun.

Purba Champaran (3968 sq kms, *Motihari*): Turkaulia; Areraj; Kalyanpur; Dhaka; Ghorasahan; Harsidhi; Chiraiya; Madhuban; Kesariya; Chakia (Pipra); Sagauli; Raxaul; Pakridayal; Ramgarhwa; Adapur; Paharpur; Patahi; Mehsi; Nakardei.

Purnia (7943 sq kms): Forbesganj; Araria; Banmankhi; Raniganj; Narpatganj; Kirtianandnagar; Dhamdaha; Thakurganj; Jakihat; Kochadhamin; Qasba; Baisi; Bahadurganj; Amaur; Pothia; Rupauli; Palasi; Bhargama; Kishanganj; Barhara; Dighalbank; Baisa; Bhawanipur; Sikti; Kursakatta; Terhagachh.

Ranchi (18,266 sq kms): Kanke; Gumla; Burmu; Simdega; Tamar—I; Silli; Bero; Ratu; Angora; Nankum; Sonahatu; Khunti; Karra; Ghaghra; Sisai; Jaldega; Kurdeg; Thethaitanagar; Bano; Lohardaga; Mandar; Torpa; Dumri; Erki (Tamar—II); Bundu; Murhu; Palkot; Kuru; Basia; Chanho; Raidih; Ormanjhi; Kamdara; Kolebira; Senha; Verno; Lapung; Chainpur; Pisko; Bishunpur; Bhandra; Rania; Bolba.

Rohtas (7213 sq kms, *Sasaram*): Dehri; Kargahar; Dinara; Bhabua; Ramgarh; Bikramganj; Narakat; Nasriganj; Nokha; Mohania; Dawath; Bhagwanpur; Chainpur; Shivsagar; Kudra; Rahtab; Durgawati; Chand; Chenari; Nawhatta; Adhaura.

Saharsa (5900 sq kms): Kahara; Supaul; Kishanganj; Madhepura; Singheshwar; Sour Bazar; Tribeniganj; Raghopur; Murliganj; Chhatapur; Simri Bakhtiyarpur; Kumar Khand; Sonbarsa; Kishanpur; Salkhua; Chausa; Pipra; Alamnagar; Mahishi; Basantpur; Nirmali; Nauhatta; Marauna.

Samastipur (2904 sq kms): Warisnagar; Rusera; Hasanpur; Dalsingh Sarai; Kalyanpur; Bibhutipur; Ujiarpur; Sarairanjan; Tajpur Morwa; Pusa; Patori; Singia; Mohiuddinnagar.

Santhal Pargana (14,206 sq kms, *Dumka*): Deoghar; Rajmahal; Jamtara; Godda; Madhupur; Meherma; Pakaur; Maheshpur; Nala; Pathargama; Borio; Poreyahat; Jarmundi; Ramgarh; Narayanpur; Mahagama; Mohanpur; Kundabit; Sarath; Barharwa; Jama; Boarijor; Palojori; Sahibganj; Sarwan; Barhait; Masalia; Saraiyahat; Shikarapara; Karon; Panishwar; Pakuria; Litipara; Taljhari; Hiranpur; Pathna; Kathikund; Sundar Pahari; Amarpara; Gopikandar.

Saran (2,641 sq kms, *Chhapra*): Baniapur; Masrakh; Jalalpur; Manjhi; Taraiya; Dariapur; Marhaura; Garkha; Sonepur; Ekmsa; Parsa; Amnaur; Dighwara; Revelganj.

Sitamarhi (2643 sq kms): Dumra; Belsand; Runisaidpur; Parihar; Nanpur; Bathnaha; Pupri; Sonbarsa; Riga; Sheohar; Bajpatti; Piprarhi; Majorganj; Sursand; Bairgania.

Singhbhum (13,440 sq kms, *Chaibasa*): Golmuri-cum-Jugsalai; Chakradharpur; Potka; Adityapur (Gamhria); Baharagora; Patamda; Chakulia; Govindpur; Mushabani; Ghatsila; Manoharpur; Noamundi (Barajamda); Sonua; Ichagarh; Nimdih; Chandil Jagannathpur; Seraikela; Goilkera; Jhinkpani; Daalbhumgarh; Kharsawan; Kumardungi; Khuntpani; Bangaon; Dumaria; Kuchai; Majhgaon; Manjhari; Tonto; Tatanagar.

Siwan (2219 sq kms): Barharia; Hussainganj; Mairwa; Pachrukhi; Goriakothi; Basantpur; Bhagwanpurhat; Duraundha; Maharajganj; Andar; Raghunathpur; Darauli; Siswan; Gothini.

Vaishali (2036 sq kms, *Hajipur*): Lalganj; Mahua; Goraul; Patepur; Jandaha; Vaishali; Bidupur; Sahdei Buzurg (Desari); Raghopur; Mahnar.

Note: 38 districts in 1985. Saharsa was divided into Madhepura and Saharsa; Ranchi into Lohardaga, Gumla, Ranchi; Santhal Pargana into Deoghar, Godda, Sahibganj, Santhal Pargana; Bhagalpur into Khagaria, Bhagalpur.

GUJARAT (196,024 sq kms, 19 districts, 183 taluks)

Ahmadabad (8707 sq kms): Daskroi; Viramgam; Dholka; Dhandhuka; Dehgam; Sanand.

Amreli (6760 sq kms): Kunkavav-Vadia; Kodinar; Rajula; Dhari; Lathi; Babra; Jafarabad; Lilia; Khambha.

Banas Kantha (12,703 sq kms, *Palanpur*): Disa; Tharad; Kankrej (*Shiheri*); Diyodar; Dhanera; Vadgam; Vav; Danta; Radhanpur; Santalpur (*Varahi*).

Bharuch (9038 sq kms): Nandod (*Rajpipla*); Jambusar; Ankleshwar; Jhagadiya; Valia; Dediapada; Amod; Sagbara; Hansot.

Gandhinagar (649 sq kms)

Bhavnagar (11,155 sq kms): Mahuva; Kundla (*Savarkundla*); Talaja; Palitana; Sihor; Gadhada; Gariadhar; Vallabhipur; Botad.

Jamnagar (14,125 sq kms): Khambhalia; Kalavad; Kalyanpur; Jamjodhpur; Okhamandal (*Dwarka*); Bhanvad; Lalpur; Jodiya; Dhrol.

Junagadh (10,607 sq kms): Porbandar; Patan Veraval; Una; Mangral; Keshod; Manavadar; Visavadar; Malia; Vanthali; Talala; Kutiyana; Ranavav; Bhesan; Mendarda.

Kachchh (45,652 sq kms, *Bhuj*): Anjar; Rapar; Mandvi; Nakhatrana; Bhachau; Abdasa (*Naliya*); Mundra.

Kheda (7194 sq kms): Anand; Nadiad; Borsad; Kapadvanj; Petlad; Khambhat; Thasra; Mehmadabad; Balasinor; Matar.

Mahesana (9027 sq kms): Vijapur; Patan; Sidhpur; Kheralu; Kalol; Kadi; Chanasma; Visnagar; Sami; Harij.

Panch Mahals (8866 sq kms, *Godhra*): Santrampur; Dohad; Devgad Bariya; Lunavada; Limkheda; Jhalod; Kalol; Shehera; Halol; Jambughoda.

Rajkot (11,203 sq kms): Morvi; Gondal; Jetpur; Upleta; Jasdan; Dhoraji; Wankaner; Paddhari; Jamkandorna; Maliya; Kotda Sangani; Lodhika.

Sabar Kantha (7390 sq kms, *Himatnagar*): Idar; Prantij; Modasa; Bayad; Bhiloda; Khedbrahma; Maghraj; Malpur; Vijaynagar.

Surat (7657 sq kms): Chorasi; Vyara; Mangrol; Bardoli; Songarh; Mandvi; Olpad; Kamrej; Mahuva; Nizar; Palsana; Valod; Velachcha.

Surendranagar (10,489 sq kms): Wadhwan; Limbdi; Dasada; Dhrangadhra; Chotila; Halvad; Muli; Sayla; Lakhtar.

The Dangs (1764 sq kms, *Ahwa*)

Vadodara (7794 sq kms): Chhota Udaipur; Padra; Savli; Jetpur Pavi; Dabhoi; Sankheda; Karjan; Vaghodia; Nasvadi; Sinor; Tilakwada.

Valsad (5244 sq kms): Navsari; Dharampur; Chikhli; Pardi; Gandevi; Bansda; Umargam.

HARYANA (44,212 sq kms, 12 districts, 41 taluks)

Ambala (3832 sq kms): Jagadhri; Narayangarh, Kalka.

Bhiwani (5099 sq kms): Charkhi Dadri (Dadri); Bawani Khera; Loharu.

Faridabad (2150 sq kms): Ballabgarh; Palwal.

Gurgaon (2716 sq kms): Firozpur-Jhirka; Nuh.

Hisar (6315 sq kms): Hansi; Fatehabad; Tohana.

Jind (3306 sq kms): Narwana; Safidon.

Karnal (3740 sq kms): Panipat; Asandh.

Kurukshetra (3740 sq kms): Thanesal; Kaithal; Guhla; Pehowa.

Mahendragarh (3010 sq kms, *Narnaul*): Rewari; Mahendragarh; Bawal.

Rohtak (3841 sq kms): Jhajjar; Bahadurgarh; Maham.

Sirsa (4276 sq kms): Dabwali.

Sonipat (2206 sq kms): Gohana.

HIMACHAL PRADESH 55,673 sq kms, 12 districts, 74 taluks)

Bilaspur (1167 sq kms): Ghumarwin; Naina Devi.

Chamba (6528 sq kms): Bhattiyat; Chaurah; Saluni; Bharmour; Sihunta; Pangi.

Hamirpur (1118 sq kms): Bhoranj; Nadaun; Barsar; Tira Sujanpur.

Kangra (5739 sq kms, *Dharmsala*): Kangra; Palampur; Dera Gopipur; Nurpur; Lambagraon; Fatehpur; Indora; Khundian.

Kinnaur (6401 sq kms, *Kalpa*): Nichar; Sangla; Moorang; Puh; Hangrang.

Kullu (5503 sq kms): Nermand; Banjar; Ani.

Lahul & Spiti (13,835 sq kms, *Keylang*): Spiti (*Kaja*); Udaipur.

Mandi (3950 sq kms): Sarkaghat; Sundar Nagar; Jogindar Nagar; Karsog; Chachyot; Chachyot (Thunag); Sandhol; Bali Chowki; Lad Bharol.

Shimla (5131 sq kms, *Kasumpti*): Rohru; Theog; Rampur; Kumharsain; Nerua; Kotkhai; Jubbal; Suoni; Chaupal; Nankhari.

Sirmaur (2825 sq kms, *Nahan*): Paonta Sahib; Renuka; Shilai; Pachhad; Rajgarh.

Solan (1936 sq kms): Nalagarh; Kasauli; Arki; Kandaghat; Ramshahr.

Una (1540 sq kms): Amb; Haroli; Bangana.

JAMMU & KASHMIR (222,236 sq kms, 14 districts, 50 taluks)

Anantnag (3984 sq kms): Kulgam; Doru; Pahalgam; Bijbiara.

Badgam (1371 sq kms): Chadura; Beerwah.

Baramula (4588 sq kms): Sopur; Sonewari; Bandipora; Uri; Tangmarg.

Doda (11,691 sq kms): Kishtwar; Ramban; Bhadarwah.

Jammu (3097 sq kms): Samba; Akhnur; Rambir Singh Pora; Bishna.

Kargil (14,036 sq kms): Zanskar.

Kathua (2651 sq kms): Hiranagar; Bilaur; Basoli.

Kupwara (2379 sq kms): Handwara; Karnah.

Ladakh (82,665 sq kms *Leh*).

Pulwama (1398 sq kms): Shupiyan; Tral.

Punch (1674 sq kms, *Haveli*): Mendhar.

Rajauri (2630 sq kms): Budhal; Naushahra; Kalakote; Sunderbani.

Srinagar (2228 sq kms): Gandarbal.

Udhampur (4550 sq kms): Ramnagar; Gool Gulab Garh; Reasi; Chineni.

Note: The above districts and taluks are according to Census 1981 which was restricted to areas that were not occupied by Pakistan and China. According to the Survey of India the districts in Jammu & Kashmir are as follows: Baramula (Census 1981) = Kashmir North (Survey 1985); Pulwama, Srinagar, Badgam and Anantnag = Kashmir South; Rajauri = Riasi; Udhampur and Doda = Udhampur; Kargil and Ladakh = Ladakh; Kupwara = Muzaffarabad.

KARNATAKA (191,791 sq kms, 19 districts, 177 taluks)

Bangalore (8005 sq kms): Bangalore North; Bangalore South; Kanakapura; Channapatna; Magadi; Hoskote; Dod Ballapur; Devanhalli; Ramanagaram; Anekal; Nelamangala.

Belgaum (13,415 sq kms): Chikodi; Gokak; Athni; Hukeri; Sampgaon; Parasgad; Raybag; Khanapur; Ramdurg.

Bellary (9885 sq kms): Hospet; Harpanahalli; Kudligi; Siruguppa; Sandur; Huvvina Hadagalli; Hagari Bommanahalli.

Bidar (5448 sq kms): Basauakalyan; Homnabad; Bhalki; Aurad.

Bijapur (17,069 sq kms): Jamkhandi; Indi; Sindgi; Basavana Bagevadi; Hungund; Badami; Muddebihal; Bagalkot; Mudhol; Bilgi.

Chikmagalur (7201 sq kms): Kadur; Tarikere; Mudigere; Koppa; Narasimharajapura; Sringeri.

Chitradurga (10,852 sq kms): Davangere; Challakere; Hiriyur; Harihar; Hosdurga; Holalkere; Jagalur; Molakalmuru.

Dakshin Kannad (8441 sq kms, *Mangalore*): Udupi; Coondapoor; Bantval; Karkal; Puttur; Beltangadi; Sulya.

Dharwad (13,738 sq kms): Hubli; Gadag; Ranibennur; Ron; Haveri; Hangal; Hirekerur; Shirhatti; Navalgund; Kundgol; Shiggaon; Savanur; Kalghatgi; Byadgi; Mundargi; Nargund.

Gulbarga (16,224 sq kms): Chitapur; Shorapur; Yadgir; Aland; Shahpur; Chincholi; Jevargi; Sedam; Afzalpur.

Hassan (6814 sq kms): Arsikere; Channarayapatna; Arkalgud; Belur; Holenarsipur; Manjarabad; Alur.

Kodagu (4102 sq kms, *Madikeri (Meroore)*): Virarajendrapet; Somvarpet.

Kolar (8223 sq kms): Bangarapet; Gauribidanur; Chintamani; Mulbagal; Malur; Srinivaspur; Sidlaghatta; Chikballapur; Bagepalli; Gudibanda.

Mandya (4961 sq kms): Maddur; Malvalli; Krishnarajpet; Nagamangala; Pandavapura; Shrirangapattana.

Mysore (11,954 sq kms): Chamrajnagar; Nanjan Gud; Kollegal; Tirumakudal-Narsipur; Krishnarajanagara; Hunsur; Heggadadevan Kote; Gundlupet; Piriyapatna; Yelandur.

Raichur (14,017 sq kms): Gangawati; Sindhnur; Manvi; Lingsugur; Koppal; Yelbarga; Devadurga; Kushtagi.

Shimoga (10,553 sq kms): Bhadravati; Channagiri; Honnali; Sagar; Shikarpur; Sorab; Tirthahalli; Hosanagara.

Tumkur (10,598 sq kms): Sira; Madhugiri; Kunigal; Gubbi; Pavagada; Tiptur; Chiknayakanhalli; Turuvekere; Koratagere.

Uttar Kannad (10,291 sq kms, *Karwar*): Sirsi; Haliyal; Honavar; Kumta; Bhatkal; Siddapur; Ankola; Yellapur; Mundgod; Supa.

KERALA (38,863 sq kms, 12 districts, 62 taluks)

Alleppey (1883 sq kms): Shertallai; Ambalapulai; Karthigappally (*Haripad*); Tiruvalla; Mavekikara; Changannur; Kuttanad (*Moncombu*).

Cannanore (4958 sq kms): Tellicherry; Talipparamba; Kasaragod; Hosdurg.

Ernakulam (2408 sq kms): Kanayannur; Alwaye; Kunnathunadu; Parur; Muvattupuzha; Kothamangalam.

Idukki (5061 sq kms, *Painavu*): Udumbanshola; Thodupulai; Devikolam; Pirmed.

Kottayam (2204 sq kms): Meenachil; Changanacherry; Vaikom; Kanjirapally.

Kozhikode (2345 sq kms): Quilandy; Badagara.

Malappuram (3548 sq kms, *Manjeri*): Ernad; Tirur; Peritalmanna; Ponnani.

Palghat (4480 sq kms): Ottappalam; Chittur; Alathur; Mannarkkat.

Quilon (4620 sq kms): Kottarakara; Pathanamthitta; Karunagappally; Pathanapuram; Kunnathur.

Trichur (3032 sq kms): Mukundapuram (*Irinjalakuda*); Talappilly (*Vadakkanchery*); Chavakkad; Kodungallur.

Trivandrum (2192 sq kms): Neyyattinkara; Chirayinkil (*Attingal*); Nedumangad.

Wayanad (2132 sq kms, *Kalpetta*): Sultan's Battery; Vayittiri; Manantavadi.

Note: 14 districts in 1985. Cannanore was divided into Kasaragod, Cannanore; Quilon into Pathanthitta, Quilon.

MADHYA PRADESH (443,446 sq kms, 45 districts, 195 taluks)

Balaghat (9229 sq kms): Wara Seoni; Baihar.

Bastar (39,114 sq kms, *Jagdalpur*): Kondagaon; Kanker; Narainpur; Dantewara; Konta; Bijapur; Bhanupratappur.

Betul (10,043 sq kms): Multai; Bhonadehi.

Bhind (4459 sq kms): Lahar; Mahgawan; Gohad.

Bhopal (2772 sq kms): Huzur; Berasia.

Bilaspur (19,897 sq kms): Katghora; Mungeli; Janjgir; Sakti.

Chhatarpur (8687 sq kms): Bijawar; Laundi.

Chhindwara (11,815 sq kms): Sausar; Amarwara.

Damoh (7306 sq kms): Hatta.

Datia (2038 sq kms): Seondha.

Dewas (7020 sq kms): Sonkach; Bagli; Kannod; Khategaon.

Dhar (8153 sq kms): Manawar; Kukshi; Sardarpur; Badnawar.

Durg (8537 sq kms): Bemetara; Sanjari Balod.

East Nimar (10,779 sq kms, *Khandwa*): Burhanpur; Harsud.

Guna (11,065 sq kms): Ashoknagar; Mungaoli; Raghogarh; Chachavra.

Gwalior (5214 sq kms): Gird; Pichhore; Bhander.

Hoshangabad (10,037 sq kms): Harda Khas; Sohagpur; Seoni Malwa.

Indore (3898 sq kms): Mhow; Depalpur; Sanwer.

Jabalpur (10,160 sq kms): Murwara; Sihora; Patan.

Jhabua (6782 sq kms): Alirajpur; Thandla; Jobat; Petlawad.

Mandla (13,269 sq kms): Dindori; Niwas.

Mandsaur (9791 sq kms): Nimach; Sitamau; Manasa; Jawad; Garot; Malhargarh; Bhanpura.

Morena (11,594 sq kms): Ambah; Jaura; Sheopur; Sabalgarh; Bijaipur.

Narsimhapur (5133 sq kms): Gadarwara.

Panna (7135 sq kms): Pawai; Ajaigarh.

Raigarh (12,924 sq kms): Jashpur Nagar; Udaipur; Sarangarh; Gharghoda.

Raipur (21,258 sq kms): Baloda Bazar; Mahasamund; Dhamtari; Bindranawagarh.

Raisen (8466 sq kms): Bareily; Goharganj; Udaipura; Begamganj; Silvani; Gairatganj.

Rajgarh (6154 sq kms): Khilchipur; Narsinghgarh; Biaora; Sarangpur.

Rajnandgaon (11,127 sq kms): Khairagarh; Kawardha.

Ratlam (4861 sq kms): Jaora; Alot; Sailana.

Rewa (6314 sq kms): Huzur (Rewa); Mauganj; Sirmour; Teonthar.

Sagar (10,252 sq kms): Rehli; Khurai; Banda.

Satna (7502 sq kms): Raghurajnagar; Nagod; Amarpatan; Maihar.

Sehore (6578 sq kms): Ashta; Nasrullahganj; Budni; Ichhawar.

Seoni (8758 sq kms): Lakhnadon.

Shahdol (14,028 sq kms): Sohagpur; Bandhavgarh; Beohari; Pushpa Rajgarh.

Shajapur (6196 sq kms): Shujalpur; Agar; Susner.

Shivpuri (10,278 sq kms): Pichor; Karera; Kolaras; Pohri.

Sidhi (10,526 sq kms): Gopadbanas; Deosar; Singrauli.

Sarguja (22,337 sq kms, *Ambikapur*): Surajpur; Palma; Manendragarh; Baikunthpur; Samri; Bharatpur.

Tikamgarh (5048 sq kms): Jatara; Nivari.

Ujjain (6091 sq kms): Khacharod; Barnagar; Tarana; Mehidpur.

Vidisha (7371 sq kms): Basoda; Sironj; Korwai; Lateri.

West Nimar (13,450 sq kms, *Khargon*): Sendhwa; Rajpur; Bhikangaon; Barwah; Maheshwar; Kasrawad.

MAHARASHTRA (307,690 sq kms, 26 districts, 236 taluks)

Ahmadnagar (17,048 sq kms): Kopargaon; Shrirampur; Sangamner; Nevasa; Rahuri; Shrigonda; Parner; Akola; Pathardi; Karjat; Shevgaon; Jamkhede.

Akola (10,575 sq kms): Washim; Akot; Murtazapur; Mangrulpir; Balapur.

Amravati (12,212 sq kms): Achalpur; Morsi; Chandur; Daryapur; Melghat.

Aurangabad (16,305 sq kms): Jalna; Ambad; Sillod; Kanad; Paithan; Vaijapur; Bhokardan; Gangapur; Jafferabad; Khuldabad; Soegaon.

Bhandara (9213 sq kms): Gondiya; Sakoli.

Beed (11,085 sq kms): Ambejogai; Manjlegaon; Kaij; Georai; Ashti; Patoda.

Buldana (9661 sq kms): Chikhli; Mehekar; Malkapur; Khamgaon; Jalgaon.

Chandrapur (25,923 sq kms): Garhchiroli; Warora; Brahmapuri; Sironcha; Rajura.

Dhule (13,150 sq kms): Sakri; Shahada; Sindkheda; Shirpur; Nandurbar; Nawapur; Akkalkuwa; Taloda; Akrani (*Dhadgaon*).

Greater Bombay (603 sq kms).

Jalgaon (11,765 sq kms): Bhusawal; Chalisgaon; Erandol; Amalner; Raver; Jamner; Yawal; Chopda; Pachora; Parola; Bhadgaon; Edlabad.

Kolhapur (8047 sq kms): Karvir; Hatkanangale; Shirol; Kagal; Gadhinglaj; Ponhela; Radhanagari; Shahuwadi; Chandgad; Bhudargad; Ajra; Bavda.

Nagpur (9931 sq kms): Ramtek; Umred; Katol; Saoner.

Nanded (10,502 sq kms): Kandahar; Biloli; Hadgaon; Kinwat; Mukher; Bhokar; Deglur.

Nashik (15,530 sq kms): Malegaon; Niphad; Baglan; Sinnar; Nandgaon; Igatpuri; Dindori; Kalwan; Yeola; Chandvad; Peint; Surgana.

Osmanabad (14,210 sq kms): Latur; Udgir; Ahmadpur; Nilanga; Umarga; Ausa; Kallam; Tulzapur; Parenda; Bhum.

Parbhani (12,561 sq kms): Hingoli; Pathri; Gangakher; Basmath; Kalamnuri; Jintur; Partur.

Pune (15,642 sq kms): Haveli; Baramati; Junnar; Khed; Daund; Mawal (*Vadgaon*); Shirur; Ambegaon (*Ghodegaon*); Purandhar (*Sasvad*); Bhor; Mulshi (*Baud*); Velhe; Indapur.

Raigad (7148 sq kms, *Alibag*): Panvel; Mangaon; Mohad; Karjat; Pen; Roha; Khalapur; Uran; Shrivardhan; Murud; Mhasla; Poladpur; Sudhagad.

Ratnagiri (13,054 sq kms): Chiplun; Sangameshwar; Rajapur; Sawantwadi; Khed; Dapoli; Kudal; Kankavli; Malwan; Devgad; Guhagar; Lanja; Vengurla; Mandangad.

Sangli (8572 sq kms): Miraj; Walwa (*Islampur*); Tasgaon; Khanapur; Jat; Shirala; Kavathe Mahankal; Atpadi.

Satara (10,484 sq kms): Karad; Patan; Phaltan; Khatav; Koregaon; Man (*Dohiwadi*); Wai; Jaoli; Khandala; Mahabaleshwar.

Solapur (15,017 sq kms,): Malsiras; Barsi; Pandharpur; Akkalkot; Madha; Sangole; Karmala; Mohol; Solapur South (*Mandrup*); Mangalvedha.

Thane (9558 sq kms): Kalyan; Ulhasnagar; Bhiwandi; Vasai; Palghat; Dahanu; Shahapur; Murbad; Jawhar; Vada; Talasari; Mokhada.

Wardha (6310 sq kms): Hinganghat; Arvi.

Yavatmal (13,584 sq kms): Darwha; Pusad; Kelapur (*Pandhar Kawaoo*); Wani.

Note: 30 districts in 1985. Ratnagiri was divided into North Ratnagiri, Sindhudurg; Osmanabad into Latur, Osmanabad; Chandrapur into Garhchiroli, Chandrapur; Aurangabad into Jalna, Aurangabad.

MANIPUR (22,327 sq kms, 6 districts, 27 taluks)

Manipur Central (2238 sq kms, *Imphal*): Imphal West; Thoubal; Imphal East; Bishnupur; Jiribam.

Manipur East (4544 sq kms, *Ukhrul*): Ukhrul Central; Ukhrul North (*Chingal*); Phunguar Phaisat (*Phunguar*); Kamjong Chassad; Ukhrul South (*Kasom Khullon*).

Manipur North (3271 sq kms, *Senapati*): Mao-Maram (*Tadubi*); Sadar Hills West (*Kangpokpi*); Sadar Hills East (*Salkut*).

Manipur South (4570 sq kms, *Churachandpur*): Tipaimukh (*Parbung*); Thanlon; Churachandpur North (*Henglep*); Singhgat.

Manipur West (4391 sq kms, *Tamenglong*): Nungba; Tamenglong West; Tamenglong North.

Tengnoupal (3313 sq kms, *Chandel*): Chakpikarong.

Note: 8 districts in 1985. Central Manipur was divided into Imphal, Thoubal, Bishnupur. The names of Manipur North East, West and South have been changed to: Senapati, Ukhrul, Tamenglong and Churachandpur.

MEGHALAYA (24,429 sq kms, 5 districts, 34 taluks)

East Garo Hills (2603 sq kms, *Williamnagar*): Resubelpare; Dambu Rongjeng; Songsak; Samanda.

East Khasi Hills (5196 sq kms, *Shillong*): Mylliem; Bhoi Area; Nongpoh; Shella Bhollaganj; Py Nursla; Mawsynram; Mawphlang; Mawryngkneng; Mawkynrew.

Jaintia Hills (3819 sq kms, *Jowai*): Laskein; Thadpaskein (*Jowai*); Khliehriat; Amlarem.

West Garo Hills (5564 sq kms, *Tura*): Selsella; Dadenggiri; Betasing; Zikzak; Dalu; Chokpot; Dambukaga; Resubelpara (West); Rongohugir.

West Khasi Hills (5247 sq kms, *Nongstoin*): Mawkyrwat; Mairang; Mawshynrut.

NAGALAND (16,579 sq kms, 7 districts, 77 taluks)

Kohima (4041 sq kms): Dimpur Sadar; Dimapur; Chiephobozou; Ghaspani; Jakhama; Jaluke; Tseminyu; Nihokhu; Zubza; Dhansiripar; Pughoboto; Pedi; Ghatthashi; Peren; Tening; Neong.

Mokokchung (1615 sq kms): Ongpangkong; Tuli; Kabulong; Alongkima; Chuchuyimlang; Mangkulengba; Changtongya; Longchem.

Mon (1786 sq kms): Mon Sadar; Champang; Naginimare; Phomching; Chare; Tizit.

Phek (2026 sq kms): Pfutsero; Phek Sadar; Chazouba; Chizami; Meluri; Chetheba; Khezakenoma; Phokhungri.

Tuensang (4228 sq kms): Longleng; Noklak; Kiphire Sadar; Tuensang Sadar; Tobu; Pungro; Shamatorr; Chare; Noksen; Monyakshu; Thonoknyo; Chessore; Seyochung; Tamlu; Sitimi; Longkkim.

Wokha (1628 sq kms): Wokha Sadar; Bhandari; Aitepyong; Sungro; Baghty; Chukitong; Lotsu.

Zunheboto (1255 sq kms): Satakha; Atoizu; Aghunato; Zunheboto Sadar; Suruhoto; Aguto; Akuloto; V.K.; Satoi.

ORISSA (155,707 sq kms, 13 districts, 114 taluks)

Balangir (8913 sq kms): Titlagarh; Patnagarh; Rampur; Sonapur; Birmaharajpur.

Baleshwar (6311 sq kms): Bhadrak; Soro; Jaleswar; Basta; Dhamnagar; Chandbali; Basudebpur; Nilagiri.

Cuttack (11,142 sq kms): Salepur; Jagatsinghapur; Darpan; Kujang; Marsaghai; Jajapur; Sukinda; Kendrapara; Binjharpur; Niali; Banki; Patamundai; Rajkanika; Narsinghpur; Athagarh; Aali; Badamba; Rajnagar; Tigiria.

Dhenkanal (10,827 sq kms): Kamakhyanagar; Anugul; Talcher; Athmallik; Hindol; Chhendipada; Palalahara.

Ganjam (12,531 sq kms, *Chhatrapur*): Brahmapur; Ghumsur; Kodala; Asika; Parlakhemandi; Digapahandi; Khallikote; R. Udayagiri; Chikiti; Surada.

Kalahandi (11,772 sq kms, *Bhawanipatna*): Dharamgarh; Jayapatna; Lanjigarh; Nuaparha; Khariat.

Kendujhargarh (8303 sq kms): Anandapur; Barabil; Telkoi; Champua.

Koraput (26,961 sq kms): Umarkote; Malkangiri; Rayagarha; Nabarangapur; Jaypur; Gunupur; Nandapur; Bissamcuttack; Borigumma; Kotaparh; Motu; Kodinga; Kashipur; Machhkund.

Mayurbhanj (10,418 sq kms, *Baripada*): Rairangpur; Karanjia; Udala; Betnoti.

Phulbani (11,119 sq kms): Baligurharna; Baudh; G. Udayagiri; Kantamal.

Puri (10,182 sq kms): Bhubaneshwar; Nimaparha; Nayagarh; Khordha; Banapur; Pipili; Khandaparha; Ranapur; Dasapalla; Krushnaprasad.

Sambalpur (17,516 sq kms): Padmapur; Bargarh; Jharsuguda; Attabira; Debagarh; Kochinda; Redhakhol.

Sundargarh (9712 sq kms): Panposh; Banei; Rajagangapur; Hemagiri.

PUNJAB (50,362 sq kms, 12 districts, 45 taluks)

Amritsar (5087 sq kms): Tarn Taran; Ajnala; Patti; Baba Bakala.

Bathinda (5551 sq kms): Mansa; Rampura Phul; Talwandi Sabo.

Faridkot (5740 sq kms): Muktsar; Moga.

Gurdaspur (3562 sq kms): Batala; Pathankot.

Ferozpur (5874 sq kms): Fazilka; Zira.

Hoshiarpur (3881 sq kms): Dasua; Garhshankar; Balachaur.

Jalandhar (3401 sq kms): Nawa Shahr; Phillaur; Nakodar.

Kapurthala ((1633 sq kms): Phagwara; Sultanpur.

Ludhiana (3857 sq kms): Jagraon; Khanna; Samrala.

Patiala (4584 sq kms): Rajpura; Nabha; Fatehgar; Samana.

Rupnagar (2085 sq kms): Kharar; Anandpur Sahib.

Sangrur (5107 sq kms): Malerkotla; Barnala; Sunam.

RAJASTHAN (342,239 sq kms, 26 districts, 199 taluks)

Ajmer (8481 sq kms): Beawar; Kishangarh; Kekri; Sarwar.

Alwar (8380 sq kms): Lachhmangarh; Bahror; Rajgarh; Kishangarhbas; Mandawar; Tijara; Bansur; Ramgarh; Thanagazi.

Banswara (5037 sq kms): Ghatal; Bagidora; Kushalgarh; Garhi.

Barmer (28,387 sq kms): Chauhtan; Pachpadra; Siwan; Shiv.

Bharatpur (8100 sq kms): Dholpur; Kaman; Wer; Bayana; Rupbas; Dig; Pahri; Nagar; Baseri; Nadba; Rajakhera.

Bhilwara (10,455 sq kms): Mandalgarh; Asind; Mandal; Jahazpur; Shahpura; Kotri; Sahara; Hurda; Banera; Raipur.

Bikaner (27,244 sq kms): Nokha; Lunkaransar; Kolayat.

Bundi (5550 sq kms): Keshoraipatan; Hindoli; Naenwa.

Chittaurgarh (10,856 sq kms): Pratapgarh; Begun; Kapasan; Nimbahera; Gangrar; Chhoti Sadri; Bari Sadri; Bhadesar; Dungla; Rashmi.

Churu (16,830 sq kms): Sujangarh; Rajgarh; Sardarshahr; Ratangarh; Shri Dungargarh; Taranagar.

Dungarpur (3770 sq kms): Sagwara; Aspur.

Ganganagar (20,634 sq kms): Nohar; Suratgarh; Hanumangarh; Anupgarh; Bhadra; Raisinghnagar; Padampur; Karanpur; Sadulsahar; Tibi; Sangaria.

Jaipur (19,527 sq kms): Amer; Phulera; Viratnagar; Jamwa Ramgarh; Idausa; Kotputli; Bhesawa; Lalsot; Sanganer; Basi; Sikrai; Dudu; Chaksu; Phagi.

Jaisalmer (38,401 sq kms): Pokaran.

Jalor (10,640 sq kms): Bhinmal; Sanchore; Ahor.

Jhalawar (6219 sq kms): Jhalrapatan; Aklera; Pirawa; Khanpur; Gangdhar; Pachpahar.

Jhunjhunun (5928 sq kms): Udaipur; Khetri; Chirawa.

Jodhpur (22,850 sq kms): Bilara; Osian; Phalodi; Shergarh.

Kota (12,436 sq kms): Ladpura; Ramganj Mandi; Pipalda; Mangrol; Sangod; Baran; Digod; Chhipabarod; Atru; Kishanganj; Chhabra; Shahbad.

Nagaur (17,718 sq kms): Parbatsar; Merta; Didwana; Nawa; Degana; Jayal; Ladnu.

Pali (12,387 sq kms): Bali; Desuri; Marwar Junction; Sojat; Jaitaran; Raipur.

Sawai Madhopur (10,527 sq kms): Karauli; Hindaun; Gangapur; Todabhim; Bonali; Mahwa; Sapotra; Bamanwas; Nandahti; Khandar.

Sikar (7732 sq kms): Sri Madhopur; Neem Kathana; Dauta Ramgarh; Lachhmangarh; Fatehpur.

Sirohi (5136 sq kms): Pindwara; Abu Road; Reodar; Sheoganj.

Tonk (7194 sq kms): Malpura; Deoli; Nawal; Todaraisingh; Uniaro.

Udaipur (17,279 sq kms): Vallabhnagar; Nathdwara; Kherwara; Mavli; Sarada; Salumbar; Dariawad; Rajsamand; Jhadol; Gogunda; Kotra; Kumbhalgarh; Bhim; Railmagra; Amet; Devgarh.

Note: 27 districts in 1985. Bharatpur was divided into Daulpur and Bharatpur.

SIKKIM (7096 sq kms, 4 districts, 4 taluks)

East Sikkim (954 sq kms, *Gangtok*).

North Sikkim (4226 sq kms, *Mangan*).

South Sikkim (750 sq kms, *Namchi*).

West Sikkim (1166 sq kms, *Gesing*).

TAMIL NADU (130,058 sq kms, 16 districts, 159 taluks)

Chengalpattu (7863 sq kms, *Kanchipuram*): Saidapet; Sriperumbudur; Chengalpattu; Madurantakam; Tiruvallur; Ponneri; Tiruttani; Pallipattu; Gummidipundi; Uttiramerur; Uttukkottai.

Coimbatore (7469 sq kms): Palladam; Pollachi; Udumalaippettai; Avanashi; Mettupalaiyam.

Dharmapuri (9622 sq kms): Krishnagiri; Harur; Denkanikota; Hosur; Palakkodu; Uttangarai; Pennagaram.

Kanniyakumari (1684 sq kms, *Nagercoil*): Kalkulam; Vilavankod; Agastiswaram; Thovala.

Madras (170 sq kms).

Madurai (12,624 sq kms): Madurai South; Dindigul; Uttamapalaiyam; Periyakulam; Nilakkottai; Palani; Usilampatti; Tirumangalam; Vedasandur; Melur; Madurai North; Nattam; Kodaikanal.

Nilgiri (2549 sq kms, *Udagamandalam*): Coonoor; Gudalur; Kotagiri.

North Arcot (12,268 sq kms, *Vellore*): Gudiyattam; Vaniyambadi; Tiruvannamalai; Arakkonam; Tiruppattur; Polur; Chengam; Walajapet; Vandavasi; Cheyyar; Arani; Arcot.

Periyar (8209 sq kms, *Erode*): Dharapuram; Bhavani; Gopichettipalaiyam; Perundurai; Satyamangalam.

Pudukkottai (4661 sq kms): Kulattur; Tirumayam; Alangudi; Arantangi; Avadaiarkovil.

Ramanathapuram (12,590 sq kms): Rajapalaiyam; Karaikkudi; Sivakasi; Aruppukkottai; Virudhunagar; Srivilliputtur; Paramakkudi; Devakottai; Rameswaram; Kilakkarai; Sattur; Sivaganga; Tiruppattur; Ilaiyankudi; Manamadurai; Tondi; Tiruppuvanam; Mandapam; Kamudi; Singampunari; Anaiyur; Pallapatti.

Salem (8650 sq kms): Namakkal; Attur; Tiruchengodu; Sankaridurg; Omalur; Mettur; Rasipuram; Yercaud.

South Arcot (10,895 sq kms, *Cuddalore*): Kallakkurichchi; Villupuram; Tindivanam; Panruti; Chidambaram; Gingee; Tirukkoyilur; Vriddhachalam; Ulundurpet; Kattumannar Koil; Tittakudi.

Thanjavur (8280 sq kms): Mayiladuthurai; Mannargudi; Nannilam; Kumbakonam; Pattukkottai; Papanasam; Sirkazhi; Orattanadu; Nagappattinam; Tirutturaippundi; Vedaranniyam; Thiruvidaimarudur; Thiruvarur; Thiruvaiyaru; Peravurani.

Tiruchchirappalli (11,095 sq kms): Karur; Perambalur; Lalgudi; Kulittalai; Udaiyarpalaiyam (*Jayamkondacholapuram*); Ariyalur; Manapparai; Musiri; Turaiyur.

Tirunelveli (11,429 sq kms): Tenkasi; Nanguneri; Ambasamudram; Tuticorin; Sankarankovil; Tiruchchendur; Kovilpatti; Srivaikuntam; Villattikulam; Sivagiri; Ottappidaram; Sattankulam; Shencottah.

Note: 18 districts in 1985. Ramanathapuram was divided into Kamarajar, Pasumpon Muthuramalingam and Ramanathapuram.

TRIPURA (10,486 sq kms, 3 districts, 10 taluks)

North Tripura (3872 sq kms, *Dharmanagar*): Dharmanagar; Kamalpur.

South Tripura (3581 sq kms, *Udaipur*): Amarpur Sub. Div.; Belonia Sub. Div.; Sabrum Sub. Div.

West Tripura (3033 sq kms, *Agartala*): Khowai Sub. Div.; Sonamura Sub. Div.

UTTAR PRADESH (294,411 sq kms, 56 districts, 248 taluks)

Agra (4805 sq kms): Firozabad; Etmadpur; Kiraoli (*Fatehpur Sikri*); Kheragarh; Bah; Fatehabad.

Aligarh (5019 sq kms): Koil; Hathras; Atrauli; Khair; Sikandra Rao; Iglas.

Allahabad (7261 sq kms): Chail; Handia; Karchana; Meja; Soraon; Phulpur; Manjhanpur; Sirathu.

Almora (5385 sq kms): Ranikhet; Bageshwar.

Azamgarh (5740 sq kms): Muhammadabad; Phulpur; Ghosi; Lalganj; Sagri.

Bahraich (6877 sq kms): Nanpara; Kaisarganj; Bhinga.

Ballia (3189 sq kms): Rasra; Bansdih.

Banda (7624 sq kms): Baberu; Karwi; Narani; Mau.

Bara Banki (4401 sq kms): Ram Sanehighat; Fatehpur; Nawabganj; Haidargarh.

Bareilly (4120 sq kms): Baheri; Aonla. Faridpur; Nawabganj.

Basti (7228 sq kms): Khalilabad; Domariaganj; Haraiya; Naugarh; Bansi.

Bijnor (4848 sq kms): Dhampur; Nagina; Najibabad.

Budaun (5168 sq kms): Bisauli; Dataganj; Sahaswan; Gunnaur.

Bulandshahr (4352 sq kms): Anupshahr; Khurja; Sikandarabad.

Chamoli (9125 sq kms, *Gopeshwar*): Karnaprayag; Okhimath; Joshimath.

Dehra Dun (3088 sq kms): Chakrata.

Deoria (5445 sq kms): Padrauna; Salempur; Hata.

Etah (4446 sq kms): Kasganj; Aliganj; Jalesar.

Etawah (4326 sq kms): Auraiya; Bharthana; Bidhuna.

Faizabad (4511 sq kms): Akbarpur; Tanda; Bikapur.

Farrukhabad (4274 sq kms, *Fatehgarh*): Farrukhabad; Chhibramau; Kannauj; Kaimganj.

Fatehpur (4152 sq kms): Bindki; Khaga.

Garhwal (5440 sq kms, *Pauri*): Lansdowne; Kotdwara.

Ghaziabad (2590 sq kms): Hapur; Dadri; Garhmuktesar.

Ghazipur (3377 sq kms): Saidpur; Muhammadabad; Zamania.

Gonda (7352 sq kms): Balrampur; Utraula; Tarabganj.

Gorakhpur (6272 sq kms): Maharaj Ganj; Bansgaon; Pharenda.

Hamirpur (7165 sq kms): Rath; Maudaha; Kulpahar; Mahoba; Charkhari.

Hardoi (5986 sq kms): Sandila; Bilgram; Shahabad.

Jalaun (4565 sq kms, *Orai*): Jalaun; Konch; Kalpi.

Jaunpur (4038 sq kms): Shahganj; Mariahu; Machhlishahr; Kerakat.

Jhansi (5024 sq kms): Mauranipur; Moth; Garautha.

Kanpur (6176 sq kms): Ghatampur; Bilhaur; Derapur; Akbarpur; Bhognipur.

Kheri (7680 sq kms, *Lakhimpur*): Mohamdi; Nighasan; Dhaurhra.

Lalitpur (5039 sq kms): Mahrauni; Talbahat.

Lucknow (2528 sq kms): Malihabad; Mohanlalganj.

Mainpuri (4343 sq kms): Bhongaon; Shikohabad; Jasrana; Karhal.

Mathura (3811 sq kms): Sadabad; Mat; Chhata.

Meerut (3911 sq kms): Baghpat; Sardhana; Mawana.

Mirzapur (11,310 sq kms): Robertsganj; Chunhar; Dudhi.

Moradabad (5967 sq kms): Sambhal; Amroha; Bilari; Hasanpur; Thakurdwara.

Muzaffarnagar (4176 sq kms): Kairana; Jansath; Budhana.

Naini Tal (6794 sq kms): Kashipur; Kichha; Haldwani; Khatima; Sitarganj.

Pilibhit (3499 sq kms): Bisalpur; Puranpur.

Pithoragarh (8856 sq kms): Champawat; Didihat; Dharchula; Munsyari.

Pratapgarh (3717 sq kms, *Bela*): Kunda; Patti.

Rae Bareli (4609 sq kms): Dalmau; Salon; Maharajganj.

Rampur (2367 sq kms): Suar; Milak; Shahabad; Bilaspur.

Saharanpur (5595 sq kms): Roorkee; Deoband; Nakur.

Shahjahanpur (4575 sq kms): Pawayan; Tilhar; Jalalabad.

Sitapur (5743 sq kms): Misrikh; Biswan; Mahmudabad; Sidhauli.

Sultanpur (4436 sq kms): Kadipur; Musafirkhana; Amethi.

Tehri Garhwal (4421 sq kms, *Narendranagar*): Devaprayag; Pratapnagar; Tehri.

Unnao (4558 sq kms): Purwa; Hasanganj; Safipur.

Uttarkashi (8016 sq kms): Dunda; Bhatwari; Rajgarhi; Purola.

Varanasi (5091 sq kms): Chandauli; Gyanpur; Chakia.

Note: 57 districts in 1985. Kanpur was divided into Kanpur (Urban) and Kanpur (Rural).

WEST BENGAL (88,752 sq kms, 16 districts, 307 taluks)

Bankura (6882 sq kms): Raipur; Onda; Bishnupur; Borjora; Chhatna; Khatra; Ganga Jalghati; Sonamukhi; Kotalpur; Patrasaer; Indas; Indpur; Joypur; Saltora; Taldangra; Simlapal; Ranibandh; Mejia.

Barddhaman (7024 sq kms): Durgapur; Memari; Kalna; Katoya; Asansol; Purbasthali; Rayna; Kulti; Andal; Ketugram; Bhatar; Mangal Kot; Ausgram; Galsi; Jamuria;

Jamalpurganj; Monteswar; Raniganj; Hirapur; Khandaghosh; Kaksa; Bud Bud; Barabani; Salanpur; Faridpur; Chittaranjan; New Township; Coke Oven.

Birbhum (4545 sq kms, *Siuri*): Rampurhat; Nalhati; Murarai; Mayureswar; Bolpur; Samthia; Nanur; Dubrajpur; Labpur; Khoyrasole; Ilambazar; Muhammadbazar; Rajnagar.

Calcutta (104 sq kms).

Darjiling (3149 sq kms): Shiliguri; Kalimpang; Phansidewa; Karsiyang; Naksalbari; Kharibari; Jore Bungalow; Rangli Rangliot; Pulbazar; Gorubathan; Mirik; Sukhiapokhri.

Haora (1467 sq kms): Haora (M.C.); Uluberia; Amta; Shampur; Bagna; Bally; Domjur; Sankrail; Jagatballabhpur; Panchla; Udaynarayanpur; Bawria; Liluah.

Hugli (3149 sq kms, *Chunchura*): Shrirampur; Khanakul; Chanditala; Arambag; Dhaniakhali; Goghat; Pandua; Singur; Uttarpara; Haripal; Bhadreswar; Balagarh; Magra; Jangipara; Tarakeswar; Pursurah; Chandannagar; Polba; Dadpur.

Jalpaiguri (6227 sq kms): Alipurduar; Mal; Mainaguri; Rajganj; Falakata; Kalchini; Banarhat; Dhubgari; Kumargram; Mitiali; Birpara; Nagrakara; Madarihat.

Koch Bihar (3387 sq kms): Dinhata; Matabhanga; Tufanganj; Sital Kuchi; Mekliganj; Haldibari; Sitai.

Maldah (3733 sq kms, *Ingraj Bazar*): Kalia Chak; Ratua; Harishchandrapur; Kharba; Gajol; Manik Chak; Habibpur; Maldah; Bamangola.

Medinipur (14,081 sq kms): Panskura; Kanthi; Nandigram; Daspur; Tamluk; Mahishadal; Bhagabanpur; Kharagpur Town; Egra; Binpur; Pataspur; Kharagpur; Kesapur; Garhbeta; Chandrakona; Ghatal; Ramnagar; Debra; Sabang; Khejuri; Sutabatal; Dantan; Jhargram; Gopiballabhpur; Mayna; Belda; Goaltor; Pingla; Salbani; Narayangarh; Keshiyari; Nayagram; Sankrail; Mohanpur; Haldia; Durgachak; Digha.

Murshidabad (5324 sq kms, *Baharampur*): Beldanga; Raghunathganj; Bharatpur; Suti; Raninagar; Domkal; Bhagwangola; Khargram; Barwan; Kandi; Shamsherganj; Sagardighi; Lalgola; Hariharpara; Jalangi; Farrakka; Nawda; Nabagram; Murshidabad; Jiaganj.

Nadia (3927 sq kms, *Krishnanagar*): Ranaghat; Chakdaha; Karimpur; Tehata; Nakasipara; Shantipur; Kaliganj; Nabadwip; Chapra; Hanskhali; Haringhata; Kalyani; Krishnaganj.

Puruliya (6259 sq kms): Jhaida; Manbazar; Raghunathpur; Kashipur; Barabazar; Para; Puncha; Hura; Arsa; Balarampur; Baghmundi; Jaipur; Nituria; Banduan; Santuri; Santaldih.

Twenty-Four Parganas (14,136 sq kms, *Alipur*): Behala; Barasat; Habra; Khardaha; Jagatdal; Magrahat; Bangaon; Basirhat; Budge-Budge; Bishnupur; Dum Dum; Jaynagar; Kasba; Canning; Bhangar; Mathurapur; Titagarh; Sonarpur; Belgharia; Baruipur; Diamond Harbour; Bijpur; Pathar Pratima; Maheshtola; Gaighata; Garden Reach; Baduria; Sandeshkhali; Deganga; Basanti; Naihati; Baranagar; Gosaba; Kulpi; Kakdwip; Airport; Rajarhat; Phalta; Hasnabad; Swarupnagar; Noapara; Kultali; Bagdah; Harua; Mandir Bazar; Sagore; Amdanga; Hingalganj; Minakhan; Namkhana; Metiabruz; Salt Lake; Barakpur; Nimta; Regent Park; Jadabpur; Tiljala Lake.

West Dinajpur (5358 sq kms, *Balurghat*): Raiganj; Itahar; Islampur; Goalpokhar; Karandighi; Kaliyaganj; Gangarampur; Tapan; Banshihari; Chopra; Kushmandi; Chakalia; Kumarganj; Hemtabad; Hilli.

ANDAMAN & NICOBAR ISLANDS (8249 sq kms, 2 districts, 7 taluks)

Andamans (6408 sq kms, *Port Blair*): Ferrargunj; Rangat; Diglipur; Mayabunder.

Nicobars (1841 sq kms, *Car Nicobar*): Nancowry.

ARUNACHAL PRADESH (83,743 sq kms, 9 districts, 97 taluks)

Dibang Valley (13,029 sq kms, *Anini*): Roing; Dambuk; Desali; Etalin; Anelin.

East Kameng (4134 sq kms, *Seppa*): Chayengtajo; Bameng; Pipu-Dipu; Khenewa; Pakke-Deshang; Lada; Seijosa.

East Siang (6512 sq kms, *Pasighat*): Mariyang; Mebo; Nari; Boleng; Pangin; Yingkiong; Karko.

Lohit (11,402 sq kms, *Tezu*): Namsai; Lekang; Chowkham; Hauling; Wakro; Hawai; Chaglongam; Goiliang; Walong; Kibithoo.

Lower Subansiri (13,010 sq kms, *Ziro*): Old Itanagar; Nyapin; Palin; Raga; Tali; New Itanagar; Sagalee; Balijan; Damin; Koloriang; Doimukh; Kimin; Mengio; Sarli.

Tirap (7024 sq kms, *Khonsa*): Miao; Diyun; Changlang; Niausa; Pongchou; Bordumsa; Namsang; Laju; Kanubari; Nampong; Wakka; Vijoynagar; Manmao.

Upper Subansiri (7032 sq kms, *Daporijo*): Dumporijo; Taliha; Giba; Siyum; Nacho; Limeking; Taksing.

West Kameng (9594 sq kms, *Bomdila*): Dirang; Tawang; Kalaktang; Lumla; Mukto; Nafra; Thrizino; Bhalukpong; Zemithang; Thingbu.

West Siang (12,006 sq kms, *Along*): Basar; Rumgong; Likabali; Darak; Tirbin; Tuting; Liromoba; Kaying; Mechuka; Monigong; Gensi; Payum; Tato; Singa; Gelling.

Note: 10 districts in 1985. Tawang, a new district for which boundaries are not available, was also formed.

CHANDIGARH (114 sq kms, 1 district, 1 taluk).

DADRA & NAGAR HAVELI (491 sq kms, 1 district, 1 taluk).

Dadra & Nagar Haveli (491 sq kms, *Silvassa*).

DELHI (1483 sq kms, 1 district, 1 taluk).

GOA, DAMAN & DIU (3814 sq kms, 3 districts, 13 taluks)

Daman (72 sq kms).

Diu (40 sq kms).

Goa (3702 sq kms, *Panaji*): Salcete; Bardez; Ponda; Marmagao; Bicholim; Pernem; Sanguem; Quepem; Satari; Konkan.

LAKSHADWEEP (32 sq kms, 1 district, 4 taluks)

Lakshadweep (32 sq kms, *Kavaratti*): Amindivi; Andrott (Androth); Minicoy.

MIZORAM (21,081 sq kms, 3 districts, 22 taluks)

Aizawl (12,588 sq kms): Tlangnuam; Khawzawl; Thingdawl; East Lungdar; Lokicherra; Serchhip; Ngopa; Thingsulthliah; Darlawn; West Phaileng; Aibawk; Reiek.

Chhimtuipui (3957 sq kms, *Saiha*): Tuipang; Lawngtlai; Chawngte; Sansau.

Lunglei (4536 sq kms): Lungsen; Hnahthial; West Bunghmun.

PONDICHERRY (492 sq kms, 4 districts, 15 taluks)

Karaikal (160 sq kms): Tirunallar; Tirumalairayan Pattinam; Kottucherry; Neravy; Nedungadu.

Mahe (9 sq kms).

Pondicherry (293 sq kms): Ozhukarai; Villianur; Mannadipet; Bahour; Ariankuppam; Nettapakkam.

Yanam (30 sq kms).

PARLIAMENTARY CONSTITUENCIES

India is a secular democratic republic with a parliamentary form of government elected on the basis of universal adult franchise at the Centre (Parliament) and in the states (Assemblies). It is the world's largest democracy (235 million votes were polled in the 1984 parliamentary elections).

The central legislature (Parliament) has two houses — the Lok Sabha (the House of the People) and the Rajya Sabha (the Council of States). The Constitution provides that the Lok Sabha shall consist of not more than 545 Members of Parliament (in 1986 there were 544). The number of representatives elected by each state to the Lok Sabha depends on the population of the state. Between 500,000 and 750,000 voters form one unit or constituency in a state, and the number of representatives a state may elect to the Lok Sabha is equal to the number of constituencies in that state. Uttar Pradesh has 85 constituencies, being the most populous state, but Madhya Pradesh, which is the largest in area, has 40.

The Rajya Sabha shall consist of not more than 250 members (its strength in 1986 was 244). Twelve members are nominated by the President to represent the arts, sciences, literature and social service. The remaining members are elected by members of the State Legislative Assemblies. The number of representatives from each state is determined in accordance with a system of proportional representation (thus Uttar Pradesh has 34 representatives and Tripura 1) by means of a single transferable vote. The representatives of the union territories are chosen in such manner as Parliament may prescribe.

The President, the executive head of the Union, is elected by an electoral college consisting of the members of both the houses of the Central Legislature as well as the Legislative Assemblies of the states in accordance with a system of proportional representation by means of a single transferable vote.

The Prime Minister is the leader of the majority party in the Lok Sabha. Members of the Lok Sabha have a tenure of five years, after the lapse of which the house is dissolved and elections held. The Rajya Sabha, however, is a permanent house, though one-third of its members retire every second year. All members are elected for a term of six years.

Every state has a Legislative Assembly and a few, namely, Bihar, Jammu & Kashmir, Karnataka, Maharashtra and Uttar Pradesh, also have Legislative Councils (an upper house). Parliament may create or dissolve the Legislative Council in any state if the proposal is supported by the Legislative Assembly concerned.

Every state elects its own Legislative Assembly. The number of Members (MLAs) of each assembly is fixed by Parliament, which divides each state into constituencies, each having a population of no more than 45,000. There is a certain per cent of seats reserved for Scheduled Castes and Scheduled Tribes (see page 27) in every state. The term of an Assembly is normally five years.

Legislative Councils are generally composed of members elected by members of the Legislative Assembly, by local self-governing bodies, by registered university graduates and registered secondary school and university teachers, and members nominated by the governor for their special contribution to the arts, sciences, the co-operative movement or social service. The term of each member is six years, though one-third have to retire every second year, thus ensuring that the house is never dissolved.

The Governor, the executive head of the state, is appointed by the President of the Union. The leader of the majority party in the Legislative Assembly becomes the Chief Minister.

The union territories are administered by the President through Lieutenant Governors, except in Chandigarh, where the administrator is designated Chief Commissioner. Some of the union territories (Arunachal Pradesh, Goa and Pondicherry) have Legislative Assemblies and councils of ministers. Delhi has a Metropolitan Council and an Executive Council. The Andaman and Nicobar Islands have a Pradesh Council and Councillors appointed from among the members of the Pradesh Council.

India has a multi-party system. In 1986 there were seven national political parties with an all-India base and 27 state parties (regional parties each recognized by less than four states of the Union).

In the Parliamentary elections held in December 1984, the Congress(I) Party polled the highest proportion of votes ever (49.97 per cent of the total votes polled) and won 403 of the 513 seats contested. (Elections were not held in 14 centres in Assam, 13 in Punjab and 1 each in Himachal Pradesh and Jammu & Kashmir.) Ten states, Andhra Pradesh, Assam, Jammu & Kashmir, Karnataka, Kerala, Punjab, Sikkim, Tamil Nadu, Tripura, and West Bengal, have non-Congress(I) governments (at the end of March 1987).

Both the parliamentary and assembly elections in India are based on adult suffrage; all Indian citizens who are 21 years of age are eligible to vote, provided they are not insolvent, insane or convicted of any serious offence. (There were 397.8 million voters in the electoral rolls at the beginning of 1985.)

The Election Commission, a constitutional authority, set up in pursuance of Article 324(1) of the Constitution, supervises the preparation of electoral rolls and controls and directs the conduct of elections, which are held by secret ballot.

Sources: 1. *India 1985*, Ministry of Information & Broadcasting.
2. *Delimitation of Parliamentary and Assembly Constituencies Order, 1976*, Election Commission.

Table 4   PARTY POSITION IN ASSEMBLIES

	Year[1]	Total seats	Bharatiya Janata Party	Communist Party of India	Communist Party of India (Marxist)	Indian Congress (Socialist)	Congress (I)	Janata Party	Lok Dal[2]	Regional parties	Others[3]
States											
Andhra Pradesh	1985	294	8	11	11	—	49	2	—	202[4]	11
Assam	1985	126	—	—	2	4	25	—	—	64[5]	31
Bihar	1985	324	12	12	1	1	192	11	38	—	57
Gujarat	1985	182	11	—	—	—	149	14	—	—	8
Haryana	1982	90	6	—	—	—	36	1	31	—	16
Himachal Pradesh	1985	68	7	—	—	—	55	—	1	—	5
Jammu & Kashmir	1987	76	2	—	—	—	24	—	—	36[6]	14
Karnataka	1985	224	2	4	2	—	66	139	—	—	11
Kerala	1987	140	—	16	36	6	33	7	1	—	41
Madhya Pradesh	1985	320	58	—	—	1	250	5	—	—	6
Maharashtra	1985	288	16	2	2	54	162	20	—	—	32
Manipur	1983	60	—	1	—	—	27	4	—	—	28
Meghalaya	1983	60	—	—	—	—	25	—	—	16[7]	19
Nagaland	1983	60	—	—	—	—	36	—	—	—	24
Orissa	1985	147	1	1	—	—	117	20	—	—	8
Punjab	1985	117	4	1	—	—	32	1	—	73[8]	6
Rajasthan	1985	200	38	—	1	—	113	10	27	—	11
Sikkim	1985	32	—	—	—	—	1	—	—	30[9]	1
Tamil Nadu	1984	234	—	2	5	—	62	3	—	133[10]	29
Tripura	1983	60	—	—	37	—	12	—	—	—	11
Uttar Pradesh	1985	425	16	6	2	4	266	19	85	—	27
West Bengal	1987	294	—	11	187	—	40	—	—	—	56
Union Territories											
Andaman & Nicobar		—	—	—	—	—	—	—	—	—	—
Arunachal Pradesh	1983	30	1	—	—	—	21	—	—	4[11]	4
Chandigarh		—	—	—	—	—	—	—	—	—	—
Dadra & Nagar Haveli		—	—	—	—	—	—	—	—	—	—
Delhi	1983	56	19	—	—	—	34	1	2	—	—
Goa, Daman & Diu	1983	30	—	—	—	—	18	—	—	8[12]	4
Lakshadweep		—	—	—	—	—	—	—	—	—	—
Mizoram	1983	33	—	—	—	—	8	—	—	24[13]	1
Pondicherry	1985	30	—	—	—	—	15	2	—	6[10]	7
INDIA		4,000	201	67	286	70	1,868	259	185	596	468

Note: [1]Election year   [2] Dalit Mazdoor Kisan Party   [3] Including countermanded seats & Independent members.
[4] Telugu Desam, [5] Assam Gana Parishad, [6] National Conference, [7] All Party Hill Leaders Conference, [8] Akali Dal, [9] Sikkim Sangram Parishad, [10] All India Anna Dravida Munnetra Kazhagam, [11] People's Party of Arunachal Pradesh, [12] Maharashtrawadi Gomantak Party, [13] People's Conference.
Given above is the position of political parties in the various state and union territory assemblies after the last territory-wide election held in each. The date of the election is given beside the name of the territory. It should be noted that the union territories of Andaman & Nicobar, Chandigarh, Dadra & Nagar Haveli and Lakshadweep do not have assembly elections and are administered by a Lt. Governor/Administrator appointed by the Centre.

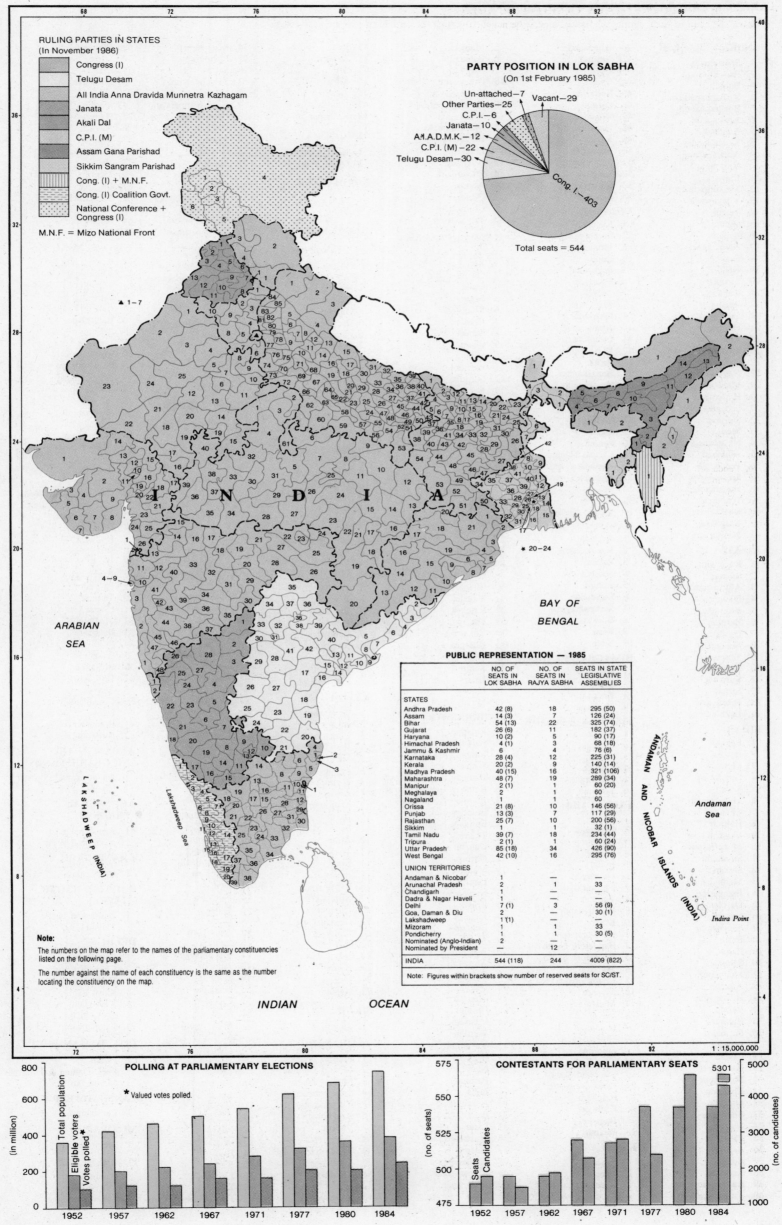

RULING PARTIES IN STATES
(In November 1986)

- Congress (I)
- Telugu Desam
- All India Anna Dravida Munnetra Kazhagam
- Janata
- Akali Dal
- C.P.I. (M)
- Assam Gana Parishad
- Sikkim Sangram Parishad
- Cong. (I) + M.N.F.
- Cong. (I) Coalition Govt.
- National Conference + Congress (I)

M.N.F. = Mizo National Front

PARTY POSITION IN LOK SABHA
(On 1st February 1985)

- Un-attached—7
- Other Parties—25
- Vacant—29
- C.P.I.—6
- Janata—10
- A.I.A.D.M.K.—12
- C.P.I. (M)—22
- Telugu Desam—30
- Cong. I.—403

Total seats = 544

PUBLIC REPRESENTATION — 1985

	NO. OF SEATS IN LOK SABHA	NO. OF SEATS IN RAJYA SABHA	SEATS IN STATE LEGISLATIVE ASSEMBLIES
STATES			
Andhra Pradesh	42 (8)	18	295 (50)
Assam	14 (3)	7	126 (24)
Bihar	54 (13)	22	325 (74)
Gujarat	26 (6)	11	182 (37)
Haryana	10 (2)	5	90 (17)
Himachal Pradesh	4 (1)	3	68 (18)
Jammu & Kashmir	6	4	76 (6)
Karnataka	28 (4)	12	225 (31)
Kerala	20 (2)	9	140 (14)
Madhya Pradesh	40 (15)	16	321 (106)
Maharashtra	48 (7)	19	289 (34)
Manipur	2 (1)	1	60 (20)
Meghalaya	2	1	60
Nagaland	1	1	60
Orissa	21 (8)	10	146 (56)
Punjab	13 (3)	7	117 (29)
Rajasthan	25 (7)	10	200 (56)
Sikkim	1	1	32 (1)
Tamil Nadu	39 (7)	18	234 (44)
Tripura	2 (1)	1	60 (24)
Uttar Pradesh	85 (18)	34	426 (90)
West Bengal	42 (10)	16	295 (76)
UNION TERRITORIES			
Andaman & Nicobar	1	—	—
Arunachal Pradesh	2	1	33
Chandigarh	1	—	—
Dadra & Nagar Haveli	1	—	—
Delhi	7 (1)	3	56 (9)
Goa, Daman & Diu	2	—	30 (1)
Lakshadweep	1 (1)	—	—
Mizoram	1	1	33
Pondicherry	1	1	30 (5)
Nominated (Anglo-Indian)	2	—	—
Nominated by President	—	12	—
INDIA	544 (118)	244	4009 (822)

Note: Figures within brackets show number of reserved seats for SC/ST.

ARABIAN SEA

INDIA

BAY OF BENGAL

LAKSHADWEEP (INDIA)

Lakshadweep Sea

ANDAMAN AND NICOBAR ISLANDS (INDIA)

Andaman Sea

Indira Point

INDIAN OCEAN

Note:

The numbers on the map refer to the names of the parliamentary constituencies listed on the following page.

The number against the name of each constituency is the same as the number locating the constituency on the map.

1 : 15,000,000

POLLING AT PARLIAMENTARY ELECTIONS

* Valued votes polled.

- Total population
- Eligible voters
- Votes polled

(in million)

1952 1957 1962 1967 1971 1977 1980 1984

CONTESTANTS FOR PARLIAMENTARY SEATS

5301

- Seats
- Candidates

(no. of seats) (no. of candidates)

1952 1957 1962 1967 1971 1977 1980 1984

LIST OF PARLIAMENTARY CONSTITUENCIES

Note: The name of each constituency is the same as the number locating the constituency on the map (page 13).

ANDHRA PRADESH
1. Srikakulam
2. Parvatipuram
3. Bobbili
4. Vishakhapatnam
5. Bhadrachalam
6. Anakapalle
7. Kakinada
8. Rajahmundry
9. Amalapuram
10. Narasapur
11. Eluru
12. Machilipatnam
13. Vijayawada
14. Tenali
15. Guntur
16. Bapatla
17. Narasarapet
18. Ongole
19. Nellore
20. Tirupati
21. Chittoor
22. Rajampet
23. Cuddapah
24. Hindupur
25. Anantapur
26. Kurnool
27. Nandyal
28. Nagarkurnool
29. Mahbubnagar
30. Hyderabad
31. Secunderabad
32. Siddipet
33. Medak
34. Nizamabad
35. Adilabad
36. Peddapalli
37. Karimnagar
38. Hanamkonda
39. Warangal
40. Khammam
41. Nalgonda
42. Mirialguda

ASSAM
1. Karimganj
2. Silchar
3. Autonomous District
4. Dhuburi
5. Kokrajhar
6. Barpeta
7. Guwahati
8. Mangaldai
9. Tezpur
10. Nowgong
11. Kaliabor
12. Jorhat
13. Dibrugarh
14. Lakhimpur

BIHAR
1. Bagaha
2. Bettiah
3. Mothari
4. Gopalganj
5. Siwan
6. Maharajganj
7. Chhapra
8. Hajipur
9. Vaishali
10. Muzaffarpur
11. Sitamarhi
12. Sheohar
13. Madhubani
14. Jhanjharpur
15. Darbhanga
16. Rosera
17. Samastipur
18. Barhi
19. Balia
20. Saharsa
21. Madhepura
22. Araria
23. Kishanganj
24. Purnia
25. Katihar
26. Rajmahal
27. Dumka
28. Godda
29. Banka
30. Bhagalpur
31. Khagaria
32. Munger
33. Begusarai
34. Nalanda
35. Patna
36. Ara
37. Buxar
38. Sasaram
39. Bikramganj
40. Aurangabad
41. Jahanabad
42. Nawada
43. Gaya
44. Chatra
45. Kodarma
46. Giridih
47. Dhanbad
48. Hazaribagh
49. Ranchi
50. Jamshedpur
51. Singhbhum
52. Khunti
53. Lohardaga
54. Palamu

GUJARAT
1. Kachchh
2. Surendranagar
3. Jamnagar
4. Rajkot
5. Porbandar
6. Junagadh
7. Amreli
8. Bhavnagar
9. Dhandhuka
10. Ahmadabad
11. Gandhinagar
12. Mahesana
13. Patan
14. Banaskantha
15. Sabarkantha
16. Kapadvanj
17. Dohad
18. Godhra
19. Kaira
20. Anand
21. Chhota Udepur
22. Vadodara (Baroda)
23. Bharuch (Broach)
24. Surat
25. Mandvi
26. Valsad (Bulsar)

HARYANA
1. Ambala
2. Kurukshetra
3. Karnal
4. Sonipet
5. Rohtak
6. Faridabad
7. Mahendragarh
8. Bhiwani
9. Hisar
10. Sirsa

HIMACHAL PRADESH
1. Shimla
2. Mandi
3. Kangra
4. Hamirpur

JAMMU & KASHMIR
1. Baramula
2. Srinagar
3. Anantnag
4. Ladakh
5. Udhampur
6. Jammu

KARNATAKA
1. Bidar
2. Gulbarga
3. Raichur
4. Koppal
5. Bellary
6. Davangere
7. Chitradurga
8. Tumkur
9. Chikballapur
10. Kolar
11. Kanakapura
12. Bangalore North
13. Bangalore South
14. Mandya
15. Chamrajnagar
16. Mysore
17. Mangalore
18. Udupi
19. Hassan
20. Chikmagalur
21. Shimoga
22. Kanara
23. Dharwad South
24. Dharwad North
25. Belgaum
26. Chikkodi
27. Bagalkot
28. Bijapur

KERALA
1. Kasaragod
2. Cannanore
3. Badagara
4. Kozhikode (Calicut)
5. Manjeri
6. Ponnani
7. Palghat
8. Ottapalam
9. Trichur
10. Mukundapuram
11. Ernakulam
12. Muvattupuzha
13. Kottayam
14. Idukki
15. Alleppey
16. Mavelikara
17. Adoor
18. Quilon
19. Chirayinkil
20. Trivandrum

MADHYA PRADESH
1. Morena
2. Bhind
3. Gwalior
4. Guna
5. Sagar
6. Khajuraho
7. Damoh
8. Satna
9. Rewa
10. Sidhi
11. Shahdol
12. Surguja
13. Raigarh
14. Janjgir
15. Bilaspur
16. Sarangarh
17. Raipur
18. Mahasamund
19. Kanker
20. Bastar
21. Durg
22. Raj Nandgaon
23. Balaghat
24. Mandla
25. Jabalpur
26. Seoni
27. Chhindwara
28. Betul
29. Hoshangabad
30. Bhopal
31. Vidisha
32. Rajgarh
33. Shajapur
34. Khandwa
35. Khargon
36. Dhar
37. Indore
38. Ujjain
39. Jhabua
40. Mandsaur

MAHARASHTRA
1. Rajapur
2. Ratnagiri
3. Raigad (Kolaba)
4. Bombay South
5. Bombay South Central
6. Bombay North Central
7. Bombay North East
8. Bombay North West
9. Bombay North
10. Thane
11. Dahanu
12. Nashik
13. Malegaon
14. Dhule
15. Nandurbar
16. Erandol
17. Jalgaon
18. Buldana
19. Akola
20. Washim
21. Amravati
22. Ramtek
23. Nagpur
24. Bhandara
25. Chimur
26. Chandrapur
27. Wardha
28. Yavatmal
29. Hingoli
30. Nanded
31. Parbhani
32. Jalna
33. Aurangabad
34. Beed (Bhir)
35. Latur
36. Osmanabad
37. Solapur
38. Pandharpur
39. Ahmadnagar
40. Kopargaon
41. Khed
42. Pune
43. Baramati
44. Satara
45. Karad
46. Sangli
47. Ichalkaranji
48. Kolhapur

MANIPUR
1. Inner Manipur
2. Outer Manipur

MEGHALAYA
1. Shillong
2. Tura

NAGALAND
1. Nagaland

ORISSA
1. Mayurbhanj
2. Baleshwar
3. Bhoadrak
4. Jajpur
5. Kendrapara
6. Cuttack
7. Jagatsinghpur
8. Puri
9. Bhubaneshwar
10. Aska
11. Brahmapur
12. Koraput
13. Nowrangpur
14. Kalahandi
15. Phulbani
16. Balangir
17. Sambalpur
18. Deogarh
19. Dhenkanal
20. Sundargarh
21. Kendujhar

PUNJAB
1. Gurdaspur
2. Amritsar
3. Tarn Taran
4. Jalandhar
5. Phillaur
6. Hoshiarpur
7. Rupnagar
8. Patiala
9. Ludhiana
10. Sangrur
11. Bathinda
12. Faridkot
13. Ferozpur

RAJASTHAN
1. Ganganagar
2. Bikaner
3. Churu
4. Jhunjhunun
5. Sikar
6. Jaipur
7. Dausa
8. Alwar
9. Bharatpur
10. Bayana
11. Sawai Madhopur
12. Ajmer
13. Tonk
14. Kota
15. Jhalawar
16. Banswara
17. Salumber
18. Udaipur
19. Chittaurgarh
20. Bhilwara
21. Pali
22. Jalor
23. Barmer
24. Jodhpur
25. Nagaur

SIKKIM
1. Sikkim

TAMIL NADU
1. Madras North
2. Madras Central
3. Madras South
4. Sriperumbudur
5. Chengalpattu
6. Arakkonam
7. Vellore
8. Tiruppattur
9. Vandavasi
10. Tindivanam
11. Cuddalore
12. Chidambaram
13. Dharmapuri
14. Krishnagiri
15. Rasipuram
16. Salem
17. Tiruchengodu
18. Nilgiri
19. Gobichettipalayam
20. Coimbatore
21. Pollachi
22. Palani
23. Dindigul
24. Madurai
25. Periyakulam
26. Karur
27. Tiruchchirappalli
28. Perambalur
29. Mayiladuthurai
30. Nagappattinam
31. Thanjavur
32. Pudukkotai
33. Sivaganga
34. Ramanathapuram
35. Sivakasi
36. Tirunelveli
37. Tenkasi
38. Tiruchchendur
39. Nagercoil

TRIPURA
1. Tripura West
2. Tripura East

UTTAR PRADESH
1. Tehri Garwal
2. Garhwal
3. Almora
4. Naini Tal
5. Bijnor
6. Amroha
7. Moradabad
8. Rampur
9. Sambhal
10. Budaun
11. Aonla
12. Bareilly
13. Pilibhit
14. Shahjahanpur
15. Kheri
16. Shahabad
17. Sitapur
18. Misrikh
19. Hardoi
20. Lucknow
21. Mohanlalganj
22. Unnao
23. Rae Bareli
24. Pratapgarh
25. Amethi
26. Sultanpur
27. Akbarpur
28. Faizabad
29. Bara Banki
30. Kaiserganj
31. Bahraich
32. Balrampur
33. Gonda
34. Basti
35. Domariaganj
36. Khalilabad
37. Bansgaon
38. Gorakhpur
39. Maharajganj
40. Padrauna
41. Deoria
42. Salempur
43. Ballia
44. Ghosi
45. Azamgarh
46. Lalganj
47. Machhlishahr
48. Jaunpur
49. Saidpur
50. Ghazipur
51. Chandauli
52. Varanasi
53. Robertsganj
54. Mirzapur
55. Phulphur
56. Allahabad
57. Chail
58. Fatehpur
59. Banda
60. Hamirpur
61. Jhansi
62. Jalaun
63. Ghatampur
64. Bilhaur
65. Kanpur
66. Etawah
67. Kannauj
68. Farrukhabad
69. Mainpuri
70. Jalesar
71. Etah
72. Firozabad
73. Agra
74. Mathura
75. Hathras
76. Aligarh
77. Khurja
78. Bulandshahr
79. Hapur
80. Meerut
81. Baghpat
82. Muzaffarnagar
83. Kairana
84. Saharanpur
85. Haridwar

WEST BENGAL
1. Koch Bihar
2. Alipurduars
3. Jalpaiguri
4. Darjiling
5. Raiganj
6. Balurghat
7. Maldah
8. Jangipur
9. Murshidabad
10. Baharampur
11. Krishnanagar
12. Navadwip
13. Barasat
14. Basirhat
15. Joynagar
16. Mathurapur
17. Diamond Harbour
18. Jadavpur
19. Barakpur
20. Dum Dum
21. Calcutta North West
22. Calcutta North East
23. Calcutta South
24. Haora
25. Uluberia
26. Serampore
27. Hugli
28. Arambagh
29. Panskura
30. Tamluk
31. Contai
32. Medinipur
33. Jhargram
34. Puruliya
35. Bankura
36. Bishnupur
37. Durgapur
38. Asansol
39. Barddhaman
40. Katwa
41. Bolpur
42. Birbhum

ANDAMAN AND NICOBAR ISLANDS
1. Andaman and Nicobar Islands

ARUNACHAL PRADESH
1. Arunachal East
2. Arunachal West

CHANDIGARH
1. Chandigarh

DADRA & NAGAR HAVELI
1. Dadra and Nagar Haveli

DELHI
1. East Delhi
2. Chandni Chowk
3. Delhi Sadar
4. Karol Bagh
5. New Delhi
6. Outer Delhi
7. South Delhi

GOA, DAMAN & DIU
1. Panaji (incl. Daman & Diu)
2. Marmagao

LAKSHADWEEP
1. Lakshadweep

MIZORAM
1. Mizoram

PONDICHERRY
1. Pondicherry (incl. Karaikal, Yanam & Mahe)

14

The People

This section comprises 41 maps & 54 charts. It offers a broad but detailed picture of Indian society and relates development with the overwhelming number of persons in every part of the country.

Note:

In this section, most of the data used is based on Census 1981 when there were 412 districts. The maps, however, show the districts as they existed at the end of 1985, when there were 439 districts (437 districts for which boundaries were available are shown on the maps). The change in the number of districts has been due to the division of 1981 districts into 2 or more smaller districts or the combination of some small districts into larger districts (as in Jammu & Kashmir). However, the 1981 data have been used to colour the 1985 configuration of districts on the maps. Such shading makes hardly any difference in cases where the shaded information on the maps is based on per cent or ratio values, such as density. However, in maps where absolute numbers are colour-shaded — pages 17, 51, 53, 55, 57 — there are likely to be differences in the data shown for the relevant districts and it is advisable that, in these cases, the districts be viewed in their 1981 configuration. Reference may be made to pages 8-11 to find which districts have been divided or enlarged and to obtain a picture of the 1981 configuration.

POPULATION

India, with 2.5 per cent of the total world area, has nearly 16 per cent of the world's population. It is, after China (983 million in 1981), the second most populous country in the world. In 1981 India had a population of 685 million. In 1986, India's estimated population was about 760 million.

The distribution of population within India varies widely. The most populous state is Uttar Pradesh, which is followed by Bihar, Maharashtra, West Bengal and Andhra Pradesh in that order. Kerala, the most densely populated state, however, ranks 12th. All the union territories together have less than the population of the 15th ranking state, Haryana.

The world's five most populous countries (China, India, USSR, USA and Indonesia) account for 51 per cent of the total world population. Uttar Pradesh alone has a population almost equal to that of Japan (the 7th most populated country) and more than that of Bangladesh, Nigeria or Pakistan, which follow Japan in terms of population. In fact, the majority of the states in India have populations greater than most of the nations in the world. Haryana, for instance, has a population greater than 127 countries, two-thirds of the nations in the world.

While the major administrative units of India themselves present a complex pattern of population distribution, the district distribution is still more complex. Most districts with low populations are in the north-east and north-west (in Sikkim, Arunachal Pradesh and Manipur; and in northern Himachal Pradesh and Jammu & Kashmir respectively). The districts in these regions generally have a population of 100,000-200,000, only two or three having 200,000-400,000 people. On the other hand, districts in the peninsula and in the Indo-Gangetic Plain have populations ranging between 800,000 and 3,000,000. The most populated districts, with populations between 6,000,000 and 11,000,000, are Greater Bombay, Delhi and 24 Parganas and Medinipur in West Bengal. According to the Census of India 1981, there were 39 districts with populations between 3,000,000 and 6,000,000. Of these districts, 18 were in the peninsula and the rest in the Indo-Gangetic Plain. With official policy being to make districts smaller, these 39 districts had become 48 in number by the end of 1985 and many of them may now be out of this population classification. But this only increases the number of districts with populations ranging from 1 to 3 million. The majority of districts in the country — 211 out of 439 (206 out of 412 in 1981) — are in the 1 to 3 million category.

There are wide and complex population differences of academic and societal importance in India, which it is difficult to demonstrate in a single map. Some of these differences are therefore presented in the maps that follow.

Note: All Census data are dated 1981 and refer to the 412 districts that then existed. The map shows the districts as they existed at the end of 1985, when there were 439 districts (437 districts for which boundaries were available are shown on the maps). Where 1981 districts have been broken up into two or more smaller districts or combined into larger districts (as in Jammu & Kashmir) the 1981 data is used to shade the 1985 configurations. Reference may be made to pages 8-11 to find which districts have been divided or enlarged. In the text, which refers to the districts as they existed in 1985, the 1981 status is given within brackets.

Sources: 1. *Primary Census Abstract Part II B(i)*, Census of India 1981.
2. World Bank's *World Development Report 1983*, Oxford University Press.
3. *Report of the Expert Committee on Population*, New Delhi, 1986.

POPULATION GROWTH

Table 5

	1971[1]			1981[1]			1986[2] Total (in m)	1990[2] Total (in m)	1995[2] Total (in m)	2000[2] Total (in m)
	Male	Female	Total	Male	Female	Total				
States										
Andhra Pradesh	22,008,663	21,494,045	43,502,708	27,108,922	26,440,751	53,549,673	58.8	63.2	68.3	72.7
Assam	7,885,064	7,072,478	14,957,542	10,467,461[2]	9,429,382[2]	19,896,843[2]	22.4	24.5	27.3	30.1
Bihar	28,846,944	27,506,425	56,353,369	35,930,560	33,984,174	69,914,734	77.7	84.7	94.2	103.9
Gujarat	13,802,494	12,894,981	26,697,475	17,552,640	16,533,159	34,085,799	37.7	40.4	43.3	45.9
Haryana	5,377,258	4,659,550	10,036,808	6,909,938	6,012,680	12,922,618	14.8	16.2	17.6	18.7
Himachal Pradesh	1,766,957	1,693,477	3,460,434	2,169,931	2,110,887	4,280,818	4.7	5.0	5.4	5.7
Jammu & Kashmir	2,458,315	2,158,317	4,616,632	3,164,660	2,822,729	5,987,389	6.7	7.3	8.0	8.7
Karnataka	14,971,900	14,327,114	29,299,014	18,922,627	18,213,087	37,135,714	41.4	44.6	48.3	51.5
Kerala	10,587,851	10,759,524	21,347,375	12,527,767	12,925,913	25,453,680	27.8	29.7	31.8	33.8
Madhya Pradesh	21,455,334	20,198,785	41,654,119	26,886,305	25,292,539	52,178,844	58.2	63.1	69.1	74.3
Maharashtra	26,116,351	24,295,884	50,412,235	32,415,126	30,369,045	62,784,171	69.7	74.2	79.7	85.7
Manipur	541,675	531,078	1,072,753	721,006	699,947	1,420,953	1.6	1.7	1.9	2.1
Meghalaya	520,967	490,732	1,011,699	683,710	652,109	1,335,819	1.5	1.7	1.9	2.1
Nagaland	276,084	240,365	516,449	415,910	359,020	774,930	0.9	1.1	1.3	1.5
Orissa	11,041,083	10,903,532	21,944,615	13,309,786	13,060,485	26,370,271	28.8	30.9	33.5	35.9
Punjab	7,266,515	6,284,545	13,551,060	8,937,210	7,851,705	16,788,915	18.5	19.6	20.5	21.8
Rajasthan	13,484,383	12,281,423	25,765,806	17,854,154	16,407,708	34,261,862	39.2	43.5	49.2	54.9
Sikkim	—	—	—	172,440	143,945	316,385	0.4	0.4	0.5	0.6
Tamil Nadu	20,828,021	20,371,147	41,199,168	24,487,624	23,920,453	48,408,077	52.6	55.7	59.2	62.4
Tripura	801,126	755,216	1,556,342	1,054,846	998,212	2,053,058	2.3	2.5	2.8	3.0
Uttar Pradesh	47,016,421	41,324,723	88,341,144	58,819,276	52,042,737	110,862,013	123.1	133.7	147.9	162.7
West Bengal	23,435,987	20,876,024	44,312,011	28,560,901	26,019,746	54,580,647	60.3	64.8	70.2	74.9
Union Territories										
Andaman & Nicobar	70,027	45,106	115,133	107,261	81,480	188,741	0.3	0.3	0.4	0.4
Arunachal Pradesh	251,231	216,280	467,511	339,322	292,517	631,839	0.7	0.8	0.9	1.0
Chandigarh	147,080	110,171	257,251	255,278	196,332	451,610	0.6	0.7	0.9	1.2
Dadra & Nagar Haveli	36,964	37,206	74,170	52,515	51,161	103,676	0.2	0.2	0.2	0.2
Delhi	2,257,515	1,808,183	4,065,698	3,440,081	2,780,325	6,220,406	7.6	8.9	10.7	12.8
Goa, Daman & Diu	431,214	426,557	857,771	548,450	538,280	1,086,730	1.3	1.4	1.5	1.6
Lakshadweep	16,078	15,732	31,810	20,377	19,872	40,249	0.04	0.05	0.05	0.05
Mizoram	—	—	—	257,239	236,518	493,757	0.6	0.7	0.8	0.9
Pondicherry	237,112	234,595	471,707	304,561	299,910	604,471	0.7	0.7	0.8	0.8
INDIA	283,936,614	264,013,195	547,949,809	354,397,884[2]	339,786,808[2]	685,184,692[2]	761.1	822.0	898.2	972.0

Note: 1971 data for Assam includes Mizoram. Sikkim was a protectorate of India till 1975, when it became a state of the Indian Union.

[1] Census
[2] Estimated

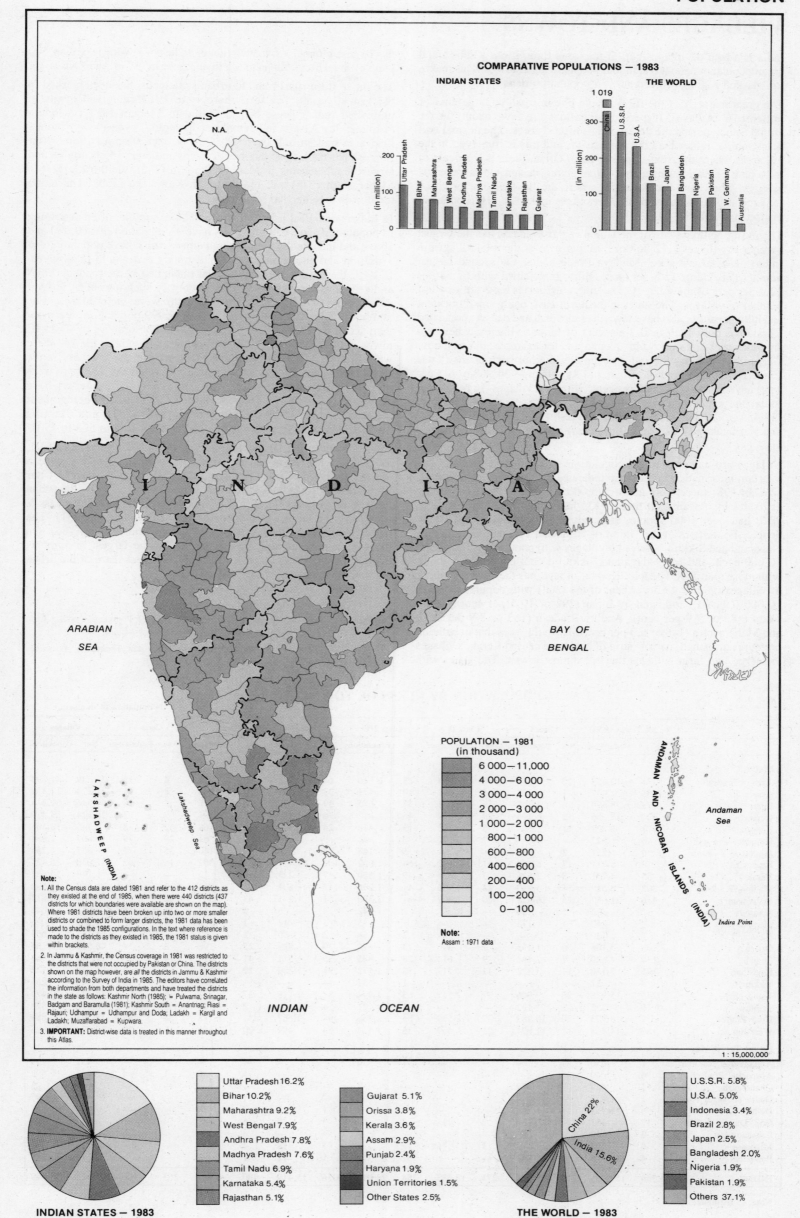

COMPARATIVE POPULATIONS — 1983

INDIAN STATES

(in million)

Uttar Pradesh, Bihar, Maharashtra, West Bengal, Andhra Pradesh, Madhya Pradesh, Tamil Nadu, Karnataka, Rajasthan, Gujarat

THE WORLD

1 019

(in million)

China, U.S.S.R., U.S.A., Brazil, Japan, Bangladesh, Nigeria, Pakistan, W. Germany, Australia

N.A.

I N D I A

ARABIAN SEA

BAY OF BENGAL

LAKSHADWEEP (INDIA)

Lakshadweep Sea

ANDAMAN AND NICOBAR ISLANDS (INDIA)

Andaman Sea

Indira Point

POPULATION — 1981
(in thousand)

6 000 – 11,000
4 000 – 6 000
3 000 – 4 000
2 000 – 3 000
1 000 – 2 000
800 – 1 000
600 – 800
400 – 600
200 – 400
100 – 200
0 – 100

Note:
Assam : 1971 data

INDIAN OCEAN

Note:

1. All the Census data are dated 1981 and refer to the 412 districts as they existed at the end of 1985, when there were 440 districts (437 districts for which boundaries were available are shown on the map). Where 1981 districts have been broken up into two or more smaller districts or combined to form larger districts, the 1981 data has been used to shade the 1985 configurations. In the text where reference is made to the districts as they existed in 1985, the 1981 status is given within brackets.

2. In Jammu & Kashmir, the Census coverage in 1981 was restricted to the districts that were not occupied by Pakistan or China. The districts shown on the map however, are *all* the districts in Jammu & Kashmir according to the Survey of India in 1985. The editors have correlated the information from both departments and have treated the districts in the state as follows: Kashmir North (1985); = Pulwama, Srinagar, Badgam and Baramulla (1981); Kashmir South = Anantnag; Riasi = Rajauri; Udhampur = Udhampur and Doda; Ladakh = Kargil and Ladakh; Muzaffarabad = Kupwara.

3. **IMPORTANT:** District-wise data is treated in this manner throughout this Atlas.

1 : 15,000,000

Uttar Pradesh 16.2%	Gujarat 5.1%	U.S.S.R. 5.8%
Bihar 10.2%	Orissa 3.8%	U.S.A. 5.0%
Maharashtra 9.2%	Kerala 3.6%	Indonesia 3.4%
West Bengal 7.9%	Assam 2.9%	Brazil 2.8%
Andhra Pradesh 7.8%	Punjab 2.4%	Japan 2.5%
Madhya Pradesh 7.6%	Haryana 1.9%	Bangladesh 2.0%
Tamil Nadu 6.9%	Union Territories 1.5%	Nigeria 1.9%
Karnataka 5.4%	Other States 2.5%	Pakistan 1.9%
Rajasthan 5.1%		Others 37.1%

China 22%

India 15.6%

INDIAN STATES — 1983
(total population: 733.2 millions)

THE WORLD — 1983
(total population: 4700 millions)

VILLAGES AND TOWNS

India is a land of villages but, at the same time, it has a substantial urban population, about one in four people living in towns. This makes for marked regional differences in population density.

The great majority of the rural people live in small or large nuclear settlements. Areas of dispersed habitations are few, mainly in the Himalayan zone and the desert and semi-arid areas. The normal unit is the *hamlet* rather than the *homestead*, and this is true even in the hill areas where the tribals predominate. Often, there is a process of 'bunching', with a few hamlets together getting designated as 'revenue' villages. There are also instances where the 'mother', or original, settlements give rise to hamlets, for example in the Kolli Hills of Salem district in Tamil Nadu.

The most populous state of the Union, Uttar Pradesh, has the largest number of villages (112,568 or 20.2 per cent of the total villages in India). The largest state, Madhya Pradesh, has the second largest number (71,352 or 12.8 per cent). Bihar ranks third with 67,546 or 12.1 per cent of the villages. The union territories have the smallest number of villages. The states in the north-east, being small in extent and hilly in nature, also have few villages. Small and densely populated Kerala has only 1219 or 0.2 per cent of India's villages, whereas a single district in Tamil Nadu, Thanjavur, for example, has as many. Generally speaking, the states occupying the northern plains (Uttar Pradesh, Madhya Pradesh, Bihar and Rajasthan — 34,968 or 6.3 per cent) have the bulk, over 50 per cent, of India's villages. In peninsular India, the states of Orissa (46,553 or 8.4 per cent), Maharashtra (39,354 or 7.1 per cent), Andhra Pradesh (27,379 or 4.9 per cent) and Karnataka (27,028 or 4.9 per cent) have the largest number of villages.

Villages are categorized by population sizes into six classes. The overlain information on the map indicates the number of villages by size. In 1981 there were about 575,000 villages in India (including estimates for Assam). Of these, 1835 had over 10,000 people each, 7228 had over 5000 people, and about 282,000 had less than 500 people. Four states of the Union — Himachal Pradesh, Meghalaya, Nagaland and Sikkim — have no villages with populations exceeding 10,000 each, and Meghalaya and Sikkim do not even have villages with more than 5,000 people. Kerala, in fact, has the largest number of villages (905 or 49.3 per cent of the total) with populations over 10,000 (Class I), followed by Bihar (192 or 10.5 per cent), Tamil Nadu (182 or 9.9 per cent), Andhra Pradesh (163 or 8.9 per cent) and Maharashtra (116 or 6.3 per cent). It would seem that population pressures in a small, narrow strip of land have caused Kerala's villages to merge into large villages that are almost towns. The states with the largest number of small villages (Class VI, with less than 500 people) are Uttar Pradesh, Madhya Pradesh, Bihar and Orissa.

The move from rural India to urban settlements has been marked in the last 80 years. In 1901, there were 1917 cities and towns in undivided India. These had increased to 3366 in 1981 (including estimates for Assam and Jammu & Kashmir), even though India's land area had shrunk by nearly 20 per cent after partition. Uttar Pradesh has the largest number of towns and cities (659), the second largest number being in Madhya Pradesh (303), the third largest in Maharashtra (276) and the fourth in Karnataka (250). The union territories together have 45 towns.

In Indian statistical usage, towns and cities are classified according to population size. Class I towns have a population of 100,000 and above and include the metropolitan cities, which have populations of 1 million and above and are administrative districts in themselves. In 1981, there were 219 Class I towns (including 12 metropolitan cities and estimates for Assam) as against only 25 such towns in 1901, of which only two (Bombay and Calcutta) were metropolises. The number of Class II towns with 50,000-99,999 population, are more than six times the 44 there were in 1901. The number of Class III towns, with 20,000 to 49,999 population, is five times the 144 there were in 1901. The largest number of towns in any category is the Class IV towns, 1076, only two-and-a-half times the 427 there were in 1901. In the case of small towns, however, their numbers have decreased, Class V towns (population 5,000 to 9,999) being about the same number (780) as there were in 1901 (771); and Class VI towns (population less than 5000) 263 as compared to 503 in 1901.

In all the states the largest number of towns belong to Classes III, IV and V. Class III towns are largest in number in Andhra Pradesh (88 towns), while Class IV towns are the largest in number in Uttar Pradesh (194), Madhya Pradesh (113), Karnataka (100), Rajasthan (98), Maharashtra (91), Tamil Nadu (82) and Gujarat (76). Class V towns are the largest in number only in Uttar Pradesh (231). The largest number of Class VI towns is also in Uttar Pradesh (82). On the whole, medium-sized towns are by far the largest in number, though it is the Class I towns that hold the largest share of the urban population in many states.

Sources: 1. *Statistical Abstract India 1984*, Ministry of Planning. Quoting from Registrar General, India, Ministry of Home Affairs.

2. *Highlights from the 1981 Census*, Hindustan Thompson Associates Limited.

Table 6

POPULATION BY CLASS OF TOWN — 1981

(Population in thousand)

	All Towns (I-VI)		Class I		Class II		Class III		Class IV		Class V		Class VI		Villages	
	No.	Population	No.	Population	No.	Population	No.	Population	No.	Population	No.	Population	No.	Population	No.	Population
States																
Andhra Pradesh	234	12,458	20	6,688	30	2,014	88	2,610	64	912	28	218	4	15	27,379	41,092
Assam¹	65	2,047	1	307	4	491	10	471	23	491	19	246	8	41	21,274	17,850
Bihar	179	8,699	16	4,708	19	1,251	57	1,675	59	874	23	176	5	15	67,546	61,216
Gujarat	220	10,556	13	6,114	23	1,534	46	1,412	76	1,080	53	394	9	22	18,114	23,530
Haryana	77	2,822	11	1,598	5	305	13	418	24	328	22	164	2	8	6,745	10,101
Himachal Pradesh	46	327	—	—	1	70	2	41	5	73	9	64	29	79	16,807	3,954
Jammu & Kashmir¹	56	1,251	2	824	—	—	5	148	5	66	19	140	25	73	6,477	4,736
Karnataka	250	10,711	17	6,277	11	692	64	1,902	100	1,471	42	307	16	62	27,028	26,425
Kerala	85	4,771	8	2,535	7	454	49	1,520	17	228	4	33	—	—	1,219	20,683
Madhya Pradesh	303	10,589	14	4,960	28	1,905	41	1,296	113	1,601	104	812	3	13	71,352	41,590
Maharashtra	276	21,967	25	16,528	20	1,307	81	2,390	91	1,341	43	347	16	55	39,354	40,817
Manipur	32	373	1	156	—	—	2	42	4	55	9	62	16	58	2,035	1,048
Meghalaya	7	240	1	173	—	—	1	35	1	13	1	6	3	12	4,902	1,096
Nagaland	7	120	—	—	—	—	2	68	2	30	3	22	—	—	1,112	655
Orissa	103	3,106	6	1,293	7	396	23	678	39	529	25	196	3	14	46,553	23,264
Punjab	134	4,620	7	2,144	9	614	28	985	35	511	41	311	14	56	12,342	12,169
Rajasthan	195	7,140	11	3,322	11	717	52	1,572	98	1,339	22	186	1	4	34,968	27,122
Sikkim	8	51	—	—	—	—	1	37	—	—	—	—	7	14	440	265
Tamil Nadu	245	15,928	20	9,906	37	2,547	63	1,994	82	1,178	37	281	6	22	15,831	32,476
Tripura	10	225	1	132	—	—	1	21	4	52	2	14	2	7	856	1,828
Uttar Pradesh	659	19,973	30	10,283	37	2,539	85	2,466	194	2,667	231	1,728	82	299	112,568	90,889
West Bengal	130	14,433	12	11,091	21	1,557	36	1,113	34	499	20	151	7	23	38,024	34,608
Union Territories																
Andaman & Nicobar	1	50	—	—	—	—	1	50	—	—	—	—	—	—	491	139
Arunachal Pradesh	6	40	—	—	—	—	—	—	—	—	5	36	1	4	3,257	592
Chandigarh	1	429	1	429	—	—	—	—	—	—	—	—	—	—	24	23
Dadra & Nagar Haveli	1	7	—	—	—	—	—	—	—	—	1	7	—	—	70	97
Delhi	6	5,753	1	5,714	—	—	—	—	1	13	4	26	—	—	214	467
Goa, Daman & Diu	17	351	—	—	3	211	2	47	3	39	5	41	4	13	412	736
Lakshadweep	3	19	—	—	—	—	—	—	—	—	3	19	—	—	7	21
Mizoram	6	123	—	—	1	76	—	—	1	18	4	29	—	—	721	371
Pondicherry	4	316	1	251	—	—	1	43	1	12	1	10	—	—	291	288
INDIA¹	3,366	159,486	219	95,424	274	18,683	754	23,033	1,076	15,419	780	6,027	263	900	578,413	525,699

Note: ¹ Estimated.

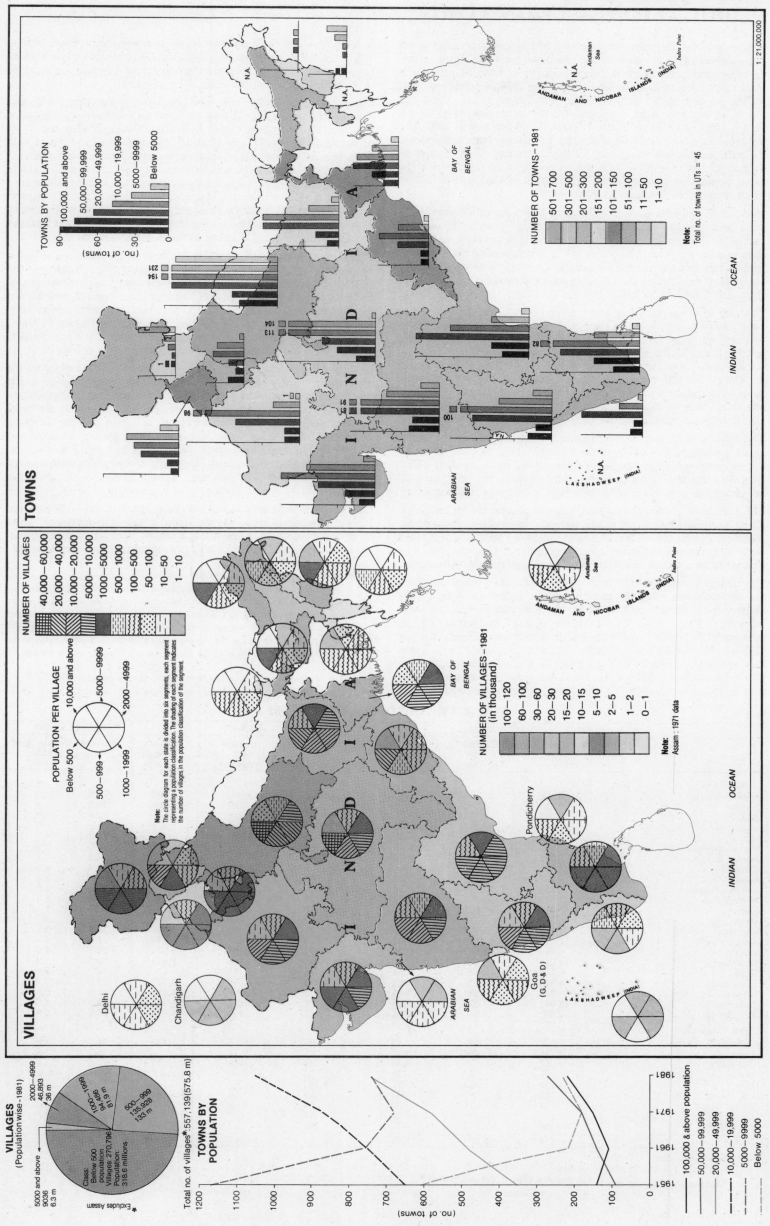

TOWNS

TOWNS BY POPULATION

(no. of towns)

100,000 and above
50,000—99,999
20,000—49,999
10,000—19,999
5000—9999
Below 5000

NUMBER OF TOWNS—1981

501—700
301—500
201—300
151—200
101—150
51—100
11—50
1—10

Note:
Total no. of towns in UTs = 45

VILLAGES

NUMBER OF VILLAGES

40,000—60,000
20,000—40,000
10,000—20,000
5000—10,000
1000—5000
500—1000
100—500
50—100
10—50
1—10

POPULATION PER VILLAGE

Below 500
500—999
1000—1999
2000—4999
5000—9999
10,000 and above

Note:
The circle diagram for each state is divided into six segments, each segment representing a population classification. The shading of each segment indicates the number of villages in the population classification of the segment.

NUMBER OF VILLAGES—1981
(in thousand)

100—120
60—100
30—60
20—30
15—20
10—15
5—10
2—5
1—2
0—1

Note:
Assam : 1971 data

Delhi
Chandigarh
Pondicherry
Goa (G., D & D)

VILLAGES
(Population wise-1981)

2000—4999
46,893
36 m

1000—1999
94,496
81.9 m

5000 and above
9036
6.3 m

Below 500
270,796
Villages
318.6 millions
Population

500—999
135,928
133 m

*Excludes Assam

Total no. of villages* 557,139 (575.8 m)

TOWNS BY POPULATION

(no. of towns)

100,000 & above population
50,000—99,999
20,000—49,999
10,000—19,999
5000—9999
Below 5000

1951   1961   1971   1981

POPULATION — URBANIZATION

While the rural and urban components of the Indian population are both continuously growing, there has been, over the years, a discernible drop in the share of village population, from 82 per cent in 1961 (about 360 million) to 80.1 per cent in 1971 (439 million), and to 76.6 per cent (525.5 million) in the 1981 Census return. The urban share, on the other hand, has been steadily increasing, from 18 per cent in 1961 to 19.9 per cent in 1971 and 23.3 per cent in 1981. But the absolute increase of both population groups has been tremendous, the rural population increasing 79 million during 1961-71 and 69 million during 1971-81 — 148 million people in 20 years — and the urban population growing from 78.9 million in 1961 to 109 million in 1971 and 159.7 million in 1981.

In the eight decades between 1901 and 1981, the urban population increased at a rate of 22.8 per cent per decade, while the average rate of increase in the rural population was 10.9 per cent, and in the total population about 17 per cent. The urban population grew six-fold in this period, from 25.9 million in 1901, whereas the rural population, though also increasing, grew only two-and-a-quarter-fold from 238.3 million in 1901.

The rate of urbanization against the total population, however, has been comparatively low. In 1901, the share of urban population to the total was only 10.8 per cent, in 1951 it was 17.3 per cent, and in 1981 it was 23.3 per cent. This was less than the average for Asia, Africa and Latin America. On the other hand, the total urban population of 160 million is much more than that of any other developing country except China (208 million) and of many developed countries, such as Japan (92.9 million), and about the same as the USSR (169 million) and USA (177 million).

The latest statistics, however, indicate a disproportionately higher urban growth rate — 46.4 per cent in 1971-1981 in contrast to the 25 per cent overall growth rate. Following the rural population shift typical of the developing countries, India should, by 1990, have 27 per cent of its population (225 million people) living in urban areas.

The three most important urbanized states at present are Maharashtra, Tamil Nadu and Gujarat. Karnataka, however, has a larger urban population than Gujarat but ranks only fourth. Madhya Pradesh, which ranks twelfth, has nearly the same urban population as Gujarat; and Uttar Pradesh ranks 15th with almost the same urban population as Maharashtra.

The districts with 100 per cent or very high urban populations are the metropolitan districts and the urban union territories. The metropolises, India's million cities, are Calcutta (9.2 m), Greater Bombay (8.2 m), Delhi (5.7 m), Madras (4.3 m), Bangalore (2.9 m), Ahmadabad (2.5 m), Hyderabad (2.5 m), Pune (1.7 m), Kanpur (1.6 m), Nagpur (1.3 m), Jaipur (1 m) and Lucknow (1 m). These metropolises account for 25 per cent of the Class I urban population. Of the four major metropolises, Delhi has grown the most — 57 per cent — in the last decade. But the fastest growing metropolitan city

is Bangalore with a 77 per cent increase in population from 1971 to 1981. While Lucknow has increased least (24 per cent) in population, most metropolitan cities have been registering a population increase of around 45 per cent between 1971 and 1981.

The districts with 100 per cent rural population are Wayanad in Kerala, The Dangs in Gujarat, Lahul & Spiti and Kinnaur in Himachal Pradesh, East Kamang, Upper Subansiri, Dibang Valley and Tirap in Arunachal Pradesh, and Phek in Nagaland. Other districts with predominantly rural and marginal urban populations (ratios of 11:1 to 30:1), are essentially hill districts: Malappuram and Idukki in Kerala, North Ratnagiri and Sindhudurg (together known as Ratnagiri in 1981) in Maharashtra, the Dadra & Nagar Haveli union territory, Bastar district in Madhya Pradesh, the three most north-eastern districts in Orissa, the Himalayan foothills districts in Uttar Pradesh and Bihar and several districts in north-eastern India.

That there is substantial urban population in much of India is revealed by the fact that the ratio of rural to urban population in most of the districts of peninsular India, and the west, north and north-west of India (Gujarat, interior Rajasthan, western Uttar Pradesh, and interior Madhya Pradesh) ranges from 2:1 to 5:1. This includes almost entirely the developed states of Tamil Nadu, Karnataka, Andhra Pradesh, Gujarat and a major part of Maharashtra, where only a few districts in each state have a rural-urban ratio going up to 6:1 to 10:1.

If India were divided into two halves along the 80°E longitude, the western half is far more urbanized than the eastern, which is the reverse of the population density pattern. But the eastern districts are beginning to catch up; it is in the eastern half that there are more districts showing a 'one-step increase' in urban content, with six districts showing a 'two-step increase' (Tamenglong, Senapati and Ukhrul in Manipur, Pithorgarh and Unnao in Uttar Pradesh, and Tikamgarh in Madhya Pradesh), and three north-eastern districts even registering a 'three-step increase' (Lower Subansiri of Arunachal Pradesh, and Mon and Tuensang of Nagaland).

Only seven districts have shown a decrease in urbanization: West Khasi Hills in Meghalaya alone has registered a 'two-step decrease', while three border districts in Jammu & Kashmir, Banaskantha in Gujarat and Idukki and Kottayam in Kerala have registered a 'one-step decrease'.

Note: All Census data are dated 1981 and refer to the 412 districts that then existed. The maps show the districts as they existed at the end of 1985, when there were 439 districts (437 districts for which boundaries were available are shown on the maps). Where 1981 districts have been broken up into two or more smaller districts or combined into larger districts (as in Jammu & Kashmir) the 1981 data are used to shade the 1985 configurations. In the text, which refers to the districts as they existed in 1985, the 1981 status is given within brackets.

Source: 1. *Primary Census Abstract Part II B(i)*, Census of India 1981.
2. World Bank's *World Development Report 1983*, Oxford University Press.

Table 7

RURAL-URBAN POPULATION — 1981

	RURAL			URBAN			URBANIZATION[2]	
	Male (in thousand)	Female (in thousand)	Density (Persons/ sq km)	Male (in thousand)	Female (in thousand)	Density (Persons/ sq km)	Per cent	Growth Rate[3] 1971-81 (%)
States								
Andhra Pradesh	20,698	20,364	152	6,411	6,076	3,086	23.3	4.0
Assam[1]	9,310	8,540	228	1,158	889	4,705	10.3	1.5
Bihar	31,170	30,025	359	4,760	3,958	2,726	12.5	2.5
Gujarat	11,987	11,497	123	5,566	5,035	2,225	31.1	3.0
Haryana	5,381	4,714	232	1,529	1,298	3,702	21.9	4.2
Himachal Pradesh	1,988	1,966	71	181	144	1,535	7.6	0.6
Jammu & Kashmir	2,492	2,234	21	672	588	2,146	21.0	2.4
Karnataka	13,352	13,054	140	5,570	5,519	2,913	28.9	4.6
Kerala	10,167	10,515	558	2,360	2,411	2,668	18.8	2.6
Madhya Pradesh	21,266	20,326	95	5,620	4,966	2,170	20.3	4.0
Maharashtra	20,527	20,263	135	11,887	10,106	3,735	35.0	3.8
Manipur	530	515	47	190	185	2,478	26.4	13.2
Meghalaya	557	537	49	127	114	2,846	18.0	3.5
Nagaland	345	309	40	71	49	1,106	15.5	5.5
Orissa	11,636	11,623	152	1,673	1,437	1,359	11.8	3.4
Punjab	6,444	5,697	247	2,492	2,155	3,875	27.7	4.0
Rajasthan	14,013	13,038	80	3,841	3,369	1,603	20.9	3.6
Sikkim	142	123	N.A.	30	21	N.A.	16.2	6.9
Tamil Nadu	16,334	16,121	261	8,153	7,798	2,722	33.0	2.7
Tripura	939	888	175	115	110	4,186	11.0	0.6
Uttar Pradesh	48,041	42,921	314	10,778	9,121	4,363	18.0	4.0
West Bengal	20,617	19,516	466	7,943	6,503	5,460	26.5	1.8
Union Territories								
Andaman & Nicobar	78	61	17	29	20	3,520	26.4	3.6
Arunachal Pradesh	314	276	N.A.	25	16	N.A.	6.3	2.6
Chandigarh	17	11	630	238	184	6,190	93.6	3.1
Dadra & Nagar Haveli	49	47	200	4	3	1,032	6.7	N.A.
Delhi	250	202	507	3,190	2,578	9,745	92.8	3.1
Goa, Daman & Diu	365	369	203	183	168	1,824	32.5	6.1
Lakshadweep	11	11	1,010	10	9	1,757	46.3	N.A.
Mizoram	192	179	18	64	57	382	25.2	N.A.
Pondicherry	146	142	736	159	157	3,160	52.3	10.3
INDIA[1]	269,364	256,093	165	85,034	74,693	3,000	23.3	3.4

Note: [1] Estimated data
[2] Urban population in state population
[3] Urbanization growth rate = $\dfrac{\text{urbanization 1981} - \text{urbanization 1971}}{\text{urbanization 1971}}$ %

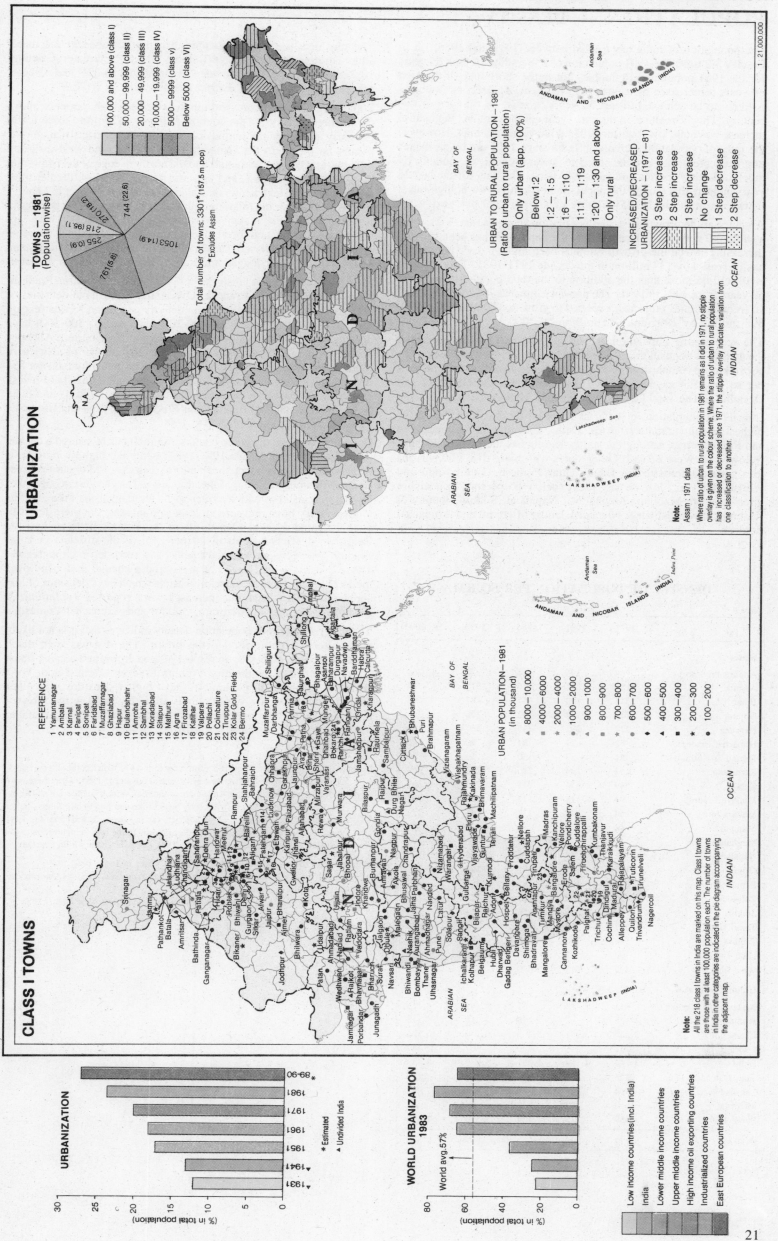

URBANIZATION

TOWNS — 1981
(Populationwise)

100.000 and above (class I)
50.000 – 99.999 (class II)
20.000 – 49.999 (class III)
10.000 – 19.999 (class IV)
5000 – 9999 (class V)
Below 5000 (class VI)

1053 (14.9)
744 (22.6)
270 (18.2)
218 (95.1)
255 (0.9)
761 (5.8)

Total number of towns: 3301* (157.5 m pop)
*Excludes Assam

URBAN TO RURAL POPULATION — 1981
(Ratio of urban to rural population)

Only urban (app. 100%)
Below 1:2
1:2 – 1:5
1:6 – 1:10
1:11 – 1:19
1:20 – 1:30 and above
Only rural

INCREASED/DECREASED URBANIZATION — (1971–81)

3 Step increase
2 Step increase
1 Step increase
No change
1 Step decrease
2 Step decrease

Note:
Assam : 1971 data

Where ratio of urban to rural population in 1981 remains as it did in 1971, no stipple overlay is given on the colour scheme. Where the ratio of urban to rural population has increased or decreased since 1971, the stipple overlay indicates variation from one classification to another.

CLASS I TOWNS

REFERENCE

1 Yamunanagar
2 Ambala
3 Karnal
4 Panipat
5 Sonipat
6 Faridabad
7 Muzaffarnagar
8 Ghaziabad
9 Hapur
10 Bulandshahr
11 Amroha
12 Sambhal
13 Moradabad
14 Sitapur
15 Mathura
16 Agra
17 Firozabad
18 Valparai
19 Pollachi
20 Coimbatore
21 Tiruppur
22 Kolar Gold Fields
23 Karaikkudi
24 Bermo

URBAN POPULATION — 1981
(in thousand)

8000–10,000
4000–6000
2000–4000
1000–2000
900–1000
800–900
700–800
600–700
500–600
400–500
300–400
200–300
100–200

Note:
All the 218 class I towns in India are marked on this map. Class I towns are those with at least 100,000 population each. The number of towns in India in other categories are indicated in the pie diagram accompanying the adjacent map.

URBANIZATION

30
25
20
15
10
5
0
(% in total population)

*89–90
1981
1971
1961
1951
1941
1931

* Estimated
▲ Undivided India

WORLD URBANIZATION 1983

80
60
40
20
0
(% in total population)

World avg. 57%

Low income countries (incl. India)
India
Lower middle income countries
Upper middle income countries
High income oil exporting countries
Industrialized countries
East European countries

POPULATION — DENSITY

The population of India has trebled between 1901 and 1986, i.e. it is today 3.2 times the 1901 population of 238 million. It took 65 years for the 1901 population to double, but in the short span of the next 20 years it increased by a further 75 per cent. The second doubling is likely to be achieved in just 35 years, that is, by the turn of this century. This is because, in the early years of the century, the natural increase was quite low, but since 1951 it has been very high. However, the rates of population increase in India since 1951 are by no means unprecedented; they are exceeded by several countries, notably by much of Latin America and parts of Africa. The data for the last 25 years indicate that though the population certainly grew at high rates, it was not at the high 1951-61 rate of 2.2 per cent per year.

A glance at the growth of actual population suggests that, except for a slight fall between 1911 and 1921, the population has been steadily increasing. Since the growth rate has been high, India's population has increased by 137 million in the decade 1971-1981. This increase is 14 million more than the addition to the total population over the 50 years from 1901 to 1951. The present estimated 761 million (1986) is, from the trends projected, expected to grow to 822 million by 1990, 898 million by 1995 and 972 million by 2000.

The accelerating population growth in India has steadily increased the density of population in the country, and in 1981 it was 208 persons per sq km. In comparison, Asia as a whole has a density of 95 persons per sq km, Europe 98, Africa 16, the USA 15, the USSR 12 and Oceania 3. The world average is only 33 persons per sq km.

Density of population in India, as revealed by the 1981 Census, is highest in four union territories: Delhi, Chandigarh, Lakshadweep and Pondicherry. Among the states, Kerala has the highest density, followed by West Bengal. Other states with more than 300 persons to a square kilometre are Bihar, Uttar Pradesh, Tamil Nadu and Punjab. The very low density areas (below 65 persons to a square kilometre) are Manipur, Meghalaya, Nagaland, Sikkim, Jammu & Kashmir, the Andaman & Nicobar Islands, Mizoram and Arunachal Pradesh.

The city densities are far greater than the densities of the major administrative units: Greater Bombay, 13,671; Greater Calcutta, 10,788; Delhi, 10,595; Greater Hyderabad, 10,325 and Greater Madras, 7,499. Calcutta district has a density of 31,779.

Population densities from medium to extremely high, about or above the average density for the country (208 per sq km) are to be found in nearly all the mainly humid, coastal lowland districts fringing the peninsular block. In the Gangetic plain, they extend westward into the semi-arid country and south into the northern edge of the peninsular block. In the north-east, and extreme north, however, the densities are far lower (50 and below) than elsewhere. Bastar district in Madhya Pradesh, Kachchh in Gujarat and Jaisalmer, Bikaner, and Barmer in Rajasthan also have very low population densities. These are attributable to hilly and forested terrain (north-eastern states, extreme north and Bastar district), desert and semi-arid climates (Rajasthan districts), and water-logged and marshy tracts in Kachchh, Gujarat.

Most of the interior districts of Karnataka, Andhra Pradesh, Maharashtra, Madhya Pradesh, Orissa and Bihar have densities of 100 to 200 persons. All districts in densely populated Kerala (except hilly Idukki with a density of 100-200 and Palghat, 300-500) have densities of more than 500, with three districts (Trivandrum, Alleppey and Ernakulam) in the 1000-2000 category. A number of districts of Madhya Pradesh, two districts in Orissa (Koraput, Phulbani), a few districts in Gujarat, Jammu & Kashmir, Rajasthan, Himachal Pradesh and the north-east have densities of 50-100. The foothills and plain districts in the states of Bihar, Uttar Pradesh, Haryana and Himachal Pradesh have densities of 300 to 1000.

Only one district in the entire country (Shimla) has registered a decline in density over the decade 1971-1981; while the highest increase is registered (a 'two-step increase') in three districts of Manipur (together Central Manipur in 1981). Mahendragarh in Haryana is the only other district to register a 'two-step increase'. As many as half the districts in the country have registered a 'one-step increase' in their densities over the same decade. Two-thirds of the Punjab districts, half the Rajasthan districts and about 40 per cent of the districts in Uttar Pradesh, West Bengal and Karnataka have recorded such increases. Those districts which have not registered a change in density over the 1971 levels are those interior districts which have densities of 200 and below. No substantial increases were reported by Jammu & Kashmir, Meghalaya, Mizoram, Sikkim and Arunachal Pradesh.

The general pattern of population density indicates that it is not likely to be altered significantly in the future. The districts with much productive land, especially in respect of food crops, and notably much of the rice land of the country (Andhra Pradesh, Tamil Nadu, West Bengal), are likely to continue to have higher densities than other districts, while the low density areas are likely to be in the forested hills and the interior, semi-arid areas (parts of Karnataka, Maharashtra, and Andhra Pradesh).

Note: All Census data are dated 1981 and refer to the 412 districts that then existed. The maps show the districts as they existed at the end of 1985, when there were 439 districts (437 districts for which boundaries were available are shown on the maps). Where 1981 districts have been broken up into two or more smaller districts or combined into larger districts (as in Jammu & Kashmir) the 1981 data are used to shade the 1985 configurations. In the text, which refers to the districts as they existed in 1985, the 1981 status is given within brackets.

Sources: 1. *Primary Census Abstract II B(i)*, Census of India 1981.
2. *Family Welfare Programme in India, Year book 1983-84*, Ministry of Health & Family Welfare.
3. *Report of the Expert Committee on Population*, New Delhi, 1986.

DENSITY OF POPULATION PER SQ.KM.

Table 8

	1971	1981	1986[1]	1990[1]	2000[1]
States					
Andhra Pradesh	158	195	214	230	264
Assam	150[2]	254[1]	286	313	384
Bihar	324	402	447	487	597
Gujarat	136	174	192	206	234
Haryana	227	292	335	367	423
Himachal Pradesh	62	77	84	90	102
Jammu & Kashmir	21	27	30	33	39
Karnataka	153	194	216	233	269
Kerala	549	655	715	763	869
Madhya Pradesh	94	118	131	142	168
Maharashtra	164	204	227	241	279
Manipur	48	64	72	76	94
Meghalaya	45	60	67	76	94
Nagaland	31	47	54	66	90
Orissa	141	167	185	198	231
Punjab	269	333	367	389	433
Rajasthan	75	100	115	127	160
Sikkim	30	45	56	63	85
Tamil Nadu	317	372	404	428	480
Tripura	148	196	219	238	286
Uttar Pradesh	300	377	418	454	553
West Bengal	499	615	679	730	843
Union Territories					
Andaman & Nicobar	14	23	37	43	55
Arunachal Pradesh	6	8	8	10	12
Chandigarh	2,257	3,961	5,263	6,140	10,526
Dadra & Nagar Haveli	151	211	286	367	449
Delhi	2,742	4,194	5,067	5,933	8,533
Goa, Daman & Diu	225	285	342	368	421
Lakshadweep	994	1,258	1,375	1,438	1,688
Mizoram	—	23	28	33	43
Pondicherry	959	1,229	1,429	1,510	1,633
INDIA	177	208[1]	232	251	296

Note: [1] Estimated
[2] Includes Mizoram

22

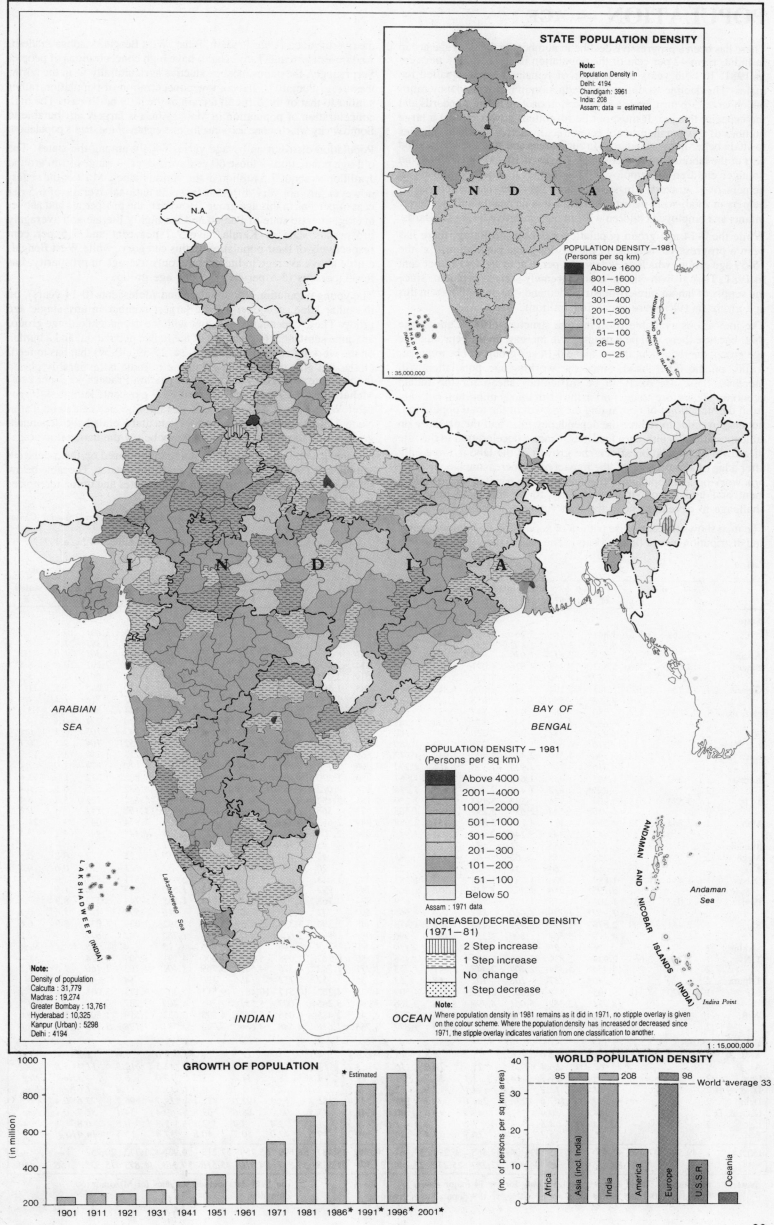

STATE POPULATION DENSITY

Note:
Population Density in
Delhi : 4194
Chandigarh : 3961
India : 208
Assam; data = estimated

INDIA

POPULATION DENSITY —1981
(Persons per sq km)

Above 1600
801 — 1600
401 — 800
301 — 400
201 — 300
101 — 200
51 — 100
26 — 50
0 — 25

1 : 35,000,000

N.A.

I   N   D   I   A

ARABIAN
SEA

BAY OF
BENGAL

POPULATION DENSITY — 1981
(Persons per sq km)

Above 4000
2001 — 4000
1001 — 2000
501 — 1000
301 — 500
201 — 300
101 — 200
51 — 100
Below 50

Assam : 1971 data

INCREASED/DECREASED DENSITY
(1971—81)

2 Step increase
1 Step increase
No change
1 Step decrease

Note:
Where population density in 1981 remains as it did in 1971, no stipple overlay is given
on the colour scheme. Where the population density has increased or decreased since
1971, the stipple overlay indicates variation from one classification to another.

LAKSHADWEEP (INDIA)

Lakshadweep
Sea

ANDAMAN AND NICOBAR ISLANDS (INDIA)

Andaman
Sea

Indira Point

Note:
Density of population
Calcutta : 31,779
Madras : 19,274
Greater Bombay : 13,761
Hyderabad : 10,325
Kanpur (Urban) : 5298
Delhi : 4194

INDIAN   OCEAN

1 : 15,000,000

GROWTH OF POPULATION

* Estimated

(in million)

1901 1911 1921 1931 1941 1951 1961 1971 1981 1986* 1991* 1996* 2001*

WORLD POPULATION DENSITY

(no. of persons per sq km area)

95   208   98   World average 33

Africa   Asia (incl. India)   India   America   Europe   U.S.S.R.   Oceania

POPULATION — AGE

There has been a progressive decline in numbers in the 0-14 age group in India, from 42 per cent of the population in 1971 to 39.6 per cent in 1981. In both years the number of females almost equalled the males. The decline would mean a reduction in the dependency ratio, which would, in turn, have a beneficial impact on living standards and the capacity to save. It must not be forgotten, however, that a large number of this population are economically active and help families to attain better living standards. In rural areas and small towns, a good part of the labour force is between the ages of 10 and 14. Often, even younger children are employed in agricultural activities, mostly as helpers (*e.g.* grazers). In small towns they are also employed as helpers in small-scale industries. Hence it is debatable whether fewer infants and employed children will, in fact, improve living standards.

While the 0-14 age group population has been declining, there has been a progressive increase in the working-age population, *i.e.* the 15-59 age group, which grew from 52 per cent in 1971 to 54 per cent in 1981. This growth will pose an extremely serious problem, since the supply of labour already outstrips demand (the unemployed in this age group in 1981 were about 167 million).

The indications available from the age structure (1981 Census) are that, because there will be a progressive increase in the reproductive age group, the reduction, if any, in the 0-14 age group will be minimal, while, on the other hand, since the working-age population will increase, there are likely to be difficulties ahead for the Indian economy. Therefore, reduction in the birth rate is important not only from the standpoint of restraining the growth of the total population, but also in order to reduce the dependency ratio and the pressure on educational and health-care facilities. Since a deceleration in the rate of growth of population affects the growth of the labour force only after a lag of about 15 years, the impact of an increasing labour force or a working-age population, together with the backlog of unemployment and underemployment, is likely to be the most important challenge to development in the near future.

The map shows the state-wise pattern of population distribution (base) and distribution by age (located bars). The plains, the deltaic areas and the riverine tracts (Uttar Pradesh, Bihar, West Bengal, Madhya Pradesh, Andhra Pradesh and Tamil Nadu) have high concentrations of people. Surprisingly, the plains most productive agriculturally — in the north-west (*e.g.* Punjab) — have a low concentration of population rather similar to that of the difficult terrain of the hilly north-east. The high concentration of population in Maharashtra is largely attributable to Bombay city which alone accounts for one-eighth of the state's population.

Population distribution by age varies widely among the states. The old-age population — those 60 years and over — ranges from around 1 million to about 7.5 million in the various states. Most of the states, however, are not very different from the national average of 6.5 per cent of persons in this age group. However, the prosperous and above-average literate state of Punjab and the highly literate and averagely prosperous state of Kerala have 7.7 per cent and 7.5 per cent respectively of their population in this category, while West Bengal, a little above average in literacy and only average in prosperity, has fewer persons (5.5 per cent) in this age group.

The young population, *i.e.* children and adolescents (0-14 years), on the other hand, accounts for the largest number in any single age group. These dependants, together with the dependent old-age group, are quite substantial (46 per cent of the Indian population) and a burden on the working-age population (15-24, 25-39, 40-59) that has to feed, clothe and give them shelter. No state in India is favourably placed in this context (Orissa 46.0 per cent, Andhra Pradesh 45.3 per cent, Maharashtra 44.7 per cent, Gujarat 44.7 per cent, Punjab 44.6 per cent, West Bengal 44.0 per cent, Kerala 42.5 per cent, and Tamil Nadu 41.4 per cent), though Tamil Nadu and Kerala have dependent group populations that are substantially below the national average.

The age group data presented in the map are based on the 5 per cent sampling of the 1981 Census data relevant to 14 states. The table below shows population by age group for *all* the states and union territories in 1971 and for 14 major states in 1981.

Source: 1. *Report & Tables Based on 5 per cent Sample Data, Part II Special*, Census of India 1981.
2. *Social & Cultural Tables, Part II - c(ii)*, Census of India 1971.

Table 9

POPULATION BY AGE AND SEX

(in thousand)

State	#	0-4 M	0-4 F	5-9 M	5-9 F	10-14 M	10-14 F	15-19 M	15-19 F	20-24 M	20-24 F	25-39 M	25-39 F	40-59 M	40-59 F	60 & above M	60 & above F	Age not stated M	Age not stated F
States																			
A.P.	1	3,269	3,244	3,841	3,786	3,367	3,183	2,489	2,345	2,131	2,232	5,495	5,505	4,741	4,336	1,765	1,801	10.8	7.9
	2			6,202*	6,178	2,692	2,536	1,891	1,780	1,680	1,756	5,694	5,657	3,758	3,359	1,388	1,391	1.9	1.3
Assam¹	2			2,509*	2,517	1,030	954	661	612	581	553	1,568	1,337	1,130	796	404	301	1.0	1.5
Bihar	1	4,653	4,700	5,533	5,213	4,876	4,071	3,209	2,736	2,634	2,814	6,960	6,994	5,627	5,149	2,429	2,301	10.8	6.8
	2			8,757*	8,360	3,747	3,130	2,373	2,117	2,029	2,272	5,678	5,810	4,604	4,148	1,656	1,667	2.0	3.3
Gujarat	1	2,154	2,080	2,368	2,194	2,345	2,078	1,880	1,691	1,685	1,690	3,500	3,270	2,652	2,463	1,120	1,055	3.5	1.7
	2			4,098*	3,850	1,888	1,658	1,347	1,183	1,122	1,091	2,626	2,511	2,033	1,882	685	718	1.3	1.0
Haryana	1	896	828	986	867	974	847	777	627	638	565	1,385	1,081	1,108	845	444	350	4.8	3.0
	2			1,683*	1,493	791	672	534	445	401	370	861	824	756	623	350	231	0.5	0.7
H.P.	2			491*	481	233	220	161	162	130	140	322	340	286	245	143	105	0.1	0.1
J & K	2			710*	682	310	278	215	180	189	169	498	461	382	282	152	104	0.1	0.1
Kar.	1	2,352	2,288	2,596	2,589	2,447	2,400	1,913	1,772	1,669	1,588	3,794	3,670	3,033	2,652	1,217	1,240	13.2	12.7
	2			4,316*	4,263	1,935	1,920	1,405	1,272	1,198	1,156	2,821	2,844	2,378	1,996	910	874	2.0	1.4
Kerala	1	1,392	1,348	1,485	1,439	1,644	1,604	1,443	1,567	1,250	1,370	2,463	2,597	1,954	1,976	896	1,022	1.3	2.6
	2			2,899*	2,832	1,446	1,418	1,127	1,210	978	1,012	1,868	2,005	1,635	1,585	632	695	0.5	0.4
M.P.	1	3,573	3,518	3,904	3,756	3,597	3,197	2,584	2,259	2,159	2,157	5,162	4,866	4,264	3,809	1,637	1,725	5.4	5.1
	2			6,719*	6,543	2,650	2,291	1,716	1,503	1,527	1,568	4,344	4,107	3,328	2,945	1,168	1,241	1.3	0.8
Mah.	1	3,854	3,672	4,289	4,127	4,224	3,899	3,151	2,688	2,849	2,663	6,609	6,350	5,436	4,899	1,974	2,047	29.2	24.3
	2			7,413*	7,243	3,246	2,938	2,335	1,942	2,094	2,008	5,352	5,130	4,227	3,589	1,441	1,442	4.2	3.5
Manipur	2			157*	156	71	71	51	53	47	45	101	100	79	71	32	32	—	—
Megh.	2			158*	158	62	62	46	45	40	42	110	100	77	59	25	21	0.2	0.1
Nag.	2			68*	66	31	30	26	23	27	20	62	50	41	32	22	15	0.6	0.7
Orissa	1	1,551	1,582	1,887	1,906	1,785	1,740	1,307	1,282	1,082	1,043	2,618	2,594	2,263	2,034	811	871	3.9	2.6
	2			3,195*	3,276	1,473	1,349	916	864	744	792	2,260	2,289	1,812	1,646	635	686	1.9	1.6
Punjab	1	1,037	951	1,103	975	1,122	981	1,004	879	874	768	1,701	1,534	1,344	1,185	744	572	8.9	7.1
	2			1,957*	1,739	1,020	877	774	665	602	532	1,243	1,135	1,069	916	597	416	3.6	0.1
Raj.	1	2,441	2,399	2,678	2,471	2,437	2,158	1,775	1,473	1,448	1,396	3,330	3,075	2,707	2,392	1,030	1,037	8.9	6.6
	2			4,179*	3,864	1,800	1,537	1,190	996	981	993	2,609	2,432	2,016	1,780	742	677	0.9	1.2
Sikkim	2			26*	27	13	14	11	10	10	8	30	20	16	13	4	3	Neg.	Neg.
T.N.	1	2,743	2,665	2,939	2,839	2,907	2,801	2,437	2,466	2,150	2,167	5,098	5,267	4,606	4,191	1,609	1,524	—	—
	2			5,456*	5,354	2,404	2,347	1,841	1,753	1,771	1,763	4,376	4,589	3,769	3,401	1,203	1,160	1.8	1.5
Tripura	2			240*	237	108	102	68	63	56	57	152	148	120	98	54	48	—	—
U.P.	1	7,670	7,312	9,105	7,838	8,005	6,412	5,529	4,309	4,335	4,262	10,517	10,481	9,511	7,307	4,123	3,008	47.1	31.2
	2			13,724*	12,381	6,061	4,798	3,978	3,206	3,246	3,285	9,007	8,551	7,705	6,409	3,287	2,690	3.7	3.5
W.B.	1	3,136	3,141	3,741	3,645	3,773	3,562	2,976	2,774	2,745	2,485	6,055	5,240	4,593	3,653	1,519	1,504	22.8	15.6
	2			6,580*	6,630	3,072	2,725	2,213	1,877	1,768	1,618	4,848	4,086	3,743	2,731	1,178	1,170	28.0	15.0
Union Territories																			
A. & N.	2			16.3*	16.0	6.1	5.3	4.8	3.8	8.5	4.7	25.1	12.0	9.7	4.0	1.8	1.2	—	—
Ar. P.	2			67.3*	66.4	23.5	21.7	16.0	16.7	23.2	17.2	71.8	53.9	56.3	42.5	12.5	11.4	3.8	3.7
Cha.	2			33.3*	29.7	14.4	12.1	15.7	11.3	19.1	14.3	48.1	34.4	27.7	17.0	6.0	4.3	Neg.	Neg.
D. & N.H.	2			12.0*	12.2	4.8	4.6	2.9	2.9	2.5	3.0	7.3	7.2	6.0	5.5	1.4	1.6	Neg.	Neg.
Delhi	2			559.4*	507.0	270.4	234.0	234.3	185.1	231.4	182.1	513.5	392.0	351.3	229.7	96.7	77.6	0.4	0.5
G.D. & D.	2			114.6*	110.4	51.9	49.9	43.6	40.0	41.1	35.5	92.1	88.6	63.8	69.4	24.1	32.7	0.1	Neg.
Lak.	2			4.9*	4.6	1.9	1.6	1.4	1.6	1.3	1.5	3.4	3.4	2.3	2.3	0.8	0.8	—	—
Pon.	2			66.4*	65.2	27.9	26.9	20.4	20.5	19.2	19.5	47.0	50.7	40.8	37.4	15.5	14.4	Neg.	Neg.
INDIA²	1	49,587	47,498	47,430	44,715	43,304	39,465	36,295	33,164	30,089	28,482	68,690	65,759	57,213	50,935	21,773	20,757	—	—
	2			82,416*	79,145	36,492	32,282	25,222	22,246	21,573	21,528	55,897	54,010	45,518	39,013	16,873	15,825	56	60

Note: 1. 1981 data — These data are available only for the 14 major states.
2. 1971 data (*0-4 and 5-9 breakup not available, so 0-9 population is given)

¹ The 1971 data include statistics for Mizoram too.
² Estimated.

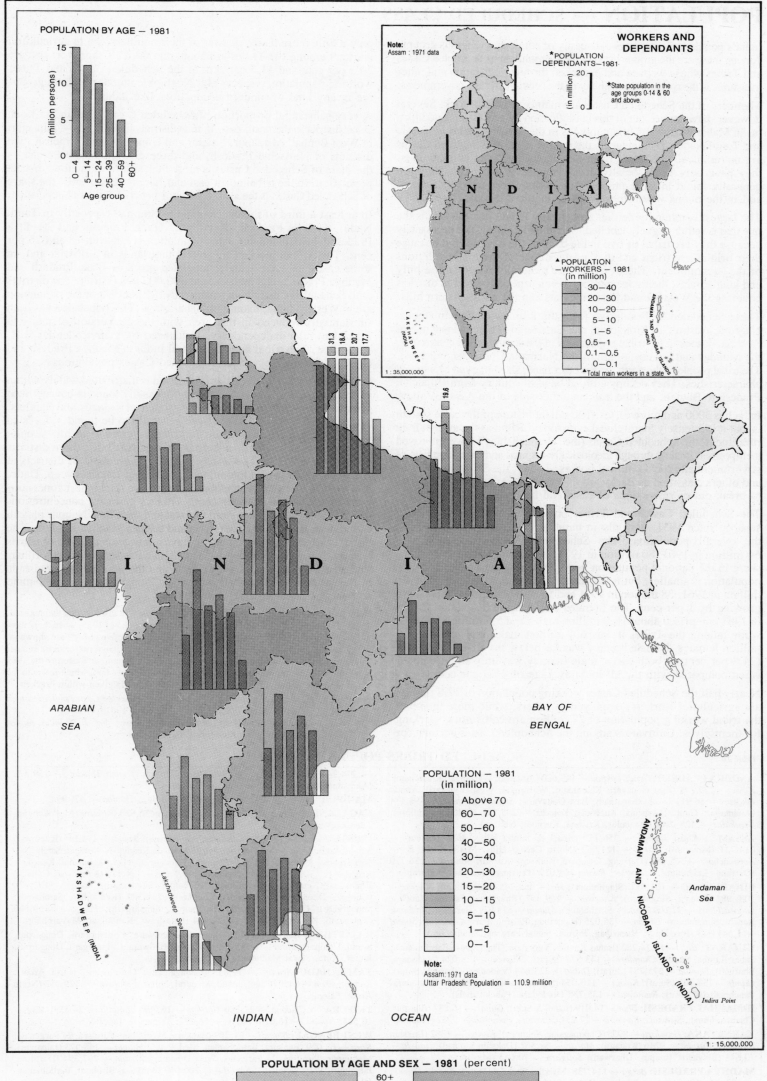

POPULATION BY AGE — 1981

(million persons)

15
10
5

0—4 5—14 15—24 25—39 40—59 60+

Age group

WORKERS AND
DEPENDANTS

Note:
Assam : 1971 data

*POPULATION
—DEPENDANTS—1981

(in million)

20

*State population in the
age groups 0-14 & 60
and above.

I N D I A

POPULATION
—WORKERS — 1981
(in million)

30—40
20—30
10—20
5—10
1—5
0.5—1
0.1—0.5
0—0.1

▲Total main workers in a state

1 : 35,000,000

31.3 18.4 20.7 17.7

19.6

I   N   D   I   A

ARABIAN
SEA

BAY OF
BENGAL

Lakshadweep
Sea

Lakshadweep
(INDIA)

ANDAMAN
AND
NICOBAR
ISLANDS
(INDIA)

Andaman
Sea

Indira Point

POPULATION — 1981
(in million)

Above 70
60—70
50—60
40—50
30—40
20—30
15—20
10—15
5—10
1—5
0—1

Note:
Assam: 1971 data
Uttar Pradesh: Population = 110.9 million

INDIAN   OCEAN

1 : 15,000,000

POPULATION BY AGE AND SEX — 1981 (per cent)

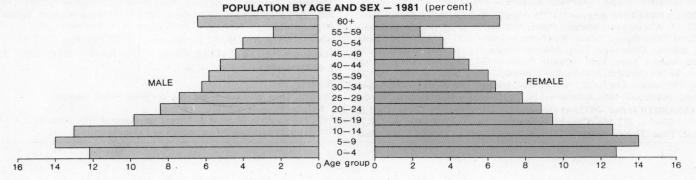

60+
55—59
50—54
45—49
40—44
35—39
30—34
25—29
20—24
15—19
10—14
5—9
0—4

MALE   FEMALE

16 14 12 10 8 6 4 2 0 2 4 6 8 10 12 14 16

Age group

POPULATION — SCHEDULED CLASS

India's population comprises a diversity of ethnic and religious groups. Among these are the groups listed in the Constitution as Scheduled Castes and Tribes whose welfare and progress, through legislation and other measures, is the special responsibility of the central and state governments.

Members of the Scheduled Castes are mainly of Hindu stock. Several however, have moved out of this classification by professing Buddhism (as in Maharashtra) or Christianity (as in the southern states of Kerala and Tamil Nadu). By and large, the members of the Scheduled Castes are not ethnically different from neighbouring Hindu communities; they have only been socially disadvantaged in the past. They are generally found in large numbers in coastal areas (e.g. fishermen) and in the plains where they work as landless labourers.

The largest community neither specifically Hindu nor Muslim is the one that is rather vaguely labelled 'Scheduled Tribes', not necessarily because they are tribal or live in the back of the beyond, but because their origins are tribal and they profess none of the major religions of the subcontinent. These 'tribes' are generally found in the hilly and jungle tracts, their concentration being especially great in districts covering the Western and Eastern Ghats and the north-eastern hills.

These Scheduled Tribes derive from the four major human stocks: Negroid, Australoid, Mongoloid and Caucasoid. There is evidence of Australoid stock in the tribal populations of the south and the centre (e.g. the Mundas and Santals, south of the Narmada-Chota Nagpur line). The tribal people of the north, on the other hand, show marked Mongoloid characteristics. They occupy a broad band of country from Himachal Pradesh to Bhutan, and the hills on either side of the Assam Valley.

India has 3000 and more castes. The salient feature of the caste system is that it is intensely hierarchical even today. Rigid as the system is in rural India, the Scheduled Castes (the 'Untouchables') have improved their position, mainly through ideologically-inspired administrative fiats of government and local political bodies. Today, the Scheduled Castes and others classified as Backward Classes have greater opportunities to break out of restrictions of caste and become upwardly mobile.

The Scheduled Castes and Tribes population has been increasing steadily since 1951, not only in numbers but also in proportion to the overall population. The Scheduled Castes population was 64 million in 1961, 80 million in 1971 and 105 million in 1981, their share in the national population growing by 1.1 per cent. The tribal population is smaller, but it has increased at about the same rate: 30 million in 1961, 38 million in 1971 and 52 million in 1981, their share growing by 1 per cent. The literates amongst the Scheduled Castes in 1981 comprised about 16.9 million males and 5.5 million females, while among the Tribes it was 6.4 million males and just about 2 million females. Female literacy was far below the all-India average (24.8 per cent) in both cases; male literacy was only slightly better when compared with the all-India average of 46.9 per cent.

Nearly half the Scheduled Castes working population of 37.8 million are agricultural workers, employed by others, while more than half the tribal working population of 22 million are cultivators, working for themselves. Cultivators among the Scheduled Castes account for

only a little more than a quarter of their numbers, while agricultural workers among the Tribes are about one third. 'Other workers' are 20.3 per cent and 11.5 per cent of the Scheduled Castes and Tribes working population respectively. Negligible numbers in both groups are in any kind of modern occupations like industry.

A very substantial proportion of Scheduled Castes is found in Koch Bihar district in the north-east. In nine districts in the north — Jalpaiguri in West Bengal, Mirzapur, Sitapur and Unnao in Uttar Pradesh, two districts of Himachal Pradesh, and three districts of the Punjab — the share of Scheduled Castes is more than a third of the total district population. In the tribal north-east and Jammu & Kashmir, the share of Scheduled Castes is the lowest, below 1 per cent of district population.

In at least a third of the districts in the country, especially in Tamil Nadu, Andhra Pradesh, Rajasthan, Uttar Pradesh and Madhya Pradesh, the Scheduled Castes population is between 15 and 25 per cent. This would amount to several thousands in a district and, in some cases, in such heavily populated areas as Uttar Pradesh and Madhya Pradesh, can even exceed 300,000 in a district. The districts with Scheduled Castes comprising only 0-5 per cent of the population are in Western India, mainly in Maharashtra. This population has been substantially reduced in this area with the conversion of many to Buddhism after independence, influenced by their leader Dr B.R. Ambedkar. The tribal belts of Central India in Madhya Pradesh and Orissa also have very small Scheduled Castes populations.

Whereas the Scheduled Castes population is distributed widely within a particular district, the Tribes are more usually found in concentrated groups, occupying in some cases a stretch of villages and hamlets. These concentrations of Tribes are generally to be found in forested hills and ranges. The Tribes are thus often referred to as 'hill' people. The districts in the north-eastern Assam-Burma hills, three districts in Madhya Pradesh, The Dangs in Gujarat, a couple of districts in Himachal Pradesh and the union territories of Lakshadweep, Dadra & Nagar Haveli and the Nicobar Islands have the highest concentrations of tribals (60-100 per cent). Other significant concentrations (40-60 percent) are in southern Gujarat and the Chota Nagpur plateau of southern Bihar, northern Orissa and eastern Madhya Pradesh. The districts around the hill areas also have a fair concentration of tribals because of tribal migrants. Most of Uttar Pradesh, northern Bihar, southern Tamil Nadu, coastal Kerala and the Saurashtran peninsula in Gujarat have minute tribal populations. Punjab, Haryana and Jammu & Kashmir have no tribal populations.

Note: All Census data are dated 1981 and refer to the 412 districts that then existed. The maps show the districts as they existed at the end of 1985, when there were 439 districts (437 districts for which boundaries were available are shown on the maps). Where 1981 districts have been broken up into two or more smaller districts or combined into larger districts (as in Jammu & Kashmir) the 1981 data are used to shade the 1985 configurations. In the text, which refers to the districts as they existed in 1985, the 1981 status is given within brackets.

Sources: 1. *General Population and Population of Scheduled Castes and Scheduled Tribes, Paper 2 of 1984*, Census of India 1981.
2. *Basic Statistics Relating to the Indian Economy, Vol. I: All India, August 1986*, Centre for Monitoring Indian Economy.

Table 10   SCHEDULED TRIBES POPULATION — 1971

ANDHRA PRADESH: *Gond/Naikpod* — 143,680 (Adilabad, Warrangal, Khammam); *Koya* — 220,146 (East Godavari, Khammam, Warrangal, West Godavari); *Konda Dhoras* — 86,911 (Vishakhapatnam, East Godavari); *Sugalis/Lambadis* — 96,174 (Anantapur, Guntur, Krishna, Kurnool); *Yenadis* — 205,381 (Chittoor, Nellore); *Yerukulas* — 128,024 (Anantapur, Krishna, Kurnool, Nellore).

ASSAM [1] : *Khasi Jaintia* — 356,155 (Khasi & Jaintia hills); *Mizo (Lushai)* — 214,721 (Mizoram); *Mikir* — 121,082 (North Cachar Hills, Mikir Hills); *Boro/Borokachari* — 345,983 (Darrang, Goalpara, Kamrup); *Kachari/Sonwal* — 256,720 (Darrang, Lakhimpur, Sibsagar); *Rabha* 108,029 (Darrang, Goalpara, Kamrup).

BIHAR: *Bhumij* — 101,057 (Singhbhum); *Ho* — 454,745 (Singhbhum); *Kharia* — 108,983 (Ranchi, Singhbum); *Kharwar* — 109,357 (Bhagalpur, Palamu, Ranchi); *Lohara/Lohru* — 92,609 (Ranchi, Singhbhum); *Munda* — 628,931 (Hazaribag, Palamu, Ranchi, Singhbhum); *Oraon* — 735,025 (Hazaribag, Palamu, Purnia, Ranchi); *Santal* — 1,541,345 (Bhagalpur, Hazaribag, Purnia, Santal Pargana, Singhbhum).

GUJARAT: *Bhil* — 1,124,282 (Banas Kantha, Vadodara, The Dangs, Panchmahals, Sabar Kantha, Surat); *Chandhri* — 143,576 (Surat); *Dhanka* — 128,024 (Vadodara, Bharuch); *Dhodia* — 275,787 (Surat); *Dubla* — 323,644 (Vadodara, Bharuch, Surat); *Gamit* — 158,703 (Surat); *Kokna* — 110,054 (Surat); *Naikda* — 108,024 (Vadodara, Panchmahals, Surat); *Rathawa* — 135,730 (Vadodara, Panchmahals).

HIMACHAL PRADESH: *Bhot* — 14,019 (Lahul & Spiti); *Gujjar* — 16,887 (Bilaspur, Chamba, Mandi, Sirmaur); *Konaura* — 27,251 (Kinnaur); *Pangwala* — 7722 (Chamba).

KARNATAKA: *Hasalaru* — 8905 (Chikmagalur, Shimoga); *Koraga* — 6382 (Dakshin Kannad); *Kuruba* — 9246 (Kodagu); *Marati* — 38,562 (Dakshin Kannad); *Naikda* — 70,648 (Belgaum, Bijapur, Dharwad); *Soligaru* — 10,653 (Mysore).

MADHYA PRADESH: *Baiga* — 114,738 (Mandla, Shahdol, Sidhi); *Bharia-Bhumia* — 122,902 (Chhindwara, Jabalpur, Panna, Shahdol, Surguja); *Bhils & Bhilalas* — 1,221,565 (Dhar, Jhabua, Ratlam, West Nimar); *Gond* — 2,857,515 (Balaghat, Bastar, Betul, Bilaspur, Chhindwara, Durg, Jabalpur, Mandla, Narsimhapur, Raigarh, Raipur, Raisen, Shahdol, Seoni, Sidhi, Surguja); *Halba* — 130,123 (Bastar, Durg, Raipur); *Kol* — 386,009 (Jabalpur, Mandla, Rewa, Shahdol, Satna, Sidhi); *Korku* — 154,245 (Betul, Chhindwara, East Nimar, Hoshangabad, Surguja); *Oraon* — 277,640 (Raigarh, Surguja); *Sahariya/Soharia/Saharia* — 174,320 (Guna, Gwalior, Morena, Vidisha).

MAHARASHTRA: *Bhil* — 575,041 (Ahmadnagar, Dhule, Nashik); *Gamit* — 102,321 (Dhule); *Gond* — 272,564 (Chandrapur, Nanded, Yavatmal); *Kathodi* — 138,730 (Raigad, Pune, Thane); *Kokna* — 212,836 (Dhule, Nashik, Thane); *Koli Mahadev*

— 274,244 (Ahmadnagar, Nashik, Pune, Thane); *Thakur/Thakar* — 159,372 (Ahmadnagar, Raigad, Nashik, Pune, Thane).

MANIPUR [2] : *Mao* — 28,810; *Tangkhul* — 43,943; *Thadon* — 47, 994.

NAGALAND: *Dimasa* — 2376 (Kohima); *Naga* — 36,820 (Kohima, Mokokchung, Tuensang).

ORISSA: *Bathudi* — 104,542 (Kendujhar, Mayurbhanj); *Bhumij* — 116,184 (Baleshwar, Mayurbhanj); *Gond* — 446,704 (Balangir, Kalahandi, Kendujhar, Koraput, Sambalpur, Sundargarh); *Khond* — 818,847 (Phulbani, Balangir, Ganjam, Kalahandi, Koraput); *Kisan* — 125,709 (Sambalpur, Sundargarh); *Kolha* — 194,111 (Baleshwar, Cuttack, Dhenkanal, Kendujhar, Mayurbhanj, Sambalpur, Sundargarh); *Lodha* — 192,078 (Cuttack, Dhenkanal, Kalahandi); *Munda* — 221,309 (Kendujhar, Sambalpur, Sundargarh); *Oraon* — 128,058 (Sambalpur, Sundargarh); *Pareja* — 159,866 (Kalahandi, Koraput); *Santal* — 574,964 (Baleshwar, Kendujhar, Mayurbhanj).

RAJASTHAN: *Bhil* — 906,705 (Banswara, Bhilwara, Chittaurgarh, Dungarpur, Sirohi, Udaipur); *Mina* — 1,155,399 (Bhartpur, Bhilwara, Chittaurgarh, Dungarpur, Jaipur, Kota, Sawai Madhopur, Tonk, Udaipur).

TAMIL NADU: *Irular* — 79,835 (Chengalpattu, Coimbatore, North Arcot); *Kattunayakan* — 6459 (Chengalpattu, Madurai, Nilgiri); *Malayali* — 129,952 (North Arcot, Salem).

TRIPURA [2] : *Chakma* — 22,386 *Haltam* — 16,298; *Jamatia* — 24,359; *Mag* — 10,524; *Noatia* — 16,010; *Riang* — 56,597; *Tripuri* — 186,799.

WEST BENGAL: *Bhumij* — 91,289 (Bankura, Medinipur, Puruliya, 24 Parganas); *Kora* — 62,029 (Bankura, Barddhaman, Medinipur); *Munda* — 145,475 (Jalpaiguri, Medinipur, Puruliya, 24 Parganas, West Dinajpur); *Oraon* — 297,394 (Darjiling, Jalpaiguri, West Dinajpur); *Santal* — 1,200,019 (Bankura, Birbhum, Barddhaman, Jalpaiguri, Maldah, Medinipur, 24 Parganas, West Dinajpur).

ANDAMAN & NICOBAR: *Nicobarese* — 13,903 (Nicobar).

DADRA & NAGAR HAVELI: *Dhodia* — 7887 *Kokna* — 7611; *Varli* — 32,494.

LAKSHADWEEP [3] : *Total* — 24,108.

Note:

Principal areas inhabited by the tribe are given within brackets.
[1] Includes Meghalaya & Mizoram.
[2] ST Population in Manipur & Tripura are scattered throughout the state.
[3] No specific name for ST population in Lakshadweep.

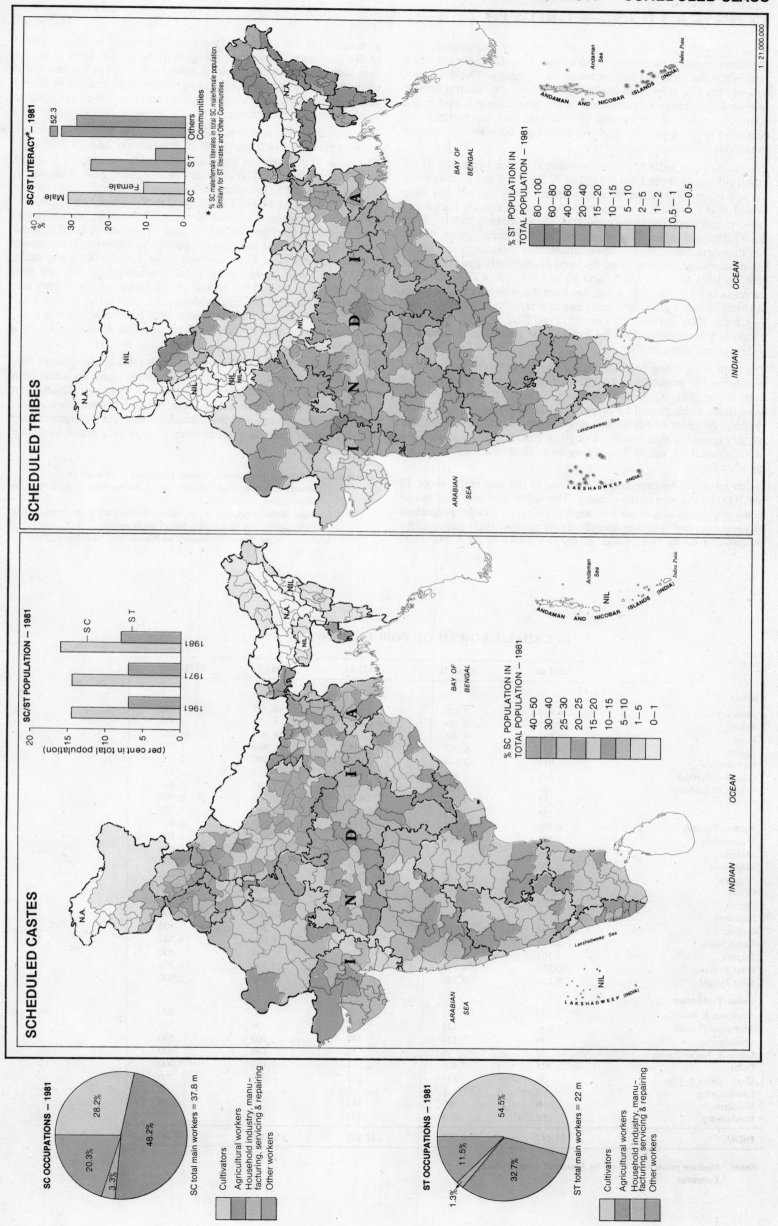

SCHEDULED TRIBES

SC/ST LITERACY* — 1981

52.3

Male

Female

SC   ST   Others
          Communities

* % SC male/female literates in total SC male/female population.
Similarly for ST literates and Other Communities.

% ST POPULATION IN
TOTAL POPULATION — 1981

80—100
60—80
40—60
20—40
15—20
10—15
5—10
2—5
1—2
0.5—1
0—0.5

BAY OF BENGAL

ANDAMAN AND NICOBAR ISLANDS (INDIA)

Andaman Sea

Indira Point

INDIAN OCEAN

ARABIAN SEA

Lakshadweep Sea

LAKSHADWEEP (INDIA)

1 : 21,000,000

SCHEDULED CASTES

SC/ST POPULATION — 1981

(per cent in total population)

SC

ST

1961   1971   1981

% SC POPULATION IN
TOTAL POPULATION — 1981

40—50
30—40
25—30
20—25
15—20
10—15
5—10
1—5
0—1

BAY OF BENGAL

ANDAMAN AND NICOBAR ISLANDS (INDIA)

Andaman Sea

Indira Point

NIL

INDIAN OCEAN

ARABIAN SEA

Lakshadweep Sea

LAKSHADWEEP (INDIA)   NIL

SC OCCUPATIONS — 1981

48.2%
3.3%
20.3%
28.2%

SC total main workers = 37.8 m

Cultivators
Agricultural workers
Household industry, manu-
facturing, servicing & repairing
Other workers

ST OCCUPATIONS — 1981

54.5%
1.3%
11.5%
32.7%

ST total main workers = 22 m

Cultivators
Agricultural workers
Household industry, manu-
facturing, servicing & repairing
Other workers

POPULATION — GROWTH

India's biggest problem is its ever increasing population. The population growth can be explained by the sharp decline in the death rate, while the birth rate has decreased only slightly. The difference between the two rates is normally referred to as the 'natural growth rate' and the change in population in any given place is due to the combined effect of the natural growth rate and the increase (or decrease) in net migration (*i.e.* the difference between the rates of in-migration and out-migration).

The birth rate in India has been declining from the beginning of the century (from 45.8 per 1000 in 1901 to 39.9 in 1951 and 37.2 in 1981, *i.e.* a 8.6 per 1000 fall in 80 years). In the years immediately after World War I, India had a high birth rate and an equally high death rate and correspondingly the population increased only slowly. After the 1920s, with general improvement in food supply, new drugs, health programmes, control of communicable diseases, and reduced infant mortality, the death rate declined dramatically (from 44.4 per 1000 in 1901 to 27.4 in 1951 and 15 in 1981, *i.e.* a 29.4 per 1000 decrease in 80 years). The result has been the widening gap between the birth rate and death rate, resulting in a rapid natural growth rate — 1.4 per 1000 in 1901; 12.5 in 1951; and 22.2 in 1981. The trend portends greater difficulties for India in the future, as the absolute increases are considerable.

The improving living conditions since the 1920s have also led to an increasing life expectancy. In 1921, an Indian could expect to live to be 20; in 1941, 32; in 1961, 41; and in 1981, 51. Life expectancy is expected to be around 59 in 1990 and about 65 by the turn of the century. In terms of life expectancy at birth, males have a slightly higher expectancy than females, though by 1990 female life expectancy is anticipated to be 59.7 years against 58.6 years of male life expectancy.

Death rates can be expected to continue to fall and reach about 10 per 1000 by the turn of the century. Thereafter they will tend to rise as the population as a whole will begin to comprise a larger proportion of ageing people. The age-specific death rates of 1980 indicate, for example, a death rate in age group 0-4 years of 41.8 per 1000 population; in the 60-64 age group of 31.2; in the 65-69 age group of 50; and in the 70-plus age group of 91.6. For the same year, the data indicate slightly higher death rates for females of 0-4 years, but slightly lower death rates in the age group over 50 years. For all age groups and in males and females, urban areas have lower age-specific death rates than rural areas.

The natural growth rate of population (1983) has been lowest in two union territories: 14.7 per 1000 in Pondicherry and 13.7 in Goa, Daman & Diu. In eleven states and one union territory, it is above the national average (21.4) — Bihar, 24.2; Assam, 22.4; Meghalaya, 21.7; Gujarat, 22.5; Haryana, 26.9; Himachal Pradesh, 22.3; Jammu & Kashmir, 22.8; Madhya Pradesh, 24; Rajasthan, 26.5; Sikkim, 23.6; Uttar Pradesh, 22.7 and Andaman & Nicobar islands, 24.8. The highest natural growth rate is seen in Rajasthan and Haryana, but in terms of actual numbers the growth in Rajasthan is enormous compared to Haryana. Three southern states, Kerala (18.2), Karnataka (19.5) and Tamil Nadu (16.2), show less natural growth rate than the northern states and the central Indian states, of which only one is below the national average (Maharashtra, 20.5). But even in these states, such increases in absolute terms can be several millions each, whereas in small states, such as Meghalaya and Assam, it is only a few thousands.

In the whole of India, rural birth and death rates are far greater than urban rates, one exception being Uttar Pradesh, where the urban birth rate (32.8) is slightly higher than the rural birth rate (29.6). But whether in rural India or urban India, the gap between death rates and birth rates is large and the natural growth rate will reduce only slowly, unless there is a remarkable change in attitude to family planning.

Sources: 1. *Population Data Sheet*, Madras Communication Institute.
2. *Seventh Five Year Plan 1985-90, Vol. II*, Government of India Planning Commission.
3. *Basic Statistics relating to the Indian Economy vol. I: All India, August 1986*, Centre for Monitoring Indian Economy.
4. *Report of the Expert Committee on Population*, New Delhi, 1986.

DECADAL GROWTH OF POPULATION

Table 11

(in thousand)

	1951-61	1961-71	1971-81	1981-91[1]	1991-2001[1]
States					
Andhra Pradesh	4,868	7,520	10,047	10,700	9,300
Assam	2,808	3,788	5,272[2]	5,100	5,500
Bihar	7,665	9,906	13,562	16,700	19,200
Gujarat	4,370	6,065	7,388	6,900	5,500
Haryana	1,917	2,446	2,886	3,600	2,400
Himachal Pradesh	426	648	821	800	700
Jammu & Kashmir	307	1,056	1,370	1,500	1,400
Karnataka	4,185	5,712	7,837	8,200	6,700
Kerala	3,355	4,443	4,107	4,600	4,100
Madhya Pradesh	6,300	9,282	10,525	12,100	10,800
Maharashtra	7,552	10,858	12,372	12,500	11,500
Manipur	202	293	348	400	400
Meghalaya	163	243	324	300	500
Nagaland	96	147	259	400	400
Orissa	2,903	4,396	4,425	5,100	4,800
Punjab	1,975	2,416	3,238	3,000	2,300
Rajasthan	4,185	5,610	8,496	10,300	11,300
Sikkim	24[2]	48[2]	106[2]	200	100
Tamil Nadu	3,568	7,512	7,209	8,000	6,700
Tripura	503	414	497	500	500
Uttar Pradesh	10,535	14,586	22,521	25,700	29,100
West Bengal	8,626	9,386	10,269	11,300	9,900
Union Territories					
Andaman & Nicobar	33	51	74	100	100
Arunachal Pradesh	N.A.	131	164	200	200
Chandigarh	96	137	195	200	500
Dadra & Nagar Haveli	17	16	30	100	100
Delhi	915	1,407	2,154	3,100	4,000
Goa, Daman & Diu	31	231	229	300	200
Lakshadweep	3	8	8	10	10
Mizoram	70[2]	66[2]	162[2]	200	200
Pondicherry	52	103	132	100	100
INDIA	78,147	108,925	137,025[2]	152,000	148,900

Note: [1] Medium projections as on 1st March of Census years.
[2] Estimated.

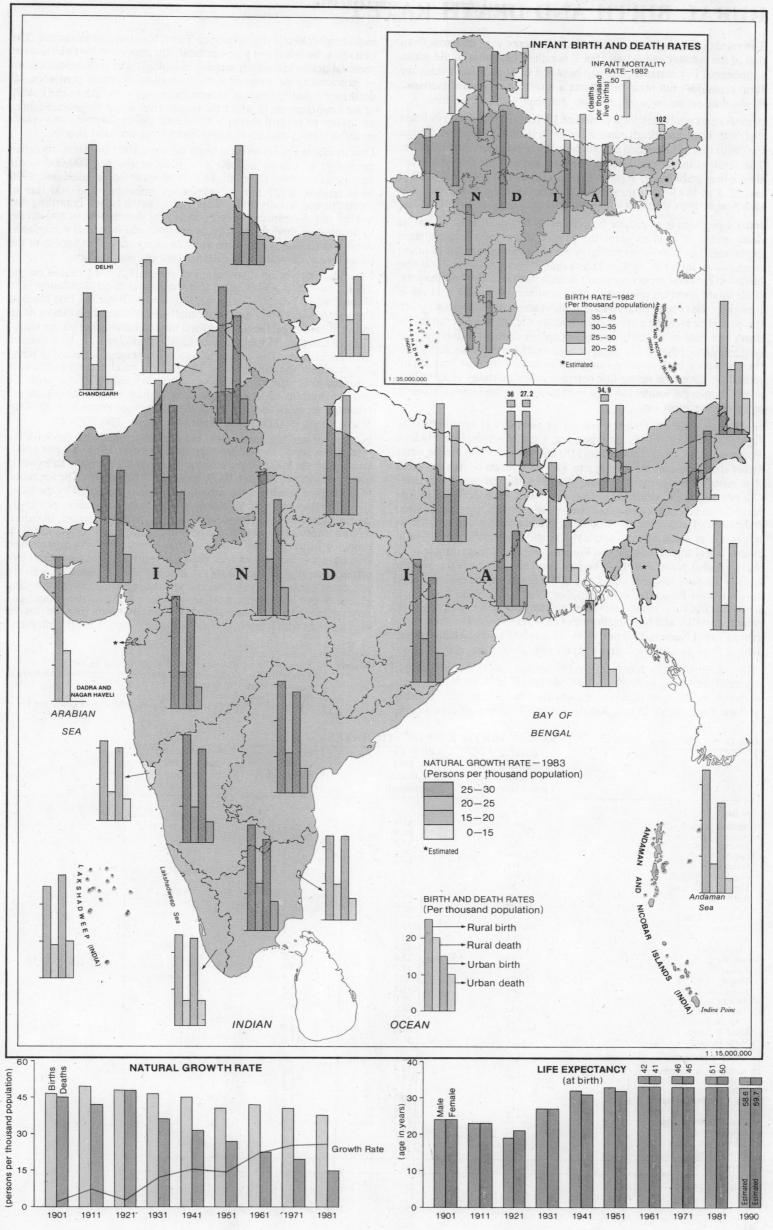

INFANT BIRTH AND DEATH RATES

INFANT MORTALITY RATE—1982
(deaths per thousand live births)

50
0

102

I N D I A

BIRTH RATE—1982
(Per thousand population)

35—45
30—35
25—30
20—25

* Estimated

1 : 35,000,000

LAKSHADWEEP (INDIA)

ANDAMAN AND NICOBAR ISLANDS

DELHI

CHANDIGARH

I N D I A

36 27.2   34.9

DADRA AND NAGAR HAVELI

ARABIAN SEA

BAY OF BENGAL

LAKSHADWEEP (INDIA)

Lakshadweep Sea

NATURAL GROWTH RATE—1983
(Persons per thousand population)

25—30
20—25
15—20
0—15

* Estimated

BIRTH AND DEATH RATES
(Per thousand population)

20 — Rural birth
   — Rural death
10 — Urban birth
   — Urban death
0

ANDAMAN AND NICOBAR ISLANDS (INDIA)

Andaman Sea

Indira Point

INDIAN   OCEAN

1 : 15,000,000

NATURAL GROWTH RATE

(persons per thousand population)

60
45
30
15
0

Births
Deaths

Growth Rate

1901 1911 1921 1931 1941 1951 1961 1971 1981

LIFE EXPECTANCY
(at birth)

(age in years)

40
30
20
10
0

Male
Female

42 41   46 45   51 50   58.6 59.7

Estimated Estimated

1901 1911 1921 1931 1941 1951 1961 1971 1981 1990

RURAL BIRTH AND DEATH RATES

The rural demographic experience in India is not very different from that of the country as a whole, but it is rather dissimilar to the urban experience. For instance, the gap between birth and death rates for rural areas does not seem to indicate a lowering of natural increase, in absolute terms, whereas the gap between urban rates does.

Annual rural vital rates for a period of 13 years (1970-1983) indicate that both birth and death rates have been declining slowly, though the death rate has declined at a slightly faster rate than the birth rate. The result is that the difference between the two for the period has shown near stability: 21.6 per 1000 in 1970, 20 in 1974, 19.4 in 1978 and 22.3 in 1983. This means that the convergence of the two may take longer than three or four generations.

India's problem is to reduce birth rates faster, or at least as fast as death rates, if it is to stem the consequences of the population explosion. The high infant mortality rate (the annual number of infant deaths — an infant being a child less than a year old — per 1000 live births) in the rural areas may result in high birth rates for two reasons. If an infant survives and the mother continues to breast feed, as is inevitably the case in rural India, lactation amenorrhoea will lengthen the period before the next conception; on the other hand, an infant death can lead to an early fresh conception due to the interruption of lactation and the earlier onset of renewed fecundity. Secondly, for the parents, a high fertility rate is an insurance against high infant mortality. Given the strong link between infant mortality and fertility, a reduction in the former can lead to a reduction in the latter, ultimately lowering the birth rate.

In three states of the Union, rural infant mortality is very high. The figures according to 1982 estimates are: Uttar Pradesh, 156; Madhya Pradesh, 145; Orissa, 139. Bihar (116) and Gujarat (120) are the other states above the national average of 114 rural infant deaths per 1000 live births. Kerala, the only state where both rural infant mortality (32 per 1000) and death rates (6.6 per 1000) have declined to low levels, is in advance of all other states, even though, by conventional indicators of economic development, Kerala cannot claim to be as advanced as some others, notably Punjab, Maharashtra and Tamil Nadu. Rural infant mortality in Punjab is 82, in Maharashtra it is 77 and in Tamil Nadu it is 97. The rate for Tamil Nadu is the highest among the four southern states and in the northern and eastern states (except West Bengal) it is even higher. Yet Tamil Nadu has shown a low level of rural natural increase, mainly because of relatively high infant mortality and high death rates (13.4 per 1000). The rural birth rate in Tamil Nadu is 29.2 per 1000. This is true also of Maharashtra, with a birth rate of 31.3 and a death rate of 9.9 per 1000.

The infant mortality rate is generally divided into two components: neonatal, referring to deaths within one month after birth, and post-neonatal, referring to deaths of infants after the first month. In terms of these rates, highly literate Kerala is naturally better off than others,

including economically developed Tamil Nadu and Maharashtra. The two rates, neonatal and post-neonatal, are important for two reasons: neonatal deaths are mainly a result of biological or endogenous causes — prematurity or congenital malformation; whereas post-neonatal deaths are mainly a result of exogenous causes — the nature and quality of the environments in which the infants live. Since exogenous causes are easier to control through immediate policy interventions, post-neonatal deaths can be reduced faster than neonatal deaths.

Data available (1978) on rural death rates per 100,000 infants by broad categories of causes indicate that 1248 deaths per 100,000 occur because of maternal ill-health, 1622 from water-borne infections, 1980 from tetanus, 2707 due to respiratory infections and 500 due to mosquito-borne infections. This raises basic issues regarding the quality and assurance of water supply, the distribution of and access to health care, and the quality of environment provided for infants. A system of health care does exist in every state, but access to the programmes is limited by distance and economic status.

From the population distribution shown in Table 12 (based on the 1981 Census), it will be seen that the largest rural population is in Uttar Pradesh (about 91 million), followed by Bihar (61 m), Madhya Pradesh (about 42 m), Andhra Pradesh and Maharashtra (about 41 m each), West Bengal (about 40 m) and the rest (among the bigger states) all have less than 33 million each. Small states obviously have smaller rural populations. This pattern is, to a certain extent, reflected in births and deaths in the same states. The highest rural natural increase is recorded in the largest rural states, Uttar Pradesh and Bihar with 2.1 and 1.5 million additions. The next highest natural increases are recorded in West Bengal (920,000), Madhya Pradesh (900,000), Maharashtra (800,000) and Andhra Pradesh (800,000).

In terms of national death rate, India is fast closing the gap with the rest of the world (world average death rate in 1981: 11 per 1000; India: 12.5 per 1000), but the birth rate gap is still considerable (world average birth rate: 29 per 1000; India: 36.2 per 1000). The approach to be adopted to reduce this natural increase, as suggested by the facts, should be two-pronged: an attack on rural infant mortality, especially deaths caused by exogenous forces relating to environment, to reduce overall mortality rates in the country; and a concerted effort at reducing births by rural family welfare (birth control) programmes. Demographic factors, such as the age at marriage, the number of live births, literacy of the mothers, income of the parents, delivery practices, the level of personal and social hygiene, and breast feeding practices are all important determinants of infant mortality, and improvement in all these aspects is necessary to provide for an environment which might in due course help reduce rural birth rates.

Sources:
1. *Primary Census Abstract, Part II B (i)*, Census of India.
2. *Family Welfare Programme in India Year Book 1983-84*, Ministry of Health & Family Welfare.
3. *Registrar General, India*.
4. World Bank's *World Development Report 1983*, Oxford University Press.

Table 12   **RURAL BIRTH AND DEATH RATES**

	Rural Population — 1981 (in thousand)	Birth Rate — 1983 (Per thousand population)	Death Rate — 1983 (Per thousand population)	Infant Mortality Rate[1] — 1982 (Per thousand live births)
States				
Andhra Pradesh	41,062	31.5	11.2	86
Assam	17,850[1]	34.9	12.1	103
Bihar	61,196	37.7	13.5	116
Gujarat	23,484	35.2	12.8	120
Haryana	10,095	37.9	9.8	100
Himachal Pradesh	3,960	33.2	10.5	70
Jammu & Kashmir	4,726	33.2	9.1	74
Karnataka	26,406	29.8	10.5	71
Kerala	20,682	24.9	6.7	32
Madhya Pradesh	41,592	40.1	15.9	145
Maharashtra	40,791	31.4	10.4	77
Manipur	1,045	30.0	7.3	N.A.
Meghalaya	1,094	32.8	9.2	N.A.
Nagaland	656	23.8	8.0	N.A.
Orissa	23,260	33.7	12.5	139
Punjab	12,141	30.7	10.3	82
Rajasthan	27,051	41.5	14.4	105
Sikkim	265	36.0	12.2	N.A.
Tamil Nadu	32,456	29.0	13.4	97
Tripura	1,827	24.5	7.8	N.A.
Uttar Pradesh	90,963	29.6	16.9	156
West Bengal	40,134	36.0	11.6	93
Union Territories				
Andaman & Nicobar	139	34.4	8.6	N.A.
Arunachal Pradesh	590	36.7	18.7	N.A.
Chandigarh	29	27.4	6.7	N.A.
Dadra & Nagar Haveli	97	40.1	14.0	N.A.
Delhi	452	33.1	8.4	N.A.
Goa, Daman & Diu	735	21.8	8.1	N.A.
Lakshadweep	22	24.6	9.4	N.A.
Mizoram	372	N.A.	N.A.	N.A.
Pondicherry	288	23.0	10.5	N.A.
INDIA	525,457[1]	35.3	13.0	114

Note: [1] Estimated

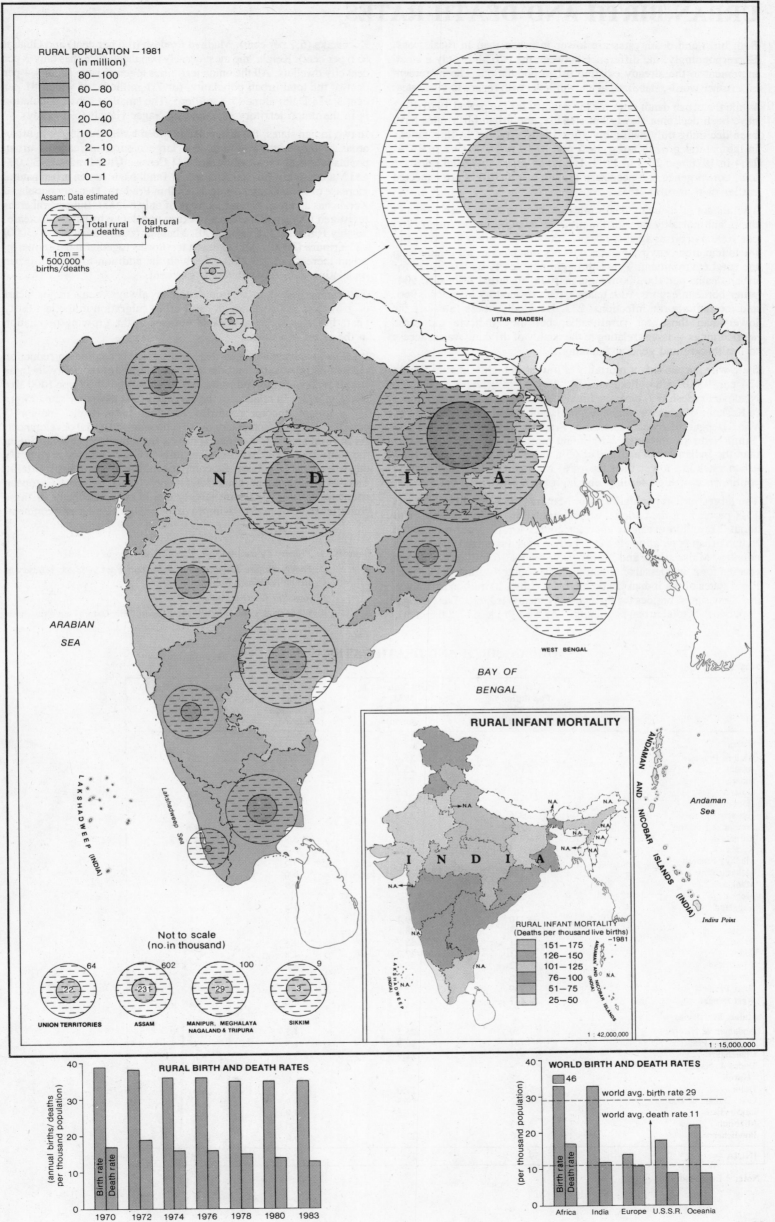

RURAL POPULATION — 1981
(in million)

80 — 100
60 — 80
40 — 60
20 — 40
10 — 20
2 — 10
1 — 2
0 — 1

Assam: Data estimated

Total rural deaths — Total rural births

1 cm = 500,000 births/deaths

UTTAR PRADESH

WEST BENGAL

ARABIAN SEA

BAY OF BENGAL

Andaman Sea

RURAL INFANT MORTALITY

I N D I A

RURAL INFANT MORTALITY
(Deaths per thousand live births)
-1981

151 — 175
126 — 150
101 — 125
76 — 100
51 — 75
25 — 50

N.A.

ANDAMAN AND NICOBAR ISLANDS (INDIA)

Indira Point

LAKSHADWEEP (INDIA)

1 : 42,000,000

Not to scale
(no. in thousand)

64   602   100   9

22   231   29   3

UNION TERRITORIES   ASSAM   MANIPUR, MEGHALAYA NAGALAND & TRIPURA   SIKKIM

1 : 15,000,000

RURAL BIRTH AND DEATH RATES

(annual births/ deaths per thousand population)

Birth rate   Death rate

1970   1972   1974   1976   1978   1980   1983

WORLD BIRTH AND DEATH RATES

(per thousand population)

46

world avg. birth rate 29

world avg. death rate 11

Birth rate   Death rate

Africa   India   Europe   U.S.S.R.   Oceania

URBAN BIRTH AND DEATH RATES

Both birth and death rates are lower in urban than in rural areas. Correspondingly, the difference between the two adds only a small increment to the already existing urban population. The increment is, in other words, stabilized at a lower level with regard to vital rates.

While the urban death rates for the country for the period 1970-1983 have been declining slowly but steadily, the urban birth rates have been declining only negligibly after 1972, resulting in fluctuations in the natural growth rates: 19.5 per 1000 in 1970, 19.2 in 1974, 18.4 in 1978 and 20.3 in 1983. However, since both vital rates are low, convergence and near zero growth of population are expected earlier than in rural areas.

The impact of reducing death rates in urban areas is clearly seen in the infant mortality rate for the urban population of the country (65 per 1000 live births against 114 per 1000 in rural areas in 1982). But the Indian urban environment is not very much better for infants than the rural environment. This is brought out clearly by the causes for urban deaths per 100,000 infants in 1978: maternal ill-health, 934; water-borne infections, 939; tetanus, 530; respiratory infections, 966; and mosquito-borne infections, 233. Decidedly, these rates are far lower than those for rural India, but they indicate the same shortcomings — issues relating to the quality of drinking water, access to health care and general sanitation.

State-wise, urban infant mortality is lowest in highly literate Kerala (24 per 1000). Uttar Pradesh (99), Gujarat (89), Assam (72) and Madhya Pradesh (79) have the highest rates. Besides Kerala, Jammu & Kashmir, Karnataka, Maharashtra, Bihar, Punjab, Andhra Pradesh, West Bengal, Haryana, Himachal Pradesh, Rajasthan, Orissa and Tamil Nadu are the states with infant mortality rates that are lower than the Indian urban average of 65 per 1000. Although these 1982 urban rates are lower than the rural rates, they still make for the continuance of high fertility and high birth rates.

The largest urban population is in Maharashtra (22 million); this is 13.8 per cent of the national urban population. Uttar Pradesh, with about 20 million urban dwellers, accounts for 12.5 per cent of the Indian urban population (1981). The two most urbanized states in the country, Maharashtra and Tamil Nadu (about 16 million), together account for a little less than a quarter of the country's urban population. Eight states of the Indian Union have more than 10 million urbanites, and these states, besides the three already mentioned, are: West Bengal (9 per cent of total urban population), Andhra Pradesh (7.8 per cent),

Karnataka (6.7 per cent), Madhya Pradesh (6.6 per cent) and Gujarat (6.6 per cent). Kerala, the most densely populated state, has only 3 per cent city dwellers. All the union territories together have about 4.4 per cent of the total urban population (or 7.1 million), of which 81 per cent are in Delhi alone (5.8 million). The smallest urban population is in the union territory of Dadra & Nagar Haveli (6914 only).

In two Indian states, the difference between births and deaths in urban areas is very high, resulting in a large natural increase of urban population being recorded by the 1981 Census: Uttar Pradesh, 430,000 and Maharashtra, 370,000. Urbanized Tamil Nadu has an urban natural increase of 260,000, the same as Andhra Pradesh. Densely populated Kerala has an urban natural increase of only 80,000. But several areas registered very small increases of urban population by natural increase, notably Himachal Pradesh (4000), Manipur (8000), Meghalaya (3000) and Tripura (2000). All the union territories together have shown an urban increase of 124,000, of which the addition to Delhi's urban population alone constitutes 84 per cent.

Natural increase in urban population has always been a major factor of growth in India, although rural-urban migration in the last thirty years has also added a sizeable number to the growing population in the cities and towns.

It must, however, be noted that, while there is continuing reduction in both national birth and death rates, the infant mortality in India is still high as compared to the world's average of 59 per 1000 live births in 1983. In relation to the average infant mortality rate (75 per 1000) for low-income countries, of which India is one, India's 93 per 1000 is a sad commentary on its economic and social development. As in rural areas, post-neonatal deaths form the larger part of infant mortality rates in urban areas because deaths in infancy are mainly due to exogenous causes. By comparison, the urban infant mortality rates of industrialized countries (10 per 1000) and East European countries (30 per 1000) are largely due to neonatal deaths as these countries have been able to improve the quality of their environment.

Sources: 1. *Primary Census Abstract, Part II B (i)*, Census of India.

2. *Family Welfare Programme in India Year Book 1983-84*, Ministry of Health & Family Welfare.

3. *Registrar General, India*.

4. World Bank's *World Development Report 1983*, Oxford University Press.

URBAN BIRTH AND DEATH RATES

Table 13

	Urban Population — 1981 (in thousand)	Birth Rate — 1983 (Per thousand population)	Death Rate — 1983 (Per thousand population)	Infant Mortality Rate[1] — 1982 (Per thousand live births)
States				
Andhra Pradesh	12,488	27.7	6.9	50
Assam	2,047[1]	23.7	6.7	72
Bihar	8,719	32.1	7.4	60
Gujarat	10,602	31.3	8.7	89
Haryana	2,827	29.7	6.5	62
Himachal Pradesh	326	22.2	6.3	42
Jammu & Kashmir	1,260	25.2	6.8	43
Karnataka	10,730	25.8	5.9	47
Kerala	4,771	24.6	6.7	24
Madhya Pradesh	10,586	31.7	8.5	79
Maharashtra	21,994	26.2	6.6	55
Manipur	375	23.9	5.9	N.A.
Meghalaya	241	17.2	4.4	N.A.
Nagaland	120	18.6	1.4	N.A.
Orissa	3,110	29.1	8.7	64
Punjab	4,648	28.8	7.3	53
Rajasthan	7,211	33.7	9.8	60
Sikkim	51	27.2	4.8	N.A.
Tamil Nadu	15,952	25.6	8.2	51
Tripura	226	16.6	4.9	N.A.
Uttar Pradesh	19,899	32.8	10.4	99
West Bengal	14,447	21.3	6.6	52
Union Territories				
Andaman & Nicobar	50	25.5	4.2	N.A.
Arunachal Pradesh	41	19.9	1.5	N.A.
Chandigarh	423	21.9	3.5	N.A.
Dadra & Nagar Haveli	7	N.A.	N.A.	N.A.
Delhi	5,768	27.0	7.0	N.A.
Goa, Daman & Diu	352	19.5	5.9	N.A.
Lakshadweep	19	27.8	9.9	N.A.
Mizoram	122	N.A.	N.A.	N.A.
Pondicherry	316	22.6	6.6	N.A.
INDIA	159,727[1]	28.0	7.7	65

Note: [1] Estimated

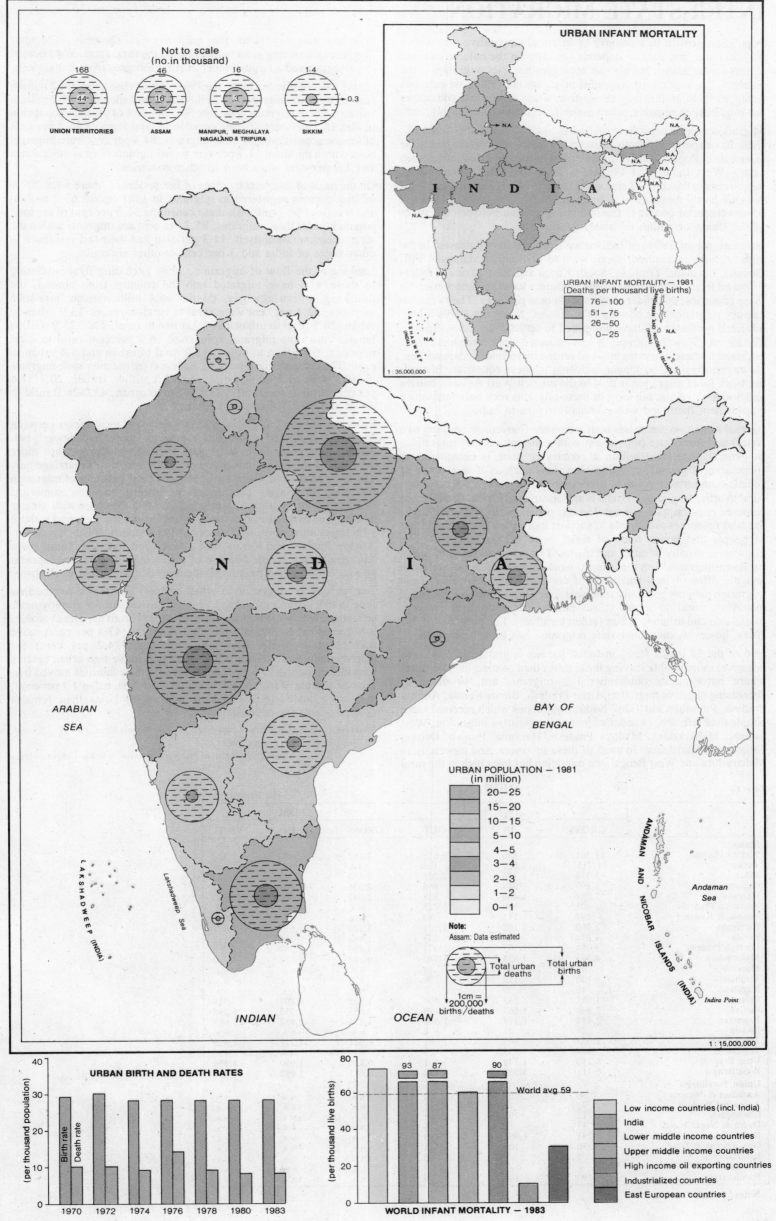

Not to scale
(no. in thousand)

168	46	16	1.4
44	16	3	0.3
UNION TERRITORIES	ASSAM	MANIPUR, MEGHALAYA NAGALAND & TRIPURA	SIKKIM

URBAN INFANT MORTALITY

I N D I A

URBAN INFANT MORTALITY — 1981
(Deaths per thousand live births)

- 76 — 100
- 51 — 75
- 26 — 50
- 0 — 25

1 : 35,000,000

LAKSHADWEEP (INDIA)

ANDAMAN AND NICOBAR ISLANDS (INDIA)

I N D I A

ARABIAN SEA

BAY OF BENGAL

URBAN POPULATION — 1981
(in million)

- 20 — 25
- 15 — 20
- 10 — 15
- 5 — 10
- 4 — 5
- 3 — 4
- 2 — 3
- 1 — 2
- 0 — 1

Note:
Assam: Data estimated

Total urban deaths → Total urban births
1cm = 200,000 births/deaths

LAKSHADWEEP (INDIA)

Lakshadweep Sea

ANDAMAN AND NICOBAR ISLANDS (INDIA)

Andaman Sea

Indira Point

INDIAN   OCEAN

1 : 15,000,000

URBAN BIRTH AND DEATH RATES

(per thousand population)

Birth rate / Death rate

1970 1972 1974 1976 1978 1980 1983

WORLD INFANT MORTALITY — 1983

(per thousand live births)

93 87 90

World avg 59

- Low income countries (incl. India)
- India
- Lower middle income countries
- Upper middle income countries
- High income oil exporting countries
- Industrialized countries
- East European countries

INTER-STATE MIGRATION

Population growth in a country or in an administrative unit within it — in India, the state — depends not only on the natural increase (births minus deaths) but also on net migration. Net migration is the number of in-migrants (immigrants) minus the number of out-migrants (emigrants). Migration may be positive, when in-migrants outnumber out-migrants, or negative, when out-migrants outnumber in-migrants.

Migration may be broadly classified as 'international' and 'internal'. India has experienced both in the past, labour migration being the main cause. Populations of Indian origin in regions or countries such as the West Indies and Fiji are a result of international labour movements in the colonial era, while those in North America and West Asia are partly the result of a 'brain drain'. A corresponding process of movement of people is the internal migration between the states of the Union or within the states themselves.

International migration of Indians has occurred in two phases. In the 19th century, indentured labour went to Sri Lanka, Malaysia, Fiji, Guyana, Surinam, Trinidad, South Africa and East Africa. Traders followed them. While most of the labourers have become settlers in these countries, the traders have sought new pastures. The twentieth century migration has been in two streams. Since the 1970s highly educated professionals have emigrated to settle in the USA, Britain, Canada and Western Europe. A less-educated work force had earlier migrated to these countries and had settled there since Independence. However, they are no longer welcome in these countries. Instead, the work force migration is now to the oil-rich West Asian countries and South-east Asia, but most of these migrants seek only temporary employment there and will eventually return to India.

Internal migration comprises both inter-state (movement between two states) and intra-state (movement within the same state) migration. In general, migration within a country or state is considered as 'population dispersal' or adjustment, inasmuch as it is a result of social, economic or political processes. For instance, the dominant rural to urban migration in India is an adjustment effected to overcome resource constraints in the rural areas; pressure on land being high, the land resources are unable to support more than a certain number of people and hence some of them move out in search of jobs elsewhere, usually to cities and towns. The urban nodes also attract rural out-migrants with the offer to work in new economic activities and the offer of new standards of comfort and life-styles. These migration patterns generally represent four types: rural to rural (rural turnover), rural to urban (rural push), urban to rural (reverse migration) and urban to urban (urban turnover). For purposes of this Atlas, however, only inter-state migration has been considered.

In 6 of the 14 major states in India, the net migration in 1981 was negative, more people leaving these states then coming in. The states where out-migrants outnumbered in-migrants are, in order of decreasing negative migration, Uttar Pradesh, Bihar, Kerala, Andhra Pradesh, Rajasthan and Tamil Nadu. Eight states which received more people than left, are, in order of increasing positive migration, West Bengal, Maharashtra, Madhya Pradesh, Haryana, Punjab, Orissa, Gujarat and Karnataka. In most of these instances, and especially in Maharashtra and West Bengal, the migration has been high in the rural to urban category, with the million cities (Bombay, Calcutta, Bangalore) attracting great numbers from the rural areas. West Bengal has also received a large number of rural refugees from Bangladesh.

On the basis of place of birth, the 1981 Census registered 204.2 million persons as being migrants, 62 million of them males and 142.2 million females. These migrants constitute 30.7 per cent of the total population (males 18.1 per cent of male population, and females 44.2 per cent of female population). Of these migrants, 84.4 per cent were migrants born within the state, 11.7 per cent were migrants born in other states and 3.9 per cent were born in other countries.

On the basis of migrants by place of last residence, there were 207.9 million persons registered as migrants in 1981 (males 63.2 million and females 144.7 million); these constitute 31.3 per cent of the total population. Of these migrants, 85.7 per cent are migrants within the state of enumeration itself, 11.3 per cent had their last residence in other states of India and 3 per cent in other countries.

Looking at the flow of migration, while excluding figures relating to those who have migrated into the country from abroad, the following pattern emerges. Of the 46.4 million male intra-state migrants, 52.4 per cent were rural to rural migrants, 13.9 urban to urban, 26.5 rural to urban and 7.2 urban to rural. The 125.9 million female intra-state migrants comprised 76.9 per cent rural to rural migrants, 6.7 urban to urban, 11.0 rural to urban and 5.4 urban to rural. The comparative figures for male and female inter-state migrants in 1981 were 11.5 million male and 12.5 million female, 20.7/37.6 per cent rural to rural, 30.5/28.1 urban to urban, 42.0/26.0 rural to urban and 60.7/7.8 urban to rural.

From the numbers migrating and the pattern of movements between states, differentiated in terms of number as 'more' and 'fewer', two principal impressions may be gathered: some migrants are more attracted by job opportunities, others move because of marriage ties. The large number of women migrants is a clear indicator of migration following marriage. The migration pattern and the numerical differentials of migration clearly indicate that the states with greater attraction in terms of employment, education etc. are preferred to those with lesser attractions, for instance, Uttar Pradesh, Bihar, Rajasthan, Andhra Pradesh and Kerala. But a developed state like Tamil Nadu has been losing people because of a slowing down of development or saturation caused by a larger-than-needed, qualified work force.

The reasons for migration to urban centres, ascertained for the first time in the 1981 Census, have been categorized under employment, education, family move, marriage and others. From the data available, it becomes apparent that more male migrants (43.1 per cent) move into cities or towns than female migrants (4.2 per cent) for employment; more females (32.5 per cent) move into urban centres than male migrants (27.3 per cent) because their families moved into them or because of marriage (females 46.6 per cent, males 1.1 per cent); and more males (6.0 per cent) migrate to cities than females (2.4 per cent) for education.

Sources: 1. *Report and Tables based on 5 per cent Sample Data, Part II — Special*, Census of India 1981.
2. *Geographic Distribution of Internal Migration in India*, Census of India 1971.

Table 14

MIGRANTS

(in thousand)

	1971			1981		
	GROSS	IN	OUT	GROSS	IN	OUT
States						
Andhra Pradesh	11,361	746	10,615	2,192	870	1,322
Assam	1,715¹	1,522¹	193¹	N.A.	N.A.	N.A.
Bihar	3,378	1,176	2,201	3,885	1,348	2,537
Gujarat	1,809	944	865	2,336	1,304	1,032
Haryana	2,244	1,406	838	2,842	1,686	1,156
Himachal Pradesh	403	180	223	N.A.	N.A.	N.A.
Jammu & Kashmir	192	111	80	N.A.	N.A.	N.A.
Karnataka	2,309	1,188	1,121	3,128	1,637	1,491
Kerala	1,214	271	943	1,468	356	1,112
Madhya Pradesh	3,084	2,107	977	3,877	2,369	1,508
Maharashtra	4,765	3,625	1,140	6,123	4,633	1,490
Manipur	48	37	11	N.A.	N.A.	N.A.
Meghalaya	146	117	29	N.A.	N.A.	N.A.
Nagaland	46	39	7	N.A.	N.A.	N.A.
Orissa	1,081	587	494	1,335	807	528
Punjab	2,860	1,667	1,193	3,053	1,757	1,296
Rajasthan	2,484	1,118	1,366	3,135	1,452	1,683
Sikkim	29	21	8	N.A.	N.A.	N.A.
Tamil Nadu	2,053	969	1,083	2,621	1,212	1,409
Tripura	584	555	29	N.A.	N.A.	N.A.
Uttar Pradesh	5,210	1,749	3,461	6,836	1,980	4,856
West Bengal	6,111	5,332	780	6,673	5,692	981
Union Territories						
Andaman & Nicobar	60	57	2	N.A.	N.A.	N.A.
Arunachal Pradesh	72	70	2	N.A.	N.A.	N.A.
Chandigarh	212	192	103	N.A.	N.A.	N.A.
Dadra & Nagar Haveli	17	12	5	N.A.	N.A.	N.A.
Delhi	2,241	1,961	280	N.A.	N.A.	N.A.
Goa, Daman & Diu	230	109	121	N.A.	N.A.	N.A.
Lakshadweep	2	1.8	0.2	N.A.	N.A.	N.A.
Mizoram	—	—	—	N.A.	N.A.	N.A.
Pondicherry	190	104	86	N.A.	N.A.	N.A.

Note: ¹ Includes Mizoram

34

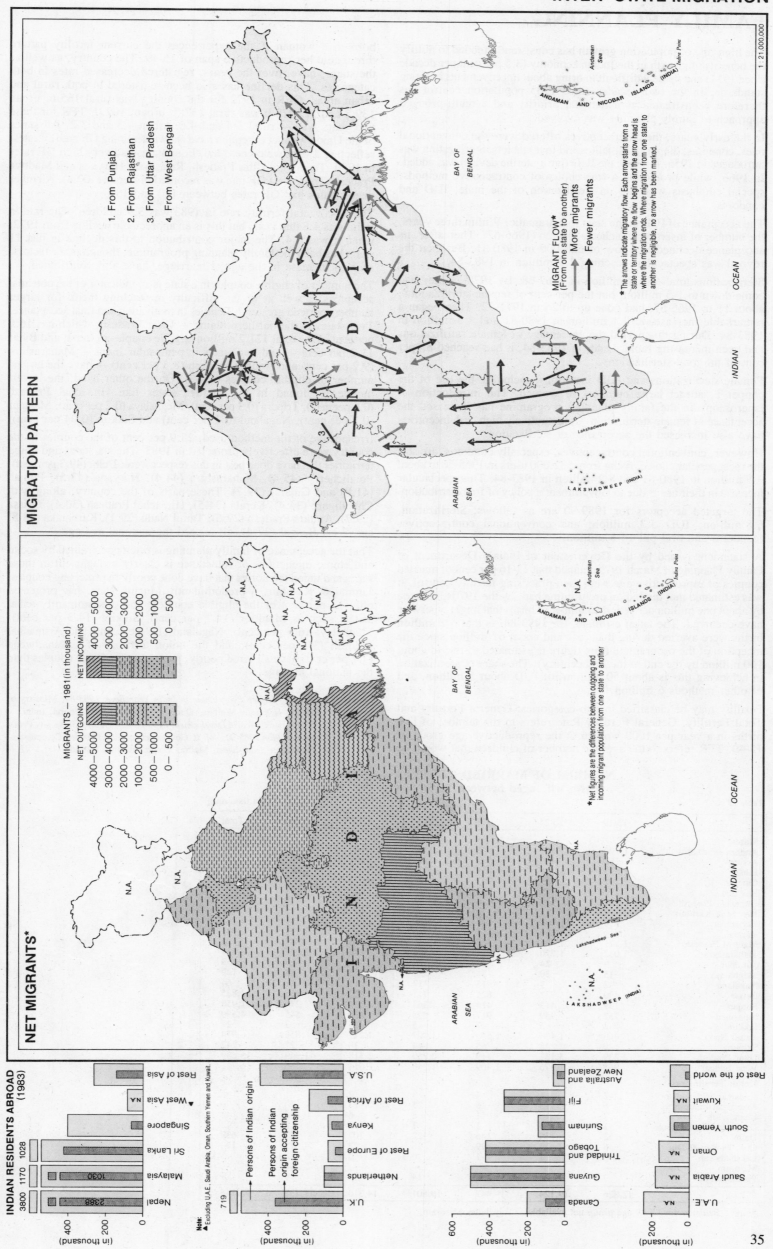

MIGRATION PATTERN

1. From Punjab
2. From Rajasthan
3. From Uttar Pradesh
4. From West Bengal

MIGRANT FLOW*
(From one state to another)
More migrants
Fewer migrants

* The arrows indicate migratory flow. Each arrow starts from a state or territory where the flow begins and the arrow heads is where the migration ends. Where migration from one state to another is negligible, no arrow has been marked.

1 : 21.000.000

NET MIGRANTS*

MIGRANTS — 1981 (in thousand)

NET OUTGOING	NET INCOMING
4000 – 5000	4000 – 5000
3000 – 4000	3000 – 4000
2000 – 3000	2000 – 3000
1000 – 2000	1000 – 2000
500 – 1000	500 – 1000
0 – 500	0 – 500

*Net figures are the differences between outgoing and incoming migrant population from one state to another

INDIAN RESIDENTS ABROAD (1983)

Rest of Asia
West Asia ▲ N.A.
Singapore
Sri Lanka   1028
Malaysia   1170   1030
Nepal   3800   2388

(in thousand) 400 200 0

Note:
▲ Excluding U.A.E. Saudi Arabia, Oman, Southern Yemen and Kuwait.

U.S.A.
Rest of Africa
Kenya
Rest of Europe
Netherlands
U.K.   719

Persons of Indian origin

Persons of Indian origin accepting foreign citizenship

(in thousand) 400 200 0

Australia and New Zealand
Fiji
Surinam
Trinidad and Tobago
Guyana
Canada

(in thousand) 600 400 200 0

Rest of the world
Kuwait   N.A
South Yemen
Oman   N.A
Saudi Arabia N.A
U.A.E.   N.A

(in thousand) 200 0

35

FAMILY PLANNING

The high rate of population growth has consistently tended to nullify the benefits of growth in the Indian economy (3.5 per cent per decade since 1951) and made it difficult to bring about improvements in living standards. In the two decades since 1965 population control has therefore been treated as a national priority and a multi-pronged approach to family planning was evolved.

In the early years, the contraceptives offered were the conventional ones: condoms, diaphragms, jellies and foam tablets. Sterilization was introduced in 1956. In 1965, the IUD (intra-uterine device) was added. In 1966, while promoting a free choice of contraceptive methods, special emphasis was laid on sterilization of the male, IUD and condoms.

The acceptance of IUD at the start was dramatic. Within three years, the number of insertions reached a million (1966-67). Thereafter, its acceptance declined to a low of 628,000 in 1980-81. But then the revival was spectacular, reaching 2.1 million in 1983-84.

Sterilizations totalled 1.8 million in 1967-68. By 1970-71 they had come down to 1.3 million, but the per cent of female sterilizations, about 11 in 1966-67, had gone up to 26 in 1971-72. Then began a remarkable increase, to 2.1 million in 1980-81 and 4.5 million in 1983-84. During this period the percentage of female sterilizations had been increasing rapidly and, in 1983-84, it had reached 85 per cent of the total sterilizations.

Female sterilizations and IUD insertions, including the use of the copper-T, attract large numbers of women. The introduction of laparoscopy in the family planning programme has increased the acceptance of female sterilization. Comparatively high cash incentives have also increased the acceptance of sterilization.

However, conventional contraceptives, especially condoms, are still the most popular choice, rising from 582,000 users in 1965-66 to about 3.8 million in 1980-81 and 8.2 million in 1983-84. This spectacular increase in their use is due to Government's policy of free distribution.

The targeted acceptors for 1989-90 are as follows: Sterilization, 6.8 million; IUD 5.3 million; and conventional contraceptives (condom) and oral pill 14.5 million.

A statement issued by the Government of India's Department of Family Planning in March 1975 claimed that 12-16 per cent of married couples of reproductive age were then practising fertility control. It was estimated that the Indian programme had, by the 1970s, reduced by about one million per year the number of births that might otherwise have occurred. The latest estimate, for 1983-84, is that 6.2 million births were averted during that year and about 61 million since the inception of the programme (this figure is estimated to rise to about 100 million by the end of the 21st century). The share of sterilization in achieving this is about 50 million, of IUD about 5 million, and of other methods 6 million.

Fertility may be classified in two categories: General Fertility and Total Fertility. General Fertility Rate refers to the number of live births in a year per 1000 women in the reproductive age group of 15-49. TFR refers to the average number of children that would be born to a woman if she experiences the current fertility pattern throughout her reproductive span of 15-49. The country, as well as the states, have, over the years, registered decreased rates in both categories. This decline has also been registered in both rural and urban areas. GFR in 1972 for the country was rural 165.6, urban 139.8 and in 1978 it was rural 137.3, urban: 102.0. TFR for these two years was — rural 5.4 and 4.6, urban 4.3 and 3.2. All states of the Union without exception have shown declining GF and TF rates, reflecting a very welcome trend. Four states had very high GF rates in 1980. They were Uttar Pradesh, Rajasthan, Haryana and Madhya Pradesh. The lowest GF rate was recorded in Kerala (100.4). All other major states had GF rates between 110 and 150.

The world total fertility rate in 1983 was 3.9 children. The rate in India was 4.8 that year, but this is an appreciable decline from 1972, when it was 6.4. The major contribution to this decline in just 11 years has been the family planning programme, though other factors, such as increase in the age at marriage, have also contributed.

The number of eligible couples in a state is an indicator of the potential acceptors as well as of the difficulty in reaching them, for larger numbers portend greater difficulties in motivation and final acceptance. Two states in the northern plains — Uttar Pradesh, which has 16.9 per cent of the total 121.2 million eligible couples in India, and Bihar 11.2 per cent — and two states in peninsular India — Maharashtra (9.2 per cent) and Andhra Pradesh (8.3 per cent) — have the largest number of eligible couples (1983). On the other hand, the lowest number is found in the smaller states like Himachal Pradesh (0.6 per cent), Tripura (0.3 per cent), Meghalaya (0.2 per cent), Manipur (0.2 per cent), Nagaland (0.1 per cent) and Sikkim (0.04 per cent).

Irrespective of the method used, 25.9 per cent of the couples in the country were effectively protected in 1983. The six states and union territories that have done best in this respect were: Delhi (49.3 per cent), Pondicherry (45.4), Maharashtra (44.4), Haryana (43.5), Punjab (41.8) and Gujarat (39.7). These parts of the country, along with Chandigarh (34.4), Kerala (34.5), Himachal Pradesh (30.4), Orissa (29.7), Andhra Pradesh (29.6), Tamil Nadu (29.1), Karnataka (28.8) and West Bengal (27.4), all exceed the national average.

That the acceptance of family planning is often constrained by social and ethnic injunctions or reluctance is clearly brought out in those states and union territories that have done poorly in protecting couples. Jammu & Kashmir, a predominantly Muslim state, has protected only 13 per cent of the eligible couples; and predominantly tribal states, such as Manipur (14.7 per cent), Sikkim (14.3 per cent), Meghalaya (6.6 per cent), Nagaland (1.9 per cent) and Arunachal Pradesh (3.1 per cent), and the union territory of Lakshadweep (7.7 per cent) have all fared poorly, protecting far fewer couples than the national average.

Sources: 1. *Family Welfare Programme in India Year Book 1983-84*, Ministry of Health & Family Welfare. Quoting from Registrar General, India.
2. World Bank's *World Development Report 1985*, Oxford University Press.
3. *Seventh Plan 1985-90, Vol. II*, Government of India Planning Commission.
4. *Population Data Sheet*, Madras Communication Institute.

NUMBER OF MARRIED COUPLES — 1971
(with wife aged between 15 & 44 years[1])

Table 15 (in thousand)

	15-19	20-24	25-29	30-34	35-39	40-44	Total
States							
Andhra Pradesh	1,190	1,627	1,638	1,341	1,148	900	7,846
Assam[2]	262	460	509	387	341	237	2,197
Bihar	1,613	2,164	2,055	1,921	1,738	1,267	10,558
Gujarat	467	966	907	811	673	542	4,266
Haryana	271	348	305	267	232	198	1,620
Himachal Pradesh	81	125	124	104	93	71	598
Jammu & Kashmir	89	151	170	142	131	95	778
Karnataka	630	1,004	1,034	850	728	551	4,797
Kerala	219	649	618	540	562	391	2,981
Madhya Pradesh	1,171	1,496	1,506	1,337	1,076	847	7,433
Maharashtra	1,032	1,770	1,854	1,574	1,375	1,027	8,631
Manipur	9	24	29	29	25	20	138
Meghalaya	13	30	38	27	23	18	149
Nagaland	2	9	15	13	13	11	64
Orissa	490	731	825	711	622	456	3,835
Punjab	149	413	414	363	324	276	1,938
Rajasthan	752	959	917	808	620	545	4,599
Sikkim	2	5	7	5	5	3	27
Tamil Nadu	469	1,430	1,669	1,307	1,243	856	6,974
Tripura	29	48	55	44	38	27	242
Uttar Pradesh	2,339	3,131	3,119	2,789	2,318	1,949	15,639
West Bengal	965	1,376	1,409	1,240	1,014	736	6,722
Union Territories							
Andaman & Nicobar	2	4	4	3	2	1	16
Arunachal Pradesh	6	13	18	15	13	10	76
Chandigarh	2	10	11	8	6	5	42
Dadra & Nagar Haveli	1	3	3	2	2	2	12
Delhi	52	135	146	122	107	77	639
Goa, Daman & Diu	5	22	29	26	22	15	120
Lakshadweep	0.7	1	1	0.9	0.9	0.6	5.1
Mizoram	—	—	—	—	—	—	—
Pondicherry	7	16	18	14	13	9	77
INDIA	12,326	19,124	19,447	16,801	14,310	11,137	93,145

Note: [1] Statistics for 45-49 age group not available   [2] Includes Mizoram.

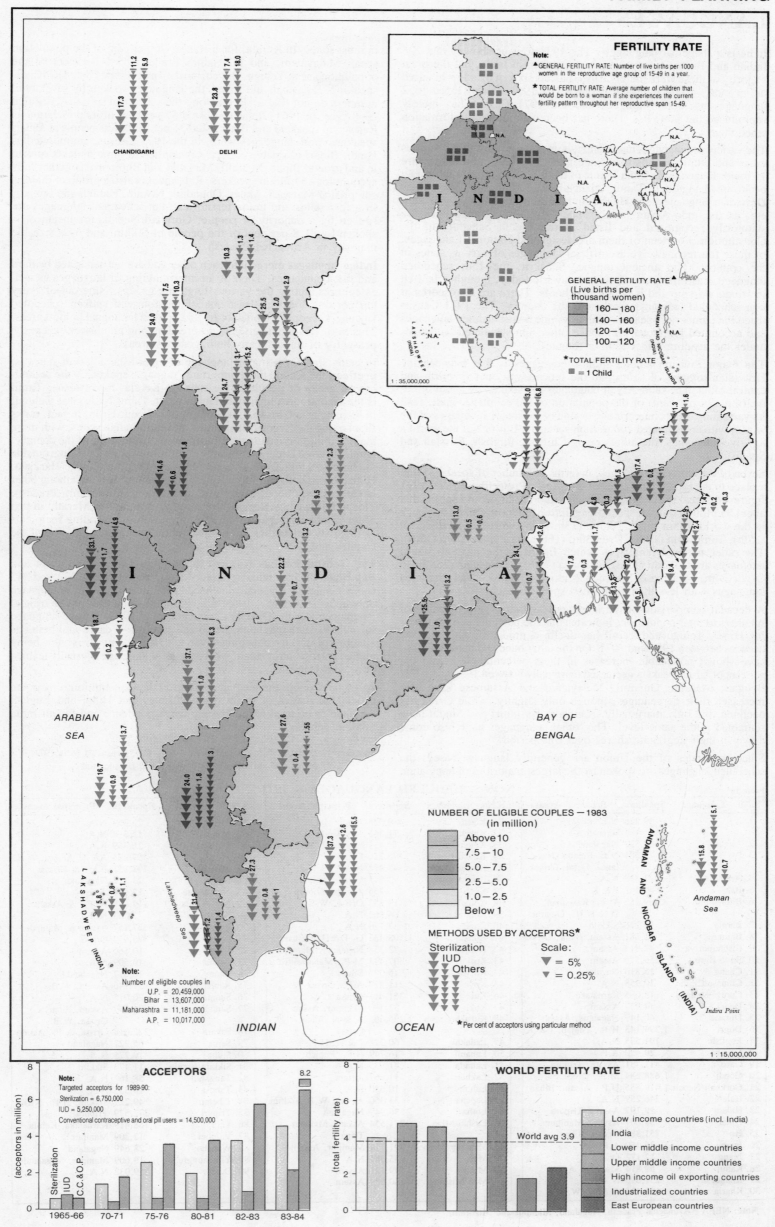

FERTILITY RATE

Note:
▲ GENERAL FERTILITY RATE: Number of live births per 1000 women in the reproductive age group of 15-49 in a year.

★ TOTAL FERTILITY RATE: Average number of children that would be born to a woman if she experiences the current fertility pattern throughout her reproductive span 15-49.

GENERAL FERTILITY RATE ▲
(Live births per thousand women)
- 160—180
- 140—160
- 120—140
- 100—120

★ TOTAL FERTILITY RATE
■ = 1 Child

1 : 35,000,000

CHANDIGARH

DELHI

ARABIAN SEA

BAY OF BENGAL

I N D I A

LAKSHADWEEP (INDIA)

Lakshadweep Sea

ANDAMAN AND NICOBAR ISLANDS (INDIA)

Andaman Sea

Indira Point

INDIAN OCEAN

NUMBER OF ELIGIBLE COUPLES — 1983
(in million)
- Above 10
- 7.5 — 10
- 5.0 — 7.5
- 2.5 — 5.0
- 1.0 — 2.5
- Below 1

METHODS USED BY ACCEPTORS★
Sterilization
IUD
Others

Scale:
▼ = 5%
▾ = 0.25%

★ Per cent of acceptors using particular method

Note:
Number of eligible couples in
U.P. = 20,459,000
Bihar = 13,607,000
Maharashtra = 11,181,000
A.P. = 10,017,000

1 : 15,000,000

ACCEPTORS

Note:
Targeted acceptors for 1989-90:
Sterilization = 6,750,000
IUD = 5,250,000
Conventional contraceptive and oral pill users = 14,500,000

(acceptors in million)

Sterilization
IUD
C.C.&O.P.

8.2

1965-66 | 70-71 | 75-76 | 80-81 | 82-83 | 83-84

WORLD FERTILITY RATE

(total fertility rate)

World avg 3.9

- Low income countries (incl. India)
- India
- Lower middle income countries
- Upper middle income countries
- High income oil exporting countries
- Industrialized countries
- East European countries

LANGUAGES

India is a 'babel of languages'. The 1931 Census showed that 225 Indian and Burmese languages were spoken in India, but the great majority of these were merely tribal dialects spoken by a few hundred or, at most, a few thousand people. The 1961 Census listed 1652 languages as mother tongues and the 1971 Census has merely reproduced the same list. There has been no additional information since then.

The variety of languages has been built up through the ages by various races and ethnic groups that made their way into India. There are 15 major languages in India. Some of these are closely allied and may be grouped together: Tamil, Malayalam, Telugu and Kannada as Dravidian languages; and Hindi, Punjabi, Bengali, Gujarati, Marathi, etc. as the Indo-Aryan family. The fifteen major languages are statutorily recognized and listed in the Eighth Schedule of the Constitution. Fourteen of them are designated National Languages; of these fourteen, twelve are official languages of various states in the country. The ancient tongue, Sanskrit, the 15th scheduled language, has little place in day-to-day life today, though it is used in Hindu worship and is taught in schools. There are also scores of non-scheduled languages spoken in the country. The 1971 Census listed the number of speakers of eighty-nine non-scheduled languages and accounted for those who spoke other non-scheduled languages under the heading 'other mother tongues'.

The major division among India's languages is that between the Dravidian tongues of south and the Indo-Aryan ones of north and central India. The Indo-Aryan languages are spoken by about three-quarters (74 per cent) of the population of the country. Hindi, with its variations, ranks numerically as one of the major languages of the world, with an estimated 200 million speakers. It is one of the world's top five languages, the others being Chinese, English, Russian and Spanish.

Seven of the national languages, in terms of number of speakers, find a place in the top 20 languages of the world. Projecting official 1971 figures for speakers of these languages into 1982 it would appear that about 55 million Indians speak Bengali (the 6th most spoken language in the world), 55m speak Telugu (13th in the world), Marathi 53m (15th), Tamil 45m (19th), Urdu 36m (16th) and Punjabi 20m (17th). The other national languages which figure in the next 20 world languages are Gujarati (33m), Kannada (28m), Malayalam (26m) and Oriya (24m). Assamese, Sindhi, Kashmiri and Sanskrit are scheduled languages with less than 10 million speakers each.

A decadal comparison of the languages of India, by proportion of speakers to total population, indicates that the percentage of speakers for Hindi, Telugu and Tamil has declined gradually over the two decades between 1951 and 1971. On the other hand, Punjabi and Urdu have shown noticeable increases in their percentages though the per cent of Urdu speakers decreased marginally between 1961 and 1971. Bengali, Marathi, Gujarati, Malayalam and Assamese have also increased their percentages, though only slightly, while Oriya has declined, but only marginally. Kannada, Kashmiri and Sindhi have remained at the same level. The 'other' languages have also come down in their aggregate shares over this period.

Since the states of the Union are generally language-based, the scheduled languages are spoken by the largest majorities of population in most states. In Kerala, for instance, 96 per cent of the population speak Malayalam, and in Andhra Pradesh 85.4 per cent of the population speak Telugu. But curiously, in two states listed as Hindi-speaking, Rajasthan and Bihar, the languages spoken by substantial numbers do not find a place among the scheduled or non-scheduled languages. In 1961, Rajasthan listed 57 per cent of its population as Rajasthani speakers and Bihar listed 36 per cent of its people as Bihari speakers. Both languages were, in the 1971 Census, enumerated as Hindi. Bihari is a name collectively applied to three dialects spoken in and around Bihar: Maithili, Magadhi and Bhojpuri. Together they account for 16 million speakers. Rajasthani is also a group of dialects comprising Marwari, Malvi, Dhundari, Niwadi, Haranti and Banjari. In a few states and union territories, non-scheduled languages are spoken by a majority of people. Gorkhali/Nepali, for instance, is spoken by 81.8 per cent of the people of Sikkim and by a sizeable majority in Arunachal Pradesh.

Indian languages merge into each other and are not separated by hard and fast boundaries. There is an intermingling of languages as well as cultures among the various linguistic groups. Nine out of every ten persons speak at least one of the fourteen spoken scheduled languages. People are often bi-lingual or even tri-lingual in most areas. This bi-or tri-lingualism is widespread among all classes except the peasantry of linguistically homogeneous areas.

In many states, the major language of one of the adjacent states is, in effect, the second most important language, spoken by the second-largest group of people in the state. In Kerala, for instance, Tamil is the second most important language; in Tamil Nadu, it is Telugu; in Karnataka and in Andhra Pradesh, it is Urdu. Urdu, in fact, ranks first among the 'second' languages spoken in Indian states, with more than 28 million speakers distributed over various parts of the country. Hindi follows Urdu with nearly 25 million speakers living outside the home region (Uttar Pradesh, Madhya Pradesh, Bihar, Haryana, Himachal Pradesh and Rajasthan). Telugu minorities lead among other languages with about 10 million speakers outside Andhra Pradesh. This is followed by Bengali, Rajasthani, Punjabi and Marathi in that order. Maharashtra and Delhi are, at present, the leading recipients of linguistic minorities, as Bombay is the commercial capital of the country and Delhi is the national capital.

The Constitution has recognized Hindi in Devanagari script and English as the official languages of the Union (Articles 343 and 345 et seq.). English, originally intended to continue as an official language only up to 26 January, 1965, has continued to be used as the official language under the Official Language Act of 1963. A three-language formula — state language, Hindi and English — came into being as a compromise to changing over to Hindi; but it is not being implemented in most states, where the local language is usually teamed only with English.

As matters stand, the languages listed in the Constitution remain the official languages of the respective states and Hindi and English continue to be used for inter-state correspondence and for all-India links in general.

Sources: 1. *Social & Cultural Tables, Part II — C(i)*, Census of India 1971.
2. *India 1985*, Ministry of Information and Broadcasting.

Table 16

NON-SCHEDULED LANGUAGES — 1971

S.No	Language	Speakers	Principal Areas
1.	Adi	99,228	N.E. Region, Rajasthan
2.	Angami	68,522	Nagaland
3.	Ao	75,381	N.E. Region (Nagaland), Delhi, Orissa
4.	Arabic/Arbi	23,318	N.A.
5.	Balti	40,142	J & K
6.	Bhili/Bilodi	3,399,285	M.P., Rajasthan, D & N.H., Gujarat
7.	Bhotia	33,226	Sikkim
8.	Bhumiji	51,681	Assam, Bihar, Orissa
9.	Bishnupuriya	43,813	Assam
10.	Bodo/Boro	556,576	Assam
11.	Chang	15,816	Nagaland
12.	Chinese/Chini	10,958	N.A.
13.	Coorgi/Kodagu	72,085	Karnataka
14.	Deori	14,937	Assam
15.	Dimasa	40,149	Nagaland, Assam
16.	Dogri	1,299,143	H.P., Punjab, J & K
17.	English	191,595	N.A.
18.	Gadaba	20,420	A.P.
19.	Garo	411,731	Assam, Tripura
20.	Gondi	1,688,284	M.P., Orissa
21.	Gorkhali/Nepali	1,419,835	U.P., Assam, Bihar
22.	Halabi	346,259	N.A.
23.	Halam	19,197	Assam, Tripura
24.	Hmar	38,207	Assam, Manipur
25.	Ho	751,389	Orissa, Bihar
26.	Jatapu	36,450	A.P.
27.	Juang	12,172	Orissa
28.	Kabui	50,814	Manipur, Assam
29.	Khandeshi	251,896	N.A.
30.	Kharia	191,421	A & N, W.B.
31.	Khasi	479,028	Assam
32.	Khezha	11,363	Nagaland
33.	Khiemnungan	14,414	Nagaland, Manipur
34.	Khond/Kondh	196,316	Orissa
35.	Kinnauri	45,472	N.A.
36.	Kisan	73,847	Orissa, Assam
37.	Koch	14,256	Assam, Tripura
38.	Koda/Kora	14,333	Orissa, W.B.
39.	Kolami	66,868	N.A.
40.	Konda	33,720	N.A.
41.	Konkani	1,508,432	G, D & D, Karnataka
42.	Konyak	72,338	Nagaland, Manipur
43.	Korku	307,434	M.P., Maharashtra
44.	Korwa	15,097	Bihar
45.	Koya	211,877	A.P., Orissa
46.	Kui	351,017	Orissa
47.	Kuki	32,560	Manipur, Assam
48.	Kurukh/Oraon	1,235,665	A & N, Bihar, Orissa
49.	Ladakhi	60,272	J & K, H.P.
50.	Lahauli	16,749	H.P., Punjab
51.	Lahnda	41,935	N.A.
52.	Lakher	11,867	Assam
53.	Lalung	10,650	N.A.
54.	Lepcha	33,360	Assam, W.B.,Sikkim
55.	Lotha	36,949	Nagaland
56.	Lushai/Mizo	271,554	Assam, Manipur, Tripura
57.	Manipuri/ Meithei	791,714	Manipur, Assam
58.	Mao	35,381	N.A.
59.	Mikir	199,121	Assam
60.	Miri/Mishing	180,684	Assam
61.	Mishmi	22,354	N.A.
62.	Mogh	12,378	N.A.
63.	Monpa	26,369	N.A.
64.	Munda	309,293	A & N, Assam, W.B.
65.	Mundari	771,253	Bihar & Orissa
66.	Naga	23,115	Nagaland
67.	Nicobarese	17,971	A & N
68.	Nissi/Dafla	114,678	NEFA & Assam
69.	Nocte	25,263	NEFA
70.	Paite	27,157	Assam & Manipur
71.	Parji	73,912	Orissa
72.	Pawi	10,560	Assam
73.	Persian	10,509	N.A.
74.	Phom	18,017	Nagaland
75.	Rabha	51,146	Assam, W.B.
76.	Sangtam	20,015	N.A.
77.	Santali	3,786,899	Assam, Bihar, Orissa, W.B.
78.	Savara	222,018	Orissa, A.P., Assam
79.	Sema	65,227	Nagaland
80.	Shina	10,275	N.A.
81.	Sikkim/Bhotia	11,173	Sikkim
82.	Tangkhul	58,167	N.A.
83.	Tangsa	13,333	N.A.
84.	Tibetan	49,221	Sikkim, H.P., NEFA
85.	Tripuri	372,579	Tripura
86.	Tulu	1,158,419	Karnataka, Kerala
87.	Vaiphei	12,209	Manipur
88.	Wancho	28,649	Nagaland
89.	Yimchungre	19,609	Manipur, Nagaland
90.	Others	359,032	N.A.

Note: NEFA = Arunachal Pradesh, Nagaland, Manipur and Mizoram.

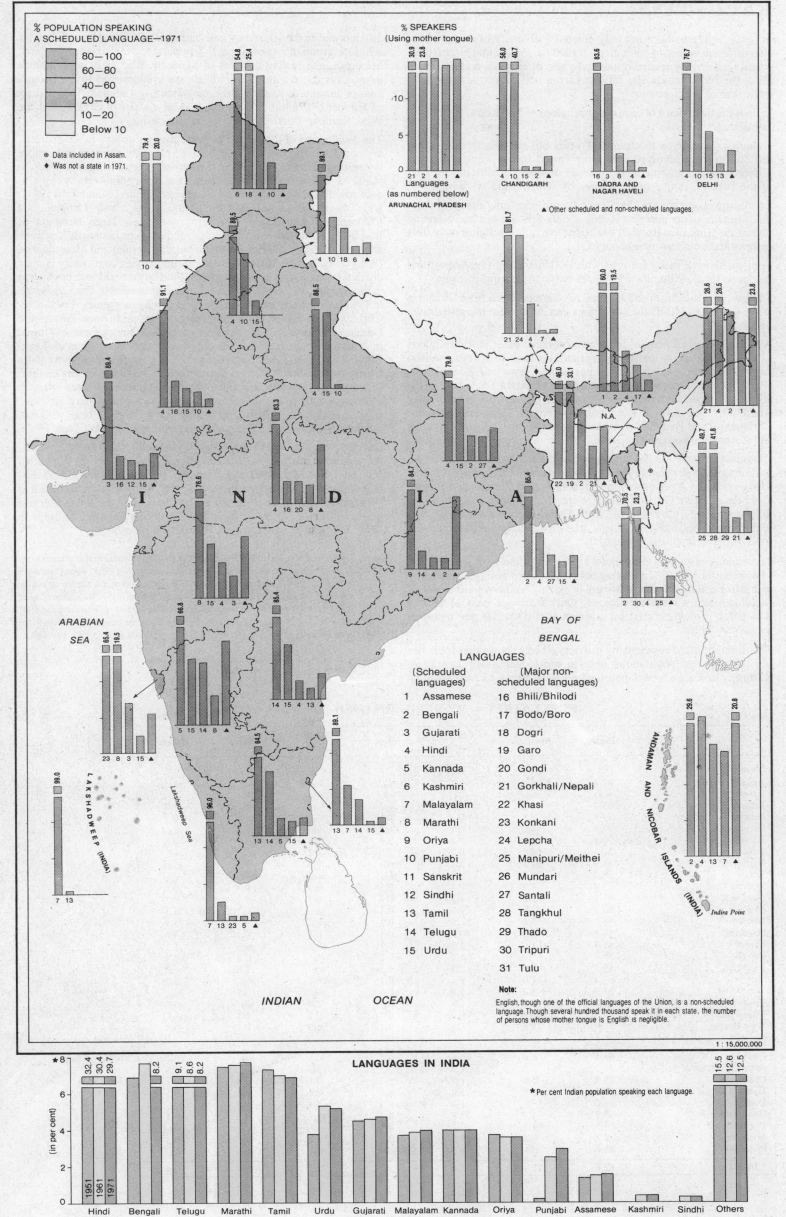

% POPULATION SPEAKING
A SCHEDULED LANGUAGE—1971

80—100
60—80
40—60
20—40
10—20
Below 10

⊛ Data included in Assam.
◆ Was not a state in 1971.

% SPEAKERS
(Using mother tongue)

Languages
(as numbered below)

ARUNACHAL PRADESH   CHANDIGARH   DADRA AND NAGAR HAVELI   DELHI

▲ Other scheduled and non-scheduled languages.

ARABIAN
SEA

BAY OF
BENGAL

LAKSHADWEEP (INDIA)

Lakshadweep Sea

ANDAMAN AND NICOBAR ISLANDS (INDIA)

Indira Point

LANGUAGES

(Scheduled languages)	(Major non-scheduled languages)
1 Assamese	16 Bhili/Bhilodi
2 Bengali	17 Bodo/Boro
3 Gujarati	18 Dogri
4 Hindi	19 Garo
5 Kannada	20 Gondi
6 Kashmiri	21 Gorkhali/Nepali
7 Malayalam	22 Khasi
8 Marathi	23 Konkani
9 Oriya	24 Lepcha
10 Punjabi	25 Manipuri/Meithei
11 Sanskrit	26 Mundari
12 Sindhi	27 Santali
13 Tamil	28 Tangkhul
14 Telugu	29 Thado
15 Urdu	30 Tripuri
	31 Tulu

Note:
English, though one of the official languages of the Union, is a non-scheduled language. Though several hundred thousand speak it in each state, the number of persons whose mother tongue is English is negligible.

INDIAN   OCEAN

1 : 15,000,000

LANGUAGES IN INDIA

★ Per cent Indian population speaking each language.

(in per cent)

Hindi Bengali Telugu Marathi Tamil Urdu Gujarati Malayalam Kannada Oriya Punjabi Assamese Kashmiri Sindhi Others

RELIGIONS

In India, religions have not only created a diversity of cultures, but also different forms of settlement, a plethora of architectural styles, a variety of activity-related rituals and a host of religious edifices which make the Indian landscape different from one area to another, and from that of other countries.

India is the birth-place of four major religions — Hinduism, Buddhism, Jainism and Sikhism.

Hinduism, a religion thousands of years old and whose origins are difficult to trace, evolved out of the varied Indian ways of life, so different and yet in some intangible way unified. 'Hindustan', the land of the Hindus, is one of the names by which India is known.

Buddhism, as a religion, has few adherents in India today, though the influence of its precepts is still widely felt and has had a lasting impact on Hinduism itself. It has become a major religion of India's eastern and southern neighbours.

Jainism, sharing many features with Buddhism and founded around the same time as Buddhism, has its major following in India.

Sikhism, like Buddhism and Jainism, has derived much from Hinduism while developing its distinct doctrines and values, but has also drawn on Islam.

A land of religious people, India has embraced other world religions as well. Christianity and Islam in India go back to almost the first days of their prophets. And the persecuted Jews and Zoroastrians (Parsis) found sanctuary here. However troubled the times have been, all these religions have coexisted. India is perhaps the only country where we see four major world religions (Hinduism, Islam, Christianity and Buddhism), three lesser religions (Jainism, Sikhism and Zoroastrianism) and hundreds of primitive religions coexisting.

Hinduism accounts for the largest part of the Indian population, 549.8 million or 82.6 per cent (1981 Census). Islam has 75.5 million Indian followers, or 11.4 per cent of the population, and Christianity (Roman Catholics, Eastern Orthodox and Protestants) a 2.4 per cent following. Sikhism, mainly prevalent in North India, accounts for 2 per cent of the population, while Buddhism, Jainism and other religions account for 1.6 per cent of the population.

The Hindus of India were about 12 per cent of the world population of 4588 million in 1982, ranking below those professing Christianity and about equal with the followers of Islam, but above adherents of Buddhism and all other religions. Only 0.01 per cent of India's population profess no religion, a substantially smaller per cent than the world figure.

The distribution of religions by districts clearly shows that India has Muslim-majority populations only in most districts of Jammu & Kashmir, in Kerala's Malappuram and West Bengal's Murshidabad districts and in the islands of Lakshadweep. Christians are the major religious group throughout sparsely populated Meghalaya, Nagaland, Mizoram, and in five districts of Manipur. The Sikhs predominate in the districts of Punjab. Buddhists are predominant in the Ladakh area of Jammu & Kashmir, the neighbouring Lahul & Spiti district of Himachal Pradesh, the North Sikkim district of Sikkim and the West Kameng district of Arunachal Pradesh.

The distribution of the second-largest religious groups in the districts, as seen from the overlay in the map, presents an altogether more interesting pattern. Hindus are the second-largest religious group in most of the Christian north-east, the exceptions being Hindu-dominated Imphal, Thoubal and Bishnupur (Central Manipur in 1981), with their substantial Muslim populations, and Lunglei and Chhimtuipui districts in Mizoram, with their large numbers of Buddhists, in most of the Sikh Punjab and Muslim Jammu & Kashmir, and in the Muslim districts of Malappuram (Kerala) and Murshidabad (West Bengal). The Muslims are the second largest community in most districts of India, just as they are the second largest religious population in the country. In Kerala, however, in its eight southern districts (seven in 1981), Christians are the second-largest religious group. Elsewhere, Christians are the second-largest religious group in once-Portuguese Goa, and in several of the districts of the Coromandel coast — three in Andhra Pradesh, once-French Pondicherry and three in Tamil Nadu. Two other Tamil Nadu districts in the interior also have a large Christian population. Other districts which have Christians as their second-largest group are those in Central India (Chota Nagpur Plateau), southern Orissa and southern Assam where there are large tribal populations. Christianity came to all these areas with the missionaries, long after it had established itself in Kerala and Tamil Nadu.

Buddhists are the second-largest group in Sikkim's southern districts, the extreme north-east districts and in one district in Himachal Pradesh. Nine Maharashtra districts also register substantial Buddhist populations, post-independence converts who were once listed as 'Scheduled Castes and Tribes'. The Jains are the second-largest group in three districts of Rajasthan bordering Gujarat, Lalitpur in Uttar Pradesh and Bilaspur in Himachal Pradesh.

Note: All Census data are dated 1981 and refer to the 412 districts that then existed. The map shows the districts as they existed at the end of 1985, when there were 439 districts (437 districts for which boundaries were available are shown on the maps. Where 1981 districts have been broken up into two or more smaller districts or combined into larger districts (as in Jammu & Kashmir) the 1981 data are used to shade the 1985 configurations. In the text, which refers to the districts as they existed in 1985, the 1981 status is given within brackets.

Source: *Household Population by Religion of Head of Household, Paper 4 of 1984*, census of India 1981.

Table 17

POPULATION — BY RELIGION (1981)

(in per cent)

	Hindus	Muslims	Christians	Sikhs	Buddhists	Jains	Others
States							
Andhra Pradesh	88.8	8.5	2.7	Neg.	Neg.	Neg.	Neg.
Assam	N.A.	N.A.	N.A.	N.A.	N.A.	N.A.	N.A.
Bihar	83.0	14.1	1.1	0.1	Neg.	Neg.	1.7
Gujarat	89.5	8.5	0.4	0.1	Neg.	1.4	0.1
Haryana	89.4	4.0	0.1	6.2	Neg.	0.3	Neg.
Himachal Pradesh	95.8	1.6	0.1	1.2	1.2	Neg.	0.1
Jammu & Kashmir	32.2	64.2	0.2	2.2	1.2	Neg.	Neg.
Karnataka	85.9	11.1	2.1	Neg.	0.1	0.8	Neg.
Kerala	58.2	21.2	20.6	Neg.	Neg.	Neg.	Neg.
Madhya Pradesh	93.0	4.8	0.7	0.3	0.1	0.8	0.3
Maharashtra	81.4	9.2	1.3	0.2	6.3	1.5	0.1
Manipur	60.0	7.0	29.7	0.1	Neg.	0.1	3.1
Meghalaya	18.0	3.1	52.6	0.1	0.2	0.1	25.9
Nagaland	14.4	1.5	80.2	0.1	0.1	0.1	3.6
Orissa	95.4	1.6	1.8	0.1	Neg.	Neg.	1.1
Punjab	36.9	1.0	1.1	60.7	Neg.	0.2	0.1
Rajasthan	89.3	7.3	0.1	1.5	Neg.	1.8	Neg.
Sikkim	67.3	1.0	2.2	0.1	28.7	Neg.	0.7
Tamil Nadu	88.9	5.2	5.8	Neg.	Neg.	0.1	Neg.
Tripura	89.3	6.8	1.2	Neg.	2.7	Neg.	Neg.
Uttar Pradesh	83.3	15.9	0.2	0.4	0.1	0.1	Neg.
West Bengal	77.0	21.5	0.6	0.1	0.3	0.1	0.4
Union Territories							
Andaman & Nicobar	64.5	8.6	25.6	0.5	0.1	Neg.	0.7
Arunachal Pradesh	29.2	0.8	4.3	0.2	13.7	Neg.	51.8
Chandigarh	75.3	2.0	1.0	21.1	0.1	0.4	0.1
Dadra & Nagar Haveli	95.6	1.9	1.9	Neg.	0.2	0.3	0.1
Delhi	83.6	7.8	1.0	6.3	0.1	1.2	Neg.
Goa, Daman & Diu	65.9	4.5	29.3	0.1	Neg.	0.1	0.1
Lakshadweep	4.5	94.8	0.7	—	—	—	Neg.
Mizoram	7.1	0.5	83.8	0.1	8.2	Neg.	0.3
Pondicherry	85.6	6.1	8.3	Neg.	Neg.	0.1	Neg.
INDIA	82.6	11.4	2.4	2.0	0.7	0.5	0.4

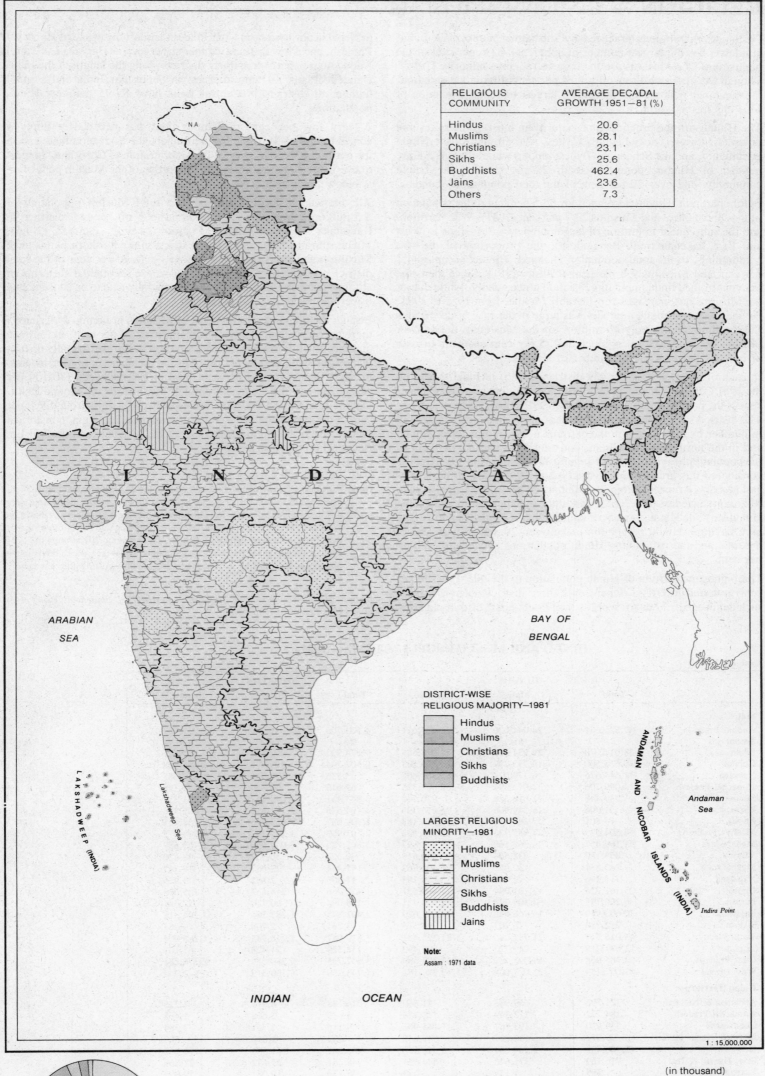

RELIGIOUS COMMUNITY	AVERAGE DECADAL GROWTH 1951—81 (%)
Hindus	20.6
Muslims	28.1
Christians	23.1
Sikhs	25.6
Buddhists	462.4
Jains	23.6
Others	7.3

ARABIAN SEA

BAY OF BENGAL

Lakshadweep Sea

L A K S H A D W E E P (INDIA)

I N D I A

A N D A M A N   A N D   N I C O B A R   I S L A N D S   (INDIA)

Andaman Sea

Indira Point

DISTRICT-WISE
RELIGIOUS MAJORITY—1981

Hindus
Muslims
Christians
Sikhs
Buddhists

LARGEST RELIGIOUS
MINORITY—1981

Hindus
Muslims
Christians
Sikhs
Buddhists
Jains

Note:
Assam : 1971 data

INDIAN   OCEAN

1 : 15,000,000

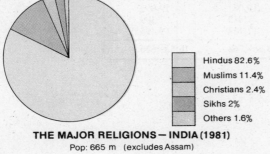

Hindus 82.6%
Muslims 11.4%
Christians 2.4%
Sikhs 2%
Others 1.6%

THE MAJOR RELIGIONS — INDIA (1981)
Pop: 665 m   (excludes Assam)

		(in thousand)
RELIGIOUS COMMUNITY	1951-61	1961-71	1971-81*
Hindus	62,800	86,900	113,360
Muslims	11,500	14,500	15,342
Christians	2 300	3 500	3 615
Sikhs	1 600	2 600	3 304
Buddhists	3 000	600	900
Jains	400	600	800
Others	−3 500	300	300

*Estimated

RELIGION — HINDUISM AND ISLAM

Of the seven religions practised by substantial numbers in India, Hinduism has the largest number of adherents, 82.6 per cent of the population. The Muslims are the biggest religious minority (1981). Though Muslims constitute only 11.4 per cent of India's population, their number makes them the fourth largest national population of Muslims in the world.

The Hindus are the preponderant majority in most of the states and union territories. The Scheduled Castes, who are part of the Hindu community, and the Scheduled Tribes, among whom there is a large number of Hindus, comprise about 28 per cent of the Hindu community and over 20 per cent of the total population of India.

Before partition, Hindus accounted for 65.5 per cent of the population of the subcontinent and Muslims 23.7 per cent (1941). With partition, and the subsequent migration of large numbers of Muslims to West and East Pakistan (now Bangladesh), the proportion of the two communities in the total population changed: Hindus became 84.1 per cent and Muslims 9.8 per cent (Census 1951). Since then, the per cent of the Hindu population has fallen marginally, while that of the Muslims has increased considerably. While, from 1951 to 1981, the increase in absolute numbers was large in the case of the Hindus (246 million against only 40 million for the Muslims), the number of Muslims increased by as much as 113 per cent against a growth of only 81 per cent in the number of Hindus.

In each of about 300 of the 439 districts (1985) in India (based on the 1981 Census), Hindus predominate, constituting more than 60 per cent of the population (at the time of the 1981 Census, India comprised 412 districts. Division and re-alignment of districts made these 439 in number by 1985). It is in the extreme north (Jammu & Kashmir) and in the north-east that Hindus constitute less than 20 per cent of the population. Five districts of Jammu & Kashmir (the 1981 Census considered this area as nine districts) and 19 districts in the north-east (Assam, Arunachal Pradesh, Manipur, Nagaland, Meghalaya and Mizoram) have less than 20 per cent of the total population professing Hinduism. The Hindus are also a minority in the Punjab, four districts here having a Hindu population of between 20 and 30 per cent of the total and the rest having Hindu populations between 30 and 60 per cent.

The highest proportion of Hindu population to the total (95-100 per cent) is found in Orissa (where only three districts do not register such dominance), 17 districts of Madhya Pradesh, six districts adjacent to Nepal in north-western Uttar Pradesh, almost the whole of Himachal Pradesh, and a few districts in other states such as Haryana and Tamil Nadu. A stretch of some thirty districts along the length of the Indo-Gangetic Plain, as many districts in peninsular India and a small number of districts in Central India have 85-95 per cent Hindu populations.

A very low per cent of Hindus does not necessarily imply a concentration of Muslims. In the Punjab, the Sikhs dominate and in the north-east there is a very high concentration of Christians. In most of the districts in these areas, the proportion of the Muslim population is below 5 per cent.

All districts in Tamil Nadu (with a total Muslim population of 2.5 million), three districts in Kerala (5.4 m), several districts in Karnataka (4.1m), Andhra Pradesh (4.5m), Gujarat (2.9m), Maharashtra (5.8m), and a few districts in the northern plains have Muslim populations that constitute less than 10 per cent of the total district population. Nevertheless, these are substantial in terms of absolute numbers. Also, the Muslims generally tend to be concentrated in certain pockets within a district.

Besides the preponderant numbers of Muslims in Jammu & Kashmir, there are substantial numbers of them (over 20 per cent) in fourteen districts of Uttar Pradesh, in Malappuram in Kerala (the only district in the South with a 60-80 per cent Muslim population), in Murshidabad in West Bengal (50-60 per cent) and in Lakshadweep (80-100 per cent). There is also a concentration of Muslims in and around Delhi. Delhi's total of 482,000 Muslims is, however, only around 7.5 per cent of the population. This concentration owes its origin to historical factors, Delhi having been the seat of the Mughals and the Muslim rulers who preceded them.

Note: All Census data are dated 1981 and refer to the 412 districts that then existed. The maps show the districts as they existed at the end of 1985, when there were 439 districts (437 districts for which boundaries were available are shown on the maps). Where 1981 districts have been broken up into two or more smaller districts or combined into larger districts (as in Jammu & Kashmir) the 1981 data are used to shade the 1985 configurations. In the text, which refers to the districts as they existed in 1985, the 1981 status is given within brackets.

Source: *Household Population by Religion of Head of Household, Paper 4 of 1984*, Census of India 1981.

HINDU AND MUSLIM POPULATIONS — 1981

Table 18

	HINDUS			MUSLIMS		
	Total	Male	Female	Total	Male	Female
States						
Andhra Pradesh	47,525,681	24,044,070	23,481,611	4,533,700	2,311,700	2,221,737
Assam	N.A.	N.A.	N.A.	N.A.	N.A.	N.A.
Bihar	58,011,070	29,931,565	28,079,505	9,874,993	4,990,368	4,884,625
Gujarat	30,518,500	15,733,297	14,785,203	2,907,744	1,485,976	1,421,768
Haryana	11,547,676	6,179,071	5,368,605	523,536	279,971	243,565
Himachal Pradesh	4,099,706	2,072,391	2,027,315	69,613	38,415	31,198
Jammu & Kashmir	1,930,448	1,010,328	920,120	3,843,451	2,042,113	1,801,338
Karnataka	31,906	16,249,996	15,656,797	4,104,616	2,101,616	2,002,896
Kerala	14,801	7,267,159	7,534,188	5,409,687	2,657,226	2,752,461
Madhya Pradesh	48,504,575	24,985,733	23,518,842	2,501,919	1,297,247	1,204,672
Maharashtra	51,109,457	26,363,560	24,745,897	5,805,785	3,042,096	2,763,689
Manipur	853,180	432,564	420,616	99,327	50,507	48,820
Meghalaya	240,831	132,062	108,769	41,434	22,087	19,347
Nagaland	111,266	72,081	39,185	11,806	7,881	3,925
Orissa	25,161,725	12,701,094	12,460,631	422,266	216,633	205,633
Punjab	6,200,195	3,309,724	2,890,471	168,094	90,710	77,384
Rajasthan	30,603,970	15,966,950	14,637,020	2,492,145	1,287,145	1,204,318
Sikkim	212,780	116,207	96,573	3,241	2,708	533
Tamil Nadu	43,016,546	21,791,085	21,225,461	2,519,947	1,256,945	1,263,002
Tripura	1,834,218	941,722	892,496	138,529	71,428	67,101
Uttar Pradesh	92,365,968	49,093,429	43,272,539	17,657,735	9,280,854	8,376,881
West Bengal	42,007,159	22,027,008	19,980,151	11,743,259	6,100,517	5,642,742
Union Territories						
Andaman & Nicobar	121,793	69,933	51,860	16,188	8,973	7,215
Arunachal Pradesh	184,732	112,349	72,383	5,073	3,608	1,465
Chandigarh	339,920	193,967	145,953	9,115	5,731	3,384
Dadra & Nagar Haveli	99,072	50,027	49,045	1,932	1,072	860
Delhi	5,200,432	2,884,054	2,316,378	481,802	271,534	210,268
Goa, Daman & Diu	716,169	371,646	344,523	48,461	25,592	22,869
Lakshadweep	1,799	1,141	658	38,173	19,081	19,092
Mizoram	35,245	25,660	9,585	2,205	1,812	393
Pondicherry	517,288	263,069	254,159	36,663	17,168	19,495
INDIA	549,779,481	284,392,942	265,386,539	75,512,439	38,989,763	36,522,676

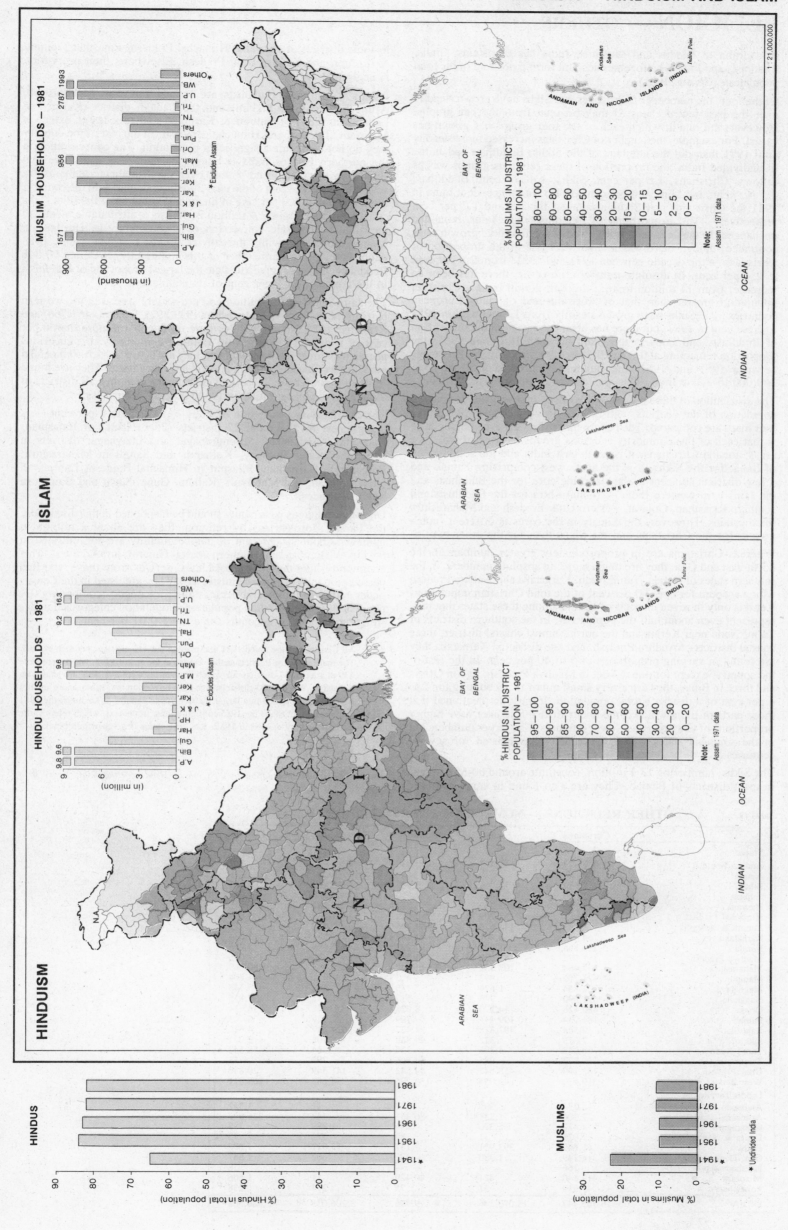

ISLAM

MUSLIM HOUSEHOLDS — 1981
(in thousand)

* Excludes Assam

% MUSLIMS IN DISTRICT
POPULATION — 1981

80—100
60—80
50—60
40—50
30—40
20—30
15—20
10—15
5—10
2—5
0—2

Note:
Assam : 1971 data

HINDUISM

HINDU HOUSEHOLDS — 1981
(in million)

* Excludes Assam

% HINDUS IN DISTRICT
POPULATION — 1981

95—100
90—95
85—90
80—85
70—80
60—70
50—60
40—50
30—40
20—30
0—20

Note:
Assam : 1971 data

HINDUS
(% Hindus in total population)

* Undivided India

MUSLIMS
(% Muslims in total population)

* Undivided India

1:21,000,000

RELIGION — OTHERS

In addition to Hindus and Muslims, India has Christians, Sikhs, Buddhists and Jains who together account for 5.6 per cent of the total population (Census 1981).

Of the four, the numbers of Christians and Sikhs have grown steadily over the past few decades as may be seen from the bar graphs. However, the rate at which each of the four groups has grown has varied. For example, the number of Christians increased more steadily until 1971 than did the numbers of the Sikhs, Buddhists and Jains. In undivided India, the proportion of these religious groups was as follows: Christians, 1.6 per cent; Sikhs, 1.4 per cent; Buddhists, 0.1 per cent; and Jains, 0.4 per cent (1941). In independent India in 1951 the Christians and Sikhs accounted for 2.3 and 1.7 per cent respectively. The proportions of Buddhists and Jains remained unchanged. In the decades that followed, the differential growth rates continued, though there was a slight decrease in the proportion of Christians in the decade between 1971 and 1981: from 2.6 per cent to 2.4 per cent. In absolute numbers, however, there has been an increase, from 14 million to 16.2 million, giving India a Christian population greater than that of Australia and of many European countries. The Sikhs have grown steadily from 1.7 per cent in 1951 to 2 per cent in 1981, but there has been no change in the proportion of Buddhists and Jains in the years between 1961 and 1981, the proportion remaining at 0.7 and 0.5 per cent respectively. However, between 1951 and 1961, the Buddhists increased 8.6 fold from 360,000 to 3.1 million. This was due to causes referred to later.

The distribution of these religious minorities, in proportion to the total population of the districts, will be seen to vary considerably if the four maps are compared. The pattern that strikes the eye at first glance is that each of these minority religious groups has one or two core areas: for the Christians, it is north-east India and the southern tip of India; for the Sikhs, it is the north-west comprising Punjab and a few districts adjacent to the state; the core for the Buddhists and the Jains is in western India — Maharashtra for the Buddhists and southern Rajasthan, Gujarat, western Uttar Pradesh and Maharashtra for the Jains. However, the density in the cores is different, much greater for the Christians and the Sikhs than for the other two.

Whereas Christians are in proportionately greater numbers in the north-east and Goa, they are much more, in absolute numbers, in the southern states of Kerala, Tamil Nadu, Karnataka and Andhra Pradesh which account for over 60 per cent of the total Christian population. Yet it is only in seven districts of Kerala among these states that they represent even about half the population. In the southern districts of Tamil Nadu near Kerala and the northernmost coastal district, three coastal districts of Andhra Pradesh, and one district in Karnataka they are found in varying proportions of 5 to 40 per cent. In the rest of the country, except for one district in Madhya Pradesh, two in Orissa and three in Bihar, they are a very small minority accounting for 1 to 4 per cent of a district's total population. It must be mentioned that those parts of India which came under British rule later have higher proportions of Christians but fewer numbers. The larger numbers are in the southern states where Christianity has existed for several centuries.

The Sikhs, numbering 13.1 million, constitute around 60-65 per cent in most districts of Punjab. They are also found in small numbers in some districts of Haryana, Himachal Pradesh, Rajasthan, Jammu & Kashmir and western Uttar Pradesh. Elsewhere, their proportion is negligible.

India's 4.7 million Buddhists are largely found in Maharashtra, Arunachal Pradesh, Sikkim, and the Ladakh district (Kargil and Ladakh in 1981) of Jammu & Kashmir, and to a lesser extent in Mizoram, Tripura, and Himachal Pradesh. In the rest of the country they do not constitute a significant population. The concentration in the northern part of India is mainly because it is near here that Buddhism originated and spread over the Himalaya. Maharashtra accounts for 85 per cent of the total Buddhist population of the country, though it is only 6.3 per cent of the total population of the state. This concentration of some 3.9 million Buddhists is attributable mostly to the almost wholesale conversion of a community of Harijans in Maharashtra, following the advice of their leader, Dr. B.R. Ambedkar, himself a member of this community. Sikkim's 90,848 Buddhists, however, give this state the largest proportion of Buddhists in the country, 28.7 per cent of its population.

The Jains number 3.2 million and are widely spread in the western parts of the country — Maharashtra (939,392), Gujarat (467,768) and Rajasthan (624,317) — but nowhere do they form more than 6 per cent of a district's population. Being a community of industrious people with great financial acumen, they are active in such commercial activities as finance, banking and the jewellery trade. They are found in small numbers as a close-knit community in almost all districts of the country, concentrating in the towns to do business.

The district-wise proportion of Jains — more than 2 per cent — is the greatest in 11 of the 27 districts (26 in 1981) of Rajasthan; Kachchh, Ahmadabad, Surendranagar and Bhavnagar districts in Gujarat; Greater Bombay, Kolhapur and Sangli in Maharashtra; Belgaum in Karnataka; Bilaspur in Himachal Pradesh; Lalitpur in Uttar Pradesh; and Mandsaur, Ratlam, Guna, Sagar and Damoh in Madhya Pradesh.

One other religious community should be mentioned in this discussion, the Parsis. Zoroastrians by religion, they are about a million in number. About 90 per cent of this community are concentrated in the city of Bombay and southern coastal Gujarat, around Surat. This community, like the West Coast Jews, derives from those who fled persecution in West Asia centuries ago. They are listed in the Census under 'other religions and persuasions', a category comprising 0.4 per cent of the total Indian population. Population categorized under 'religion not stated' accounts for a mere 0.01 per cent.

Note: All Census data are dated 1981 and refer to the 412 districts that then existed. The maps show the districts as they existed at the end of 1985, when there were 439 districts (437 districts for which boundaries were available are shown on the maps). Where 1981 districts have been broken up into two or more smaller districts or combined into larger districts (as in Jammu & Kashmir) the 1981 data are used to shade the 1985 configurations. In the text, which refers to the districts as they existed in 1985, the 1981 status is given within brackets.

Source: *Household Population by Religion of Head of Household, Paper 4 of 1984*, Census of India 1981.

Table 19   **OTHER RELIGIONS — NUMBER OF FOLLOWERS (1981)**

	Christians	Sikhs	Buddhists	Jains	Others
States					
Andhra Pradesh	1,433,327	16,222	12,930	18,642	9,171
Assam	N.A.	N.A.	N.A.	N.A.	N.A.
Bihar	740,186	77,704	3,003	27,613	1,180,165
Gujarat	132,703	22,438	7,550	467,768	29,096
Haryana	12,215	802,230	761	35,482	718
Himachal Pradesh	3,954	52,209	52,629	1,046	1,661
Jammu & Kashmir	8,481	133,675	69,706	1,576	52
Karnataka	764,449	6,401	42,147	297,974	13,334
Kerala	5,223,865	1,295	223	3,605	2,255
Madhya Pradesh	351,972	143,020	75,312	444,960	157,086
Maharashtra	795,464	107,255	3,946,149	939,392	80,669
Manipur	421,702	992	473	975	44,304
Meghalaya	702,854	1,674	2,739	542	345,745
Nagaland	621,590	743	517	1,153	27,855
Orissa	480,426	14,270	8,028	6,642	276,914
Punjab	184,934	10,199,141	799	27,049	8,703
Rajasthan	39,568	492,818	4,427	624,317	4,617
Sikkim	7,015	322	90,848	108	2,071
Tamil Nadu	2,798,048	4,395	735	49,564	18,842
Tripura	24,872	285	54,806	297	51
Uttar Pradesh	162,199	458,647	54,542	141,549	20,880
West Bengal	319,670	49,054	156,296	38,663	265,071
Union Territories					
Andaman & Nicobar	48,274	991	127	11	1,357
Arunachal Pradesh	27,306	1,231	86,483	42	326,972
Chandigarh	4,470	95,370	454	1,889	392
Dadra & Nagar Haveli	2,025	11	189	372	75
Delhi	61,609	393,921	7,117	73,917	1,608
Goa, Daman & Diu	318,249	1,380	302	602	1,567
Lakshadweep	266	—	—	—	8
Mizoram	413,840	421	40,429	11	1,606
Pondicherry	49,914	31	75	277	238
INDIA	16,165,447	13,078,146	4,719,796	3,206,038	2,826,502

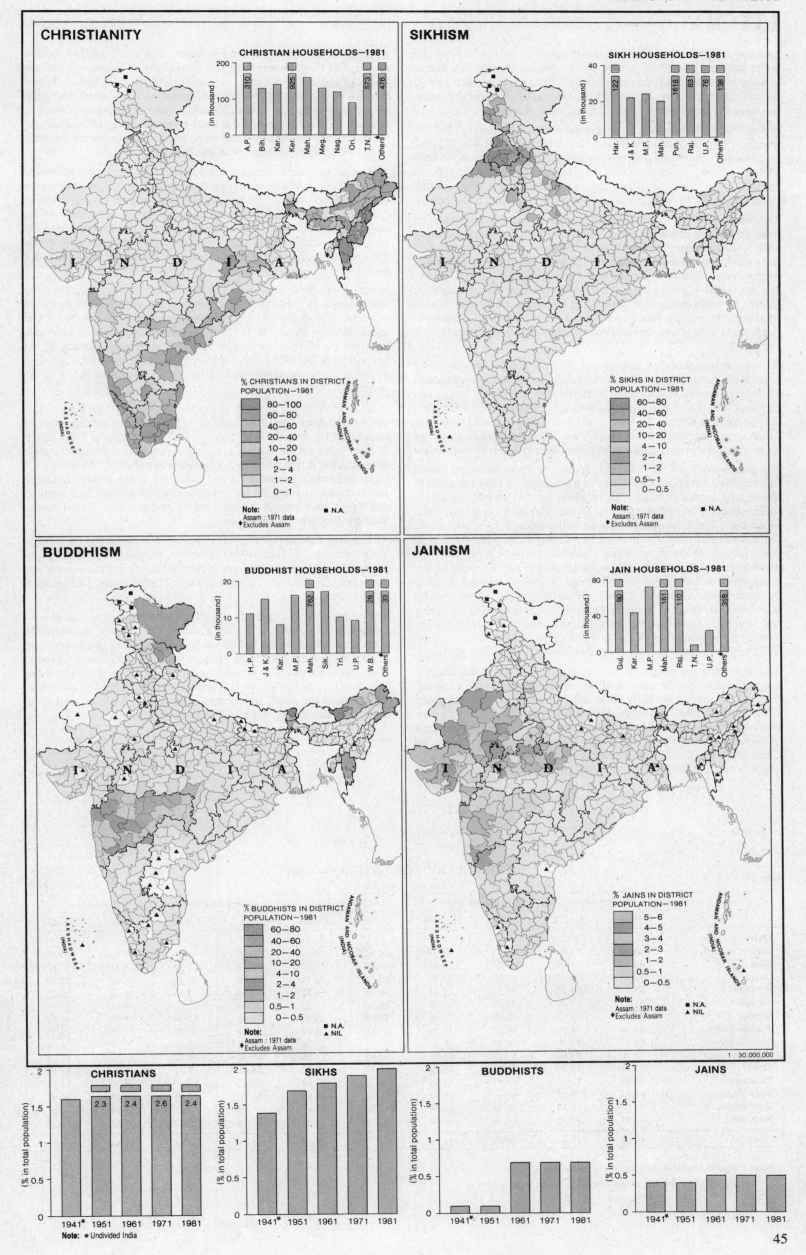

CHRISTIANITY

CHRISTIAN HOUSEHOLDS—1981
(in thousand)

Values shown: A.P. 310, Ker. 925, T.N. 573, Others 476
States: A.P., Bih., Kar., Ker., Mah., Meg., Nag., Ori., T.N., Others

% CHRISTIANS IN DISTRICT POPULATION—1981
- 80—100
- 60—80
- 40—60
- 20—40
- 10—20
- 4—10
- 2—4
- 1—2
- 0—1

Note:
Assam : 1971 data
◆ Excludes Assam
■ N.A.

LAKSHADWEEP (INDIA)
ANDAMAN AND NICOBAR ISLANDS (INDIA)

SIKHISM

SIKH HOUSEHOLDS—1981
(in thousand)

Values shown: Har. 122, Pun. 1618, Raj. 83, U.P. 76, Others 138
States: Har., J & K., M.P., Mah., Pun., Raj., U.P., Others

% SIKHS IN DISTRICT POPULATION—1981
- 60—80
- 40—60
- 20—40
- 10—20
- 4—10
- 2—4
- 1—2
- 0.5—1
- 0—0.5

Note:
Assam : 1971 data
◆ Excludes Assam
■ N.A.

LAKSHADWEEP (INDIA)
ANDAMAN AND NICOBAR ISLANDS (INDIA)

BUDDHISM

BUDDHIST HOUSEHOLDS—1981
(in thousand)

Values shown: Mah. 762, W.B. 28, Others 33
States: H.P., J & K., Kar., M.P., Mah., Sik., Tri., U.P., W.B., Others

% BUDDHISTS IN DISTRICT POPULATION—1981
- 60—80
- 40—60
- 20—40
- 10—20
- 4—10
- 2—4
- 1—2
- 0.5—1
- 0—0.5

Note:
Assam : 1971 data
◆ Excludes Assam
■ N.A.
▲ NIL

LAKSHADWEEP (INDIA)
ANDAMAN AND NICOBAR ISLANDS (INDIA)

JAINISM

JAIN HOUSEHOLDS—1981
(in thousand)

Values shown: Guj. 90, M.P. 161, Mah. 110, Others 359
States: Guj., Kar., M.P., Mah., Raj., T.N., U.P., Others

% JAINS IN DISTRICT POPULATION—1981
- 5—6
- 4—5
- 3—4
- 2—3
- 1—2
- 0.5—1
- 0—0.5

Note:
Assam : 1971 data
◆ Excludes Assam
■ N.A.
▲ NIL

LAKSHADWEEP (INDIA)
ANDAMAN AND NICOBAR ISLANDS (INDIA)

1 : 30,000,000

CHRISTIANS
(% in total population)
1941*: 1.6, 1951: 2.3, 1961: 2.4, 1971: 2.6, 1981: 2.4

SIKHS
(% in total population)
1941*, 1951, 1961, 1971, 1981

BUDDHISTS
(% in total population)
1941*, 1951, 1961, 1971, 1981

JAINS
(% in total population)
1941*, 1951, 1961, 1971, 1981

Note: * Undivided India

45

LITERACY — I

Literacy is generally defined as the ability of a person to read and write with understanding in any one language. Despite this liberal concept of literacy, this is an area in which India has done very poorly; though the percentage of literacy has been increasing over the years, the total number of illiterates has also been increasing.

Since 1901 there has been a continuous increase in literacy. But the greatest progress was made in just three decades, 1931-41, 1951-61 and 1971-81, there being approximately a 7 per cent increase in each of these periods. In absolute numbers, 1971-81 produced the greatest increase in literates, from 161 million to 240 million. But, correspondingly, the illiterates also increased by about 38 million in this period.

Despite the progress made, when compared with other countries, even other developing countries, India is near the bottom of literacy ranking with just 36 per cent literacy in 1981. Europe tops the list with 92.9 per cent literacy while Africa is at the very bottom with just about 34 per cent; in between are Eurasia — comprising Cyprus, Turkey, and the USSR — with a literacy rate of 83 per cent, South America, despite having a number of developing countries, ranking high with 82.5 per cent literacy, North and Central America with 77.8 per cent, Australasia with 70 per cent and Asia (including India) with only 53.7 per cent.

It is generally accepted that countries with predominantly urban populations have very high levels of literacy. But in India, even though the urban population (including Assam) is comparatively large (160 million), it is a small percentage of the total population (23.3 per cent) and so national literacy levels are still low. Even in India, however, urban areas have high literacy rates. Easy access to schools is a major factor that helps in the attainment of higher literacy levels in these areas.

An examination of the literacy levels in India becomes more meaningful if the educational levels of the literates in 1981 are studied. Those with formal primary education top the list, accounting for 76 million people, while those without formal education are almost equal in number, about 74 million. Those with middle school education, i.e. 6-8 years of formal instruction, are 42.4 million, while those with matriculation, or higher secondary education, i.e. 9-12 years of formal schooling, are about 37.9 million. Graduates and others with higher education, including professional and technical education from collegiate institutions, are a mere 9.4 million, while those literates with certificates that are not quite equal to collegiate degrees, including technical and non-technical diploma holders, are still fewer, amounting to only a little over a million. Yet, ironically, the country has the third largest technically qualified manpower pool in the world.

The inter-state variations in the per cent of literates in the country is considerable. Kerala is the most literate state in India with a literacy level of 70 per cent. It is followed by seven union territories: Chandigarh, Delhi, Mizoram, Goa, Daman & Diu, Lakshadweep, Pondicherry and the Andaman & Nicobar Islands. The high per cent of literates (over 60 per cent) in the case of the union territories of Chandigarh and Delhi is attributable to the fact that both are virtually major urban agglomerations with high concentrations of educational

institutions. In the other union territories, their small size has enabled the Central Government to invest adequately in increasing literacy levels to at least 50 per cent. Maharashtra and Tamil Nadu, the two states with the highest levels of urbanization, rank ninth and tenth respectively in literacy levels. The largest state in the country, Madhya Pradesh, ranks 24th, while the most populous state, Uttar Pradesh, ranks 25th among the states and union territories. Arunachal Pradesh, in the extreme north-east, is at the bottom of the list.

A look at how administrative units have, over the years, improved literacy levels within their respective borders shows that two union territories have made the largest leap forward in the most recent Census decade, Arunachal Pradesh bettering its 1971 level by 86 per cent, and Dadra & Nagar Haveli by 78.5 per cent. Among the states, the greatest improvement was in Nagaland (improving on the 1971 level by 56.7 per cent), followed by Jammu & Kashmir (44.2 per cent).

Several states and union territories with high literacy rates in 1971 have, however, shown little improvement, e.g. Kerala (only 14 per cent), Tamil Nadu (16.5 per cent), and Delhi (10 per cent). The states with low literacy in 1971, on the other hand, made considerable progress by 1981 — Jammu & Kashmir increasing literacy by 44 per cent over the 1971 level, Uttar Pradesh by 31 per cent, Rajasthan by 30 per cent, Madhya Pradesh by 27 per cent, and Bihar by 24 per cent.

In no state or union territory is the female literacy rate higher than the male literacy rate. The disparity is lowest in the smaller units of Mizoram (3.3 per cent), Meghalaya (7.7 per cent) and Chandigarh (9.5 per cent). Of the states, Kerala has done best in eradicating female illiteracy, the difference now being only 9.5 per cent. The difference is generally high in other states and territories, being greatest in Orissa and Haryana — about 26 per cent, nearly that amount in Rajasthan and Bihar, and around 24 per cent in Maharashtra, Tamil Nadu, Madhya Pradesh and Manipur.

All the states and territories have improved on their female literacy levels between 1971 and 1981. However, as many as five states, all major ones, made little progress in female literacy in the same period; they are Uttar Pradesh, Andhra Pradesh, Tamil Nadu, Maharashtra and Gujarat.

It is because these disparities continue that women continue to hold a subordinate position in Indian society. The country has certainly made great strides in the last 80 years towards reducing this inequity. In 1901 there were 1466 male literates for every 100 female literates. This difference has been steadily reduced until, in 1981, there were 202 male literates for every 100 female literates. There is, however, a great need for doing still better, both on this count as well as in improving the overall literacy levels of the country.

Sources: 1. *Primary Census Abstract, Part II B (i),* Census of India 1981.
2. *Family Welfare Programme in India Year Book 1983-84,* Ministry of Health & Family Welfare. Quoting from *Key Population Statistics Based on 5 per cent Sample Data, Paper 2 of 1983,* Census of India 1981.

Table 20

LITERATES BY AGE GROUPS — 1981[1]

(in per cent)

	5 and above		10 and above		15 and above		35 and above		All ages	
	Male	Female	Male	Female	Male	Female	Male	Female	Male	Female
States										
Andhra Pradesh	44.64	23.25	46.53	22.93	44.32	20.03	36.97	11.57	39.26	20.39
Bihar	43.76	15.79	47.23	15.86	44.85	13.17	36.42	7.32	38.11	13.62
Gujarat	62.05	36.95	65.26	37.30	62.76	33.16	52.26	20.48	54.44	32.30
Haryana	55.36	25.80	58.57	25.81	54.43	21.58	36.94	9.21	48.18	22.25
Karnataka	55.73	35.32	58.69	31.81	56.90	28.23	48.83	16.01	48.81	27.71
Kerala	84.67	73.39	87.72	74.58	85.98	70.79	77.16	51.36	75.26	65.73
Madhya Pradesh	45.54	18.03	49.65	18.59	47.63	15.88	36.44	7.77	39.49	15.53
Maharashtra	66.72	39.57	68.89	39.45	67.62	34.56	55.91	18.70	58.79	34.79
Orissa	53.30	24.02	57.38	24.46	55.98	21.16	49.18	10.48	47.10	21.12
Punjab	53.31	38.28	54.63	37.38	50.71	32.42	35.61	12.72	47.16	33.69
Rajasthan	42.04	13.36	45.97	13.78	43.01	12.03	30.87	6.17	38.30	11.42
Tamil Nadu	65.61	39.38	68.03	38.74	65.99	34.65	56.93	19.66	58.26	34.99
Uttar Pradesh	44.56	16.32	48.42	16.68	45.36	13.92	33.06	6.93	38.76	14.04
West Bengal	56.91	34.39	61.35	36.18	61.15	33.25	56.66	20.87	50.67	30.25
INDIA[2]	53.48	28.47	56.99	28.99	54.84	25.68	44.61	14.44	46.89	24.82

Note: [1] Figures available only for the 14 major states.

[2] Excludes Assam, but includes other states & union territories.

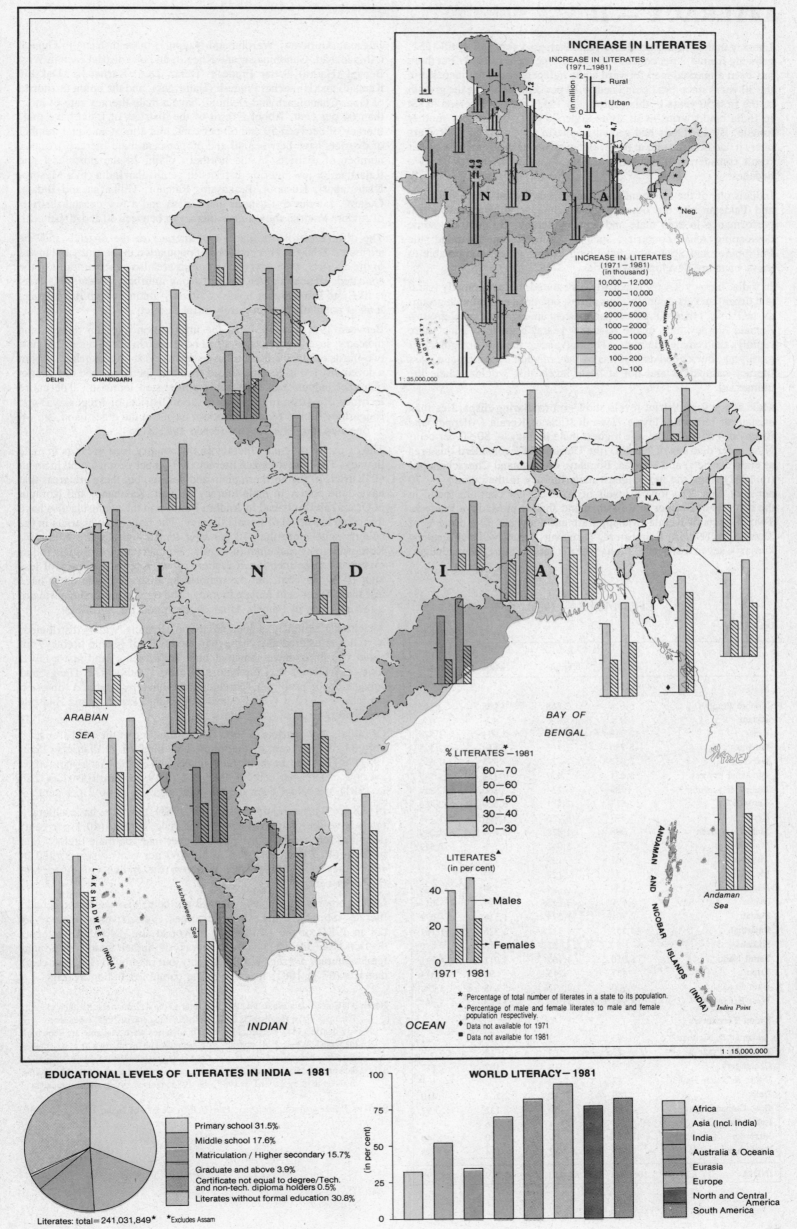

INCREASE IN LITERATES

INCREASE IN LITERATES
(1971—1981)

(in million)
2 → Rural
0 → Urban

DELHI

INCREASE IN LITERATES
(1971 — 1981)
(in thousand)

10,000 — 12,000
7000 — 10,000
5000 — 7000
2000 — 5000
1000 — 2000
500 — 1000
200 — 500
100 — 200
0 — 100

ANDAMAN AND NICOBAR ISLANDS (INDIA)

*Neg.

1 : 35,000,000

DELHI   CHANDIGARH

I   N   D   I   A

ARABIAN
SEA

BAY OF
BENGAL

N.A.

LAKSHADWEEP (INDIA)

Lakshadweep Sea

% LITERATES *—1981

60—70
50—60
40—50
30—40
20—30

LITERATES ▲
(in per cent)

40

20 → Males

→ Females

0
1971 1981

ANDAMAN AND NICOBAR ISLANDS (INDIA)

Andaman
Sea

Indira Point

* Percentage of total number of literates in a state to its population.
▲ Percentage of male and female literates to male and female population respectively.
♦ Data not available for 1971
■ Data not available for 1981

INDIAN   OCEAN

1 : 15,000,000

EDUCATIONAL LEVELS OF LITERATES IN INDIA — 1981

Primary school 31.5%
Middle school 17.6%
Matriculation / Higher secondary 15.7%
Graduate and above 3.9%
Certificate not equal to degree/Tech. and non-tech. diploma holders 0.5%
Literates without formal education 30.8%

Literates: total = 241,031,849*   *Excludes Assam

WORLD LITERACY — 1981

100

75

50

(in per cent)
25

0

Africa
Asia (Incl. India)
India
Australia & Oceania
Eurasia
Europe
North and Central America
South America

LITERACY — II

Literacy in the pre-independence years increased very slowly till 1951, growing from 5.2 per cent in 1901 to 16.7 per cent in 1951. But there has been a marked leap forward since independence, the increase in the 30 years since 1951 being nearly 19 per cent, doubling the growth of the first 50 years of this century. In 1951, only one person in six in India could write, with some understanding, his or her mother tongue. By 1961 this had grown to one in four. In 1981, 20 years later, it had become one in three. But even this improvement is a poor result considering the volume of educational effort in the last two decades.

India is one of the six countries (Bangladesh, Egypt, Nepal, Nigeria and Pakistan are the other five) which have shown the poorest performance in both male and female literacy. By contrast, some developing Asian countries, such as South Korea, Thailand, the Philippines and Sri Lanka, have high literacy rates, comparable to those of many developed countries.

In India, literacy is very unequally distributed, both regionally (rural and urban) and communally. It is almost universal among such groups as the Parsis, but is still almost non-existent among some tribal groups. Female literacy has, in the past few years, been increasing more rapidly than male literacy. The advance, though irregular, is gratifying, but a good deal of it is by way of adult education and mass literacy campaigns and not at basic levels that would include the unmarried.

Male literacy at district levels shows more glaring disparities than at state or territorial levels. Two districts of Kerala (Alleppey and Kottayam) have, by far, the highest male literacy — 80-90 per cent. Nine other districts of Kerala, (the Census 1981 considerd this area as eight districts) and Madras, Bombay, Calcutta and Churachandpur (in 1981) show the next highest male literacy in the country — 70 per cent plus. The lowest levels of 10 to 20 per cent are found in the districts of Barmer in Rajasthan and Jhabua in Madhya Pradesh. Two districts of Kerala (Malappuram and Palghat), five districts of Tamil Nadu (Nilgiri, Coimbatore, Thanjavur, Tirunelveli and Kanniyakumari), seven districts of Maharashtra (Satara, Pune, Ahmadnagar,

Jalgaon, Amravati, Wardha and Nagpur), three districts in Gujarat (Ahmadabad, Gandhinagar and Kheda), and one district each in West Bengal (Haora), Uttar Pradesh (Dehra Dun) Karnataka (Dakshin Kannad) and Himachal Pradesh (Hamirpur), and the union territories of Goa, Chandigarh and Delhi all have a male literacy rate of more than 60 per cent. About a third of the districts of India have male literacy of between 50 and 60 per cent, and almost an equal number of districts have between 40 and 50 per cent male literacy. A large number of districts in the northern plain, in the north-east and Rajasthan, a few interior districts in peninsular India (like Mysore, Dharmapuri, Kurnool, Prakasam, Raichur, Gulbarga and Bidar), Gujarat, Jammu & Kashmir and Orissa and a few coastal districts of Andhra Pradesh show a male literacy of between 30 and 40 per cent.

The differences in the pattern of literacy in the districts may be attributed to large concentrations of population in the cities and towns of the districts, for such concentrations are also concentrations of the educated. Hence, a given district strong in urban population content is likely to show higher levels of literacy amongst both its male and female populations than purely rural districts.

Between 1971 and 1981, as the information overlain on the map indicates, several districts have had no change in male literacy. Almost two-thirds of India's districts have shown a 'one-step increase'. About a dozen districts have experienced 'two-step increases' and one district (Churachandpur) has even had a 'three-step increase'. The entire territory of Mizoram, the two western districts of Meghalaya, and Rajgarh district in Madhya Pradesh have, on the other hand, shown a 'one-step decrease' in the decade 1971-81.

The disparity in female literacy in the country is as great as in male literacy. The lowest female literacy of 0-10 per cent is found in about 60 districts of the northern plain and deserts, but these are areas that have done poorly in male literacy as well. Kalahandi and Koraput in Orissa, and Adilabad in Andhra Pradesh in the peninsula also have this same low level of female literacy. The only three districts in the country with more than 70 per cent female literacy are Alleppey, Kottayam and Ernakulam in Kerala. No district in peninsular India, except the three mentioned earlier, shows a female literacy of less than 10-20 per cent, but the majority of districts in the north have less than 20 per cent female literacy. The exceptions in the north are a few districts in Punjab, Uttar Pradesh and the north-east.

As with male literacy, a good number of districts, widely distributed, have not registered any changes in the levels of female literacy, but about two-thirds have shown a 'one-step increase' and just a few a 'two-step increase' (Coimbatore, Thane, Dhule, The Dangs and Junagadh, for example). Three districts have registered a 'one-step decrease'; they are West Godavari in Andhra Pradesh, and Bilaspur and Sehore in Madhya Pradesh.

Of India's 439 districts in 1985 (412 at the time of the 1981 Census), only 34 (7.7 per cent of the total), including all 14 districts (12 in 1981) in Kerala, have literacy levels above the Asian combined average of 53.7 per cent. In fact, 263 (59.9 per cent) districts (258 in 1981) are below even the all-India average of 36.2 per cent.

In 118 (26.9 per cent) districts (110 in 1981), the males have a literacy rate above the combined average of Asia; in 176 (40.1 per cent) districts (166 in 1981), the rate is higher than the male literacy rate in India (46.7 per cent); and in 303 (69 per cent) districts (289 in 1981) the male literacy rate is above the Indian combined rate (36.2 per cent).

On the other hand, in only 21 (4.8 per cent) districts is the female literacy rate above the Asian combined average; these include 13 (11 in 1981) of the 14 districts in Kerala. In 154 (35.1 per cent) districts (147 in 1981), female literacy is higher than the all-India female literacy average (24.9 per cent), but in only 70 (15.9 per cent) districts (67 in 1981) it is above the combined Indian average.

LITERATES — 1981

Table 21

(in thousand)

	RURAL		URBAN	
	Male	Female	Male	Female
States				
Andhra Pradesh	6,674	2,868	3,968	2,525
Assam	N.A.	N.A.	N.A.	N.A.
Bihar	10,718	3,053	2,974	1,576
Gujarat	5,736	2,766	3,819	2,575
Haryana	2,338	724	993	615
Himachal Pradesh	1,021	577	133	87
Jammu & Kashmir	789	234	360	214
Karnataka	5,617	2,581	3,620	2,465
Kerala	7,537	6,756	1,890	1,741
Madhya Pradesh	6,998	1,828	3,619	2,099
Maharashtra	10,521	5,042	8,536	5,522
Manipur	262	129	123	74
Meghalaya	172	129	87	67
Nagaland	159	94	49	28
Orissa	5,179	2,145	1,090	614
Punjab	2,701	1,574	1,514	1,072
Rajasthan	4,155	712	2,326	1,161
Sikkim	57	22	18	9
Tamil Nadu	8,356	4,160	5,911	4,211
Tripura	453	245	92	74
Uttar Pradesh	16,900	4,075	5,898	3,232
West Bengal	8,985	4,306	5,488	3,565
Union Territories				
Andaman & Nicobar	43	22	21	12
Arunachal Pradesh	83	27	15	6
Chandigarh	9	4	167	112
Dadra & Nagar Haveli	17	7	2	1
Delhi	150	65	2,202	1,410
Goa, Daman & Diu	228	159	132	97
Lakshadweep	7	4	6	4
Mizoram	116	89	50	40
Pondicherry	87	52	114	85
INDIA	106,066	44,450	55,219	35,295

Note: All Census data are dated 1981 and refer to the 412 districts that then existed. The maps show the districts as they existed at the end of 1985, when there were 439 districts (437 districts for which boundaries were available are shown on the maps). Where 1981 districts have been broken up into two or more smaller districts or combined into larger districts (as in Jammu & Kashmir) the 1981 data are used to shade the 1985 configurations. In the text, which refers to the districts as they existed in 1985, the 1981 status is given within brackets.

Source: *Primary Census Abstract, Part II B(i)*, Census of India, 1981.

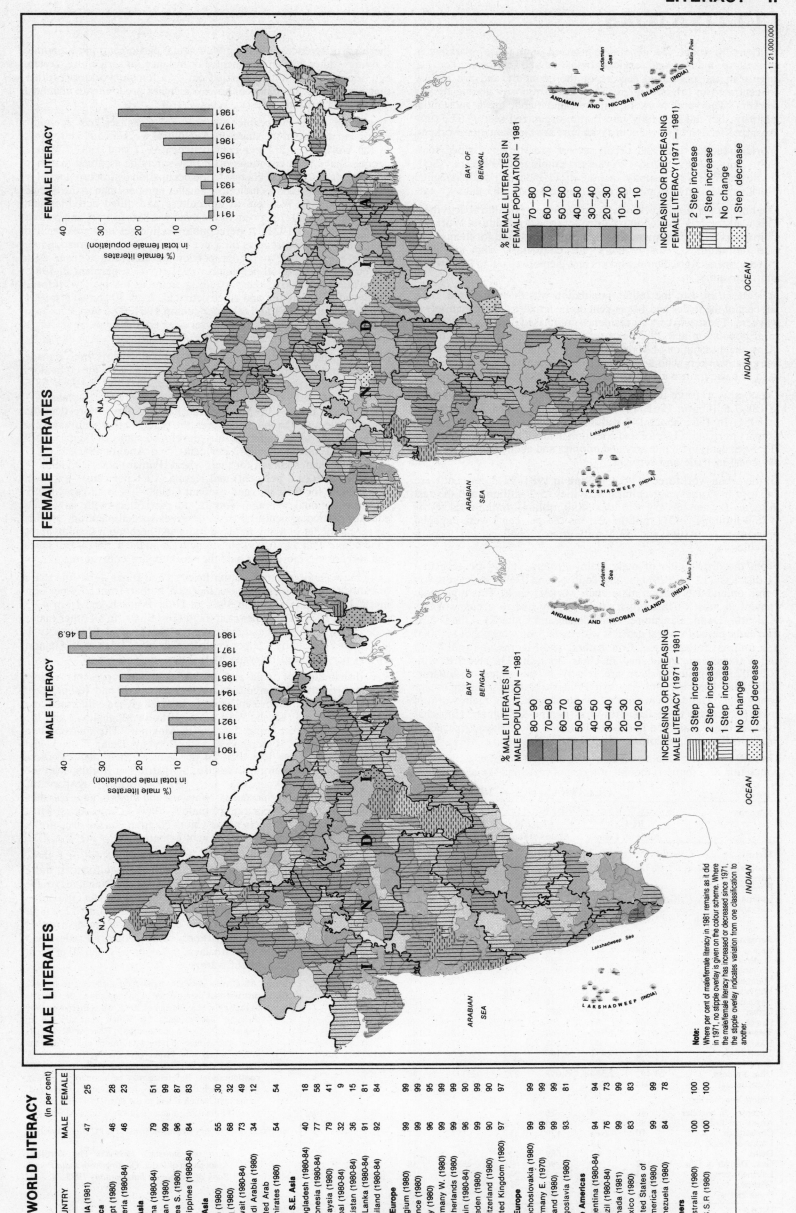

1 : 21,000,000

FEMALE LITERATES

FEMALE LITERACY

% female literates in total female population

1911 1921 1931 1941 1951 1961 1971 1981

% FEMALE LITERATES IN
FEMALE POPULATION –1981

| 70–80 |
| 60–70 |
| 50–60 |
| 40–50 |
| 30–40 |
| 20–30 |
| 10–20 |
| 0–10 |

INCREASING OR DECREASING
FEMALE LITERACY (1971 – 1981)

2 Step increase
1 Step increase
No change
1 Step decrease

BAY OF
BENGAL

Andaman
Sea

ANDAMAN AND NICOBAR ISLANDS (INDIA)

Indira Point

INDIAN OCEAN

ARABIAN
SEA

Lakshadweep Sea

LAKSHADWEEP (INDIA)

MALE LITERATES

MALE LITERACY

% male literates in total male population

46.9

1901 1911 1921 1931 1941 1951 1961 1971 1981

% MALE LITERATES IN
MALE POPULATION –1981

| 80–90 |
| 70–80 |
| 60–70 |
| 50–60 |
| 40–50 |
| 30–40 |
| 20–30 |
| 10–20 |

INCREASING OR DECREASING
MALE LITERACY (1971 – 1981)

3 Step increase
2 Step increase
1 Step increase
No change
1 Step decrease

BAY OF
BENGAL

Andaman
Sea

ANDAMAN AND NICOBAR ISLANDS (INDIA)

Indira Point

INDIAN OCEAN

ARABIAN
SEA

Lakshadweep Sea

LAKSHADWEEP (INDIA)

Note:
Where per cent of male/female literacy in 1981 remains as it did
in 1971, no stipple overlay is given on the colour scheme. Where
the male/female literacy has increased or decreased since 1971,
the stipple overlay indicates variation from one classification to
another.

WORLD LITERACY
(in per cent)

COUNTRY	MALE	FEMALE
INDIA (1981)	47	25
Africa		
Egypt (1980)	46	28
Nigeria (1980-84)	46	23
E. Asia		
China (1980-84)	79	51
Japan (1980)	99	99
Korea S. (1980)	96	87
Philippines (1980-84)	84	83
W. Asia		
Iran (1980)	55	30
Iraq (1980)	68	32
Kuwait (1980-84)	73	49
Saudi Arabia (1980)	34	12
United Arab Emirates (1980)	54	54
S & S.E. Asia		
Bangladesh (1980-84)	40	18
Indonesia (1980-84)	77	58
Malaysia (1980)	79	41
Nepal (1980-84)	32	9
Pakistan (1980-84)	36	15
Sri Lanka (1980-84)	91	81
Thailand (1980-84)	92	84
W. Europe		
Belgium (1980)	99	99
France (1980)	99	99
Italy (1980)	96	95
Germany W. (1980)	99	99
Netherlands (1980)	99	99
Spain (1980-84)	96	90
Sweden (1980)	99	99
Switzerland (1980)	90	90
United Kingdom (1980)	97	97
E. Europe		
Czechoslovakia (1980)	99	99
Germany E. (1970)	99	99
Poland (1980)	99	99
Yugoslavia (1980)	93	81
The Americas		
Argentina (1980-84)	94	94
Brazil (1980-84)	76	73
Canada (1981)	99	99
Mexico (1980)	83	83
United States of America (1980)	99	99
Venezuela (1980)	84	78
Others		
Australia (1980)	100	100
U.S.S.R (1980)	100	100

49

CULTIVATORS

Workers in India are generally grouped into those working in agriculture, industry and services. Industrial and service workers are largely in the organized sectors of the economy and are mainly concentrated in urban centres. They are a minority among India's workers. The rural workers, on the other hand, are a substantial majority, but fall generally into the un-organized sector. The only organized section of workers in agriculture are the plantation workers.

Workers in the rural and urban sectors are further divided into: Cultivators, Agricultural Workers, Employed Workers, Self-employed Workers (usually in household industry) and Marginal Workers, who, by definition, work for less than 183 days a year.

The bulk of the rural workforce is engaged in agricultural activities. This workforce is divided into Cultivators and Agricultural Workers. Cultivators work their own, leased or mortgaged land for themselves (see definition on map); there are cultivators who are even sharecroppers. Agricultural workers hire themselves out to landowners and cultivators.

In 1981, a third of the Indian population was the workforce, 51.6 per cent of the males and 14 per cent of the females calling themselves workers. These workers by residence comprised 34.8 per cent of the rural population and 29.2 per cent of the urban population. In the maps, the distribution of Main Workers (the total workforce excluding marginal workers who are unemployed for most part of the year) is used as basic information and the overlays record cultivators by sex.

The total workforce in 1951 was about 139.5 million; this went up to 188.7 million in 1961, 180.3 million in 1971, and 222.5 million in 1981. In 1981 about 148 million workers were engaged in the primary agricultural sector and 5 million in allied activities, 30 million (13.5 per cent) in all industry and mining, and about 40 million (17.8 per cent) in trade and services.

Of the urban workforce of 46.1 million in 1981, 11.3 per cent were engaged in primary agricultural activities (5.2 million) and the rest in secondary and tertiary activities (40.9 million). In the rural sector (176.4 million), 142.9 million persons (81 per cent) were engaged in primary activities and 33.5 million in secondary and tertiary activities.

From the distribution of male main workers, it will be seen that 21 districts (19 in 1981) in the country and Delhi have between 1 million and 1.6 million male main workers each. This high male content in most of these districts is largely due to urban workers (Calcutta, Delhi, Bangalore, Ahmadabad and Cuttack). This is also true in the case of several districts with male main workers of 800,000 to a million: for example, Coimbatore, and Madras in Tamil Nadu and Varanasi and Allahabad in Uttar Pradesh. The districts with 200-400,000 male workers are mostly in the states of Gujarat, Rajasthan, Haryana and Madhya Pradesh and most of them are contiguous. The districts with a low number of male main workers are in the extreme north and north-west (Jammu & Kashmir, Himachal Pradesh, Rajasthan) and north-east (Sikkim, Meghalaya, Arunachal Pradesh, Tripura, Mizoram, Manipur and Nagaland).

The distribution of main workers normally reflects the underlying population distribution, though in terms of male and female workers there are differences reflecting their actual participation in economic activities rather than employment. For instance, in agriculture, certain activities are traditionally carried out by a female workforce rather than a male one. On balance, however, males predominate in almost all working activities.

A general trend noticeable over several Censuses has been the continued decline in the share of agricultural workers and cultivators in the workforce (2.2 per cent between 1971 and 1981). Where declines have been encountered, it may safely be assumed that they are a result of a general change in the occupational structure and they must not be construed as indicating smaller numbers than in the decade ending in 1971. Workers in agriculture and allied activities and cultivators were 71.7 per cent of the main workers in 1901 and peaked at 76 per cent in 1921. A perceptible decline set in thereafter and their share was recorded as 68.8 per cent in 1981. A corresponding decline is reflected in the proportion of cultivators, from 50.6 per cent in 1901, to 50 per cent in 1951, to 41.6 per cent in 1981. Cultivators also have, thus, a falling share in the total workforce employed in agriculture and allied activities: from 70 per cent in 1901 to 69 per cent in 1951 and 60.6 per cent in 1981. The trend has been just the opposite in industrial and services employment, revealing a changing employment equation.

In India, as mentioned earlier, 41.6 per cent of the workforce is made up of cultivators. Among the major states, in Rajasthan the cultivators constitute 61.6 per cent of the workforce, in Uttar Pradesh 58.5 per cent, Madhya Pradesh 52 per cent, Orissa 47 per cent, Haryana 44.7 per cent and Bihar 43.6 per cent, all having shares more than the national average. On the other hand, the percentage of cultivators in the workforce in the agriculturally developed states of Punjab (35.9 per cent), Tamil Nadu (29.2 per cent), and Andhra Pradesh (32.7 per cent) — states known for their wheat (Punjab) and rice yields — West Bengal (29.8 per cent) and Kerala (13.1 per cent) — known respectively for their jute and coconut yields — is in all cases lower than the national average. Kerala, in particular, with its strong educational and agricultural base, deserves special mention, for this sta te is a model in many fields in India and here too it demonstrates good output with a low working strength. In all the north-eastern states and territories, more than half the workers are cultivators.

Only in two major states — Uttar Pradesh and Bihar — has the share of cultivators shown an increase, marginal at that, from 57.4 per cent in 1971 to 58.5 per cent in 1981 in Uttar Pradesh and from 43.3 per cent in 1971 to 43.6 per cent in 1981 in Bihar. In all other major states the share has come down over the decade 1971-81, notably in the case of Kerala by 4.7 per cent, Punjab by 6.7 per cent, Gujarat by 5.6 per cent and Haryana by 4.7 per cent.

The distribution of male and female cultivators presents widely different patterns. Male cultivators are between 80 and 100 per cent in seven districts: Jhabua in Madhya Pradesh, Kheri, Bahraich and Hardoi in Uttar Pradesh; West Khasi Hills in Meghalaya; and the Senapati and Tamenglong districts in Manipur. The metropolitan districts of Madras, Hyderabad, Greater Bombay, Delhi, Calcutta and Chandigarh all have less than 20 per cent of their male workers categorized as cultivators. Likewise, all districts in Kerala, with the exception of five (Quilon, Kottayam, Pathanamitta, Wayanad and Idukki), have less than 20 per cent of male workers as cultivators. Most districts in the Chota Nagpur Plateau area are in the 40-60 per cent category.

The districts in the north and north-east show higher concentrations of female cultivators, more than 60 per cent in most areas (Nagaland is the only state in the country where female cultivators outnumber the male — 143,152 against 123,089). Kerala stands out, along with many districts of West Bengal, Bihar, coastal Andhra Pradesh and northern Karnataka, in having less than 20 per cent female cultivators.

Note: All Census data are dated 1981 and refer to the 412 districts that then existed. The map shows the districts as they existed at the end of 1985, when there were 439 districts (437 districts for which boundaries were available are shown on the maps). Where 1981 districts have been broken up into two or more smaller districts or combined into larger districts (as in Jammu & Kashmir) the 1981 data are used to shade the 1985 configurations. Reference may be made to pages 8-11 to find which districts have been divided or enlarged. In the text, which refers to the districts as they existed in 1985, the 1981 status is given within brackets.

Sources: 1. *Primary Census Abstract, Part II B(i)*, Census of India 1981.
2. *Fertiliser Statistics 1984-85*, The Fertiliser Association of India. Quoting from *India Pocket Book of Economic Information, 1973-74*.

Table 22

CULTIVATORS — 1981

(in thousand)

| | MAIN WORKERS | | | | | CULTIVATORS | | Total |
| | RURAL | | URBAN | | | RURAL | | (including |
	Male	Female	Male	Female	Total	Male	Female	urban)
States								
Andhra Pradesh	12,327	6,506	3,159	637	22,629	5,557	1,673	7,408
Assam	N.A.	N.A.	N.A.	N.A.	N.A.	N.A.	N.A.	N.A.
Bihar	15,589	2,912	2,087	165	20,753	8,094	778	9,042
Gujarat	6,386	1,547	2,775	276	10,984	3,458	535	4,115
Haryana	2,617	230	764	52	3,664	1,452	137	1,637
Himachal Pradesh	979	381	97	14	1,471	643	352	1,002
Jammu & Kashmir	1,317	137	334	30	1,819	901	103	1,034
Karnataka	7,532	2,908	2,667	543	13,650	4,167	825	5,222
Kerala	4,188	1,417	953	233	6,792	780	78	887
Madhya Pradesh	11,759	5,239	2,630	413	20,041	7,553	2,637	10,414
Maharashtra	11,055	6,361	5,964	921	24,302	5,567	2,769	8,536
Manipur	253	200	78	42	573	186	141	365
Meghalaya	303	199	60	18	580	209	152	363
Nagaland	179	147	36	5	368	122	142	266
Orissa	6,411	1,287	826	110	8,635	3,649	388	4,053
Punjab	3,458	98	1,291	80	4,928	1,688	10	1,767
Rajasthan	7,149	1,380	1,764	150	10,442	5,242	995	6,431
Sikkim	80	47	18	3	147	49	40	89
Tamil Nadu	9,532	4,491	4,145	859	19,026	4,177	1,190	5,559
Tripura	468	80	51	9	609	228	34	263
Uttar Pradesh	24,492	2,534	5,098	272	32,397	17,159	1,329	18,958
West Bengal	10,044	1,209	3,869	303	15,424	4,306	219	4,591
Union Territories								
Andaman & Nicobar	42	3	16	1	63	9	0.5	10.2
Arunachal Pradesh	180	117	15	2	313	113	109	223
Chandigarh	10	Neg	129	17	157	1	Neg	2
Dadra & Nagar Haveli	27	13	2	1	42	16	10	26
Delhi	117	12	1,689	169	1,986	25	2	35
Goa, Daman & Diu	162	60	90	21	332	38	20	63
Lakshadweep	3	1	3	Neg	8	—	—	—
Mizoram	100	66	30	11	206	75	62	146
Pondicherry	71	20	69	13	173	13	1	16
INDIA	**136,831**	**39,603**	**40,713**	**5,370**	**222,517**	**75,476**	**14,682**	**92,523**

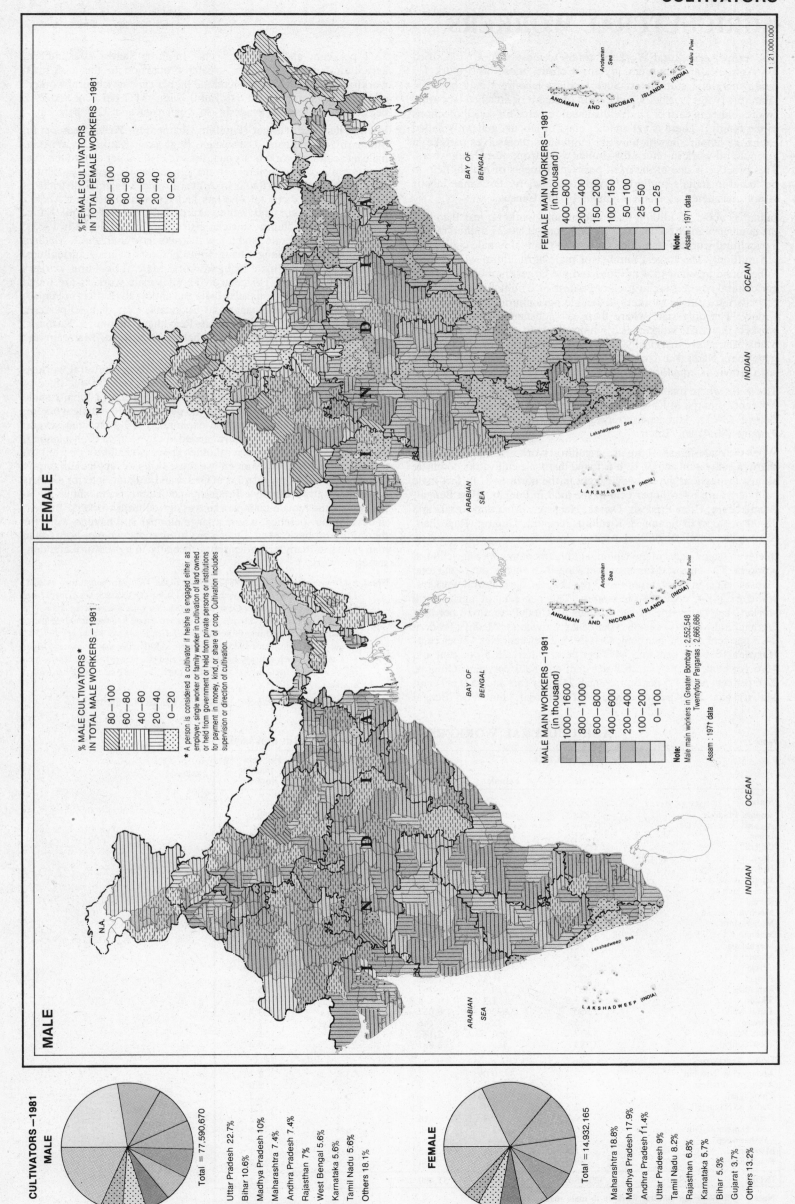

FEMALE

% FEMALE CULTIVATORS
IN TOTAL FEMALE WORKERS —1981

80—100
60—80
40—60
20—40
0—20

FEMALE MAIN WORKERS —1981
(in thousand)

400—800
200—400
150—200
100—150
50—100
25—50
0—25

Note:
Assam : 1971 data

MALE

% MALE CULTIVATORS *
IN TOTAL MALE WORKERS — 1981

80—100
60—80
40—60
20—40
0—20

* A person is considered a cultivator if he/she is engaged either as
employer, single worker or family worker in cultivation of land owned
or held from government or held from private persons or institutions
for payment in money, kind, or share of crop. Cultivation includes
supervision or direction of cultivation.

MALE MAIN WORKERS — 1981
(in thousand)

1000—1600
800—1000
600—800
400—600
200—400
100—200
0—100

Note:
Male main workers in Greater Bombay : 2,552,548
Twentyfour Parganas : 2,666,686
Assam : 1971 data

CULTIVATORS —1981
MALE

Total = 77,590,670

Uttar Pradesh 22.7%
Bihar 10.6%
Madhya Pradesh 10%
Maharashtra 7.4%
Andhra Pradesh 7.4%
Rajasthan 7%
West Bengal 5.6%
Karnataka 5.6%
Tamil Nadu 5.6%
Others 18.1%

FEMALE

Total = 14,932,165

Maharashtra 18.8%
Madhya Pradesh 17.9%
Andhra Pradesh 11.4%
Uttar Pradesh 9%
Tamil Nadu 8.2%
Rajasthan 6.8%
Karnataka 5.7%
Bihar 5.3%
Gujarat 3.7%
Others 13.2%

51

AGRICULTURAL WORKERS

The term 'Agricultural Workers' refers to those who are classified as 'Agricultural Labourers' by the Census, those who work for landowners and cultivators. Such landless labourers may be wage labourers (wage is usually in the form of cash), or attached labourers (wages either in cash or kind with annual rewards for social functions in the family). There is yet another class of labour, generally called exchange labour, in which owner-cultivators themselves partake in agricultural work in return for similar work from other landowners. This practice is one of the most positive elements of mutual help to be found in India's tradition-bound rural society. Exchange labour is not considered as landless labour by the Census.

India has about 35 million male agricultural workers, less than half the number of male cultivators (77.6 million). It has 21 million female agricultural workers, 6 million more than the total female cultivators (15 million). The largest number of male agricultural workers are to be found in Bihar (5.4 million) and the largest number of female agricultural workers are in Andhra Pradesh (4.2 million). But the ratio of male agricultural workers to female agricultural workers varies widely. The only state where there are substantially more female workers than male workers is Manipur. Though Andhra Pradesh and Maharashtra have marginally more female workers, they are, together with Tamil Nadu, Karnataka, Goa and Madhya Pradesh, states where such labour is equally shared.

The states where male agricultural labour dominates are Punjab (95.9 per cent), Jammu & Kashmir (94.2 per cent), Haryana (89.5), West Bengal (84.7), Himachal Pradesh (83.1), Uttar Pradesh (80.9), Tripura (80.4) and Bihar (73.5).

When the male-female share in agricultural work is compared to their share in cultivation activities, it is found that male cultivators dominate almost throughout the country except in the north-east. In fact male cultivators are over 90 per cent of the total in Punjab, West Bengal, Pondicherry, Uttar Pradesh, Orissa, Haryana, Bihar and Kerala and 75-90 per cent in Jammu & Kashmir, Tripura, Gujarat, Rajasthan, Karnataka, Tamil Nadu and Andhra Pradesh.

In eleven major states of the Union, cultivators outnumber agricultural workers. The largest difference is in Rajasthan where, out of the total main workers in the state, only 7.3 per cent are agricultural workers, while 61.1 per cent are cultivators. The per cent of agricultural workers and cultivators in other states is: Uttar Pradesh 16 per cent (agricultural workers) and 58.5 per cent (cultivators); Madhya Pradesh 24.2 per cent and 52 per cent; Orissa 19.2 per cent and 47 per cent; Haryana 16.1 per cent and 44.7 per cent; Bihar, 35.5 per cent and 43.6 per cent; Karnataka 26.8 per cent and 38.3 per cent; Gujarat, 22.6 per cent and 37.5 per cent; Punjab 21.1 per cent and 35.9 per cent; Maharashtra 26.6 per cent and 35.1 per cent; and West Bengal

25.2 per cent and 29.8 per cent. In three states, all southern, agricultural workers form a greater proportion of the total main workforce than cultivators: Kerala, 28.2 per cent agricultural workers and 13.1 per cent cultivators; Tamil Nadu, 31.7 per cent and 29.2 per cent; and Andhra Pradesh, 36.8 per cent and 32.7 per cent.

In eight districts of Bihar (Paschim Champaran, Purba Champaran, Sitamarhi, Madhubani, Darbhanga, Begusarai, Katihar and Purnia) male agricultural workers account for some 40-60 per cent of the total male main workers, whereas in the peninsula, the same per cent of total male workers is found in Amravati and Yavatmal districts of Maharashtra. Six coastal districts and Kurnool district of Andhra Pradesh, Thanjavur and Kanniyakumari districts of Tamil Nadu, Karaikal of Pondicherry, thirteen districts in Bihar (eleven in 1981), four districts of Kerala, and a few districts in Maharashtra, Madhya Pradesh and West Bengal have 30-40 per cent of male agricultural workers of the total male main workers. Male agricultural workers accounting for 20-30 per cent of the total male workers are found in several districts of peninsular India and almost all districts in Punjab, except Kapurthala, Rupnagar and Ludhiana. Less than 10 per cent are found in almost all districts in Rajasthan, Jammu & Kashmir, Himachal Pradesh, Manipur, Meghalaya, Nagaland, Mizoram and Arunachal Pradesh.

In five districts (four in 1981) of Bihar, female agricultural workers account for 80-100 per cent of the female workers. In one-third of the districts of peninsular India and several districts in Bihar female agricultural workers account for 60-80 per cent of all female workers. This higher contribution may be explained by the fact that several agricultural activities are usually carried out by women (transplanting, weeding, harvesting and several other lesser agricultural operations) and in landless families women use these skills to supplement family income. Almost all the districts in Orissa and Gujarat, interior districts of Maharashtra, Madhya Pradesh and Tamil Nadu show 40-60 per cent of the female main workers as agricultural workers. Indeed, in developing societies, where a large number still have to work for the minimum basic needs, there are often more women-hours than man-hours in many activities, but especially in agriculture and construction-work.

Note: All Census data are dated 1981 and refer to the 412 districts that then existed. The map shows the districts as they existed at the end of 1985, when there were 439 districts (437 districts for which boundaries were available are shown on the maps). Where 1981 districts have been broken up into two or more smaller districts or combined into larger districts (as in Jammu & Kashmir) the 1981 data are used to shade the 1985 configurations. Reference may be made to pages 8-11 to find which districts have been divided or enlarged. In the text, which refers to the districts as they existed in 1985, the 1981 status is given within brackets.

Source: *Primary Census Abstract, Part II B(i),* Census of India 1981.

AGRICULTURAL WORKERS — 1981

Table 23   (in thousand)

| | RURAL | | Total (including urban) | Per cent male workers[1] | Per cent female workers[2] |
	Male	Female			
States					
Andhra Pradesh	3,878	4,035	8,325	26.5	59.0
Assam	N.A.	N.A.	N.A.	N.A.	N.A.
Bihar	5,251	1,911	7,367	30.6	63.3
Gujarat	1,515	835	2,488	17.6	48.0
Haryana	501	59	590	15.6	22.0
Himachal Pradesh	32	6.6	40	3.1	1.7
Jammu & Kashmir	51.6	3.3	63	3.6	2.1
Karnataka	1,778	1,600	3,655	19.0	49.7
Kerala	1,134	685	1,917	23.3	43.5
Madhya Pradesh	2,448	2,218	4,857	17.8	40.6
Maharashtra	2,954	3,166	6,470	18.5	45.6
Manipur	7.8	14	28.6	3.3	7.3
Meghalaya	32.8	23	57.9	9.4	10.9
Nagaland	2.1	0.4	2.9	1.2	0.3
Orissa	1,583	736	2,397	22.6	54.0
Punjab	972	41.8	1,092	22.0	25.3
Rajasthan	483	229	765	5.8	15.7
Sikkim	3.3	1.5	4.9	3.4	3.0
Tamil Nadu	2,950	2,697	6,037	23.2	53.4
Tripura	115	28	146	22.5	32.0
Uttar Pradesh	3,877	964	5,177	14.2	35.0
West Bengal	3,193	585	3,891	23.7	39.4
Union Territories					
Andaman & Nicobar	2.2	0.09	2.3	3.8	2.3
Arunachal Pradesh	5.2	2.3	7.7	2.8	2.0
Chandigarh	0.4	0.06	0.8	0.6	0.4
Dadra & Nagar Haveli	2.5	1.9	4.6	8.8	15.1
Delhi	9.7	1.7	16	0.7	1.2
Goa, Daman & Diu	16	13	32	6.9	18.5
Lakshadweep	—	—	—	—	—
Mizoram	1.4	0.7	5	2.7	2.2
Pondicherry	30	16	54	25.6	56.2
INDIA	32,835	19,878	55,499	19.6	46.2

Note [1] % Male agricultural workers in male main workers.
[2] % Female agricultural workers in female main workers.

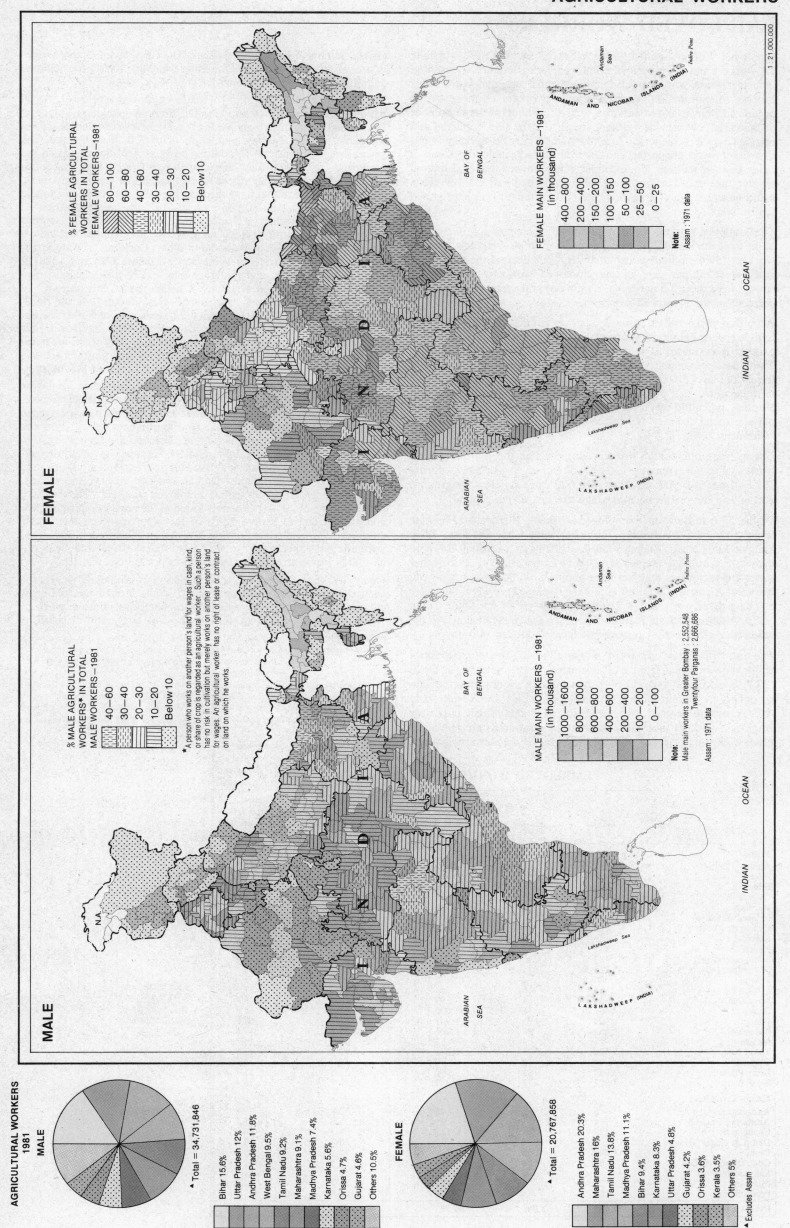

FEMALE

MALE

EMPLOYED WORKERS

Employed Workers, called 'Other Workers' by the Census, include all workers other than those employed in agriculture, household industry (those who are self-employed) and marginal work. This process of elimination leaves employed workers as those in the organized industrial and services sectors, although a large number of industrial and service workers are also in the self-employed and marginal workers categories. In addition, there may be several millions of workers who are in the unorganized sectors, including those in industries and services. For example, it is estimated that 11.2 million workers are children, but it has been suggested that child-workers could be as many as 44 million, the bulk of them thus unaccounted for by the Census.

The employed workers, 66.8 million in all, accounted for 30 per cent of the total workers (222.5 million) in 1981. Of these, male employed workers (59.6 million) constituted 26.8 per cent of the total main workers (89.2 per cent of the employed workers) and females constituted a mere 3.2 per cent (7.2 million) of the total workers (10.8 per cent of the employed workers).

According to the 1981 Census, 5.3 per cent of the total workers (11.2 million) were below 15 years of age. Of these, about 1.7 million children were stated to be in the organized sectors of the economy. These are, presumably, child workers in organizations that strictly follow the laws governing child labour. Some 16.2 million workers in the country (7.2 per cent of the total) were in the age group of 60 plus, indicating that a substantial number of people do not enjoy old-age social security and hence have to work beyond the age of retirement.

The average number of workers employed in the mines — these workers also fall within the category of employed workers — during 1979-80 was 740,000. Of these, 310,000 worked below the ground, 206,000 in the open-cast mines and the rest above ground.

Of the 59.6 million male employed workers, Maharashtra has the largest number, 7.7 million. Together with Uttar Pradesh, West Bengal, Tamil Nadu, Andhra Pradesh and Gujarat it accounts for more than half of all male employed workers in the country; all the other states and territories together have about 25 million male employed workers. As for the 7.2 million female employed workers, the states that have the largest share (4.7 million totally) are Maharashtra, Tamil Nadu, Andhra Pradesh, Kerala, Karnataka and West Bengal All the other states and territories together have about 2.5 million.

The ratio of urban employed workers in all categories to rural employed workers is highest in such urban union territories as Delhi (20.7:1) and Chandigarh (17.5:1). Amongst the major states, the ratio is highest in Maharashtra (2.4:1), Gujarat (2:1) and Tamil Nadu (1.7:1), the three most urbanized states. States which have less employed workers in urban areas than in rural areas are Himachal Pradesh, Kerala, Tripura, Sikkim, Manipur, Meghalaya, Nagaland,

Orissa, Jammu & Kashmir and Bihar. The economy of all these states is strongly rural-based except for Bihar (mining), Kerala (plantations) and Himachal Pradesh (orchards).

The very high concentration of male employed workers, 80-100 per cent, in Nilgiri, Delhi, Greater Bombay, Calcutta, Madras, Hyderabad, Chandigarh and Mahe (Pondicherry) districts is understandable. Nilgiri, a tiny plantation district in Tamil Nadu, has several major industries in addition to large tea plantations in the organized sector. The other districts have a relatively high concentration of industries and service activities and a low rural component, if any. The districts that have the next highest concentration of male employed workers, 60-80 per cent, are half the coastal districts of Kerala, Goa, Pondicherry, Thane, Ahmadabad and Haora, the interior districts of Bangalore in Karnataka, Indore and Bhopal in Madhya Pradesh, Dhanbad — a coal district in Bihar; Darjiling and Jalpaiguri, — tea plantation districts of West Bengal; Dehra Dun and Lahul & Spiti in North India; and the Andaman and Nicobar Islands. The districts around industrial towns (Chengalpattu, Trivandrum, 24 Parganas, Raigad, Valsad, Gurgaon, Faridabad etc) have the third-highest concentration, for they are in the hinterland of such metropolises and capitals as Madras, Trivandrum, Calcutta, Bombay and Delhi. The districts with very low numbers of male employed workers are found in Uttar Pradesh and Bihar (seven in the Himalayan foothills and one in the interior) and the north-east (West Khasi Hills of Meghalaya).

The pattern of distribution of female employed workers is somewhat similar, though not in terms of intensity. The Nilgiris, Delhi, Greater Bombay, Calcutta, Madras, Hyderabad and Chandigarh districts again have the highest concentrations, 80-100 per cent. Kozhikode and Idukki in Kerala, Kodagu — the coffee plantation district in Karnataka — Haora, Darjiling, Jalpaiguri and 24 Parganas in West Bengal, Bareilly in Uttar Pradesh and five districts in Punjab, Mahe, the Andaman & Nicobar Islands, Daman and Diu all have the second-highest concentration of female employed workers. The medium concentration of 20-40 per cent is found in the north-western interior and around Delhi. In two-thirds of the other districts, the female employed workers concentration is less than 10 per cent, and in the northern part of peninsular India, Uttar Pradesh, Bihar and the no th-east, this low concentration is found in contiguous districts.

Generally, then, the pattern of distribution of employed workers is such that most coastal districts and districts around metropolises, capitals and seaports have relatively higher concentrations of these workers than the interior, non-industrial or non-urban districts.

Note: All Census data are dated 1981 and refer to the 412 districts that then existed. The map shows the districts as they existed at the end of 1985, when there were 439 districts (437 districts for which boundaries were available are shown on the maps). Where 1981 districts have been broken up into two or more smaller districts or combined into larger districts (as in Jammu & Kashmir) the 1981 data are used to shade the 1985 configurations. Reference may be made to pages 8-11 to find which districts have been divided or enlarged. In the text, which refers to the districts as they existed in 1985, the 1981 status is given within brackets.

Source: *Primary Census Abstract, Part II B (i),* Census of India 1981.

Table 24

EMPLOYED WORKERS — 1981
(in thousand)

	URBAN		Total (including Rural)	% Male employed workers in total male main workers	% Female employed workers in total female main workers
	Male	Female			
States					
Andhra Pradesh	2,639	348	5,832	32.1	12.0
Assam	N.A.	N.A.	N.A.	N.A.	N.A.
Bihar	1,692	111	3,849	20.3	8.4
Gujarat	2,502	206	4,114	41.0	19.4
Haryana	664	45	1334	37.2	26.3
Himachal Pradesh	89	12	402	34.5	8.0
Jammu & Kashmir	274	21	625	35.4	23.7
Karnataka	2,184	328	4,214	34.9	19.0
Kerala	841	184	3,736	58.6	43.8
Madhya Pradesh	2,193	246	4,063	25.0	8.2
Maharashtra	5,427	673	8,675	45.2	13.5
Manipur	46	12	125	30.7	9.4
Meghalaya	56	16	154	32.0	17.6
Nagaland	34	4	98	41.4	5.4
Orissa	682	77	1,900	23.2	15.6
Punjab	1,104	73	1,941	38.5	63.9
Rajasthan	1,463	98	2,906	30.1	14.8
Sikkim	17	3	52	45.1	16.7
Tamil Nadu	3,528	531	6,531	40.9	17.5
Tripura	47	9	190	32.0	27.2
Uttar Pradesh	3,947	182	7,062	22.8	11.7
West Bengal	3,577	271	6,399	41.8	38.4
Union Territories					
Andaman & Nicobar	16	1.4	48	77.0	75.8
Arunachal Pradesh	14	1	81	38.6	5.2
Chandigarh	127	17	153	97.3	98.7
Dadra & Nagar Haveli	1.4	0.2	11	30.4	10.8
Delhi	1,649	165	1,729	95.8	95.6
Goa, Daman & Diu	84	18	226	73.5	51.0
Lakshadweep	3.4	0.3	6.9	91.9	51.7
Mizoram	23	6	54	35.2	10.5
Pondicherry	60	9	98	61.8	37.2
INDIA	34,985	3,667	66,783	33.6	16.0

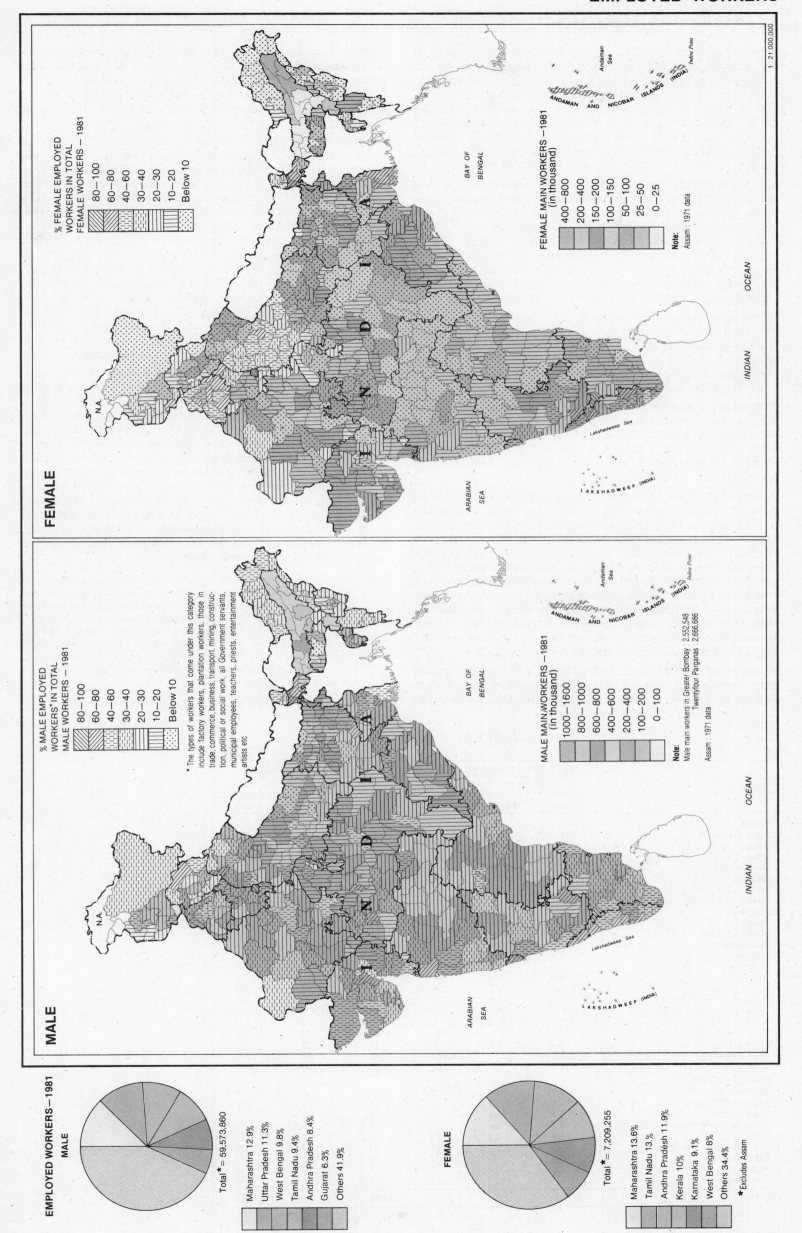

1 : 21,000,000

FEMALE

% FEMALE EMPLOYED
WORKERS IN TOTAL
FEMALE WORKERS – 1981

80—100
60—80
40—60
30—40
20—30
10—20
Below 10

FEMALE MAIN WORKERS –1981
(in thousand)

400—800
200—400
150—200
100—150
50—100
25—50
0—25

Note:
Assam : 1971 data

BAY OF
BENGAL

ARABIAN
SEA

INDIAN
OCEAN

Lakshadweep Sea

LAKSHADWEEP (INDIA)

ANDAMAN AND NICOBAR ISLANDS (INDIA)

Andaman
Sea

Indira Point

N.A.

MALE

% MALE EMPLOYED
WORKERS* IN TOTAL
MALE WORKERS – 1981

80—100
60—80
40—60
30—40
20—30
10—20
Below 10

* The types of workers that come under this category
include factory workers, plantation workers, those in
trade, commerce, business, transport, mining, construction, political or social work, all Government servants,
municipal employees, teachers, priests, entertainment
artists etc

MALE MAIN WORKERS – 1981
(in thousand)

1000—1600
800—1000
600—800
400—600
200—400
100—200
0—100

Note:
Male main workers in Greater Bombay : 2,552,548
Twenty four Parganas : 2,666,686
Assam : 1971 data

BAY OF
BENGAL

ARABIAN
SEA

INDIAN
OCEAN

Lakshadweep Sea

LAKSHADWEEP (INDIA)

ANDAMAN AND NICOBAR ISLANDS (INDIA)

Andaman
Sea

Indira Point

N.A.

EMPLOYED WORKERS—1981
MALE

Total* = 59,573,860

Maharashtra 12.9%
Uttar Pradesh 11.3%
West Bengal 9.8%
Tamil Nadu 9.4%
Andhra Pradesh 8.4%
Gujarat 6.3%
Others 41.9%

FEMALE

Total* = 7,209,255

Maharashtra 13.6%
Tamil Nadu 13.%
Andhra Pradesh 11.9%
Kerala 10%
Karnataka 9.1%
West Bengal 8%
Others 34.4%

*Excludes Assam

55

OTHER WORKERS

Other Workers, as classified in this volume, include all those who are, according to the Indian Census, Self-employed (main workers in household industry) or are Marginal Workers (not counted among main workers). The self-employed in the country amounted to 7.7 million in 1981, of which 5.6 million were males and the rest females. On the other hand, marginal workers, who work less than 183 days a year for want of employment, numbered a staggering 22.1 million. Together, they accounted for about 30 million of the country's workers, a little more than 4 per cent of the total population.

Of the 5.6 million male self-employed, a little more than half (57.4 per cent) are in five states: Uttar Pradesh, Andhra Pradesh, Tamil Nadu, Madhya Pradesh and Maharashtra. Of the 2.1 million female self-employed, nearly 56 per cent are also in just four states: Andhra Pradesh, Tamil Nadu, Karnataka and Madhya Pradesh. Andhra Pradesh, Tamil Nadu and Madhya Pradesh, which have done well in male self-employment, have also done well in female self-employment, whereas most other states that have done well in one (male/female) have done rather poorly in the other.

Of the 22 million marginal workers, males account for a mere 16 per cent (3.5 million). That over 18 million women are marginal workers is understandable. With household chores occupying most of her time, the female worker in a family has little time to spare, yet she has to supplement the family income, so part-time work, or work for part of the year, is acceptable to her.

Five states together have more than half the male marginal workers (55 per cent), with Kerala having the largest number, followed by West Bengal, Maharashtra, Bihar and Uttar Pradesh. Again, five states together account for a little more than half the female marginal workers: Madhya Pradesh, Maharashtra, Rajasthan, Andhra Pradesh and Gujarat.

Almost a fifth of the districts in the country, including a third of the peninsular districts, registered more than a million main workers each in 1981. With the exception of four districts (Kanniyakumari, Madras, Dharmapuri and Pudukkottai) in Tamil Nadu and eleven districts in Andhra Pradesh, all the other districts in these two states registered more than a million main workers each. On the other hand, only three districts in Karnataka (Bangalore, Dharwad and Belgaum) registered such a concentration. Such a high concentration is also found in seven districts in Maharashtra, six districts in West Bengal, ten districts in Bihar (four in 1981), seven districts (six in 1981) in Uttar Pradesh, two districts in Madhya Pradesh, and one district each in Orissa, Gujarat and Rajasthan and in the union territory of Delhi.

Another third of the districts registered between 500,000 and a million main workers each. The rest of the districts had low concentrations of main workers. Most of the districts with 100,000 - 500,000 main

workers each are found in the north; districts with such low concentrations are found in the south in only 15 districts. Still lower concentrations of main workers, of less than 50,000 are found only in the north-eastern states and in the north, in Ladakh (Kargil and Ladakh in 1981) and Lahul & Spiti, and Kinnaur in Himachal Pradesh.

A high proportion of self-employed to total workers (15-25 per cent) is found in Dakshin Kannad in Karnataka, Imphal, Thoubal and Bishnupur in Manipur (all three were part of Central Manipur in 1981), Nicobar in the Andaman & Nicobar Islands, and Chhindwara and Seoni in Madhya Pradesh. The second-highest concentration (10-15 per cent) is found in Tirunelveli in Tamil Nadu, Nizamabad in Andhra Pradesh, Bhandara in Maharashtra, Lakshadweep, Varanasi in Uttar Pradesh, and Kashmir North (Baramulla in 1981), in Jammu & Kashmir. In the rest of the districts in the country, a less than 10 per cent concentration has been registered, with the very lowest, 0-1 per cent, in Idukki and Wayanad in Kerala, Kodagu in Karnataka, Nilgiri in Tamil Nadu, Ladakh (Kargil and Ladakh in 1981) in Jammu & Kashmir, Chamba, Lahul & Spiti, Kullu and Shimla in Himachal Pradesh, Tehri-Garhwal, Garhwal and Kheri in Uttar Pradesh, most districts in north-east India including the whole of Arunachal Pradesh, and Dadra & Nagar Haveli, Andaman and Chandigarh.

As for marginal workers, the highest concentration — 60-80 persons per 100 main workers — is in Muzaffarabad, Punch and Riasi in Jammu & Kashmir, and Dungarpur in Rajasthan. The next highest concentration is also found around these three districts, in Kashmir North (Baramulla in 1981), Jammu, and Kathua in Jammu & Kashmir, Hamirpur in Himachal Pradesh and Banswara in Rajasthan. The rest of the country has concentrations of less than 40 marginal workers for every 100 main workers. With the exception of a few districts with concentrations of 15 to 30, all the rest in the peninsula have less than 15 marginal workers for every 100 main workers. About two-thirds of the districts in Uttar Pradesh have less than five persons for every 100 main workers doing marginal labour. In the whole of India, about hundred districts have this same low percentage of marginal workers, indicative of greater opportunities for work.

NOTE: All Census data are dated 1981 and refer to the 412 districts that then existed. The map shows the districts as they existed at the end of 1985, when there were 439 districts (437 districts for which boundaries were available are shown on the maps). Where 1981 districts have been broken up into two or more smaller districts or combined into larger districts (as in Jammu & Kashmir) the 1981 data are used to shade the 1985 configurations. Reference may be made to pages 8-11 to find which districts have been divided or enlarged. In the text, which refers to the districts as they existed in 1985, the 1981 status is given within brackets.

Source: *Primary Census Abstract, Part II B (i),* Census of India 1981.

Table 25   **OTHER WORKERS — 1981**

(in thousand)

	SELF-EMPLOYED WORKERS			MARGINAL WORKERS		
	Rural		Total (incl. Urban)	Rural		Total (incl. Urban)
	Male	Female		Male	Female	
States						
Andhra Pradesh	557	289	1,064	131	1,645	1,877
Assam	N.A.	N.A.	N.A.	N.A.	N.A.	N.A.
Bihar	345	76	495	331	1,486	1,864
Gujarat	155	28	267	108	1,540	1,718
Haryana	68	5	103	65	394	424
Himachal Pradesh	22	2.6	27	64	275	343
Jammu & Kashmir	52	13	96	110	692	831
Karnataka	214	155	560	103	1,095	1,293
Kerala	103	114	251	411	447	980
Madhya Pradesh	351	167	706	230	2,033	2,356
Maharashtra	270	116	621	313	1,917	2,416
Manipur	3.4	34.6	55	3.8	23	40
Meghalaya	1.8	1.8	4.9	5.5	27	33
Nagaland	0.4	0.3	1.5	2.6	2.4	5.4
Orissa	179	72	206	184	1,164	1,387
Punjab	75	5.6	118	51	295	360
Rajasthan	206	28	340	163	1,878	2,100
Sikkim	0.8	0.2	1.6	1	4.1	5.4
Tamil Nadu	337	197	898	145	919	1,172
Tripura	5.9	2	8.7	14	38	54
Uttar Pradesh	669	95	1,200	246	1,348	1,655
West Bengal	304	95	543	380	527	1,042
Union Territories						
Andaman & Nicobar	1.3	0.4	1.8	2	4.6	6.9
Arunachal Pradesh	0.6	0.1	1	4	14.9	19
Chandigarh	0.1	Neg	0.9	Neg	0.6	1
Dadra & Nagar Haveli	0.2	Neg	0.3	0.5	7.5	8.4
Delhi	1.9	0.2	33	2.1	5.4	15.8
Goa, Daman & Diu	6.5	2.2	11	11	33	50
Lakshadweep	0.5	0.4	1.1	0.8	0.5	1.9
Mizoram	0.5	0.3	1.8	4.6	10.8	18
Pondicherry	1	0.4	4	1.8	5.3	10.6
INDIA	3,932	1,502	7,711	3,090	17,785	22,088

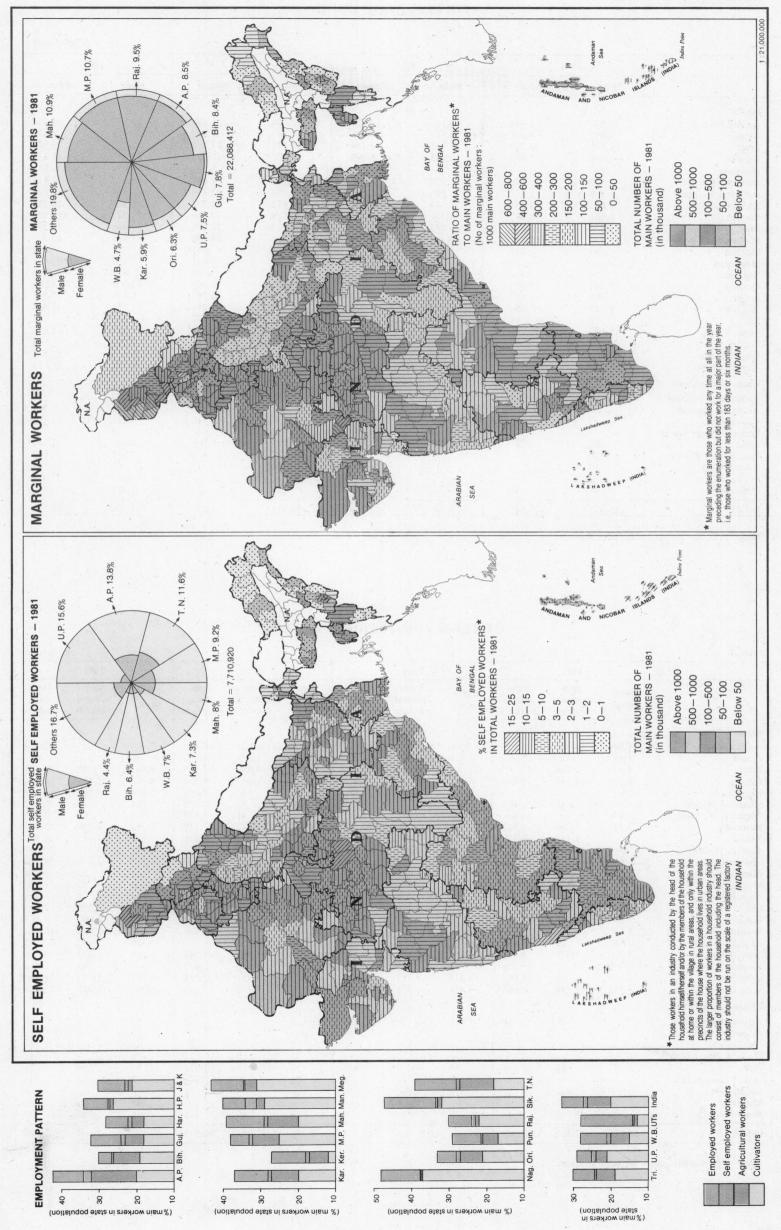

MARGINAL WORKERS

Total marginal workers in state

MARGINAL WORKERS — 1981

Male
Female

Mah. 10.9%
M.P. 10.7%
Raj. 9.5%
A.P. 8.5%
Bih. 8.4%
Gui. 7.8%
Total = 22,088,412
U.P. 7.5%
Ori. 6.3%
Kar. 5.9%
W.B. 4.7%
Others 19.8%

RATIO OF MARGINAL WORKERS
TO MAIN WORKERS — 1981
(No of marginal workers :
1000 main workers)

600—800
400—600
300—400
200—300
150—200
100—150
50—100
0—50

TOTAL NUMBER OF
MAIN WORKERS — 1981
(in thousand)

Above 1000
500—1000
100—500
50—100
Below 50

BAY OF
BENGAL

ANDAMAN AND NICOBAR ISLANDS (INDIA)
Andaman Sea
Indira Point

ARABIAN
SEA

LAKSHADWEEP (INDIA)
Lakshadweep Sea

INDIAN OCEAN

*Marginal workers are those who worked any time at all in the year
preceding the enumeration but did not work for a major part of the year,
i.e., those who worked for less than 183 days or six months.

1 : 21,000,000

SELF EMPLOYED WORKERS

Total self employed
workers in state

SELF EMPLOYED WORKERS — 1981

Male
Female

U.P. 15.6%
A.P. 13.8%
T. N. 11.6%
M.P. 9.2%
Mah. 8%
Total = 7,710,920
Kar. 7.3%
W.B. 7%
Bih. 6.4%
Raj. 4.4%
Others 16.7%

% SELF EMPLOYED WORKERS*
IN TOTAL WORKERS — 1981

15—25
10—15
5—10
3—5
2—3
1—2
0—1

TOTAL NUMBER OF
MAIN WORKERS — 1981
(in thousand)

Above 1000
500—1000
100—500
50—100
Below 50

BAY OF
BENGAL

ANDAMAN AND NICOBAR ISLANDS (INDIA)
Andaman Sea
Indira Point

ARABIAN
SEA

LAKSHADWEEP (INDIA)
Lakshadweep Sea

INDIAN OCEAN

*Those workers in an industry conducted by the head of the
household himself/herself and/or by the members of the household
at home or within the village in rural areas, and only within the
precincts of the house where the household lives in urban areas.
The larger proportion of workers in a household industry should
consist of members of the household including the head. The
industry should not be run on the scale of a registered factory.

EMPLOYMENT PATTERN

A.P. Bih. Gui. Har. H.P. J & K

Kar. Ker. M.P. Mah. Man. Meg.

Nag. Ori. Pun. Raj. Sik. T. N.

Tri. U.P. W.B. UTs India

(% main workers in state population)

Employed workers
Self employed workers
Agricultural workers
Cultivators

57

POPULATION — 1951-81

Table I   (in thousand)

States	1951 Male	1951 Female	1961 Male	1961 Female	1971 Male	1971 Female	1981 Male	1981 Female
States								
Andhra Pradesh	15,670	15,444	18,161	17,821	22,008	21,494	27,109	26,440
Assam	4,812	4,231	6,519	5,722	7,885	7,072	10,468	9,429
Bihar	19,491	19,294	23,301	23,154	28,846	27,506	35,930	33,984
Gujarat	8,331	7,930	10,633	9,999	13,802	12,895	17,552	16,533
Harayana	—	—	—	—	5,377	4,695	6,909	6,013
Himachal Pradesh	579	529	702	648	1,767	1,693	2,170	2,110
Jammu & Kashmir	1,736	1,517	1,896	1,664	2,458	2,158	3,164	2,823
Karnataka	9,866	9,535	12,040	11,545	14,972	14,372	18,922	18,213
Kerala	6,681	6,867	8,361	8,541	10,588	10,759	12,527	12,926
Madhya Pradesh	13,255	12,816	16,578	15,794	21,455	20,198	26,886	25,292
Maharashtra	16,490	15,512	20,428	19,124	26,116	24,296	32,415	30,369
Manipur	283	293	387	393	541	531	721	699
Meghalaya	—	—	—	—	520	491	683	652
Nagaland	—	—	—	—	276	240	415	359
Orissa	7,242	7,403	8,770	8,778	11,041	10,903	13,309	13,060
Punjab	8,681	7,453	10,891	9,415	7,266	6,285	8,637	7,851
Rajasthan	8,313	7,656	10,564	9,591	13,484	12,281	17,854	16,408
Sikkim	72	65	85	77	—	—	172	144
Tamil Nadu	15,003	15,115	16,910	16,775	20,828	20,371	24,487	23,920
Tripura	335	303	591	550	801	755	1,055	998
Uttar Pradesh	33,098	30,116	38,634	35,112	47,016	41,324	58,819	52,043
West Bengal	14,105	12,194	18,599	16,327	23,436	20,876	28,560	26,019
Union Territories								
Andaman & Nicobar	19	12	39	24	70	45	107	81
Arunachal Pradesh	—	—	—	—	251	216	339	292
Chandigarh	—	—	—	—	147	110	255	196
Dadra & Nagar Haveli	—	—	—	—	37	37	52	51
Delhi	986	757	1,489	1,169	2,257	1,808	3,440	2,780
Goa, Daman & Diu	—	—	—	—	431	426	548	538
Lakshadweep	10	10	11	12	16	15	20	19
Mizoram	—	—	—	—	—	—	257	236
Pondicherry	—	—	—	—	237	234	304	299
INDIA	183,333	173,545	225,600	212,244	283,936	264,013	354,398	330,786

Note: 1971 data for Assam includes Mizoram.

WORKING POPULATION — 1951-81

Table II   (in thousand)

	1951 Culti-vators	1951 Agri-cultural workers	1951 Total main workers	1961 Culti-vators	1961 Agri-cultural workers	1961 Total main workers	1971 Culti-vators	1971 Agri-cultural workers	1971 Total main workers	1981 Culti-vators	1981 Agri-cultural workers	1981 Total main workers
States												
Andhra Pradesh	4,048	3,849	11,527	7,486	5,336	18,663	5,794	6,828	18,005	7,407	8,325	22,629
Assam	2,415	129	3,917	3,516	189	5,356	2,410	405	4,240	N.A.	N.A.	N.A.
Bihar	8,147	3,238	13,565	10,361	4,418	19,234	7,579	6,806	17,488	9,042	7,366	20,753
Gujarat	2,929	1,196	6,532	4,519	1,252	8,474	3,619	1,887	8,395	4,114	2,488	10,984
Haryana	—	—	—	—	—	—	1,302	430	2,653	1,636	590	3,663
Himachal Pradesh	571	16	653	670	11	805	903	53	1,278	1,001	40	1,471
Jammu & Kashmir	N.A.	N.A.	N.A.	1,153	18	1,523	889	41	1,373	1,033	63	1,818
Karnataka	3,251	1,296	6,612	5,806	1,761	10,726	4,072	2,717	10,179	5,222	3,655	13,650
Kerala	1,012	1,114	4,359	1,178	978	5,630	1,106	1,908	6,216	887	1,917	6,791
Madhya Pradesh	5,849	4,329	12,875	10,611	2,815	16,929	8,084	4,062	15,295	10,414	4,857	20,041
Maharashtra	5,463	4,423	14,534	8,737	4,510	18,948	6,537	5,393	18,390	8,535	6,470	24,301
Manipur	184	0.7	298	234	2	357	248	13	370	364	28	573
Meghalaya	—	—	—	—	—	—	308	44	446	363	57	580
Nagaland	—	—	—	—	—	—	203	3	262	266	3	368
Orissa	2,839	1,009	5,470	4,353	1,303	7,661	3,368	1,937	6,850	4,053	2,396	8,635
Punjab	3,321	691	6,049	3,996	543	7,101	1,665	786	3,912	1,767	1,092	4,927
Rajasthan	5,170	596	7,875	7,005	393	9,583	5,225	749	8,048	6,431	764	10,442
Sikkim	33	38	79	92	2	103	—	—	—	88	5	147
Tamil Nadu	3,319	1,949	8,771	6,457	2,828	15,351	4,607	4,490	14,741	5,559	6,037	19,026
Tripura	168	23	266	32	13	280	235	86	432	263	146	608
Uttar Pradesh	17,945	2,017	26,863	18,428	3,261	28,850	15,697	5,453	27,334	18,957	5,177	32,396
West Bengal	3,218	1,388	9,066	4,458	1,771	11,580	3,954	3,272	12,368	4,590	3,891	15,424
Union Territories												
Andaman & Nicobar	1	0.1	18	6	0.3	31	6	2	45	10	2	62
Arunachal Pradesh	—	—	—	—	—	—	211	5	269	223	8	313
Chandigarh	—	—	—	—	—	—	2	1	85	2	0.8	156
Dadra & Nagar Haveli	—	—	—	—	—	—	25	5	34	26	4	42
Delhi	47	14	633	55	7	854	32	15	1,228	34	16	1,986
Goa, Daman & Diu	—	—	—	—	—	—	65	40	271	62	32	332
Lakshadweep	0.7	Neg.	11	0.1	Neg.	12	Neg.	—	8	—	—	8
Mizoram[1]	—	—	—	—	—	—	—	—	—	145	5	206
Pondicherry	—	—	—	—	—	—	16	46	141	15	54	173
INDIA	69,938	27,326	139,983	99,461	31,438	188,218	78,176	47,489	180,373	92,522	55,499	222,516

Note: 1971 data for Assam includes Mizoram.

The Climate

This section comprises 10 maps & 5 charts. It provides information on the factors generally beyond human control that must be considered when selecting developmental projects for implementation in a diverse land.

Note:

There are 220 meteorological stations in India. Of these, 20 are located on the maps in this section. These 20 are not only major meteorological stations in the country but they are also sufficiently scattered around the country to give a reasonable picture of the climate prevailing in the various parts of India. Data from the 220 stations are, however, provided at the end of this section. The district in which each station is, is indicated by a number which corresponds with the district number on the transparent overlay provided with this atlas.

There are no averages of state rainfall and temperature. We have, however, attempted to give a state-wise picture in the inset maps in the section by computing the average rainfall and temperature from the data available from the various meteorological stations in each state.

ANNUAL RAINFALL

The Indian economy depends almost entirely on the bounty of the monsoons which bring the seasonal rains essential for agriculture, irrigation and power production. Although considerable areas of the country are provided for by irrigation from major and minor projects, nearly 80 per cent of the cropped area still depends on the seasonal rainfall. Hydro-electric power, which is about 35 per cent of the power produced in the country for agriculture, industry and domestic use, is also dependent on such rainfall.

India has two monsoons, the south-west and the north-east. The south-west monsoon (June to September) accounts for 70 per cent of the annual rainfall. The north-east (October to December) is important only for Tamil Nadu, which receives about 50 per cent of its rainfall from this monsoon. It also brings winter rainfall to Punjab.

The south-west monsoon blows with great regularity and its dates of onset and withdrawal have been determined with a fair degree of accuracy. It sets in along the Kerala coast by 1 June, extends to Bombay by 10 June, and covers the whole country by the beginning of July. The north-east monsoon, an extension of the retreating monsoon, usually sets in by November. Whereas the south-west monsoon is dependable, with almost no case of *total* failure being reported from any part of the country, even in years of the worst drought, the north-east monsoon is less dependable and failures for several years at a stretch have led to distress from time to time.

The average annual rainfall of India is 125 cms, the highest for a land of such size anywhere in the world. But this is highly variable in space and time, with the heaviest rains occurring over the north-eastern states and along the west coast, Cherrapunji in Meghalaya receiving a record rainfall of nearly 900 cms a year. In contrast, some districts in south-west Rajasthan receive hardly 15 cms a year. In the extreme south of the country, Kanniyakumari and Tirunelveli districts get less than 30 cms during the south-west monsoon, while heavy rains (200 cms and more) lash the Kerala coast nearby.

Variability in time is also significant. This may be seen in the histograms of monthly rainfall received at selected meteorological stations in the country during a calendar year. In most stations, the monthly variability is quite high. The long-term variability, however, shows both years of plenty as well as scanty rainfall. There are, in fact, years when the monsoon in plentiful in the early season (June-July) and fails in the late season (August-September) and *vice versa*, causing concern to farmers in several parts of the country. The formation and movement of low-pressure areas, or depressions, largely determine the activity of the monsoons and account for their vagaries.

The spatial variations shown by the isohyets, the lines joining places with equal rainfall, indicate the differences in annual receipt of rainfall in different parts of the country. The areas of high rainfall include the west coast from Trivandrum, in the south, to Bombay in the north, receiving 200-400 cms with some localized areas in between receiving rainfall of 400-800 cms. Almost the whole of Assam, Nagaland

Meghalaya, Mizoram, Arunachal Pradesh and Sikkim and parts of Manipur, Tripura and the north-eastern tip of West Bengal also receive 200 cms and more, with isolated pockets getting more than 400 cms and one unique area, Cherrapunji, as mentioned earlier, receiving nearly 900 cms of rain a year. Meghalaya and Arunachal Pradesh are the wettest parts of the country, but Assam and Kerala have the most rainfall among the bigger states. In contrast, the areas receiving rainfall of less than 40 cms a year include desert areas that extend as far south as Kachchh and most of Jammu & Kashmir. Rajasthan has the least rainfall amongst the major states. Between these two extremes lies the major part of India, with rainfall receipts of 40-200 cms.

The areas east of the 400 cm line along the west coast show rapidly decreasing rainfall averages over the narrow width of the Western Ghats, while the great land mass east of the Ghats is a rain-shadow area (40-80 cms). There is, however, a tremendous difference in the economic impact of rainfall between the moderately heavy rainfall region of the eastern slopes of the Western Ghats (the Nilgiris, Anamalais and Kodagu in the south), the northern Tamil Nadu coast, areas in eastern and central India spread over northern Uttar Pradesh, eastern Madhya Pradesh, southern Bihar, West Bengal, Orissa and parts of Andhra Pradesh on the one hand (rainfall of 120-200 cms) and the whole of the Deccan Plateau, the plains of southern Punjab and the whole of Gujarat on the other (rainfall of 40-100 cms).

The variations in rainfall are due to two important factors: the distance from the sea and the rain-shadow effect. The variations at different times shown by the histograms for selected stations emphasize the role these factors play in the rainfall India gets. The dominance of summer rainfall throughout almost the whole of India may be noted, followed by dry conditions that exist throughout winter. There is a sudden increase in rainfall from May to July and a decrease during September and October, the exception being Madras, which reflects the rainfall south-eastern Andhra Pradesh and eastern Tamil Nadu gets. Madras shows a peak in October-November, mainly as a result of rainfall blown in by Bay of Bengal cyclones, while Srinagar demonstrates similar conditions, but due to winter cyclones from the Mediterranean.

The rainfall distribution in the interior of India is mostly controlled by the rain-shadow effect; note the difference in rainfall receipts between the western slopes of the Western Ghats and the areas lying to the east — the rainfall decreases with distance from the sea. A reverse of this occurs in the north where the rainfall declines westwards with distance from West Bengal. The orographic, or relief, effect on rainfall is also seen in the distribution of rainfall, for the windward sides of the hills and mountains (Western Ghats, north-eastern hills and the Himalayan foothills) receive more rain than other parts of the country.

Source: *Statistical Abstract India, 1984,* Ministry of Planning. Quoting from India Meteorological Department.

ACTUAL ANNUAL RAINFALL

Table 26   (millimetre)

Sub-divisions	1951	1956	1961	1966	1971
Andaman & Nicobar Islands	3,274.8[1]	3,031.2[1]	4,362.4[1]	2,927.7[1]	2,982.1[1]
Assam (North)	2,357.5	2,565.3	1,783.3	2,175.5[1]	2,172.6[1]
Assam (South)	2,955.7	3,115.5	3,008.4	2,860.7[1]	2,267.4[1]
Sub-Himalayan West Bengal	2,692.0	3,103.3	2,714.6	2,368.4	2,899.2[1]
Gangetic West Bengal	1,188.3	1,693.1	1,247.0	966.4	2,166.1[1]
Orissa	1,403.6	1,849.9	2,014.7[1]	1,283.6[1]	1,678.2[1]
Bihar Plateau	1,083.5	1,595.1	1,552.2	926.6[1]	1,975.1[1]
Bihar Plain	900.2	1,490.3	1,179.1	684.0[1]	1,616.0[1]
East Uttar Pradesh	760.4	1,378.3	1,163.5	704.3	1,356.3
West Uttar Pradesh	764.0	1,087.5	1,257.0	893.0	1,191.9
Punjab	480.2	793.6	755.1	518.4	646.6
Haryana	325.4	639.3	771.1	763.3	637.1
Himachal Pradesh	1,532.3	1,665.4	1,649.8	1,390.7	1,628.8[1]
Jammu & Kashmir	782.8	1,300.1	1,282.2	1,269.9	343.0[1]
East Rajasthan	366.3	870.1	926.2	453.2	860.2
West Rajasthan	210.0	437.4	412.2	275.7	292.2
East Madhya Pradesh	1,281.0	1,506.0	1,870.1	984.8	1,510.3
West Madhya Pradesh	758.4	1,189.8	1,489.6	698.4	1,138.5
Gujarat	489.8	1,332.7	982.4	806.7	877.8
Saurashtra & Kachchh (Gujarat)	327.4	816.5	773.7	381.8	525.1
Konkan (coastal region of Maharashtra)	2,765.1	3,625.3	3,864.1	2,191.1	2,879.2
Madhya Maharashtra	861.6	1,260.7	1,036.2	868.7	808.0
Marathwada (south Maharashtra)	686.4	994.0	920.8	754.6	704.1
Vidarbha (north Maharashtra)	1,037.2	1,039.8	1,355.8	960.1	830.8
Coastal Andhra Pradesh	1,076.1	1,354.4	1,032.1	1,002.5	892.3[1]
Telengana (north-west Andhra Pradesh)	900.4	1,225.4	1,092.5	858.8	734.9[1]
Rayalaseema (south-west Andhra Pradesh)	492.4	864.8	553.7	785.9	597.1[1]
Tamil Nadu	929.2	973.3	922.5	1,155.3	1,034.2[1]
Coastal Karnataka	3,114.0	3,284.2	5,604.1	3,012.3	3,395.5[1]
North interior Karnataka	697.0	974.9	772.2	762.7	688.9[1]
South interior Karnataka	1,165.1	1,651.0	1,693.7	1,128.7	774.2[1]
Kerala	2,328.4	2,723.0	4,003.6	2,558.2	3,069.6
Lakshadweep	N.A.	N.A.	1,836.2[1]	1,675.9[1]	1,928.8[1]

Note: [1]Figures are based on observatory data.

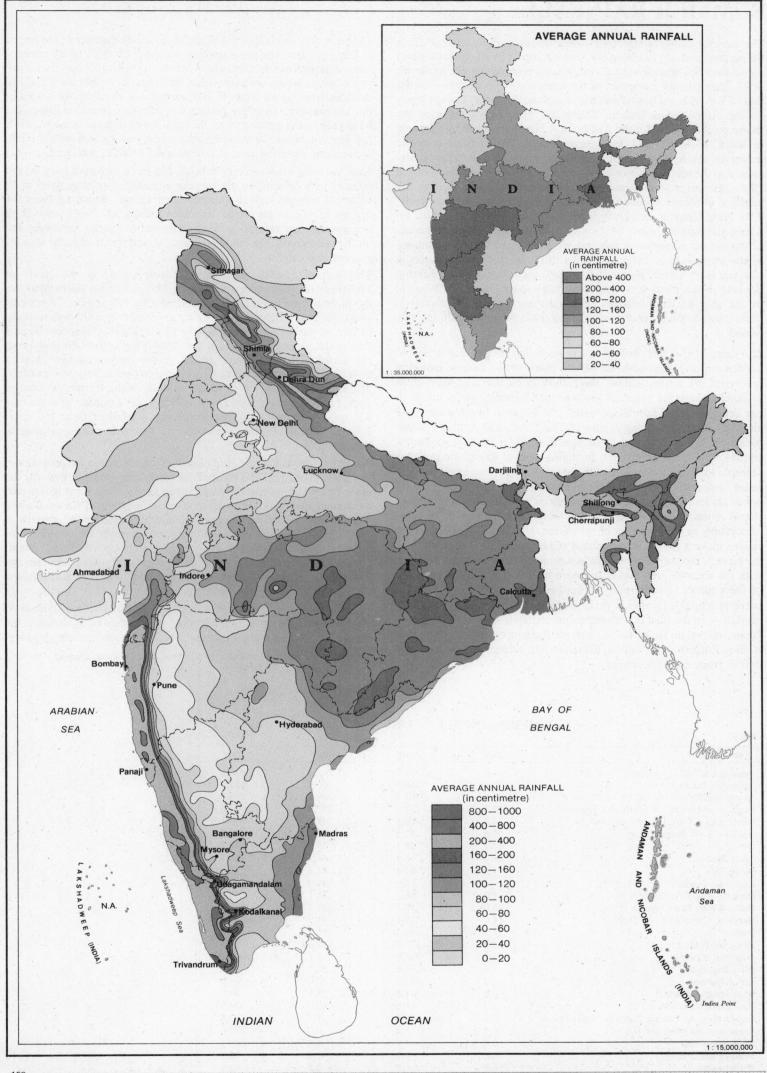

AVERAGE ANNUAL RAINFALL

AVERAGE ANNUAL
RAINFALL
(in centimetre)

Above 400
200 — 400
160 — 200
120 — 160
100 — 120
80 — 100
60 — 80
40 — 60
20 — 40

1 : 35,000,000

Srinagar

Shimla

Dehra Dun

New Delhi

Lucknow

Darjiling

Shillong

Cherrapunji

Ahmadabad INDORE I N D I A Calcutta

Indore

Bombay

Pune

ARABIAN
SEA

BAY OF
BENGAL

Panaji

Hyderabad

Bangalore Madras

Mysore

L A K S H A D W E E P   S e a

LAKSHADWEEP
(INDIA)

N.A.

Udagamandalam

Kodaikanal

ANDAMAN
AND
NICOBAR
ISLANDS
(INDIA)

Andaman
Sea

AVERAGE ANNUAL RAINFALL
(in centimetre)

800 — 1000
400 — 800
200 — 400
160 — 200
120 — 160
100 — 120
80 — 100
60 — 80
40 — 60
20 — 40
0 — 20

Trivandrum

Indira Point

INDIAN   OCEAN

1 : 15,000,000

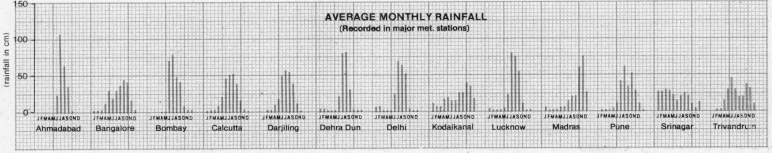

AVERAGE MONTHLY RAINFALL
(Recorded in major met. stations)

(rainfall in cm)

Ahmadabad Bangalore Bombay Calcutta Darjiling Dehra Dun Delhi Kodaikanal Lucknow Madras Pune Srinagar Trivandrum

SUMMER RAINFALL

The summer rains, in effect the south-west monsoon, blow up from the equator and, on reaching the southern tip of India, bifurcate into an Arabian Sea branch and a Bay of Bengal branch. For the Andaman & Nicobar Islands, the onset of the monsoon is connected with the Bay of Bengal branch, whereas for peninsular India the onset depends on the Arabian Sea branch. Therefore, the date of onset of the monsoon is different for the Andaman & Nicobar islands and the rest of India: 20 May and 1 June, respectively. With the onset of the monsoon, the Arabian Sea branch brings heavy rainfall to the coastal areas and the western slopes as it is forced to ascend over the Western Ghats. However, the intensity of this rainfall is conditioned by the angle of incidence the monsoon currents make with the mountains. This branch moves quickly up the west coast and reaches Goa on 5 June and Bombay, north of it, on 10 June. The Bay of Bengal branch of the monsoon moves as fast, but gets deflected by the mountains in the north-east of India and, thus, the onset of the monsoon in Calcutta is on 8 June. By the first of July, the entire country, barring pockets in the north-west and the south-eastern coast of India, is in the grip of the monsoon. Even in years when the monsoon 'fails', some rain at least does fall; failure is never total during this period.

The rains of summer, however, are such that there are discernible variations in the areas covered. This is linked to two factors: humidity and relief. To a great extent, the pattern of rainfall and the pattern of humidity are the same. A low summer humidity of less than 70 per cent, affecting rainfall adversely, is found in two pockets, one along the east coast, taking in parts of Tamil Nadu and Andhra Pradesh (this stretch receives most of its rain from the north-east monsoon) and the other in the north-west, including parts of Rajasthan (the Thar Desert), Punjab and Haryana. On the other hand, wherever there are relief features in the path of the monsoon winds, the cloud-bearing winds are forced to ascend and give copious rainfall. Thus, the western slopes of the Western Ghats and the foothills of the Himalaya, most importantly in north-east India, where the Himalayan-Arakan Yoma ranges make a *funnel* of mountains (Cherrapunji is at the mouth of the funnel), receive plenty of rain in summer. These areas also coincide with the areas of high humidity (more than 90 per cent). In the rest of the country, the humidity is between 65 and 90 per cent and, correspondingly, rainfall is variable. Where the rain-shadow affects rainfall — in the land that stretches immediately east of the Western Ghats, taking in Tamil Nadu, interior Karnataka, Maharashtra and Andhra Pradesh — the rainfall decreases, the further east the monsoon moves from the high ranges.

The monsoon winds and, for that matter, winds in general, are linked to low pressure. The pressure distribution over much of the country generally represents conditions of low pressure, but there are nevertheless great contrasts within the country: the pressure steadily declines from south to north and north-west, reaching its lowest in the Thar desert. This low pressure is so effective that it draws the winds across the equator, from the high-pressure areas in the southern hemisphere, but while benefiting the country as a whole with variable amounts of rain, the area it hovers over receives the least.

How variable even summer rainfall can be in different parts of the country may be seen by the varying amounts of rain received at 20 scattered meteorological stations. Cherrapunji, which receives the highest rainfall in the world, averages around 900 cms a year. High average summer rainfall is received by stations in the north-east and on the west coast. On the other hand, inland and east coast stations receive low rainfall.

The highest rainfall is along the west coast, in the states of Maharashtra, Karnataka and Kerala. Mangalore (in Karnataka) has the highest rainfall on the west coast during this season. North-east India (Assam, Meghalaya and Tripura) receives equally high rainfall. In fact, no part of north-eastern India gets a rainfall of less than 80 cms during the summer and the narrow zone lying in the rain-shadow of the Khasi hills in Assam, which is the only area here getting 80-120 cms, is considered as receiving comparatively low rainfall. The rainfall is equally high in the foothills of the Himalaya. Central India, and the area around Dandakaranya, get a rainfall of more than 120 cms, with isolated areas receiving rainfall of over 160 cms. Rainfall decreases westwards along the Gangetic plain and towards the south.

Along the west coast, the rainfall is high mainly due to the orographic, or relief, effect. But the rainfall decreases very sharply towards the east. The whole of the Deccan Plateau, accounting for a major part of Maharashtra, Andhra Pradesh, almost the whole of Karnataka and Tamil Nadu, experiences very low rainfall (less than 40 cms) with only a few hilly areas relieving this arid monotony with a rainfall of 80 cms. Coimbatore and the south-eastern tip of Tamil Nadu, consisting of the coastal areas of the districts of Ramanathapuram and Tirunelveli, share with west Rajasthan, and Jammu & Kashmir the distinction of having the lowest rainfall in the entire country.

Indian agriculture depends so much on the rainfall of the south-west monsoon that the onset and advances of summer rains make all the difference between a good year and a bad one, economically speaking.

Source: *Statistical Abstract India, 1984,* Ministry of Planning. Quoting from India Meteorological Department.

SUMMER RAINFALL — 1983

Table 27   (millimetre)

Sub-divisions	March	April	May	June
Andaman & Nicobar Islands	4.6	19.3	242.6	380.0
Arunachal Pradesh	179.1	268.5	200.9	582.2
Assam and Meghalaya	79.8	163.1	289.8	396.3
Nagaland, Mizoram, Manipur and Tripura	245.7	281.7	292.6	327.3
Sub-Himalayan West Bengal and Sikkim	34.4	80.5	359.3	549.5
Gangetic West Bengal	55.6	52.3	129.3	196.7
Orissa	27.5	58.1	63.5	148.9
Bihar Plateau	18.2	48.1	84.9	150.5
Bihar Plain	14.5	42.4	82.4	112.1
East Uttar Pradesh	0.4	24.0	39.0	105.6
Plain of West Uttar Pradesh	4.7	68.4	36.1	78.9
Hills of West Uttar Pradesh	85.0	159.5	131.7	168.9
Haryana, Chandigarh and Delhi	14.0	135.3	44.9	46.5
Punjab	17.9	146.4	51.7	36.8
Himachal Pradesh	123.2	177.2	126.6	85.8
Jammu & Kashmir	307.7	187.3	92.1	47.3
West Rajasthan	3.0	37.6	37.1	37.6
East Rajasthan	0.2	37.7	40.0	64.0
West Madhya Pradesh	0.0	111.5	15.1	98.1
East Madhya Pradesh	1.7	17.9	40.0	134.9
Gujarat Region, Daman, Dadra & Nagar Haveli	0.0	17.6	1.3	117.8
Saurashtra, Kachchh and Diu	0.0	5.4	0.0	322.6
Konkan and Goa	0.0	1.1	1.3	610.1
Madhya Maharashtra	0.0	0.1	1.7	98.3
Marathwada	4.5	0.0	0.5	62.5
Vidarbha	4.8	4.7	5.3	153.1
Coastal Andhra Pradesh	0.0	8.3	32.0	150.5
Telengana	1.3	6.1	36.6	165.6
Rayalaseema	5.8	8.2	75.9	78.9
Tamil Nadu and Pondicherry	0.6	3.8	75.3	38.9
Coastal Karnataka	0.0	0.4	38.1	504.0
North interior Karnataka	0.0	0.1	35.3	197.2
South interior Karnataka	0.0	3.1	80.7	158.3
Kerala	0.0	44.7	71.1	301.8
Lakshadweep	0.0	0.1	113.5	181.6

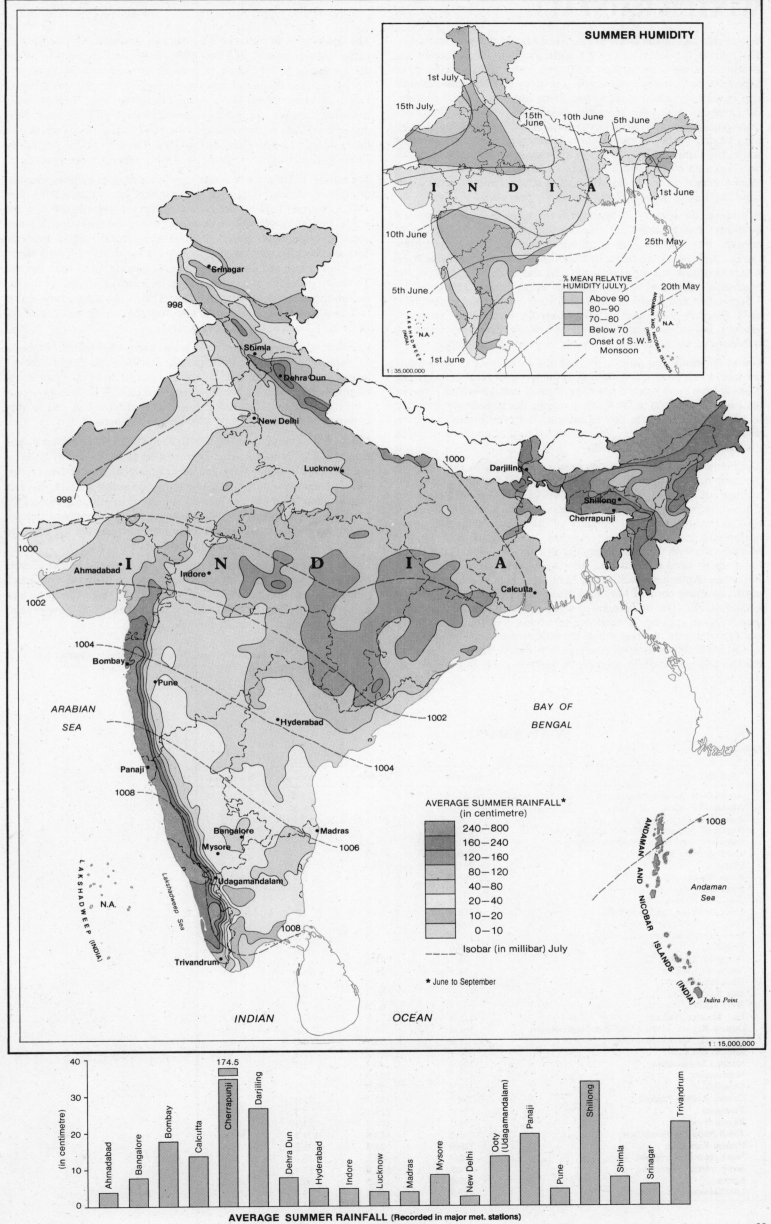

SUMMER HUMIDITY

1st July
15th July
15th June
10th June
5th June
10th June
25th May
5th June
20th May
1st June
I N D I A
1st June

% MEAN RELATIVE HUMIDITY (JULY)
Above 90
80—90
70—80
Below 70
Onset of S.W. Monsoon

1 : 35,000,000

LAKSHADWEEP (INDIA) N.A.

ANDAMAN AND NICOBAR ISLANDS (INDIA) N.A.

Srinagar
998
Shimla
Dehra Dun
New Delhi
Lucknow
1000
Darjiling
Shillong
Cherrapunji
998
1000
Ahmadabad
Indore
I N D I A
Calcutta
1002
1004
Bombay
Pune
ARABIAN SEA
Hyderabad
1002
BAY OF BENGAL
Panaji
1008
1004
Bangalore
Madras
Mysore
1006
Udagamandalam
LAKSHADWEEP Sea
N.A.
LAKSHADWEEP (INDIA)
1008
Trivandrum

ANDAMAN AND NICOBAR ISLANDS (INDIA)
1008
Andaman Sea
Indira Point

AVERAGE SUMMER RAINFALL*
(in centimetre)
240—800
160—240
120—160
80—120
40—80
20—40
10—20
0—10

Isobar (in millibar) July

* June to September

INDIAN OCEAN

1 : 15,000,000

AVERAGE SUMMER RAINFALL (Recorded in major met. stations)

(in centimetre)

40 — 174.5 (Cherrapunji)
30
20
10
0

Ahmadabad, Bangalore, Bombay, Calcutta, Cherrapunji, Darjiling, Dehra Dun, Hyderabad, Indore, Lucknow, Madras, Mysore, New Delhi, Ooty (Udagamandalam), Panaji, Pune, Shillong, Shimla, Srinagar, Trivandrum

WINTER RAINFALL

Winter rainfall in India is due to the retreat of the south-west monsoon. This retreat is gradual across the north and west of the country, a striking contrast to the sudden burst with which it arrives. Much of September is fairly dry in the north; it is sticky and hot, with a distinct rise in temperature in many areas. Then begins a gradual cooling. By October, there may be storms related to troughs in the upper troposphere westerlies sometimes interacting violently with a trough (or troughs) in the retreating upper troposphere easterlies. Farther south, the rains (the north-east, or retreating, monsoon) sometimes lash down on south-eastern India in violent thunderstorms, but are more common as wider disturbances in which easterly depressions from the Bay of Bengal move across the peninsula.

In a sense, the winter is a reversal of the summer winds, rains and humidity, though there are marked differences in the pattern of rainfall. For instance, the north-west continues to get the lowest rainfall, being now the centre of a high-pressure area from which the winds blow towards the east, south-east and south. The onset of the north-east monsoon in the extreme north-west of India is by 1 September, and it is in the north and west (near Delhi and Bombay) by 1 October. The onset of this monsoon, because the retreat is gradual, lasts longer and the whole of India is, therefore, under the influence of the north-east monsoon only by 15 December.

Unlike the south-west monsoon winds, the north-east monsoon winds are land winds and, hence, their effect in terms of rainfall is limited. Rainfall from these winds is low over much of India, though it is higher and fairly widespread in the southern region, due to gathering moisture as it swings in from the Bay of Bengal. But even in the south, the winter rainfall is relatively low compared to the summer rainfall in the area, except in Tamil Nadu.

The humidity at this time of the year is generally low (less than 70 per cent) in the north-west, central India and the Deccan. The lowest average, of less than 50 per cent, occurs in the Saurashtra region of Gujarat, in Himachal Pradesh and in a part of the Marathwada region of Maharashtra. Relative humidity of 50-70 per cent occurs over northern Karnataka, western Andhra Pradesh and the rest of Marathwada with this zone extending as a narrow belt across Central India up to central Bihar and also westwards, covering most of Rajasthan. An isolated belt in this humidity range also occurs in the north, covering parts of Jammu & Kashmir, Himachal Pradesh and Uttar Pradesh. The occurrence of a zone of high humidity of more than 80 per cent along the north-western borders of India in Kashmir and Punjab is directly traceable to the influence of the Mediterranean winds. A broad belt along the coast of Tamil Nadu and Andhra Pradesh also has a humidity of more than 80 per cent during this period.

The distribution of pressure is not very pronounced in winter, unlike in the summer months, and the difference between the north-west and the south-east is very small. But then, low pressures in the Bay of Bengal develop frequently and cause a series of depressions that give rise to cyclones. These cyclones are often disastrous, their recurved maritime courses commonly causing terrible havoc in densely populated deltaic regions, both by wind and, often, floods that follow the build-up of a storm surge in the Bay. The heavy rains from these easterly disturbances are associated with upper air divergences.

The rainfall distribution in winter shows that most of India experiences drought conditions: Rajasthan, for instance, gets a rainfall of as low as 2.5 cms and below. The rainfall increases towards the east and south, rainfall of 40-80 cms occurring along the east (Coromandel) coast in Tamil Nadu and in south-eastern coastal Andhra Pradesh. The whole of the east coast, spread over the states of Tamil Nadu, Andhra Pradesh and Orissa, gets an average rainfall of over 10 cms. The highest rainfall is recorded along the coast of Thanjavur district in Tamil Nadu, a little over 80 cms. This is the main rainy season in Tamil Nadu. High rainfall is also recorded along the crest of the Western Ghats (near Idukki in Kerala).

Rainfall is generally very low over parts of Madhya Pradesh. Maharashtra, Gujarat, Uttar Pradesh, Haryana and Punjab, though in the foothills of the Himalaya in northern Himachal Pradesh the rainfall is somewhat higher (20 cms). This rainfall is from the strong winds which blow into India from the Mediterranean and is very important for the cultivation of wheat in the area.

Much of the east coast rainfall is from cyclones that develop during the retreating monsoon. The south-east tip of India, Tamil Nadu, is the only part of the country that gets rainfall in October-November, the north-east monsoon continuing to remain active. In fact, this rain comes when the north-east monsoon is most developed: that is, the north-east monsoon winds, while recurving over the Bay of Bengal, absorb a certain amount of moisture that, on reaching the southern Coromandel coast, condenses to produce rain.

In looking at the winter rainfall pattern, no aspect of winter climate is of greater human import than the incidence of flood over much of the east coast and drought over much of the rest of India, which causes great loss to life and property and makes life in these regions difficult during this time of the year.

Source: *Statistical Abstract India, 1984*, Ministry of Planning. Quoting from India Meteorological Department.

WINTER RAINFALL — 1983

Table 28   (millimetre)

Sub-divisions	September	October	November	December
Andaman & Nicobar Islands	467.9	264.6	297.7	90.8
Arunachal Pradesh	653.5	151.9	9.5	39.9
Assam and Meghalaya	466.7	205.6	7.9	34.8
Nagaland, Mizoram, Manipur and Tripura	177.5	177.1	23.2	45.9
Sub-Himalayan West Bengal and Sikkim	541.3	74.3	0.8	18.1
Gangetic West Bengal	230.8	175.8	21.8	17.0
Orissa	297.4	103.3	4.8	8.0
Bihar Plateau	283.7	112.2	0.0	22.5
Bihar Plain	222.0	72.9	0.0	15.7
East Uttar Pradesh	371.5	105.3	0.0	18.2
Plain of West Uttar Pradesh	284.4	56.8	0.0	13.9
Hills of West Uttar Pradesh	390.9	47.1	0.0	8.1
Haryana, Chandigarh and Delhi	106.8	12.8	0.0	9.6
Punjab	60.3	14.1	0.0	2.3
Himachal Pradesh	206.1	45.8	1.6	10.7
Jammu & Kashmir	83.3	26.2	14.8	20.1
West Rajasthan	40.0	16.5	0.0	0.3
East Rajasthan	124.9	39.2	0.0	0.9
West Madhya Pradesh	314.9	60.3	7.3	1.2
East Madhya Pradesh	316.8	71.6	0.3	9.8
Gujarat Region, Daman, Dadra & Nagar Haveli	161.4	84.6	0.0	0.0
Saurashtra, Kachchh and Diu	79.6	22.7	0.0	0.0
Konkan and Goa	636.1	127.2	12.3	3.2
Madhya Maharashtra	232.9	46.2	1.7	4.5
Marathwada	422.7	102.9	0.5	8.4
Vidarbha	316.5	59.0	0.1	14.3
Coastal Andhra Pradesh	259.5	192.7	38.5	22.6
Telengana	307.4	229.1	3.5	10.7
Rayalaseema	294.5	104.7	23.1	66.5
Tamil Nadu and Pondicherry	143.5	142.3	139.5	362.6
Coastal Karnataka	779.0	124.8	36.6	32.5
North interior Karnataka	240.3	83.2	16.0	17.7
South interior Karnataka	134.5	116.0	15.4	43.7
Kerala	429.0	124.5	200.3	86.5
Lakshadweep	247.9	68.8	47.0	53.8

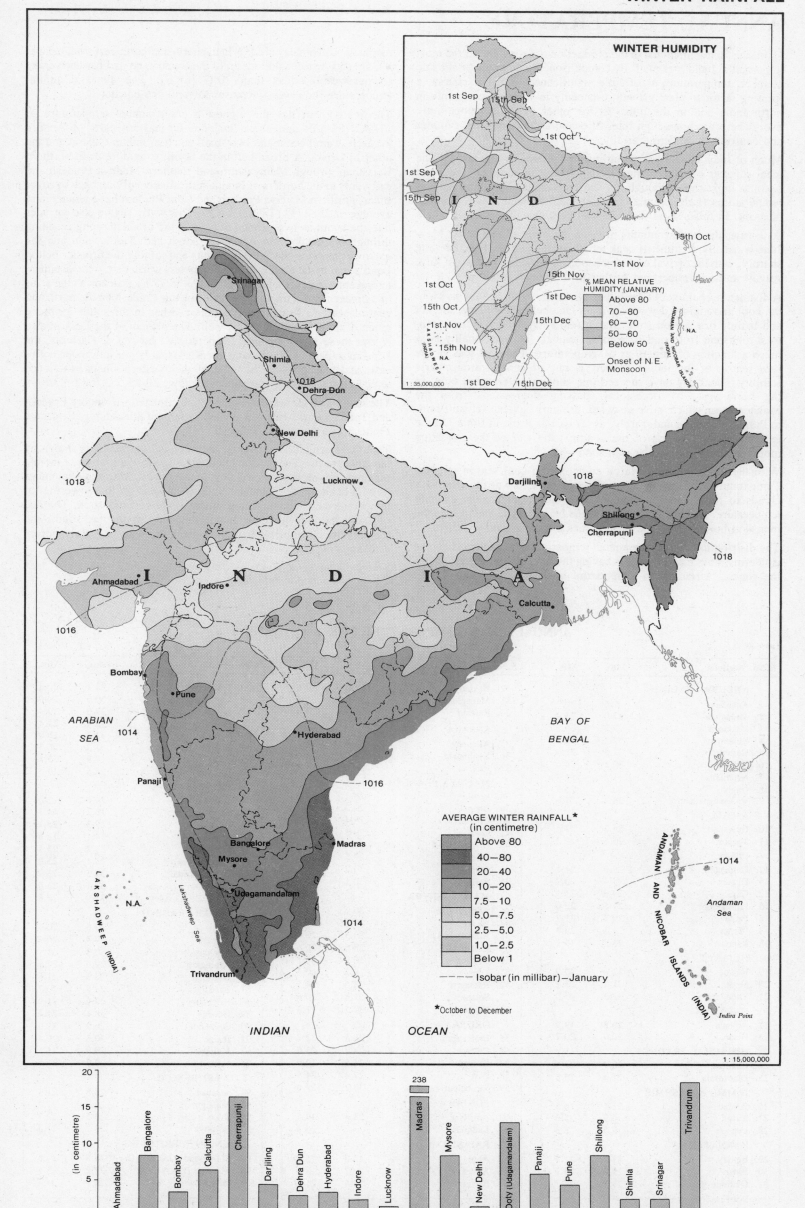

WINTER RAINFALL

WINTER HUMIDITY

1st Sep
15th Sep
1st Oct
1st Sep
15th Sep
I N D I A

15th Oct

1st Nov

1st Oct
1st Dec
15th Oct
15th Nov
1st Nov
15th Nov
1st Dec
15th Dec

L A K S H A D W E E P
(INDIA)
N.A.

1st Dec
15th Dec

ANDAMAN AND NICOBAR ISLANDS
N.A.

% MEAN RELATIVE HUMIDITY (JANUARY)
Above 80
70—80
60—70
50—60
Below 50
—— Onset of N.E. Monsoon

1 : 35,000,000

Srinagar

Shimla

1018
Dehra Dun

New Delhi

Lucknow

Darjiling
1018

Shillong
Cherrapunji
1018

1018

Ahmadabad I Indore N D I A

Calcutta

1016

Bombay

Pune

ARABIAN SEA
1014

Panaji

Hyderabad

BAY OF BENGAL

1016

Bangalore Madras

Mysore

ANDAMAN AND NICOBAR ISLANDS (INDIA)
1014

Udagamandalam

L A K S H A D W E E P (INDIA)
N.A.

Lakshadweep Sea

1014

Andaman Sea

**AVERAGE WINTER RAINFALL ★
(in centimetre)**
Above 80
40—80
20—40
10—20
7.5—10
5.0—7.5
2.5—5.0
1.0—2.5
Below 1

—— Isobar (in millibar)—January

★October to December

Trivandrum

Indira Point

INDIAN   OCEAN

1 : 15,000,000

AVERAGE WINTER RAINFALL (Recorded in major met. stations)

(in centimetre)

20 — 15 — 10 — 5 —

Ahmadabad, Bangalore, Bombay, Calcutta, Cherrapunji, Darjiling, Dehra Dun, Hyderabad, Indore, Lucknow, Madras (238), Mysore, New Delhi, Ooty (Udagamandalam), Panaji, Pune, Shillong, Shimla, Srinagar, Trivandrum

ANNUAL TEMPERATURE

Of all the factors involved in the production of crops, by far the most important in India is rainfall. By comparison, the other major climatic element, temperature, is of little significance. It is, of course, a limiting factor to plant growth, especially in winter, and mainly in north India and in the hills. At the other extreme, temperature everywhere promotes high rates of evaporation during the summer and causes crop-affecting aridity.

Much of India enjoys a pleasantly warm winter, comparable to the ideal summer of many temperate lands. But in summer, India can become unbearable, though, in the mountains, elevation reduces temperatures (Leh, Shimla, Udagamandalam) and makes summer pleasant. In this season, the effect of continentality is clear.

As a rule, the hottest months precede the onset of the rains. May-June is usually the time of peak temperatures, which then fall quite sharply, with the arrival of the monsoon, to a modest plateau in July, August and September and further in the winter months.

A distinctive character of the Indian climate is its three-fold division: the cool and mainly dry winter from October to February, the hot and mainly dry season from March into early June and the south-west monsoon from mid-June to September. This seasonality is as much a function of rainfall as of temperature. The cool season, the *rabi* season of the Indian farmer, is one of general atmospheric stability. Clear dry days succeed one another, interrupted only, in the north-west, by occasional, passing depressions from the Mediterranean which move eastward, bringing light but valuable rains to Punjab, Haryana and, rarely, as far east as Patna in Bihar. At the extreme south-east end of the country, the *end* of the retreating monsoon delays the dry season somewhat.

With the passing of the March equinox, the cool season is over. Temperatures rise day by day and people look forward to afternoon clouds to reduce the build-up of heat and moderate the evening temperature. In April, thunder-showers bring respite and in June the monsoon bursts. With the rains comes more pleasant weather.

The distribution of average annual temperature, as indicated by the isotherms, lines that join places having the same temperatures, shows the general variations in temperature in the country. Barring the mountainous Himalayan areas in the north and north-east (Meghalaya, Sikkim and Arunachal Pradesh) of the country, no part has an average temperature of less than 20°C (68°F). The average annual temperature, however, increases towards the equator.

The region that has the highest average annual temperature of 27.5°C (81.5°F) and above includes the eastern parts of Andhra Pradesh, Tamil Nadu and two small patches, one in Gujarat and the other in Orissa. A broad belt in the north, extending from northern Rajasthan through Maharashtra and southern Madhya Pradesh and eastwards to the borders of Bangladesh, finally curving back to cover Bihar, northern Madhya Pradesh and Uttar Pradesh has a temperature average of 25°-27°C (77° to 81.5°F). This warm region also intrudes into the south as two prongs of a fork, the broader prong reaching through Karnataka into Andhra Pradesh and Tamil Nadu and the narrower hugging the Western Ghats and joining the broader belt at Land's End in the South. A marginally less warm belt, with an annual temperature average of 22.5°-25°C (73° to 77°F), extends as the arms of a pincer, from Punjab into western Uttar Pradesh on the north and central Madhya Pradesh and southern Bihar in the south, gripping much of the Gangetic Plain as well. The plateau of the Deccan also enjoys these lower average temperatures. Jammu & Kashmir, the Himalayan parts of North India, the Khasi and Jaintia hills in the north-east, and places along the crest of the Western Ghats experience temperatures that are comparatively low: 20-22.5°C (68.5-73°F).

The states with the warmest year-round climates are Andhra Pradesh and Tamil Nadu, while it is coolest in Jammu & Kashmir, Himachal Pradesh and Arunachal Pradesh.

That there are clearly seasonal and regional variations is shown by the histograms. The stations in the Gangetic Plain show a double maxima, pre-monsoonal and post-monsoonal in May and September (*e.g.* Lucknow & Begampet). This is due to the lowering of temperature during the period of the south-west monsoon. The rest of India exhibits a single maximum and, in the far south, the temperature is uniformly high throughout the year, (*e.g.* Trivandrum & Mangalore).

Source: *Climatological Tables of Observatories in India (1931-1960)*, India Meteorological Department.

ANNUAL MEAN TEMPERATURE FOR 100 STATIONS

Table 29

(in °Celsius)

S.No. Stations	Max.	Min.	S.No. Stations	Max.	Min.	S.No. Stations	Max.	Min.
ANDHRA PRADESH			32. Hassan	25.6	19.4	65. Bikaner	32.1	20.3
1. Anantapur	31.6	21.7	33. Mangalore	28.3	24.8	66. Ganganagar	30.6	20.0
2. Cuddapah	32.0	24.4	34. Raichur	31.7	22.0	67. Jaipur	29.6	19.4
3. Kakinada	29.2	24.9	**KERALA**			68. Jaisalmer	32.4	22.4
4. Khammam	32.2	23.6	35. Alleppey	28.7	25.5	69. Jodhpur	32.1	20.0
5. Kurnool	32.3	22.6	36. Kozhikode	29.0	25.6	70. Kota	31.2	20.3
6. Nellore	30.5	24.6	37. Palghat	29.6	23.8	71. Sikar	29.9	22.1
7. Nizamabad	31.2	21.4	**MADHYA PRADESH**			72. Udaipur	28.6	19.5
8. Ongole	30.3	25.7	38. Betul	28.2	19.1	**TAMIL NADU**		
9. Vishakhapatnam	28.4	25.6	39. Bhopal	29.2	18.9	73. Coimbatore	29.0	22.0
ASSAM			40. Chhindwara	28.4	19.2	74. Coonoor	18.2	14.3
10. Guwahati	26.2	22.5	41. Gwalior	30.3	20.0	75. Cuddalore	29.3	25.1
11. Tezpur	25.5	22.2	42. Hoshangabad	30.7	20.9	76. Kodaikanal	14.1	12.6
BIHAR			43. Jabalpur	29.5	20.6	77. Madurai	31.2	23.5
12. Daltenganj	29.0	21.1	44. Raipur	30.2	21.4	78. Nagapattinam	29.1	25.0
13. Dhanbad	25.4	20.2	45. Ratlam	30.1	20.6	79. Palayamkottai	30.5	23.9
14. Gaya	29.6	21.1	**MAHARASHTRA**			80. Salem	31.1	23.0
15. Jamshedpur	29.5	22.0	46. Ahmadnagar	30.1	19.8	81. Tiruchchirappalli	31.7	23.8
16. Patna	28.8	21.7	47. Akola	32.1	21.2	**UTTAR PRADESH**		
17. Purnia	27.3	22.4	48. Aurangabad	30.5	20.4	82. Agra	30.6	20.7
18. Ranchi	25.8	19.2	49. Kolhapur	28.6	20.6	83. Aligarh	29.7	20.5
GUJARAT			50. Nagpur	30.9	21.5	84. Allahabad	30.3	21.1
19. Rajkot	32.0	20.9	51. Pune	29.2	20.3	85. Azamgarh	29.0	22.9
20. Surat	31.1	23.7	52. Ratnagiri	28.5	24.5	86. Bahraich	28.8	21.6
21. Vadodara	32.3	23.0	53. Solapur	31.8	20.8	87. Bareilly	28.8	20.7
HARYANA			54. Yavatmal	30.5	21.3	88. Fatehpur	29.9	21.9
22. Ambala	28.8	19.9	**ORISSA**			89. Gonda	28.4	21.4
23. Hisar	30.0	20.7	55. Brahmapur	28.3	22.8	90. Gorakhpur	26.9	21.9
HIMACHAL PRADESH			56. Cuttack	29.2	23.6	91. Jhansi	30.5	20.4
24. Chamba	30.7	21.5	57. Koraput	24.8	19.0	92. Mainpuri	29.9	21.5
25. Dharmsala	21.4	15.5	58. Puri	27.4	24.5	93. Varanasi	29.9	21.5
JAMMU & KASHMIR			59. Sambalpur	30.2	22.6	**WEST BENGAL**		
26. Jammu	21.7	16.3	**PUNJAB**			94. Asansol	28.8	21.9
27. Kargil	6.9	3.0	60. Amritsar	28.4	19.5	95. Jalpaiguri	26.7	22.2
28. Leh	9.5	2.8	61. Ludhiana	29.3	20.3	96. Koch Bihar	26.5	22.7
KARNATAKA			**RAJASTHAN**			97. Puruliya	28.7	21.3
29. Bellary	31.2	21.5	62. Abu	22.8	15.8	**UNION TERRITORIES**		
30. Bidar	29.2	20.3	63. Ajmer	29.2	18.6	98. Marmagao (Goa D.&D.)	27.8	24.8
31. Chitradurga	28.7	20.7	64. Barmer	32.6	21.5	99. Minicoy (Lak.)	28.2	25.1
						100. Port Blair (A. & N.)	26.3	24.3

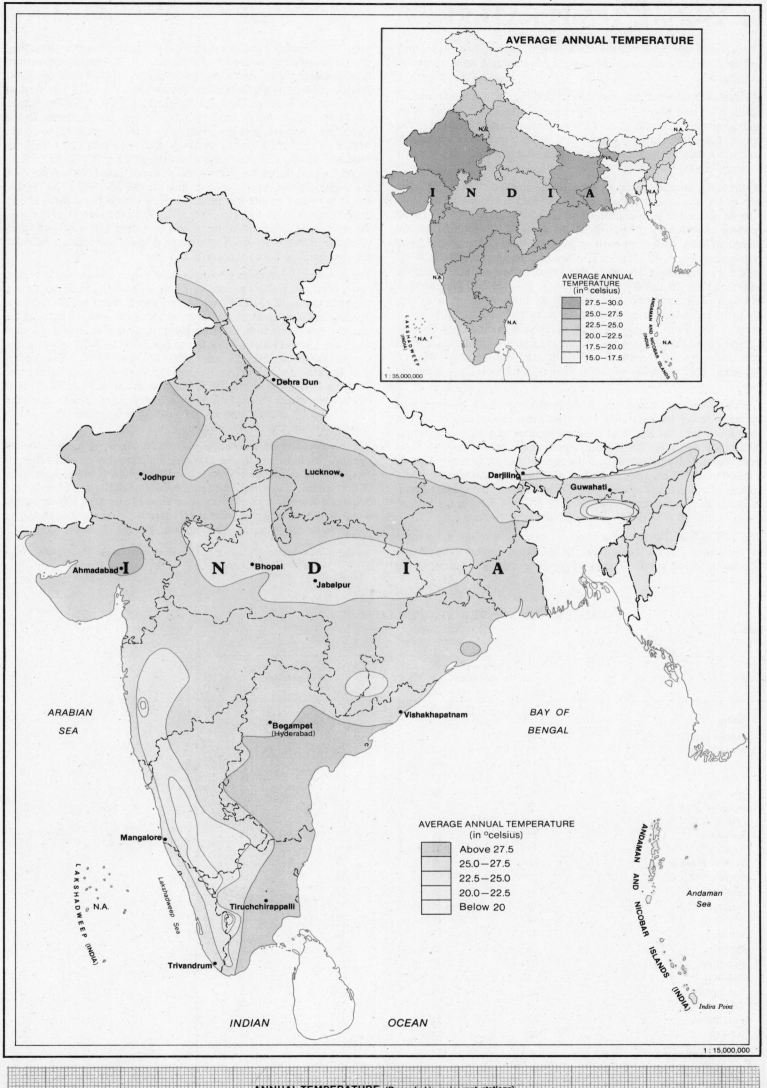

AVERAGE ANNUAL TEMPERATURE

I N D I A

N.A.

AVERAGE ANNUAL
TEMPERATURE
(in° celsius)

- 27.5 — 30.0
- 25.0 — 27.5
- 22.5 — 25.0
- 20.0 — 22.5
- 17.5 — 20.0
- 15.0 — 17.5

1 : 35,000,000

LAKSHADWEEP (INDIA) N.A.

N.A.

ANDAMAN AND NICOBAR (INDIA) N.A.

Dehra Dun

Jodhpur

Lucknow

Darjiling

Guwahati

I N D I A

Ahmadabad

Bhopal

Jabalpur

ARABIAN
SEA

Begampet
(Hyderabad)

Vishakhapatnam

BAY OF
BENGAL

Mangalore

ANDAMAN AND NICOBAR ISLANDS (INDIA)

Andaman
Sea

LAKSHADWEEP (INDIA)

Lakshadweep Sea

N.A.

AVERAGE ANNUAL TEMPERATURE
(in °celsius)

- Above 27.5
- 25.0 — 27.5
- 22.5 — 25.0
- 20.0 — 22.5
- Below 20

Tiruchchirappalli

Trivandrum

Indira Point

INDIAN   OCEAN

1 : 15,000,000

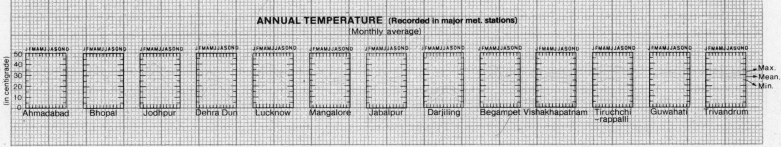

ANNUAL TEMPERATURE (Recorded in major met. stations)
(Monthly average)

(in centigrade)

Ahmadabad | Bhopal | Jodhpur | Dehra Dun | Lucknow | Mangalore | Jabalpur | Darjiling | Begampet | Vishakhapatnam | Tiruchchi-rappalli | Guwahati | Trivandrum

Max.
Mean.
Min.

SEASONAL TEMPERATURE

The advance and retreat of the sun's vertical rays across the country and the resultant monsoon winds have a great bearing on temperature conditions in India.

The Indian climate may conveniently be divided into a summer season, lasting from mid-March to mid-September, and a winter season, from mid-September to mid-March. The two seasons can be further divided into a hot, dry weather season and a hot, wet weather season (summer); and a retreating monsoon season and a cool, dry weather season (winter).

As the sun's vertical rays move towards the north, the hot dry weather sets in and peninsular India south of the Satpura range heats up to a mean maximum temperature of 40°C (104°F) in April. The entire country experiences a high mean of about 32°C (89.5°F) by the middle of May, with some parts in the north-west and in central India registering over 45°C (113°F). Record temperatures of 47°C (116.5°F) and above are often registered in several places during May and June (Ahmadabad, New Delhi, Bikaner and Nagpur, for example). The heat continues to increase northwards in June and July, but in the peninsula the temperatures begin to decline with the monsoon rains in May-June. In July, August and September the rains provide a respite from the heat (the hot, wet weather season); peninsular India gets cooler by 3-6°C (appx 37-43°F) in June and the north-western dry areas register a fall of 2-3°C (appx 35-38°F) in July after the rains.

The cool weather (December to mid-March) is generally characterized by much lower temperatures, but even these are relatively high in the south where the length of the cool weather season is also shorter than in the north. As the Himalaya protect the great Gangetic Plain from the cold, dry polar winds, India generally enjoys a much warmer climate in winter than countries like the USA and China where cold northerly winds freely sweep southwards. January, the coldest month in India, has a mean temperature of 12.5°C (54.5°F) in the Punjab plain — and 25.5°C (78°F) in southern Tamil Nadu.

The 21°C (70°F) isotherm for January, running through the middle of the country from the estuary of the Tapi to the delta of the Mahanadi, divides India into two regions in the cool season: in the north, the weather becomes progressively colder towards the north, from a mean temperature of 17.5°C (63.5°F) in central and eastern India (from Gujarat, across the subcontinent, through southern Uttar Pradesh, Bihar, West Bengal and Assam) to 15°C (59°F) in most of Rajasthan and western Uttar Pradesh and 12.5°C (54.5°F) in Punjab and Haryana. Jammu & Kashmir has very low temperatures. In the south, the peninsula gets progressively warmer from north to south, with 21°C (69.8°F) in the north of the peninsula (Maharashtra, southern Madhya Pradesh and Orissa) and 25.5°C (77.9°F) in the south. Coastal Karnataka, Kerala and southern Tamil Nadu experience the highest temperatures during this season 25-30°C (77 - 86°F). But it is cooler on the Karnataka plateau where Bangalore, at 916 metres, records around 21°C (69.8°F). With the retreat of the sun, the temperature rises towards the east and stays that way until June, when the monsoon sets in. The average range of temperature between the extreme south and the north in winter is 20°C (68°F).

The summer distribution of temperature is much more uniform; the entire country swelters in varying degrees of heat. The highest mean temperature is 35°C (95°F) while the lowest mean is 20°C (68°F). The highest temperatures are found in the north-western parts of the country, extending from the Thar desert through Punjab and Haryana and into Uttar Pradesh. But the major part of India experiences temperatures between 25 and 30°C (77 and 86°F). The lowest temperatures are, surprisingly, registered in the south, along the Western Ghats, extending from Maharashtra to Karnataka. This is due to the effect of elevation. The lowest temperature recorded 15.7°C (60°F) at this time of the year in this region has been in Udagamandalam in Tamil Nadu.

The temperatures for both the seasons at scattered meteorological stations, as shown in the histograms, indicate the general effect of elevations, with centres at greater heights registering low temperatures (Cherrapunji, Darjiling, Udagamandalam, Shillong, Shimla and Srinagar) while low level areas generally register high temperatures, whether they are in the interior (Hyderabad, Indore, Lucknow) or along the coast (Madras, Trivandrum, Panaji).

Source: *Climatological Tables of Observatories in India (1931-1960),* India Meteorological Department.

SUMMER AND WINTER TEMPERATURES

Table 30

(in °Celsius)

	Summer Temperature		Winter Temperature	
	Max.	Min.	Max.	Min.
States				
Andhra Pradesh	41	20	32	13
Assam	35	18	26	17
Bihar	47	20	28	4
Gujarat	41	27	29	14
Haryana	46	35	14	2
Himachal Pradesh	33	14	15	0
Jammu & Kashmir	N.A.	N.A.	N.A.	N.A.
Karnataka	35	26	25	14
Kerala	35	21	N.A.	N.A.
Madhya Pradesh	48	22	23	4
Maharashtra	39	22	34	12
Manipur	29	14	25	7
Meghalaya	25	15	16	4
Nagaland	18	29	24	11
Orissa	49	27	16	5
Punjab	45	35	14	0
Rajasthan	45	17	32	7
Sikkim	N.A.	N.A.	N.A.	N.A.
Tamil Nadu	43	18	N.A.	N.A.
Tripura	35	24	27	13
Uttar Pradesh	45	11	32	2
West Bengal	40	24	26	7
Union Territories				
Andaman & Nicobar	33	22	31	22
Arunachal Pradesh	N.A.	N.A.	N.A.	N.A.
Chandigarh	43	35	14	7
Dadra & Nagar Haveli	As in Gujarat & Maharashtra			
Delhi	45	35	20	7
Goa, Daman & Diu	32	21	N.A.	N.A.
Lakshadweep	N.A.	N.A.	N.A.	N.A.
Mizoram	29	18	24	11
Pondicherry	As in Tamil Nadu			

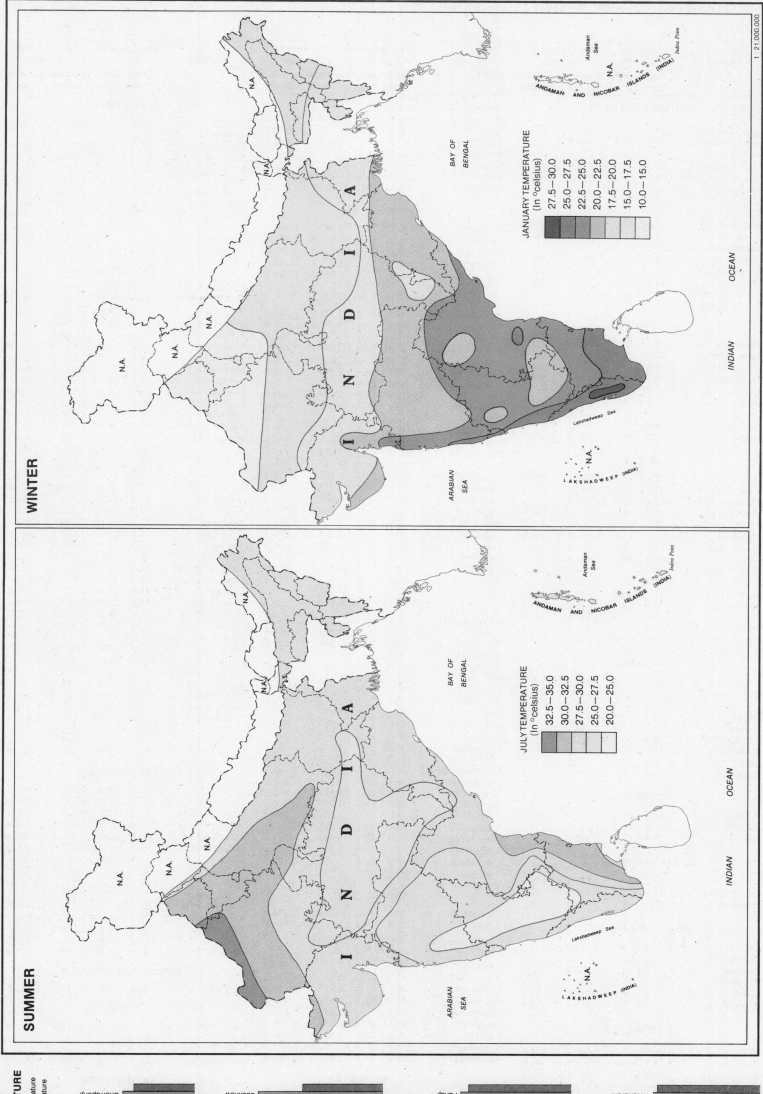

WINTER

JANUARY TEMPERATURE
(In °celsius)
27.5 – 30.0
25.0 – 27.5
22.5 – 25.0
20.0 – 22.5
17.5 – 20.0
15.0 – 17.5
10.0 – 15.0

1 : 21,000,000

BAY OF BENGAL

ARABIAN SEA

INDIAN OCEAN

I N D I A

ANDAMAN AND NICOBAR ISLANDS (INDIA)

Andaman Sea

N.A.

Indira Point

LAKSHADWEEP (INDIA)

N.A.

Lakshadweep Sea

SUMMER

JULY TEMPERATURE
(In °celsius)
32.5 – 35.0
30.0 – 32.5
27.5 – 30.0
25.0 – 27.5
20.0 – 25.0

BAY OF BENGAL

ARABIAN SEA

INDIAN OCEAN

I N D I A

ANDAMAN AND NICOBAR ISLANDS (INDIA)

Andaman Sea

N.A.

Indira Point

LAKSHADWEEP (INDIA)

N.A.

Lakshadweep Sea

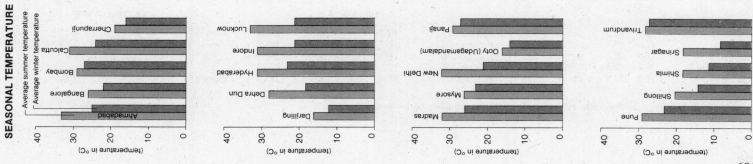

SEASONAL TEMPERATURE

Average summer temperature
Average winter temperature

Cherrapunji
Calcutta
Bombay
Bangalore
Ahmadabad

Lucknow
Indore
Hyderabad
Dehra Dun
Darjiling

Panaji
Ooty (Udagamandalam)
New Delhi
Mysore
Madras

Trivandrum
Srinagar
Shimla
Shillong
Pune

(temperature in °C)

METEOROLOGICAL REPORT (1931-60 AVERAGE)

Table III

Station	Temperature (in °C) Daily Max.	Daily Min.	Total Rainfall (in mm) In wettest month	In driest month
ANDHRA PRADESH				
Anantapur(2)	33.3	21.9	777.9	233.2
Arogy Varam(3)	33.2	19.8	955.5	473.2
Cuddapah(4)	34.5	24.0	1,273.6	299.0
Gannavaram(10)	33.5	23.5	1,488.0	676.4
Hanamkonda(22)	33.1	22.3	1,615.7	384.3
Hyderabad(7)	31.7	20.0	1,430.8	455.2
Kakinada(5)	31.7	23.8	1,825.5	407.9
Kalingapatam(19)	30.9	23.4	1,718.1	611.6
Khammam(9)	34.2	22.9	1,341.8	530.9
Kurnool(11)	34.0	21.1	1,095.5	280.7
Machilipatnam(10)	31.7	24.0	1,765.0	541.8
Nellore(15)	34.0	24.4	1,638.0	295.9
Nizamabad(16)	33.1	20.9	1,722.9	495.1
Ongole(17)	32.8	24.3	1,357.9	588.8
Rentachintala(6)	34.7	23.2	1,009.6	165.8
Vishakhapatnam(20)	31.0	23.5	1,442.0	473.2
ASSAM				
Dhubri(2)	28.3	20.4	3,669.3	1,703.6
Dibrugarh(3)	27.7	18.7	3,300.2	2,165.0
Guwahati(6)	29.5	19.7	2,121.1	874.0
Guwahati(6)*	29.3	19.4	2,121.1	1,568.2
Lumding(11)	29.9	18.3	1,723.1	947.9
Nowgong(12)	32.6	18.1	1,868.4	344.9
Sibsagar(14)	28.3	19.2	3,209.3	1,744.0
Silchar(15)	29.9	19.9	5,598.0	1,939.0
Tezpur(16)	29.1	19.6	2,882.9	1,334.3
BIHAR				
Bhagalpur(3)	31.6	21.0	1,594.9	837.7
Chaibasa(35)	32.3	20.6	1,906.8	740.4
Daltonganj(23)	31.8	19.0	1,856.5	622.3
Darbhanga(5)	30.8	19.7	1,942.1	799.9
Dehri(29)	32.4	21.0	1,718.6	567.7
Dhanbad(7)	31.1	20.4	1,781.7	869.2
Dumka(33)	31.6	20.0	1,993.4	83.5
Gaya(8)	32.3	20.1	1,892.8	588.0
Hazaribag(13)	29.4	18.2	2,146.0	874.2
Jamshedpur(36)	32.4	20.7	1,862.7	745.5
Jamui(19)	32.2	20.6	1,350.4	671.6
Motihari(26)	30.6	18.7	2,020.5	254.8
Patna(25)	31.6	20.8	1,959.4	642.4
Purnia(27)	30.8	19.0	2,338.3	691.4
Ranchi(28)	29.5	18.0	2,106.4	978.7
GUJARAT				
Ahmadabad(1)	34.2	20.5	1,997.5	119.9
Bharuch(4)	34.9	21.1	1,663.7	425.5
Bhavnagar(6)	33.4	21.0	1,438.0	147.1
Bhuj(9)	33.0	20.1	1,770.0	21.8
Dwarka(7)	29.2	22.9	1,080.0	26.2
Jamnagar(7)*	32.1	20.0	1,119.6	78.2
Rajkot(13)	33.9	19.6	2,050.3	154.2
Surat(15)	33.6	21.9	2,284.0	363.0
Veraval(8)	30.6	21.4	1,327.2	69.1
HARYANA				
Ambala(1)	31.8	17.5	2,047.7	347.7
Hisar(5)	32.8	17.4	1,048.0	158.0
Karnal(7)	30.9	17.7	1,150.9	383.9
HIMACHAL PRADESH				
Chamba(2)	33.2	21.5	1,785.1	1,178.1
Dharmsala(4)	23.6	15.5	4,721.3	2,219.2
Shimla(9)	17.1	10.1	2,786.6	992.6
JAMMU & KASHMIR				
Dras(8)	8.6	−4.7	1,273.0	357.4
Gulmarg(6)	N.A.	N.A.	N.A.	N.A.
Jammu(4)	30.0	18.7	1,964.7	646.4
Kargil(8)	14.5	3.3	636.5	85.1
Leh(8)	12.4	−1.4	231.1	25.4
Srinagar(6)	19.5	7.2	1,291.6	398.5
KARNATAKA				
Balehonnur(6)	26.9	17.5	3,665.7	1,975.6
Bangalore(1)	28.8	18.4	1,348.5	544.3
Bangalore(1)*	29.6	18.6	1,087.1	542.9
Belgaum(2)	30.1	18.0	2,847.5	730.0
Bellary(3)	32.9	22.2	949.4	208.3
Bidar(4)	31.1	20.7	1,837.9	437.9
Bijapur(5)	32.5	20.3	991.4	243.3
Chitradurga(7)	30.8	20.1	1,059.7	295.1
Gadag(9)	31.7	20.1	1,041.0	345.4
Gulbarga(10)	33.4	21.0	1,431.8	357.9
Hassan(11)	28.5	17.9	1,294.4	510.0
Honavar(19)	30.9	22.9	5,730.0	2,292.9
Mangalore(8)	30.5	23.7	4,703.1	2,269.7
Mercara(12)	24.1	16.5	5,768.6	2,146.8
Mysore(15)	29.6	19.2	1,295.9	450.6
Raichur(16)	33.3	22.4	1,270.5	269.0
Shimoga(17)	30.6	19.2	1,189.7	673.9
KERALA				
Alleppey(1)	30.7	23.9	3,839.2	2,462.0
Calicut(7)	30.9	23.7	4,623.8	1,887.2
Fort Cochin(3)	29.8	24.4	4,215.4	2,047.0
Palghat(9)	32.3	23.4	2,907.1	1,507.7
Trivandrum(13)	30.7	23.5	3,035.6	1,029.2
MADHYA PRADESH				
Ambikapur(41)	30.2	17.4	2,338.8	1,082.0
Betul(3)	30.7	17.9	1,662.2	708.9
Bhopal(5)	31.5	18.5	1,779.5	508.8
Chhindwara(8)	30.6	18.2	1,809.8	487.2
Guna(15)	31.8	17.6	1,812.8	675.4
Gwalior(16)	32.5	18.8	1,514.3	224.0
Hoshangabad(17)	32.4	20.0	2,045.7	587.8
Indore(18)	31.3	17.5	1,743.2	400.1
Jabalpur(19)	32.1	18.3	2,407.2	774.4
Jagdalpur(2)	31.2	18.9	2,395.2	881.6
Kanker(2)	31.6	19.9	1,975.6	879.1
Khandwa(14)	33.6	19.6	1,517.7	225.5
Mandla(21)	31.7	17.1	1,648.7	1,152.9
Nimach(22)	31.5	18.6	1,751.1	305.1
Pachmarhi(17)	26.9	15.9	3,365.5	1,002.5
Pendra(6)	29.8	18.6	1,937.0	892.2
Raigarh(26)	33.5	21.5	2,046.7	987.3
Raipur(27)	32.6	21.2	2,181.4	688.9
Ratlam(31)	31.9	19.2	1,512.1	563.1
Sagar(33)	30.9	19.5	2,006.1	592.3
Satna(34)	32.1	18.7	1,682.5	674.6
Seoni(36)	30.8	18.5	2,127.2	620.8
Sheopur(23)	32.9	18.5	1,320.0	671.8
Umaria(37)	31.4	18.0	2,124.2	712.2
MAHARASHTRA				
Ahmadnagar(1)	32.2	18.5	1,119.4	287.3
Akola(2)	34.0	20.4	1,683.0	306.8
Alibag(23)	30.0	22.6	3,396.7	720.6
Amraoti(3)	N.A.	N.A.	N.A.	N.A.
Aurangabad(4)	32.4	19.7	1,171.8	265.5
Bombay(11)	31.0	23.6	3,481.6	848.6
Bombay(11)*	31.6	21.9	3,784.9	1,814.1
Buldana(7)	30.9	20.5	1,272.5	529.6
Chanda(8)	33.4	20.8	2,114.3	577.3
Devgarh(26)	30.5	23.9	3,367.5	1,908.8
Gondia(5)	32.7	20.9	1,668.0	986.0
Jalgaon(12)	34.6	19.9	1,351.8	205.7
Kolhapur(14)	31.2	19.2	1,400.8	805.9
Mahabaleshwar(25)	24.1	16.1	10,221.2	3,545.1
Malegaon(18)	33.5	18.6	976.3	205.0
Nagpur(16)	33.5	20.3	1,931.4	364.7
Parbhani(21)	33.7	19.9	1,496.3	273.3
Pune(22)	32.0	18.2	1,242.3	268.5
Ratnagiri(19)	30.8	23.1	4,093.2	1,406.9
Sironcha(10)	33.9	22.3	1,960.5	247.4
Solapur(27)	33.7	20.5	1,239.5	325.4
Vengurla(26)	31.2	22.4	3,955.8	2,583.3
Yavatmal(30)	32.8	21.0	1,745.7	720.6
MEGHALAYA				
Cherrapunji(2)	20.6	14.3	15,706.6	7,178.0
Tura(4)	28.4	19.4	4,133.1	2,816.8
ORISSA				
Angul(4)	32.5	21.5	1,831.3	756.1
Baleshwar(2)	31.5	22.0	2,957.1	1,076.5
Chandbali(2)	33.5	22.1	2,335.3	866.5
Cuttack(8)	32.9	22.5	2,363.8	884.8
Gopalpur(5)	33.3	23.0	2,040.9	655.1
Koraput(8)	28.1	17.8	2,157.7	1,395.2
Puri(11)	30.0	24.1	1,984.0	599.9
Sambalpur(12)	32.9	20.8	2,307.8	934.7
Titlagarh(1)	33.0	21.3	2,024.5	1,086.6
PUNJAB				
Amritsar(1)	30.5	15.9	1,013.5	257.8
Ferozpur(5)	31.3	16.7	820.9	361.2
Ludhiana(9)	31.9	17.3	1,402.1	236.5
Pathankot(4)	29.8	16.6	2,065.8	986.8
Patiala(10)	31.4	17.6	1,204.2	335.8
RAJASTHAN				
Abu(25)	24.9	16.5	3,990.5	290.1
Ajmer(1)	31.2	18.3	1,227.0	149.3
Barmer(4)	34.0	22.2	895.3	128.8
Bikaner(7)	33.6	18.2	770.9	29.0
Ganganagar(13)	32.9	17.0	640.6	123.4
Jaipur(14)	31.7	18.4	1,403.9	120.1
Jaisalmer(15)	34.0	19.0	453.6	104.2
Jhalawar(17)	32.6	18.9	1,698.2	445.0
Jodhpur(19)	33.6	19.8	1,176.5	24.4
Kota(2)	33.3	20.8	1,593.1	171.7
Sikar(24)	31.9	17.3	631.9	269.2
Udaipur(27)	31.1	17.5	1,222.8	252.0
TAMILNADU				
Coimbatore(2)	31.1	21.7	1,059.2	323.6
Coimbatore(2)*	31.7	21.2	880.4	243.6
Coonoor(8)	21.2	12.6	2,239.0	1,130.3
Cuddalore(15)	32.1	24.2	2,443.5	537.0
Kallakkurichchi(15)	33.9	23.6	1,360.2	567.2
Kodaikanal(7)	17.9	10.6	2,354.6	1,184.1
Madras(1)*	32.9	24.3	2,134.9	522.2
Madurai(7)	33.9	24.0	1,436.6	410.5
Nagapattinam(16)	31.9	25.2	2,196.3	603.5
Palayankottai(18)	33.7	24.9	1,310.9	326.6
Pamban(13)	30.6	25.7	1,772.2	410.5
Salem(14)	33.5	22.6	1,532.1	527.3
Tiruchchirappalli(17)	33.7	24.0	1,324.1	511.6
Udagamandalam(8)	19.5	9.1	1,926.6	903.2
Vellore(9)	33.1	22.8	1,701.5	468.4
UTTAR PRADESH				
Agra(1)	32.3	19.0	1,202.2	276.6
Aligarh(2)	31.9	18.7	1,342.9	204.5
Allahabad(3)	32.4	19.8	1,935.5	516.1
Azamgarh(5)	32.3	19.6	1,604.5	637.1
Bahraich(6)	31.5	19.1	2,442.7	469.1
Banda(8)	32.9	19.9	1,342.5	793.9
Bareilly(10)	31.5	18.9	1,974.4	492.0
Dehra Dun(16)	27.8	15.8	3,118.9	1,152.4
Fatehpur(22)	32.4	19.5	1,360.7	537.2
Gonda(26)	31.5	18.8	2,222.0	405.6
Gorakhpur(27)	31.4	20.0	2,455.2	649.7
Hardoi(29)	31.9	19.0	1,845.9	543.3
Jhansi(32)	32.8	19.7	1,495.8	314.7
Kanpur(34)*	32.1	19.3	1,567.2	420.6
Kheri(35)	31.5	19.0	1,517.4	747.7
Lucknow(37)	32.3	19.4	1,866.7	424.2
Lucknow(37)*	31.9	18.5	1,529.7	545.5
Mainpuri(38)	32.8	18.7	1,476.7	299.8
Meerut(40)	31.2	18.3	1,520.4	302.3
Mukteswar(44)	17.6	9.5	2,254.3	522.9
Mussoorie(16)	18.0	10.4	3,284.7	1,475.7
Orai(30)	32.6	19.1	1,724.6	737.9
Roorkee(50)	30.2	17.0	2,299.7	511.6
Varnasi(57)	32.2	19.8	2,108.5	445.8
WEST BENGAL				
Asansol(2)	32.0	20.8	2,135.1	963.7
Barddhaman(2)	31.7	21.4	2,257.5	886.0
Baghdogra(5)	29.7	18.3	4,429.3	2,379.8
Baharampur(12)	31.9	20.8	2,136.9	774.9
Calcutta(4)	31.8	22.1	2,501.4	909.1
Calcutta(4)*	31.4	21.4	2,625.6	866.7
Contai(11)	30.7	22.6	2,853.9	1,300.5
Darjiling(5)	16.4	10.2	4,024.4	2,271.3
Jalpaiguri(8)	28.8	19.3	4,292.3	1,719.6
Kalimpong(5)	21.5	14.5	3,102.4	1,777.7
Koch Bihar(9)	29.7	19.4	6,534.1	2,112.3
Krishna Nagar(13)	32.4	20.6	2,295.7	884.9
Maldah(10)	30.6	20.2	2,306.1	858.9
Medinipur(11)	32.2	21.5	2,479.8	761.2
Puruliya(14)	31.9	21.0	1,570.1	927.6
ANDAMAN & NICOBAR				
Port Blair(1)	29.6	3.5	4,027.1	2,131.6
DELHI				
New Delhi(1)	31.7	18.8	1,533.1	262.9
GOA				
Marmagao(3)	29.5	23.7	3,500.4	1,843.5
LAKSHADWEEP				
Amini(1)	31.1	25.4	2,550.4	909.8
Minicoy(1)	29.9	24.6	2,247.6	1,050.0
MIZORAM				
Aizawl(1)	23.9	16.3	2,846.1	1,583.5

Note: 1. * Airport   2. Bracketed numbers refer to district numbers on loose leaf key map.

The Natural Resources

This section comprises 23 maps & 26 charts. It assesses the potential of land, water, forest and mineral resources and offers information on the extent exploitation is possible.

Note:

Page 73: The detailed map on soil distribution is accompanied by a generalized inset map. This inset map reflects a general cropping pattern and broadly indicates the crop (or crops) best suited to particular areas. It should be noted that the soil distribution in these areas has been generalized for the purpose of this map.

Page 83: The potential of wasteland in India has only been realised in the 1980s and a Wasteland Development Board set up only in 1985.

Page 85: The natural regions mentioned on the map are a broad socio-political geographic breakdown of the physiographic divisions indicated on pages 2 and 3.

Page 91: The detailed map of the geological systems in India is accompanied by a generalized inset map. This inset map broadly indicates the rock types to be found in particular areas. It should be noted that the geological systems in these areas have been generalized for the purpose of this map.

SOILS

India is primarily an agricultural country, with nearly 70 per cent of its population either directly or indirectly dependent on agriculture. The aspects of the physical environment most affecting the cultivator are climate, soil and slope. Soil is the major factor influencing the extent and intensity of cultivation. For example, the presence of a vast tract of regur, black soils with moisture-retentive characteristics, in the northern Deccan, has given rise to a concentration of cotton cultivation there.

Fundamental to an understanding of the soils that the Indian farmers cultivate is some knowledge of the country's geology and morphology. India consists of three basic structural features (see page 91): the geologically ancient and now stable 'shield' of peninsular India; the relatively young mountain chains of the Himalaya, still a region of considerable geological instability; and, between them, the extensive alluvium-filled depression occupied by the three great river systems — the Indus, Ganga and Brahmaputra. Through aeons of geological time, the peninsular shield poured sediments into a broad, deepening trough. To it was added vast quantities of coarse material and river-borne alluvium from the Himalaya. Numerous streams have also deposited alluvium from the Western and Eastern Ghats on to the coastal areas.

In terms of soils, what lies beneath the surface concerns the farmer as much as what he sees as he ploughs. Characteristically, alluvium, in all but the most humid parts of India, contains *kankar*, concretions of lime that accumulate through the fluctuation of water in the soil. This older alluvium is generally darker and of a more clayey structure than younger alluvium. In certain areas, particularly in Assam, Meghalaya and West Bengal, the older alluvium has undergone intensive leaching, producing lateritic soils which are 'droughty' as far as field crops are concerned.

Indian soils are commonly classified into 'zonal' and 'azonal' soils; regur is a zonal soil, while alluvium is an azonal soil. Among the zonal soils, two groups that stand out are the red and black soils. Alluvium and terai soils are the important azonal soils.

The soils of India are normally categorized into 13 soil types. Alluvial soils occupy by far the largest area (43 per cent of the area), followed by black soils (13.7 per cent) and red (sandy and loamy) soils (10.2 per cent). Thus, two-thirds of the total area of the country is under these three soils. Laterite, red and yellow, and desert soils occupy almost one-sixth of the area of the country. Mountain soils, terai soils, submontane and grey and brown soils occupy a considerable area, (9.1 per cent). Other soils that occupy numerous pockets in small areas account for just about 7.8 per cent of the total area.

From the map it is clear that two soil groups are important in peninsular India: red soils and their variants, and black soils. There are three distinct subtypes of red soils: red sandy soils (a large part of the southern states of Tamil Nadu, Karnataka and Andhra Pradesh, and portions of Orissa, Madhya Pradesh and Bihar); red loamy soils (a long stretch along the Western Ghats and some pockets in the foothills in the interior); and red and yellow soils (a large part of Orissa, Bihar, Madhya Pradesh, Uttar Pradesh and Rajasthan). Black soil occurs for the most part in one vast block in the western half of the northern Deccan. This soil is derived from the trap rocks (volcanic origin), is clayey in structure and responds well when irrigated. The black soils are spread over north-western Andhra Pradesh, northern Karnataka, western Madhya Pradesh, Maharashtra and across the Gulf of Khambhat into Saurashtra in Gujarat. Mixed red and black soil is found along the southern border of the Krishna-Tungabhadra valley, an area noted for cotton cultivation. In Tamil Nadu, patches of this red and black soil occur in Coimbatore, Madurai and Tirunelveli districts. Isolated patches of this soil type are also found in the black soil and red soil belts of peninsular India. Unlike the northern black soils, these soils are derived from ferruginous gneiss and schistose rocks under semi-arid conditions.

Laterites are a result of leaching of surface soils under very humid conditions and were originally identified by Buchanan in the old Malabar (northern Kerala). It is surfaced with a hard crust, and is found all along the coast of Kerala and Karnataka, in south-west India. It is also seen in Orissa, along the south-western border of West Bengal and in south-eastern Tamil Nadu, besides in patches in the states of the north-east.

Desert soils occur in central and western Rajasthan, an area that is part of the Thar desert. They cover about 150,000 square kilometres. When irrigated, these soils are highly suitable for agriculture, though there is always the danger of their becoming saline. Hilly soils of the Himalayan region cover an area of 3.4 million hectares. In the eastern Himalaya, the soil is loamy, rich in humus and is suitable for tea cultivation. The glacial, terai and skeletal soils develop under swamp conditions. Terai soils are found along the foothills of the Himalaya and skeletal soils along the coast of West Bengal and Andhra Pradesh.

Alluvial soils are one of the most important soils in terms of economic importance. They occupy a broad belt between the foothills of the Himalaya and the peninsular plateau and extend from Rajasthan through Punjab, Haryana, Uttar Pradesh and Bihar to West Bengal and the Assam Valley. These soils are the result of the deposits brought by the Ganges and its tributaries. Alluvial soils are also found along the east coast, in the deltas of the rivers Mahanadi, Godavari, Krishna and Kaveri. Likewise, there are scattered alluvial deposits along the west coast.

The cropping pattern of the country is greatly influenced by the soils and the elements of the physical environment. Farmers in areas with the thinner black soils have, for example, to adhere to the more usual practice of *kharif* cropping, during the rainy season. Setting aside soil regions of the country with more or less extreme differences, the country may be divided for agricultural purposes into three regional types: the soils of the plateau country, poor in nutrients but, where thick enough, responding to irrigation; the lava-covered Deccan, which carries some remarkably rich and productive soils; and the alluvial soils, very fertile from the point of view of agriculture.

Source: *A study on Wastelands of India including saline, alkaline and waterlogged lands and their reclamation measures.* Report of the Committee on Natural Resources (Government of India Planning Commission).

Table 31   CROPS AND FAVOURABLE SOIL TYPES

Crops	Favourable Soil Types
Rice	: Alluvial friable loams with sub-soil of clay.
Wheat	: Clay loam soils or fertile silt.
Maize	: Sandy, deep and well-watered soil.
Jowar	: Black soil, mixed red and black soil.
Bajra	: Sandy loams, shallow soils.
Gram	: Alluvial soil.
Coconut	: Sandy soil that is loose and porous, along the coast. Alluvial flats exposed to mild sea breezes.
Groundnut	: Light sandy soils which are friable.
Sugarcane	: Rich alluvial or lava soil.
Tea	: Light and friable loam with porous sub-soil which will allow water to percolate. Sandy loam best. Iron in the soil beneficial. Stagnant water harmful, so mountain slopes preferred.
Coffee	: Weathered volcanic soil (deep loamy soil formed from lava) on well-drained hillsides from 450-1800 metres. There should be humus in the soil.
Cocoa	: Well-drained deep and porous soil.
Tobacco	: Sandy loams with sandy clay soils.
Rubber	: Alluvial soil from which virgin forest has been cleared.
Cotton	: Light limestone soil or black lava soil. The Deccan black lava soil (regur) has the quality of retaining moisture.
Jute	: Alluvial soil found in the flood plains and deltas of rivers.
Flax	: Rich alluvial soil.
Pepper	: Alluvial soil and heavy rainfall.
Cardamom	: Well-drained soil, rich in humus.
Turmeric	: Sandy loams.

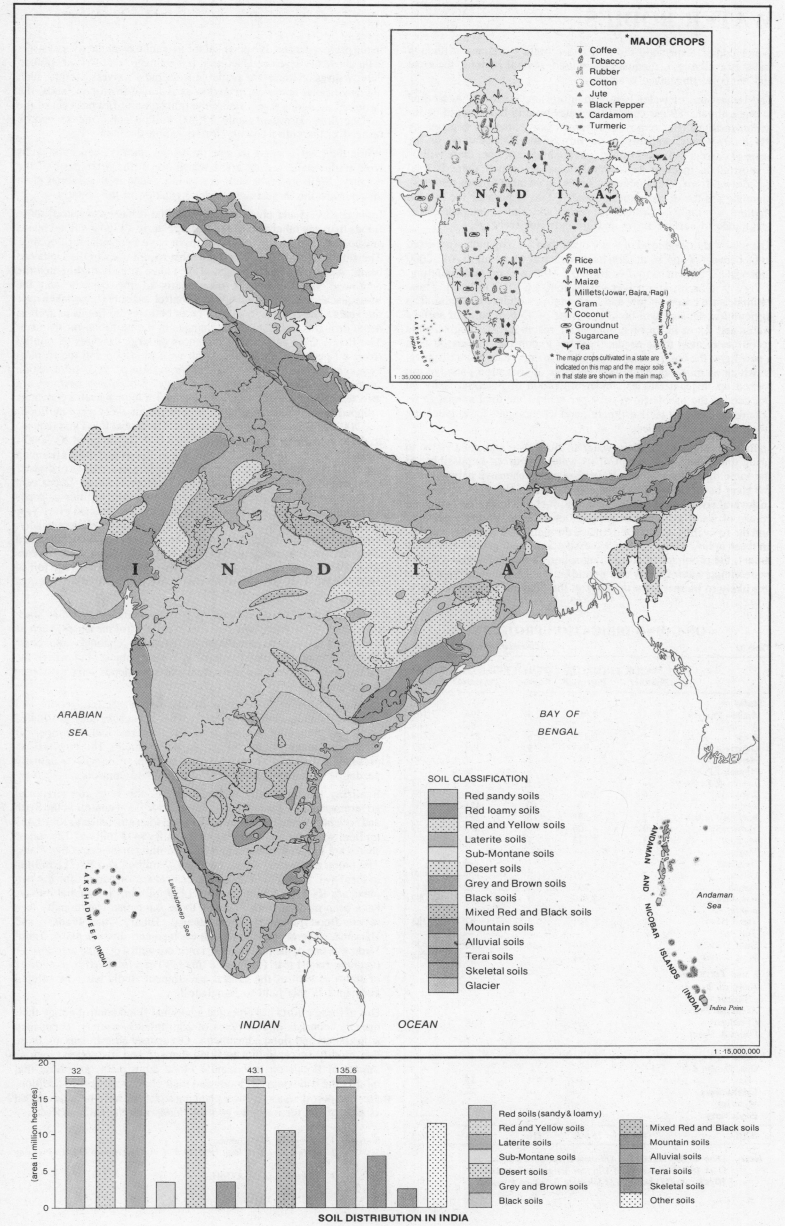

MAJOR CROPS

- Coffee
- Tobacco
- Rubber
- Cotton
- Jute
- Black Pepper
- Cardamom
- Turmeric

- Rice
- Wheat
- Maize
- Millets (Jowar, Bajra, Ragi)
- Gram
- Coconut
- Groundnut
- Sugarcane
- Tea

* The major crops cultivated in a state are indicated on this map and the major soils in that state are shown in the main map.

1 : 35,000,000

INDIA

ARABIAN SEA

BAY OF BENGAL

ANDAMAN AND NICOBAR ISLANDS (INDIA)

Andaman Sea

LAKSHADWEEP (INDIA)

Lakshadweep Sea

SOIL CLASSIFICATION

- Red sandy soils
- Red loamy soils
- Red and Yellow soils
- Laterite soils
- Sub-Montane soils
- Desert soils
- Grey and Brown soils
- Black soils
- Mixed Red and Black soils
- Mountain soils
- Alluvial soils
- Terai soils
- Skeletal soils
- Glacier

INDIAN OCEAN

Indira Point

1 : 15,000,000

SOIL DISTRIBUTION IN INDIA

(area in million hectares)

32   43.1   135.6

- Red soils (sandy & loamy)
- Red and Yellow soils
- Laterite soils
- Sub-Montane soils
- Desert soils
- Grey and Brown soils
- Black soils
- Mixed Red and Black soils
- Mountain soils
- Alluvial soils
- Terai soils
- Skeletal soils
- Other soils

WATER BODIES

Water bodies, especially the rivers, are India's lifelines. Although water is available in adequate amounts in several regions, the areal and temporal imbalance is severe.

The largest use of water is in agriculture and industry. According to one estimate, 89 per cent of the water used in India would be for *withdrawal* use (*i.e.* removal of water from its natural courses) by AD 2000. Worse, well before the turn of the 20th century, on the basis of current estimates of water availability and use, there will be a shortfall in meeting agricultural demands even if there is total withdrawal from *all* available resources. In no other area of resources planning are the uncertainties so many. This is because the rainfall rhythm, so highly variable in space and time, determines the magnitude of surface flows and groundwater recharge.

The total withdrawal need of water other than for irrigation is expected to be between 8 and 16 million hectare metres (m ha m) in A.D. 2000 (domestic: 2 - 2.5 m ha m; livestock: 0.5 - 0.6 m ha m; manufacturing: 1.7 - 3.9 m ha m; and power generation: 3.6 - 9.3 m ha m. These estimates are based on low and high annual withdrawal needs). The agricultural demand will be of the order of 45.6 m ha m of surface water and 21 m ha m of groundwater resources. To augment the resources to meet these needs, India has to consider three aspects in a new light: the cost of delayed decisions in inter-state water disputes, in which as much as 20 million hectares of irrigation potential are locked up in legal tangles; under-utilization of potential (only 30 per cent of the usable flow, or 9 per cent of the total annual flow, is now utilized — a 1981 estimate); and technologies for augmenting water resources supplies.

India cannot afford to allow water to run waste and must strive to bring under control as much of its water resources as possible. At the same time, it must work towards a comprehensive national plan for river basin flows. As things are, the total volume of water used in several river basins exceeds the total surface flows, indicating the re-use of water through the exploitation of groundwater reservoirs and the re-cycling of surface drainage through canal and tank systems. In these areas, where there is excessive use (most project command areas), the problems of controlling salinity and waterlogging and of maintaining water quality at a non-toxic level for downstream users are likely to increase, particularly in the drier parts of the country.

In the present decade, the potential for irrigation from the river systems is far above the developed facilities. It is estimated that 167,500 million cubic metres of water are available from India's rivers, despite such intensive canal and well irrigation as is found in, for instance, the basins of the Indus river, where the utilization is 140 per cent of the surface flow. However, only 55,500 million cubic metres can be utilized in the entire country for irrigating the land.

While the river systems irrigate the whole country, tanks and wells have traditionally served areas which are in the drier parts of the country. Wells and tubewells are mainly found in the alluvial plain and nurture the largest area under irrigation in the country.

India is so vast and physically diverse that it has experienced more floods than any other part of the world. Nearly all India's flood plains and coastal areas have flood problems or have a potential for flooding. The floods of 1977, 1978 and the most recent ones in the Godavari basin, Assam and West Bengal (1986) have already highlighted the continued vulnerability of many parts of the country and the inadequacies of flood estimation and control measures undertaken over the years. The country regularly faces both severe floods as well as acute droughts because of the changes in the monsoon mechanism. The floods are a result of occurrences of large excesses of rainfall (over 80 per cent) during the south-west monsoon and the tropical cyclonic storms in the pre- and post-monsoon periods. Information from 1953 indicates that, on an average, 7.8 million hectares are affected by floods every year, of which 3.3 million hectares constitute cropped area. About 24 million people are hit every year by floods and 700,000 houses a year are damaged. The annual flood loss reduced to the wholesale price index of 1977-78 is in the range of Rs 3000 - 4000 million and the annual deaths due to floods and their aftermath are in the region of 50,000 - 100,000 (according to a long-term disaster study made in 1977-78). The damage to crops is about 72 per cent of the total damage, the rest being to such public utilities as roads and communication lines. The fury of the floods unleashed every year in different parts of the country kills upwards of 1500 Indians annually. Statistics indicate that the death toll exacted by floods has declined from the average annual high of 4052 recorded during the five years from 1975 to 1979, but it is also true that the annual average toll of 1852 during 1980-84 was somewhat more than double the average of 902 recorded during 1971-74.

A national programme of flood control was launched in 1954. Since then, Rs 9760 million have been spent on control measures, such as construction of new embankments, drainage channels and other structural controls. Flood control is a state subject, but where the calamity is of exceptional severity, the centre extends some assistance to the states.

Despite efforts at controlling floods and protecting people and property, floods continue to occur. This is because they are linked with the prevailing human land use systems and improper or inadequate measures for mitigating the problem. The emphasis at present is on structural (engineering) controls while what is actually needed is a change in non-structural (human) controls.

Realizing the immensity of the problem, the state and territorial governments have made large allocations for flood control in the Sixth and Seventh Plans. The outlay in the Sixth Plan amounted to Rs 10,451 million, while the Seventh Plan outlay is Rs 9474 million. The states' shares are Rs 8274 million and Rs 7264 million in the respective Plans. The union territories' shares are Rs 427 million and Rs 711 million in these two Five Year Plans. The central sector outlays for the two plans are Rs 1750 million and Rs 1499 million. Substantial outlays have been made in both Plans by those states that traditionally face severe flood problems: West Bengal, Bihar, Uttar Pradesh and Haryana. Several states have also evolved anti-disaster plans. Tamil Nadu was one of the first states to come out with a detailed anti-disaster (cyclone-relief) plan (1974) and this has been followed by a number of states as well as the central government. India has now offered assistance in this field to Bangladesh.

Out of these efforts has emerged a national flood control policy that, broadly, consists of two types of adjustments, namely, occupancy adjustments and flood adjustments. Occupancy adjustments are those that modify susceptibility to flood damage and disruption, such as impact of floods on individuals. Flood adjustments are those that modify the behaviour of floods and their characteristics. In addition, pre- and post-disaster physical planning strategies have been developed as part of the measures to control floods and their after-effects.

ON-GOING IRRIGATION PROJECTS[1]

Table 32   (Potential in thousand hectares)

	MAJOR PROJECTS Number	Potential[2]	OTHER PROJECTS Number	Potential[3]	TOTAL Potential
States					
Andhra Pradesh	4	1,968	4	194	2,162
Assam	—	—	2	68	68
Bihar	8	3,171	8	277	3,448
Gujarat	5	953	9	369	1,322
Haryana	3	923	5	257	1,180
Himachal Pradesh	—	—	—	—	—
Jammu & Kashmir	—	—	1	53	53
Karnataka	6	1,467	7	316	1,783
Kerala	—	—	10	454	454
Madhya Pradesh	8	1,881	18	738	2,619
Maharashtra	10	1,405	25	1,124	2,529
Manipur	3	1,173	2	100	1,273
Meghalaya	—	—	—	—	—
Nagaland	—	—	—	—	—
Orissa	3	767	3	173	940
Punjab	3	835	1	12	847
Rajasthan	6	2,960	3	129	3,089
Sikkim	—	—	—	—	—
Tamil Nadu	1	101	1	82	183
Tripura	—	—	—	—	—
Uttar Pradesh	10	4,844	16	576	5,420
West Bengal	4	1,548	—	—	1,548
Union Territories					
Andaman & Nicobar	—	—	—	—	—
Arunachal Pradesh	—	—	—	—	—
Chandigarh	—	—	—	—	—
Dadra & Nagar Haveli	—	—	—	—	—
Delhi	—	—	—	—	—
Goa, Daman & Diu	—	—	1	15	15
Lakshadweep	—	—	—	—	—
Mizoram	—	—	—	—	—
Pondicherry	—	—	—	—	—
INDIA	74	23,996	116	4,934	28,933

Note: [1] Over 10,000 hectares of Ultimate Irrigation Potential
    [2] Over 100,000 hectares of Ultimate Irrigation Potential.
    [3] 10,000-100,000 hectares of Ultimate Irrigation Potential.

Sources: 1. *Commerce 1975*, Bombay.
        2. *Seventh Five Year Plan 1985-90, Vol: II*, Government of India Planning Commission.
        3. *National Atlas*, NATMO.

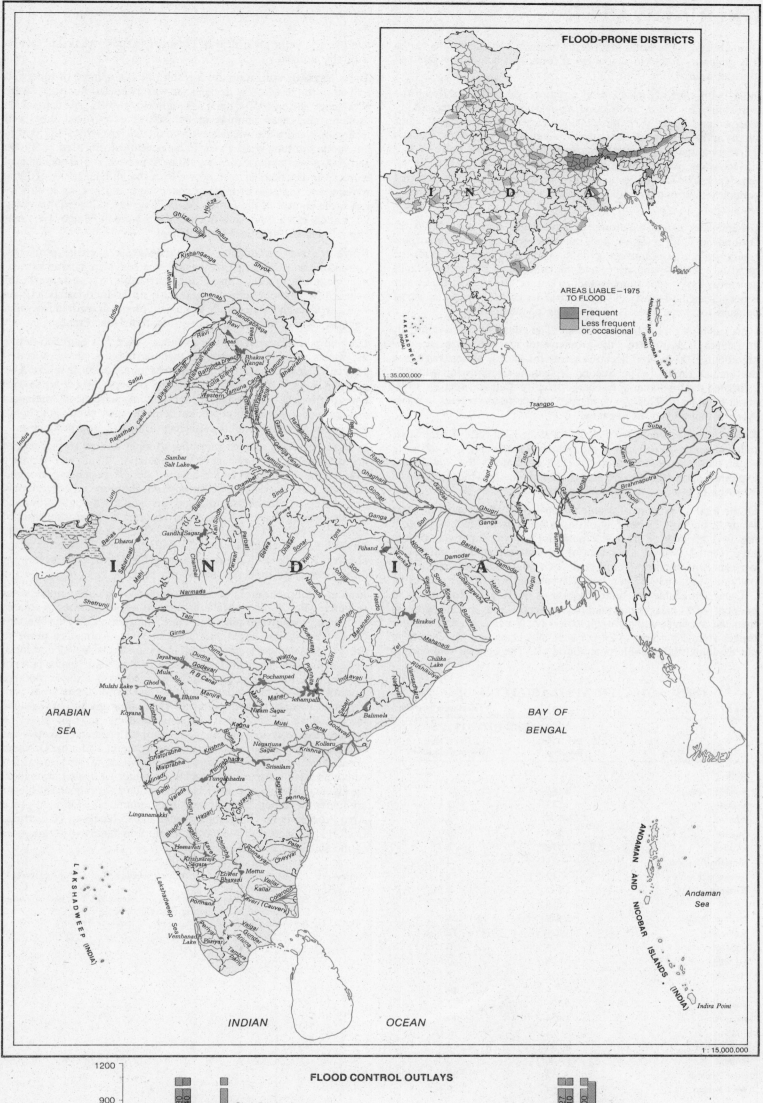

FLOOD-PRONE DISTRICTS

I N D I A

AREAS LIABLE—1975
TO FLOOD

Frequent

Less frequent
or occasional

35,000,000

LAKSHADWEEP (INDIA)

ANDAMAN AND NICOBAR ISLANDS (INDIA)

I N D I A

ARABIAN
SEA

BAY OF
BENGAL

ANDAMAN AND NICOBAR ISLANDS (INDIA)

Andaman
Sea

Indira Point

LAKSHADWEEP (INDIA)

INDIAN   OCEAN

1 : 15,000,000

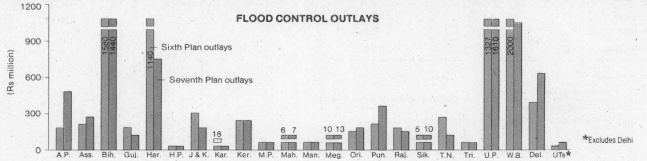

FLOOD CONTROL OUTLAYS

Sixth Plan outlays

Seventh Plan outlays

*Excludes Delhi

A.P. Ass. Bih. Guj. Har. H.P. J & K. Kar. Ker. M.P. Mah. Man. Meg. Ori. Pun. Raj. Sik. T.N. Tri. U.P. W.B. Del. UTs*

GROUNDWATER

Groundwater is the water that occurs below the surface of the earth. It is generally found in a geological formation commonly referred to as an *aquifer*.

India, with a mainly agricultural economy, depends heavily on the availability of water for a good year. Rainfall, characterized by seasonal and annual variations, is not a very reliable source of water supply in most parts of the country. Availability of groundwater is therefore a major asset that can greatly influence agriculture. While surface water potential has been used to the full in several states, the same cannot be said of groundwater. Nevertheless, the droughts that periodical failures of rainfall bring, place a great strain on groundwater resources.

Groundwater offers a number of advantages over surface water. Groundwater, for instance, does not suffer seepage losses as surface reservoirs do. Loss through evaporation is also minimal. With the rapid increase in population, and the consequent pressure on land, new areas have been brought under irrigation. But such surface water is not sufficient to satisfy the agricultural demand for water. Industrial demand for water has also been rising phenomenally.

India has high groundwater potential. The guesstimates have it that, at a depth of 300 metres, the groundwater reserves of the country are about 3700 million hectare-metres (m ha m), almost ten times the annual rainfall. The Second Irrigation Commission in 1972 estimated the total annual re-charge from rainfall and seepage to be about 67 m ha m. This can be increased with improvements in water systems that add to the recharge.

Recent investigations have indicated that the hard rock of peninsular India — nearly 70 per cent of the country's land mass — holds more water than was assumed earlier. The estimate is that the annual exploitable potential may be around 42 m ha m, of which only about 10 m ha m is being exploited at present.

Of the total 42 m ha m of usable potential, 21.9 per cent is in Uttar Pradesh, but only 29 per cent of this is being used. Madhya Pradesh has 14.1 per cent of the usable potential but only 8 per cent of it is used. Punjab, with 3.1 per cent of the usable potential, utilizes 73 per cent of it, whereas Tamil Nadu, with 6.4 per cent of the usable potential, uses only 37 per cent. Haryana utilizes 70 per cent of its 2.1 per cent usable potential. Delhi uses the largest proportion (88 per cent) of its meagre potential (0.6 per cent). Eleven states use less than the average national exploitation (23.7 per cent) and this use ranges from 1 per cent to 24 per cent. Nagaland and Manipur have not exploited their usable potential at all. The general inference is

that there is yet a great deal of potential in most states that can be fruitfully utilized.

These levels of utilization are a result of development of tube well irrigation in the last few decades. It was estimated that there were 5000 tube wells in 1950-51 and 3.9 million dug wells. The nationwide droughts and near famines in the sixties encouraged tube well technology and there was an average annual growth of 172,000 in the number of tube wells a year in the seventies compared to 50,000 and 2000 a year in the sixties and fifties respectively. The phenomenal increase in the number of tube wells in the alluvial tracts of India is related to the ease of tapping water there and placing it directly under the control of the farmer. A 1977 survey indicated that about 90 per cent of the tube wells in the country were in Punjab, Haryana, Uttar Pradesh, Bihar and West Bengal.

However, over-exploitation of groundwater can create problems. Groundwater withdrawal consistently in excess of re-charge has led, in some areas, to a lowering of the water table. Six such areas, for instance, were reported in a recent study of the Indo-Gangetic Plain — Kapurthala and Maler Kotla in Punjab, Mahendragarh in Haryana, Baghpat, Alipur and Saharanpur in western Uttar Pradesh.

Drought is a recurring disaster in India. Often it is easier to define its effects after the occurrence than during the period when it gradually, almost imperceptibly, becomes more and more severe. One problem is the relative nature of the term 'drought'. There are several types of drought — atmospheric, agricultural, permanent, contingent and seasonal — agricultural drought being the most common in India. It is a condition of rainfall deficiency affecting crop production.

In India, the hard core of recurring drought affects 16 per cent of the total area and 12 per cent of the population. In all, as of January 1986, some 90 districts in the country were experiencing drought. Although these districts are designated drought-prone, in general, only some blocks in each of these districts are regularly drought-prone and not the whole district. It has been reported that, on the basis of correlation between several parameters, three types of drought could be identified, covering about 1.55 million square kilometres in the country — extreme drought accounting for 12 per cent, severe for 42 per cent, and moderate for 46 per cent.

In all drought-prone areas, there is a compulsion to utilize groundwater resources. For instance, Maharashtra as a whole qualifies as an area with high groundwater potential. Here, the areas east of the Western Ghats (rain-shadow areas) *i.e.* Deccan Maharashtra, are prone to drought. It is in these areas that well irrigation has been well developed (irrigated sugarcane is the crop). Western Rajasthan, major parts of Gujarat, interior Karnataka, Tamil Nadu and the interior of southern Andhra Pradesh fall under this zone and in all these areas well irrigation assumes great importance. This, unfortunately, has resulted in groundwater depletion in many districts in this zone. A comprehensive system of management for efficient and effective use of groundwater has now become a necessity there. Under the Drought-prone Area Programme, efforts are being made to work towards a comprehensive water management and monitoring system throughout the country. Twenty-one districts in the country with about 131 blocks are under the Desert Development Programme. The Drought-prone Area Programme is being implemented in 615 blocks of 90 districts. In all, 111 districts with 746 blocks benefit from these two programmes of the state and central governments.

Sources: 1. *Human Response to Agricultural Drought: A Systems Approach, 1984*, Kumaran.
2. *Basic Statistics Relating to the Indian Economy, Vol. 2: States, September 1986*, Centre for Monitoring Indian Economy.
3. *Economic Times, 1-12-1985*, Times of India.

GROUNDWATER AVAILABILITY[1] — 1981

Table 33   (Hectare-metre/year)

	USABLE POTENTIAL	
	Per Taluk (by Area)[2]	Per Taluk (by population)[2]
States		
Andhra Pradesh	4,164	21
Assam	543	2.2
Bihar	9,597	2.6
Gujarat	1,895	10.9
Haryana	776	2.7
Himachal Pradesh	75	0.7
Jammu & Kashmir	42.5	1.6
Karnataka	1,112	5.7
Kerala	329	0.5
Madhya Pradesh	2,549	21.7
Maharashtra	3,321	12.9
Manipur	9	0.1
Meghalaya	38.7	0.5
Nagaland	14	0.3
Orissa	1,574	9.3
Punjab	1,171	3.5
Rajasthan	1,054	10.5
Sikkim	N.A	N.A.
Tamil Nadu	3,143	8.4
Tripura	56.2	0.3
Uttar Pradesh	7,712	20.5
West Bengal	5,754	9.4
Union Territories (Total)[3]	0.8	0.01

Note: [1] Estimated

[2] $\dfrac{\text{Total area/population of a state}}{\text{No. of taluks in the state}}$ = Average taluk area/taluk population

[3] Only collective data is available for union territories.

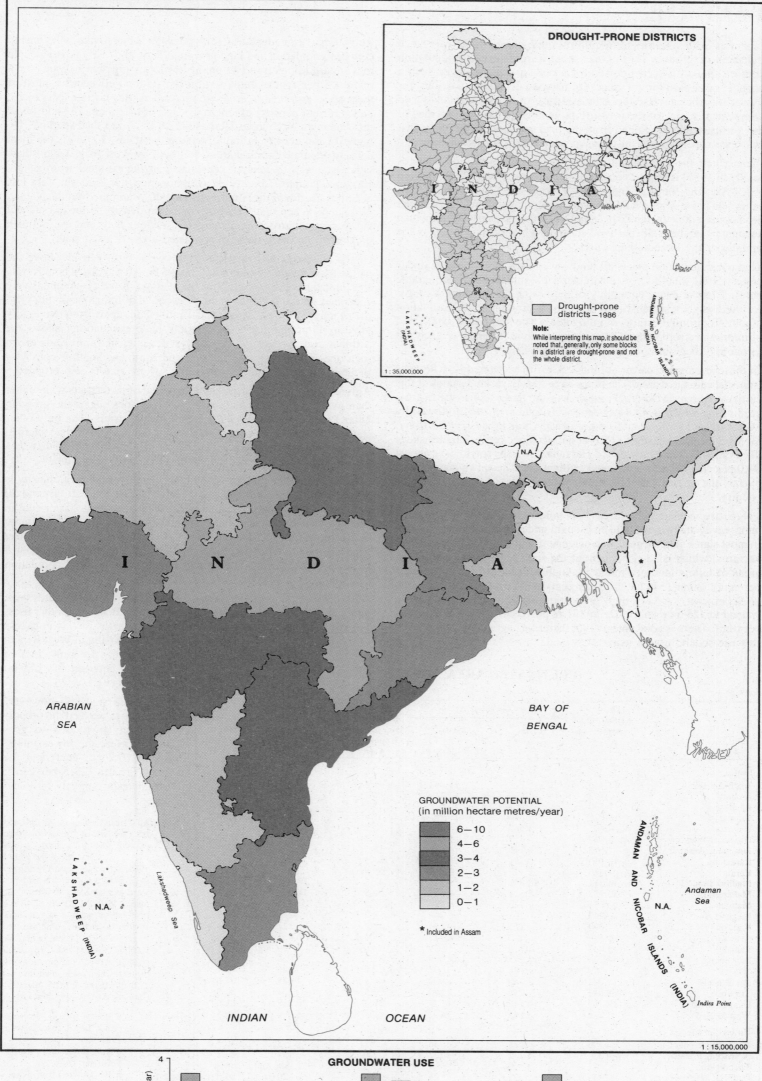

DROUGHT-PRONE DISTRICTS

I N D I A

Drought-prone
districts—1986

Note:
While interpreting this map, it should be
noted that, generally, only some blocks
in a district are drought-prone and not
the whole district.

1 : 35,000,000

N.A.

I N D I A

ARABIAN
SEA

BAY OF
BENGAL

LAKSHADWEEP
(INDIA)

Lakshadweep
Sea

N.A.

GROUNDWATER POTENTIAL
(in million hectare metres/year)

6—10
4—6
3—4
2—3
1—2
0—1

* Included in Assam

ANDAMAN
AND
NICOBAR
ISLANDS
(INDIA)

Andaman
Sea

N.A.

Indira Point

INDIAN   OCEAN

1 : 15,000,000

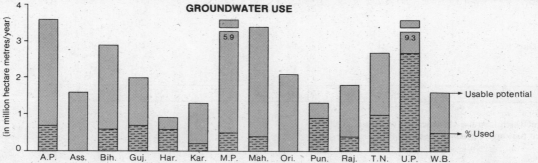

GROUNDWATER USE

(in million hectare metres/year)

5.9

9.3

Usable potential

% Used

A.P. Ass. Bih. Guj. Har. Kar. M.P. Mah. Ori. Pun. Raj. T.N. U.P. W.B.

LAND USE

India has made a mark in agricultural production in the years since independence, but is yet to make a mark in the area of population control, though it seems poised for a demographic change. Such a change is essential for the success of national economic planning, for food self-sufficiency and for better standards of living. And this should come about soon, for the pressure on land is increasing. An idea of this pressure may be gained from the land use data for the states that has been superimposed in the map over rural population density.

Almost half the arable land of the world (1371.6 million hectares totally in 1983) is in four countries, namely the USSR (227.5 million ha), USA (187.9 m ha), India (164.9 m ha) and China (97.5 m ha). India has more arable land than more populous China, yet enough has not been done on the farm front. Nevertheless, agriculture, which accounts for about 48 per cent of India's national income, is the kingpin of the national economy.

That agriculture dominates all land use is clear from the area under crops in 1981-82 (43.2 per cent or about 142 million hectares of arable land). There is every possibility of increasing the area under crops, for there is yet a sizeable area that can be brought under cultivation (about 40 million hectares of cultivable waste and fallow land), and of improving crop production, with productivity-oriented science and technology beginning to take root in the villages.

Of the total geographical area of India, 328.7 m hectares, 142 m hectares is the net cultivated area. The states have widely varying shares of this cultivated area. Madhya Pradesh has the largest net sown area, 18.8 m ha, which is 42.4 per cent of its total area. It is followed by Maharashtra (18.3 m ha; 59.6 per cent), Uttar Pradesh (17.3 m ha; 58.8 per cent), Rajasthan (15.6 m ha; 45.6 per cent). Smaller states like Punjab (4.2 m ha sown) and Haryana (3.7 m ha sown) have almost 84.0 per cent of their land under cultivation. These two states have, by far, the largest proportion of net sown area in any state in the country.

There are considerable areas of fallow land (being cultivated temporarily and which could be brought under permanent cultivation) in most states. The largest such acreage in India is in Rajasthan, 4 m hectares, which is 11.8 per cent of the state area. Other significant areas of fallow land are found in Andhra Pradesh (3.4 m ha; 12.4 per cent), Bihar (3 m ha; 17.2 per cent) and Tamil Nadu (2.2 m ha; 16.9 per cent). A very high proportion of fallow land is found in Nagaland (25 per cent), Delhi (21.6 per cent) and Mizoram (19 per cent). There is good potential for increasing agricultural production in these states through using such land.

Rural population densities provide some clue to the relationship between population and land use. India's rural density is 165 people to a square kilometre. The union territories together account for a mere 0.5 per cent of India's rural population yet, because most of them are constricted in area, they have densities far in excess of the national rural density, except in the case of Mizoram (18/sq km) and the Andaman & Nicobar Islands (17/sq km). Lakshadweep has the highest rural density in India. Among the states, Kerala is the most densely populated in rural areas, followed by West Bengal and Bihar. Other states with greater densities than the national average are Haryana, Punjab, Tamil Nadu, Tripura and Uttar Pradesh. Very low densities are found in Himachal Pradesh, Manipur, Meghalaya and Nagaland. When district-wise population density is looked at, and cultivation patterns are studied, it is obvious that land use is much greater in the more densely populated areas of each state.

Of the country's 439 districts, 209 have a rural density above the national average (199 out of 412 in 1981). All the districts in Kerala and Punjab have densities above the average. Kerala is an intensively cultivated state and Punjab is the wheat bowl of India (maximum yield per hectare). In West Bengal and Tamil Nadu, apart from two totally urbanized metropolises, all districts have populations above the national average density. The highest densities in Tamil Nadu are in such rice-rich districts as Thanjavur and North and South Arcot. West Bengal is a major jute-growing area, where rice is also cropped.

Agriculturally rich Haryana has 83 per cent of its districts with higher than average rural population density. So does sugarcane-growing Uttar Pradesh which has 79 per cent such districts. The only other state with over 50 per cent of its districts having higher than average rural population density is Bihar (89 per cent), which, curiously, has a poor agricultural record but leads the county in mining.

In other states, the district-wise higher than average rural density is 50 per cent in Manipur, 43 per cent in Andhra Pradesh, around 30 per cent in Himachal Pradesh and Karnataka, around 20 per cent in Maharashtra, Orissa and Rajasthan and around 15 per cent in Jammu & Kashmir (36 per cent in 1981). India's largest state, Madhya Pradesh, population-wise, is comparatively well off, only about 5 per cent of its districts having higher than average rural population densities, but it is poorly off agriculturally.

With the exception of Orissa and Bihar all the 15 major states have larger areas under conventional agricultural use than under other uses. For instance, in Punjab and Haryana, the net area sown is in excess of 80 per cent and in West Bengal it is about 63 per cent. In five states it is between 50 and 60 per cent.

In the smaller states, however, forest areas far exceed the proportion of net sown area to geographical area, the lowest being Manipur (6.2 per cent) and the highest Tripura (23.5 per cent). These states are mostly hilly and forested.

Note: All Census data are dated 1981 and refer to the 412 districts that then existed. The maps show the districts as they existed at the end of 1985, when there were 439 districts (437 districts for which boundaries were available are shown on the maps). Where 1981 districts have been broken up into two or more smaller districts or combined into larger districts (as in Jammu & Kashmir) the 1981 data are used to shade the 1985 configurations. In the text, which refers to the districts as they existed in 1985, the 1981 status is given within brackets.

Sources: 1. *Primary Census Abstract, Part II B(i)*, Census of India 1981.
2. *Fertiliser Statistics 1984-85*, Fertiliser Association of India. Quoting from Directorate of Economics & Statistics, Ministry of Agriculture and *1984 FAO Production Year Book, Vol. 38*, FAO, Rome.

CULTIVATED AREA[1] — 1981-82

Table 34

(Area in thousand hectares)

| | Total Reporting Area[2] | Cultivated Area | | % Net Cultivated Area to Total Reporting Area | Area under Food Grain | % Area under Food Grain to Gross Cropped Area |
		Net	Gross[3]			
States						
Andhra Pradesh	27,440	11,325	13,047	41.3	9,222	70.7
Assam	7,852	2,696	3,439	34.3	2,505	72.8
Bihar	17,330	7,861	10,628	45.4	9,555	89.9
Gujarat	18,826	9,670	10,903	51.4	4,743	43.5
Haryana	4,405	3,660	5,826	83.1	4,342	74.5
Himachal Pradesh	3,089	573	949	18.5	856	90.2
Jammu & Kashmir	4,675	716	978	15.3	834	85.3
Karnataka	19,050	10,391	11,228	54.5	7,502	66.8
Kerala	3,885	2,170	2,905	55.9	846	29.1
Madhya Pradesh	44,211	18,841	21,756	42.6	17,882	82.2
Maharashtra	30,758	18,314	20,386	59.5	14,219	69.7
Manipur	2,211	140	240	6.3	178	74.2
Meghalaya	2,249	193	203	8.6	134	66.0
Nagaland	1,099	153	164	13.9	127	77.4
Orissa	15,540	6,130	8,743	39.4	6,632	75.9
Punjab	5,033	4,210	6,929	83.6	4,996	72.1
Rajasthan	34,234	15,577	18,596	45.5	13,026	70.0
Sikkim	719	86	92	12.0	66	71.7
Tamil Nadu	13,002	5,740	6,909	44.1	4,628	67.0
Tripura	1,048	246	380	23.5	304	80.0
Uttar Pradesh	29,708	17,288	24,773	58.2	20,218	81.6
West Bengal	8,846	5,565	7,402	62.9	5,986	80.5
Union Territories						
Andaman & Nicobar	790	33	36	4.2	13	36.1
Arunachal Pradesh	5,530	112	152	2.0	134	88.2'
Chandigarh	11	3	4	27.3	N.A.	N.A.
Dadra & Nagar Haveli	49	23	25	46.9	20	80.0
Delhi	147	56	84	38.1	69	82.1
Goa, Daman & Diu	371	132	142	35.6	60	42.3
Lakshadweep	3	3	3	100.0	N.A.	N.A.
Mizoram	2,102	65	68	3.1	33	48.5
Pondicherry	47	30	51	63.8	40	78.4
INDIA	304,280	142,002	177,041	46.7	129,138	72.9

Note: [1] Provisional.
[2] Area for which land use classification of area is available.
[3] Net area plus area cultivated more than once.

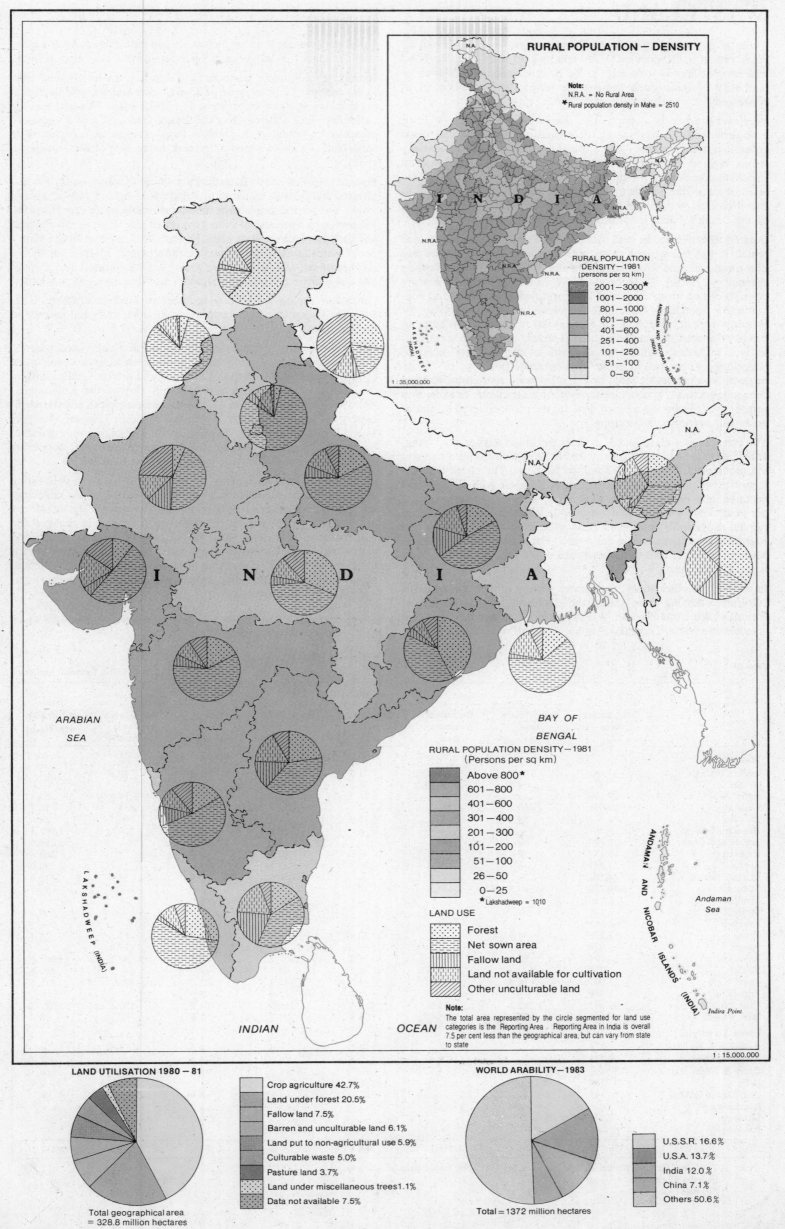

RURAL POPULATION — DENSITY

Note:
N.R.A. = No Rural Area
*Rural population density in Mahe = 2510

RURAL POPULATION
DENSITY—1981
(persons per sq km)

2001—3000*
1001—2000
801—1000
601—800
401—600
251—400
101—250
51—100
0—50

1 : 35,000,000

ARABIAN
SEA

BAY OF
BENGAL

RURAL POPULATION DENSITY—1981
(Persons per sq km)

Above 800*
601—800
401—600
301—400
201—300
101—200
51—100
26—50
0—25
*Lakshadweep = 1010

LAND USE

Forest
Net sown area
Fallow land
Land not available for cultivation
Other unculturable land

Note:
The total area represented by the circle segmented for land use categories is the Reporting Area . Reporting Area in India is overall 7.5 per cent less than the geographical area, but can vary from state to state

1 : 15,000,000

INDIAN   OCEAN

LAKSHADWEEP (INDIA)

ANDAMAN AND NICOBAR ISLANDS (INDIA)

Andaman
Sea

Indira Point

LAND UTILISATION 1980 – 81

Crop agriculture 42.7%
Land under forest 20.5%
Fallow land 7.5%
Barren and unculturable land 6.1%
Land put to non-agricultural use 5.9%
Culturable waste 5.0%
Pasture land 3.7%
Land under miscellaneous trees1.1%
Data not available 7.5%

Total geographical area
= 328.8 million hectares

WORLD ARABILITY—1983

U.S.S.R. 16.6%
U.S.A. 13.7%
India 12.0%
China 7.1%
Others 50.6%

Total = 1372 million hectares

WASTELAND

There is an increasing awareness and concern in India about the environment and the extent of its relationship with land, man and technology. This has resulted in the establishment of a Wasteland Authority of India, with the express purpose of development of wasteland for social and economic uses.

Whatever the reason — population pressure, intensive use of land, crop unsuitability, overgrazing, deforestation, excessive fuel wood collection, or soil erosion — the cause of land degradation is essentially its use beyond its suitability and capability. Land degradation is an insidious process; by the time a community becomes aware of the seriousness of the problem, it often no longer has the capacity to mobilize itself to take remedial measures. In India, 0.5 - 1.0 per cent of the country's area is turning into wasteland every year.

Wasteland is defined as land that is at present lying unused; or land which is not being used to its optimum potential due to various constraints; or land which cannot be used (National Remote Sensing Agency — NRSA — Hyderabad). Wasteland in India therefore consists of two broad classes of land: culturable wasteland and non-culturable wasteland. Culturable wasteland is capable of, or has the potential for, development for agricultural or pastoral purposes or can be afforested. It is not being used at present due to such constraints as lack of water, salinity or alkalinity of the soil, soil erosion, waterlogging, an unfavourable physiographic position, or human neglect. Non-culturable land, on the other hand, is barren land and cannot be put to any productive use, either for agriculture or to develop forest cover. Examples of such land are the snow-covered or glacial areas and barren rock outcrops.

The extent of wasteland in the 23 states and union territories of India, according to the latest estimate, by the NRSA, which has been studying the problem in depth, is 53.3 million hectares. The states with the largest areas of wasteland are Jammu & Kashmir with 13.3 million hectares, and Rajasthan with 12.9 m ha. They comprise nearly 50 per cent of the total national wasteland. Whereas almost the entire Jammu & Kashmir wasteland cannot be cultivated, most of the Rajasthan wasteland can be cultivated. The wasteland in Jammu & Kashmir is 60 per cent of the total area of the state and in Rajasthan it is 37.7 per cent.

In most of the other states, the greater portion of their wasteland is culturable. Among these states, those with the largest amount of wasteland are Uttar Pradesh, 4.3 m ha, of which about 65 per cent is culturable waste, Gujarat 3.3 m ha, Andhra Pradesh 2.4 m ha and

Himachal Pradesh 2 m ha (almost all non-culturable). All the other states account for only about 28 per cent of the wasteland in India.

Rajasthan's culturable wasteland (11.9 m ha), the largest such area in the country, is the expanse of desert, semi-arid and arid lands that occupy almost the entire western part of the state. This will become culturable once schemes like the Indira Gandhi Canal Project are completed. Jammu & Kashmir's huge acreage of non-culturable wasteland is snow-covered or glacial, being part of the Himalayan chain.

Gujarat ranks second to Rajasthan in culturable waste, most of it salt-affected and perhaps a quarter of it undulating upland with or without scrub. The second-largest area of non-culturable waste is in Himachal Pradesh, again because of snow-covered or glacial areas. Sikkim also has such snow-covered or glacial areas, 27.5 per cent of the state's geographical area being unsuitable for any use. A large number of patches of salt-affected land are found in Uttar Pradesh and Gujarat and this is attributed to over-irrigation resulting in salt accumulation.

Gullied or ravinous land predominates in Madhya Pradesh, Uttar Pradesh, Rajasthan, Maharashtra and in small scattered patches in the south.

The *jhum* or forest blanks (*jhum* is the shifting cultivation practice prevalent in the Assam hills and is often referred to as *kumri* in the south) occupy parts of the Western and Eastern Ghats (shifting cultivation is generally confined to hill slopes: the Kollaimalai, Kalrayan, and Javadi hills in Tamil Nadu are examples), and the north-eastern hill areas (in Assam, Arunachal Pradesh, Tripura, Manipur, Meghalaya and Mizoram). In these areas, forests are often felled or burnt for cultivation. When the crop is reaped, the wandering cultivator moves on, leaving behind ravaged land.

The National Remote Sensing Agency, which has been delineating areas of wasteland through satellite imageries and ground checking, has suggested a national wasteland monitoring system. The indications are that this system will be flexible enough to include changes that will become necessary with the increasing awareness of land degradation and alternatives suggested. This monitoring system for the moment focuses on four types of wasteland which are subject to rapid changes: namely, saline areas, waterlogged areas, the ravinous areas and the forest blanks.

Source: *Mapping of Wastelands in India from Satellite Imagery 1980-82*, National Remote Sensing Agency.

WASTELAND IN INDIA — 1980-82

Table 35

(in thousand hectares)

| | Geographical area | Culturable | | | | | | Non-culturable wasteland |
		Salt-affected land	Gullied or ravinous land	Water-logged or marshy land	Undulating upland with or without scrub	Jhum or forest blanks	Sandy area (coastal or desert)	
States								
Andhra Pradesh	27,682	66	206.8	71.3	1,667.6	11.2	40.4	404
Assam	7,852	6.1	30.3	36.2	118.6	—	—	—
Bihar	17,388	42.2	63.7	128.7	578.9	69.2	19.9	105.1
Gujarat[1]	19,598	2,060.1	316	—	837.8	—	38	58.7
Haryana	4,422	69.4	—	25.9	12.4	—	166.1	59.6
Himachal Pradesh	5,567	—	—	—	45.9	38.4	—	1,957.4
Jammu & Kashmir	22,224	—	—	6.1	11.3	68.4	—	13,280.1
Karnataka	19,177	32.4	78.9	—	1,182.4	79.9	1.3	268.8
Kerala	3,887	—	—	22.3	27.4	113.1	23.7	39.9
Madhya Pradesh	44,284	—	1,247.2	—	1,056.6	289.8	6.9	311.4
Maharashtra	30,776	90.3	430.5	—	2,324.5	45.5	—	90.7
Manipur	2,236	—	—	—	12	308.9	—	—
Meghalaya	2,249	—	—	—	—	158.9	—	9
Nagaland	1,653	—	—	—	—	158.6	—	—
Orissa	15,578	4.9	—	20.4	905.9	196.2	18.3	8.4
Punjab[2]	5,036	123.1	—	45	4	2.2	177.8	—
Rajasthan	34,221	62.9	915.9	28.4	1,067.7	14.4	9,789.2	1,061.6
Sikkim	730	—	—	—	69.3	—	—	200.4
Tamil Nadu	13,007	16.8	62.6	44.5	545.7	123.5	106.7	111.5
Tripura	1,048	—	—	—	17.8	75.9	—	—
Uttar Pradesh	29,441	1,282.3	995.8	220.4	116.5	61.2	130.1	1,511.7
West Bengal	8,785	51	5	234.3	245.7	8.2	1.7	18
Union Territories								
Andaman & Nicobar	829	—	—	2	6.4	4.4	2.5	—
Arunachal Pradesh	8,358	—	0.4	0.4	14.6	244.1	—	951.2
Dadra & Nagar Haveli	N.A.	N.A.	N.A.	N.A.	N.A.	N.A.	N.A.	N.A.
Delhi	149	0.6	—	0.9	3	—	—	9.7
Goa, Daman & Diu	381	—	—	—	8	0.4	3.8	—
Lakshadweep	N.A.	N.A.	N.A.	N.A.	N.A.	N.A.	N.A.	N.A.
Mizoram	2,109	—	—	—	—	210.6	—	—
Pondicherry	48	1.2	—	—	0.4	—	4	—
INDIA[3]	328,715	3,909.3	4,353.1	886.8	10,880.4	2,283	10,530.4	20,457.2

Note: [1] Total area includes Daman & Diu, but excludes 1,471,800 ha. of Rann of Kachchh.
[2] Includes Chandigarh.
[3] Estimated.

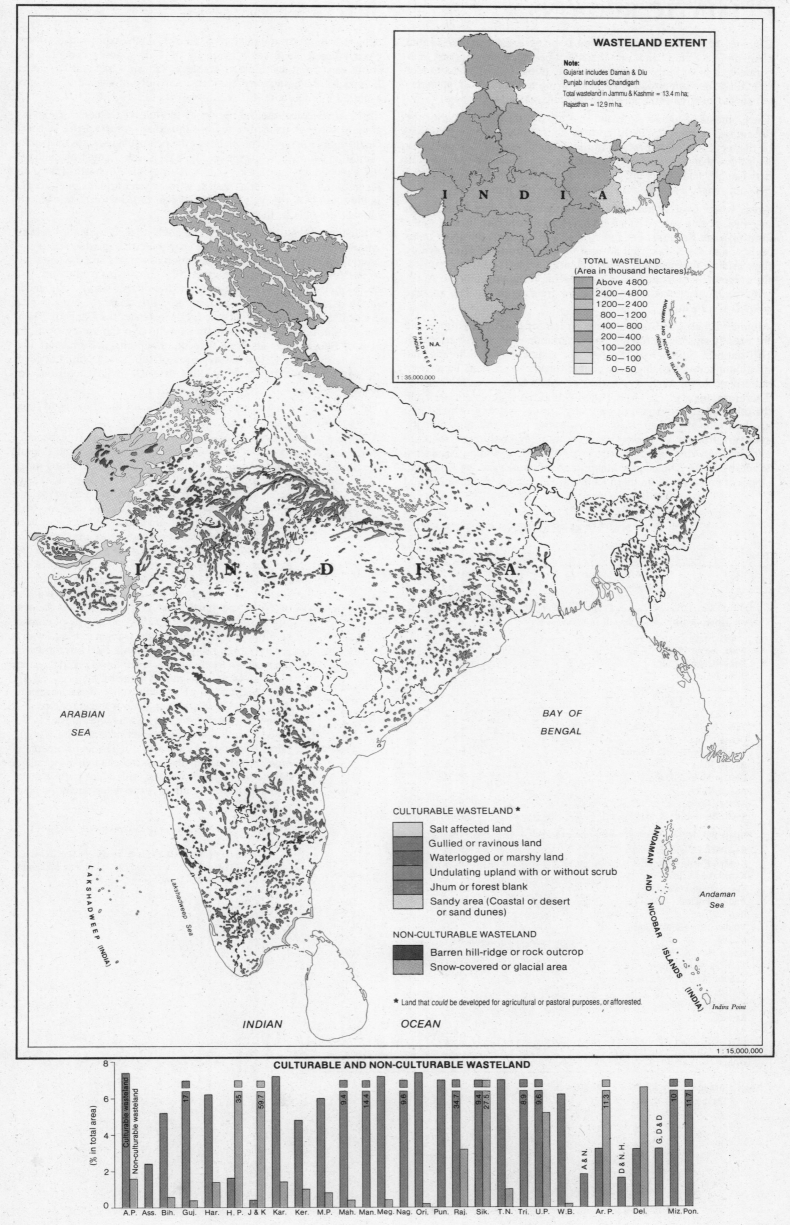

WASTELAND EXTENT

Note:
Gujarat includes Daman & Diu
Punjab includes Chandigarh
Total wasteland in Jammu & Kashmir = 13.4 m ha;
Rajasthan = 12.9 m ha.

I N D I A

TOTAL WASTELAND
(Area in thousand hectares)
- Above 4800
- 2400 — 4800
- 1200 — 2400
- 800 — 1200
- 400 — 800
- 200 — 400
- 100 — 200
- 50 — 100
- 0 — 50

LAKSHADWEEP (INDIA)

N.A.

ANDAMAN AND NICOBAR ISLANDS (INDIA)

1 : 35,000,000

ARABIAN SEA

I N D I A

BAY OF BENGAL

LAKSHADWEEP (INDIA)

Lakshadweep Sea

ANDAMAN AND NICOBAR ISLANDS (INDIA)

Andaman Sea

Indira Point

CULTURABLE WASTELAND *
- Salt affected land
- Gullied or ravinous land
- Waterlogged or marshy land
- Undulating upland with or without scrub
- Jhum or forest blank
- Sandy area (Coastal or desert or sand dunes)

NON-CULTURABLE WASTELAND
- Barren hill-ridge or rock outcrop
- Snow-covered or glacial area

* Land that *could* be developed for agricultural or pastoral purposes, or afforested.

INDIAN   OCEAN

1 : 15,000,000

CULTURABLE AND NON-CULTURABLE WASTELAND

(% in total area)

Culturable wasteland
Non-culturable wasteland

A.P. Ass. Bih. Guj. Har. H.P. J & K Kar. Ker. M.P. Mah. Man. Meg. Nag. Ori. Pun. Raj. Sik. T.N. Tri. U.P. W.B.   Ar. P.   Del.   Miz. Pon.

A.N.   D & N.H.   G, D & D

CLIMATIC REGIONS

Climates (the term refers to regions) bring together the essential elements of weather conditions that affect people, their work and their comfort: temperature, and the amount, duration and incidence of rainfall. More importantly, climates bring together those elements which are of greatest concern to the cultivators of the land.

The elements of temperature and rainfall are, thus, the basis for the differences in lifestyles in different regions. Four climatic types may be distinguished for the whole of the country: *Humid* — those areas with more than 1000 mm of rainfall a year; *Sub-humid* — those with between 500 and 1000 mm; *Semi-arid* — those with between 250 and 500 mm; and *Arid* — with rainfall below 250 mm a year.

Areas where average temperatures are below freezing for a month or more, posing severe constraints on agriculture, are limited to the Himalaya and Kashmir. The inner regions of Kashmir, the upper Indus Valley, and the Karakoram are distinguished from the Himalaya by their aridity; for example, in the Leh area, with its cold dry climate, the minimum temperature is below freezing from November to April, with the January maximum also below freezing point. The average rainfall is 116 mm.

Altitude has an important bearing on temperature, but latitude and the equally significant topographical situation cannot be overlooked. Hill stations, Shimla and Darjiling for instance, record an absolute minimum below freezing point, but air drainage prevents them from suffering cold for long, despite their elevation being over 2100 metres Therefore, these places are not classified as having cold climates. On the other hand, the length of the rainy season — during which there are two relevant regimes, one the summer rains and the other the summer drought — gives rise to some regional types that are agriculturally distinct. The extreme north-east, including Arunachal Pradesh, Assam, Nagaland, Manipur and Tripura, and the windward slopes of the Ghats in southern Kerala are all categorized under humid climate with maximum rainfall in summer. At the same time, northern coastal Tamil Nadu has a tendency towards summer drought. The entire west coast and most of Madhya Pradesh, Orissa, Bihar, West Bengal and the eastern and Himalayan foothills of Uttar Pradesh have a humid climate with 4-7 wet months.

Three areas constitute the region of sub-humid climate. Occupying about a third of India, the region immediately east of the Western Ghats but in the northern reaches of the peninsula receives maximum rainfall in summer, while the region in the south, in the rain-shadow, has a tendency to summer drought. A small area in north-west India receives rains in summer as well as winter. Semi-arid climate prevails in the south of north-west India (Rajasthan and Gujarat) while an arid climate prevails in the Thar desert.

While the cultivator (in most of India) looks to the clouds for the rains which will enable him to plant his crops, the soil in which he plants them is itself a product of the interaction of climate and the materials of the land surface. The least affected by climate is the alluvium, but crops on it still depend on the rains and the surface waters impounded in the dams. Besides soil and rain there are certain features which are of local agricultural relevance. For instance, the Gangetic Plain, despite the funnelling of the monsoon through the valley from the Bay of Bengal, gets less rain than the mountains and plateaux that flank it, so much so that a small area of sub-humid climate occurs in western Bihar and eastern Uttar Pradesh and is notorious for famines.

The map simplifies these divisions and presents the climates of India in six distinct types.

A *tropical wet* climate is characteristic of the west coast, especially in Kerala, Karnataka and Goa, where both temperature and rainfall are high. A *tropical wet and dry* climate is mainly found along much of the east coast and in the interior of the northern peninsula, with a tongue of such climate penetrating southwards into Maharashtra and Karnataka. In this region, both temperature and rainfall are important for land-related activities, with rainfall perhaps more important, for it does not greatly alter crop rhythms. There are two patches of *semi-arid* areas, one in the south immediately east of the Western Ghats where the rain-shadow has the effect of causing drought, and the other in the north-west circling the desert, *arid* climate type. The *humid sub-tropical* type of climate is found in the entire Indo-Gangetic Plain which receives high rainfall. The rainfall declines towards the west where the waters from the Gangetic system make up for it. This makes it one of the most densely populated and most agriculturally worked areas of the world. There are humid sub-tropical regions also in the extreme north-east, where the rainfall is the highest, and in a small pocket of land with a slight rain-shadow (the Assam Valley). The undifferentiated *highlands* stretching the length of the Indian Himalaya have cold climates but are arid. Here temperature, because of the altitudes, is the determinant of plant growth.

Sources: 1. *India, Resources and Development, 1978:* B.L.C. Johnson.
2. *Basic Statistics Relating to the Indian Economy, Vol.2; States, September 1985,* Centre for Monitoring Indian Economy.

NATURAL REGIONS OF INDIA

Table 36

Region	Area thousand sq. kms.	Density — 1981 (persons per sq. kms.)
Population density above the all-India average		
Lakshadweep	0.03	1,333
West Bengal Plain	60.9	696
Kerala Coastal Plain	38.9	655
Bihar Plain	94.1	556
Konkan Coastal Lowland	30.9	494
Uttar Pradesh Plain	202.6	487
West Bengal Upland	24.7	452
Tamil Nadu Coastal Plain	69.0	418
Orissa Coast	27.6	355
Tamil Nadu Hills & Upland	61.4	328
Punjab Plain	89.7	316
Gujarat Plain & Dangs	57.0	312
Assam Valley	63.3	304[2]
Andhra Coast	93.0	255
South Bihar Hills & Plateaux	79.8	221
Population density below the all-India average		
Rajasthan Hills & Plateaux	153.6	205
Konkan-Kerala Transition	22.4	199
Karnataka Deccan	173.1	195
Maharashtra Deccan	277.3	172
Andhra Pradesh	182.2	164
Bundelkhand	66.4	162
Vindhya Ranges & Plateaux	75.0	134
Malwa	80.5	133
Orissa Hills & Plateaux	128.1	129
Kachchh & Kathiawar	139.1	188
Central Madhya Pradesh Plateaux	262.3	107
West Himalaya[1]	329.0	73
North-eastern Ranges	108.3	62
Rajasthan Plain	195.0	56
Andamans	8.2	23
East Himalaya	93.9	21
INDIA	3,287.3[1]	208[2]

Note: [1] Includes areas occupied by Pakistan & China.
[2] Estimated.

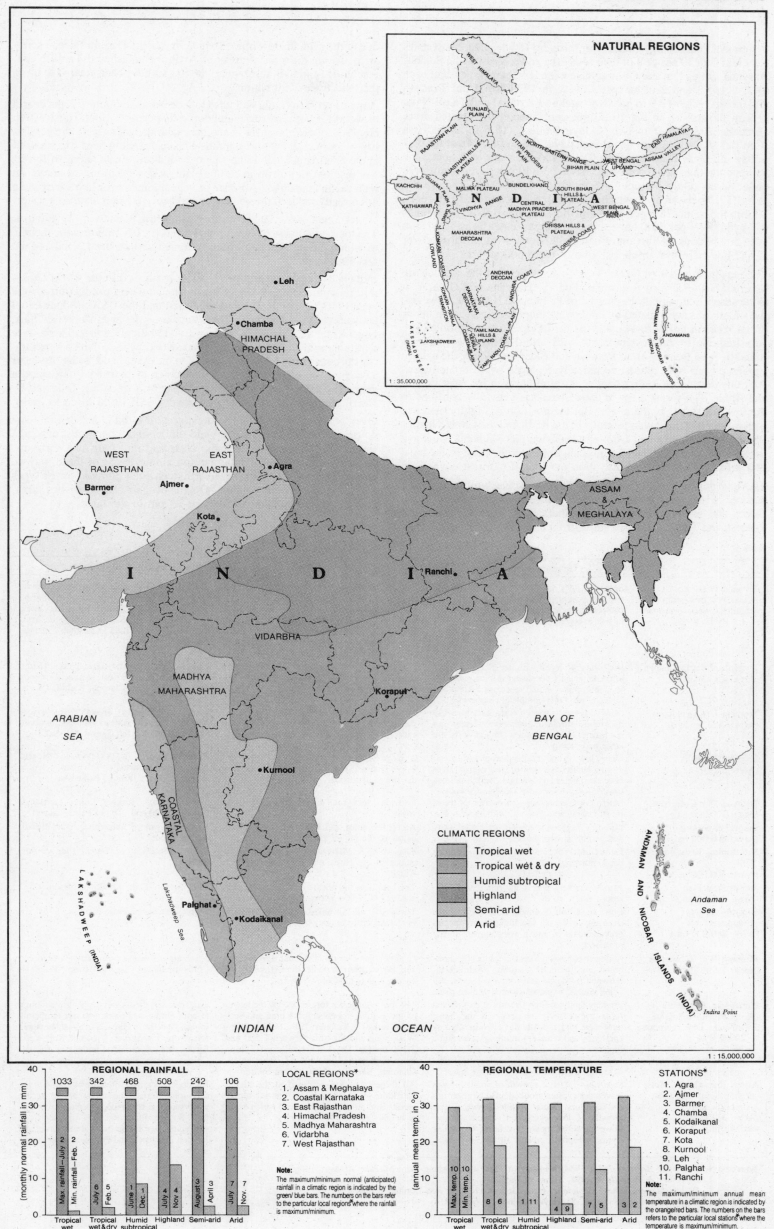

FORESTS

In terms of land ecology, Indian forest cover is inadequate. Most states have less than 30 per cent of their total area under forest while Punjab, Haryana, Rajasthan and Gujarat have even less than 10 per cent each (0.2, 0.16, 3 and 2 m ha respectively in 1979-80). Uttar Pradesh (5.1 m ha), Bihar (2.9 m ha), Karnataka (3.8 m ha) and Tamil Nadu (2.2 m ha) are all in the 10-20 per cent bracket of forested area. Himachal Pradesh (2 m ha) Madhya Pradesh (15.5 m ha), Orissa (6 m ha), Assam (2.8 m ha), Meghalaya (0.8 m ha), Mizoram (0.7 m ha) and Goa, Daman & Diu (0.1 m ha) are somewhat above the national average (22.8 per cent), having 30-40 per cent of their area under forest. Only Manipur 1.5 m ha (67.6 per cent), Tripura 0.6 m ha (57.2 per cent), Arunachal Pradesh 5.2 m ha (61.5 per cent), Sikkim 0.5 m ha (76.8 per cent) and the Andaman & Nicobar Islands 0.7 m ha (89.8 per cent) have substantial areas under forest. To Delhi goes the dubious distinction of being a territory without forest cover at all, but then it is an almost wholly metropolitan area.

The reasons for lack of forest cover in many parts of the country are several. In the case of Rajasthan, Punjab, Haryana and Gujarat, it is the absence of rainfall. In the case of Punjab and Haryana, however, the situation is accentuated by land being increasingly brought under the plough (in northern Punjab, the forests along the foothills have vanished due to deforestation and cultivation). This is a condition found in West Bengal, Uttar Pradesh and Bihar too, where the natural vegetation has been almost completely replaced by cultivated area. And this despite the rather copious rainfall enjoyed by West Bengal and Bihar. The forest areas in these states are generally confined to the low hill slopes in the north. But the position is slightly better in Bihar, with its Himalayan foothills in the north and the Chota Nagpur Plateau in the south; and Uttar Pradesh, with its Kumaon Himalaya in the north and the Chambal ravines in the south. Both states have two areas unsuitable for cultivation but ideal for forest cover.

The low per cent of forest area in Jammu & Kashmir is due to a number of reasons, the important ones being low rainfall, steep barren slopes and snow covered peaks. Maharashtra, Karnataka, Andhra Pradesh and Tamil Nadu also suffer from lack of forest cover, mainly

due to the rain-shadow effect (which makes the Deccan Plateau semi-arid). Along the coastal areas of Andhra Pradesh and Tamil Nadu, where the rainfall is substantial, the forests have been cleared to bring the land under cultivation.

A major problem India will have to face in the near future is the dearth of wood for both industrial and domestic purposes (fuel, construction etc). Such a dearth ushers in another major danger as well: the possibility of an overall drop in rainfall. This has already been experienced in the Nilgiris in South India where the indiscriminate felling of forests has resulted in a drop in rainfall and, when the rains do come, an increase in landslides. Absence of vegetational cover in most of India has already reduced the soil cover and hence the percolation of water.

Government is the largest timber producing agency in the country. The total production of wood is estimated at 21.5 million cubic metres, of which 7.6 million cubic metres constitute industrial wood and the rest fuel wood.

India is by far the largest producer of hardwood in the world (203.4 million cubic metres). The average annual world production for the year 1980-82 was 1695 m cu m. Brazil and the USA produce equal amounts (153 m cu m each). These three countries and Indonesia, China, Nigeria, USSR, Malaysia, Tanzania and Philippines together produce a little more than 1000 m cu m of hardwood, while all the other countries produce only 694.95 m cu m. On the other hand, India produces 17.5 m cu m of the world's 437 m cu m of industrial hardwood a year. The USA is the largest producer (74.3 m cu m), followed by Brazil (35 m cu m) and Malaysia and USSR (30.6 m cu m each).

Over 16 per cent of the total forest area in India is inaccessible and 45 per cent of this area could yield timber and other forest produce. Forests in India may be broadly classified into four major *groups:* tropical, sub-tropical, temperate and alpine. Each of these may be further divided into forest *types.* Sixteen such climatic forest types are recognized in the country. Details of their characteristics, area, and the species found in them are set out in the table.

Note: m cu m = million cubic metres.

Table 37

FOREST TYPES IN INDIA

Forest Types	Characteristics	Area	Some Important Species
Tropical Evergreen (Area: Wet evergreen — 4.5 m. ha; Semi-evergreen — 1.9 m. ha.)	Both are lofty, dense forests with a large number of species and numerous epiphytes; climbers heavy in semi-evergreen, though few in wet evergreen forests.	Wet and semi-evergreen forests found in the Andamans. Wet evergreen also along western face of the W.Ghats and in a strip running south-west from Upper Assam through Cachar. Semi-evergreen occurs along the west coast, in Assam, Orissa and the lower slopes of the Himalaya.	Mesua, White Cedar, Hopea and Bamboo occur in both semi- and wet evergreen forests. Specific to wet evergreen are Jamun, Canes. Semi-evergreen: Kadam, Irul, Laurel, Rosewood, Haldu, Indian Chestnut, Champa, Mango.
Tropical Moist Deciduous (Area: 23.3 m. ha.)	Multi-layered forest with irregular top storey of predominantly deciduous species, a definite second storey of mixed species and a shrubby undergrowth fairly complete; climbers heavy.	Andamans, moister parts of Uttar Pradesh, Madhya Pradesh, Gujarat, Maharashtra, Karnataka and Kerala.	Padauk, Badam, Kokko, Teak, Laurel, Haldu, Rosewood, Bijasal, Irul, Amla, Common Bamboo, Sal, Pula, Jamun, Mahul, Canes.
Tropical Dry Deciduous (Area: 29.2 m. ha.)	Multi-layered forest, almost entirely deciduous; shrubs and bamboos present, but not luxuriant; climbers few, though some are large and woody.	Irregular wide strip running north-south from the foot of the Himalaya to Cape Comorin except in Rajasthan, W. Ghats and W. Bengal.	Teak, Axlewood, Bijasal, Rosewood, Amaltas, Palas, Haldu, Common Bamboo, Red Sanders, Laurel, Satinwood, Sal.
Tropical Thorn (Area: 5.2 m. ha.)	Open, low, pronouncedly xerophytic forest, thorny, leguminous species predominate; trees have short boles and low branches; ill-defined lower storey of smaller trees and shrubs.	A large strip in south Punjab, Rajasthan, Upper Gangetic Plain, the Deccan Plateau and the lower peninsular India.	Khair, Reunjha, Axlewood, Neem, Sandalwood, Nirmali, Dhaman, *Acacia senegal*, Khejra, Kanju, Palas, Ak.
Tropical Dry Evergreen (Area: 700,000 ha.)	Single complete canopy of evergreen trees, mostly coriaceous leaved, with short boles.	Restricted to a small area along the Konkan coast.	Khirni, Jamun, Kokko, Ritha, Tamarind, Neem, Machkund, Toddy Palm, Canes.
Littoral & Swamp (Area: 600,000 ha.)	Mainly evergreen species of varying density and height, always associated with wetness.	Along the coast and swamp forests in the deltas of the bigger rivers.	Sundri, *Bruguiera, Sonneratia,* Agar, Bhendi, Keora, *Nipa.*
Sub-tropical Broad-leaved Hill (Area: 300,000 ha.)	Luxuriant forests, evergreen species predominating.	Khasi, Nilgiri, Mahabaleshwar and the lower slopes of the Himalaya in W. Bengal and Assam.	Jamun, Machilus, *Melissma, Elaeocarpus, Celtis.*
Sub-tropical Pine (Area: 3.7 m. ha.)	Almost entirely pure Chir pine, no underwood, few shrubs.	The length of the north-west Himalaya between 1,000-1,800 metres except in Kashmir.	Chir, Jamun, Oak, Rhododendron.
Sub-tropical Dry Evergreen (Area: 200,000 ha.)	Low, scrub forest; small, stunted evergreen trees appear, shrubs including thorny species prevalent; herbs and grasses appear during monsoon.	The Bhabar, the Siwalik range and the western Himalaya up to about 1000 metres.	Olive, *Acacia modesta, Pistacia.*
Montane Wet Temperate (Area: 1.6 m. ha.)	A closed evergreen forest, trees mostly short-boled and branchy attaining wide gerth, crowned by dense and rounded leaves, red when young, branches clothed with mosses.	Higher hills of Tamil Nadu and Kerala from 150 metres upwards and eastern Himalaya — on the higher hills of W. Bengal.	*Machilus, Cinnamomum, Litsea,* Magnolia, Chilauni, Indian Chestnut, Birch, Plum.
Himalayan Temperate (Area: Moist Temperate — 2.7 m ha.; Dry Temperate — 200,000 ha.)	Predominantly coniferous forests; mosses and ferns grow freely on trees in the moist temperate forest, while there are hardly any epiphytes and climbers in the dry temperate forests.	The moist temperate forests occupy the length of the Himalaya between the pine and sub-alpine forests in Kashmir, H.P., Punjab, U.P., W. Bengal and Sikkim, between 1,500 and 3,300 metres. Dry temperate forests: along inner range of the Himalaya in Ladakh, Lahul, Chamba, Bashahr, Garhwal and Sikkim.	Oak, Deodar, Celtis and Mapla are common to both Himalayan moist and dry temperate forests. Specific to Himalayan moist temperate forests are: Fir, Spruce, Chestnut, Kail, Yew, Birch; and to the dry temperate forests: Chilgoza, Ash, Parrotia and Olive.
Alpine (Sub-alpine, Moist Alpine Scrub and Dry Alpine Scrub) (Area: 3,000 ha., total for all)	*Sub Alpine:* Dense growth of small, broad-leaved trees, often crooked; also large shrubs and conifers. *Moist Alpine Scrub:* Dense growth of low evergreen, alpine shrubs, flowering herbs, mosses and ferns. *Dry alpine scrub:* xerophytic, dwarf shrubs.	All three types are confined to the higher elevations of the Himalaya, the sub-alpine forests adjoining the alpine scrub which occur at heights above 3000 metres	Rhododendron and Birch are common to the sub-alpine forest and moist alpine scrub; honeysuckle to both types of alpine scrub. Fir, Kail, Spruce, Plum and Yew are specific to sub-alpine; Berberis to moist alpine scrub; and Juniper, Artemesia and Potentilla to dry alpine scrub.

Sources: 1. *Statistical Abstract India 1984,* Ministry of Planning. Quoting from Ministry of Agriculture.
2. *Directory & Year Book 1984,* Times of India.
3. *Philips Modern School Economic Atlas,* George Philip & Son Limited, London.
4. *Forests and Forestry,* K.P. Sagreiya, National Book Trust.

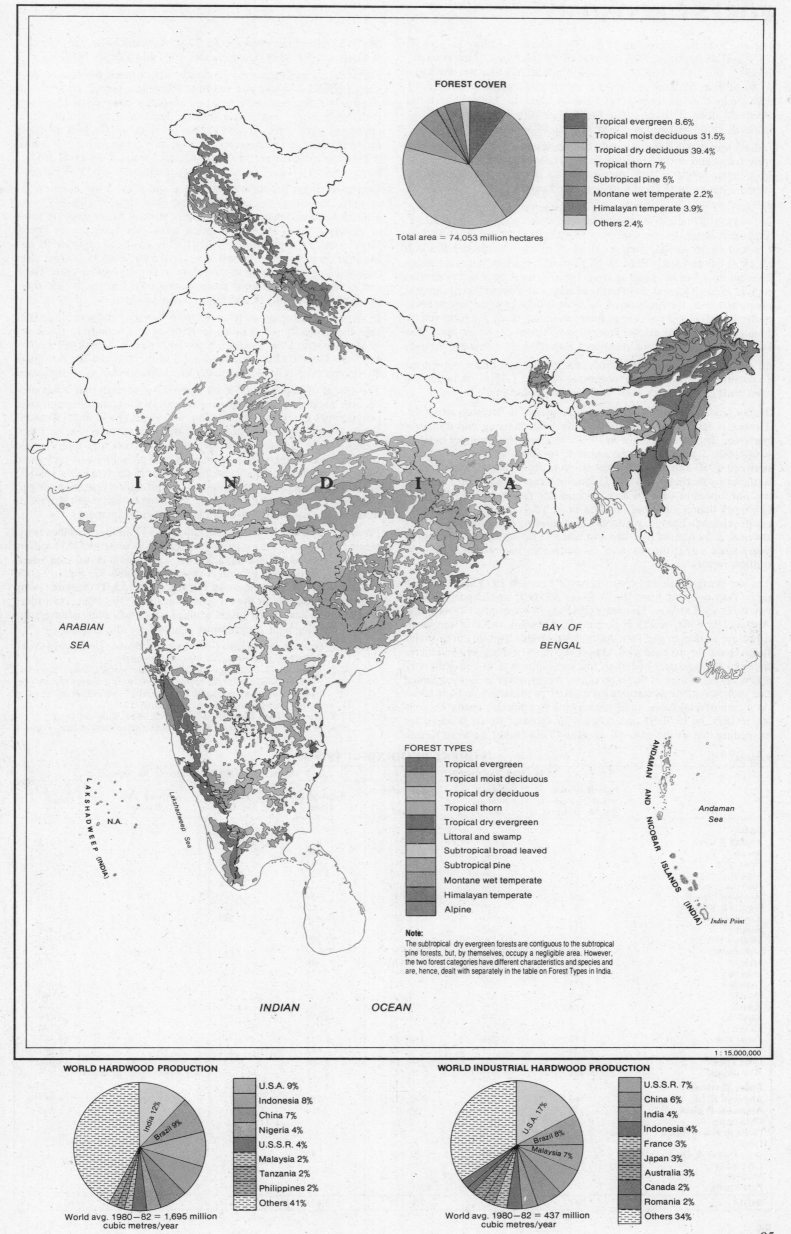

FOREST COVER

	Tropical evergreen 8.6%
	Tropical moist deciduous 31.5%
	Tropical dry deciduous 39.4%
	Tropical thorn 7%
	Subtropical pine 5%
	Montane wet temperate 2.2%
	Himalayan temperate 3.9%
	Others 2.4%

Total area = 74.053 million hectares

I N D I A

ARABIAN SEA

BAY OF BENGAL

LAKSHADWEEP

Lakshadweep Sea

N.A.

LAKSHADWEEP (INDIA)

ANDAMAN AND NICOBAR ISLANDS (INDIA)

Andaman Sea

Indira Point

FOREST TYPES

	Tropical evergreen
	Tropical moist deciduous
	Tropical dry deciduous
	Tropical thorn
	Tropical dry evergreen
	Littoral and swamp
	Subtropical broad leaved
	Subtropical pine
	Montane wet temperate
	Himalayan temperate
	Alpine

Note:
The subtropical dry evergreen forests are contiguous to the subtropical pine forests, but, by themselves, occupy a negligible area. However, the two forest categories have different characteristics and species and are, hence, dealt with separately in the table on Forest Types in India.

INDIAN OCEAN

1 : 15,000,000

WORLD HARDWOOD PRODUCTION

India 12%

Brazil 9%

	U.S.A. 9%
	Indonesia 8%
	China 7%
	Nigeria 4%
	U.S.S.R. 4%
	Malaysia 2%
	Tanzania 2%
	Philippines 2%
	Others 41%

World avg. 1980—82 = 1,695 million cubic metres/year

WORLD INDUSTRIAL HARDWOOD PRODUCTION

U.S.A. 17%

Brazil 8%

Malaysia 7%

	U.S.S.R. 7%
	China 6%
	India 4%
	Indonesia 4%
	France 3%
	Japan 3%
	Australia 3%
	Canada 2%
	Romania 2%
	Others 34%

World avg. 1980—82 = 437 million cubic metres/year

FOREST ECONOMY

Forests in India, according to a 1983 estimate, occupy an area of 75.1 million hectares, 22.8 per cent of the total area of the country. In 1951, the area was about 73.5 million hectares and this dwindled to 69 million hectares in 1960-61. A sustained effort increased the acreage to 75 million hectares in 1964-65 and this forest pattern has persisted since. The per capita forest area in India is a meagre 0.11 hectares compared to the world per capita area of 1.08 hectares.

India's forests are very unevenly distributed being located mostly in hilly tracts and remote interior areas. Their production yield is also low, being only 0.5 cubic metres per hectare a year as against the world average of 2.1 cubic metres per hectare (1980).

In 1981-82 there were 74.7 million hectares of forest area in India, with Madhya Pradesh having by far the largest acreage, 15.54 m ha. The other large acreages were in Maharashtra (6.4 m ha), Andhra Pradesh (6.36 m ha), Orissa (about 6 m ha), Arunachal Pradesh (5.15 m ha) and Uttar Pradesh (5.12 m ha). Each of the other states had less than 5 m ha, ranging from 130,000 ha (Goa, Daman & Diu) to 3.83 m ha (Karnataka). In terms of state and territorial geographical areas, however, the Andaman & Nicobar Islands have the largest areas under forest, 80-90 per cent of their total area. They are followed by Manipur, Sikkim, Arunachal Pradesh and Tripura (50-80 per cent). The lowest forest acreages are in Gujarat, Rajasthan, Haryana and Punjab.

Of the country's total forest area, 68.9 m ha are under the Forest Department, 2 m ha under corporate bodies, 1.96 m ha under the civil authorities and the rest with private individuals.

Today, about 3.5 million persons are engaged in forestry activities. Timber is felled for agricultural, industrial, housing and domestic purposes. The forest revenue in 1979-80 , about 56 rupees per hectare, was about 2 per cent of the national revenue. In the thirty years between 1950 and 1980, the net revenue from forests has shown a tremendous increase — from 152.2 million rupees in 1951-52 to 428.2 million rupees in 1965-66 and 1395 million rupees in 1979-80. This is a more than a nine-fold increase in 30 years and the trend, based on provisional figures available, continues to be alarming, as this revenue is increasing by leaps and bounds; for instance, in the three years since 1977, the revenue has shown an increase of some 350 million rupees.

The per hectare value of forest produce between 1973 and 1980 has more than doubled from Rs 23 to Rs 56. But, in comparison with countries such as West Germany (Rs 565), Switzerland (Rs 494) and Austria (Rs 336), it was rock-bottom. This low value of output to a hectare is due to the fact that a considerable part of 'area under forest' today is just bare with, at best, a couple of trees here and there, the poor management of Indian forests resulting in low productivity and widespread theft of forest products. It is however well-documented that, whenever private farmers undertake tree plantation on their farms on a commercial basis, their net income to a hectare easily exceeds Rs 10,000. In 1979-80, the states with value of forest produce far exceeding that of the national average of Rs 56 per ha were Kerala

(Rs 335), Himachal Pradesh (Rs 276), Karnataka (Rs 236), Jammu & Kashmir (Rs 153). Haryana (Rs 130) and Punjab (Rs 125).

Forest produce in India is of two kinds: Major forest produce (mainly teak, deodar, sal, sissoo, chir and kail), comprising timber and firewood; and minor forest produce consisting of such commercially important items as bamboo, cane, gum, resins, dyes, tans, lac, fibres, floss, medicinal plants, fodder, and grass. The total production of major forest produce is estimated at 20.3 million cubic metres, of which 8.2 million cubic metres constitute industrial wood. Fuel wood production is about 12.0 million cubic metres, including wood for charcoal.

Export of forest products (both major and minor) has been on the increase, touching over nine hundred million rupees in 1976-77. But imports have also been very high, only slightly lower than the value of exports. Many states and union territories have set up Forest Development Corporations. By 1978-79, 16 such corporations had been set up and were expected to achieve a target of 35,000 hectares of plantation and 13 million man-days of employment a year. These corporations have attracted much institutional finance to help them promote sale of timber and other forest products.

In most states, the gross revenue from forests has been far greater than the expenditure on forest activities, most notably in the states of Maharashtra, Uttar Pradesh, Karnataka, Kerala, Orissa and Andhra Pradesh. Gujarat, Punjab, Haryana, Tamil Nadu and the Andaman & Nicobar Islands have had greater expenditures than gross revenues.

The extent of depletion of India's forests is so high that it has now become necessary to implement two special Government-sponsored programmes to reverse the situation: afforestation and social forestry. While afforestation aims at planting trees to increase general forest wealth, social forestry aims at augmenting fuel and minor wood resources for farmers by encouraging new plantations of fuel wood trees and small trees suitable for farm needs. In the five years 1980-81 to 1984-85, the country has seen massive plantation operations as part of 'Save Forests' and 'Plant Trees' movements. But much still needs to be done, especially by way of forestry management.

From a modest afforestation programme of 847 million seedlings being planted in 1980-81, afforestation has steadily increased to 2533 million seedlings planted in 1984-85, a 200 per cent growth in just four years. A further 150 per cent growth is expected by 1986-87. Equally good progress has been made in social forestry, with 153,000 hectares being planted in 1986-87. In the on-going 7th Five Year Plan, both these programmes have been given a place of significance amongst the several development programmes planned for the country.

Sources: 1. *Statistical Abstract India 1984*, Ministry of Planning. Quoting from Central Forestry Commission, Ministry of Agriculture.
2. *Basic Statistics Relating to the Indian Economy, Vol.2: States, September 1985 and 1986,*Centre for Monitoring Indian Economy. Quoting from Interim Report on Production Forestry Man - Made Forests issued by the National Commission on Agriculture, 1972.
3. *Times of India Directory & Year Book 1984*, Times of India.
4. *Seventh Five Year Plan 1985-90, Vol.II*, Government of India Planning Commission.

Table 38

FOREST PRODUCE — 1979-80

	MAJOR PRODUCE			MINOR PRODUCE (Value in Rs. thousand)			
	Pulp & Match-wood (in thousand cu. metres)	Firewood (in thousand cu. metres)	Total Value (in Rs. thousand)	Bamboo & Cane	Fodder and Grass	Gums and Resins	Others
States							
Andhra Pradesh	192	2,035	162,081	34,976	—	—	54,213
Assam	457	895	64,872	1,880	59	—	12,314
Bihar	380	429	94,442	13,033	433	60	73,203
Gujarat	167	199	186,567	3,509	4,607	269	1,772
Haryana	38	178	16,019	232	516	567	279
Himachal Pradesh	464	158	231,042	485	1,601	22,938	2,370
Jammu & Kashmir	587	150	447,797	640	—	180,000	21,050
Karnataka	1,183	1,238	712,167	20,427	385	24	63,013
Kerala	435	363	362,383	—	—	—	5,904
Madhya Pradesh	1,085	2,289	213,190	92,200	—	—	350,800
Maharashtra	320	1,466	319,882	10,979	3,885	2,866	43,758
Manipur	19	90	1,470	46	—	—	872
Meghalaya	43	1	1,782	32	Neg.	44	1,081
Nagaland	33	26	3,742	4	—	—	1,316
Orissa	469	677	179,490	21,092	783	410	66,046
Punjab	88	40	27,143	101	3,477	649	1,269
Rajasthan	4	209	44,944	14,574	892	2,583	23,456
Sikkim	8	15	1,257	5	20	—	2,588
Tamil Nadu	12	288	186,323	1,721	1,337	—	68,365
Tripura	39	133	7,343	483	152	—	1,093
Uttar Pradesh	962	2,359	344,284	4,904	6,829	50,252	54,026
West Bengal	226	477	56,817	—	—	—	—
Union Territories							
Andaman & Nicobar	162	56	45,617	282	—	57	1,087
Arunachal Pradesh	138	20	35,057	729	—	—	1,896
Chandigarh	—	—	—	—	—	—	—
Dadra & Nagar Haveli	N.A.	4	581	—	—	—	—
Delhi	—	—	—	—	—	—	—
Goa, Daman & Diu	47	125	21,689	80	—	—	—
Lakshadweep	—	—	—	—	—	—	—
Mizoram	3	Neg.	163	19	—	—	57
Pondicherry	—	—	—	—	—	—	—
INDIA	7,561	13,920	3,768,147	222,433	24,976	260,719	851,828

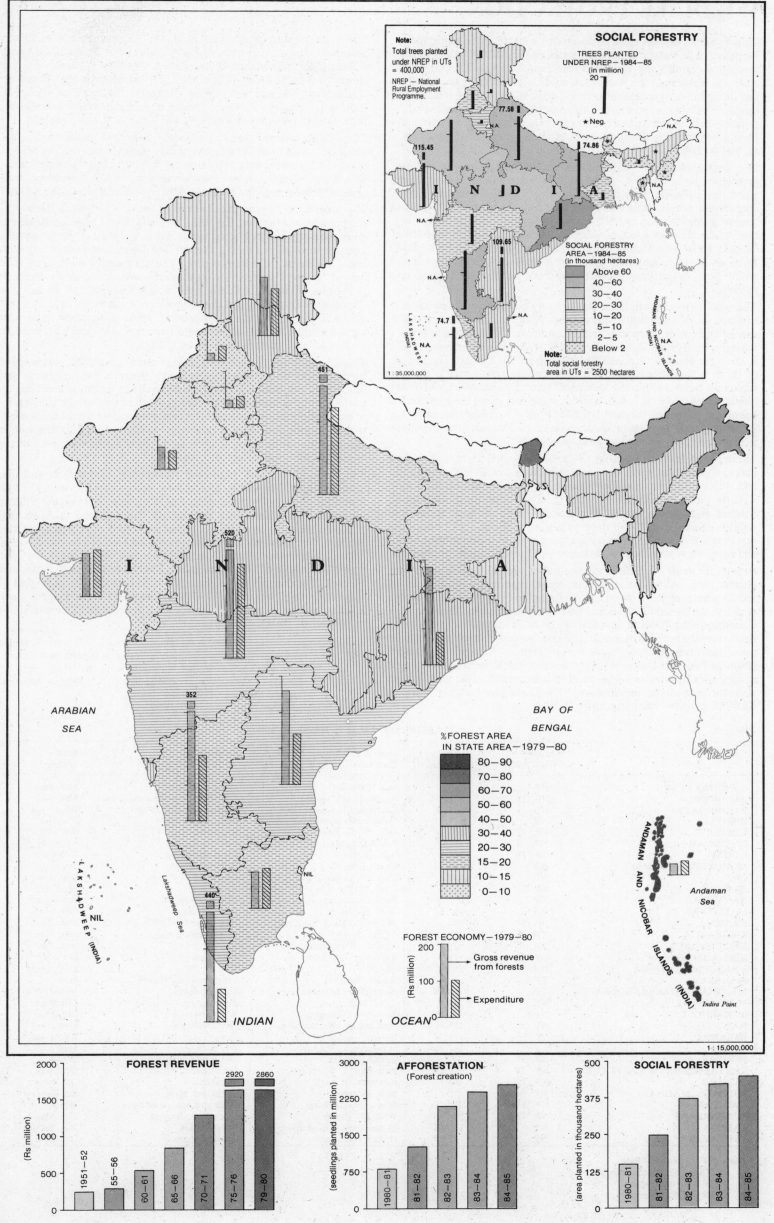

WILDLIFE PROTECTION

India is one of the foremost protectors of wildlife in the world today. To preserve its vanishing species, it has given statutory status to such protection through the Wildlife Protection Act of 1972. Out of this Act have developed the numerous wildlife reserves in the country. But though the organized destruction of wildlife — once indulged in by the former Indian princes and British officers, often in collaboration — has ended, poachers and herdsmen, who persistently penetrate the forests to graze their cattle, are now taking a toll of the country's animal wealth.

The Indian Board of Wildlife has listed 36 species which need protection: lion, wild ass, pangolin, brow-antlered deer, swamp deer, musk deer, gharial, four-horned antelope, clouded leopard, Nilgiri stag, lesser panda, Kashmir stag, wild buffalo, rhinoceros, Nilgiri langur, golden langur, gazelle, markhor, spotted linsang, pigmy hog, blackbuck, snow leopard, golden cat, marble cat, hunting leopard or cheetah (now considered extinct), dugong, great Indian bustard, Jerdon's courser, mountain quail, pink-headed duck, white-winged wood duck, tragopan, crocodile, leathery turtle, water lizard and python.

The wildlife reserves in India may broadly be classified in two types: national parks and wildlife sanctuaries. Wildlife sanctuaries have the special purpose of preserving animals and birds. National parks protect the entire ecosystem. A special category of animal sanctuaries in India is the tiger reserve, a consequence of the 1970 Project Tiger. One of these is the Sundarbans, a unique swamp forest that is the last remaining bastion of the Royal Bengal Tiger. This sanctuary and others like it, developed to save the once fast-vanishing tiger, have succeeded beyond the wildest dreams of the Project's founders; now the increasing tiger population in these reserves does not have the space it needs and this is causing fresh problems.

There are in India, at present, 54 national parks, including marine parks, high altitude parks and parks in protected areas in the Andaman and Nicobar Islands. Of these parks, 26 may be considered major wildlife parks or wildlife systems in urban areas (e.g. Guindy in Madras and Bannirgatta in Bangalore). There are also about 160 wildlife sanctuaries. Of all these protected areas only 147 important reserves are shown in the accompanying map. All the national parks and tiger reserves as well as the major wildlife sanctuaries are located.

Among the reserves located are several of special significance. The Asiatic lion, one of the rarest and most important wild animals in India, where alone it survives, is found in only two parts of the country: the famed Gir National Park in Gujarat and the lesser known Chandraprabha Wildlife Sanctuary in Uttar Pradesh, where Gir lions have recently been introduced. The one-horned rhinoceros, another once vanishing species, is now protected in Assam's Kaziranga National Park and Manas Wildlife Sanctuary. But the species is much sought after by poachers who covet its horn, much in demand in East Asia for aphrodisiac preparations. The Dachigam National Park in Kashmir protects the hangul (the Kashmir stag).

The country's tiger projects are in Manas (Assam), Palamau (Bihar), Simlipal (Orissa), Corbett National Park (Uttar Pradesh), Dhakna Kolkaz (Melghat, Maharashtra), Kanha (Madhya Pradesh), Periyar (Kerala), Ranthambhor (Rajasthan), Sariska (Rajasthan), Bandipur (Karnataka), Sundarbans (West Bengal), Baxa (West Bengal), Indrawati (Madhya Pradesh), Nagarjuna Sagar (Andhra Pradesh), Nam Dapha (Arunachal Pradesh) and Dudwa (Uttar Pradesh).

The wildlife reserves are fairly widely dispersed, providing for the needs of animal lovers in all parts of the country and providing protection to a very varied animal population. It is also clear that some states, Kerala for example, are better endowed with wildlife than others, but, generally speaking, the north is richer in animals than the south.

The proportion of area under forest in India is largest in the north-eastern states (Manipur, Tripura, Meghalaya and Nagaland), where more than half the total geographical area is forested. But the states here are so small that their protected forest areas compare poorly with areas allocated for national parks and sanctuaries elsewhere. The largest areas reserved for sanctuaries are in Madhya Pradesh (1.5 m ha), Andhra Pradesh (0.9 m ha), Gujarat (0.7 m ha) and Karnataka (0.6 m ha). But the proportion of sanctuary area to forest area is highest in Gujarat (35 per cent). Most of the other states with a good wildlife conservation record have sanctuary areas between 10 to 20 per cent of their total forest area.

Though late, efforts are now being taken by the Department of Environment, Government of India to preserve the ecosystem and biotic life. The Man And Biosphere (MAB) programme launched by the Government aims at conserving as much of the biological diversity of the country as possible and forms part of the international scheme to set up a global network of biosphere reserves. The idea of biosphere reserves was initiated by UNESCO in 1973-74 and the first reserve in the Nilgiris was established in 1986.

The Indian National Man and Biosphere Committee and the Environmental Research Committee recommended in 1979 the following 12 areas for development as biosphere reserves: the Nilgiris in Tamil Nadu, Karnataka and Kerala; Nam Dapha in Arunachal Pradesh; Nanda Devi and Uttarakhand (or Valley of Flowers) in Uttar Pradesh; North Islands of the Andamans; the Gulf of Mannar in Tamil Nadu; Kaziranga in Assam; the Sundarbans in West Bengal; the Thar desert in Rajasthan; Manas in Assam; Kanha in Madhya Pradesh; and Nokrek (Tura range) in Meghalaya. Project documents in respect of four (Nilgiris, Nanda Devi, Uttarakhand and Nam Dapha) have already been prepared while surveys are in progress (October 1986) in respect of the other areas.

Sources: 1. *Basic Statistics Relating to the Indian Economy, Vol. 2: States, September 1986*, Centre for Monitoring Indian Economy. Quoting from Reply to Unstarred Question No.6209 in the Lok Sabha on 13 May 1985.
2. *Glimpses of India, Occasional Monograph - 2*, National Atlas and Thematic Mapping Organisation.
3. *Hindustan Times, 28.8.1986*, Delhi.

Table 39

MAJOR WILDLIFE RESERVES

Reserve	Status NP/WS/TP/BS	Area (in Sq. km.)	State & Location	Forest cover	Best Season
BANDHAVAGARH	NP	105	M.P. (Shahdol)	Moist deciduous	Nov.-June
BANDIPUR	NP, TP	874.2	Karnataka (Mysore)	Low elevation mixed deciduous	Mar.-Aug.
BANNIRGATTA	NP	104.2	Karnataka (Bangalore)	Tropical dry deciduous	Whole year
BORIVLI	NP	679.9	Maharashtra (Greater Bombay)	Moist deciduous	Jan.-Mar.
CORBETT	NP, TP	525	U.P. (Naini Tal)	Dry deciduous	Nov.-May
DACHIGAM	NP	N.A.	J & K (Srinagar)	Sub-Himalayan	Apr.-Nov.
DUDWA	NP, TP	500	U.P. (Kheri)	Moist deciduous	Nov.-May
ERAVIKULAM RAJMALLAY	NP	97	Kerala (Idukki)	Semi-evergreen	Oct.-Apr.
GIR	NP	140.4	Gujarat (Junagadh)	Dry deciduous plain	Dec.-June
GUINDY	NP	2.8	T.N. (Madras City)	Arable	Whole Year
HARARIBAG	NP	186.25	Bihar (Hazaribag)	Mixed hilltop Sal	Oct.-June
JALDAPARA	WS	115.53	W.B. (Jalpaiguri)	Sub-Himalayan	Dec.-May
KANHA	NP, TP	940	M.P. (Mandla & Balaghat)	Mixed Sal & Meadows	Mar.-June
KAZIRANGA	NP	430	Assam (Jorhat)	Magnificent trees	Feb.-May
KEIBUL LAMJAO	NP	25	Manipur (Churachandpur)	Arable	Nov.-Feb.
KEOLADEO GHANA	NP, BS	29	Rajasthan (Bharatpur)	Deciduous	Oct.-Feb.
KHANGCHENDZONGA	NP	850	Sikkim	Pine	May-Aug.
MANAS	TP	80	Assam (Barpeta)	Sub-Himalayan riverine	Feb.
MELGHAT (DHAKNANKOLKAZ)	TP	381.58	Maharashtra (Amravati)	Dry deciduous	Apr. & May
MUDUMALAI	WS	321	T.N. (Nilgiri)	Mixed deciduous hill	Feb.-June
NAGARHOLE	NP	571.55	Karnataka (Kodagu)	Wet evergreen	Oct.-Mar.
NAWEGAON	NP	133.88	Maharashtra (Bhandara)	Dry deciduous	May
PALAMAU	TP	979.27	Bihar (Daltenganj)	Dry deciduous	Whole year
PENCH	NP	257.26	Maharashtra (Nagpur)	Dry deciduous	May
PERIYAR	NP, TP	777	Kerala (Idukki)	Green hilltop sholas	Oct.-Apr.
RANTHAMBHOR (S. MADHOPUR)	NP, TP	392.2	Rajasthan (S. Madhopur)	Deciduous	Oct.-June
SARISKA	NP, TP	195	Rajasthan (Alwar)	Deciduous	Oct.-June
SHIVPURI (MADHAV)	NP	156	M.P. (Shivpuri)	Deciduous plain	Jan.-June
SIMLIPAL	NP, TP	303	Orissa (Mayurbhanj)	Wide streches of sal trees	Winter
SUNDARBANS	NP, TP	2,585	W.B. (24 Parganas)	Lihoral and Swamp	Sep.-May
TADOBA	NP	116.55	Maharashtra (Chandrapur)	Natural Teak	May & June
VELVADAR	NP	17.83	Gujarat (Bhavnagar)	Tropical thorn	Oct.-June

Note: NP — National Park
WS — Wildlife Sanctuary
TP — Tiger Project
BS — Bird Sanctuary

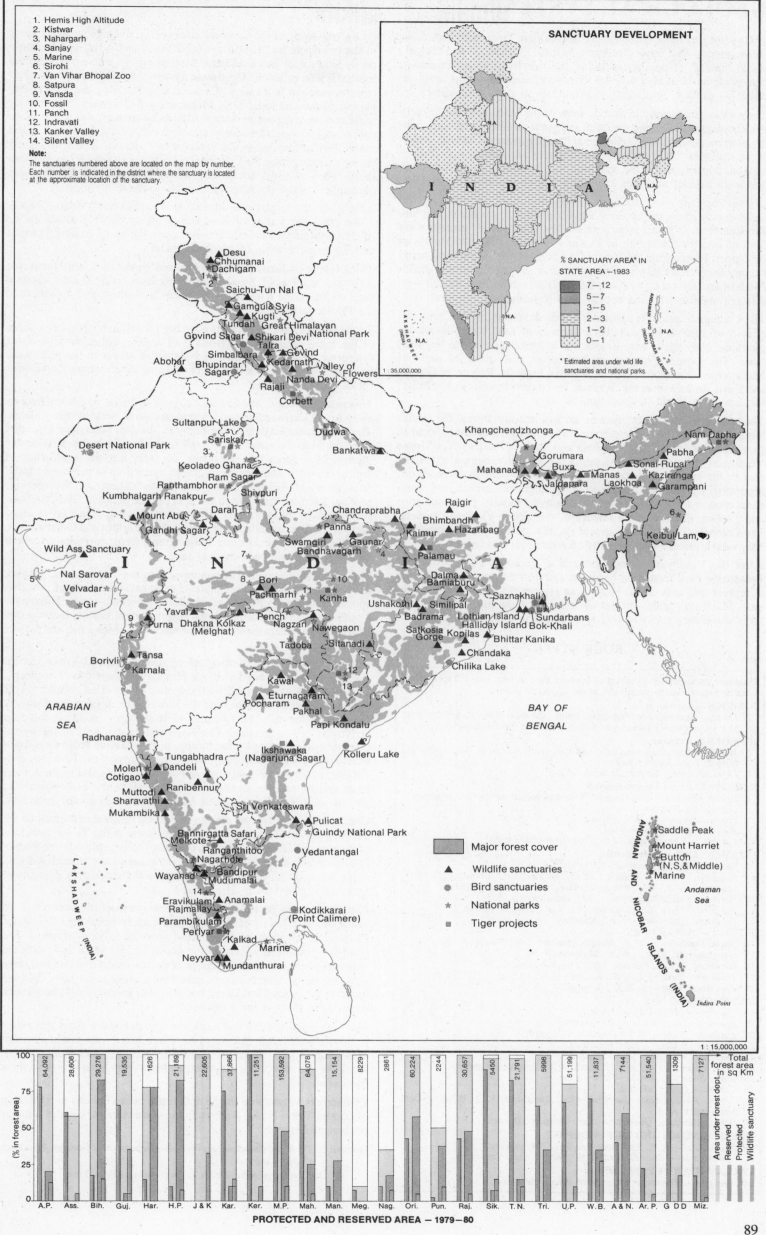

GEOLOGICAL SYSTEMS

The geology of India spans the entire Geologic Time Scale (see table below maps). The three structural features of India — the ancient, stable 'shield' of peninsular India, the relatively young mountain chains of the Himalaya and the extensive, alluvium-filled northern plains — exhibit significant developments in geological history.

The rocks of the peninsula include some of the oldest known, the early Pre-Cambrian formations. Granites and gneisses are widespread; there are also metamorphozed sedimentary and igneous rocks with which India's mineral belts are associated. There are several occurrences of iron, copper, lead, zinc, chromium, manganese, gold, limestone and mica besides abrasive, refractive and radio-active minerals.

In the Pre-Cambrian formations may be found remnants of the sedimentary rocks of the early Pre-Cambrian or Archaean and late Pre-Cambrian or Proterozoic periods. The Proterozoic formations are divided into the Cuddapah or Lower Proterozoic and the Vindhyan or Upper Proterozoic periods. Old rocks of the Archaean period occupy almost the entire southern peninsula and stretch across Andhra Pradesh and Orissa up to south Bihar, from where they stretch west through the Bundelkhand region, into Rajasthan and Gujarat.

The Cuddapah system is particularly well-developed in the Cuddapah district of Andhra Pradesh. Rich deposits of sandstone, limestone, quartzite, slaty shale, baryte, asbestos and soapstone are to be found here. The Lower Proterozoic formations also occur in parts of Maharashtra, Karnataka, Orissa, Madhya Pradesh, Rajasthan, Gujarat and Bihar. In Rajasthan they account for copper deposits and in Bihar, for limestone. Lead is also found in some places.

The younger Vindhyan system occurs mainly along the entire Vindhyan range: Lower Vindhyan rocks, mainly limestone, occur in parts of Rajasthan, Andhra Pradesh and Madhya Pradesh. The Upper Vindhyans — the north-west slopes of the Aravalli range and extending east to the edge of the Indo-Gangetic Plain — are in three series, separated by diamond-bearing conglomerates, which are best-developed in the Panna district of Madhya Pradesh. Pink sandstone and gypsum deposits are a feature of the Upper Vindhyan system. Pre-Cambrian rocks also appear in the north-east, forming the Meghalaya plateau and the hills of Arunachal Pradesh.

After the Vindhyan period, there is a 'geologic gap': no Lower Palaeozoic rock formation has been found in the peninsula. The era is represented only in the Himalayan region of the north and north-east. Much geological interest has centred on the marine fossils that the sedimentary rocks of this age contain in abundance.

ROCK TYPES

Table 40

(Numbers correspond to numbers of symbols used in inset 'Rock Types' map. Localities mentioned in states refer to the respective districts).

IGNEOUS — Intrusives
1. *Granite, Grano-diorite, Pegmatite* — Kar.: Tumkur; U.P.: Jhansi, Lalitpur; M.P.: Chhatarpur, Tikamgarh; J. & K.: Ladakh; Ori.: Kendujhar.

IGNEOUS — Effusives
2. *Basalt, associated lavas and tuffs* — all Mah.; north-west M.P.; Kar.: Gulbarga, Bijapur; Guj.: Amreli, Rajkot.
3. *Rhyolite* — Raj.: Barmer, Jodhpur.

SEDIMENTARY — Unconsolidated
4. *Recent Alluvium* — Indo-Gangetic Plain: east coastal region; Guj.: Mahesana, Ahmadabad, Bharuch.
5. *Older alluvium* — J. & K.: Srinagar.
6. *Blown sand* — north-west Raj.
7. *Laterite* — Ker.: Palghat; T.N.: Thanjavur, Pudukkottai.

SEDIMENTARY — Consolidated
8. *Sandstone* — Raj.: Jodhpur; T.N.: South Arcot.
9. *Sandstone, Shale* — J. & K.: Punch, Rajauri; Raj.: Chittaurgarh, Sawai Madhopur; M.P.: Rewa; Manipur; Mizoram; Nagaland; Tripura.
10. *Sandstone, Shale, Limestone* — Raj.: Jaisalmer.
11. *Sandstone, Shale, Conglomerate* — J. & K.: Jammu, Kathua.
12. *Sandstone, Quartzite* — A.P.: Adilabad.
13. *Shale* — A.P.: Warangal.
14. *Shale, Sandstone* — M.P.: Shadol, Jabalpur; Guj.: Bharuch.
15. *Conglomerate, Clay* — W.B.: Medinipur.
16. *Limestone* — Raj.: Nagaur; T.N.: Tiruchchirappalli.
17. *Limestone, Shale* — J. & K.: Ladakh; M.P.: Raipur.
18. *Limestone, Slate* — J. & K.: Srinagar.

METAMORPHIC
19. *Quartzite* — Raj.: Udaipur, Alwar.
20. *Quartzite, Limestone* — U.P.: Chamoli.
21. *Quartzite, Shale* — M.P.: Gwalior, Tikamgarh.
22. *Quartzite, Schists* — Bih.: Hazaribag.
23. *Quartzite, Phyllite* — M.P.: Jabalpur.
24. *Slate, Quartzite* — J. & K.: Ladakh; A.P.: Prakasam, Guntur.
25. *Slate, Quartzite, Schists* — M.P.: Sidhi.
26. *Schists* — Guj.: Vadodara.
27. *Schists, Phyllite, Slate, Quartzite* — Raj.: Dungarpur; M.P.: Balaghat; Ori.: Sundargarh; Ar.P.
28. *Schists, Gneisses, Quartzite* — Kar.: Belgaum, Dharwad.

OTHER IGNEOUS AND METAMORPHIC
29. *Khondalite* — A.P.: East Godavari; Ker.: Trivandrum.
30. *Charnockite* — Ori.: Koraput; T.N.: Nilgiri; Ker.: Cannanore.

UNCLASSIFIED CRYSTALLINE
31. *Gneisses* — Peninsular India; South Bihar; Meghalaya.

By the beginning of the Carboniferous period, the geological activities of the peninsula had almost entirely ceased and it had settled into a stable land mass. But a different kind of crustal movement resulted in what is now called the Gondwana System. Mid-Carboniferous earth-stresses resulted in a series of sunken troughs along the present-day valleys of the Damodar-Son, Mahanadi and Godavari. Fresh water sediments were deposited in these troughs by streams, and sedimentary rocks, such as sandstone and shale with rich coal deposits are found there. Similar formations are observed in Australia, Africa and South America, supporting the Continental Drift theory which assumes that there were two primeval land units—Laurasia in the northern hemisphere and Gondwanaland in the southern hemisphere.

India's best and largest coal deposits are found in this system — mainly in the Damodar Valley of West Bengal-Bihar, the Mahanadi Valley of Orissa-Madhya Pradesh, the Godavari Valley of Andhra Pradesh and the Satpura basin of Madhya Pradesh.

Thick layers of Jurassic limestone occur in the northern Himalaya, Rajasthan and Kachchh and shales in the Lahul & Spiti area. Jurassic beds of limestone and shale also occur in Ladakh and a portion of the Pir Panjal range.

No significant mineral deposits have been found in the Mesozoic areas of the extra-peninsular or Himalayan region, although deposits of limestone, celestite, gypsum and phosphate occur in the peninsula, near the Tiruchchirappalli district of Tamil Nadu where Cretaceous rocks occur in patches.

Towards the close of the Cretaceous period, India, which had broken away from Gondwanaland and drifted eastwards, welded with Eurasia. The consequent sinking of the Tethyan Sea—that had once extended from the Mediterranean in the west to the Pacific in the east—under the weight of depositing sediments, upset the internal equilibrium of the earth's crust and numerous cracks opened up in peninsular India through which basaltic lavas poured out. These, on weathering, produced fertile black-cotton soil and bauxite deposits. The eruptions spread a layer of lava, over 3000 metres thick in places, over large parts of the Deccan (Madhya Pradesh, Maharashtra and Gujarat). The basaltic rock of the Deccan Traps (trap=step), provides excellent building stone and road-making material.

The great earth movements continued into the Tertiary era, pushing the continental masses, or 'plates', upwards. The Tethyan Sea disappeared and in its place rose the lofty mountain-chain, Himalaya. The Himalayan mountain system, rich in marine fossils, is composed of sedimentary formations of the Palaeozoic, Mesozoic and Lower Cainozoic eras.

When the Himalaya were rising, a long, narrow depression was formed along the southern border of the Himalaya, known as 'Foredeep'. This depression was filled with deposits of denuded rocks brought by the Himalayan streams, and the Lower Himalaya or the Shiwalik range was formed. This phase of Himalayan mountain-building activity is still continuing. The Naga and Lushai hills of the north-east were also formed by the folding of the massive sediments deposited in the trough formed in place of the Tethyan Sea. Tertiary deposits also occur in the Andaman & Nicobar islands and in the Kaveri basin and Ramanathapuram district in Tamil Nadu, in Pondicherry, the coastal strip of Kerala, Kachchh in Gujarat and parts of Rajasthan.

Coal, oil and natural gas are the important mineral resources of the Tertiary period. Most of India's oil fields are in the Tertiary regions of Assam and Gujarat. Tertiary deposits in the peninsula are mainly sandstone and shale with lignite. Limestone of the period is found in the Khasi-Jaintia hills of Meghalaya and India's only deposits of rock salt are in Mandi, Himachal Pradesh.

The Quarternary period of the Cainozoic era is a little over a million years old and has 'just begun', geologically speaking. The older Pleistocene division of this period saw the advent of man. The younger, 'recent' division started about 12,500 years ago, with the withdrawal of glaciation. Evidence of the ice-sheets that covered large areas of the extreme north and made the climate impossibly cold are found in the faceted boulders, moraines and grooved and planed rock surfaces in parts of the Himalaya.

The vast tract of land that separates the peninsula from the extra-peninsula is the fertile Indo-Gangetic Plain. It is a wide belt of alluvium consisting of soft, unconsolidated clay, silt and sand washed down from the Himalaya by the Ganga, Indus, Brahmaputra and their tributaries.

Quarternary deposits in the peninsula are restricted to a narrow strip of coastal Tamil Nadu, Andhra Pradesh and Orissa, besides a major portion of Gujarat.

Sources: 1. Dr. V. A. Chandrasekar, Professor, Department of Geology, Presidency College, Madras.
2. *Geology of India*, Dr. A.K. Dey, National Book Trust.

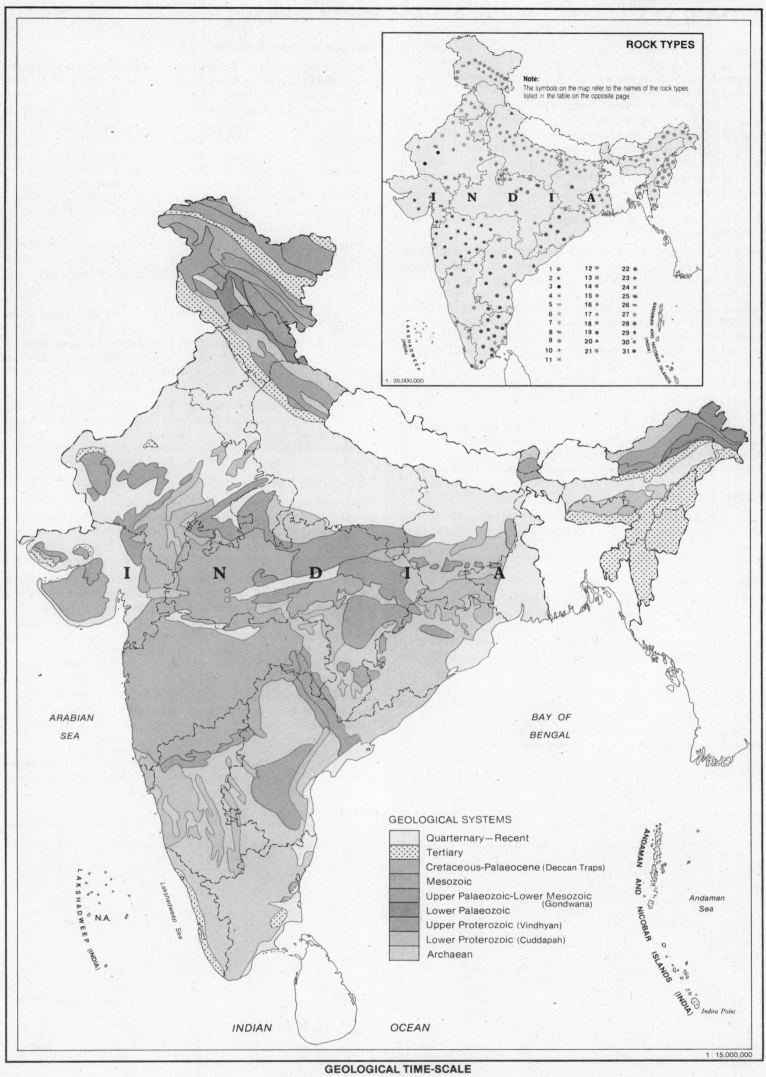

ROCK TYPES

Note:
The symbols on the map refer to the names of the rock types listed in the table on the opposite page.

I N D I A

1 : 35,000,000

LAKSHADWEEP (INDIA)

1	12	22
2	13	23
3	14	24
4	15	25
5	16	26
6	17	27
7	18	28
8	19	29
9	20	30
10	21	31
11		

ANDAMAN AND NICOBAR ISLANDS (INDIA)

I N D I A

ARABIAN SEA

BAY OF BENGAL

LAKSHADWEEP (INDIA)

N.A.

Lakshadweep Sea

ANDAMAN AND NICOBAR ISLANDS (INDIA)

Andaman Sea

Indira Point

GEOLOGICAL SYSTEMS

- Quarternary—Recent
- Tertiary
- Cretaceous-Palaeocene (Deccan Traps)
- Mesozoic
- Upper Palaeozoic-Lower Mesozoic (Gondwana)
- Lower Palaeozoic
- Upper Proterozoic (Vindhyan)
- Lower Proterozoic (Cuddapah)
- Archaean

INDIAN OCEAN

1 : 15,000,000

GEOLOGICAL TIME-SCALE

PRE-CAMBRIAN	PALAEOZOIC	MESOZOIC	CAINOZOIC	
Late Pre-Cambrian or Proterozoic	Permian (270)	Cretaceous (135)	Quarternary	Recent (0.01)
Early Pre-Cambrian or Archaean (4000)	Upper Carboniferous	Jurassic (180)		Pleistocene (1)
	Lower Carboniferous (350)	Triassic (225)	Tertiary	Mio-Pliocene (3)
	Devonian (400)			Lower Miocene (25)
	Silurian (440)			Oligocene (40)
	Ordovician (500)			Eocene (60)
	Cambrian (600)			Palaeocene (70)

Note: Numbers within brackets indicate the beginning of period in million years.

MINERALS

India's mineral resources, rich and varied, adequately provide for an industrial base, especially a ferrous-based one. But the distribution is very uneven. The Damodar valley has the largest concentration of mineral wealth (except petroleum), whereas much of the peninsular area west of a line from Mangalore to Kanpur has very little mineral wealth. East of the line are the major reserves of metallic minerals, coal, mica and many other non-metallic minerals. But sedimentary rocks in the western flank of the peninsular formation, in Gujarat, and in the east, in Assam, have substantial reserves of petroleum. Also west of this line is Rajasthan, with its reserves of many non-ferrous minerals. Outside this area, most of the states, including Jammu & Kashmir, Punjab, Haryana, Uttar Pradesh, Himachal Pradesh, Tripura, Nagaland and Gangetic West Bengal, are very poor in mineral resources.

In 1950, the production of minerals by value amounted to Rs 617 million, with mineral fuel accounting for 76 per cent, metallic minerals 22 per cent and non-metallic minerals a mere 2 per cent. But mineral fuel production took off so spectacularly in the last decade, with several new finds off-shore, that its value reached Rs 73,394 million in 1985, a 156-fold increase. Metallic and non-metallic mineral production also accelerated after the 1970s, reaching Rs 5224 million and Rs 3012 million respectively in 1985 (a nearly 37-fold and 215-fold increase in 35 years, respectively).

The total value of mineral production is Rs 81,630 million (1985), a little more than a third of it coming from mineral fuel production at off-shore Bombay High (Rs 28,630 million). Bihar (Rs 12,740 million) and Madhya Pradesh (Rs 8910 million) earn the most from the production of metallic and non-metallic minerals and coal. Rajasthan, Tamil Nadu, Karnataka, Uttar Pradesh, Kerala, Haryana, Meghalaya, Himachal Pradesh and Jammu & Kashmir all appear to be minerally deprived states, with production ranging from Rs 10 million to Rs 1510 million.

There are, according to the Geological Survey of India, 50 important mineral occurrences and about 400 major sites where these minerals occur. A few of these minerals and locations are discussed generally here and more elaborately in the two maps that follow.

Mineral Fuels

India's most important coal fields are Raniganj (West Bengal) and Jharia and Bokaro (Bihar). Lignite occurs in Neyveli (Tamil Nadu). Bihar has the largest coal reserves in the country, 37 per cent of all reserves. West Bengal has 18 per cent and Madhya Pradesh 16 per cent, mainly in Rajgarh, Chhindwara, Bilaspur and Surguja districts.

India's oldest oil-producing areas are in the valleys of the Nova, Dirling and Buri rivers in north-east Assam (Lakhimpur and Dibrugarh districts). The well in Digboi is the oldest in the country. Oil is also found in Kheda, Ahmadabad, Mahesana and Surat districts of Gujarat. But the biggest oil strikes in the country have been more recent off-shore finds, especially on the continental shelf off the coast of Maharashtra (Bombay High). The most recent major oil find (August 1986) was in the Gulf of Khambhat (off Gujarat).

Metallic (Ferrous) Minerals

The chief deposits of chromite ore are in Dhenkanal and Kendujhar districts of Orissa; North Ratnagiri district in Maharashtra; Singhbhum district in Bihar; Mysore and Hassan districts of Karnataka; West Godavari district in Andhra Pradesh; and Salem district in Tamil Nadu.

Iron ore is almost entirely concentrated in Orissa, Bihar, Madhya Pradesh, Karnataka and Goa (96 per cent of the total reserves). In Orissa, the principal deposits occur in a series of hill ranges across Sundargarh, Sambalpur, Kendujhar and Mayurbhanj districts. In Bihar, they are found in Singhbhum district. Most of the reserves in Madhya Pradesh are in 14 isolated areas in the Bailadila range, Raoghat area and near Aridongri, all in the Bastar district, and in the Dhalli-Rajhara range in Durg district.

Extensive deposits of manganese ore occur in Madhya Pradesh, Maharashtra, Orissa, Karnataka and Rajasthan. The largest reserves are in the Nagpur-Bhandara belt in Maharashtra and at Balaghat in Madhya Pradesh.

Metallic (Non-ferrous) Minerals

The major bauxite deposits occur in Gumla and Lohardaga districts of Bihar; Mopa and Pernem area in Goa; Surguja, Balaghat, Durg, Mandla, Bilaspur and Raigarh districts of Madhya Pradesh; Madurai and Salem districts of Tamil Nadu; Jamnagar and Kachchh districts of Gujarat; Chitradurga district of Karnataka; and Kalahandi and Koraput districts of Orissa.

Copper occurs in India as sulphides. It occurs in ancient crystallines as well as in younger rock formations, including the Cudapah and Bijawar plateau and the Aravalli range. Lead and zinc are found in Zawar and Rajpura Dariba in Rajasthan (Udaipur and Tonk districts). Gold occurs in the Kolar Gold Fields of Karnataka.

Non-metallic Minerals

Magnesite, an important mineral in the manufacture of basic refractory bricks for steel making, is found in Tamil Nadu. The Chalk Hills deposits of Salem district, Tamil Nadu, are the largest in India and have been under active production for several years. Deposits in the Almora and Pithoragarh districts of Uttar Pradesh are now under exploitation.

Mica, a strategic mineral, is found in Andhra Pradesh, Bihar (Ruby Mica) and Rajasthan. Investigations by the Department of Atomic Energy have revealed a number of radioactive mineral deposits. The beaches of Kerala and Tamil Nadu are rich in radioactive monazite, which contains thorium, uranium oxide, phosphate and a large percentage of rare earth oxides. Ilmenite is recovered from the beach sands of the Kanniyakumari district of Tamil Nadu and in Kerala. It is now mainly exported to Japan.

Uranium bearing ores have been discovered in the Singhbhum Thrust Belt of Bihar and in the Himalayan regions of Himachal Pradesh and Uttar Pradesh.

Sources: 1. *Minerals of India*, Meher D. N. Wadia, National Book Trust.
2. *Basic Statistics Relating to the Indian Economy, Vol. 1: All India, August 1986* and *Vol. 2: States, September 1986*, Centre for Monitoring Indian Economy.

Table 41 | VALUE OF PRODUCTION OF MINERALS — 1985 | (in Rs. million)

	Crude petroleum	Coal	Iron ore	Limestone	Lignite	Copper ore	Gold	Zinc concentrate	Chromite	Phosphorite	Manganese ore	Lead concentrate	Bauxite	Natural gas	Others
States															
A.P.	—	2,890	Neg.	220	—	Neg.	20	—	Neg.	—	Neg.	30	—	—	140
Assam	6,830	280	—	10	—	—	—	—	—	—	—	—	—	10	—
Bihar	—	11,720	420	110	—	340	20	—	—	—	Neg.	—	30	—	90
Gujarat	5,910	—	—	110	180	—	—	—	—	—	10	—	30	40	80
Haryana	—	—	—	30	—	—	—	—	—	—	—	—	—	—	—
H.P.	—	—	—	20	—	—	—	—	—	—	—	—	—	—	—
J. & K.	—	10	—	Neg.	—	—	—	—	—	—	—	—	—	—	—
Kar.	—	—	400	90	—	10	390	—	30	—	30	—	Neg.	—	30
Kerala	—	—	—	20	—	—	—	—	—	—	—	—	—	—	60
M.P.	—	7,290	680	450	—	130	—	—	—	50	130	—	90	—	90
Mah.	—	1,900	40	60	—	—	—	—	—	—	110	—	20	—	20
Megh.	—	—	—	10	—	—	—	—	—	—	—	—	—	—	—
Orissa	—	910	360	210	—	Neg.	—	—	360	—	70	50	—	—	120
Raj.	—	—	10	150	—	350	—	410	—	310	—	150	—	—	130
T.N.	—	—	—	220	1,130	—	—	—	—	—	—	—	Neg.	—	130
U.P.	—	680	—	50	—	—	—	—	—	20	—	—	Neg.	—	20
W.B.	—	4,930	Neg.	—	—	—	—	—	—	—	—	—	—	—	20
Union Territories															
Ar. P.	50	—	—	—	—	—	—	—	—	—	—	—	—	—	—
G, D. & D.	—	—	420	—	—	—	—	—	—	—	Neg.	—	Neg.	—	—
Others															
Bombay High	28,540	—	—	—	—	—	—	—	—	—	—	—	—	90	—
INDIA	41,330	30,610	2,330	1,760	1,310	830	430	410	390	380	360	230	170	140	930

Note: Dolomite production in Bihar = Rs.10 m; Madhya Pradesh = Rs.40 m; Orissa = Rs.80 m. Gypsum production in Rajasthan = Rs.30 m; Tamil Nadu = Rs.10 m. Kyanite production in Bihar = Rs.20 m; Maharashtra = Rs.10 m. Mica (crude) production in Andhra Pradesh = Rs.10 m.

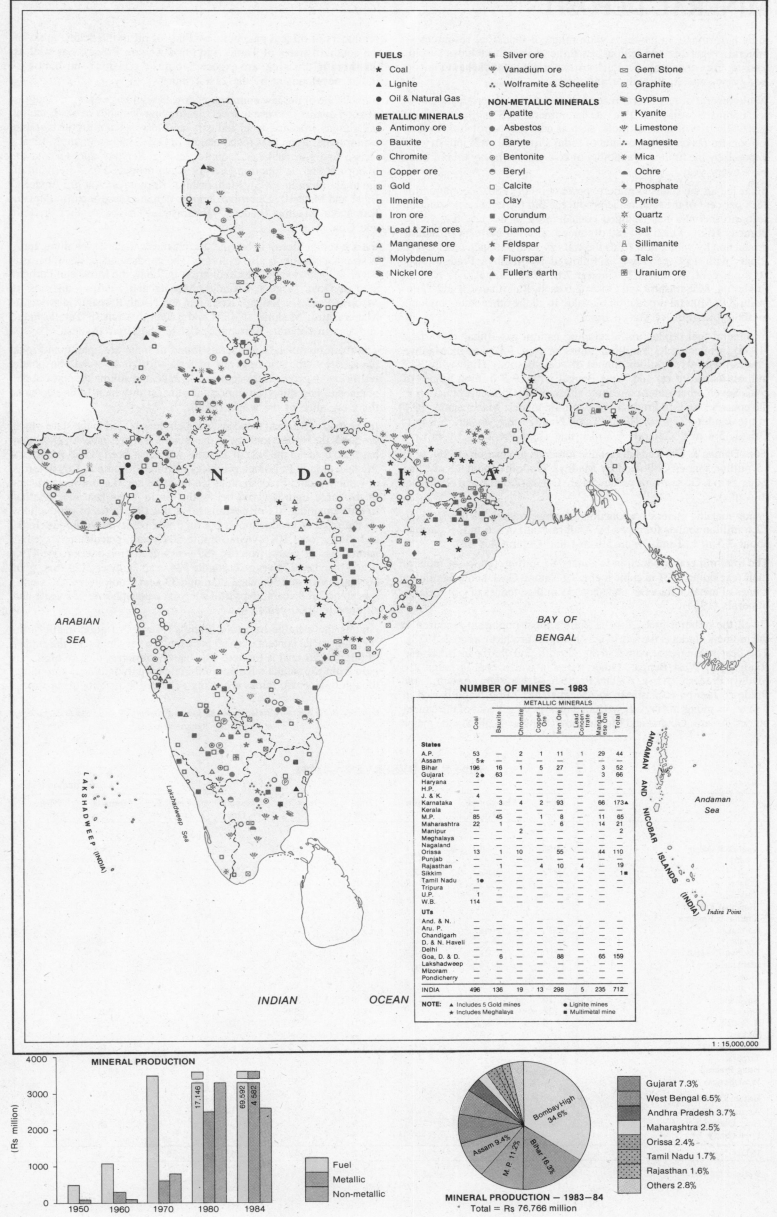

FUELS
- ★ Coal
- ▲ Lignite
- ● Oil & Natural Gas

METALLIC MINERALS
- ⊕ Antimony ore
- ○ Bauxite
- □ Chromite
- □ Copper ore
- ⊠ Gold
- ▫ Ilmenite
- ■ Iron ore
- ◆ Lead & Zinc ores
- △ Manganese ore
- ⊠ Molybdenum
- Nickel ore

- Silver ore
- Vanadium ore
- Wolframite & Scheelite

NON-METALLIC MINERALS
- ⊕ Apatite
- ● Asbestos
- ○ Baryte
- Bentonite
- Beryl
- □ Calcite
- □ Clay
- ■ Corundum
- Diamond
- ★ Feldspar
- ◆ Fluorspar
- ▲ Fuller's earth

- △ Garnet
- ⊠ Gem Stone
- ⊠ Graphite
- Gypsum
- Kyanite
- Limestone
- Magnesite
- Mica
- Ochre
- Phosphate
- Pyrite
- ☆ Quartz
- Salt
- Sillimanite
- Talc
- ⊞ Uranium ore

NUMBER OF MINES — 1983

States	Coal	METALLIC MINERALS Bauxite	Chromite	Copper Ore	Iron Ore	Lead Concentrate	Manganese Ore	Total
A.P.	53	—	2	1	11	1	29	44
Assam	5★	—	—	—	—	—	—	—
Bihar	196	16	1	5	27	—	3	52
Gujarat	2●	63	—	—	—	—	3	66
Haryana	—	—	—	—	—	—	—	—
H.P.	—	—	—	—	—	—	—	—
J. & K.	4	—	—	—	—	—	—	—
Karnataka	—	3	4	2	93	—	66	173▲
Kerala	—	—	—	—	—	—	—	—
M.P.	85	45	—	1	8	—	11	65
Maharashtra	22	1	—	—	6	—	14	21
Manipur	—	—	2	—	—	—	—	2
Meghalaya	—	—	—	—	—	—	—	—
Nagaland	—	—	—	—	—	—	—	—
Orissa	13	1	10	—	55	—	44	110
Punjab	—	—	—	—	—	—	—	—
Rajasthan	—	1	—	4	10	4	—	19
Sikkim	—	—	—	—	—	—	—	1■
Tamil Nadu	1●	—	—	—	—	—	—	—
Tripura	—	—	—	—	—	—	—	—
U.P.	1	—	—	—	—	—	—	—
W.B.	114	—	—	—	—	—	—	—
UTs								
And. & N.	—	—	—	—	—	—	—	—
Aru. P.	—	—	—	—	—	—	—	—
Chandigarh	—	—	—	—	—	—	—	—
D. & N. Haveli	—	—	—	—	—	—	—	—
Delhi	—	—	—	—	—	—	—	—
Goa, D. & D.	—	6	—	—	88	—	65	159
Lakshadweep	—	—	—	—	—	—	—	—
Mizoram	—	—	—	—	—	—	—	—
Pondicherry	—	—	—	—	—	—	—	—
INDIA	496	136	19	13	298	5	235	712

NOTE: ▲ Includes 5 Gold mines   ● Lignite mines
★ Includes Meghalaya   ■ Multimetal mine

1 : 15,000,000

ARABIAN SEA

BAY OF BENGAL

ANDAMAN AND NICOBAR ISLANDS (INDIA)

Andaman Sea

Indira Point

LAKSHADWEEP (INDIA)

Lakshadweep Sea

INDIAN OCEAN

MINERAL PRODUCTION

(Rs million)

- Fuel
- Metallic
- Non-metallic

1950 1960 1970 1980 1984

17,146
69,592
4,682

MINERAL PRODUCTION — 1983—84
Total = Rs 76,766 million

- Bombay High 34.6%
- Bihar 16.3%
- Assam 9.4%
- M.P. 11.2%
- Gujarat 7.3%
- West Bengal 6.5%
- Andhra Pradesh 3.7%
- Maharashtra 2.5%
- Orissa 2.4%
- Tamil Nadu 1.7%
- Rajasthan 1.6%
- Others 2.8%

MINERAL DEPOSITS

India is fortunate in having a wide range of industrial resources — mineral, vegetable and animal. Its mineral wealth includes surplus bauxite, mica and iron ore, which are exported; but India has to import lead, zinc, nickel, tin, silver and mercury.

While several of its mineral reserves are substantial, production does not compare well with that of the mineral-rich nations. This is especially true of mineral fuels, such as coal and petroleum. In fact, the constraints on the expansion of India's iron and steel industry are imposed by the limited availability of coal and the heavy oil bill India pays every year.

Bihar led all the states in mineral production in 1980, accounting for 28.8 per cent of the total value produced. But the mineral production scenario in India has changed considerably in the last few years. Bombay High, which in 1980 produced only 9.4 per cent of India's production by value, now leads the country with 35.7 per cent (1985). Bihar, with 15.6 per cent, is followed by Madhya Pradesh with 10.9 per cent, Assam 8.7, Gujarat 7.8, West Bengal 6.1, Andhra Pradesh 4, Maharashtra and Orissa 2.6 each, Rajasthan 1.9 and Tamil Nadu 1.8. Mineral production by value in all the other states amounts to about one-third of that in Bihar.

In mineral fuel production, excluding natural gas, Bihar ranks first (1985), producing 53.3 million tonnes of coal, followed by Madhya Pradesh with 41.3 million tonnes of coal, Bombay High with 20.7 million tonnes of oil and West Bengal with 19.3 million tonnes of coal. Seven other states contribute substantially to fuel production in the country: Andhra Pradesh, 14.6 million tonnes; Maharashtra, 10.8 m t (excluding Bombay High); Tamil Nadu, 7 m t; Assam, 5.8 m t; Orissa, 5.4 m t; Gujarat, 5.2 m t; and Uttar Pradesh, 3.6 m t.

Goa, Daman & Diu leads in metallic minerals production (1985), with 13 million tonnes, followed by Madhya Pradesh, about 11 m t and Bihar, 9 m t. Other major producers are Orissa (7.6 m t) and Karnataka (6.4 m t).

In non-metallic minerals production, Madhya Pradesh leads with about 12.6 million tonnes followed by Andhra Pradesh 7.6 m t, Rajasthan about 7.5 m t, Tamil Nadu 5.6 m t and Gujarat 4.8 m t.

The total mineral production is about 187 million tonnes of mineral fuels (excluding 3821 m cubic metres of Natural Gas), about 52 million tonnes of metallic minerals and about 55 million tonnes of non-metallic minerals (1985).

Of all the mineral fuels, coal is India's most important resource at the moment; India is the world's fifth largest producer of coal (1983). The best coking coal comes from Giridih and Jharia in Bihar and Raniganj in West Bengal. Other states in which coal is found are Madhya Pradesh, Orissa, Andhra Pradesh, Maharashtra, Assam, Uttar Pradesh, Jammu & Kashmir and Meghalaya. Tamil Nadu is rich in lignite, its mines in Neyveli producing nearly 92 per cent of the annual output. Assam, Gujarat and Maharashtra (Bombay High) are the major producers of oil and gas, but new finds of oil in the deltaic areas of the southern states of Tamil Nadu and Andhra Pradesh, as well as off-shore in this area, are expected to make a major contribution to Indian development by the 21st century.

India is one of the few countries endowed with large enough reserves of good quality iron ore to meet the growing demands of a substantial indigenous iron and steel industry and also sustain a considerable export trade. Iron ore is found mainly in Goa, Madhya Pradesh, Bihar, Orissa and Karnataka. Chromite, another ferrous mineral and an important one, is found in Orissa (which produced 85.7 per cent of the total output in 1985), Maharashtra, Karnataka, Andhra Pradesh, Bihar and Manipur. Extensive deposits of manganese occur in Punjab, Karnataka, Madhya Pradesh, Maharashtra, Orissa, Andhra Pradesh and Goa.

India is self-sufficient in bauxite, the chief raw material for aluminium. A sizeable quantity is also exported. The important deposits of bauxite occur in Palamau and Ranchi districts in Bihar; the Mopa and Pernem area in Goa; Surguja, Shahdol, Mandla and Bilaspur districts in Madhya Pradesh; Raigad, Kolhapur and North Ratnagiri districts in Maharashtra; Madurai, Nilgiri and Salem districts of Tamil Nadu; and also in Gujarat, Karnataka, Orissa and Uttar Pradesh.

The main non-metallic minerals found in India are apatite and rock phosphate — the principal raw materials for commercial phosphatic fertilizers — gypsum (for cement), kyanite and sillimanite, limestone, magnesite and mica (an important mineral in which India accounts for a big slice of the world trade).

In the map of Mineral Deposits, each circle is placed in the state which produces the highest tonnage of the mineral it represents. An exception has been made in the case of asbestos which has been placed in Andhra Pradesh though Rajasthan produces ten times its output, for, in terms of value, Andhra Pradesh ranks ten times higher. The most significant of the three figures given below the circle is the last one, relating to the remaining life of the mineral at the 1981 rates of production. At the rates of production at the time, the reserves would last India as follows: coal 1016 years, lignite 504 years, petroleum 31 years, natural gas 21 years, iron ore 439 years, lead-zinc concentrate 4795 years, bauxite 1274 years, chromite 324 years, copper 212 years, gold 26 years, asbestos 30 years, diamond 39 years, dolomite 2450 years, limestone 2288 years, gypsum 1306 years, phosphorite 236 years and sillimanite 1000 years.

Among the metallic minerals, India's reserves of good quality iron ore, with metal content over 60 per cent, are very high, but the supply of bituminous coal is limited. The bauxite reserves too are high, but gold from Karnataka's Kolar Gold Fields is a dwindling commodity, though one or two mines are being explored in neighbouring states.

Source: *Basic Statistics Relating to the Indian Economy, Vol. 2: States, September 1986*, Centre for Monitoring Indian Economy.

MINERAL RESERVES[1] — 1984

Table 42

(in million tonnes)

	Iron Ore	Manganese Ore	Chromite	Lime-stone	Dolomite	Copper Ore	Lead & Zinc Ore	Bauxite	Nickel
States									
Andhra Pradesh	257	3	—	15,971	127	9	8	480	—
Assam	50	—	—	525	—	—	—	—	—
Bihar	3,572	Neg.	5	673	35	215	—	73	—
Gujarat	—	3	—	10,792	245	8	7	90	—
Haryana	8	—	—.	52	6	15	—	—	—
Himachal Pradesh	—	—	—	959	—	—	—	—	—
Jammu & Kashmir	—	—	—	293	—	—	—	7	—
Karnataka	6,337	48	3	16,968	334	11	—	31	—
Kerala	88	—	—	42	—	—	—	16	—
Madhya Pradesh	2,470	21	—	8,218	1,426	193	—	194	—
Maharashtra	225	16	Neg.	3,485	260	3	1	102	—
Manipur	—	—	—	8	—	—	—	—	—
Meghalaya	—	—	—	6,189	—	Neg.	Neg.	—	—
Nagaland	10	—	—	337	—	—	—	—	—
Orissa	3,124	33	132	841	668	2	3	1,601	160
Rajasthan	16	Neg.	—	5,320	90	109	335	1	—
Sikkim	—	—	—	—	—	1	1	—	—
Tamil Nadu	532	—	Neg.	823	2	Neg.	1	16	—
Tripura	—	—	—	1	—	—	—	—	—
Uttar Pradesh	—	—	—	1,367	75	1	1	14	—
West Bengal	—	—	—	24	252	Neg.	3	—	—
Union Territories									
Arunachal Pradesh	—	—	—	140	426	—	—	—	—
Goa, Daman & Diu	884	3	—	129	—	—	—	28	—
Pondicherry	—	—	—	5	—	—	—	—	—
INDIA	17,573	127	135	73,202	3,946	567	360	2,635	160

Note: [1] Estimated

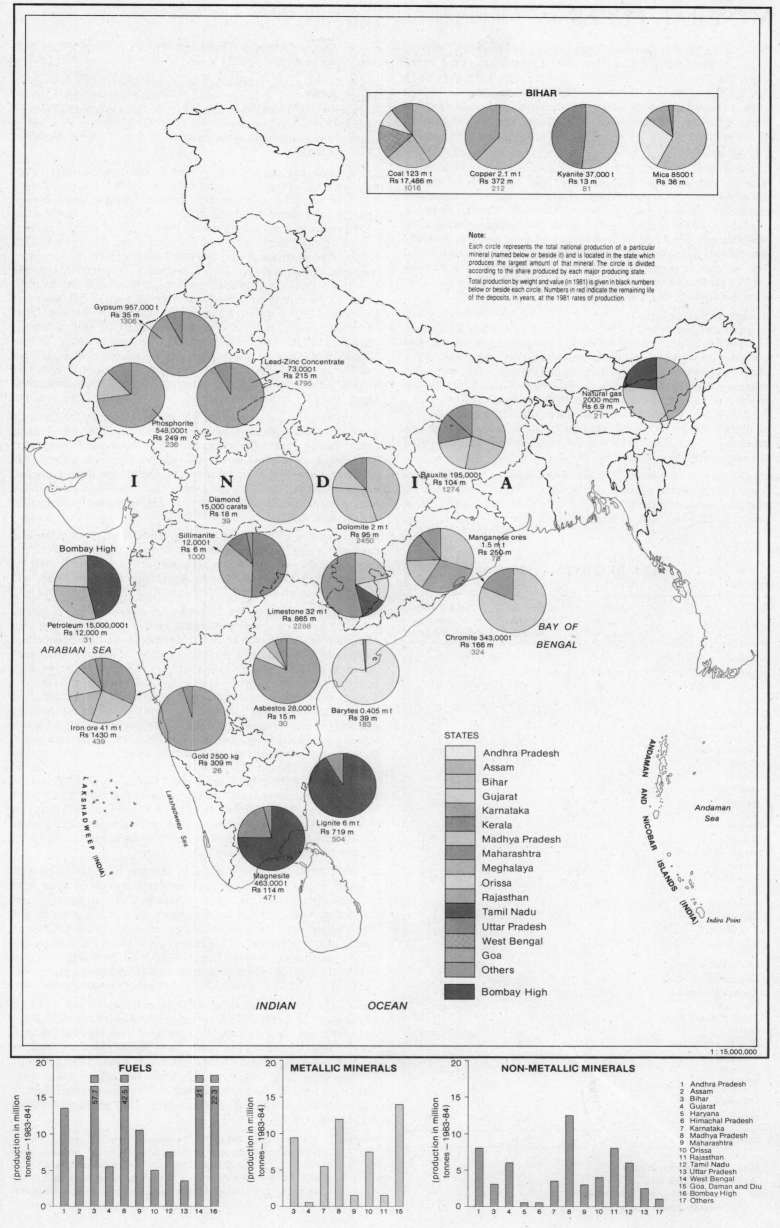

BIHAR

Coal 123 m t
Rs 17,486 m
1016

Copper 2.1 m t
Rs 372 m
212

Kyanite 37,000 t
Rs 13 m
81

Mica 8500 t
Rs 36 m

Note:

Each circle represents the total national production of a particular mineral (named below or beside it) and is located in the state which produces the largest amount of that mineral. The circle is divided according to the share produced by each major producing state.

Total production by weight and value (in 1981) is given in black numbers below or beside each circle. Numbers in red indicate the remaining life of the deposits, in years, at the 1981 rates of production.

Gypsum 957,000 t
Rs 35 m
1306

Lead-Zinc Concentrate
73,000 t
Rs 215 m
4795

Natural gas
2000 mcm
Rs 6.9 m
21

Phosphorite
548,000 t
Rs 249 m
236

I N D I A

Bauxite 195,000 t
Rs 104 m
1274

Diamond
15,000 carats
Rs 18 m
39

Bombay High

Sillimanite
12,000 t
Rs 6 m
1000

Dolomite 2 m t
Rs 95 m
2450

Manganese ores
1.5 m t
Rs 250 m
78

Petroleum 15,000,000 t
Rs 12,000 m
31

ARABIAN SEA

Limestone 32 m t
Rs 865 m
2288

Chromite 343,000 t
Rs 166 m
324

BAY OF BENGAL

Iron ore 41 m t
Rs 1430 m
439

Asbestos 28,000 t
Rs 15 m
30

Barytes 0.405 m t
Rs 39 m
183

LAKSHADWEEP (INDIA)

Lakshadweep Sea

Gold 2500 kg
Rs 309 m
26

Lignite 6 m t
Rs 719 m
504

Magnesite
463,000 t
Rs 114 m
471

STATES

- Andhra Pradesh
- Assam
- Bihar
- Gujarat
- Karnataka
- Kerala
- Madhya Pradesh
- Maharashtra
- Meghalaya
- Orissa
- Rajasthan
- Tamil Nadu
- Uttar Pradesh
- West Bengal
- Goa
- Others
- Bombay High

ANDAMAN AND NICOBAR ISLANDS (INDIA)

Andaman Sea

Indira Point

INDIAN OCEAN

1 : 15,000,000

FUELS

(production in million tonnes — 1983-84)

20
15
10
5
0

57.7 42.5 21 22.3

1 2 3 4 8 9 10 13 14 16

METALLIC MINERALS

(production in million tonnes — 1983-84)

20
15
10
5
0

3 7 8 9 10 11 15

NON-METALLIC MINERALS

(production in million tonnes — 1983-84)

20
15
10
5
0

1 3 4 5 6 7 8 9 10 11 12 13 17

1 Andhra Pradesh
2 Assam
3 Bihar
4 Gujarat
5 Haryana
6 Himachal Pradesh
7 Karnataka
8 Madhya Pradesh
9 Maharashtra
10 Orissa
11 Rajasthan
12 Tamil Nadu
13 Uttar Pradesh
14 West Bengal
15 Goa, Daman and Diu
16 Bombay High
17 Others

MINERAL FUELS

The importance of fuel in the world economy was forcefully brought to the notice of every Indian when the oil price hikes of the 1970s shook the world, hitting developing countries like India even more than the highly industrialized ones. With India determined to become a major economic power by the 21st century, its whole focus of attention in the eighties is on energy, the key to the pace of development.

The rate of consumption of energy resources is a function of the rate of population increase and the rates of development in agriculture, industry and services. For primitive man, the level of energy consumption was 100 thermal watts. When the use of firewood was discovered, the level rose to around 1000 watts. The rate stayed at that level until the continuous mining of coal, beginning about eight centuries ago, and the production of oil, beginning just over a century ago, brought an exponential increase to around 10,000 thermal watts. The increase in consumption of fossil fuel is so new that half the world's production of coal has occurred only since 1930 and half its oil consumption since 1952.

It is against this backdrop that the energy crisis of the 1970s and 1980s — a crisis of supply and price — acquires significance. Even more so when it is considered that the world is running out of oil. The fuel crisis that the world faces is due to two factors: one, the exhaustibility of the resources, especially petroleum, and the other, the distribution of these resources. Saudi Arabia has the richest of oil reserves, but at the present rate of exploitation it will get exhausted in about a hundred years. On the other hand, the position in the case of coal is slightly better. While oil distribution is confined to very few countries — the USA, USSR and the West Asian countries — coal is more widely distributed.

Among the various mineral fuels, petroleum accounts for nearly three-fourths of the world's production by value. Worldwide production of crude petroleum (1985) is about 2669 million tonnes. The USSR produces 22.3 per cent of this, USA 16 per cent, Saudi Arabia 6.5 per cent and Mexico 5.3 per cent. India's production is a little over 1 per cent of the world total.

In just twenty years, crude oil production in the country has shown a dramatic increase, from a mere 4.7 million tonnes in 1966 to a six-fold increase of 30 million tonnes in 1985. It is expected to increase by a further 4.5 million tonnes by 1989-90, There has also been a similar increase in natural gas production, from both on-shore and off-shore wells.

The internal consumption of petroleum products has, however, kept pace with the increasing on- and off-shore oil and gas production. A substantial part of it, about three-fourths, therefore, has to be met with imports from other countries, notably from West Asia. Maharashtra (6.5 million tonnes in 1984-85), Gujarat (4.7 million tonnes), Uttar Pradesh (3.8 million tonnes) and Tamil Nadu (3.5 million tonnes) consume the most petroleum products (motor spirit, kerosene, diesel, fuel oils, LPG and others). Of the approximately 39 million tonnes of petroleum products consumed in the country, about 30 per cent is consumed in the western states (Maharashtra, Gujarat and Goa) and about a quarter in the four southern states, Tamil Nadu, Kerala (1.3 million tonnes), Karnataka (1.7 million tonnes) and Andhra Pradesh (2.2 million tonnes). The states of Assam, Orissa, Nagaland, Manipur, Meghalaya, Jammu & Kashmir, Sikkim, Tripura and Himachal Pradesh consume only about 4 per cent (1.8 million tonnes) of the total, and the union territories together consume about 6 per cent (2.3 million tonnes).

This consumption, seen in tonne per 1000 persons, places the union territories in the highest position with 232 tonnes (1984-85). Gujarat (130), Punjab (111) and Maharashtra (98) top the state index list. Considered in relation to the national average (56 tonnes/1000 persons), only two other states besides these have high indices of consumption: Haryana (81) and Tamil Nadu (70). The states with the lowest indices are Bihar (25), Orissa (23), Himachal Pradesh (23) and Tripura (18).

The country is now making a major effort at oil exploration and exploitation and has begun laying a network of pipelines connecting production sites with refineries. This network is expected to web the peninsula once oil is struck in the Kaveri and Godavari basins and brought under production.

The areas of potential petroleum are four in number. Basins of proven potential for commercial production are in Assam and off-shore Maharashtra. Basins with known occurrences but where commercial exploitation is yet to begin include the whole of the east coast, both on- and off-shore, the Krishna-Godavari valley (a region of high potential), a narrow strip across western Uttar Pradesh, Haryana and Punjab, the border area of Rajasthan (desert), the Ganges delta and the Brahmaputra valley, and the seas around the Andaman & Nicobar Islands. The areas with no significant oil or gas occurrences, but which appear geologically favourable, include parts of Rajasthan and western Gujarat (the Kathiawar peninsula) and the south-west coast, both on- and off-shore. The Oil & Natural Gas Commission made a major oil find in the Khambhat Basin off Gujarat in August 1986. The wells here are estimated to have 110 million tonnes of oil besides reservoirs of natural gas. The quality of the oil here is expected to be superior to that of Bombay High. Basins with uncertain prospects include almost the whole of Uttar Pradesh, Bihar, Madhya Pradesh and parts of Maharashtra and the north-east.

Of the proven and exploited oil fields, Bombay High has the greatest potential, in both crude oil and gas reserves, though the Gandhar strike in the Khambhat basin might well surpass this. The on-shore Gujarat and Assam basins have 187 million tonnes of crude oil reserves and 100.4 billion cubic metres of natural gas reserves against 324 million tonnes of crude oil and 377.8 billion cubic metres of natural gas reserves at off-shore Bombay High (1984). In all, production potential of natural gas is 118 million cubic metres a day, including a proven and capped reserve of 53.8 million cubic metres per day in the western basin.

Coal production in India touched 149 million tonnes in 1985, of which a little more than a third (35.8 per cent or 53.4 m t) came from Bihar and a little more than a fourth (27.7 per cent) from Madhya Pradesh. Another third came from four states, West Bengal (19.3 m t), Andhra Pradesh (14.6 m t), Maharashtra (10.8 m t) and Orissa (5.4 m t). The rest came from Uttar Pradesh, Jammu & Kashmir and Assam. The target for 1989-90 has been fixed at 226 million tonnes. India's coal production is 4.7 per cent of the world's 1985 production of 3207 million tonnes.

FUEL RESERVES — 1984

Table 43

	Coal (in million tonnes)	Crude Oil (in million tonnes)	Natural Gas (in billion [1] cubic metres)
States			
Andhra Pradesh	8,505	—	—
Assam	280	99	80
Bihar	55,703	—	—
Gujarat	156[2]	88	19
Haryana	—	—	—
Himachal Pradesh	—	—	—
Jammu & Kashmir	8[2]	—	—
Karnataka	—	—	—
Kerala	—	—	—
Madhya Pradesh	23,234	—	—
Maharashtra	3,183	—	Neg.
Manipur	—	—	—
Meghalaya	509	—	—
Nagaland	12	1	Neg.[3]
Orissa	29,535	—	—
Punjab	—	—	—
Rajasthan	105[2]	—	Neg.[3]
Sikkim	—	—	—
Tamil Nadu	3,379[2]	—	—
Tripura	—	—	1
Uttar Pradesh	—	—	—
West Bengal	27,740	—	—
Union Territories			
Andaman & Nicobar	—	—	—
Arunachal Pradesh	91	—	—
Chandigarh	—	—	—
Dadra & Nagar Haveli	—	—	—
Delhi	—	—	—
Goa, Daman & Diu	—	—	—
Lakshadweep	—	—	—
Mizoram	—	—	—
Pondicherry	—	—	—
Off-shore			
Bombay High	—	324	378
INDIA	152,449[4]	511	478

Note: [1] Billion = 1000 Million
[2] Lignite reserves
[3] Less than 1 billion cubic metres
[4] Includes lignite reserves.

Sources: 1. *The Twentyninth Day:* Lester Brown.
2. *Basic Statistics Relating to the Indian Economy, Vol. 1: All India, August 1986, Vol. 2: States, September 1986 and World Economy & India's Place in It, October 1986,* Centre for Monitoring Indian Economy.
3. *Philips Modern School Economic Atlas,* George Philip & Son Limited, London.

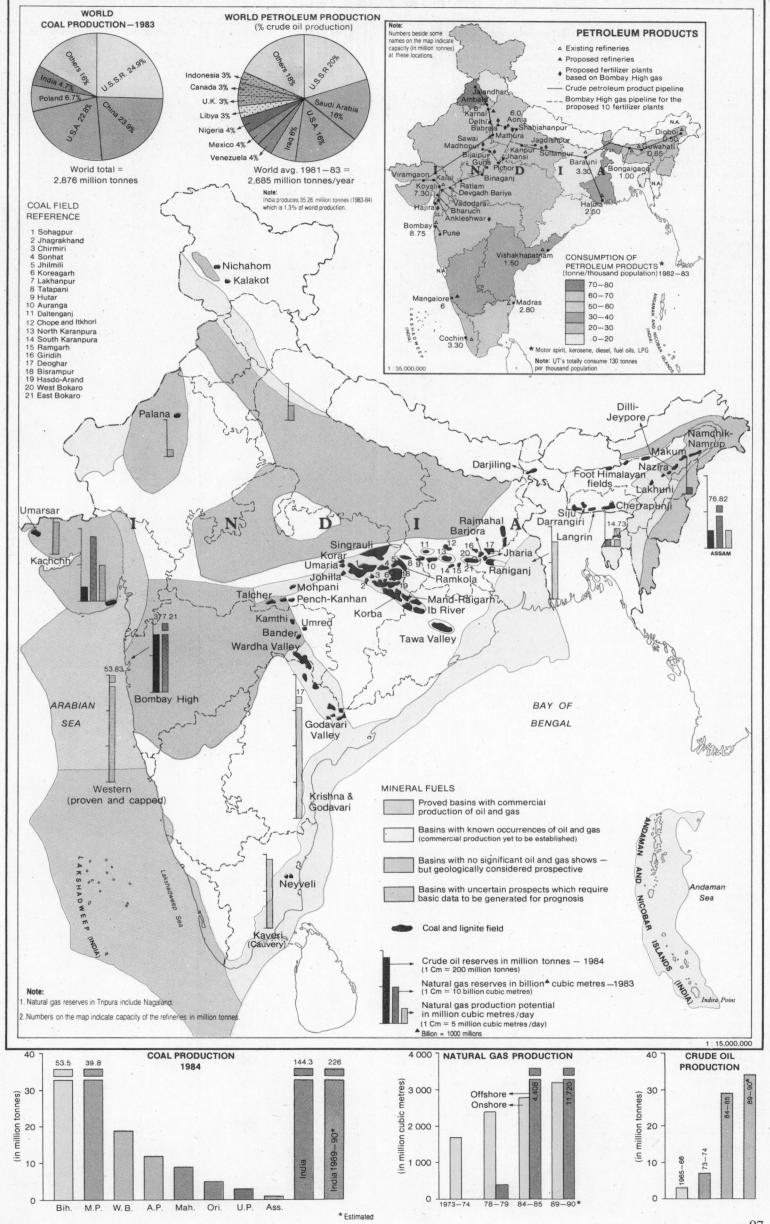

WORLD COAL PRODUCTION — 1983

Others 16%
U.S.S.R. 24.9%
India 4.7%
Poland 6.7%
U.S.A. 22.8%
China 23.9%

World total = 2,876 million tonnes

WORLD PETROLEUM PRODUCTION (% crude oil production)

Others 16%
U.S.S.R. 20%
Indonesia 3%
Canada 3%
U.K. 3%
Libya 3%
Nigeria 4%
Mexico 4%
Venezuela 4%
Iraq 9%
Saudi Arabia 16%
U.S.A. 16%

World avg. 1981—83 = 2,685 million tonnes/year

Note: India produces 35.26 million tonnes (1983-84) which is 1.3% of world production.

COAL FIELD REFERENCE

1 Sohagpur
2 Jhagrakhand
3 Chirmiri
4 Sonhat
5 Jhilmili
6 Koreagarh
7 Lakhanpur
8 Tatapani
9 Hutar
10 Auranga
11 Daltenganj
12 Chope and Itkhori
13 North Karanpura
14 South Karanpura
15 Ramgarh
16 Giridih
17 Deoghar
18 Bisrampur
19 Hasdo-Arand
20 West Bokaro
21 East Bokaro

Note: Numbers beside some names on the map indicate capacity (in million tonnes) at these locations.

PETROLEUM PRODUCTS

△ Existing refineries
▲ Proposed refineries
◆ Proposed fertilizer plants based on Bombay High gas
— Crude petroleum product pipeline
-- Bombay High gas pipeline for the proposed 10 fertilizer plants

CONSUMPTION OF PETROLEUM PRODUCTS★ (tonne/thousand population) 1982—83

70—80
60—70
50—60
30—40
20—30
0—20

★ Motor spirit, kerosene, diesel, fuel oils, LPG

Note: UT's totally consume 130 tonnes per thousand population.

1 : 35,000,000

Jalandhar
Ambala
Karnal
Delhi
Babrala
Sawai Madhopur
Bijaipur
Viramgam
Koyali 7.30
Hajira
Bombay 8.75
Pune
Aonla
Shahjahanpur
Mathura
Jagdishpur
Kanpur
Guna
Jhansi
Pichor
Binaganj
Kalol
Ratlam
Devgadh Bariya
Vadodara
Bharuch
Ankleshwar
Sultanpur
Baraini 3.30
Bongaigaon 1.00
Digboi 0.50
Guwahati 0.85
Haldia 2.50
Vishakhapatnam 1.50
Mangalore 6
Madras 2.80
Cochin 3.30

6.0

N.A.

MINERAL FUELS

Proved basins with commercial production of oil and gas

Basins with known occurrences of oil and gas (commercial production yet to be established)

Basins with no significant oil and gas shows — but geologically considered prospective

Basins with uncertain prospects which require basic data to be generated for prognosis

● Coal and lignite field

Crude oil reserves in million tonnes — 1984 (1 Cm = 200 million tonnes)

Natural gas reserves in billion ▲ cubic metres — 1983 (1 Cm = 10 billion cubic metres)

Natural gas production potential in million cubic metres/day (1 Cm = 5 million cubic metres /day)

▲ Billion = 1000 millions

1 : 15,000,000

Map labels:

Nichahom
Kalakot
Palana
Umarsar
Kachchh
Umaria
Singrauli
Korar
Johilla
Mohpani
Talcher
Pench-Kanhan
Korba
Kamthi
Bander
Wardha Valley
Umred
Bombay High
Godavari Valley
Krishna & Godavari
Neyveli
Kaveri (Cauvery)
Western (proven and capped)
Rajmahal
Barjora
Jharia
Raniganj
Ramkola
Mand-Raigarh
Ib River
Tawa Valley
Darjiling
Dilli-Jeypore
Namchik-Namrup
Makum
Nazira
Foot Himalayan fields
Lakhuni
Cherrapunji
Siju
Darrangiri
Langrin
ASSAM
Umaria

377.21
53.83
76.82
14.73

ARABIAN SEA
BAY OF BENGAL
LAKSHADWEEP (INDIA)
Lakshadweep Sea
ANDAMAN AND NICOBAR ISLANDS (INDIA)
Andaman Sea
Indira Point

Note:
1. Natural gas reserves in Tripura include Nagaland.
2. Numbers on the map indicate capacity of the refineries in million tonnes.

COAL PRODUCTION 1984

(in million tonnes)

Bih. 53.5
M.P. 39.8
W.B.
A.P.
Mah.
Ori.
U.P.
Ass.
India 144.3
India 1989—90* 226

* Estimated

NATURAL GAS PRODUCTION

(in million cubic metres)

Offshore
Onshore

1973—74
78—79
84—85 4,408
89—90* 11,720

CRUDE OIL PRODUCTION

(in million tonnes)

1965—66
73—74
84—85
89—90*

Table IV

USE OF THE LAND — 1981-82

	Total geographical area — 1981-82 (in thousand ha.)	Reporting area — 1981-82 (in thousand ha.)	Gross cropped — 1981-82 (in thousand ha.)	Area under forest 1981-82 (in thousand ha.)	Culturable wasteland 1981-82 (in thousand ha.)	Number of mines[2] 1983
States						
Andhra Pradesh	27,507	27,440	13,047	6,360	2,063	437
Assam	7,844	7,852[1]	3,439	3,080	119	6
Bihar	17,388	17,330	10,628	2,930	902	497
Gujarat	19,602	18,826	10,903	1,970	3,252	507
Haryana	4,421	4,405	5,826	160	274	11
Himachal Pradesh	5,567	3,089	949	2,100	84	25
Jammu & Kashmir	22,224	4,675	978	2,190	86	8
Karnataka	19,179	19,050	11,228	3,830	1,375	295
Kerala	3,886	3,885	2,905	1,120	186	43
Madhya Pradesh	44,305	44,211	21,756	15,540	2,600	492
Maharashtra	30,769	30,758	20,386	6,400	2,891	122
Manipur	2,233	2,221	240	1,520	321	2
Meghalaya	2,243	2,249	203	850	159	6
Nagaland	1,628	1,099	164	290	159	—
Orissa	15,571	15,540	8,743	5,990	1,146	208
Punjab	5,036	5,033	6,929	260	352	1
Rajasthan	34,224	34,234[1]	18,596	3,050	11,878	914
Sikkim	710	719[1]	92	280	69	1
Tamil Nadu	13,006	13,002	6,909	2,200	900	155
Tripura	1,049	1,048	380	600	94	—
Uttar Pradesh	29,441	29,708[1]	24,773	5,120	2,804	152
West Bengal	8,875	8,846	7,402	1,190	546	132
Union Territories						
Andaman & Nicobar	825	790	36	710	15	—
Arunachal Pradesh	8,374	5,550	152	5,150	259	—
Chandigarh	11	11	4	—	—	—
Dadra & Nagar Haveli	49	49	25	—	—	—
Delhi	148	147	84	—	5	4
Goa, Daman & Diu	381	371	142	130	12	163
Lakshadweep	3	3	3	—	—	—
Mizoram	2,108	2,102	68	1,650	211	—
Pondicherry	49	47	51	—	6	—
INDIA	328,726	304,280	177,041	74,670	32,768	4,181

Note: 1. Under verification.
2. Excludes Natural Gas and Petroleum.

EARNINGS FROM THE LAND

Table V

	Value of Agricultural output per hectare of gross cropped area — 1981-82 (in Rs.)	Value of Forest produce per hectare of forest area — 1979-80 (in Rs.)	Value of mineral output per mine[1] 1983 (in Rs. thousand)
States			
Andhra Pradesh	2,785	26	5,103
Assam	3,455	33	13,467[2]
Bihar	2,500	33	2,091
Gujarat	1,866	95	504
Haryana	2,710	130	2,087
Himachal Pradesh	2,895	276	676
Jammu & Kashmir	3,814	153	1,488
Karnataka	2,093	236	2,437
Kerala	4,510	335	1,030
Madhya Pradesh	1,186	15	13,084
Maharashtra	1,844	60	11,356
Manipur	3,727	2	14
Meghalaya	2,955	2	—
Nagaland	1,688	13	—
Orissa	1,799	27	6,899
Punjab	3,135	125	43
Rajasthan	1,205	22	1,179
Sikkim	N.A.	5	1,270
Tamil Nadu	2,609	92	7,757
Tripura	2,895	13	—
Uttar Pradesh	2,401	67	3,259
West Bengal	3,612	50	26,878
Union Territories			
Andaman & Nicobar	N.A.	N.A.	—
Arunachal Pradesh	N.A.	N.A.	—
Chandigarh	N.A.	N.A.	—
Dadra & Nagar Haveli	N.A.	N.A.	—
Delhi	N.A.	N.A.	159
Goa, Daman & Diu	N.A.	N.A.	1,606
Lakshadweep	N.A.	N.A.	—
Mizoram	N.A.	N.A.	—
Pondicherry	N.A.	N.A.	—
INDIA	2,153	56	7,096

Note: 1. Excludes Natural Gas and Petroleum.
2. Includes Meghalaya.

The Infrastructure

This section comprises 40 maps and 55 charts. It examines the facilities developed and being developed to tap human and natural resource potentials and utilize them to the maximum.

> **Note:**
>
> a) Data for the primary and secondary education maps (pp 105 and 107) is based on the Census of India's 5 per cent sampling, which is available only for the 14 major states. Data for higher education is an estimate for 1980-81 based on 1978-80 enrolment figures.
>
> b) There is some discrepancy in the three education maps due to different sources using different age-groups for their classification of education categories. The education categories themselves can be confusing. But for the purpose of this atlas, the general pattern of education in the country has been taken as ten years schooling and two years in secondary school followed by college. The years in secondary school and college have been considered as higher education, thus including all students who are 14 plus in age.
>
> The maps on printed media on p 123 broadly indicate the potential reach newspapers and magazines have in India. The first map shows the number of publications in a given town and the population they *could* reach (the towns literate population as well as the literate population of the surrounding districts being indicated). The second map shows the possible readership of each copy of a newspaper or magazine published in that state.

MINIMUM NEEDS

There is a distinctive strategy for development and improving the lot of the poor in the country in India's Five-Year Plans. This can be called a 'basic needs approach'.

Basic needs include two elements:

First, they include certain minimum requirements of a family for private consumption: adequate food, shelter and clothing are obviously included, as would be certain household equipment and furniture. Second, they include essential services provided by, and for, the community at large, such as safe drinking water, sanitation, public transport and health and educational facilities.

Minimum needs is part of the dynamic basic needs concept and is, therefore, considered here in the context of India's overall economic and social development.

To provide minimum needs, the most important consideration is the plan outlay to build the necessary infrastructure, an extremely costly affair, especially in India, where the needs are several and the people more than 700 million in number. Since the rural population is three-fourths of India's population, the greater part of any minimum needs expenditure must necessarily be spent on rural India, where very little already exists and where access is often difficult, making developmental costs still greater.

The minimum needs outlays for 1984-85 are very high in Andhra Pradesh, Uttar Pradesh, Tamil Nadu, Maharashtra and Bihar. The other states have progressively decreasing outlays of Rs 1000 million and less.

In rural areas, it appears that more villages have easier access to such minimum amenities as transport, education, water supply and electricity than to health and agricultural, including veterinary, services. This has serious implications for human health and cattle wealth (India has four cattle per person). Kerala is the lone exception to this. States that are mainly hilly and/or tribal are the worst off.

The national plan outlay for minimum needs (1985-90) is around Rs 115,500 million, to be spent on rural health, nutrition, elementary education, adult education, rural water supply, urban electrification, rural housing, urban slum housing and rural roads.

Rural health and nutrition acquire special significance under the 'Health For All by 2000' commitment and so get about a quarter (24.5 per cent) of the total outlay. Only two states in the country (Tamil Nadu, 20.8 per cent and Gujarat 22.5 per cent of the total Indian outlay on rural health and nutrition) are close to the national commitment of 9.5 per cent on rural health and 15 per cent on nutrition. The outlay in the other states is very much below the national commitment. Tamil Nadu has an ambitious Nutritious Meal Scheme with an outlay of Rs 5388 million, but Gujarat, without any such multifaceted nutrition programme, is spending even more, Rs 5955 million. The Tamil Nadu scheme has resulted in 98.7 per cent pupil enrolment in the first five classes and 72.2 per cent in the next three during the Sixth Plan — rates far exceeding the national average. It is estimated that a national scheme like Tamil Nadu's would cost Rs 40,000 million in the Seventh Plan period.

The Seventh Plan outlay on elementary and adult education amounts to about 19 per cent of the national outlay on minimum needs. Only Uttar Pradesh and West Bengal, which will spend Rs 2220 m and Rs 1950 million respectively, each have an outlay on such education of over 10 per cent of the national outlay. Adult education in all the states gets a still lower outlay than elementary education. The largest outlay will be Rs 420 million in Uttar Pradesh.

Rural water supply and electrification will have about Rs 40,000 million, or about 34 per cent of the total minimum needs outlay. Rural water supply, which gets more than 75 per cent of the outlay, is the more important need. There are about 100,000 villages (the first five Five Year Plans covered only 36,000 of the 'difficult and problem' villages) yet to be provided with safe water supply, and at least 200,000 villages do not at present have *protected* water supply. Many of these villages, however, draw unprotected water from sources close to them. Some states have set aside their entire outlays under this head for rural water supply (*e.g.* Maharashtra, Tamil Nadu, Andhra Pradesh and Haryana) in preference to further electrification.

The national outlay on housing in rural areas and urban slum localities is 7.3 per cent of the outlay. Andhra Pradesh alone will spend 29 per cent of this outlay and together with Maharashtra and Karnataka will account for more than half the national total. In all states and union territories except Maharashtra, Haryana, Tamil Nadu, Rajasthan, Punjab, West Bengal, Uttar Pradesh, Meghalaya, Mizoram and Delhi, urban slum housing gets lower allocations than rural housing. Sikkim and Andaman & Nicobar spend 100 per cent on urban slum housing. Urban slum housing, however, gets top priority in urban development. Ambitious slum clearance programmes, coupled with new housing for the displaced, are planned in Tamil Nadu and Maharashtra, the most urbanized states.

Rural roads, vital for the earlier mentioned services to rural people, gets an outlay of 15 per cent of the total outlay, with Uttar Pradesh (Rs 6500 million) spending the most. This outlay is, however, meagre considering that most rural folk are 8-15 kms from most service facilities and a fifth of India's rural population is more than 20 kms from such locations.

The indices of development of infrastructure show that the ratio of highest development to lowest in the states was 4.8 to 1 in 1972-73 and had improved to 3.7 to 1 by 1984-85. In 1972-73 six major states, Andhra Pradesh, Bihar, Assam, Orissa, Rajasthan and Madhya Pradesh, and all the minor states, were below the all-India base index of 100. By 1984-85, Karnataka had moved into the list of poor performers, while Andhra Pradesh had improved marginally and matched the all-India base. Five states maintained very high development indices over these years: Punjab, the leading state throughout, Haryana, Tamil Nadu, Kerala and West Bengal. Maharashtra, Uttar Pradesh and Gujarat were also above the all-India base index. But, over the years, the leading states have dropped from relatively high indices to lower indices, a glaring example being Tamil Nadu — from 171 in 1966-67 to 138 in 1984-85. This does not, however, mean that there has necessarily been a shrinkage of the infrastructure in the state; it is only that the rate of expansion of infrastructure has been slower than the all-India average rate of expansion. Orissa, Himachal Pradesh and Tripura are the only three states showing steadily accelerating indices.

The female-male welfare indices for the states also reflect the relative development of the states. Kerala has the highest per cent of female welfare index to male welfare index (95 per cent). Three southern states — Tamil Nadu, Karnataka and Andhra Pradesh — and three northern states — Punjab, Bihar and Uttar Pradesh — also fare well in female welfare (75-80 per cent), unlike Haryana, Orissa and Maharashtra, which have the lowest female: male welfare percentage (65-70 per cent).

Sources: 1. *Basic Statistics Relating to the Indian Economy, Vol. 2: States, September 1986*, Centre for Monitoring Indian Economy.
2. *Seventh Five Year Plan 1985-90, Vol. II*, Planning Commission.
3. *Report of the Director-General of I.L.O., 1976*, Geneva.

Table 44
BASIC AMENITIES IN VILLAGES — 1977[1]

	Villages for which data was processed	Metalled road (within 2 kms.)	Bus stop (within 2 kms.)	Primary or junior basic school (within 2 kms.)	High/higher secondary school (within 5 kms.)	Hospital (within 5 kms.)	Bank (within 5 kms.)	Electricity	Drinking water (within village)
Andhra Pradesh	25,756	15,812	12,582	23,185	14,006	8,129	9,452	13,169	23,611
Assam	23,082	7,603	11,010	21,628	16,254	12,423	6,703	2,112	22,006
Bihar	62,404	26,964	21,398	57,424	39,233	28,662	22,166	10,865	57,655
Gujarat	18,158	10,549	12,925	17,380	3,672	3,653	9,237	6,795	16,903
Haryana	6,681	6,192	4,794	6,606	5,307	2,088	5,973	6,666	6,620
Himachal Pradesh	16,275	4,575	7,420	13,627	8,963	2,381	10,535	8,264	11,212
Jammu & Kashmir	6,487	3,860	3,114	5,857	3,982	2,655	2,885	2,784	N.A.
Karnataka	27,006	19,506	15,979	25,618	13,932	13,984	12,839	17,945	24,945
Kerala	974	957	954	971	942	763	936	944	937
Madhya Pradesh	70,525	19,183	18,823	60,482	19,260	14,500	23,577	14,149	66,725
Maharashtra	35,746	15,864	16,586	33,326	11,353	8,608	12,500	19,146	34,477
Manipur	2,068	718	659	1,815	825	243	332	194	1,718
Meghalaya	4,599	1,019	917	3,846	1,089	419	349	215	4,030
Nagaland	963	253	167	946	207	161	66	268	761
Orissa	41,656	12,138	11,018	36,778	22,757	9,818	10,097	7,698	35,932
Punjab	12,162	11,512	8,577	12,035	10,225	6,117	7,701	11,941	12,088
Rajasthan	32,430	10,183	13,400	26,375	11,347	10,309	17,764	7,762	31,292
Sikkim	215	92	56	194	55	38	31	45	162
Tamil Nadu	14,875	11,705	10,753	14,145	9,947	3,835	6,558	14,061	14,149
Tripura	863	341	225	714	355	176	258	126	785
Uttar Pradesh	112,547	52,750	35,182	102,181	65,637	52,346	55,103	34,788	106,413
West Bengal	38,227	22,137	17,122	36,927	30,555	13,797	18,716	6,132	34,175
INDIA[2]	558,519	255,578	224,860	503,337	116,842	198,163	222,734	186,322	517,244

Note: [1] Based on 'provisional' data available only for the states.   [2] Including union territories for which break-up figures are not available.

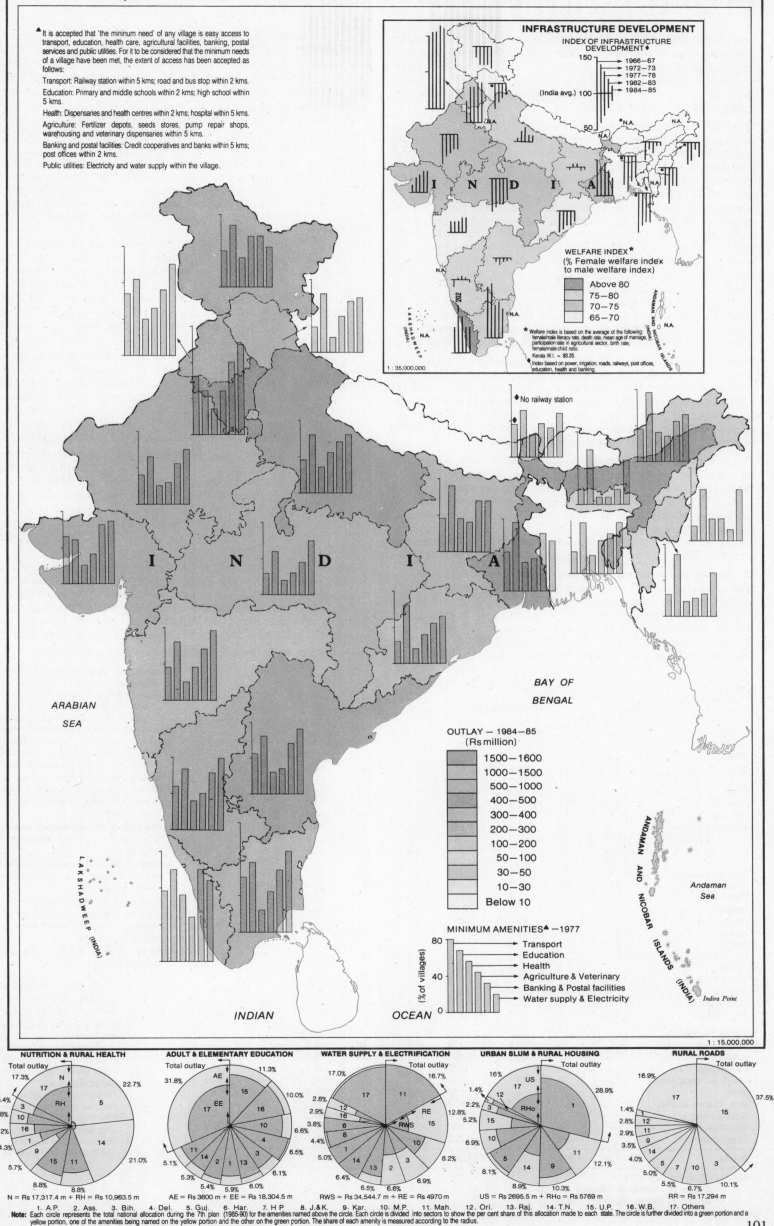

It is accepted that 'the minimum need' of any village is easy access to transport, education, health care, agricultural facilities, banking, postal services and public utilities. For it to be considered that the minimum needs of a village have been met, the extent of access has been accepted as follows:

Transport: Railway station within 5 kms; road and bus stop within 2 kms.

Education: Primary and middle schools within 2 kms; high school within 5 kms.

Health: Dispensaries and health centres within 2 kms; hospital within 5 kms.

Agriculture: Fertilizer depots, seeds stores, pump repair shops, warehousing and veterinary dispensaries within 5 kms.

Banking and postal facilities: Credit cooperatives and banks within 5 kms; post offices within 2 kms.

Public utilities: Electricity and water supply within the village.

INFRASTRUCTURE DEVELOPMENT

INDEX OF INFRASTRUCTURE DEVELOPMENT ◆

150

1966—67
1972—73
1977—78
1982—83
1984—85

(India avg.) 100

50

N.A.

I N D I A

WELFARE INDEX ∗
(% Female welfare index to male welfare index)

Above 80
75—80
70—75
65—70

∗ Welfare index is based on the average of the following: female/male literacy rate, death rate, mean of marriage, participation rate in agricultural sector, birth rate, female/male child ratio.
Kerala W.I. = 93.35.
◆ Index based on power, irrigation, roads, railways, post offices, education, health and banking.

1 : 35,000,000

◆ No railway station

LAKSHADWEEP (INDIA)

ARABIAN SEA

I N D I A

BAY OF BENGAL

OUTLAY — 1984—85
(Rs million)

1500—1600
1000—1500
500—1000
400—500
300—400
200—300
100—200
50—100
30—50
10—30
Below 10

ANDAMAN AND NICOBAR ISLANDS (INDIA)

Andaman Sea

Indira Point

MINIMUM AMENITIES ▲ —1977

80

→ Transport
→ Education
→ Health
→ Agriculture & Veterinary
→ Banking & Postal facilities
→ Water supply & Electricity

40

(% of villages)

0

LAKSHADWEEP (INDIA)

INDIAN OCEAN

1 : 15,000,000

NUTRITION & RURAL HEALTH
Total outlay
17.3%
22.7%
3.4%
3.8%
4.2%
4.3%
5.7%
8.8%
8.8%
21.0%
N
RH
17
3
10
16
1
9
15
11
5
14
N = Rs 17,317.4 m + RH = Rs 10,963.5 m

ADULT & ELEMENTARY EDUCATION
Total outlay
31.8%
11.3%
10.0%
6.6%
6.5%
6.1%
6.0%
5.9%
5.4%
5.3%
5.1%
AE
EE
17
11
14
2
1
13
15
16
10
4
3
AE = Rs 3600 m + EE = Rs 18,304.5 m

WATER SUPPLY & ELECTRIFICATION
Total outlay
17.0%
16.7%
2.8%
2.9%
3.8%
4.4%
5.0%
6.4%
6.5%
6.6%
6.9%
8.2%
12.8%
17
12
6
8
1
14
13
2
3
10
15
RWS
RE
RWS = Rs 34,544.7 m + RE = Rs 4970 m

URBAN SLUM & RURAL HOUSING
Total outlay
16%
28.9%
1.4%
2.2%
5.2%
6.9%
8.1%
8.9%
10.3%
12.1%
US
17
12
3
15
10
5
14
9
11
1
RHo
US = Rs 2695.5 m + RHo = Rs 5769 m

RURAL ROADS
Total outlay
16.9%
37.5%
1.4%
2.8%
2.9%
3.5%
4.0%
5.0%
5.5%
6.7%
10.1%
17
12
11
9
14
5
7
10
3
15
RR = Rs 17,294 m

1. A.P. 2. Ass. 3. Bih. 4. Del. 5. Guj. 6. Har. 7. H P 8. J.&K. 9. Kar. 10. M.P. 11. Mah. 12. Ori. 13. Raj. 14. T.N. 15. U.P. 16. W.B. 17. Others

Note: Each circle represents the total national allocation during the 7th plan (1985-90) for the amenities named above the circle. Each circle is divided into sectors to show the per cent share of this allocation made to each state. The circle is further divided into a green portion and a yellow portion, one of the amenities being named on the yellow portion and the other on the green portion. The share of each amenity is measured according to the radius.

MEDICAL FACILITIES

Since September 1978, when India signed the Alma Ata declaration promising 'Health For All by 2000 AD', medicare has become a subject of great governmental concern. For instance, under the New 20-point Programme there has been, since 1982, a two-and-a-half-fold increase in doctors and hospital beds and a 6-fold increase in nurses in the government sector.

The number of medical colleges (including Veterinary Science) has increased from 399 to 500 within just one year of signing the declaration. Medical facilities of all types are mainly urban-based, even though there are something like 7250 primary health centres and 83,006 sub-centres (1985) in rural areas. This is also the case with professional personnel and beds, and hence the percentage of urbanization is used as the basic information on the map while the number of registered doctors and beds for every million population is overlain. Punjab is the only state that has a better doctor-population ratio than a bed-population ratio. Where the percentage of urbanization is greater, these ratios tend to be higher (Maharashtra, Tamil Nadu and Gujarat). Madhya Pradesh has the lowest bed- and doctor-population ratios. Generally speaking, the north, with the exception of Punjab, West Bengal and Gujarat, has lower ratios than the peninsular states where only Orissa fares poorly. Wherever the rural component is larger, the bed- and doctor-population ratios are low, pointing to the need for improved medical facilities in rural areas.

In the 32 years between 1950 and 1982, India has made tremendous progress in increasing the number of medical personnel and hospital beds in government hospitals. From 16.5 doctors and 31 beds per 100,000 population in 1950, these services grew to 25.8 doctors and 60 beds per 100,000 in 1970. In the subsequent twelve years, the increase was even more spectacular: there were 42.1 doctors and 74 beds for every 100,000 people in 1982.

This ratio varies widely from state to state in India. There is one doctor for every 765 persons in the Punjab (1984) as against 1:8605 in Madhya Pradesh. In Tamil Nadu, which ranks second, the ratio is 1:1425. In Kerala, which has done commendably in reducing birth and death rates and infant mortality, while generally improving the welfare of its people, this ratio is 1:1881. The lowest doctor-population ratios after Madhya Pradesh are in Uttar Pradesh (1:4460), Rajasthan (1:3696), Bihar (1:3446), Orissa (1:3148) and Assam (1:2480).

Kerala and other small states are best off when it comes to providing beds for their people. Kerala has a bed-to-people ratio of 1:612 (in 1985), followed by Meghalaya (1:710), Nagaland (1:798) and Sikkim (1:839). Among the bigger states the better ratios are in Maharashtra (1:771), followed by Punjab (1:847), Gujarat 1:923, West Bengal (1:1145), Karnataka (1:1190) and Tamil Nadu (1:1200). The poorest ratios are in Bihar (1:2771) and Madhya Pradesh (1:3278).

There was nearly a ten-fold increase in registered nursing personnel between 1951 and 1983, from 16,550 to 160,880. There was a corresponding increase in midwives, whose number increased 8-fold from 19,281 in 1951 to 160,904 in 1983. Maharashtra and Tamil Nadu led the country in having the maximum number of nurses and midwives: Maharashtra — 27,427 and 31,210 respectively, and Tamil Nadu, 26,407 and 30,606. But the population of these states is much larger than smaller states like Himachal Pradesh, which has the smallest number of nurses and midwives — 228 and 629.

Health is a state subject in India; it is only in the union territories that the responsibility is the centre's. But the centre guides, sponsors and supports major health programmes. Expenditure by state governments on health (medical services, family planning, public health, sanitation and water supply) has been increasing yearly. From Rs 2990 million in 1970-71 it rose to Rs 16,080 million in 1980-81 and Rs 31,370 million in 1984-85. The expenditure has always been high in the larger states, Maharashtra (Rs 3690 million) and Uttar Pradesh (Rs 3210 million), for instance, spending most in 1984-85. Other states with large expenditures are Tamil Nadu (Rs 2430 million), Andhra Pradesh (Rs 2210 million), Rajasthan (Rs 2360 million), West Bengal (Rs 2100 million) and Madhya Pradesh (Rs 2110 million). This excludes central government and private charitable expenditure.

The per capita expenditure is higher in the smaller states: Nagaland Rs 225 (1984-85), Meghalaya Rs 157, Himachal Pradesh Rs 116, Sikkim Rs 133, Jammu & Kashmir Rs 132 and Manipur Rs 93. Of the major states, the highest per capita expenditures — are in Rajasthan (Rs 58), Punjab (Rs 55), Haryana (Rs 53), Maharashtra (Rs 52), Kerala (Rs 52) and Tamil Nadu (Rs 54). For all other states, this expenditure is below Rs 40 per head, with Bihar spending only Rs 22.

The total Seventh Plan outlay on the health sector alone by the centre and all the states is expected to be about Rs 34,000 million for the five-year period. Maharashtra will spend the most (Rs 3740 million), followed by Uttar Pradesh (Rs 3008 million). Other substantial outlays are in Madhya Pradesh (Rs 1573 million). Andhra Pradesh (Rs 1642 million), Tamil Nadu (Rs 1500 million), Bihar (Rs 1464 million), West Bengal (Rs 1280 million), Karnataka (Rs 1180 million), and Delhi (Rs 1808 million).

The information in the diagrams below the map indicates the number of patients treated for various diseases and the number of those that died among them. This data relates to different years for different states, but an approximation indicates that though the maximum number have been treated for infectious and parasitic diseases, the maximum deaths have been due to diseases of the respiratory system.

The diseases have been divided into six major categories:

1. **Infectious and Parasitic diseases:** influenza, diseases of the teeth and gums, tuberculosis (non-respiratory), syphillis, gonococcal infections, cholera, whooping cough, meningococcal infection, plague and leprosy, acute poliomyelitis, measles, rabies, malaria, filariasis, ancylostoma and other infectious and parasitic diseases.
2. **Diseases of the circulatory system:** chronic rheumatic heart disease, hypertensive heart diseases, ischaemic heart disease.
3. **Diseases of the respiratory system:** Lobar pneumonia and tuberculosis (respiratory).
4. **Diseases of the nervous and muscular system:** diseases of the central nervous system and tetanus.
5. **Diseases of the digestive and reproductive system:** diseases of the digestive system, liver diseases, infectious hepatitis.
6. **Other diseases** are classified in a separate category.

Table 45

MEDICAL FACILITIES[1] — 1985

States	Doctors — 1984 number	HOSPITALS				DISPENSARIES			
		Number	of which Rural (%)	Beds Number	of which Rural (%)	Number	of which Rural (%)	Beds Number	of which Rural (%)
Andhra Pradesh	26,464	612	26.8	35,911	10.3	786	69.2	277	61.7
Assam	8,630	121	33.9	11,897	26.9	455	93.4	132	36.4
Bihar	21,621	226	8.4	22,574	3.1	1,002	98.4	4,926	98.1
Gujarat	17,257	914	8.3	34,071	9.1	1,930	59.5	6,013	38.2
Haryana	—	87	9.2	7,538	5.7	234	54.3	672	56.1
Himachal Pradesh	—	66	43.9	4,128	25.1	236	90.3	273	88.3
Jammu & Kashmir	3,027	35	5.7	3,943	1.5	648	95.8	185	97.8
Karnataka	23,470	236	19.9	30,842	9.0	1,554	82.3	3,198	41.4
Kerala	14,251[2]	758	78.5	43,078	50.2	750	94.7	1,699	97.1
Madhya Pradesh	6,473[3]	284	16.5	17,246	4.7	645	81.7	114	100.0
Maharashtra	36,313	1,367	12.3	85,942	8.2	6,189	15.2	2,748	23.3
Manipur	—	21	57.1	1,246	29.5	52	94.2	—	—
Meghalaya	—	13	7.7	2,065	1.5	58	98.3	48	100.0
Nagaland	—	36	80.6	1,092	56.5	76	90.8	36	100.0
Orissa	8,831	319	50.5	12,271	20.7	294	74.8	—	—
Punjab	23,272	257	43.2	14,607	21.7	1,785	88.0	6,871	91.7
Rajasthan	10,065	253	13.4	19,754	7.3	1,372	69.5	1,984	57.5
Sikkim	—	5	80.0	477	47.2	—	—	—	—
Tamil Nadu	35,644	393	28.5	42,398	9.3	674	47.3	673	57.2
Tripura	—	16	25.0	1,251	10.0	203	96.6	12	100.0
Uttar Pradesh	26,502	735	11.3	47,178	5.5	1,713	74.9	5,683	89.7
West Bengal	35,804	402	32.3	51,675	14.7	426	64.3	10	100.0
Union Territories									
Andaman & Nicobar	—	11	90.9	728	45.1	38	84.2	—	—
Arunachal Pradesh	—	23	65.2	1,008	37.1	79	96.2	120	98.3
Chandigarh	—	2	—	1,279	—	26	26.9	—	—
Dadra & Nagar Haveli	—	1	100	50	100	7	100	6	100.0
Delhi	—	67	3.0	14,656	0.4	547	16.3	—	—
Goa, Daman & Diu	—	85	36.5	2,834	24.3	31	100	—	—
Lakshadweep	—	2	100	50	100	2	100	—	—
Mizoram	—	12	58.3	868	45.6	25	100	255	100.0
Pondicherry	—	10	—	2,322	—	35	60.0	155	63.2
INDIA	297,228	7,369	26.4	514,989	13.4	21,874	58.5	36,090	70.3

Note: [1] As in January 1985
[2] Relates to Travancore Medical Council.
[3] Relates to Bhopal Medical Council.

Sources: 1. World Bank's *Word Development Report 1982*, Oxford University Press.
2. *Basic Statistics Relating to the Indian Economy, Vol. 1: All India, August 1986* and *Vol. 2: States, September 1986*, Centre for Monitoring Indian Economy.
3. *Davidson's principles and practice of medicine*, John Macleod, The English language book society.
4. *Statistical Abstract India, 1984*, Ministry of planning.

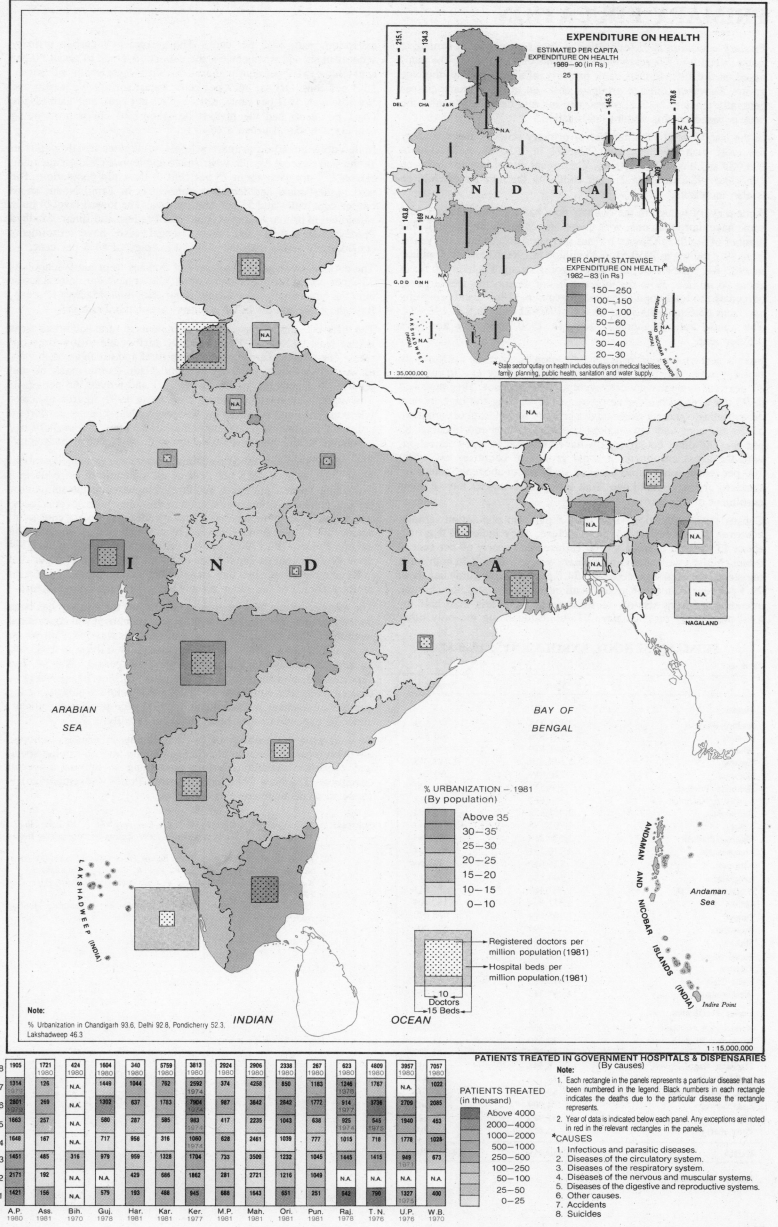

EXPENDITURE ON HEALTH

ESTIMATED PER CAPITA EXPENDITURE ON HEALTH 1989–90 (in Rs.)

PER CAPITA STATEWISE EXPENDITURE ON HEALTH* 1982–83 (in Rs.)
- 150 — 250
- 100 — 150
- 60 — 100
- 50 — 60
- 40 — 50
- 30 — 40
- 20 — 30

*State sector outlay on health includes outlays on medical facilities, family planning, public health, sanitation and water supply.

1 : 35,000,000

% URBANIZATION — 1981 (By population)
- Above 35
- 30 — 35
- 25 — 30
- 20 — 25
- 15 — 20
- 10 — 15
- 0 — 10

Registered doctors per million population (1981)

Hospital beds per million population (1981)

10 Doctors
15 Beds

Note:
% Urbanization in Chandigarh 93.6, Delhi 92.8, Pondicherry 52.3, Lakshadweep 46.3

ARABIAN SEA

BAY OF BENGAL

Andaman Sea

LAKSHADWEEP (INDIA)

ANDAMAN AND NICOBAR ISLANDS (INDIA)

Indira Point

INDIAN OCEAN

1 : 15,000,000

PATIENTS TREATED IN GOVERNMENT HOSPITALS & DISPENSARIES (By causes)

PATIENTS TREATED (in thousand)
- Above 4000
- 2000 — 4000
- 1000 — 2000
- 500 — 1000
- 250 — 500
- 100 — 250
- 50 — 100
- 25 — 50
- 0 — 25

Note:
1. Each rectangle in the panels represents a particular disease that has been numbered in the legend. Black numbers in each rectangle indicates the deaths due to the particular disease the rectangle represents.
2. Year of data is indicated below each panel. Any exceptions are noted in red in the relevant rectangles in the panels.

*CAUSES
1. Infectious and parasitic diseases.
2. Diseases of the circulatory system.
3. Diseases of the respiratory system.
4. Diseases of the nervous and muscular systems.
5. Diseases of the digestive and reproductive systems.
6. Other causes.
7. Accidents
8. Suicides

*Cause	A.P. 1980	Ass. 1981	Bih. 1970	Guj. 1978	Har. 1981	Kar. 1981	Ker. 1977	M.P. 1981	Mah. 1981	Ori. 1981	Pun. 1981	Raj. 1978	T.N. 1976	U.P. 1976	W.B. 1970
8	1905 1980	1721 1980	424 1980	1604 1980	340 1980	5759 1980	3813 1980	2924 1980	2906 1980	2338 1980	267 1980	623 1980	4809 1980	3957 1980	7057 1980
7	1314 1979	126	N.A.	1449	1044	762	2592 1974	374	4258	850	1183	1246 1976	1767	N.A.	1022
6	2901 1979	269	N.A.	1302	637	1783	7904 1974	987	3842	2842	1772	914 1977	9736	2709	2085
5	1663	257	N.A.	580	287	585	983 1974	417	2235	1043	638	925 1975	545 1975	1940	453
4	1648	167	N.A.	717	956	316	1060 1974	628	2461	1039	777	1015	718	1778	1028
3	1451	485	316	979	959	1328	1704	733	3509	1232	1045	1445	1415	949 1971	673
2	2171	192	N.A.	N.A.	429	666	1862	281	2721	1216	1049	N.A.	N.A.	N.A.	N.A.
1	1421	156	N.A.	579	193	488	945	688	1643	651	251	542 1975	790	1327 1975	400

PRIMARY EDUCATION

Primary education is the education a child receives in its formative years, when it is 5-9 years old and attends classes I to V. The map, based on the 1981 Census data, projects information about this age group. However, the age group of children in the primary classes generally tends to be 5-11 years, and all data that accompanies the map is based on this wider classification.

In the last 35 years, the number of primary schools in the country has more than doubled, from 209,671 in 1950-51 to 509,143 in 1983-84 and it is still increasing. The annual rate of increase between 1950 and 1984 has been 2.7 per cent. But the number of children is also increasing!

Almost every village in the country now has a primary school, and some have more than one, with towns and cities having several. The number of pupils in Classes I-V has increased greatly over the years, from 19.2 million in 1950-51 to 35 million in 1960-61 and 81.1 million in 1983-84. By 1989-90, this school-going strength is likely to reach about 96 million. With the increase in enrolment, the school-going percentage to total population in age group 6-11 has also been going up, from 42.6 in 1950-51 to 83.1 in 1980-81 and 91.8 in 1983-84. The annual rate of increase between 1950 and 1984 has been 4.5 per cent.

India's performance in primary education is catching up fast with world norms. In 1982, India's primary school enrolment was 87 per cent as against a world average of 95 per cent. This increased to 92 per cent within the next two years. Barring the high-income oil-exporting countries which have a primary school level of enrolment of 76 per cent and low-income countries where enrolment is 85 per cent, all other countries now have universal primary education, with the gross enrolment in some groups of countries exceeding 100 per cent as some pupils enrolled are below or above the countries' standard primary school age. East European countries have a gross enrolment of 104 per cent!

Census (1981) statistics based on a 5 per cent population sample, however, do not reflect this rosy picture. They indicated that only about 33 per cent of India's rural children and nearly 60 per cent of urban children were attending primary schools. Kerala in both cases registered the largest percentages: 74.3 per cent from rural and about 79.1 per cent from urban areas. In all other major states, the children attending primary schools in rural areas were less than half the total number of rural children (Tamil Nadu being the only other

exception, with 54.2 per cent). The enrolment in urban primary schools in six states was below the national average of about 60 per cent (Bihar, 51.2 per cent; Gujarat, 57.8 per cent; Madhya Pradesh, 55.3 per cent; Orissa, 56.9 per cent; Rajasthan, 49.1 per cent and West Bengal, 56.1 per cent). Kerala (79.1 per cent) and Tamil Nadu (70.1 per cent) had the highest percentage of children attending primary schools in urban areas.

Male pupils attending primary schools outnumber female pupils in all the states except Kerala, where male and female children attending classes I-V are each about 75 per cent of the child population. The next highest male and female enrolment is in Tamil Nadu, about 64 per cent males and 55 per cent females. The lowest level of male enrolment in primary schools is in Uttar Pradesh and Bihar. Madhya Pradesh, Rajasthan and West Bengal also have enrolments considerably lower than the national average of 44.3 per cent.

The national average of enrolment of females in primary schools is 32.2 per cent. Five major states of the Union have enrolments lower than the female national average: Bihar, Haryana, Madhya Pradesh, Rajasthan (the lowest in the country) and Uttar Pradesh.

The differences in primary school enrolments in rural and urban areas is also least in Kerala. This is mainly attributable to low drop-out rates. The next-best enrolment in both rural and urban areas in male as well as in female pupils is in Tamil Nadu (rural: male, 60 and female, 48.1 per cent; urban: male, 72.1 and female, 68 per cent). This state is doing much better than these 1981 figures indicate, following the introduction of the Nutritious Meal scheme; the current policy of feeding daily 200,000 children between 2-9 years of age is believed to be responsible for the lowest drop-out rate in the country.

The expenditure in 1980-81 on primary education amounted to about Rs 15,283 million, 48.5 per cent of all expenditure on education (including Sports and Youth Welfare), emphasizing the status now being accorded to primary education in the country. In several states, a substantial sum of money has been expended on primary education, notably Rs 1715 m or 49.4 per cent of state expenditure on education in Uttar Pradesh, Rs 1790 m or 46.4 per cent in Maharashtra, Rs 1174 m or 49.9 per cent in Tamil Nadu and Rs 1149.4 m or 54.5 per cent in Kerala; but it was Bihar that spent the highest amount (73.5 per cent or Rs 1509 m) of its outlay on education in the primary sector.

The number of teachers in primary schools in the country has been on the increase, keeping pace with student enrolment and educational institutions. In 1950-51 the strength of teachers was 537,918, which increased to 1.35 million in 1980-81 and 1.39 million in 1983-84, an annual increase of 2.9 per cent for the period 1950-84. But experience shows that while enrolment is still not 100 per cent as desired under the commitment to universal primary education, the number of teachers is inadequate, even for the present enrolment, although expenditure has been going up steadily.

To overcome the problem of low enrolments in primary schools, attention must be given to alleviating poverty and improving social and economic conditions before attempting to impose universal education. The New National Education Policy (Seventh Plan) is expected to do better on this count.

Sources: 1. *Basic Statistics Relating to the Indian Economy, Vol. 1: All India, August 1986* and *Vol. 2: States, September 1986,* Centre for Monitoring Indian Economy.
2. World Bank's *World Development Report 1986,* Oxford University Press.
3. *Seventh Five Year Plan 1985-90, Vol. II,* Planning Commission.
4. *Key Population Statistics Based on 5 per cent Sample Data, Paper 2 of 1982,* Census of India, 1981.
5. *Statistical Abstract India 1982* and *1984,* Ministry of Planning. Quoting from Ministry of Education & Social Welfare.

PRIMARY SCHOOL ENROLMENT[1] : 1980-81

Table 46

	Male	Female
States		
Andhra Pradesh	3,141,226	2,821,509
Assam	1,158,806	843,628
Bihar	4,661,879	1,886,668
Gujarat	2,619,470	1,785,218
Haryana	760,019	393,767
Himachal Pradesh	206,571	223,432
Jammu & Kashmir	350,682	196,888
Karnataka	2,462,459	1,879,092
Kerala	1,707,195	1,921,540
Madhya Pradesh	3,532,218	1,719,580
Maharashtra	4,770,331	3,623,128
Manipur	116,612	92,066
Meghalaya	104,732	95,413
Nagaland	97,460	77,913
Orissa	1,632,604	1,112,104
Punjab	1,145,694	936,070
Rajasthan	2,132,994	672,309
Sikkim	24,715	17,071
Tamil Nadu	3,489,913	2,844,117
Tripura	168,866	123,680
Uttar Pradesh	6,842,232	2,961,561
West Bengal	4,057,313	2,403,260
Union Territories		
Andaman & Nicobar	14,835	12,356
Arunachal Pradesh	39,340	18,921
Chandigarh	16,397	14,303
Dadra & Nagar Haveli	8,448	5,311
Delhi	353,151	354,648
Goa, Daman & Diu	82,475	71,638
Lakshadweep	4,009	3,442
Mizoram	31,949	29,627
Pondicherry	44,874	34,809
INDIA	45,780,119	28,585,706

Note: [1] Estimated

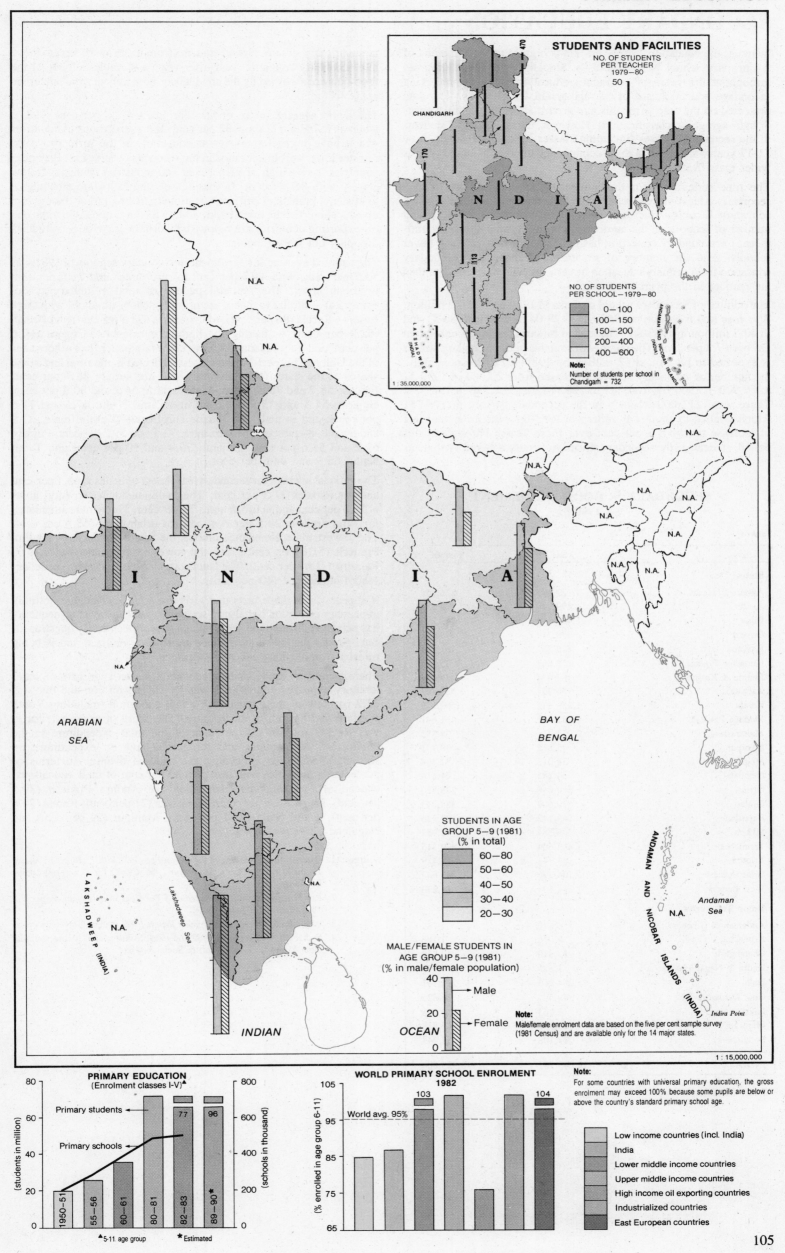

STUDENTS AND FACILITIES

NO. OF STUDENTS
PER TEACHER
1979—80

50

0

470

CHANDIGARH

170

I N D I A

113

NO. OF STUDENTS
PER SCHOOL—1979—80

- 0—100
- 100—150
- 150—200
- 200—400
- 400—600

Note:
Number of students per school in
Chandigarh = 732

1 : 35,000,000

N.A.

N.A.

N.A.

I N D I A

ARABIAN
SEA

BAY OF
BENGAL

LAKSHADWEEP (INDIA)

Lakshadweep Sea

N.A.

INDIAN

ANDAMAN AND NICOBAR ISLANDS (INDIA)

Andaman
Sea

N.A.

Indira Point

STUDENTS IN AGE
GROUP 5—9 (1981)
(% in total)

- 60—80
- 50—60
- 40—50
- 30—40
- 20—30

MALE/FEMALE STUDENTS IN
AGE GROUP 5—9 (1981)
(% in male/female population)

40

20

0

Male

Female

Note:
Male/female enrolment data are based on the five per cent sample survey
(1981 Census) and are available only for the 14 major states.

OCEAN

1 : 15,000,000

PRIMARY EDUCATION
(Enrolment classes I-V)▲

80

60

40

20

0

(students in million)

800

600

400

200

0

(schools in thousand)

Primary students

Primary schools

77

96

1950-51
55-56
60-61
80-81
82-83
89-90 *

▲5-11 age group   *Estimated

WORLD PRIMARY SCHOOL ENROLMENT
1982

105

103

104

World avg. 95%

95

85

75

65

(% enrolled in age group 6-11)

Note:
For some countries with universal primary education, the gross
enrolment may exceed 100% because some pupils are below or
above the country's standard primary school age.

- Low income countries (incl. India)
- India
- Lower middle income countries
- Upper middle income countries
- High income oil exporting countries
- Industrialized countries
- East European countries

SECONDARY EDUCATION

In India, the educational drop-out rate is highest towards the end of the primary school years (age 5-9). Students continue to drop-out throughout the years of secondary education. Students receiving secondary education are in the age-group 10-14 years. The data projected on the map is for this age-group, according to the Census of India age-group classifications. However, the age-group of students in the secondary classes (including classes XI and XII) tends to be 11-17 years, and all data that accompanies the map is based on this wider classification.

The time series data for the number of schools and students in secondary education repeat the general pattern of change in primary education: there is an increase with the years, and with the increasing number of schools, in the number of pupils. But the similarity ends there, for secondary education has both smaller numbers and fewer schools. And the number of students that move into secondary education from primary education has always been much smaller than the number in the primary streams.

The number of secondary school students in 1950-51 was 4.3 million. This rose to 9.6 million in 1960-61, to 28.04 million in 1980-81 and to 40.3 million in 1983-84. The annual rate of increase in enrolment has been 7.3 per cent. The number of students in the secondary stream is expected to be around 42 million by 1989-90. Correspondingly, the number of secondary schools increased from 20,884 in 1950-51 to 66,920 in 1960-61, 164,202 in 1980-81 and 181,580 in 1983-84. There were 212,000 teachers in the secondary stream in 1950-51 which swelled to about 1.9 million in 1983-84, with an annual rate of increase of about 7 per cent over these years. Throughout this period, students in the secondary stream were outnumbered by students in the primary stream. Not all students from the primary stream move into secondary education year after year; a sizeable number of the drop-outs are absorbed by the child labour pool in both rural and urban areas.

The world average for secondary enrolment is 59 per cent (1983), whereas in India it is about 33 per cent. But even groups of countries which have registered universal education in the primary years, register lower enrolment ratios in the secondary years. East European countries have a high of 91 per cent and industrial countries follow closely with 85 per cent. In these cases, the ratios are attributable to students branching off into vocational streams rather than giving up education. On the other hand, all low income, middle income and oil-exporting countries are below the world average due to students dropping out from schools.

The national average for enrolment in secondary schools in 1981 was 50.5 per cent, with 44.3 per cent in rural areas and 71.6 per cent in urban areas. The regional pattern of total secondary school enrolment is similar to that of the primary school situation, with three states — Kerala (86 per cent), Maharashtra (62.6 per cent) and Punjab (62.9 per cent) — having the highest enrolments. Gujarat (60.6 per cent) and Tamil Nadu (55.5 per cent) fare better than other states of the Union. This performance is also reflected in the rural and urban shares of the states (Kerala: rural, 85.4 and urban, 88.7 per cent; Punjab: 58.7 and 75 per cent; Maharashtra: 54.2 and 80.3 per cent; Gujarat: 54.3 and 75.9 per cent; and Tamil Nadu: 47.5 and 71.6 per cent); and in male and female enrolments (Kerala: male, 87.9 and female, 84 per cent; Maharashtra: 73.1 and 51.2 per cent; Punjab: 69.2 and 55.6 per cent; Gujarat: 70.4 and 50 per cent; and Tamil Nadu: 65.9 and 44.6 per cent).

The national average for secondary enrolment of males is 62.1 per cent and for females 37.5 per cent. The male attendance in rural areas is 57.8 per cent and in urban areas 77 per cent. The female attendance for rural areas is 29.2 per cent and for urban areas 65.6 per cent. The lowest male attendance among the major states is in Andhra Pradesh (51.6 per cent), and the lowest female attendance is in Rajasthan (18.7 per cent). The latter state also has the lowest secondary school attendance (40 per cent).

It appears from these facts that girls are at a disadvantage not only in primary education but also in secondary education. This inequality has persisted through the years and has never been really questioned. But 'gender justice' is now being more fully realized and is being translated into a new educational policy.

The expenditure on secondary education, as an aggregate of state/ territory expenses in 1980-81, was Rs 10,163 million and this was 32.3 per cent of the expenditure on all education (including Youth Welfare and Sports) in the country. The share of union territories was Rs 547 million (51 per cent of the total expenditure in the territories). Most states and territories have an expenditure on secondary education exceeding the national average in terms of percentage; the states with less than 30 per cent of their educational expenditure on secondary education are Andhra Pradesh (29.5 per cent), Bihar (16.8 per cent), Karnataka (21 per cent), Kerala (28.4 per cent), Tamil Nadu (26.1 per cent), Manipur (29 per cent) and Nagaland (20.9 per cent).

Sources: 1. *Basic Statistics Relating to the Indian Economy, Vol. 1: All India, August 1986* and *Vol. 2: States, September 1986,* Centre for Monitoring Indian Economy.

2. *Key Population Statistics Based on 5 Per Cent Sample Data, Paper 2 of 1983,* Census of India 1981.

3. World Bank's *World Development Report 1986,* Oxford University Press.

4. *Statistical Abstract India 1982* and *1984,* Ministry of Planning, Quoting from Ministry of Education & Social Welfare.

SECONDARY SCHOOL ENROLMENT[1]: — 1980-81

Table 47

	Male	Female
States		
Andhra Pradesh	1,129,692	526,759
Assam	499,940	304,332
Bihar	1,475,584	347,299
Gujarat	998,749	579,681
Haryana	476,362	161,031
Himachal Pradesh	175,067	73,883
Jammu & Kashmir	162,044	74,544
Karnataka	930,237	523,209
Kerala	1,259,143	1,169,558
Madhya Pradesh	1,387,451	451,641
Maharashtra	2,052,416	1,108,873
Manipur	45,815	30,101
Meghalaya	38,012	31,584
Nagaland	34,385	24,814
Orissa	610,048	273,820
Punjab	630,609	388,213
Rajasthan	849,142	192,726
Sikkim	6,384	2,962
Tamil Nadu	1,483,604	879,315
Tripura	54,741	39,419
Uttar Pradesh	3,104,296	790,042
West Bengal	1,284,101	684,230
Union Territories		
Andaman & Nicobar	6,825	4,763
Arunachal Pradesh	9,076	3,186
Chandigarh	13,310	10,658
Dadra & Nagar Haveli	1,625	791
Delhi	284,265	204,741
Goa, Daman & Diu	49,491	36,625
Lakshadweep	2,263	1,248
Mizoram	20,901	17,813
Pondicherry	27,108	17,743
INDIA	**19,049,792**	**8,955,604**

Note: [1] Estimated

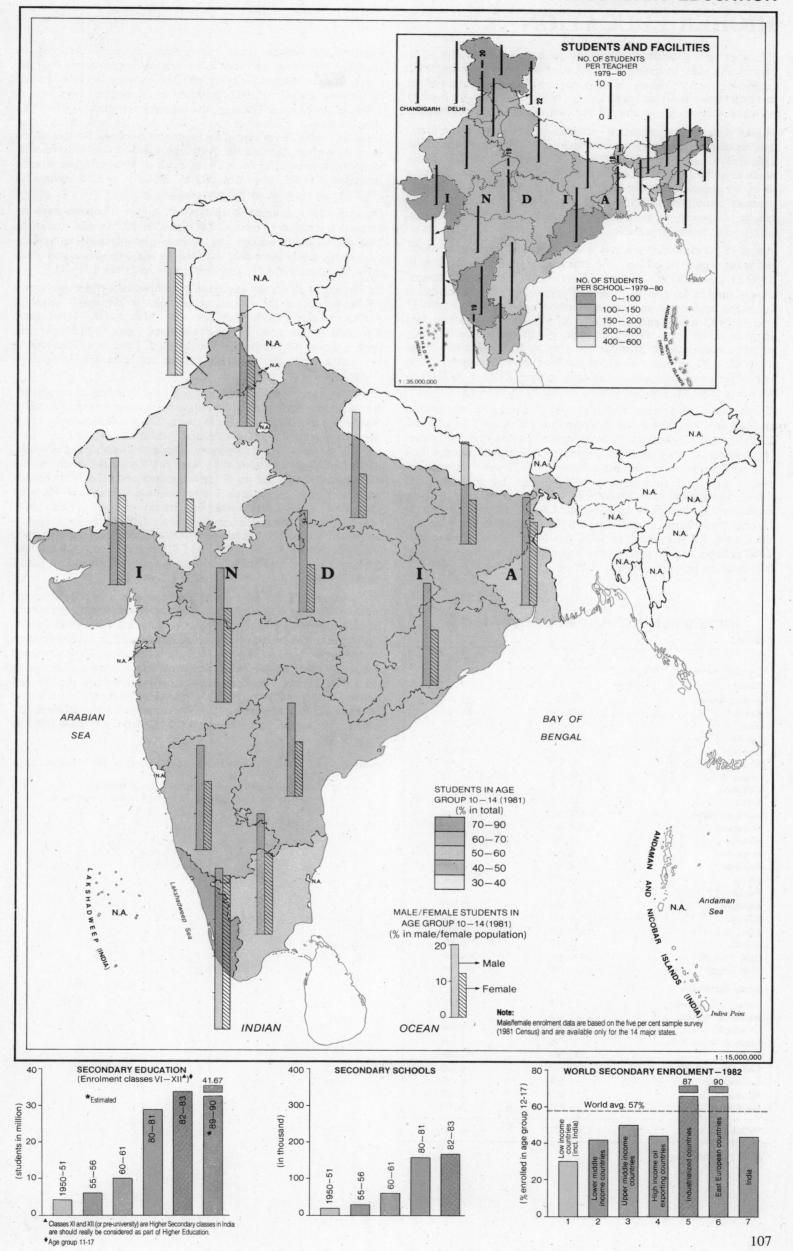

STUDENTS AND FACILITIES

NO. OF STUDENTS
PER TEACHER
1979—80

NO. OF STUDENTS
PER SCHOOL—1979—80

0—100
100—150
150—200
200—400
400—600

1 : 35,000,000

N.A.

CHANDIGARH   DELHI

I N D I A

LAKSHADWEEP
(INDIA)

ANDAMAN AND
NICOBAR ISLANDS
(INDIA)

N.A.

ARABIAN
SEA

BAY OF
BENGAL

LAKSHADWEEP (INDIA)

Lakshadweep Sea

INDIAN   OCEAN

ANDAMAN AND NICOBAR ISLANDS (INDIA)

Andaman
Sea

N.A.

Indira Point

STUDENTS IN AGE
GROUP 10—14 (1981)
(% in total)

70—90
60—70
50—60
40—50
30—40

MALE/FEMALE STUDENTS IN
AGE GROUP 10—14 (1981)
(% in male/female population)

20

10

0

Male

Female

Note:
Male/female enrolment data are based on the five per cent sample survey
(1981 Census) and are available only for the 14 major states.

1 : 15,000,000

SECONDARY EDUCATION
(Enrolment classes VI — XII▲)✦

*Estimated

(students in million)

40

30

20

10

41.67

1950—51
55—56
60—61
80—81
82—83
*89—90

▲ Classes XI and XII (or pre-university) are Higher Secondary classes in India
are should really be considered as part of Higher Education.
✦ Age group 11-17

SECONDARY SCHOOLS

(in thousand)

400

300

200

100

1950—51
55—56
60—61
80—81
82—83

WORLD SECONDARY ENROLMENT—1982

(% enrolled in age group 12-17)

80

60

40

20

World avg. 57%

87   90

1 Low income countries (incl. India)
2 Lower middle income countries
3 Upper middle income countries
4 High income oil exporting countries
5 industrialized countries
6 East European countries
7 India

HIGHER EDUCATION

Education is a concurrent subject in the Indian Constitution. The central and state governments have, over the years, invested ever increasing sums of money on it. A disproportionately large part of this expenditure is devoted to all types of higher education, the benefits of which accrue only to about 5 per cent of the Indian population.

Higher education is education provided by collegiate institutions, universities and their equivalents which falls within the scope of the University Grants Commission (UGC). It, therefore, generally covers students in the 17-plus age group in such institutions alone. Students in higher secondary schools and technical diploma - or certificate-granting institutions (the 15-17 age group) are not covered. In this chapter, however, we have estimated data for the 15-24 age group as a whole, where applicable.

About 15 per cent of all students who complete secondary education go in for higher education. The expenditure on such education in 1980-81 was Rs 4722.4 million, 15 per cent of the total educational expenditure in the country. The total expenditure on all education (including art and culture, scientific services and research) in 1970-71 was Rs 7930 million and in 1980-81, Rs 31,480 million, a four-fold increase in 10 years, even after making adjustments for differing budget classifications. This has gone up to Rs 59,800 million in 1984-85, a 90 per cent increase in just four years. During this same period per capita expenditure has gone up from Rs 15 in 1970-71 to Rs 47 in 1980-81, and has shot up to Rs 80 in 1984-85.

There were 47 universities and institutions deemed to be universities in India in 1960. These had increased to 139 in 1983-84 and had 220,177 teachers working in 5246 colleges affiliated to them. Of these, about a fifth (48,694) are in the university departments/colleges. The average growth rate of enrolment during 1970-71 was 4 per cent per annum (the total enrolment in 1979-80 was 2.7 million), with differential growth rates for various faculties. Veterinary Science registered 7 per cent growth per year and Law 6.3 per cent. Commerce had the third highest growth rate, 4.6 per cent, while Medicine had the lowest, 0.5 per cent. The UGC has in its survey revealed that 1680 colleges in the country have less than 500 students each, and 520 colleges between 100 and 200.

The inequality between sexes is again clearly discernible in collegiate education. However, this has shown gradual improvement over the years, from 14 women for every 100 men enrolled in 1950-51 to 23 in 1960-61 and 41 in 1982-83. In terms of faculties, the ratio is different. In Teacher Training, the majority are women. There are also substantial numbers of them taking Arts. In both Science and Medicine, about 28 per cent of the enrolment is female, but specialized faculties such as Engineering, Agriculture and Law have low enrolments of women. The overall enrolment of both men and women increased 10-fold, from 360,000 in 1950-51 to 3.55 million in 1983-84, an increase of 7.2 per cent per year.

Data on total enrolment in all types of higher education show an increase in enrolment from 425,000 in 1950-51 to 6.624 million in 1980-81, an almost sixteen-fold increase. The increase was continuous, being nearly three-fold over the decade 1950-61, a little over three-fold over 1960-71, and nearly two-fold over 1970-81.

Enrolment in all higher education by individuals of the age group 15-24, accounted for various percentages in the major states in 1980-81: Uttar Pradesh 8.8 per cent, Maharashtra 7 per cent, Karnataka 6.4 per cent, Andhra Pradesh and Tamil Nadu 5.7 per cent each, Gujarat 5.3 per cent, West Bengal 5.2 per cent, Rajasthan 5 per cent and all others between 2.6 per cent (Madhya Pradesh) to 4.8 per cent (Kerala).

In all major states, male enrolment is greater than female enrolment. It would seem that greater emphasis is being given to male education than to female education in these states. The greatest disparity (11.9 percentage points) is in Uttar Pradesh, which has 14.3 per cent male and 2.4 per cent female enrolment. The least disparity is in Punjab, which has 2.9 per cent male and 1.8 per cent female enrolment. Kerala does almost as well, having 5.5 per cent male and 4.2 per cent female enrolment. Madhya Pradesh (3.8 per cent and 1.3 per cent) Haryana (5.1 per cent and 2.5 per cent) and Orissa (4.7 and 1.1) also show less disparities is on this count. In most states however, the disparity is about 4 to 7 percentage points: Andhra Pradesh 8.5 per cent male and 3 per cent female, Bihar 6.4 and 1.1, Gujarat 7 and 3.4, Karnataka 9.1 and 3.4, Maharashtra 9.3 and 4.4 Rajasthan 7.8 and 1.9, Tamil Nadu 7.7 and 3.9 and West Bengal 7.2 and 3.0.

If the student/teacher ratio is considered an indicator of educational welfare in higher education, then those states with fewer students per teacher may be said to enjoy greater welfare and vice versa. On this count, Sikkim (4 students per teacher), Nagaland (6), Andhra Pradesh and Manipur (7 each), Jammu & Kashmir, Kerala, Meghalaya and Tripura (9 each), all fare better than other states. Almost all the union territories, barring Chandigarh (14), enjoy such welfare and are well below the national average of 13 students per teacher. Much worse off than the national average are Rajasthan and Uttar Pradesh (27 students per teacher each), Karnataka (24), Madhya Pradesh (20) and Gujarat (19).

Sources: 1. *Report & Tables Based on 5 per cent Sample Data, Part II Special*, Census of India.
2. *Statistical Abstract India 1984*, Ministry of Planning.
3. *Report for the Year 1983-84*, University Grants Commission.

HIGHER EDUCATION ENROLMENT —1980-81[1]

Table 48

	Male	Female
States		
Andhra Pradesh	390,650	136,744
Assam	123,382	39,516
Bihar	372,623	61,321
Gujarat	250,692	114,577
Haryana	71,765	29,920
Himachal Pradesh	15,828	5,296
Jammu & Kashmir	30,878	14,773
Karnataka	327,117	113,613
Kerala	147,533	122,293
Madhya Pradesh	179,592	57,446
Maharashtra	558,808	236,812
Manipur	13,785	7,709
Meghalaya	8,296	5,540
Nagaland	2,921	1,067
Orissa	113,233	25,170
Punjab	92,147	52,905
Rajasthan	250,254	55,849
Sikkim	437	164
Tamil Nadu	350,804	178,856
Tripura	12,055	6,349
Uttar Pradesh	1,410,968	208,639
West Bengal	411,139	157,365
Union Territories		
Andaman & Nicobar	996	659
Arunachal Pradesh	875	155
Chandigarh	19,301	9,346
Dadra & Nagar Haveli	90	64
Delhi	107,063	71,068
Goa, Daman & Diu	11,073	6,126
Lakshadweep	209	97
Mizoram	3,073	2,107
Pondicherry	6,255	3,000
INDIA	4,901,322	1,722,393

Note: [1] Estimated.

108

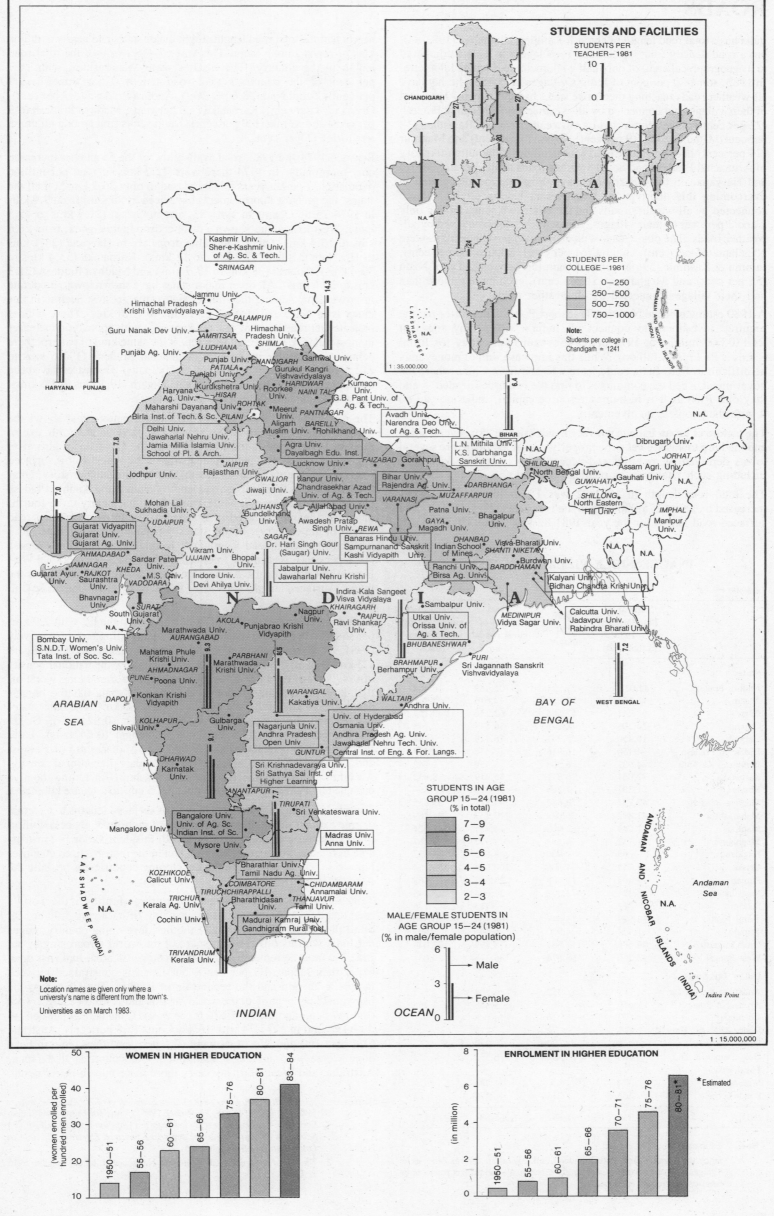

STUDENTS AND FACILITIES

STUDENTS PER
TEACHER—1981

CHANDIGARH

I N D I A

N.A.

STUDENTS PER
COLLEGE—1981

0—250
250—500
500—750
750—1000

Note:
Students per college in
Chandigarh = 1241

1 : 35,000,000

LAKSHADWEEP
(INDIA)   N.A.

ANDAMAN AND NICOBAR ISLANDS

Kashmir Univ.
Sher-e-Kashmir Univ.
of Ag. Sc. & Tech.
SRINAGAR

Jammu Univ.

Himachal Pradesh
Krishi Vishvavidyalaya
PALAMPUR

Guru Nanak Dev Univ.
AMRITSAR

Himachal
Pradesh Univ.
SHIMLA

Punjab Ag. Univ.
LUDHIANA

Punjab Univ.
Punjabi Univ.
PATIALA
CHANDIGARH

Gurukul Kangri
Vishvavidyalaya

Garhwal Univ.
HARIDWAR

Kumaon
Univ.
NAINI TAL

Kurukshetra Univ.
HISAR

Roorkee
Univ.

G.B. Pant Univ. of
Ag. & Tech.,
PANTNAGAR

Haryana
Ag. Univ.

Maharshi Dayanand Univ.
Birla Inst. of Tech.& Sc.
PILANI
ROHTAK

Meerut
Univ.

Aligarh
Muslim Univ.
BAREILLY
Rohilkhand Univ.

Avadh Univ.
Narendra Deo Univ.
of Ag. & Tech.

HARYANA   PUNJAB

Delhi Univ.
Jawaharlal Nehru Univ.
Jamia Millia Islamia Univ.
School of Pl. & Arch.

Agra Univ.
Dayalbagh Edu. Inst.
FAIZABAD   Gorakhpur

L.N. Mithila Univ.
K.S. Darbhanga
Sanskrit Univ.

Dibrugarh Univ.
JORHAT

Assam Agri. Univ.
GUWAHATI
Gauhati Univ.

Jodhpur Univ.
JAIPUR
Rajasthan Univ.

Lucknow Univ.

Kanpur Univ.
Chandrasekhar Azad
Univ. of Ag. & Tech.

Bihar Univ.
Rajendra Ag. Univ.
VARANASI   *MUZAFFARPUR*   *DARBHANGA*

North Bengal Univ.
SHILIGUBI

SHILLONG
North Eastern
Hill Univ.

IMPHAL
Manipur
Univ.

Mohan Lal
Sukhadia Univ.
UDAIPUR

GWALIOR
Jiwaji Univ.

Allahabad Univ.

JHANSI
Bundelkhand
Univ.

Awadesh Pratap
Singh Univ.
REWA

Patna Univ.

GAYA
Magadh Univ.

Bhagalpur
Univ.

Visva-Bharati Univ.
SHANTI NIKETAN

Gujarat Vidyapith
Gujarat Univ.
Gujarat Ag. Univ.
AHMADABAD

Vikram Univ.
UJJAIN

Bhopal
Univ.

SAGAR
Dr. Hari Singh Gour
(Saugar) Univ.

Banaras Hindu Univ.
Sampurnanand Sanskrit
Kashi Vidyapith   Univ.

DHANBAD
Indian School
of Mines

Burdwan Univ.
BARDDHAMAN

Gujarat Ayur.
Univ.
JAMNAGAR   *RAJKOT*
Saurashtra
Univ.

KHEDA
Sardar Patel
Univ.

M.S. Univ.
VADODARA

Indore Univ.
Devi Ahilya Univ.

Jabalpur Univ.
Jawaharlal Nehru Krishi

Ranchi Univ.
Birsa Ag. Univ.

Kalyani Univ.
Bidhan Chandra Krishi Univ.

Bhavnagar
Univ.

SURAT
South Gujarat
Univ.

N.A.

Indira Kala Sangeet
Visva Vidyalaya
KHAIRAGARH

*Sambalpur Univ.

Calcutta Univ.
Jadavpur Univ.
Rabindra Bharati Univ.

I   N   D   I   A

AKOLA

Nagpur
Univ.

Ravi Shankar
Univ.
RAIPUR

Utkal Univ.
Orissa Univ. of
Ag. & Tech.
BHUBANESHWAR

MEDINIPUR
Vidya Sagar Univ.

Marathwada
Univ.
AURANGABAD

Punjabrao Krishi
Vidyapith

Bombay Univ.
S.N.D.T. Women's Univ.
Tata Inst. of Soc. Sc.

Mahatma Phule
Krishi Univ.
AHMADNAGAR

Marathwada
Krishi Univ.
PARBHANI

Berhampur Univ.
BRAHMAPUR

PURI
Sri Jagannath Sanskrit
Vishvavidyalaya

PUNE
Poona Univ.

Konkan Krishi
Vidyapith
DAPOLI

WARANGAL
Kakatiya Univ.

ARABIAN
SEA

KOLHAPUR
Shivaji Univ.

Gulbarga
Univ.

Nagarjuna Univ.
Andhra Pradesh
Open Univ.
GUNTUR

Univ. of Hyderabad
Osmania Univ.
Andhra Pradesh Ag. Univ.
Jawaharlal Nehru Tech. Univ.
Central Inst. of Eng. & For. Langs.

WALTAIR
Andhra Univ.

BAY OF
BENGAL

WEST BENGAL

DHARWAD
Karnatak
Univ.

Sri Krishnadevaraya Univ.
Sri Sathya Sai Inst. of
Higher Learning
ANANTAPUR

N.A.

Bangalore Univ.
Univ. of Ag. Sc.
Indian Inst. of Sc.

TIRUPATI
Sri Venkateswara Univ.

Mangalore Univ.

Mysore Univ.

Madras Univ.
Anna Univ.

STUDENTS IN AGE
GROUP 15—24 (1981)
(% in total)

7—9
6—7
5—6
4—5
3—4
2—3

KOZHIKODE
Calicut Univ.

Bharathiar Univ.
Tamil Nadu Ag. Univ.

COIMBATORE
TIRUCHCHIRAPPALLI
Bharathidasan
Univ.

CHIDAMBARAM
Annamalai Univ.

THANJAVUR
Tamil Univ.

LAKSHADWEEP
(INDIA)
N.A.

TRICHUR
Kerala Ag. Univ.

Cochin Univ.

Madurai Kamraj Univ.
Gandhigram Rural Inst.

MALE/FEMALE STUDENTS IN
AGE GROUP 15—24 (1981)
(% in male/female population)

6
Male
3
Female
0

Andaman
Sea

ANDAMAN AND NICOBAR ISLANDS (INDIA)

N.A.

TRIVANDRUM
Kerala Univ.

INDIAN   OCEAN

Indira Point

Note:
Location names are given only where a
university's name is different from the town's.

Universities as on March 1983.

1 : 15,000,000

WOMEN IN HIGHER EDUCATION

(women enrolled per
hundred men enrolled)

50
40
30
20
10

1950—51   55—56   60—61   65—66   75—76   80—81   83—84

ENROLMENT IN HIGHER EDUCATION

(in million)

8
6
4
2
0

* Estimated

1950—51   55—56   60—61   65—66   70—71   75—76   80—81*

ROADS

India has a total road length of about 1.7 million kilometres (1983-84), but a road density (that is, the total road length in a state compared to its geographical area) of only half a kilometre to a square kilometre. In 1983, some 70 per cent of India's villages (415,153) did not have all-weather roads reaching them; five states had more than 70 per cent of their villages not connected by all-weather roads. Madhya Pradesh (22 per cent villages connected by all-weather roads), Rajasthan (18 per cent), Uttar Pradesh (9 per cent), Orissa (3 per cent) and Manipur (24 per cent) had very poor road accessibilities. The union territory of Arunachal Pradesh had no all-weather roads at all. Kerala, Punjab and Haryana, on the other hand, have done commendably well in overcoming this deprivation, with Kerala having all its villages connected by all-weather roads and the other two states having only 1 and 2 per cent of their villages, respectively, not connected by all-weather roads. The other states with above average connections were Nagaland (65 per cent), Gujarat (61 per cent), Sikkim (60 per cent), Jammu & Kashmir (59 per cent), Assam (56 per cent), Tamil Nadu (55 per cent) and Meghalaya (51 per cent), each having more than half their villages connected by all-weather roads.

A 1980 estimate of the National Transport Policy Committee put the required investment to connect all India's villages by roads at Rs 110,000 million at 1978 prices (Seventh Plan outlay for Rural Roads is Rs 17,294 million). While this immense sum of money may be difficult to find, India could learn a lesson from China where the government urged the communes to find the resources needed — and they did. Rich as it is in human resource capital, India too can do it with the people's involvement.

The picture of urban India is not very much brighter, for several small and medium towns are yet to be properly connected; there are several towns deprived of mobility and accessibility to service locations, including such vital ones as medical facilities.

The total road length in 1981-82 was 1.5 million kms and this increased by about 175,000 kms in the two years that followed. The surfaced road length for the year 1981-82 was about 730,000 kms,

nearly half the total road length. Eight major states had nearly a million kms of roads among them. This was 67.4 per cent of the total road length in the country. These states were Maharashtra, with 11.7 per cent of the country's total road length, Uttar Pradesh (9.9 per cent), Tamil Nadu (8.6 per cent), Andhra Pradesh (8.3 per cent), Orissa (7.8 per cent), Karnataka (7.3 per cent), Madhya Pradesh (6.9 per cent) and Kerala (6.9 per cent). Delhi ranks first among all union territories (1 per cent).

Between 1974 and 1982, road availability for the people has increased only marginally. In 1974 there were 22.2 kms of road per million population; in eight years this increased to only 26.2 kms. Of all the states, the greatest improvements have been in Sikkim, from 9.9 kms in 1974-75 to 37 kms in 1981-82, and in Orissa (21.9 kms to 45.5 kms). West Bengal has shown a distinct drop in this index, from 26.6 kms to 10.4 kms. Other noticeable drops are in Haryana (25.1 kms to 18.1 kms), Bihar (17.2 kms to 12 kms), Karnataka (33.4 kms to 30.4 kms), Gujarat (28 kms to 17.7 kms) and Madhya Pradesh (21.3 kms to 20.5 kms). All other major states have shown low to moderate improvement. Among the smaller states, the greatest decline in this index is in Nagaland, from 121.1 kms to 78.6 kms. These indices indicate that road building has not kept pace with population increases in most states. Union Territories repeat the same kind of pattern, there being a spectacular increase in Arunachal Pradesh (177.6 kms to 212.5 kms) and Mizoram (0.6 kms to 50.2 kms). Decreases have been noted in certain territories, notably Chandigarh (19.1 kms to 2.5 kms), and the Andamans (72.9 kms to 34.1 kms).

The distribution of road lengths provides a pattern that shows the southern states more favourably placed than those of the north, except Punjab, Uttar Pradesh and Madhya Pradesh. The road lengths also, in a sense, reflect the geographical size of the states. But if surfaced roads are indicators of development, that is quite another story. Uttar Pradesh has less than half its road length surfaced and Andhra Pradesh just about half its length, while Tamil Nadu has nearly three-fourths of its road length surfaced (this state has the longest surfaced road length in the country). Major states that have below average (47.3 per cent) surfaced road lengths are Kerala (22.6 per cent), Orissa (14.1 per cent), Assam (13.1 per cent), Bihar (34.3 per cent) and West Bengal (44.6 per cent). Among the union territories, Delhi has the longest surfaced length, 8669 kms, followed by Goa, Daman & Diu, 4263 kms.

As for road density, Delhi and Pondicherry top the list with 7.9 kms per square kilometre and 4.23 kms per square kilometre respectively (1981-82). Goa, Daman & Diu has a density of 2 kms and Chandigarh 1.3 kms, but all other union territories are below the national density of 0.47 kms of road per square kilometre, Mizoram being worst off (0.1 kms per sq km). Amongst the states, Kerala has the highest density, 2.74 kms, followed by Tamil Nadu 1.0 kms. The other states that exceed the national density are Punjab (0.92 kms), Orissa (0.77 kms), Assam (0.75 kms), West Bengal (0.64 kms), Uttar Pradesh (0.52 kms), Karnataka (0.59 kms), Maharashtra (0.59 kms) and Tripura (0.8 kms). Andhra Pradesh equals the national density. All other states fare poorly, Jammu & Kashmir, including the area occupied by Pakistan, having the least, 0.05 kms to a square kilometre.

The road densities and the distribution of surfaced lengths to a certain degree express the relative deprivation of mobility and accessibility. In comparison with the rest of the world, this deprivation is considerable in relation to the prevalent situation in the industrialized countries but not so bad in relation to the developing countries, especially the African ones.

Ten countries in the world accounted for three-fourths of the world's road length in 1981-82, with India ranking fifth, having 5.4 per cent of all the world's roads. Brazil, another developing country, edges out India with its 6.0 per cent share of the world's road length, but Brazil, it must be remembered, is much bigger in area. India has done better than Japan (4.7 per cent), but Japan is constrained by such factors as terrain and the separation of islands. The USA tops the list with 27 per cent of all the world's roads and is followed by France (6.4 per cent) and the USSR (6.0 per cent). Canada (3.9 per cent — it ranked 2nd in 1973-74 with 13%), China (3.8 per cent), Australia (3.4 per cent) and West Germany (2.1 per cent) may seem to have done worse in road length than India, but countries like Japan, Australia, and West Germany carry more materials on better roads.

INACCESSIBLE VILLAGES — 1983

Table 49

	Rural Population[1] — 1983 (in thousand)	Total number of villages	Total length of rural roads[2] (in km.)	Villages not connected by all-weather roads
States				
Andhra Pradesh	42,053	27,221	105,489	16,464
Assam	18,653	22,026	47,616	9,762
Bihar	63,241	67,566	62,276	46,155
Gujarat	24,208	18,275	36,844	7,148
Haryana	10,450	6,741	16,728	128
Himachal Pradesh	4.150	16,916	13,456	9,973
Jammu & Kashmir	4,932	6,503	8,439	2,662
Karnataka	27,051	26,871	85,319	18,832
Kerala	21,231	1,268	94,216	—
Madhya Pradesh	42,783	70,883	49,212	55,335
Maharashtra	41,892	36,033	119,237	24,289
Manipur	1,044	2,000	4,421	1,521
Meghalaya	1,133	4,583	3,972	2,267
Nagaland	663	960	4,853	332
Orissa	23,751	54,606	100,078	53,154
Punjab	12,443	12,188	29,582	162
Rajasthan	28,321	33,305	37,695	27,348
Sikkim	327	434	713	175
Tamil Nadu	33,029	23,047	106,193	10,390
Tripura	1,954	4,930	6,445	2,698
Uttar Pradesh	93,566	112,561	78,899	102,044
West Bengal	42,350	38,074	36,375	20,631
Union Territories				
Andaman & Nicobar	145	352	364	145
Arunachal Pradesh	650	3,463	5,555	3,463
Chandigarh	27	26	4	—
Dadra & Nagar Haveli	933	72	218	20
Delhi	420	20	239	—
Goa, Daman & Diu	788	435	6,532	33
Lakshadweep	21	N.A.	—	N.A.
Mizoram	374	N.A.	884	N.A.
Pondicherry	268	333	570	22
INDIA	541,339	591,692	1,065,684	415,153

Note: [1] Estimated

[2] Since the national and state highways link all urban centres (and some villages), all other roads except purely urban roads and project roads, have been considered as rural roads.

Sources: 1. *Basic Statistics Relating to the Indian Economy, Vol. 1: All India, August 1986 and Vol. 2: States, September 1986*, Centre for Monitoring Indian Economy. Quoting from reply to *Unstarred Question No. 274 in the Rajya Sabha* on 28 April 1982 and reply to *Unstarred Question No.1004 in Lok Sabha* on 25 March 1985.

2. *The World in Figures*, The Economist Newspaper Limited, London, 1976 & 1984.

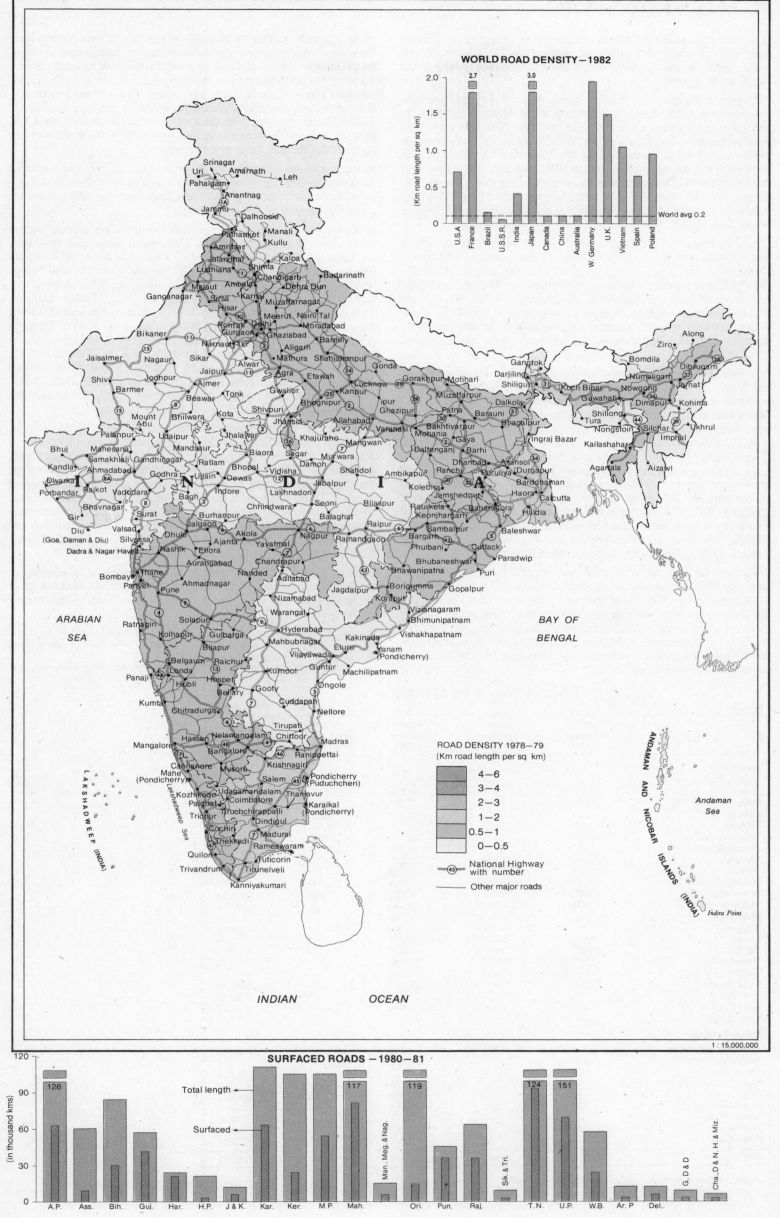

ROAD TRANSPORT

The extent of availability of, and accessibility to, road transport reflects the level of actual and potential mobility of people, goods and information. It substantially affects the location of people and the facilities they need.

Changes in the extent of roads available in a country indicate how movement potential has been changing in the country. In 1950-51, the road length in India was only 398,000 kms. This had more than quadrupled to 1.7 million kms in 1983-84. Similarly, the passenger traffic on the roads increased over ten-fold, from 23 billion (billion = 1 million million) passenger kms in 1950-51 to 270 billion passenger kms in 1978-79. The increase in goods traffic over three decades was almost 18-fold, from 6 billion tonne kms in 1950-51 to 106 billion tonne kms in 1982-83; it is estimated that it will grow to 182 billion tonne kms by AD 2000.

These increases in the movement of people and goods have been possible only with the increasing number of motor vehicles of different varieties. From a small beginning of 306,000 motor vehicles in 1950-51, their number had risen to 8.6 million in 1984-85. In all three indicators of people and goods movement, it is evident that the rate of increase has been high, though not as high as in the West. While these increases are a direct result of changing social and economic patterns, they also influence and are influenced by a changing road structure.

The number of vehicles relative to states and union territories indicate the levels of passenger and goods traffic in inter- and intra-state movement: the larger the number, the greater will be the traffic. The states with the largest number of registered motor vehicles (1982-83) are Maharashtra (1.1 m), Gujarat (515,000), West Bengal (419,000), Punjab (439,000), Uttar Pradesh (597,000), Karnataka (504,000), Tamil Nadu (408,000) and Madhya Pradesh (396,000). Delhi, the capital of the country, with large international and inter-state/internal movements centred on it, ranks second in the country with 668,000 motor vehicles. The lowest number of motor vehicles is in Arunachal Pradesh (502). The number is also comparatively small in states and territories where the terrain poses problems for traffic: Nagaland (2934), Manipur (10,500), Tripura (9300), Andaman & Nicobar Islands (2900) and Dadra & Nagar Haveli (510).

Vehicular density (the number of vehicles to a kilometre of road), based on 1981 figures, is greatest in Chandigarh, 717 vehicles per km. Compared to this, Delhi is far behind, with 30 vehicles per km. All the states and other union territories are still further behind and the national index is only 4 vehicles per kilometre. Only the states of Punjab (7 vehicles per km), Maharashtra (5), Gujarat (9) and the union territories of Pondicherry (12) and Goa, Daman & Diu (5) have more vehicles than the national index.

Motorized vehicles may be classified as vehicles for personal transport (two-wheelers, cars, jeeps), for public transport (three-wheelers, taxis, buses) and for commercial use (trucks, tractors, trailers and others). There were 3.5 million two-wheelers, 1.1 million cars and jeeps, 544,000 three-wheelers, taxis and buses, and 1.4 million trucks, tractors, trailers and other commercial vehicles on the road in 1982-83. There was in that year one personal vehicle for an estimated 148 persons, one public transport vehicle for every 1288 people and one commercial vehicle for every 492 Indians.

In India, two-wheelers are the favoured form of personal transport (all-India 1:199). Cars and jeeps are much fewer (all-India 1:576), because they are more expensive and those who can afford them are a very small part of the population.

In Arunachal Pradesh, the two-wheeler vehicle: population ratio is the lowest in the country, 1:3352. The smaller states with low ratios are Tripura (1:1311) and Nagaland (1:2564). The better-off states are Gujarat (1:115), Maharashtra (1:122), Karnataka (1:128), and Punjab (1:84) and among the union territories, Chandigarh (1:8), Delhi (1:15), Goa, Daman & Diu (1:45) and Pondicherry (1:23). The car and jeep vehicle: population ratio is very low in Madhya Pradesh (1:1079), Rajasthan (1:1071), Bihar (1:1676), Assam (1:1077), Haryana (1:1307), Orissa (1:2237) and Uttar Pradesh (1:1914). In six states and five union territories (Karnataka, 1:475; Kerala, 1:355; Maharashtra, 1:251; Punjab, 1:523; Manipur, 1:463; West Bengal, 1:341; Andaman & Nicobar, 1:482; Chandigarh, 1:42; Delhi, 1:48; Goa, Daman & Diu, 1:164 and Pondicherry, 1:105) the car and jeep vehicle: population ratio is better than the national ratio.

It is in public transport vehicles that there are wide divergences of ratio from the national vehicles: population ratio (1:1288). In Arunachal Pradesh and Nagaland, this ratio is 1:33,333, in Dadra & Nagar Haveli 1:25,000 and in Orissa 1:9860. It would appear from this that topography has a dominant influence on the public transport system. The pattern seen in the case of public carriers, such as three-wheelers, taxis and buses, repeats itself in the case of commercial vehicles as well, although the ratios are not much worse than the national ratio (1:492), except in a few states — Bihar, 1:1161; Orissa, 1:1593 and Arunachal Pradesh, 1:3141.

These indicators collectively point to an overall deprivation of mobility for people and goods in the country. However, Indian road transport operators carry in a year more materials internally than other means of transport.

Sources: 1. *Basic Statistics Relating to the Indian Economy, Vol. 1: All India, August 1986* and *Vol. 2: States, September 1986,* Centre for Monitoring Indian Economy.
2. *Handbook of Statistics 1984,* Association of Indian Engineering Industry.

ROAD & RAILWAY LENGTH

Table 50

(in kilometre)

	National highways — 1981	State highways — 1981	Other roads[1] — 1981	Railway route length — 1983	Number of regd. vehicles — 1982-83
States					
Andhra Pradesh	2,352	5,443	115,829	4,920	393,755
Assam	2,198	1,529	55,852	2,181	80,812
Bihar	2,188	4,191	77,264	5,362	214,591
Gujarat	1,424	9,158	48,915	5,635	514,748
Haryana	655	3,166	20,129	1,501	156,211
Himachal Pradesh	589	3,251	16,707	256	37,785
Jammu & Kashmir	593	688	10,516	77	44,774
Karnataka	1,983	7,813	103,007	3,024	504,252
Kerala	839	2,079	104,095	916	223,093
Madhya Pradesh	2,755	11,564	92,998	5,748	395,727
Maharashtra	2,950	19,122	158,123	5,297	1,085,118
Manipur	431	480	4,410	—	10,551
Meghalaya	462	—	4,749	—	N.A.
Nagaland	105	398	5,784	9	2,934
Orissa	1,631	2,834	115,692	1,982	78,134
Punjab	977	1,900	43,284	2,139	439,277
Rajasthan	2,521	7,273	62,685	5,614	265,627
Sikkim	62	217	832	—	N.A.
Tamil Nadu	1,867	1,814	129,045	3,894	408,592
Tripura	198	136	7,639	12	9,346
Uttar Pradesh	2,461	7,969	142,648	8,882	597,388
West Bengal	1,631	3,147	47,715	3,726	419,239
Union Territories					
Andaman & Nicobar	—	271	411	—	2,843
Arunachal Pradesh	330	—	12,433	—	502
Chandigarh	14	—	111	11	78,126
Dadra & Nagar Haveli	—	—	246	—	510
Delhi	89	—	15,701	168	668,022
Goa, Daman & Diu	224	231	7,681	79	45,407
Lakshadweep	—	—	—	—	N.A.
Mizoram	240	—	2,268	—	N.A.
Pondicherry	—	36	2,107	27	35,627
INDIA	31,769	94,680	1,419,442	61,460	6,718,539

Note: [1] Other state roads, urban roads and project roads.

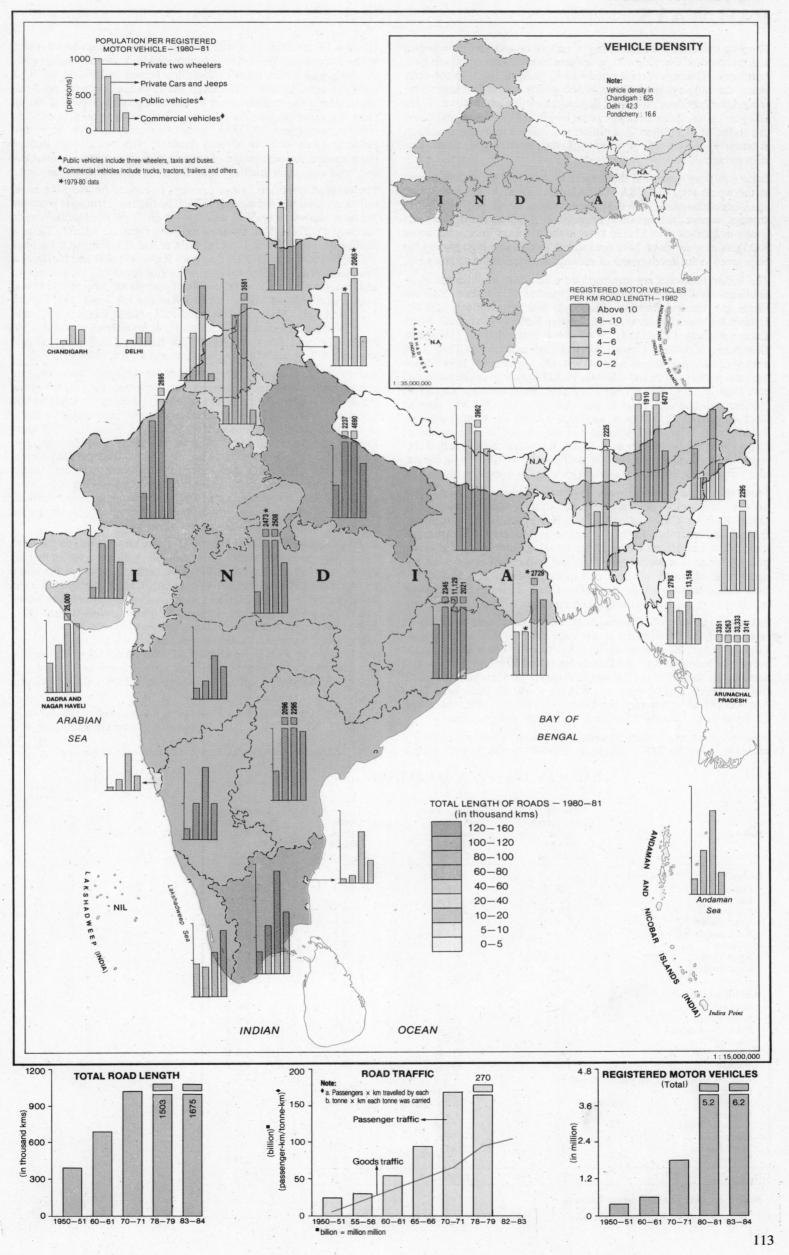

POPULATION PER REGISTERED
MOTOR VEHICLE — 1980—81

Private two wheelers
Private Cars and Jeeps
Public vehicles ▲
Commercial vehicles ♦

▲ Public vehicles include three wheelers, taxis and buses.
♦ Commercial vehicles include trucks, tractors, trailers and others.
* 1979-80 data

CHANDIGARH   DELHI

DADRA AND
NAGAR HAVELI

ARUNACHAL
PRADESH

VEHICLE DENSITY

Note:
Vehicle density in
Chandigarh : 625
Delhi : 42.3
Pondicherry : 16.6

I N D I A

N.A.

N.A.

N.A.

LAKSHADWEEP
(INDIA)

N.A.

1 : 35,000,000

REGISTERED MOTOR VEHICLES
PER KM ROAD LENGTH—1982

Above 10
8—10
6—8
4—6
2—4
0—2

I N D I A

ARABIAN
SEA

BAY OF
BENGAL

N.A.

LAKSHADWEEP (INDIA)   NIL

Lakshadweep Sea

TOTAL LENGTH OF ROADS — 1980—81
(in thousand kms)

120—160
100—120
80—100
60—80
40—60
20—40
10—20
5—10
0—5

ANDAMAN AND NICOBAR ISLANDS (INDIA)

Andaman
Sea

Indira Point

INDIAN   OCEAN

1 : 15,000,000

TOTAL ROAD LENGTH

(in thousand kms)

1503
1675

1950—51 60—61 70—71 78—79 83—84

ROAD TRAFFIC

(billion)■
(passenger-km/tonne-km)♦

Note:
♦ a. Passengers × km travelled by each
b. tonne × km each tonne was carried

270

Passenger traffic

Goods traffic

1950—51 55—56 60—61 65—66 70—71 78—79 82—83

■ billion = million million

REGISTERED MOTOR VEHICLES
(Total)

(in million)

5.2
6.2

1950—51 60—61 70—71 80—81 83—84

113

RAILWAYS

The first train in India, consisting of an engine and 14 four-wheeled teak coaches and carrying 400 passengers, made its inaugural run from Bombay to Thane, a distance of 34 kms, in April 1853. In the early years, the railways moved people and goods only a few kilometres around the Presidency towns of Bombay, Calcutta and Madras. Some 130 years later, India's railway route length is 61,500 kms and caters to a traffic of 213 billion (one billion = 1 million million) passenger-kilometres, about 15 per cent of the total world passenger-kilometrage.

India's railway system is the largest in Asia and the fourth largest in the world after the USA, USSR and Canada. It was the British who, in collaboration with the princely states, developed, largely for strategic reasons, this network, impressive by any standard. The route length in 1950-51 was 53,596 kms; in the 35 years since, only about 8000 kms of route length have been added, though very high priority has been given to the development of railways throughout the planning era.

The Indian Railways are grouped into nine zonal administrations — Southern Railway (headquarters: Madras), Western Railway (Bombay), Central Railway (Bombay), Northern Railway (Delhi), Eastern Railway (Calcutta), South Eastern Railway (Calcutta), North Eastern Railway (Gorakhpur), North-east Frontier Railway (Maligaon, Guwahati), and South Central Railway (Secunderabad). They carry over 9 million passengers a day (1984-85) and employ 1.59 million men and women to run and maintain 10,128 locomotives (steam, diesel and electric), 38,583 carriages and coaches (passenger carriages, electric multiple-unit coaches and rail cars) and 396,550 wagons (including over 30,000 tank wagons). Eleven thousand trains are scheduled daily to 7093 stations and almost one million tonnes of freight are transported per day. Indian Railways carry nearly three-fourths of the country's passenger traffic and four-fifths of its freight, thus constituting the most important means of transport to the people.

More importantly, the railways have been instrumental in building growth corridors — Bombay-Calcutta, Bombay-Ahmadabad-Delhi, Delhi-Kanpur-Calcutta, Madras-Hyderabad-Pune-Bombay, Bangalore-Hyderabad-Delhi, Coimbatore-Madras-Vishakapatnam-Calcutta — besides providing cheap travel country-wide. However, gauge-change is a serious defect in the Indian railway system and appears to inhibit growth to towns in the subordinate network. The inter-gauge transfer of goods involves delays and expense, not to mention high rates of pilferage and spoilage.

In terms of proportion of world daily railway passenger traffic, India is next only to the USSR (25 per cent or 1380 billion passenger-kilometres in 1981-82) and Japan (23 per cent). Ten countries in the world, including Japan, the USSR, India and China (11.4 per cent) account for four-fifths of the world's total passenger-kilometres. Of the 1.3 million kms of route length in the world the same year, India had 4.6 per cent, whereas the USA had 21.7 per cent, the USSR 18.1 per cent, Canada 5 per cent and China 3.8 per cent.

Revenues accruing to the Indian Railways have shown a phenomenal increase, from Rs 2630 million in 1950-51 to Rs 54,690 million in 1984-85. In 1980-81 it was only Rs 27,030 million and doubled in just three years. This revenue is garnered from the increasing traffic of goods which rose from 73.2 million tonnes in 1950-51 to 236.4 million tonnes in 1984-85, though the railways' share of road-rail goods traffic has *reduced* from 89 per cent to 65 per cent all-India. Over the same period, the passenger traffic has grown from 1284 million passengers to 3333 million passengers. But here too, proportionately less use is being made of railway transport than of other modes, the percentage of railway passenger traffic to total rail and road passenger traffic falling from 74 to around 40 per cent.

The ratio of route km: gross earnings in rupees for the nine zonal railways is, on the average, 1:1374. The highest earning is from the Western Railway (1:6502) and the lowest from the North Eastern Railway (1:274). The Central Railway ratio is 1:1230, Eastern Railway 1:1183, South Eastern Railway 1:1025, Northern Railway 1:675, South Central 1:592, Southern Railway 1:463 and North-east Frontier Railway 1:415. However, the ratios for route km: locomotive and total staff: route km, indicators of operational efficiency, reveal a somewhat different story — Central Railway 1:4.7 and 1:33, Eastern 1:2.1 and 1:53, Northern 1:6.3 and 1:21, North Eastern 1:7.3 and 1:20, North-east Frontier 1:6.9 and 1:24, Southern 1:9.8 and 1:20, South Central 1:8.9 and 1:18, South Eastern 1:5.9 and 1:37 and Western 1:7.5 and 1:20. (Approximate data for 1982-83).

Railway route lengths seen as a ratio of 100,000 population indicate considerable disparities between the different states and union territories. There is a route length of 9.03 kms for every 100,000 population in India (1984-85); the ratios in seven major states are better than this: Gujarat 16.5 kms to 100,000 people, Rajasthan 16.4 kms, Punjab 12.7 kms, Assam 11.8 kms, Haryana 11.6 kms, Madhya Pradesh 11.1 kms and Andhra Pradesh 9.3 kms. But this ratio cannot be taken as a dependable economic or development indicator, for the route lengths depend on the area of the state; several big states have extensive route lengths but comparatively low populations. Further, inter-state passenger and goods traffic far outweighs intra-state passenger and goods traffic in most areas.

Railway density (length in km to 1000 square kilometres), on the other hand, may be considered a more reliable indicator, for it implies access. In 1984-85, high densities were naturally found in the union territories of Delhi (110 kms/1000 sq kms), Chandigarh (84 kms) and Pondicherry (54 kms); but the national average was only 18.8 kms per 1000 square kilometres. Punjab (42.8 kms), West Bengal (42.3 kms), Haryana (34.1 kms), Bihar (30.8 kms), Uttar Pradesh (30.2 kms), Tamil Nadu (30.1 kms), Assam (29.6 kms), Gujarat (28.7 kms) and Kerala (23.5 kms) are all well served by internal route lengths. But the states with hilly and difficult terrain, such as Tripura (1.2 kms), Nagaland (0.5 kms) and Jammu & Kashmir (0.4 kms), are poorly served.

Sources: 1. *Basic Statistics Relating to the Indian Economy, Vol. 1: All India, August 1986* and *Vol. 2: States, September 1986,* Centre for Monitoring Indian Economy.
2. *Indian Railways Year Book 1982-83,* Ministry of Railways.

RAILWAY TRAFFIC AND REVENUE

Table 51

Railways	No. of passengers carried 1982-83 (in million)	Earnings from passengers carried 1982-83 (in Rs. million)	Goods carried 1981-82 (in m tonnes)	Earnings from goods traffic 1981-82 (in Rs. million)
BROAD GAUGE (1.676 metres)				
Central	958.2	2,116.8	62.4	3,681.9
Eastern	517.9	1,274.9	71.0	2,478.3
Northern	330.4	1,868.9	57.2	3,175.2
North Eastern	36.0	124.5	3.3	1,262.0
North-east Frontier	3.4	25.9	4.0	219.5
Southern	143.5	715.5	22.7	962.0
South Central	117.8	893.5	31.9	1,872.6
South Eastern	158.6	767.0	92.3	470.6
Western	976.3	1,466.5	33.1	2,714.8
METRE GAUGE (1.000 metres)				
Northern	67.3	293.7	8.3	240.6
North Eastern	157.0	530.5	6.7	243.9
North-east Frontier	52.8	265.8	4.8	323.5
Southern	214.6	445.5	7.5	375.2
South Central	87.2	232.7	6.3	348.6
Western	131.8	480.8	15.3	991.5
NARROW GAUGE (0.762 and 0.610 metres)				
Central	9.5	25.5	0.3	5.2
Eastern	1.4	1.2	—	—
Northern	2.9	9.6	0.2	5.3
North-east Frontier	0.1	0.3	—	—
Southern	0.4	0.6	—	—
South Eastern	15.8	40.0	1.0	12.4
Western	16.0	16.7	2.3	34.0
INDIA[1]	3,998.9	11,596.4	245.8	22,503.4

Note: [1] Excludes non-Government Railways

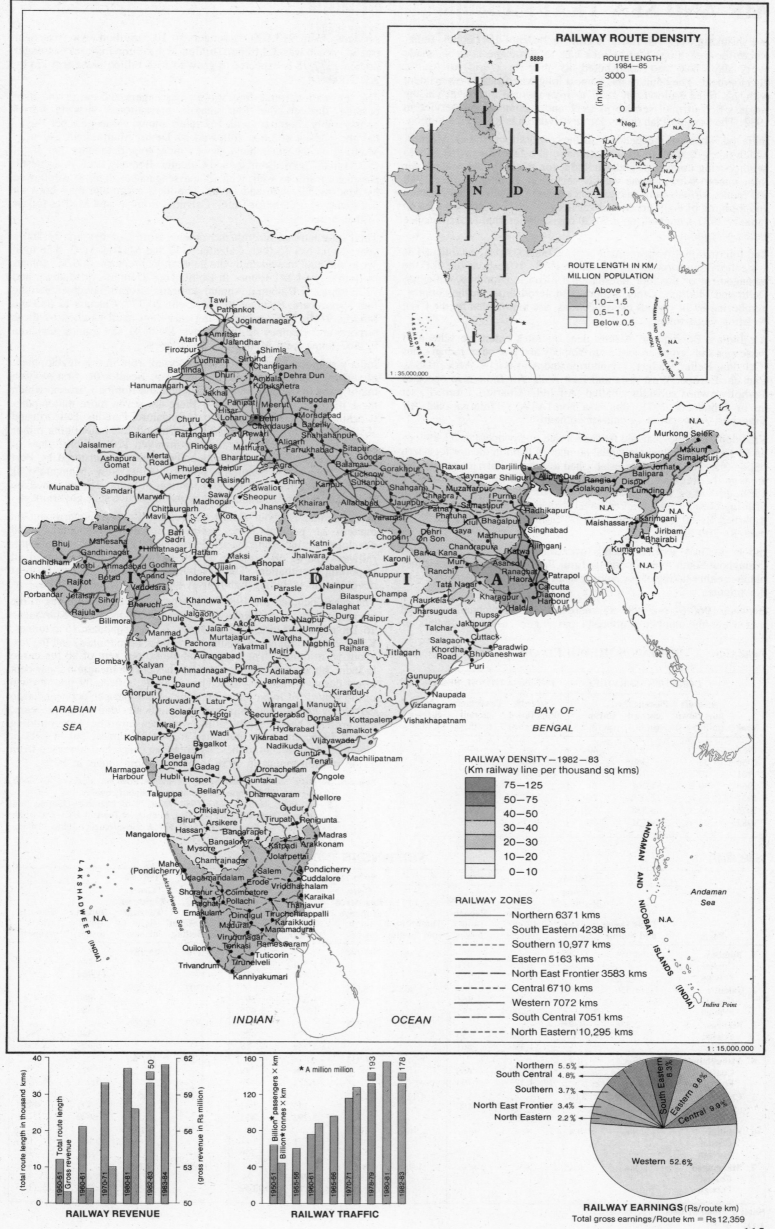

RAILWAY ROUTE DENSITY

ROUTE LENGTH
1984–85

3000

(in km)

0

*Neg.

ROUTE LENGTH IN KM/
MILLION POPULATION

Above 1.5
1.0 – 1.5
0.5 – 1.0
Below 0.5

1 : 35,000,000

ARABIAN
SEA

BAY OF
BENGAL

LAKSHADWEEP
(INDIA)

INDIAN   OCEAN

ANDAMAN AND NICOBAR ISLANDS (INDIA)

Andaman
Sea

Indira Point

RAILWAY DENSITY – 1982–83
(Km railway line per thousand sq kms)

75 – 125
50 – 75
40 – 50
30 – 40
20 – 30
10 – 20
0 – 10

RAILWAY ZONES

Northern 6371 kms
South Eastern 4238 kms
Southern 10,977 kms
Eastern 5163 kms
North East Frontier 3583 kms
Central 6710 kms
Western 7072 kms
South Central 7051 kms
North Eastern 10,295 kms

1 : 15,000,000

RAILWAY REVENUE

(total route length in thousand kms)
40
30
20
10
0

Total route length
Gross revenue

(gross revenue in Rs million)
62
59
56
53
50

50

1950-51 1960-61 1970-71 1980-81 1982-83 1983-84

RAILWAY TRAFFIC

160
120
80
40

Billion* passengers × km
Billion* tonnes × km

*A million million

193
178

1950-51 1955-56 1960-61 1965-66 1970-71 1978-79 1980-81 1982-83

RAILWAY EARNINGS (Rs/route km)
Total gross earnings/Route km = Rs 12,359

Northern 5.5%
South Central 4.8%
Southern 3.7%
North East Frontier 3.4%
North Eastern 2.2%

South Eastern 8.3%
Eastern 9.6%
Central 9.9%

Western 52.6%

AIR AND SEA TRAFFIC

In a shrinking world, air traffic is growing by leaps and bounds. India, where the transition from bullock cart to aeroplane is easily made every day, has also contributed its mite to world air traffic development. The country has come a long way from an experiment with gas filled balloons in 1877 to international air travel via the world's first airmail service in 1911 and a domestic air service in 1932. The Seventh Plan outlay for civil aviation is Rs 7578.4 million.

India today has two independent airline corporations, Indian Airlines, which takes care of all major internal and a few external flights to neighbouring countries, and Air India for international routes. Air India started with just one weekly service to London in June 1948 and Indian Airlines came into existence in 1953 with the nationalization and merging of eight private air services. These two corporations in January 1981 launched, on a 50:50 basis, an internal feeder service called Vayudoot.

The International Airports Authority of India (IAAI), established as a statutory corporation in February 1972, is responsible for the management of the four international airports (Bombay, Calcutta, Delhi and Madras). Other airports and aerodromes in the country, numbering 87 in various classifications, are maintained by the Civil Aviation Department.

Air India's Boeing and Airbus fleet operates extensive scheduled passenger and cargo services from Bombay and five other Indian cities (Calcutta, Delhi, Madras, Trivandrum and Amritsar) to Africa, USA, Canada, Europe, West Asia and East Asia. Air India also operates a wholly-owned subsidiary called 'Air India Charters Limited', set up in September 1971 (this is a non-IATA — International Air Transport Association — charter company).

Vayudoot, the five-year-old feeder service, today operates 52 stations on 61 routes (1986), a substantial number of them in the north-eastern region (the service was first established to provide links to inaccessible areas in this region). It is expected to operate from 150 stations by 1988, providing a 'third level airline' for the entire country and offering cheaper and quicker travel than surface transport. In 1985 it earned Rs 80 million and 1986-87 earnings were expected to be over Rs 200 million.

Indian Airlines service extends over the entire sub-continent and throughout South Asia. Operating from 70 stations, an Indian Airlines plane is either taking off or landing at any one of these airports every 2.5 minutes.

Between 1960-61 and 1984-85, there has been a spectacular ten-fold increase in the number of domestic passengers carried by India's air services, from 913,000 passengers to 10.3 million. Air passenger traffic, which was 3.4 billion (billion ½ 1 million million) passenger kms in 1977-78 is expected to grow to 23.6 billion passenger km by AD 2000.

The international and domestic air passengers and cargo and mail passing through the four major international airports varies considerably. Bombay, with 152 international passengers per flight, has the highest volume, followed by Delhi, Madras and Calcutta. Madras, on the other hand, books more export tonnage per flight (5.4 t) than Delhi, Bombay and Calcutta. Bombay leads in domestic passenger traffic as well, with 130 passengers per flight. It is followed by Madras, Calcutta and Delhi. The local cargo and mail booking is 2 tonnes to a flight for Delhi; Calcutta, Bombay and Madras follow in that order.

Total international aircraft movements from Bombay are 24,760 a year; Delhi has 15,080, Calcutta 5733 and Madras 3102. The total cargo and mail movement in the international sector is 106,527 tonnes in Bombay, 64,141 tonnes in Delhi, 16,870 tonnes in Madras, and 7223 tonnes in Calcutta. Annual domestic aircraft movements in the four airports are: Bombay 26,375, Delhi 20,777, Calcutta 12,885 and Madras 9310. The corresponding cargo and mail handled by these airports are: Bombay 42,985 tonnes, Delhi 41,454 tonnes, Calcutta 22,667 tonnes and Madras 14,506 tonnes.

India's seaports have played an important role in the development of several cities and towns, old and new. Besides the first modern major ports in India — Madras, Calcutta and Bombay, around which these metropolises developed — there are seven more major ports (Cochin, Kandla, Marmagao, New Mangalore, Paradip, Tuticorin and Vishakhapatnam), 21 intermediate ports and 152 functioning minor ports. The ten major ports together handled 120 million tonnes of cargo during 1985-86. Nine major ports are administered by Port Trusts, which are statutory bodies, while New Mangalore is administered directly by the Government of India. The intermediate and minor ports are administered by the respective state governments.

The National Harbour Board advises the Central Government on matters of general policy relating to port management and development, including the effects on industry, commerce, shipping, railways and so on, so as to ensure integrated consideration of all important port matters. The Seventh Five Year Plan outlay for port development including lighthouses will be Rs 12,604 million.

The cargo handled for 1985-86 has been 120 million tonnes in the ten major ports, with Bombay handling 25 million tonnes, Marmagao 16.1 m tonnes, Madras 18.2 m tonnes, Vishakhapatnam 16 m tonnes, Kandla 16.5 m tonnes, Calcutta and Haldia 12 m tonnes, and the rest about 16 m tonnes. These ports handle 90 per cent of all the cargo in India, but the rate of cargo increase has been a meagre 6.2 per cent a year during the last ten years. This, in a way, reflects the poor growth of the volume of India's foreign trade. A comparison between cargo carried by aircraft and ship, however, indicates that, over the years, air cargo has been steadily falling while sea cargo has been building up. In 1956, air cargo amounted to 44,031 tonnes, but in 1980 it amounted to only 534 tonnes, whereas sea cargo has increased from 17.43 million tonnes in 1956 to 64.8 million tonnes in 1979.

Sources: 1. *Basic Statistics Relating to the Indian Economy, Vol. 1: All India, August, 1986* Centre for Monitoring Indian Economy.
2. *Statistical Abstract India 1984*, Ministry of Planning. Quoting from Director General of Civil Aviation, Ministry of Tourism & Civil Aviation.
3. *Seventh Five Year Plan 1985-90, Vol. II*, Government of India Planning Commission.

Table 52(i) **INDIAN SCHEDULED OPERATIONS**

	DOMESTIC SERVICES			INTERNATIONAL SERVICES		
Year	Aircraft kms. flown (in m.kms)	Passengers carried (in th.)	Cargo (in th. t.)	Aircraft kms. flown (in m. kms)	Passengers carried (in th.)	Cargo carried (in th. t.)
1951	22	301	40	9.4	148	3
1956	23	368	45	14.8	191	5
1961	27.8	745	40	16.6	229	8
1966	30.8	1,261	21	19.0	228	11
1971	33.4	2,056	26	25.9	491	19
1976	43.3	3,538	39	33.6	997	42
1982	44.9	6,151	86	46.5	2,140	76

Table 52(ii) **SHIPPING IN INDIA**

Ports	1960-61				1970-71				1981-82			
	Total Steam and Sailing	Gross Registered Tonnage (in th. tonnes)	Imports (in th. tonnes)	Exports (in th. tonnes)	Total Steam and Sailing	Gross Registered Tonnage (in th. tonnes)	Imports (in th. tonnes)	Exports (in th. tonnes)	Total Steam and Sailing	Gross Registered Tonnage (in th. tonnes)	Imports (in th. tonnes)	Exports (in th. tonnes)
1. Bombay	68,054	23,998			12,581	17,862			8,778[1]	25,185[1]		
Coastal			1,223	2,011			1,364	1,793			855	3,322
Foreign			9,572	1,915			9,494	1,753			11,968	2,397
2. Calcutta	3,536	13,817			1,945	8,992			1,307	5,929		
Coastal			1,483	1,464			1,375	292			890	1,853
Foreign			4,009	2,545			1,872	2,475			5,880	1,127
3. Madras	2,336	9,495			1,817	9,835			1,853	14,506		
Coastal			768	50			178	429			340	782
Foreign			1,360	861			3,559	2,759			6,457	3,831
4. Vishakhapatnam	N.A.	N.A.			1,217	8,333			854	10,740		
Coastal			13	668			104	284			923	58
Foreign			1,373	794			2,222	6,124			3,051	6,789
5. Cochin	2,455	7,032			4,127	7,517			632	4,571		
Coastal			913	190			155	941			531	985
Foreign			708	276			3,292	451			3,672	313
6. Kandla	N.A.	N.A.	N.A.	N.A.	421	1,819			957	7,105		
Coastal							1	—			2,836	12
Foreign							—	2,156			5,971	711
7. Marmagao	N.A.	N.A.	N.A.	N.A.	1,984	8,767			1,434	12,647		
Coastal							267	—			446	—
Foreign							86	10,652			668	13,733

Note: 1. 1980-81 data.

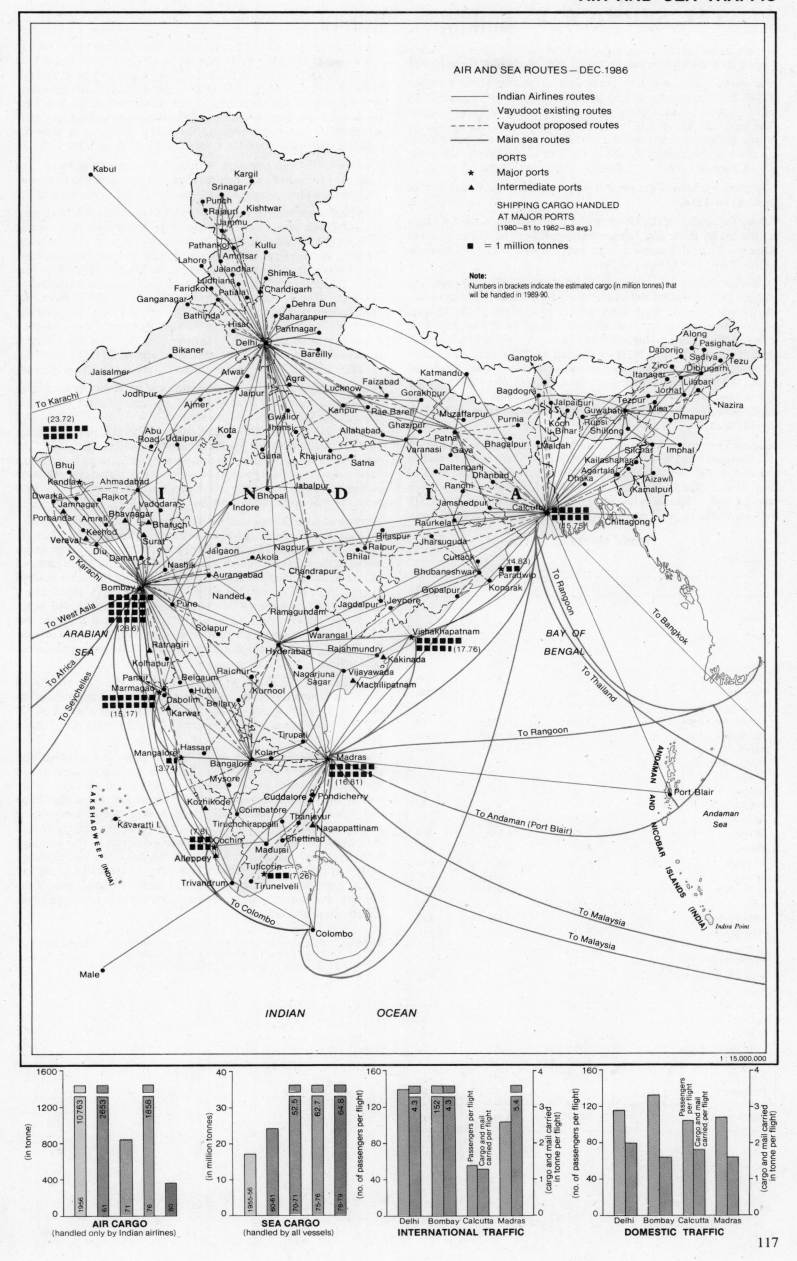

AIR AND SEA ROUTES — DEC. 1986

——————— Indian Airlines routes
——————— Vayudoot existing routes
- - - - - - Vayudoot proposed routes
══════════ Main sea routes

PORTS
★ Major ports
▲ Intermediate ports

SHIPPING CARGO HANDLED
AT MAJOR PORTS
(1980–81 to 1982–83 avg.)

■ = 1 million tonnes

Note:
Numbers in brackets indicate the estimated cargo (in million tonnes) that
will be handled in 1989-90.

1 : 15,000,000

AIR CARGO
(handled only by Indian airlines)
(in tonne)

1956 — 10,763
61 — 2,653
71
76 — 1,858
80

SEA CARGO
(handled by all vessels)
(in million tonnes)

1955-56
60-81
70-71 — 52.5
75-76 — 62.7
78-79 — 64.8

INTERNATIONAL TRAFFIC
(no. of passengers per flight) / (cargo and mail carried in tonne per flight)

Delhi — 4.3
Bombay — 152
Calcutta — 4.3
Madras — 5.4

Passengers per flight
Cargo and mail carried per flight

DOMESTIC TRAFFIC
(no. of passengers per flight) / (cargo and mail carried in tonne per flight)

Delhi, Bombay, Calcutta, Madras

Passengers per flight
Cargo and mail carried per flight

POSTAL SERVICE

India's postal service is among the world's largest. It is the second largest civil employer in the country, having over 850,000 employees.

The postal system was first introduced in India in 1766 by Robert Clive. It was meant for official purposes only. About 70 years later, in 1837, it was opened to the public. The first issue of postal stamps was made in 1852 in Sind. The first telegraph line was opened for traffic in October 1851 and the circuits were gradually extended to all parts of the country by 1867. The first airmail flight was made on 18 February 1911 by a French aviator, M. Pequet.

The statutory basis of the present postal service is the Postal Service Act, VI of 1858.

Under the control of the Ministry of Communications, the Secretary of the Indian Posts and Telegraphs Department holds the office of Director General, Posts and Telegraphs, as well as that of Chairman of the Posts and Telegraphs Board.

A 6-digit Postal Index Number Code (PIN Code), introduced in August 1972, identifies and locates every departmental delivery office. At the end of 1981, there were 140,435 post offices (125,743 rural — one for about every four villages — and 14,692 urban). This number has gone up slightly since then: in 1983-84, there were 144,719 post offices, 129,394 rural and 15,325 urban. In addition to these, there are rural mobile post offices (bicycle-based) which take care of about 2000 tribal and about 80,000 non-tribal villages (December 1981). The P & T Department also has 373 very high- and high-frequency wireless stations covering a total of 67,443 kms.

The mail is carried within the country on surface routes that cover 18.3 million kms. There were 495,853 post boxes in 1981 (out of which 421,604 were in the rural areas) and the mail handled in 1981-82 was about 9730 million pieces, of which 244 million were registered articles and 112 million inland money orders valued at Rs 12,510 million (earning the department Rs 300 million as commission).

POST OFFICES — 1983-84

Table 53

	Total	Rural	Urban
States			
Andhra Pradesh	16,403	14,682	1,721
Assam	3,321	3,074	247
Bihar	10,870	10,259	611
Gujarat	8,579	7,789	790
Haryana	2,452	2,156	296
Himachal Pradesh	2,432	2,342	90
Jammu & Kashmir	1,444	1,296	148
Karnataka	9,565	8,253	1,312
Kerala	4,734	4,066	668
Madhya Pradesh	10,489	9,591	898
Maharashtra	11,755	10,489	1,266
Manipur	555	523	32
Meghalaya	440	405	35
Nagaland	253	235	18
Orissa	7,356	6,986	550
Punjab	3,772	3,304	468
Rajasthan	9,625	8,809	816
Sikkim	126	119	7
Tamil Nadu	11,916	10,017	1,899
Tripura	615	570	45
Uttar Pradesh	18,132	16,310	1,822
West Bengal	8,116	7,098	1,018
Union Territories			
Andaman & Nicobar	80	66	14
Arunachal Pradesh	221	213	8
Chandigarh	44	7	37
Dadra & Nagar Haveli	29	29	—
Delhi	587	169	418
Goa, Daman & Diu	253	217	36
Lakshadweep	14	10	4
Mizoram	267	246	21
Pondicherry	94	64	30
INDIA	144,719	129,394	15,325

During the same years, 723,415 parcels were despatched to foreign countries and 714,586 parcels were received from abroad. Quick Mail Service (QMS) was introduced in 1975 and is available to 45 national and 410 regional centres. The QMS handles 450,000 articles a day. A Savings Bank Service is offered at 139,000 post offices and there are 45 million account holders whose total deposits at the end of 1981 amounted to Rs 78,950 million. The Postal Insurance scheme has 230,000 policies written for a value of Rs 1852 million.

There were 34,096 telegraph offices in the country at the end of 1980-81. During that year, they handled 71.4 million inland telegrams, 12.2 million phonograms and 6.7 million 'greetings telegrams'.

Despite this impressive record — and it is generally acknowledged that the Indian postal system is one of the better ones in the world — there are several thousand villages in the country without postal services. According to an unofficial estimate, in Uttar Pradesh, there are 96,000 villages without post offices, in Madhya Pradesh 61,000 and in Bihar 57,000. Even in better developed states like Maharashtra (25,000), Tamil Nadu (6000) and Gujarat (5000) there are considerable numbers of villages without post offices. But over the years, the number of post offices has been increasing; there has been a four-fold increase in 35 years (a little more than four times in rural areas and three-fold in urban areas), from around 36,000 post offices (31,000 rural and 5000 urban) in 1950-51 to about 144,000 in 1983-84.

Post office figures available for the states and union territories (March 1984) reveal that Uttar Pradesh has 18,132 post offices (16/100,000 population), followed by Andhra Pradesh 16,403 (29/100,000), Tamil Nadu 11,916 (24/100,000), Maharashtra 11,755 (18/100,000), Bihar 10,870 (15/100,000) and Madhya Pradesh 10,489 (19/100,000). The pattern of distribution generally follows the population pattern, though some union territories appear to be unfavourably placed (the Andaman & Nicobar and Lakshadweep Islands, for instance) in terms of outreach from the main population centres.

The size of population served by each post office varies considerably. In West Bengal, every 6667 persons have a post office, in Bihar it is 6250, in Assam 5882 and in Uttar Pradesh 6250. The picture seems worse in the urbanized union territories, 10,000 in Chandigarh and 11,111 in Delhi. And it seems very much rosier in Himachal Pradesh (1 post office for every 1754 people) and Mizoram (1:1851). But if accessibility is considered, the larger populations are served better by fewer offices than such places as Himachal Pradesh and Mizoram, where the people have trouble reaching their post offices. In 1984 it was estimated that for every 4762 Indians there was one post office.

The post offices in Dadra & Nagar Haveli are entirely rural. In all states and union territories, rural post offices far outnumber urban ones. With the exception of Pondicherry, there is no other territorial unit in which the number of urban post offices is greater than a fourth of all post offices.

Postal services are a very vital link in inter- and intra-national communication. The smaller the threshold (population served) and the less the average person distance to the service units, the better will be the social communication and interaction. It becomes clear that those states in India with larger thresholds than the national average are more rural and are extensive in area (the average person distance is, therefore, large), while states and territories that are small both in population and size have smaller thresholds and lower average person distances to post offices. However, in areas like Delhi and Chandigarh, where thresholds are very large, the small size of the territory, and therefore small average person distance, makes for better effectiveness and coverage.

Sources: 1. *Basic Statistics Relating to the Indian Economy, Vol. 2: States, September 1986*, Centre for Monitoring Indian Economy.
2. *Directory & Year Book 1984*, Times of India.

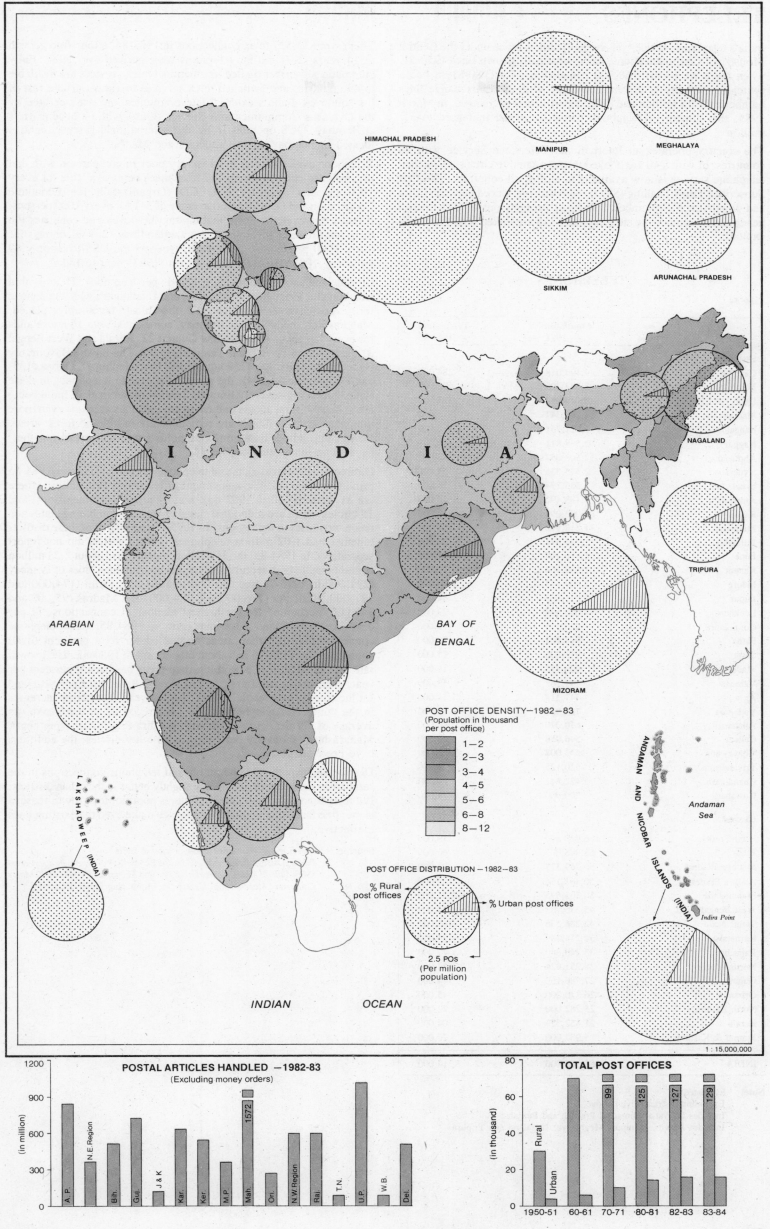

HIMACHAL PRADESH

MANIPUR

MEGHALAYA

SIKKIM

ARUNACHAL PRADESH

I N D I A

NAGALAND

ARABIAN SEA

BAY OF BENGAL

TRIPURA

MIZORAM

POST OFFICE DENSITY—1982–83
(Population in thousand per post office)

- 1—2
- 2—3
- 3—4
- 4—5
- 5—6
- 6—8
- 8—12

ANDAMAN AND NICOBAR ISLANDS (INDIA)

Andaman Sea

Indira Point

L A K S H A D W E E P (INDIA)

POST OFFICE DISTRIBUTION — 1982–83

% Rural post offices → ← → % Urban post offices

2.5 POS (Per million population)

INDIAN OCEAN

1 : 15,000,000

POSTAL ARTICLES HANDLED —1982-83
(Excluding money orders)

(in million)

1200 · 900 · 600 · 300 · 0

A. P. · N.E.Region · Bih. · Guj. · J & K · Kar. · Ker. · M.P. · Mah. 1572 · Ori. · N.W.Region · Raj. · T.N. · U.P. · W.B. · Del.

TOTAL POST OFFICES

(in thousand)

80 · 60 · 40 · 20 · 0

Rural · Urban

1950-51 · 60-61 · 70-71 99 · 80-81 125 · 82-83 127 · 83-84 129

TELEPHONES

India's telephone service, yet another public service under the Central Ministry of Communications, has dramatically grown since 1950-51, when there were only 168,000 connections. By 1980-81, it had a switching capacity of 2.8 million lines, with direct exchange lines numbering 2.15 million and 7871 departmental exchanges. In March 1985, the number of telephone connections had increased to 3.7 million.

The country operates an International Telephone Service with 44 countries, of which 40 are hooked through satellite circuits. Switched telephone service is now available to almost all countries. The first Subscriber Trunk Dialling (STD) route was introduced between Delhi and Jaipur on 8 September, 1964. At present, out of 373 district headquarters outside the state capitals, 355 are linked with their state capitals.

TELEPHONES

Table 54

	Population[1] — 1981	Telephones — 1984-85
Districts		
Calcutta	9,194,018	284,000
Bombay	8,243,405	605,000
Delhi	5,729,283	387,000
Madras	4,289,347	145,000
Bangalore	2,921,751	100,000
Nagpur	2,588,811	24,000
Ahmadabad	2,548,057	98,000
Hyderabad	2,545,836	78,000
Ernakulam	2,535,294	26,000
Amritsar	2,188,490	17,000
Rajkot	2,093,094	13,000
Ludhiana	1,818,912	21,000
Jullundur	1,734,574	14,000
Pune	1,686,109	66,000
Kanpur	1,639,064	27,000
Indore	1,409,473	23,000
Jaipur	1,015,160	44,000
Lucknow	1,007,604	29,000
Coimbatore	920,355	19,000
Patna	918,903	20,000
Surat	913,806	25,000
Madurai	907,732	16,000
Varanasi	797,162	13,000
Agra	747,318	14,000
Vadodara	744,881	22,000
Allahabad	650,070	10,000
Calicut	546,058	13,000
Vijayawada	543,008	15,000
Trivandrum	520,125	24,000
Chandigarh	422,841	24,000
Guwahati	200,000	9,000
Circles[2]		
Uttar Pradesh	106,020,800	119,000
Bihar	68,996,097	60,000
Madhya Pradesh	50,769,527	93,000
Andhra Pradesh	50,461,156	152,000
Maharashtra	50,265,675	176,000
West Bengal	45,386,982	54,000
Tamil Nadu	42,286,566	165,000
Karnataka	34,214,249	124,000
Rajasthan	33,246,840	79,000
North Western[3]	28,251,024	132,000
Gujarat	27,786,162	144,000
Orissa	26,370,000	45,000
North Eastern[4]	25,282,000	22,000
Kerala	21,852,523	104,000
Jammu & Kashmir	5,987,000	20,000
INDIA	685,185,000	3,714,000

Note: [1] Estimated.
    [2] Excluding districts given above.
    [3] Includes Haryana, Himachal Pradesh and Punjab.
    [4] Includes Assam, Manipur, Meghalaya, Nagaland and Tripura.

There were 22,527 telex connections in 1983-84, a four-fold growth in 14 years; 2.07 million telegrams were booked over telex. Fully automatic subscriber dialled international telex services are available to 49 countries and switched telex services to almost all the rest of the countries. India's external telecommunications are operated by the Overseas Communications Service (OCS) with its headquarters at Bombay. OCS operates from four international gateway centres located in Bombay, Delhi, Calcutta and Madras.

Telephone availability in India is very poor in comparison with that in developed countries, only 4.7 telephones being available for every 1000 persons against the OECD (Organization for Economic Cooperation and Development) average of 541 for every 1000 persons. It must, however, be noted that among the developed countries too there are dramatic differences: Switzerland has 1269 telephones for every 1000 persons — the highest, whereas the USSR has only 89 for every 1000 persons — the lowest (1983) next to India.

Considering another index, households per telephone, Delhi with a telephone for every four households, and Chandigarh with a telephone for every five households are the most favourably placed. Maharashtra, mainly due to Bombay, comes next with 16 households for every telephone. Then follow Gujarat 22, Kerala and West Bengal 31 each, Karnataka 33 and Tamil Nadu 35. The most unfavourably placed states are Bihar (159 households to a telephone), Orissa (135) and Uttar Pradesh (104), the people's access to telephones in these states thus being severely limited. Telephone services in India being urban biased, rural access to them is limited. This access is even more limited when the villages are remote or are in hill and forest areas. Hence, the indices based on households (rural and urban together) have to be interpreted cautiously.

Data on telephone districts and telephone circles for the year 1984-85 provide a more realistic assessment of telephone availability. There are 31 districts (30 in 1982-83), mainly urban agglomerations, and 15 circles which are individual states or regions combining states and union territories in the regions. In 1975-76, seventeen of these districts together had 1.02 million telephones (other districts did not report separately). In 1984-85, the 31 districts together had about 2.23 million connections. In both reporting years, the metropolitan cities of Bombay (271,000 in 1975-76 and 605,000 in 1984-85), Delhi (174,000 and 387,000), Calcutta (186,000 and 284,000) and Madras (95,000 and 145,000) accounted for the highest number of connections, 71 per cent in 1975-76 and 63.8 per cent in 1984-85. The telephone connections in all urban areas accounted for 53.4 per cent of all connections in 1975-76 and about 60 per cent in 1984-85. The growth rate for the eight-year period in urban areas has been 9 per cent per year, whereas in rural areas it has been 5.8 per cent per year and for the country as a whole 7.6 per cent per year. Very large increases in the number of telephones have been recorded in Karnataka (an average of 9.2 per cent per annum), Andhra Pradesh (9.1 per cent), Maharashtra (about 8.9 per cent). In all other circles, the additions have been marginal.

Despite the growth rate, the number of telephone connections in the rural areas remain poor and is only slightly better in the urban districts. But telephone communication in India is poor not for growth reasons alone; line faults due to poor maintenance make even the existing lines ineffective.

Sources: 1. *Directory & Year Book 1984*, Times of India.
    2. *Basic Statistics Relating to the Indian Economy, Vol. 1: All India, August 1986, Vol. 2: States, September 1986*, and *Economic Profiles of 40 Major Countries, March 1986*, Centre for Monitoring Indian Economy.

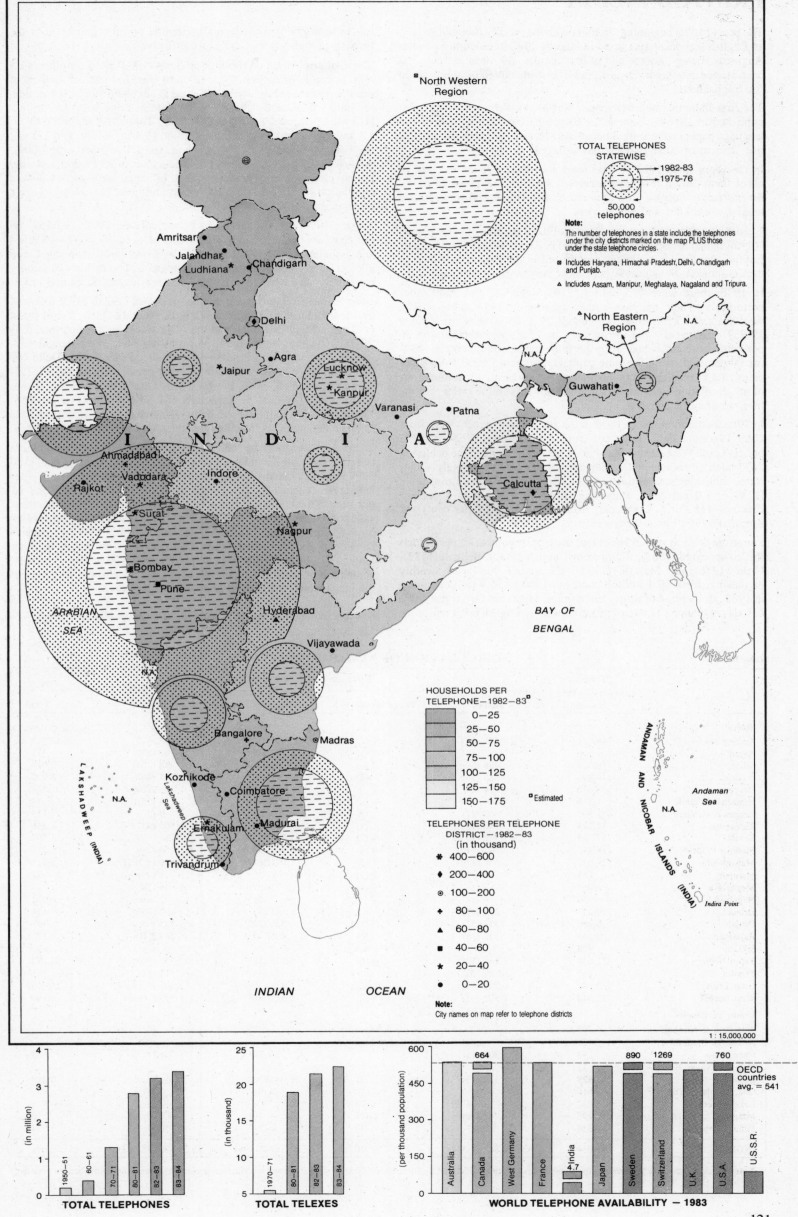

North Western Region

TOTAL TELEPHONES
STATEWISE
1982-83
1975-76
50,000
telephones

Note:
The number of telephones in a state include the telephones under the city districts marked on the map PLUS those under the state telephone circles.

⊠ Includes Haryana, Himachal Pradesh, Delhi, Chandigarh and Punjab.

△ Includes Assam, Manipur, Meghalaya, Nagaland and Tripura.

North Eastern Region

N.A.

N.A.

Amritsar
Jalandhar
Ludhiana
Chandigarh
Delhi
Agra
Jaipur
Lucknow
Kanpur
Varanasi
Patna
Guwahati

I N D I A

Ahmadabad
Vadodara
Indore
Rajkot
Surat
Nagpur
Bombay
Pune
Hyderabad
Calcutta

ARABIAN
SEA

N.A.

Vijayawada

BAY OF
BENGAL

Bangalore
Madras
Kozhikode
Coimbatore
Ernakulam
Madurai
Trivandrum

LAKSHADWEEP (INDIA)

N.A.

Lakshadweep Sea

ANDAMAN AND NICOBAR ISLANDS (INDIA)

Andaman Sea

N.A.

Indira Point

HOUSEHOLDS PER
TELEPHONE—1982-83
0—25
25—50
50—75
75—100
100—125
125—150
150—175
□ Estimated

TELEPHONES PER TELEPHONE
DISTRICT—1982-83
(in thousand)
✳ 400—600
◆ 200—400
◉ 100—200
✦ 80—100
▲ 60—80
■ 40—60
★ 20—40
● 0—20

Note:
City names on map refer to telephone districts

INDIAN OCEAN

1 : 15,000,000

TOTAL TELEPHONES

(in million)

1950—51	
60—61	
70—71	
80—81	
82—83	
83—84	

TOTAL TELEXES

(in thousand)

1970—71	
80—81	
82—83	
83—84	

WORLD TELEPHONE AVAILABILITY — 1983

(per thousand population)

Australia	Canada	West Germany	France	India	Japan	Sweden	Switzerland	U.K.	U.S.A	U.S.S.R.
	664			4.7		890	1269		760	

OECD countries avg. = 541

PRINTED MEDIA

The press had its beginnings in India with the weekly *Bengal Gazette,* an English scandal sheet started in January 1780 in Calcutta by James Augustus Hickey. By the end of that century, 19 more journals had been started, mainly in English, published and edited by Englishmen for Englishmen.

The first Indian-owned newspaper was also a *Bengal Gazette,* first published in 1816 by Gangadhar Bhattacharjee. Many more Indian language papers were started between 1818 and 1822, but the average life of most of these publications was only 5 years.

In the years since 1822 several more journals were started, but of all of them only a few have survived, such as *The Times of India, Amrita Bazar Patrika, The Statesman, The Hindu* and some Indian language weeklies and newspapers.

According to the 24th Annual Report of the Registrar of Newspapers (1980), Indian newspapers are published in 83 languages, which include English and the 15 languages recognized in the Eighth Schedule of the Constitution. More than 90 per cent of the newspapers are periodicals.

In 1970, there were 695 dailies, 3162 weeklies, and 7239 other periodicals. Their numbers gradually rose to 1423 dailies, 6122 weeklies and 13,213 other periodicals by 1983. Out of the total of 19,937 newspapers and periodicals in 1982, about 6000 (or 30 per cent) were published from the four metropolitan centres, 4000 (or 20 per cent) from the state capitals, and about 6000 from other cities with more than 100,000 population. At present, about four-fifths of all circulation is in the urban areas, for it is in the cities that the literates are concentrated.

In 1983 a little more than a third of the 20,758 newspapers were from Uttar Pradesh (2912 or 14 per cent), Maharashtra (2654 or 12.8 per cent) and West Bengal (2274 or 11 per cent). Delhi alone published 2637 such periodicals (12.7 per cent of the country's total). Other states publishing substantial numbers of periodicals were Tamil Nadu (1289 or 6.2 per cent), Rajasthan (1142 or 5.5 per cent), Andhra Pradesh (1123 or 5.4 per cent), Kerala (1085 or 5.2 per cent) and Karnataka (1019 or 4.9 per cent).

Language-wise, it was in Hindi that the largest number of periodicals (5936) appeared (1983), followed by English (3840), Bengali (1582), Urdu (1378) and Marathi (1131). Circulation has been steadily increasing, from 19.1 million copies in 1960 to 55.4 million copies in 1983, an increase of nearly three times. Hindi has the largest share of total circulation (27.9 per cent), followed by English (19.1 per cent).

But in terms of the intellectual or public opinion impact, it is the English periodicals that are more effective.

The total circulation of Hindi periodicals (1983) is 15.5 million (4.5 million copies of dailies, 4.5 million of weeklies and 6.5 million of other periodicals). English ranks second in circulation with 10.6 million (3.3 m, 1.5 m, and 5.8 m), Malayalam is third with 4.9 million (1.5 m, 1.7 m, and 1.7 m), followed by Tamil, 4.6 million (1.1 m, 2.1 m, and 1.4 m), Bengali, 3.5 million (1.1 m, 0.9 m, and 1.5 m), and Gujarati 2.8 million (1.1 m, 0.6 m, and 1.1 m). Oriya (564,000), Assamese (478,000), Sindhi (84,000) and Sanskrit (11,000) have the lowest circulations. There are 1641 bilingual newspapers with a circulation of 1.3 million, and 355 multilingual newspapers with a circulation of 250,000.

Among 19,937 newspapers and periodicals (in 1982), 13,157 (66 per cent) are owned by individuals, 3310 (16.6 per cent) by societies and associations, 979 (4.9 per cent) by firms and partnerships, and 806 (4.1 per cent) by joint stock companies. The rest are owned by the central and state governments and are located in capital cities.

Uttar Pradesh has the largest number of dailies (192 in 1983) followed by Maharashtra (177), Karnataka (121), Kerala (116), Tamil Nadu (112) and Madhya Pradesh (111). The weeklies are distributed as follows: Uttar Pradesh 1454, Maharashtra 664, West Bengal 497, Rajasthan 453, Andhra Pradesh 400, Madhya Pradesh 393, Delhi 369 and Bihar 364. The distribution of other periodicals is: Delhi 2213, Maharashtra 1813, West Bengal 1712, Uttar Pradesh 1266, Tamil Nadu 1022, Kerala 828, Karnataka 664, and Andhra Pradesh 653. There are states with just one daily (Himachal Pradesh, Meghalaya) or just a few weeklies (Manipur, Nagaland) or periodicals (Nagaland).

The majority of journals (newspapers, magazines) published in India issue from the bigger urban centres, which in 1983 together published 3289 English journals, 2927 Hindi journals and 5909 journals in other languages. Eighty-eight per cent of the English journals and 61 per cent of the Hindi journals are issued from the metropolitan cities, state capitals and other big cities. The 32 major urban centres located on the map, account for the bulk of this urban publishing: English 86 per cent and Hindi 49 per cent. Of all the urban publishing in 'other languages', these 32 cities, each of which has 100 or more journals, contribute 54 per cent, made up of 38 per cent in the language of the city and 16 per cent in some other language.

Source: *Mass Media in India 1985,* Ministry of Information & Broadcasting.

Table 55

CIRCULATION OF NEWSPAPERS — 1982

	Total Literates — 1981 (in thousand)	Circulation of Newspapers (in thousand)				
		Dailies	Tri/bi Weeklies	Weeklies	Others	Total
States						
Andhra Pradesh	16,035	695 (61)	1 (2)	929 (383)	687 (625)	2,312
Assam	N.A.	97 (7)	61 (4)	213 (74)	80 (95)	451
Bihar	18,321	498 (54)	9 (3)	1,098 (349)	439 (274)	2,044
Gujarat	14,896	888 (36)	(2)	360 (172)	675 (461)	1,923
Haryana	4,670	34 (14)	—	139 (138)	156 (220)	329
Himachal Pradesh	1,818	—	—	22 (22)	19 (55)	41
Jammu & Kashmir	1,597	43 (27)	—	67 (113)	9 (40)	119
Karnataka	14,283	748 (109)	1 (4)	599 (221)	508 (637)	1,856
Kerala	17,925	1,496 (105)	2 (2)	1,528 (139)	1,781 (797)	4,807
Madhya Pradesh	14,545	647 (110)	4 (9)	412 (380)	150 (229)	1,213
Maharashtra	29,621	2,359 (172)	77 (29)	1,911 (657)	3,970 (1,767)	8,317
Manipur	588	24 (33)	2 (2)	—	11 (48)	37
Meghalaya	455	3 (1)	1 (4)	23 (27)	19 (24)	46
Nagaland	330	—	—	2 (5)	2 (N.A.)	4
Orissa	9,027	231 (13)	—	41 (43)	192 (255)	464
Punjab	6,860	593 (37)	(1)	321 (226)	283 (366)	1,197
Rajasthan	8,354	522 (98)	(2)	401 (449)	247 (560)	1,170
Sikkim	108	N.A.	N.A.	N.A.	N.A.	N.A.
Tamil Nadu	22,638	1,130 (108)	8 (6)	2,175 (148)	2,770 (975)	6,083
Tripura	865	39 (13)	(1)	53 (43)	11 (11)	103
Uttar Pradesh	30,105	1,299 (182)	20 (20)	1,452 (1,395)	2,305 (1,202)	5,076
West Bengal	22,344	1,802 (62)	4 (7)	1,521 (481)	1,572 (1,659)	4,899
Union Territories						
Andaman & Nicobar	97	—	—	—	(8)	—
Arunachal Pradesh	131	N.A.	N.A.	N.A.	N.A.	N.A.
Chandigarh	293	305 (11)	—	56 (34)	241 (127)	602
Dadra & Nagar Haveli	28	—	—	—	—	—
Delhi	3,828	1,330 (49)	(3)	1,044 (354)	4,514 (2,052)	6,888
Goa, Daman & Diu	616	45 (9)	—	9 (9)	5 (5)	59
Lakshadweep	22	N.A.	N.A.	N.A.	N.A.	N.A.
Mizoram	296	12 (18)	(1)	4 (16)	4 (6)	20
Pondicherry	338	7 (1)	(1)	7 (7)	20 (4)	34
INDIA	241,034	14,847 (1,334)	190 (103)	14,387 (5,898)	20,670 (12,602)	50,094

Note: Figures within brackets indicate the number of publications. The total Indian figures include those of states and union territories for which break-up figures are not available.

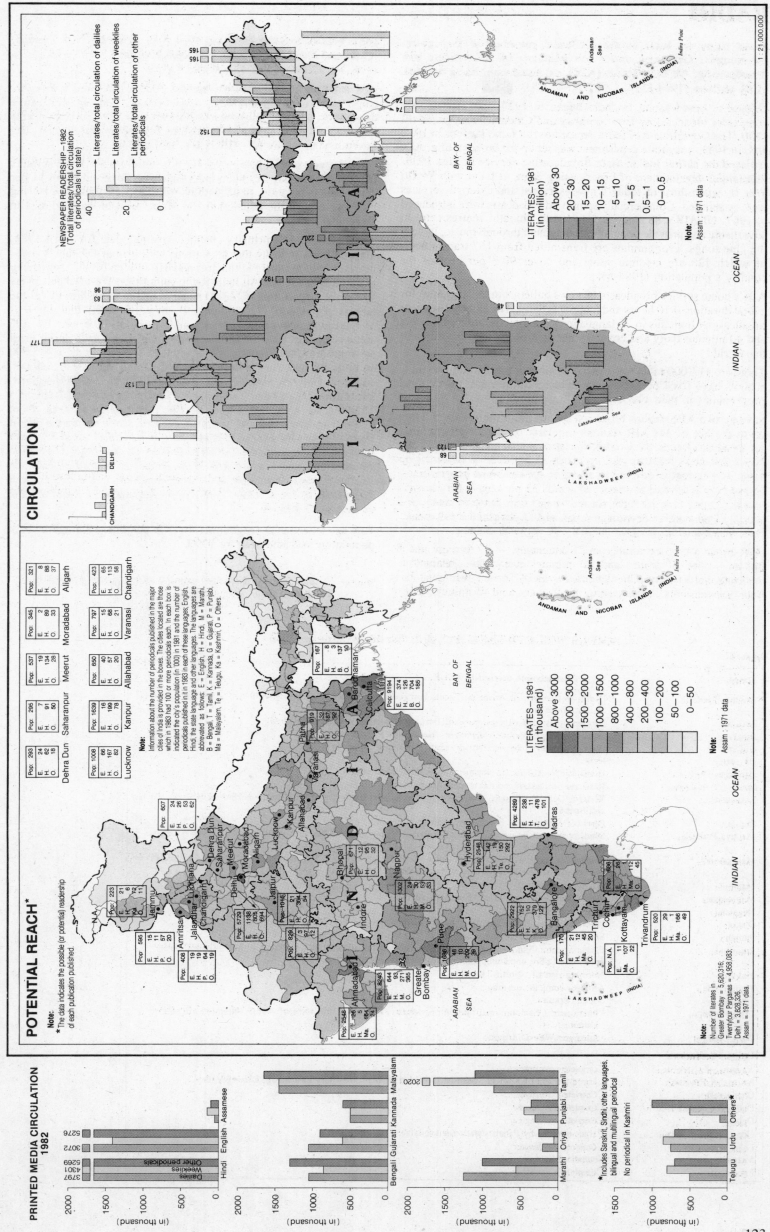

CIRCULATION

NEWSPAPER READERSHIP—1982
(Total literates/total circulation
of periodicals in state)

Literates/total circulation of dailies
Literates/total circulation of weeklies
Literates/total circulation of other
periodicals

LITERATES—1981
(in million)

Above 30
20—30
15—20
10—15
5—10
1—5
0.5—1
0—0.5

Note:
Assam: 1971 data.

POTENTIAL REACH*

Note:
* The data indicates the possible (or potential) readership
of each publication published.

Note:
Information about the number of periodicals published in the major
cities of India is provided in the boxes. The cities located are those
which in 1983 had 100 or more periodicals each. In each box is
indicated the city's population (in '000) in 1981 and the number of
periodicals published in it in 1983 in each of these languages: English,
Hindi, the state language and other languages. The languages are
abbreviated as follows: E = English, H = Hindi, M = Marathi,
B = Bengali, T = Tamil, K = Kannada, G = Gujarati, P = Punjabi,
Ma = Malayalam, Te = Telugu, Ka = Kashmiri, O = Others.

LITERATES—1981
(in thousand)

Above 3000
2000—3000
1500—2000
1000—1500
800—1000
400—800
200—400
100—200
50—100
0—50

Note:
Assam: 1971 data.

Note:
Number of literates in
Greater Bombay = 5,620,316;
Twentyfour Parganas = 4,956,083;
Delhi = 3,828,326.
Assam: 1971 data.

PRINTED MEDIA CIRCULATION
1982

Dailies
Weeklies
Other periodicals

Hindi English Assamese

Bengali Gujarati Kannada Malayalam

Marathi Oriya Punjabi Tamil

Telugu Urdu Others*

No periodical in Kashmiri
* Includes Sanskrit, other languages,
bilingual and multilingual periodical

(in thousand)

RADIO

Broadcasting in India is the exclusive preserve of the central government. Operating under the Ministry of Information and Broadcasting, All India Radio (AIR), or Akashvani, has a network of 88 stations (1984-85).

Organized broadcasting in India began in 1927 with two private companies broadcasting from Bombay and Calcutta. This lasted till 1930. The Government of India decided to run radio stations in 1932 and, in 1935, a separate department was set up for broadcasting. AIR replaced the earlier Indian State Broadcasting Service in June 1936. Substantial development of broadcasting came to India with World War II, when most of today's news services and external services were established. Commercially sponsored broadcasts were introduced in 1967. Of AIR's 88 stations, 31 Vividh Bharathi centres cater to advertising and provide popular entertainment through music, including cine songs. Programmes are transmitted from 167 transmitters, of which 128 are medium wave and cover 90.3 per cent of the country's population (1984-85).

AIR's home service broadcasts 68 news bulletins in 19 languages for a total duration of 10 hours and 8 minutes daily. The external services broadcast 64 bulletins in 25 languages for a total duration of 8 hours and 40 minutes daily and cover 79 per cent of the total land area of the world.

There are 117,000 radio sets in schools all over the country and these schools have fixed periods when children can listen to educational programmes in their classes.

Seventy-two AIR stations broadcast special programmes for children up to the age of 14. AIR stations regularly broadcast programmes for the armed forces, for women, for farmers, and on family planning. News and news features take up about 22 per cent of transmission time; 40 per cent is devoted to all other spoken-word programmes; 38 per cent is allotted to music, of which 32 per cent is for classical music, 22 per cent for light music, 18 per cent for film music and songs, 17 per cent for devotional songs, and 11 per cent for folk music. AIR also transmits programmes in 146 regional dialects.

AIR broadcasts programmes in 25 languages — 17 foreign and 8 Indian — for 57 hours and 30 minutes every day, beamed to international listeners. The General Overseas Service in English has four transmissions, broadcasting for 9 hours and 45 minutes daily and is beamed to South-east Asia, West Asia, Australia, New Zealand, Africa and Europe. The Urdu service is broadcast in three transmissions for 12 hours and 15 minutes a day.

AIR employs 17,000 persons, besides 3000 artistes and nearly 900 employees in programming, engineering, and administrative disciplines. It has 77 full-time and 200 part-time news correspondents in India, and 7 full-time and 6 part-time correspondents abroad. The Seventh Plan outlay for AIR is Rs 7000 million.

Over the years, AIR has grown from 6 stations and 18 transmitters in 1947 (on the day of Indian independence), covering about 11 per cent of the population, to 88 stations with 167 transmitters, covering 90.3 per cent of the population and 80 per cent of the land area in December 1985.

The statewise AIR primary channel coverage data for April 1984, mapped here, indicate that 88 stations with primary channels cover nearly 3 million square kilometres and 616 million people. According to this estimate by the All India Radio authorities, West Bengal covers by far the largest area (632,000 sq kms) and Maharashtra by far the largest population (72 million). It must be noted that there is overlapping coverage, area- and population-wise, by the stations of one state in another state. The states with large areal coverage are Uttar Pradesh 289,000 sq kms, Madhya Pradesh 475,500 sq kms, Maharashtra 440,000 sq kms, Andhra Pradesh 329,000 sq kms, Gujarat 300,000 sq kms, Rajasthan 231,000 sq kms, Bihar 160,000 sq kms, Tamil Nadu 152,500 sq kms and Karnataka 151,000 sq kms, Coverage in the union territories is about 214,000 sq kms and about 14 million persons, Delhi having the largest share of both (166,000 sq kms and 5 million people). Area coverage does not necessarily result in large population coverage: Population covered in Uttar Pradesh and West Bengal is 61 million each, in Andhra Pradesh 56 million, in Gujarat 58 million, in Madhya Pradesh 53 million, in Tamil Nadu 42 million, in Karnataka 32.5 million, and in Kerala 27 million.

AIR has launched a programme for 100 per cent coverage of both population and land area by 1990.

Source: 1. *Mass Media in India 1985*, Ministry of Information & Broadcasting.
2. All India Radio, Madras Office.

AREAS NOT COVERED BY ALL INDIA RADIO — 1984

Table 56

States	Districts not covered by A.I.R.
Andhra Pradesh	: North-west corner of Anantapur, south-west corner of Kurnool, north-east corner of Srikakulam, portion of East Godavari.
Assam	: Northern part of Goalpara, parts of Darrang, Nowgong and Karbi Anglong.
Bihar	: Small portion in Purnia, Singhbhum, Ranchi and Palamau.
Gujarat	: Parts of Southern and western Junagadh, south-west Jamnagar.
Haryana	: Sirsa.
Himachal Pradesh	: Lahul & Spiti, Chamba, Kinnaur, north Kangra and Kullu.
Jammu & Kashmir	: Vast area in Ladakh, northern and western districts of the state.
Karnataka	: Kodagu, Dakshin Kannad, south-west and north-west portion of Uttar Kannad, portion of Bellary, Raichur and Bijapur.
Kerala	: Parts of Malappuram, Kasaragod, parts of Cannanore.
Madhya Pradesh	: Small portion of Bilaspur, Shahdol, Rajgarh, meeting point of Narsimhapur, Raisen and Sagar, large portion of Vidisha and Shivpuri, southern corner of Morena.
Maharashtra	: Sindhudurg, parts of Kolhapur, Solapur, northern part of North Ratnagiri, parts of Satara, Central portion of Osmanabad and Beed, southern tip of Nashik.
Manipur	: Complete coverage.
Meghalaya	: West Garo Hills, portion of Khasi and Jaintia hills.
Nagaland	: Portion of Tuensang.
Orissa	: Parts of Ganjam, Phulbani, Kalahandi, Balangir and Sundargarh.
Punjab	: Small portion in Ferozpur.
Rajasthan	: Parts of central and western Jaisalmer, western corner of Barmer, western portion of Bikaner, large portion of Churu and Jhunjhunun, Sawai Madhopur, Alwar and Kota.
Sikkim	: Northern part of North Sikkim.
Tamil Nadu	: Almost complete coverage.
Tripura	: South Tripura.
Uttar Pradesh	: Pithoragarh, Chamoli, small portion of Meerut, Ghaziabad, Bulandshahr, Agra, Manipuri, Mirzapur, Lalitpur.
West Bengal	: Jalpaiguri, West Dinajpur.
Union Territories	
Andaman & Nicobar	: Complete coverage.
Arunachal Pradesh	: Large portion of Kameng, Subansiri, northern parts of West and East Siang, Dibang Valley.
Chandigarh	: Complete coverage.
Dadra & Nagar Haveli	: Complete coverage.
Delhi	: Complete coverage.
Goa, Daman & Diu	: Extent covered by Panaji station not available.
Lakshadweep	: Complete coverage.
Mizoram	: Southern Mizoram.
Pondicherry	: Complete coverage.

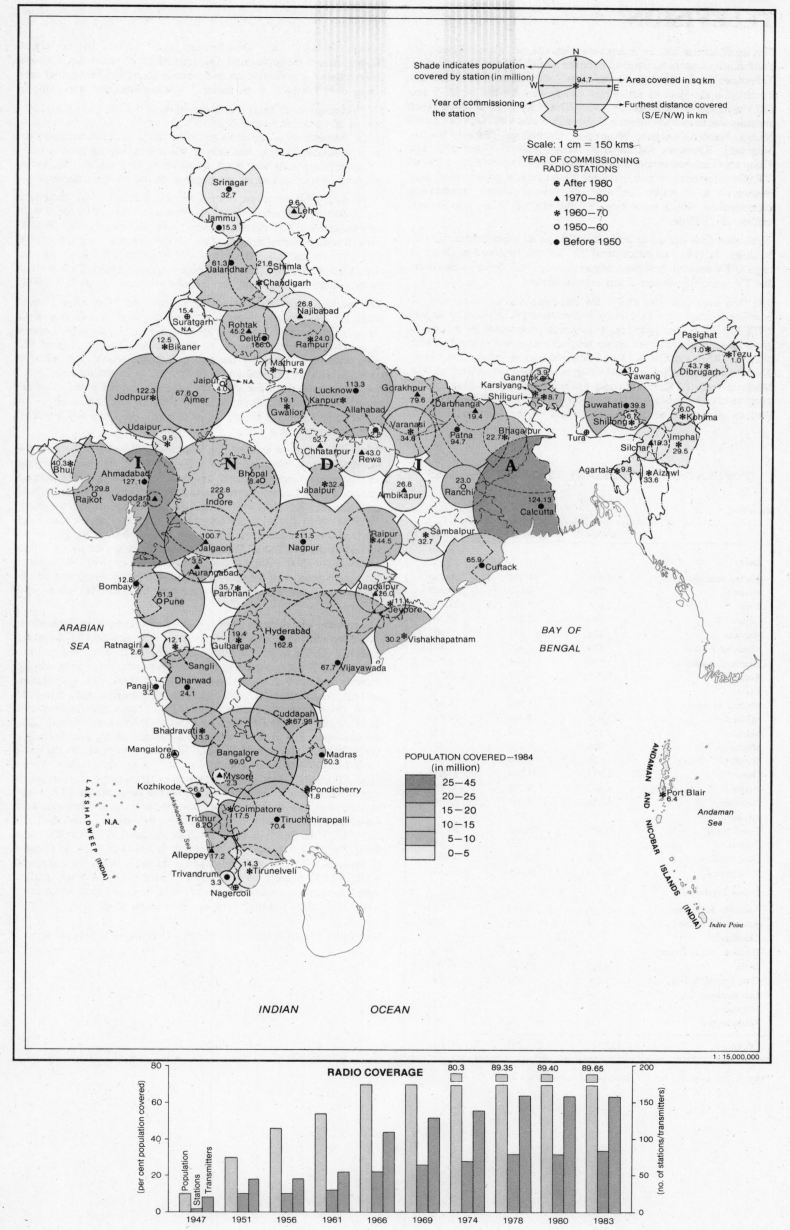

Shade indicates population
covered by station (in million)

Area covered in sq km

Year of commissioning
the station

Furthest distance covered
(S/E/N/W) in km

Scale: 1 cm = 150 kms

YEAR OF COMMISSIONING
RADIO STATIONS

⊕ After 1980
▲ 1970—80
✳ 1960—70
○ 1950—60
● Before 1950

POPULATION COVERED—1984
(in million)

25—45
20—25
15—20
10—15
5—10
0—5

ARABIAN SEA

BAY OF BENGAL

INDIAN OCEAN

ANDAMAN AND NICOBAR ISLANDS (INDIA)

Andaman Sea

Indira Point

LAKSHADWEEP (INDIA)

1 : 15,000,000

RADIO COVERAGE

(per cent population covered) — Population, Stations, Transmitters

(no. of stations/transmitters)

80.3   89.35   89.40   89.65

1947   1951   1956   1961   1966   1969   1974   1978   1980   1983

TELEVISION

The small screen has, in recent years, carved out a special place for itself in mass communication all over the world. India is no exception. Television has developed in India from 2 stations in 1970 (Delhi and Bombay) to 18 stations and Relay Centres by the end of 1979 and 181 (including 11 stations) by 1986. The relay centres added since the map was drawn are Poonch (Jammu & Kashmir), Neyveli (Tamil Nadu), Nazira (Assam), Bhagalpur, Darbhanga (Bihar), Korba, Singranti, Khandwa (Madhya Pradesh), Ukhrul (Manipur) and Kavaratti (Lakshadweep), all with low-powered transmitters. Indian TV (Doordarshan), which is owned by the central government, also registered a dramatic increase in revenue from advertising commercials, which grew from Rs 800,000 in 1976-77 to Rs 600 million in 1985-86.

Television first started in India in 1959, as an experimental service in Delhi. In 1961, an educational TV service followed in Delhi. A regular TV service, however, began only in 1965. Seven years later, the TV station at Bombay was commissioned.

The first ever exposure to TV for the rural masses came with the Satellite Instructional Television Experiment (SITE) during August 1975-July 1976. The year-long experiment, conducted by the central government, broadcast television programmes direct to rural viewers via the NASA satellite ATS-6. During the experiment, community TV sets were installed in 2400 villages in Rajasthan, Bihar, Orissa, Madhya Pradesh, Karnataka and Andhra Pradesh, and instructional programmes were beamed to them direct via the satellite. With 181 stations and relay centres now hooked to the National Satellite System (INSAT 1B), Doordarshan now reaches out to about 50 per cent of the population. The demand for TV receivers has been projected at 1.2 million black and white sets and 800,000 colour sets a year over 1986-91 by the Indian TV Manufacturers' Association.

TV, belieing earlier fears that it would wipe out the cinema houses, has, in fact, complemented them. TV, however, has, in the last few years, become an advertiser's paradise and a revenue bonanza for the government. With an audience of 400 million waiting to be tapped, TV advertising rates have been skyrocketing. It costs Rs 150,000 to sponsor a Super A special programme on the national network.

Of the 172 TV stations and centres for which detailed statistical data are available, there are today one 80 W centre, 131 one hundred watt centres, one 0.6 KW centre, fifteen 1 KW, and twenty four 10 KW centres with different ranges. The 10 KW centres have a range of 110-220 kms, the 0.6 KW centres a range of 55—110 Kms, and all others about 25 Kms. The largest population covered is 23.8 million by the Calcutta centre and the lowest by a relay centre is 10,000.

The spurt in Indian TV was a direct result of the 1982 Asian Games when 20 low-power transmitters were installed in different state capitals and towns. This paved the way for setting up relay centres in small and medium towns, so that, today, 50 per cent of the population is covered by the national transmission. But while a common television programme is now produced for the entire country, this has considerably cut short the time allotted for local, regional programmes. Second channels in the state capitals are expected to be set up to solve this problem. Some cities like Delhi and Bombay have already made a start.

While TV caters to the varying interests of the stay-at-homes, the world's biggest film industry caters exclusively to those seeking entertainment. India produces about 750 films a year, the largest number by any country in the world. The industry has a total capital investment of Rs 7000 million and provides employment to nearly 250,000. With almost every other kind of entertainment out of the reach of the common man, films have helped to fill a big gap in the social structure in India. The gross annual earnings are more or less steady at Rs 5500 million a year from theatres all over the country, but more than half of this goes to the state governments as entertainment tax. The export revenue from films in 1980 was Rs 150 million.

The number of cinema houses is pathetically short of requirements. There were 9551 cinemas in operation in 1979, including 3521 touring cinemas. This rose to 12,701 in 1985-86, 17 theatres to a million population. The states and union territories have considerably varying numbers of cinema houses and population thresholds (population served). Andhra Pradesh ranks first with 2335 theatres and a threshold of 24,754 persons to a theatre (1985). Tamil Nadu ranks second with 2153 theatres and 24,013 persons. Andhra Pradesh and Tamil Nadu are the states that produce the largest number of regional language films. Madras, the new Hollywood of India, also produces the most Hindi films (which have a countrywide audience). Maharashtra, whose capital Bombay used to churn out the most Hindi films, has the fourth largest number of cinema houses (1338) but there are 51,121 people to a theatre. The figures for other states and union territories are: Karnataka 1255 theatres (32,271 persons to a theatre), Kerala 1356 (20,206), Uttar Pradesh 898 (134,298), Assam 200 (109,000), West Bengal 672 (87,946) and Madhya Pradesh 539 (105,566). In Tripura (8 theatres with 287,500 persons to a theatre), Bihar (368 theatres with 207,065 persons to a theatre), in Mizoram and other smaller states/territories, cinema houses are very few and have to cater to large audiences. Lakshadweep has no theatre at all.

Source: *Mass Media in India 1985*, Ministry of Information & Broadcasting.

POPULATION COVERED BY TELEVISION[1] — 1985

Table 57 (in thousand)

	URBAN	RURAL	TOTAL
States			
Andhra Pradesh	8,801	11,329	20,130
Assam	704	6,682	7,386
Bihar	5,019	19,054	24,073
Gujarat	7,930	8,238	16,168
Haryana	238	671	909
Himachal Pradesh	674	1,429	2,103
Jammu & Kashmir	1,071	2,688	3,759
Karnataka	7,709	12,029	19,738
Kerala	2.352	3,841	6,193
Madhya Pradesh	6,899	11,902	11,801
Maharashtra	17,856	12,154	30,010
Manipur	155	103	258
Meghalaya	208	218	426
Nagaland	36	80	116
Orissa	1,817	11,254	13,071
Punjab	3,402	9,235	12,637
Rajasthan	4,178	7,911	12,089
Sikkim	37	66	103
Tamil Nadu	13,929	20,409	34,338
Tripura	131	200	331
Uttar Pradesh	15,888	70,041	85,929
West Bengal	14,089	38,409	52,498
Union Territories			
Andaman & Nicobar	50	157	207
Arunachal Pradesh	24	84	108
Chandigarh	—	—	—
Dadra & Nagar Haveli	—	—	—
Delhi	8,815	10,245	19,060
Goa, Daman & Diu	260	466	726
Lakshadweep	—	—	—
Mizoram	76	36	112
Pondicherry	379	356	735
INDIA	122,727	259,287	382,014

Note: [1] Data available for 174 stations and relay centres only.

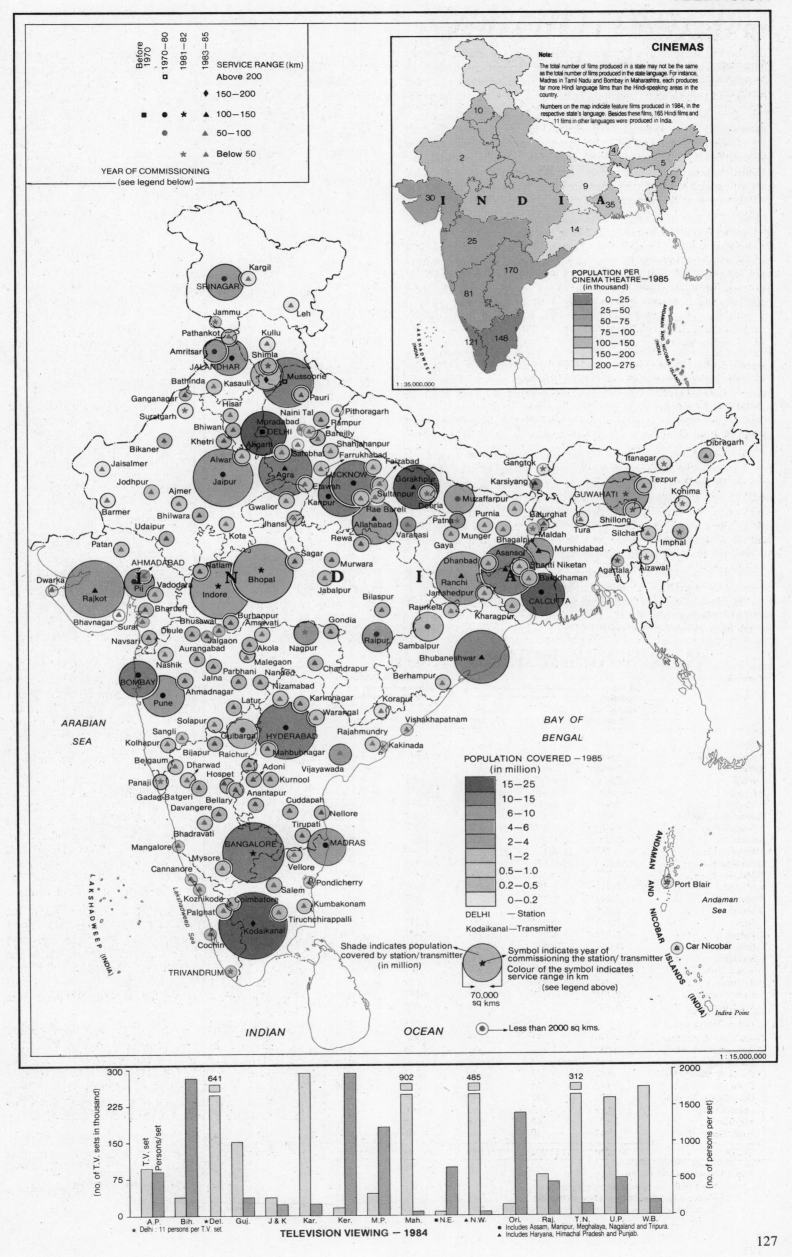

CINEMAS

Note:

The total number of films produced in a state may not be the same as the total number of films produced in the state language. For instance, Madras in Tamil Nadu and Bombay in Maharashtra, each produces far more Hindi language films than the Hindi-speaking areas in the country.

Numbers on the map indicate feature films produced in 1984, in the respective state's language. Besides these films, 165 Hindi films and 11 films in other languages were produced in India.

POPULATION PER CINEMA THEATRE—1985
(in thousand)

- 0—25
- 25—50
- 50—75
- 75—100
- 100—150
- 150—200
- 200—275

1 : 35,000,000

SERVICE RANGE (km)
- Above 200
- 150—200
- 100—150
- 50—100
- Below 50

Before 1970 · 1970—80 · 1981—82 · 1983—85

YEAR OF COMMISSIONING
(see legend below)

POPULATION COVERED — 1985
(in million)

- 15—25
- 10—15
- 6—10
- 4—6
- 2—4
- 1—2
- 0.5—1.0
- 0.2—0.5
- 0—0.2

DELHI — Station
Kodaikanal—Transmitter

Shade indicates population covered by station/transmitter (in million)

Symbol indicates year of commissioning the station/transmitter
Colour of the symbol indicates service range in km
(see legend above)

70,000 sq kms

Less than 2000 sq kms.

ARABIAN SEA

BAY OF BENGAL

INDIAN OCEAN

ANDAMAN AND NICOBAR ISLANDS (INDIA)

Andaman Sea

Port Blair

Car Nicobar

Indira Point

1 : 15,000,000

TELEVISION VIEWING — 1984

(no. of T.V. sets in thousand) / (no. of persons per set)

A.P., Bih., *Del., Guj., J & K, Kar., Ker., M.P., Mah., ■ N.E., ▲ N.W., Ori., Raj., T.N., U.P., W.B.

* Delhi : 11 persons per T.V. set.
■ Includes Assam, Manipur, Meghalaya, Nagaland and Tripura.
▲ Includes Haryana, Himachal Pradesh and Punjab.

641   902   485   312

IRRIGATED CULTIVATION

The marked seasonality of Indian rainfall and its variability in many areas make water scarce almost everywhere. Irrigation provides an artificial supply to meet such scarcity. In India, therefore, almost throughout the country, irrigation is essential for high productivity.

Though rainfall is not, by itself, a dependable source of perennial water supply, it is nevertheless the original source of supply for most irrigation works such as man-made dammed reservoirs (where water is stored during the rainy season for year-long use through canals and channels), tanks (man-made reservoirs where water is impounded against banked slopes during the rains), and wells which are dug, or bored, to tap groundwater.

These sources of irrigation serve the largest net cropped area in Uttar Pradesh (17.3 million hectares in 1981), but this is only 58.2 per cent of the state's reporting area, i.e. the area for which data regarding land use classification are available. Maximum use of this reporting area is, however, made in Punjab (83.6 per cent or 4.2 m ha) and Haryana (83.1 per cent or 3.7 m ha). Other states report as follows: Tamil Nadu (44.1 per cent or 5.7 m ha), West Bengal (63 per cent or 5.6 m ha), Bihar (45.4 per cent or 7.9 m ha) and Andhra Pradesh (41.3 per cent or 11.1 m ha).

Relative to cropped area, net irrigated area is well developed in the case of Punjab (81 per cent), Haryana (61.4 per cent), Uttar Pradesh (55.2 per cent), Tamil Nadu (47.2 per cent), Manipur (46.4 per cent), Jammu & Kashmir (42.9 per cent) and Nagaland (40.5 per cent). In all other areas, the net irrigated area is less than 40 per cent of the net cropped area. Though small in extent, Delhi (87.5 per cent) and Pondicherry (86.7 per cent) have a large part of their cropped area irrigated.

It is estimated that only about 30 per cent of the net sown area in the country lies in areas which receive a high rainfall of over 115 cms a year. Of the rest, 35.9 per cent are areas which get between 75 and 115 cms of rainfall and 34.1 per cent are areas of low rainfall. The rainfall for more than 70 per cent of the net sown area is too low to allow intensive agriculture during the main crop season. For example, sugarcane and rice require regular and sufficient water supply, which is only possible through irrigation.

The relationship of sown area to differential rainfall receipts in each of the states of the Union can help to identify areas in need of irrigation. In terms of rainfall, the country could be divided into areas of high, low, medium and inadequate rainfall. The entire sown area falls within high rainfall areas (1978) in nine states — Kerala, Orissa,

West Bengal, Manipur, Assam, Tripura, Nagaland, Meghalaya and Sikkim — and the union territory of Mizoram. On the other hand, more than three-fourths of the sown area falls in inadequate rainfall areas in Rajasthan, Karnataka and Gujarat, although Gujarat is better placed than Rajasthan in agriculture because a small proportion of Gujarat's sown area receives high rainfall. Maharashtra and Andhra Pradesh have slightly less than 75 per cent of their sown area within areas in receipt of inadequate rainfall. In Tamil Nadu, the sown area within inadequate and low and medium rainfall areas is almost equal. In Uttar Pradesh, a little less than half the sown area falls within inadequate rainfall areas and this is true of Haryana also. In Punjab, however, a little less than a quarter of the sown area is in the zone of inadequate rainfall and the rest is in low and medium rainfall areas. Surprisingly, areas that are adversely placed in respect of rainfall have done better in irrigation development and agriculture (Punjab, Tamil Nadu and Maharashtra for example), while those with plenty of it have not.

Wherever artificial irrigation has been carried out, it has the effect of reclaiming and economizing land. It reclaims land in dry areas where irrigation alone makes crop-growing possible. It saves land in places where, although there is sufficient rainfall and groundwater for agriculture, increases in yields can be economically achieved and double-cropping made possible only with irrigation. In both cases, irrigation is a measure of intensification, with the aim of achieving more than proportionately higher gross returns through increased expenditure.

Intensification of cultivation often results from cropping the same land more than once in a year. The gross sown area is, thus, the total area sown plus the area cultivated a second or, rarely, a third time. Gross irrigated area is, correspondingly, the sum of areas irrigated per crop, doubling if the land is cropped twice. Over the planning years, the gross sown area has increased from 131.9 million to 174 million hectares (1950-1984). During the same period, the net irrigated area has just about doubled, from 20.9 million to 40.8 million hectares, while the gross irrigated area has multiplied 2½ times, from 22.6 million to 56.5 million hectares. This growth, either in the net irrigated or gross irrigated area, is not as spectacular as the growth in the outlays, yet it has certainly made a large difference to productivity, resulting in a country of severe famines becoming one self-sufficient in food. What is more, India has become a food exporter as well.

One measure of irrigation intensity is the percentage of gross irrigated area to the net irrigated area. Over the years, this intensity has been more than 100 per cent, because gross area has always been larger than net area due to repeat cropping in a year: 108 per cent in 1950-51, 113.4 per cent in 1960-61, 122.8 per cent in 1970-71, 127.8 per cent in 1980-81 and 138.5 per cent in 1983-84 (this last is an estimate of the Centre for Monitoring Indian Economy). The intensity has, thus, been substantially increasing over the years, but in terms of net area the increases have tended to vary from very high to marginally more. In 1981 it was highest in Himachal Pradesh (172 per cent), followed by Punjab, Kerala and Haryana. It was above the national average of 126.3 per cent in Tripura, Maharashtra, Tamil Nadu, Orissa and Jammu & Kashmir also. In two states, Assam and Sikkim, the gross and net irrigated areas equalled one another. And it was well below the national average in Madhya Pradesh, Meghalaya, West Bengal and Nagaland. Andhra Pradesh, Uttar Pradesh, Rajasthan, Karnataka, Manipur and Gujarat also fared poorly. The union territories have an intensity of 114 per cent.

The cropped area intensity, gross cropped area to net cropped area, for 1983-84 for the country was 147.3 per cent (only 21 states reporting). Four states registered an intensity in excess of the national average: Kerala (the greatest, 345 per cent), Gujarat, Meghalaya and Karnataka. The other 17 states recorded lower than national cropped area intensity, the lowest being 109 per cent in Himachal Pradesh.

Indian plan outlays on irrigation have been steadily increasing over the years. In the very first Five Year Plan, the outlay on such projects was Rs 4460 million. This investment grew 20-fold in 30 years. It will increase even further during the Seventh Plan, the planned outlay being Rs 143,605 million, which is more than 30 times that in the first Five Year Plan, a phenomenal increase indeed.

CULTIVATED AND IRRIGATED AREA — 1981-82

Table 58

('000 hectares)

	Total Reporting Area	Cultivated Area		Irrigated Area	
		Net	Gross	Net	Gross
States					
Andhra Pradesh	27,440	11,325	13,047	3,692	4,673
Assam[1]	7,852	2,696	3,439	572[1]	572[1]
Bihar	17,330	7,861	10,628	3,001	3,582
Gujarat	18,826	9,670	10,903	2,155	2,522
Haryana	4,405	3,660	5,826	2,248	3,455
Himachal Pradesh	3,089	573	949	92	161
Jammu & Kashmir	4,675	716	978	307	396
Karnataka	19,050	10,391	11,228	1,471	1,801
Kerala	3,885	2,170	2,905	240	383
Madhya Pradesh	44,211	18,841	21,756	2,421	2,510
Maharashtra	30,758	18,314	20,386	1,927[1]	2,686[1]
Manipur[2]	2,211	140	240	65	75
Meghalaya[1]	2,249	193	203[3]	50	51
Nagaland	1,099	153	164	62	68
Orissa	15,540	6,130	8,743[3]	1,215	2,006
Punjab	5,033	4,210	6,929	3,408	5,966
Rajasthan	34,234	15,577	18,596	2,903	3,722
Sikkim[1]	719	86	92[3]	11	11
Tamil Nadu	13,002	5,740	6,909	2,709	3,425
Tripura[1]	1,048	246	380	29	38
Uttar Pradesh	29,708[4]	17,288	24,773	9,541	11,620
West Bengal	8,846[4]	5,565	7,402	1,489	1,735
Union Territories					
Andaman & Nicobar	790	33	36	N.A.	N.A.
Arunachal Pradesh[1]	5,530	112	152	24	24
Chandigarh	11	3	4	N.A.	N.A.
Dadra & Nagar Haveli	49	23	25[3]	1	1
Delhi	147	56	84	49	54
Goa, Daman & Diu	371	132	142	13	13
Lakshadweep	3	3	3	N.A.	N.A.
Mizoram[1]	2,102	65	68	8	8
Pondicherry	47	30	51	26	42
INDIA	304,280	142,002	177,041	39,729	51,601

Note: [1] Based on provisional data.
[2] Estimated
[3] Forecast data has been used in estimation
[4] Reporting area is more than the geographical area. Variations are under verification.

Sources: 1. *Fertiliser Statistics 1984-85*, Fertiliser Association of India.
2. *Basic Statistics Relating to the Indian Economy, Vol. 1: All India August 1986* and *Vol.2: States, September 1986,* Centre for Monitoring Indian Economy.
3. *Seventh Five Year Plan 1985-90, Vol. II,* Government of India Planning Commission.

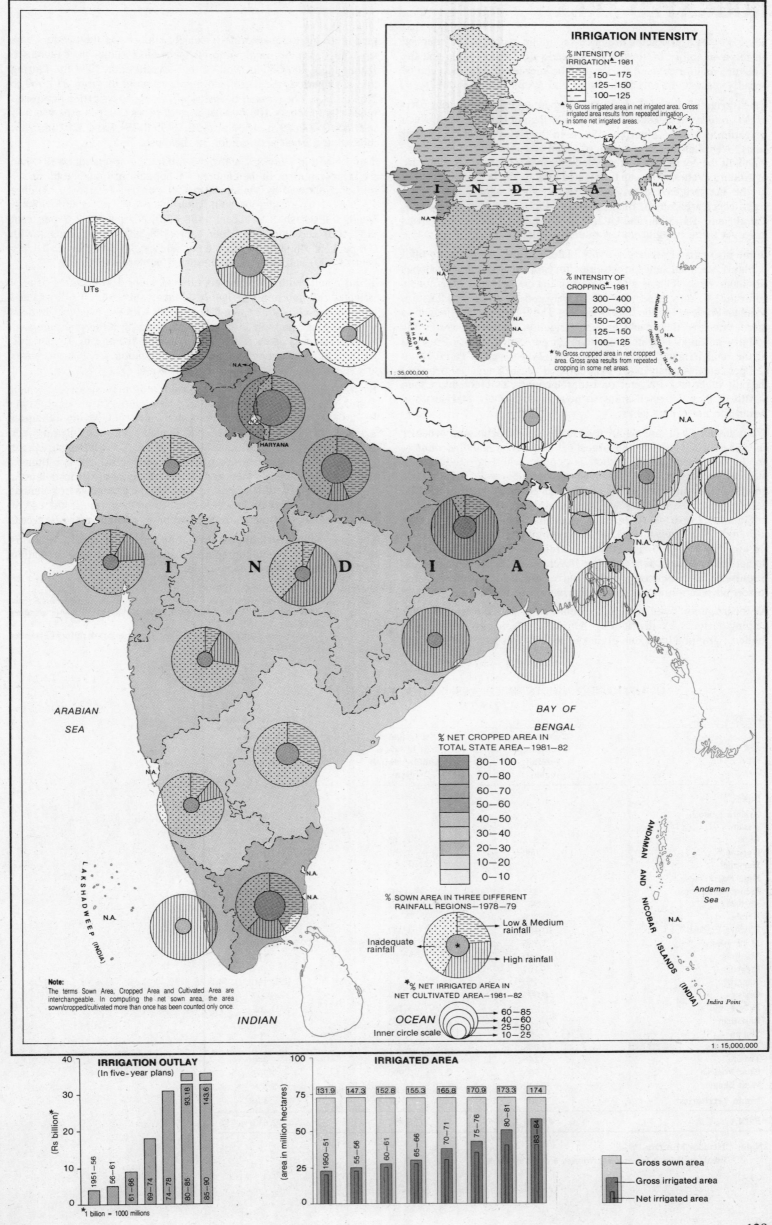

IRRIGATION INTENSITY

% INTENSITY OF IRRIGATION▲ 1981
- 150 — 175
- 125 — 150
- 100 — 125

▲ % Gross irrigated area in net irrigated area. Gross irrigated area results from repeated irrigation in some net irrigated areas.

% INTENSITY OF CROPPING* 1981
- 300 — 400
- 200 — 300
- 150 — 200
- 125 — 150
- 100 — 125

* % Gross cropped area in net cropped area. Gross area results from repeated cropping in some net areas.

1 : 35,000,000

HARYANA

N.A.

I N D I A

ARABIAN SEA

BAY OF BENGAL

% NET CROPPED AREA IN TOTAL STATE AREA — 1981-82
- 80 — 100
- 70 — 80
- 60 — 70
- 50 — 60
- 40 — 50
- 30 — 40
- 20 — 30
- 10 — 20
- 0 — 10

ANDAMAN AND NICOBAR ISLANDS (INDIA)

Andaman Sea

N.A.

LAKSHADWEEP (INDIA)

N.A.

% SOWN AREA IN THREE DIFFERENT RAINFALL REGIONS — 1978-79

Low & Medium rainfall

Inadequate rainfall

High rainfall

* % NET IRRIGATED AREA IN NET CULTIVATED AREA — 1981-82

INDIAN

OCEAN
- 60 — 85
- 40 — 60
- 25 — 50
- 10 — 25

Inner circle scale

Indira Point

Note:
The terms Sown Area, Cropped Area and Cultivated Area are interchangeable. In computing the net sown area, the area sown/cropped/cultivated more than once has been counted only once.

UTs

1 : 15,000,000

IRRIGATION OUTLAY
(In five-year plans)

(Rs billion)*

- 1951-56
- 56-61
- 61-66
- 69-74
- 74-78
- 80-85 — 93.18
- 85-90 — 143.6

*1 billion = 1000 millions

IRRIGATED AREA

(area in million hectares)

Year	Gross sown area
1950-51	131.9
55-56	147.3
60-61	152.8
65-66	155.3
70-71	165.8
75-76	170.9
80-81	173.3
83-84	174

- Gross sown area
- Gross irrigated area
- Net irrigated area

IRRIGATED AREA

Several types of irrigation are practised in the country. They depend on the availability of surface or groundwater, relief, soils and the moisture requirements of the crops. The sources of irrigation can be broadly classified as Canals, Tanks and Wells, and Tubewells.

Canal irrigation is found almost everywhere, except in the drier parts of the country. It irrigates 39 per cent of the net irrigated area. Such irrigation is a very important lifeline in the states of Orissa (66 per cent), West Bengal (64), Assam (63) and Haryana (53), where more than half the net irrigated area is fed by canals (1981-82). In the case of Andhra Pradesh, the canal irrigated area is just below 50 per cent of the net irrigated area. Other states where canals have some significance are Punjab, Bihar, Kerala, Madhya Pradesh, Karnataka, Tamil Nadu, Rajasthan and Uttar Pradesh. Large tracts in these states have yet to be brought under assured irrigation.

In the uneven and relatively rocky plateau of the peninsula, rainfall is highly seasonal, and possibilities for storing water across slopes (in tanks) exist. This is a small part of the country and, thus, tanks contribute to only 9 per cent of all irrigated area. Only in Eastern Madhya Pradesh and the interior area of Tamil Nadu is tank irrigation more extensive than canal or well irrigation. Nowhere else in the country do tanks irrigate more than 28 per cent (Andhra Pradesh) of the total irrigated area, with Kerala (24 per cent), Tamil Nadu (27 per cent), Karnataka (22 per cent) and West Bengal (20 per cent) the only other tank-irrigated parts of the country that have more than a fifth of their irrigated areas so nourished. Punjab, Haryana and Assam do not report tanks.

Wells and tubewells are mainly used in the alluvial plains and, whether energized or not, irrigate the largest net irrigated area in the country, about 46 per cent. This method is popular where groundwater is plentiful and canals few. In the past two decades, well irrigation has greatly increased in the plains of northern India. More than half the net irrigated area is under well irrigation here. Wells predominate in Gujarat (78 per cent), Maharashtra (56), Punjab (61), Rajasthan (62) and Uttar Pradesh (61), while in all other states wells account for less than half the irrigated area. Kerala, West Bengal and Assam do not report wells. Dadra & Nagar Haveli is fully dependent on wells. Because of rapid expansion of this source, canals have been relegated to second position as a source of irrigation in these areas.

As for irrigation by tubewells, the central and state governments are helping farmers by distributing pumpsets (electricity- and diesel-fuelled), granting loans and giving subsidies. This has played a large part in giving a boost to the Green Revolution. In the twenty years since 1960, the electrical pumpsets/tubewells installed have increased from a mere 200,000 in 1960-61 to 3.95 million in 1979-80. During the same period, the dieselized pumpsets went up from 230,000 to 2.65 million. The number of electrically operated irrigation pumpsets/tubewells for every 100,000 hectares of gross cropped area was 131 in 1961, 977 in 1971, 1600 in 1976, 2495 in 1984 and 3291 in 1985, indicating a great acceleration in their use.

Tamil Nadu in 1984-85, with 1.03 million pumpsets/tubewells, had the largest number in the country, 18 per cent of India's total of 5.7 million, followed by Maharashtra 15.6 per cent (891,600), Andhra Pradesh 11.6 per cent (664,800), Uttar Pradesh 9.2 per cent (526,800), Madhya Pradesh 8 per cent (458,400), Karnataka 7.6 per cent (437,100) and Punjab 6.8 per cent (388,300). West Bengal with 0.8 per cent (46,900), Orissa 0.6 per cent (31,700) and Assam 0.1 per cent (3,600) had the fewest energized pumpsets/tubewells.

Tamil Nadu, with 53 pumpsets/tubewells per 100 hectares of area irrigated through minor irrigation, Maharashtra with 49, Karnataka with 40, Kerala with 37, Andhra Pradesh with 30, Madhya Pradesh with 25, Haryana with 21 and Gujarat with 18, all have a pumpset/tubewell density greater than the national average of 16 per 100 hectares of area irrigated through minor irrigation. Low density states are Assam (1), West Bengal (3) and Orissa (3).

By 1990, India will, estimatedly, have 8 million pumpsets, of which Tamil Nadu will have 15 per cent, Maharashtra 15 per cent, Uttar Pradesh 9.4 per cent, Andhra Pradesh 8.7 per cent, Madhya Pradesh 8.4 per cent, Karnataka 8.2 per cent, Punjab 7 per cent, Bihar 5.5 per cent, Gujarat 4.9 per cent, Rajasthan 4.7 per cent, West Bengal 1.8 per cent and all union territories together 3.9 per cent. These estimates take into consideration the trends in the past in pumpset installation, as well as the difficulties due to terrain and the groundwater potential in these states. Caution has, however, to be exercised in the use of this equipment to avoid over-exploitation of groundwater resources.

The geographic distribution of irrigation sources generally conforms to the climatic and topographical conditions. But well irrigation confirms the groundwater potential.

Sources: 1. *Fertiliser Statistics 1984-85*, Fertiliser Association of India.
2. *Basic Statistics Relating to the Indian Economy, Vol. 1: All India, August 1986* and *Vol.2: States, September 1986*, Centre for Monitoring Indian Economy.
3. *Seventh Five Year Plan 1985-90, Vol. II*, Government of India Planning Commission.

CULTIVATED AREAS WITH ASSURED WATER SUPPLY
1978-79

Table 59   (in thousand hectares)

	Net sown area in high rainfall regions	Net irrigated area in low & medium rainfall regions	Net sown area with reasonably assured water supply
States			
Andhra Pradesh	—	3,655	3,655
Assam[1]	2,679	—	2,679
Bihar	6,872	1,165	8,037
Gujarat	696	1,568	2,264
Haryana	—	1,918	1,918
Himachal Pradesh	115	76	191
Jammu & Kashmir	257	267	524
Karnataka	967	1,152	2,119
Kerala	2,204	—	2,204
Madhya Pradesh	10,465	1,247	11,712
Maharashtra	3,836	1,368	5,204
Manipur	140	—	140
Meghalaya	193	—	193
Nagaland	150	—	150
Orissa	6,097	—	6,097
Punjab	—	3,262	3,262
Rajasthan	—	2,895	2,895
Sikkim	61	—	61
Tamil Nadu	1,068	2,227	3,295
Tripura	246	—	246
Uttar Pradesh	1,766	7,996	9,762
West Bengal	5,539	—	5,539
Union Territories[2]	409	61	470
INDIA	43,760	28,857	72,617

Note: [1] Includes Mizoram

[2] Break-up figures for union territories are not available.

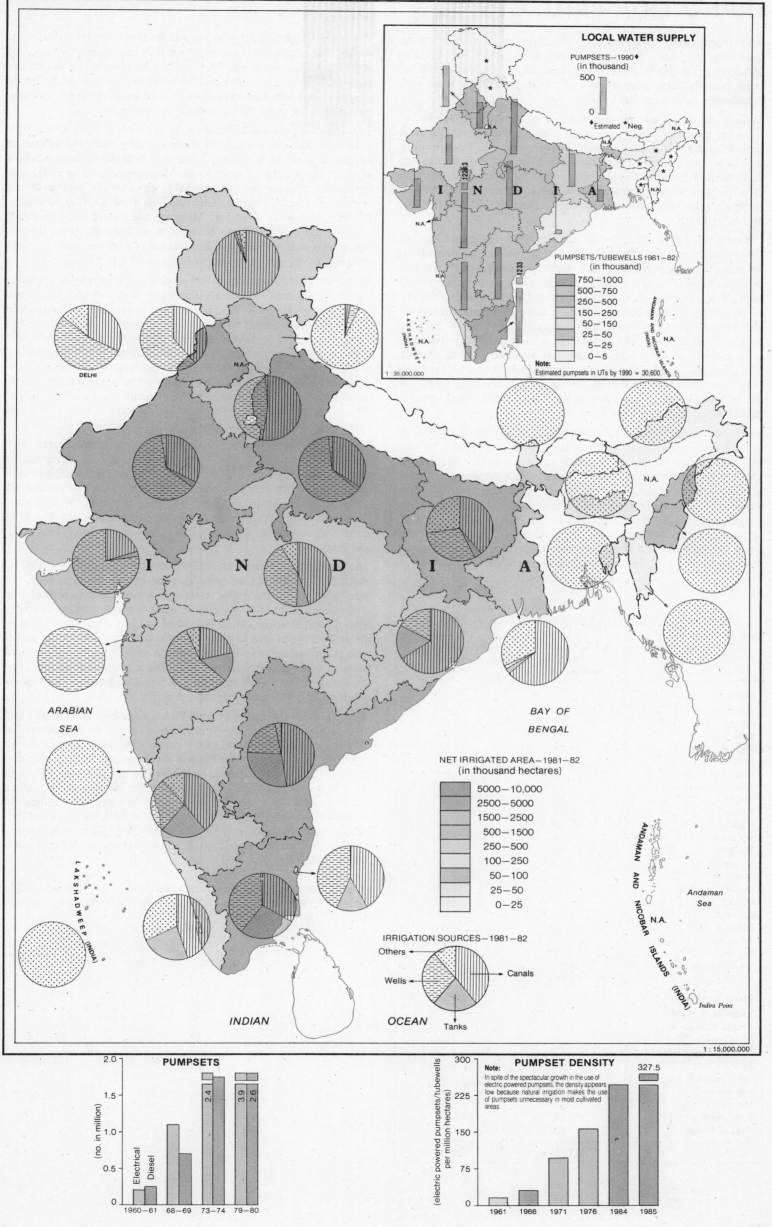

LOCAL WATER SUPPLY

PUMPSETS—1990 ◆
(in thousand)

500

0

◆ Estimated   * Neg.

PUMPSETS/TUBEWELLS 1981—82
(in thousand)

750—1000
500—750
250—500
150—250
50—150
25—50
5—25
0—5

Note:
Estimated pumpsets in UTs by 1990 = 30,600.

1 : 35,000,000

DELHI

N.A.

LAKSHADWEEP (INDIA)

ARABIAN
SEA

BAY OF
BENGAL

NET IRRIGATED AREA—1981—82
(in thousand hectares)

5000—10,000
2500—5000
1500—2500
500—1500
250—500
100—250
50—100
25—50
0—25

IRRIGATION SOURCES—1981—82

Others

Wells

Tanks

Canals

LAKSHADWEEP (INDIA)

ANDAMAN AND NICOBAR ISLANDS (INDIA)

Andaman
Sea

N.A.

Indira Point

INDIAN   OCEAN

1 : 15,000,000

PUMPSETS

2.0

1.5

1.0

0.5

(no. in million)

Electrical

Diesel

2.4   3.9 2.6

1960—61   68—69   73—74   79—80

PUMPSET DENSITY

300

225

150

75

Note:
In spite of the spectacular growth in the use of electric powered pumpsets, the density appears low because natural irrigation makes the use of pumpsets unnecessary in most cultivated areas.

327.5

(electric powered pumpsets/tubewells per million hectares)

1961   1966   1971   1976   1984   1985

IRRIGATION PROJECTS

After independence, irrigation acquired a primary position in national development plans. Throughout the development of irrigated agriculture, the major concern has been water. Much progress has been made in harnessing rivers and constructing canal systems to distribute water to the land.

India is one of the leading dam builders in the world today. India's dams are mostly multipurpose ones, *i.e.* they are designed to achieve more than the one express objective of irrigation. In the peninsula, the major rivers have been harnessed to the extent that few opportunities remain for river projects. On the other hand, to utilize effectively the flow of many of the Himalayan rivers, India will have to depend upon international cooperation, especially with Nepal, China and Bangladesh.

Between 1947 and 1982, India took up 1040 major and medium irrigation projects and has been able to create the full potential in 46 major irrigation projects, each with a culturable command area of above 10,000 hectares, and 517 medium irrigation projects, each with a culturable command area of 2000 to 10,000 hectares. Another 50 major projects have been substantially completed. Between 1950-51 and 1983-84, an additional irrigation potential of about 20.3 million hectares has been realized through major and medium projects. In all, a potential of 30.5 million hectares had been created up to 1984-85. The gross irrigated area to gross sown area, which was 17.1 per cent before the First Five Year Plan, rose to 29 per cent at the end of the Fifth Plan. The irrigation potential has more than doubled since the beginning of the Five Year Plans.

The total potential created up to 1984-85 by about 103 major projects is 15.5 million hectares, which is about 50.8 per cent of the total potential created. Of this, 14.2 million hectares are in 15 major states. Other major and medium projects account for a created potential of 15 million hectares, 49.2 per cent of the total created potential. Uttar Pradesh has a created potential of 2.7 million hectares through major irrigation projects, Bihar 1.8 million and Karnataka 1.1 million hectares. Jammu & Kashmir has the least created potential, 12,000 hectares. An additional irrigation potential of 28.9 million hectares, from some 191 projects (ongoing), is ultimately expected.

The potential for irrigation in India is far above the level developed so far. The ultimate irrigation potential in the country is estimated at 113.3 million hectares, 58.4 million hectares under major and medium irrigation schemes and 54.9 million hectares under minor irrigation. Achievement up to 1984-85 has been 67.9 million hectares, including 37.4 m ha through minor irrigation.

The irrigation potential estimated to be available in 1984-85 was 67.9 million hectares, the largest part of it in one state, Uttar Pradesh (about 28 per cent or 18.8 million hectares). In Maharashtra, Madhya Pradesh, Himachal Pradesh, Sikkim, Meghalaya, Assam and Tripura, the current irrigation potential to total cropped area is less than 20 per cent. In Kerala, Karnataka, Gujarat, Rajasthan, Orissa, Arunachal Pradesh, Nagaland, Manipur, Mizoram and the union territories it is 20-40 per cent; in Tamil Nadu, Andhra Pradesh, West Bengal, Bihar, Haryana and Jammu & Kashmir, 40-60 per cent; in Uttar Pradesh, 60-80 per cent; and in Punjab, 80-100 per cent.

In the first twenty years of Indian planning, the growth in irrigation was 100 per cent. Relatively speaking, this may be considered reasonably good, but in absolute terms it is not. The gross irrigated area at the end of the Third Plan totalled only 31 million hectares, a little less than 22 per cent of the gross cropped area. When the Fourth Plan was being formulated, it was estimated that another 40 million hectares could be brought under irrigated cultivation. However, gross irrigated area even in 1982-83 was only 52 million hectares, falling much short of the expectations of the Fourth Five Year Plan target. That is, in 16 years there had been an increase of only 21 million hectares of gross irrigated area.

In 1965 the gross sown area amounted to 155 million hectares. This had gone up to only 172.6 million hectares in 1982-83, an eleven per cent increase over 17 years. Of the gross cropped area of 172.6 million hectares, 74.7 per cent (or about 130 million hectares) were in eight states, namely Uttar Pradesh, Madhya Pradesh, Maharashtra, Rajasthan, Karnataka, Andhra Pradesh, Gujarat and Bihar. The other states with substantial gross cropped areas were Punjab, Tamil Nadu and Haryana.

The gross irrigated area in 1982-83 totalled 52.1 million hectares, about 30.1 per cent of the gross cropped area in the same year, but here the shares of the major states with a larger gross cropped area differed: from about 36 per cent in Andhra Pradesh to about 46 per cent in Uttar Pradesh, while Rajasthan had just about 20 per cent of the gross cropped area as gross irrigated area. But these states, along with other states, have indicated that a gross target of 71 million hectares could be reached by 1990. This is planned to be achieved through both major and medium (29.5 million hectares) and minor (41.5 million hectares) irrigation projects.

The investment during the Seventh Plan is scheduled to be Rs 143,605 million. But even if the target is achieved, it will be well below the ultimate irrigation potential of the country, which is now estimated at 113 million hectares. Uttar Pradesh has the largest ultimate irrigation potential (25.7 m ha), followed by Bihar (12.9 m ha), Madhya Pradesh (10.2 m ha), Andhra Pradesh (9.2 m ha), Punjab (6.6 m ha), Orissa (5.9 m ha) and Rajasthan (5.2 m ha). The gross targets for utilization set for the Seventh Plan period are substantial for Uttar Pradesh (19.6 million hectares), Bihar (6.7 m ha), Punjab (5.94 m ha), Andhra Pradesh (5.89 m ha), Rajasthan (3.8 m ha), Haryana (3.5 m ha) and Tamil Nadu (3.3 m ha).

Sources: 1. *Basic Statistics Relating to the Indian Economy, Vol. 1: All India, August 1986,* and *Vol.2: States, September 1986,* Centre for Monitoring Indian Economy.

OUTLAY ON IRRIGATION PROJECTS — 1985-90

Table 60

(in Rs. million)

	Major & medium projects	Minor projects
States		
Andhra Pradesh	11,823	1,474
Assam	1,370	1,600
Bihar	12,850	2,600
Gujarat	14,690.9	1,346
Haryana	4,185	142
Himachal Pradesh	135	540
Jammu & Kashmir	528.6	420
Karnataka	5,230	1,510
Kerala	2,800	500
Madhya Pradesh	13,759.2	4,336
Maharashtra	13,200	2,500
Manipur	600	100
Meghalaya	5.5	97
Nagaland	—	150
Orissa	5,500	1,100
Punjab	2,707.8	462
Rajasthan	6,354.6	479
Sikkim	60	100
Tamil Nadu	2,120	850
Tripura	270	150
Uttar Pradesh	14,200	5,120
West Bengal	2,080	680
Union Territories		545
Andaman & Nicobar	—	—
Arunachal Pradesh	10	—
Chandigarh	—	—
Dadra & Nagar Haveli	20	—
Delhi	1	—
Goa, Daman & Diu	540	—
Lakshadweep	—	—
Mizoram	5	—
Pondicherry	20	—
Central Sector	500	1,350
INDIA	115,555.6	28,151

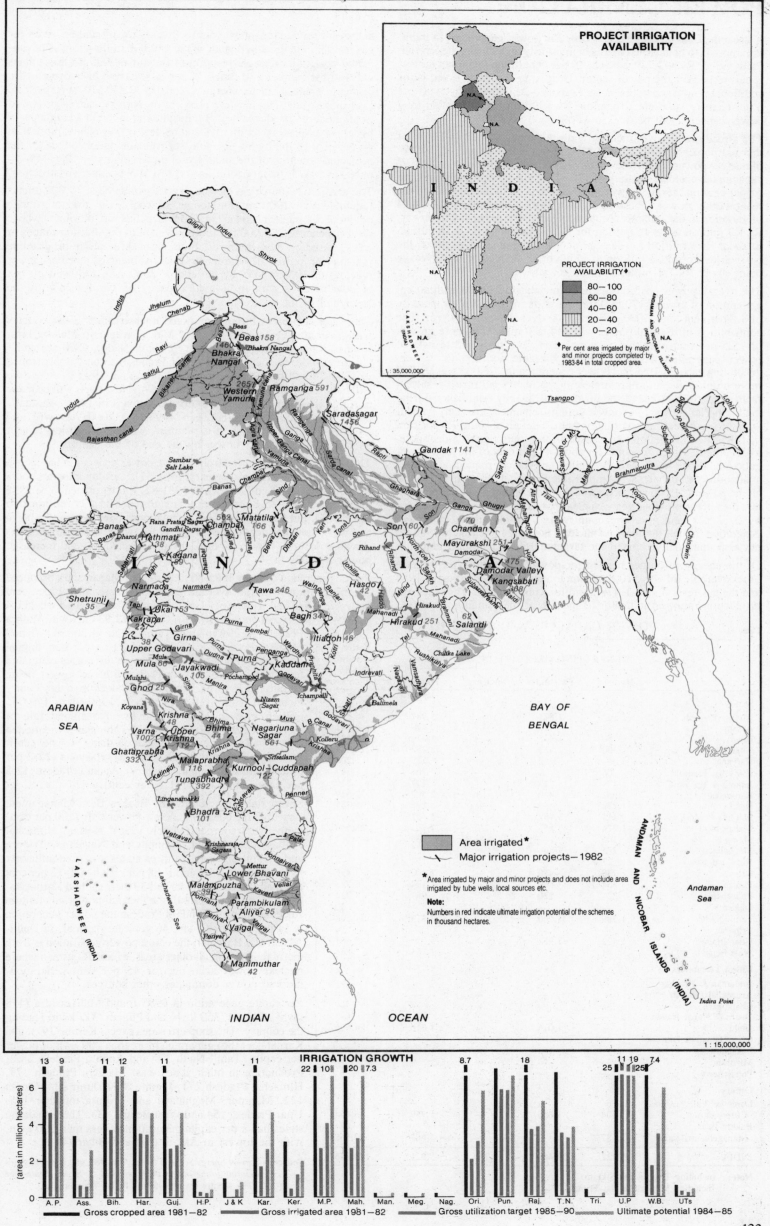

PROJECT IRRIGATION AVAILABILITY

I N D I A

1 : 35,000,000

PROJECT IRRIGATION AVAILABILITY ◆

	80–100
	60–80
	40–60
	20–40
	0–20

◆ Per cent area irrigated by major and minor projects completed by 1983-84 in total cropped area.

Beas 158
Bhakra Nangal 1460
Western Yamuna 2651
Ramganga 591
Saradasagar 1456
Gandak 1141
Ghugri 70
Chandan
Matatila 766
Rana Pratap Sagar 562
Gandhi Sagar
Chambal
Mayurakshi 251
Damodar Valley 475
Kangsabati 388
Hathmati 38
Kadana 59
Tawa 246
Hasdo 42
Narmada
Hirakud 251
Salandi 62
Shetrunji 35
Ukai 153
Kakrapar 25
Girna 38
Girna
Bagh 34
Upper Godavari
Itiadoh 46
Mula 66
Jayakwadi 105
Kaddam
Purna
Ghod 25
Krishna 48
Varna 100
Upper Krishna 112
Bhima 44
Nagarjuna Sagar 661
Ghataprabha 332
Malaprabha 116
Kurnool–Cuddapah 122
Tungabhadra 392
Bhadra 101
Krishnaraja Sagara
Lower Bhavani 79
Malampuzha 39
Parambikulam Aliyar 95
Vaigai
Manimuthar 42

ARABIAN SEA

BAY OF BENGAL

LAKSHADWEEP (INDIA)

INDIAN OCEAN

ANDAMAN AND NICOBAR ISLANDS (INDIA)

Andaman Sea

Indira Point

	Area irrigated*
	Major irrigation projects—1982

*Area irrigated by major and minor projects and does not include area irrigated by tube wells, local sources etc.

Note:
Numbers in red indicate ultimate irrigation potential of the schemes in thousand hectares.

1 : 15,000,000

IRRIGATION GROWTH

(area in million hectares)

13 9   11 12   11   11   22 10 20 7.3   8.7   18   25 11 19 7.4 25

A.P. Ass. Bih. Har. Guj. H.P. J & K Kar. Ker. M.P. Mah. Man. Meg. Nag. Ori. Pun. Raj. T.N. Tri. U.P. W.B. UTs

■ Gross cropped area 1981—82   ▧ Gross irrigated area 1981—82   ▨ Gross utilization target 1985—90   ░ Ultimate potential 1984—85

133

POWER GENERATION

A great deal has already been invested in power generation in India. Still more is to be invested in this high priority area during the Seventh Plan period. Rs 342,730 million, 19 per cent of the total plan outlay, is earmarked for the power sector. Of this, nearly 66 per cent is for additional power generation of an estimated 20,000 MW by 1990 and 6 per cent for rural electrification. By 1990 it is expected that total power capacity will be 64,736 MW.

The already huge investment provides a capacity of 29,469 MW of thermal power (1986), 15,222 MW of hydel power and 1230 MW of nuclear power. Eighty-seven mini hydel plants of less than 20 MW capacity have a total capacity of 170 MW. The investment shares are 61 per cent on thermal power, 36.7 per cent on hydel power and the rest on other power. These figures for 1985-86 indicate a great deal of variation in the amounts required to generate a megawatt of power, Rs 1.7 million to Rs 15 million for hydel plants and Rs 0.26 million to about Rs 1 million for thermal plants. Among other projects, the cost varies from Rs 5 million to a high of Rs 34 million, the average being about Rs 7.1 million per megawatt.

The overall power position in the country has improved with a total generation of around 170,000 million units in 1985-86, an increase of 8.6 per cent (or 13,000 million units) over the previous year. The contribution of thermal power was 114,000 million units (a growth of 15.5 per cent over the previous year). The sustained growth of power generation stemmed from a sizeable addition to thermal generating capacity (thermal power generation in 1984-85 was 99,000 million units) and an improvement in the plant load factor resulting from synchronization and stabilization of capacity established during the Sixth Plan period. Yet power deficits remained, unevenly spread, with the south bearing the brunt of them: a deficit of some 9.3 per cent in 1985-86. This deficit was caused in the south due to a decline in hydel output, which accounts for about 42 per cent of the total power generated in the south. Scanty rainfall and the consequent low water availability at dam sites was responsible for hydel failing in the south. Hydel power (51,000 m units) throughout the country was 5.3 per cent less than in the previous year. Nuclear power (5000 m units) increased by 22.2 per cent. For the country as a whole, the total power generation (hydel, thermal and nuclear) not only increased in 1985-86 but also marginally crossed the targeted 169,795 million units.

Of the energy resources needed for power generation, the balance of recoverable oil was 510.8 million tonnes at the beginning of 1984 and the gas reserves were 478.3 million cubic metres. The reserves/production ratio, an indicator of the life of the remaining reserves at current production rates, was 18.8 years for oil and 96 years for gas in 1983. On the other hand, of the 149,000 million tonnes of estimated coal, the mineable resources of about 60,000 million tonnes will last for another 130 years. A new assessment of hydro-electric resources estimates annual energy potential at 89,830 MW at 60 per cent load factor (load factor is the per cent of average energy produced to total energy produceable). But despite the inherent advantages of hydel power plants over thermal and nuclear power plants, more than 80 per cent of the hydro potential still remains unexploited. Of the projected potential of the small hydel power plants — 5000 MW — only 171 MW is under operation and 170 MW is under construction.

The uranium resources in the country are estimated at 70,000 tonnes (equivalent to 1900 million tonnes of coal). But the long-range potential of nuclear power in India depends on thorium, whose reserves exceed 360,000 tonnes (equivalent to 600 million tonnes of coal). Based on these reserves, India has a three-phase programme for nuclear power development: establishment of nuclear power reactors of 10,000 MW by the end of 2000 AD, followed by fast breeder reactors with an ultimate capacity of 350,000 MW by the latter half of the next century.

There are at present three operational nuclear units in the country with a cumulative capacity of 2410 MW (Tarapur in Maharashtra, Kota in Rajasthan and Kalpakkam in Tamil Nadu). Two are under construction, one at Narora (in Uttar Pradesh) to be commissioned in 1987-88 and one at Kakrapur (in Gujarat) in 1990-91. The two will have a cumulative capacity of 4070 MW. New starts will be made at Kaiga (in Karnataka) and Kota (in Rajasthan) in 1994. And units planned for commission between 1995 and 2000 AD will add 14,300 MW. Thus, India, it is planned, will have a production of 26,730 MW of nuclear power by the end of the century.

The regionwise contribution from captive power generation has shown aggregate increases over the years, from 6712 million kilowatt hours in 1978-79 to 7783 million kwh in 1981-82. It is estimated to be 8775 million kwh in 1985-86 and is expected to reach 15,577 million kwh in 1989-90. But there are regional fluctuations in production, the north and west showing increases over the years, while the north-east has shown a decline. However, they are all expected to yield increased power by 1989-90.

Power generation data for 1985-86 indicate that Maharashtra produced 28,560 million kwh (an average annual increase of 9.1 per cent between 1960 and 1985-86), Uttar Pradesh 18,567 m kwh (11.4 per cent annual growth), Gujarat 12,934 m kwh, Andhra Pradesh 16,299 m kwh, Tamil Nadu 14,308 m kwh and Madhya Pradesh 15,903 m kwh (15.1 per cent annual growth). All other states and union territories generated less than 10,000 m kwh each, with the lowest being Sikkim (30 m kwh). The power production in the country increased from 16,937 m kwh in 1960-61 to 170,045 million kwh in 1985-86, an average annual rate of increase of 9.7 per cent. Besides the states mentioned above, this rate is exceeded by Rajasthan (14.7 per cent), Jammu & Kashmir (12.7 per cent), Haryana (22.2 per cent, the maximum in the country), Andhra Pradesh (12.3 per cent), and Delhi (11.5 per cent).

Seven states (excluding the Bhakra Beas Management Board) in the country generate power only (100 per cent) from hydel sources (March 1986): Kerala, Himachal Pradesh, Meghalaya, Manipur and Nagaland, Tripura and Sikkim together. In three states it is predominantly hydel (Jammu & Kashmir: 88 per cent hydel, 12 per cent thermal; Orissa: 61 hydel and 39 thermal; and Karnataka: 83 hydel and 17 thermal, the thermal power an addition since the map). In Andhra Pradesh the power generation is 54 per cent hydel and 46 per cent thermal. In Delhi, Assam and Haryana, the entire power generation is from thermal plants. In all other areas, except Rajasthan where nuclear power accounts for 45 per cent generation, thermal power dominates other sources.

Per capita generation in 1985 found Maharashtra (418 kwh), Gujarat (350 kwh) and Punjab (317 kwh) leading the country. The southern states except Kerala (197), and Karnataka (186) generated more than 200 kwh of power per capita (Tamil Nadu 277 and Andhra Pradesh 282). Production in other states was: Madhya Pradesh 279, Himachal Pradesh 271, Jammu & Kashmir 134, Orissa 122, Manipur, Meghalaya and Tripura together 185, Uttar Pradesh 154 and West Bengal 133. The remaining states had a per capita generation of less than 100 kwh, with the lowest in Assam (39) and Bihar (44).

Table 61

POWER CAPACITY & GENERATION

	GENERATING CAPACITY IN MW — 1986				Power Generation 1985-86 (in million kwh.)
	Hydro	Thermal	Nuclear	Total	
States					
Andhra Pradesh	2,162	1,825	—	3,987	16,299
Assam	—	410	—	410	842
Bihar	150	1,425	—	1,575	3,325
Gujarat	300	2,983	—	3,283	12,934
Haryana	—	52	—	525	1,206
Himachal Pradesh	306	—	—	306	1,247
Jammu & Kashmir	174	23	—	197	870
Karnataka	2,082	420	—	2,502	7,519
Kerala	1,272	—	—	1,272	5,385
Madhya Pradesh	115	3,452	—	3,567	15,903
Maharashtra	1,326	5,555	320	7,201	28,560
Manipur	105	—	—	105	1,001[2]
Meghalaya	175	—	—	175	—
Nagaland	27[1]	—	—	27	—
Orissa	730	470	—	1,200	3,475
Punjab	334	860	—	1,194	5,763
Rajasthan	321	220	440	981	3,326
Sikkim	—	—	—	—	30
Tamil Nadu	1,389	1,960	470	3,819	14,308
Tripura	—	—	—	—	—
Uttar Pradesh	1,422	3,953	—	5,375	18,567
West Bengal	41	2,717	—	2,758	7,885
Union Territories					
Andaman & Nicobar	—	—	—	—	—
Arunachal Pradesh	—	—	—	—	—
Chandigarh	—	—	—	—	—
Dadra & Nagar Haveli	—	—	—	—	—
Delhi	—	1,016	—	1,016	4,695
Goa, Daman & Diu	—	—	—	—	—
Lakshadweep	—	—	—	—	—
Mizoram	—	—	—	—	—
Pondicherry	—	—	—	—	—
Others					
Damodar Valley Corporation	104	1,655	—	1,759	6,464
Bhakra Beas Management Board	2,687	—	—	2,687	10,568
INDIA	15,222	29,469	1,230	45,921	170,045

Note: [1] Including Tripura and Sikkim.
[2] Including Meghalaya and Tripura.

Sources: *Current Energy Scene in India, July 1986*, and *Basic Statistics Relating to the Indian Economy, Vol. 2: States, September 1986*, Centre for Monitoring Indian Economy.

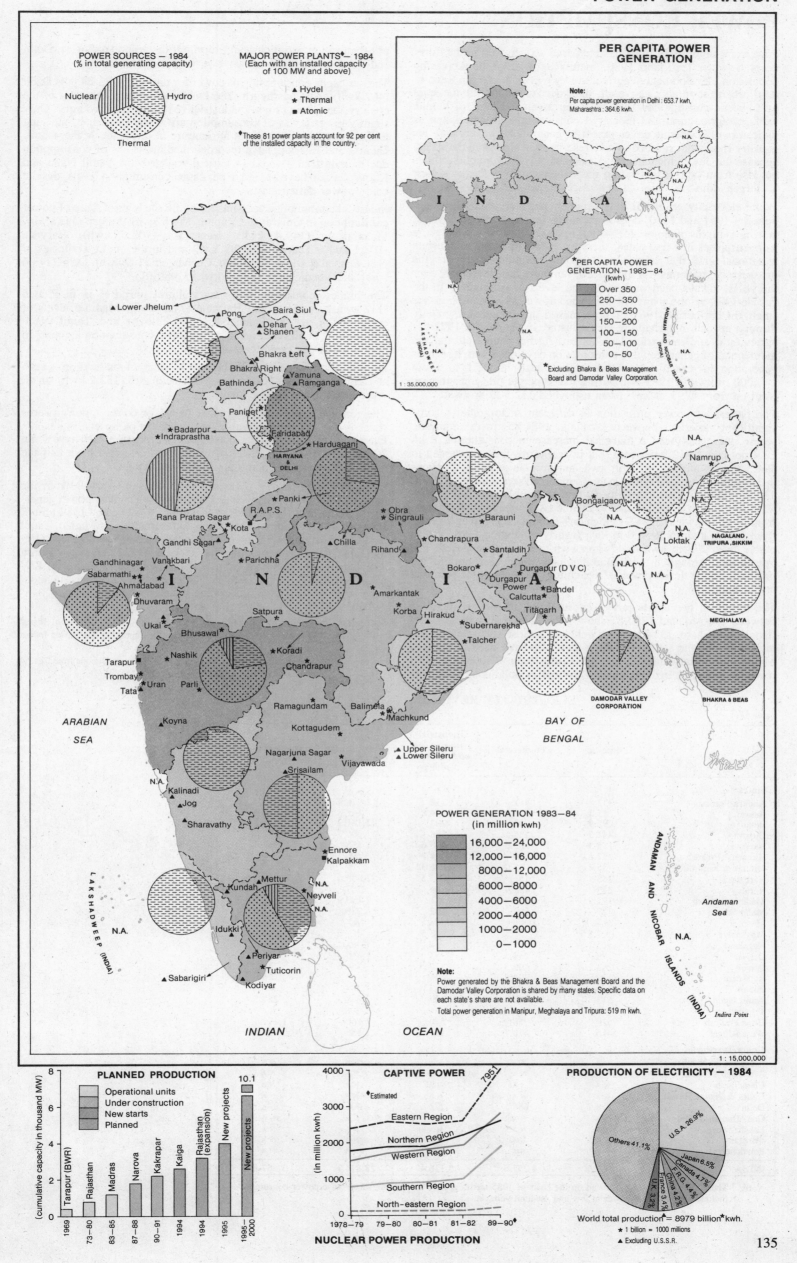

POWER SOURCES — 1984
(% in total generating capacity)

Nuclear
Hydro
Thermal

MAJOR POWER PLANTS* — 1984
(Each with an installed capacity
of 100 MW and above)

▲ Hydel
★ Thermal
■ Atomic

* These 81 power plants account for 92 per cent
of the installed capacity in the country.

PER CAPITA POWER
GENERATION

Note:
Per capita power generation in Delhi : 653.7 kwh,
Maharashtra : 364.6 kwh.

I N D I A

*PER CAPITA POWER
GENERATION — 1983—84
(kwh)

Over 350
250—350
200—250
150—200
100—150
50—100
0—50

*Excluding Bhakra & Beas Management
Board and Damodar Valley Corporation.

1 : 35,000,000

▲ Lower Jhelum
▲ Pong
▲ Baira Siul
▲ Dehar
▲ Shanen
Bhakra Left
Bhakra Right
Bathinda
★ Yamuna
▲ Ramganga
Panipet ★
★ Badarpur
★ Indraprastha
★ Faridabad
★ Harduaganj
HARYANA
& DELHI
★ Panki
Rana Pratap Sagar
R.A.P.S.
★ Kota
Gandhi Sagar ▲
Gandhinagar ★ Vanakbari
Sabarmathi ★ ★
Ahmadabad
▲ Dhuvaram
Ukai ▲
▲ Chilla
★ Parichha
★ Obra
★ Singrauli
Rihand
▲ Amarkantak
Satpura
★ Bhusawal
▲ Korba
★ Nashik
★ Koradi
Chandrapur
Tarapur ■
Trombay ★
Tata ★ ★ Uran
★ Parli
Ramagundam
Balimela ▲
Koyna ▲
Kottagudem
Nagarjuna Sagar
Srisailam
N.A.
Kalinadi ▲
▲ Jog
▲ Sharavathy
★ Ennore
■ Kalpakkam
▲ Mettur
Kundah ▲
★ Neyveli
▲ Idukki
▲ Periyar
▲ Sabarigiri
▲ Tuticorin
Kodiyar

Namrup ★
Bongaigaon ★
Loktak
NAGALAND,
TRIPURA, SIKKIM

N.A.

MEGHALAYA

Chandrapura ★
Bokaro ★
Santaldih ★
Durgapur (D V C)
Durgapur
Power ★
★ Bandel
Calcutta
Titagarh ★
★ Subernarekha
★ Talcher
Hirakud ▲
Barauni ★

DAMODAR VALLEY
CORPORATION

BHAKRA & BEAS

ARABIAN
SEA

BAY OF
BENGAL

Machkund ▲
▲ Upper Sileru
▲ Lower Sileru
Vijayawada

POWER GENERATION 1983—84
(in million kwh)

16,000—24,000
12,000—16,000
8000—12,000
6000—8000
4000—6000
2000—4000
1000—2000
0—1000

Note:
Power generated by the Bhakra & Beas Management Board and the
Damodar Valley Corporation is shared by many states. Specific data on
each state's share are not available.

Total power generation in Manipur, Meghalaya and Tripura: 519 m kwh.

ANDAMAN
AND
NICOBAR
ISLANDS
(INDIA)

Andaman
Sea

N.A.

Indira Point

LAKSHADWEEP (INDIA)
N.A.

INDIAN   OCEAN

1 : 15,000,000

PLANNED PRODUCTION

Operational units
Under construction
New starts
Planned

(cumulative capacity in thousand MW)

8
6
4
2
0

10.1

Tarapur (BWR)
Rajasthan
Madras
Narova
Kakrapar
Kaiga
Rajasthan (expansion)
New projects
New projects

1969
73—80
83—85
87—88
90—91
1994
1994
1995
1995—2000

NUCLEAR POWER PRODUCTION

CAPTIVE POWER

(in million kwh)

4000
3000
2000
1000

◆ Estimated

7951

Eastern Region
Northern Region
Western Region
Southern Region
North-eastern Region

1978—79
79—80
80—81
81—82
89—90◆

PRODUCTION OF ELECTRICITY — 1984

Others 41.1%
U.S.A. 26.9%
Japan 6.5%
Canada 4.7%
F.R.G 4.4%
China 4.2%
U.K. 3.2%
France 3.4%

World total production = 8979 billion* kwh.

★ 1 billion = 1000 millions
▲ Excluding U.S.S.R.

POWER CONSUMPTION

Power consumption in India has increased sixteen-fold in the thirty years between 1951 and 1980, mainly as a result of increasing population, development in agriculture by way of the Green Revolution, urbanization and, above all, industrialization. While the level of domestic consumption may be taken as an indicator of human welfare, agricultural and industrial consumption may be taken as indicators of economic development in general. However, data for industry should be regarded as underestimates in almost all the states, because the power for industry comes not only from public utilities but also from captive generating sets installed as standbys in almost all larger industrial units as well as public service units.

While electricity generated in the country has increased 17½ times between 1951 and 1980, consumption has increased only 16 times, leaving a surplus of almost 25,000 million kwh. However, while there were surpluses in a few states, there were deficits in others, despite a national grid that interlinks states. In addition, there are the unassessable transmission losses. Consumption has, therefore, over the years, always been well within the levels of generation. In the decade 1951-61, the growth in generation as well as consumption was small. In the period 1961-71 both picked up momentum, mainly through growth in agriculture and industry. In the decade 1971-81, agricultural and industrial development further increased and the momentum of generation and consumption developed with it. Power generation increased from 5858 million kwh in 1951 to about 140,000 m kwh in 1983-84. During the same period, consumption went up from 4793 m kwh to an estimated 112,500 m kwh.

The pattern of power utilization in different sectors shows some significant variations between 1951 and 1984-85. In the domestic sector, there has been a marginal increase in consumption from 12.4 per cent to 12.9 per cent, but in real terms these amounted to 594.3 m kwh and 14,561.8 m kwh, an increase of about 25 times in 34 years. The consumption in the commercial sector increased from 330.7 m kwh to 6660 m kwh (a 20-fold increase), in industry from 3053.1 m kwh to 64,004 m kwh (a 21-fold rise), in railways/tramways from 330.7 m kwh to 2934.9 m kwh (a 9-fold increase), in agriculture from 206 m kwh to 19,980 m kwh (nearly a hundred-fold), and in others from 278 m kwh to 4741 m kwh (a 17-fold increase). The phenomenal growth in agricultural consumption corroborates the emphasis on the development of agriculture throughout the planning era in India.

The trends in per capita power consumption in the two decades between 1960-61 and 1980-81 show great increases in the country (from 38 kwh to 132 kwh) as well as in most states and union territories. Of the states, the steep increases have been in Punjab, Maharashtra, Haryana, Gujarat and Tamil Nadu. There have been marginal increases in those states with small populations, low levels of urbanization and industrialization. Of the union territories, Delhi and Pondicherry have shown a very great increase.

The per capita consumption of power for the year 1983-84 puts Delhi (467 kwh) and Chandigarh (386 kwh) way ahead of the rest of the country, both being highly industrialized parts of the country. Only slightly lower levels of consumption are found in Punjab, attesting to the tremendous growth of agriculture in this state. Maharashtra, Gujarat and Haryana have recorded medium levels of consumption due to industrial as well as agricultural growth. Tamil Nadu and Karnataka also have moderate per capita consumption levels, despite being power deficit states.

Industrial consumption accounts for by far the highest share of power consumption in Gujarat (per capita: 154.6 kwh), Punjab (112 kwh), Maharashtra (149.4 kwh), Karnataka (105.5 kwh), Haryana (102.1 kwh), Tamil Nadu (98 kwh) and in the union territories of Goa, Daman & Diu (190.5 kwh), Chandigarh (189 kwh), Delhi (148.6 kwh) and Pondicherry (132.2 kwh) (1980-81).

The highest consumption for agricultural purposes is in Punjab (112 kwh per capita) and Haryana (74.9 kwh). Medium levels of consumption are in Pondicherry (70.2 kwh) and Tamil Nadu (49.2 kwh). In all other areas, agricultural consumption amounts to only a small per capita consumption.

Domestic consumption is the highest in urban union territories such as Delhi (125.7 kwh per capita) and Chandigarh (132.2 kwh). In all other areas it has only a small share.

The considerable agricultural share of power consumption in some states is due to the large number of energized pumpsets. States like Punjab, Haryana, Tamil Nadu and Andhra Pradesh, which have about 20 per cent of the total groundwater resources, also have half the number of energized pumpsets in the country.

In several states, large industrial units consume power from the national/regional grid as well as from their own captive power generators. Maharashtra (403 industrial units), West Bengal (384 units), Bihar (254 units), Gujarat (217 units), Uttar Pradesh (169 units), Tamil Nadu (136 units), Karnataka (117 units) and Madhya Pradesh (109 units) had 81 per cent of all the industrial units (2223 units — March 1981) that had their own captive plants with more than 100 KW capacity each. The total captive installed capacity amounted to about 9 per cent of the power generation capacity of the country.

Sources: 1. *Basic Statistics Relating to the Indian Economy, Vol. 1: All-India, August 1986* and *Vol. 2: States, September 1986*, Centre for Monitoring Indian Economy.
2. *Statistical Abstract India 1984*, Ministry of Planning. Quoting from Central Electricity Authority, Ministry of Energy.

ELECTRICITY REVENUE[1]

Table 62 (in Rs. million)

	Domestic	Commercial	Industrial power at low and medium voltage	Industrial power at high voltage	Agricultural
States					
Andhra Pradesh	332	451	238	2,076.9	95.4
Assam	55.2	29.9	144.8	141.3	3
Bihar	92.9	80	203	2,181.9	177
Gujarat	442.5	236.5	588	3,406.7	440.8
Haryana	125.5	83.6	168.6	555	332.4
Himachal Pradesh	37.6	26	44.7	*	1.3
Jammu & Kashmir	35.4	25.9	36.3	*	3.5
Karnataka	348	151	253.5	1,809.6	32.7
Kerala	273	157.9	69	557	16
Madhya Pradesh	172.9	209.9	181	2,180.9	62.9
Maharashtra	727	899	662	5,286	431.6
Manipur	N.A.	N.A.	N.A.	N.A.	N.A.
Meghalaya	7.5	7.3	18.5	N.A.	N.A.
Nagaland	N.A.	N.A.	N.A.	N.A.	N.A.
Orissa	78	54.5	53	1,456	14.2
Punjab	263.7	109.5	240	754.6	251
Rajasthan	119	106.9	664	*	226.8
Sikkim	N.A.	N.A.	N.A.	N.A.	N.A.
Tamil Nadu	513.7	490	499	3,047	284.5
Tripura	N.A.	N.A.	N.A.	N.A.	N.A.
Uttar Pradesh	526	56	530	2,160	1,171.7
West Bengal	415	426.7	251	2,855	23
Union Territories					
Andaman & Nicobar	N.A.	N.A.	N.A.	N.A.	N.A.
Arunachal Pradesh	N.A.	N.A.	N.A.	N.A.	N.A.
Chandigarh	N.A.	N.A.	N.A.	N.A.	N.A.
Dadra & Nagar Haveli	N.A.	N.A.	N.A.	N.A.	N.A.
Delhi	238.7	656	272	406	N.A.
Goa, Daman & Diu	N.A.	N.A.	N.A.	N.A.	N.A.
Lakshadweep	N.A.	N.A.	N.A.	N.A.	N.A.
Mizoram	N.A.	N.A.	N.A.	N.A.	N.A.
Pondicherry	N.A.	N.A.	N.A.	N.A.	N.A.
INDIA	4,803.6	4,257.6	5,116.4	28,874.1	3,567.8

Note: [1] These data have been calculated on the basis of 1985 tariffs applied to 1981-82 sector-wise power consumption.
* Included in industrial power at low and medium voltage.

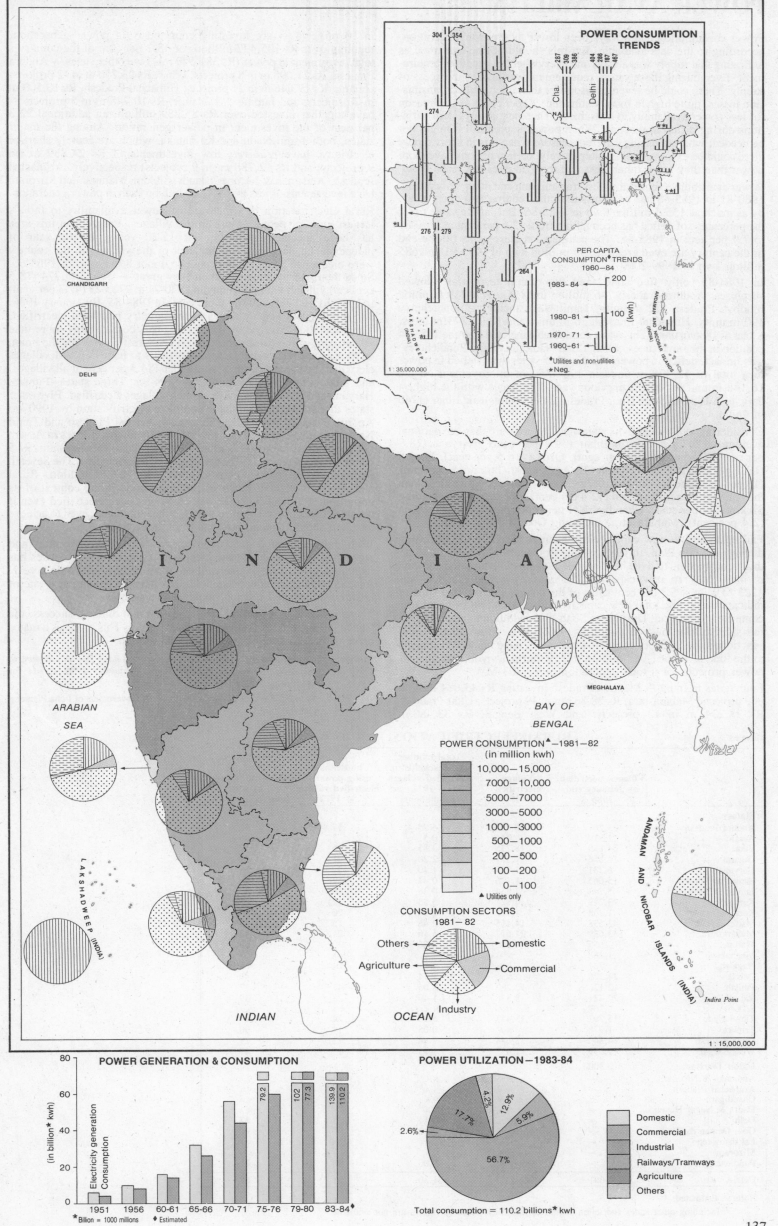

POWER CONSUMPTION TRENDS

304 354   287 404
309 286
386 187
467
Cha.   Delhi
274

262

I N D I A

276 279

264

PER CAPITA
CONSUMPTION ▲ TRENDS
1960 – 84

1983 – 84 ─────── 200
(kwh)
1980 – 81 ─────── 100

1970 – 71 ───────
1960 – 61 ─────── 0

▲ Utilities and non-utilities
* Neg.

1 : 35,000,000

CHANDIGARH

DELHI

I N D I A

ARABIAN
SEA

BAY OF
BENGAL

MEGHALAYA

ANDAMAN AND NICOBAR ISLANDS (INDIA)

LAKSHADWEEP (INDIA)

POWER CONSUMPTION ▲ —1981—82
(in million kwh)

10,000 – 15,000
7000 – 10,000
5000 – 7000
3000 – 5000
1000 – 3000
500 – 1000
200 – 500
100 – 200
0 – 100

▲ Utilities only

CONSUMPTION SECTORS
1981 – 82

Others ← → Domestic
Agriculture ← → Commercial
Industry

INDIAN   OCEAN

Indira Point

1 : 15,000,000

POWER GENERATION & CONSUMPTION

80

(in billion* kwh)

Electricity generation
Consumption

60

40

20

0
1951 1956 60-61 65-66 70-71 75-76 79-80 83-84 ◆

79.2
102
77.3
139.9
110.2

* Billion = 1000 millions   ◆ Estimated

POWER UTILIZATION — 1983-84

12.9%
4.2%
5.9%
17.7%
2.6%
56.7%

Domestic
Commercial
Industrial
Railways/Tramways
Agriculture
Others

Total consumption = 110.2 billions* kwh

137

POWER AVAILABILITY

Power consumption had always been lower than power generation, according to the data available. But this should not be construed as indicating that supply was always and everywhere in excess of requirement. Even during those years, requirement was always in excess of supply. There could be several reasons for this: unquantified transmission losses, quite high in India (estimated: 20 per cent), is one reason for less power availability to consumers. There are also regional variations during different seasons. When power is plentiful it might not be needed, whereas during different seasons when it is in short supply there could be excess demand; alternatively, certain regions have more power than they need, others less and the grid does not reach them.

Power availability (generation, less transmission and other losses) from 1980-81 to 1985-86 has been less than the requirement every year; by as much as 155.5 million kwh in 1985-86. But, at the same time, the percentage of deficit has been falling from 12.6 per cent in 1980-81 to 7.9 per cent in 1985-86. The estimated requirements till the end of the century are even greater: 295 million kwh in 1990-91 and 685 million kwh in 1999-2000.

In 1984-85, only five states and two union territories showed surpluses: Andhra Pradesh 746 million units, Gujarat 219 m units, Madhya Pradesh 422 m units, Tamil Nadu including Pondicherry 190 m units, Himachal Pradesh 36 m units and Delhi 273 m units, a total of 1886 m units. This was a considerable increase over the 1983-84 position in the same states, when Tamil Nadu (including Pondicherry) was, in fact, deficit in power. All the other states were deficit in power to a total of 12,305 m units in 1984-85, an improvement on the 16,148 m unit deficit the previous year. The states with the highest deficits in 1984-85 were Uttar Pradesh (2159 m units) and Bihar (1740 m units).

Considering 1984-85 deficits statewise, it was more than the national average of 6.7 per cent in Bihar (39.4 per cent), Haryana (28.9 per cent), Karnataka (7.2 per cent), Orissa (16.5 per cent), Punjab (19.2 per cent), Uttar Pradesh (13.2 per cent), Jammu & Kashmir (18 per cent) and Rajasthan (10.2 per cent). The deficit was less than the national average in Kerala (2.4 per cent), West Bengal, including Sikkim (2.2 per cent), Chandigarh (1 per cent), North-eastern region (2.4 per cent), Maharashtra, including Goa (4.3 per cent).

Failure to add to new capacity has been particularly disconcerting during the Sixth Plan. In 1985-86, the first year of the Seventh Plan, this trend appears to have been arrested: against the targeted addition of 4460 MW to new capacity, the capacity actually added was 4223 MW, just 5.3 per cent short of the target. The rough estimates indicate that of the total new capacity added, about 88 per cent was connected to the grid. During the Sixth Plan, addition to new capacity fell short of the lowered target of 19,666 MW (from 29,665 MW sanctioned) by 28.9 per cent. Now the states have made investments to the tune of Rs 661,050 million in 259 thermal, hydel and nuclear power projects for a total generation of 92,703 MW.

Four states in India — Madhya Pradesh investing Rs 43,084 million in 9 projects; Maharashtra, Rs 28,863 m in 19 projects; Uttar Pradesh, Rs 78,234 m in 13 projects; and West Bengal, Rs 33,480 m in 10 projects — are investing enormously in power generation, totalling over Rs 105,427 million, or 48.2 per cent of the country's total investment in power (Rs 381,529 m). Five other states — Andhra Pradesh, Rs 21,095 m in 8 projects; Bihar, Rs 23,250 m in 12 projects; Gujarat, Rs 19,646 m in 11 projects; Himachal Pradesh, Rs 10,500 m in 7 projects; and Jammu & Kashmir, Rs 10,748 m in 5 projects — have together invested over Rs 85,239 million, an additional 22.3 per cent of the investment in power generation. Among the major states, both Tamil Nadu and Karnataka, which are acutely short of electricity, have *relatively* low investments of Rs 24,869 m (in 9 projects) and Rs 12,788 m (in 9 projects) respectively. Arunachal Pradesh, Andaman & Nicobar Islands, Sikkim, Manipur and Mizoram have investments of less than Rs 100 million each in power generation.

Rural electrification is a measure of power availability to India's largest mass of people, those in the villages. Progress in this area has been spectacular in the last 15-20 years. In the year of independence, less than 0.5 per cent of the villages in the country were electrified. The number rose to about 8 per cent in 1965-66. Since then, the increase has been very great — from 106,774 (18.5 per cent) villages with electricity in 1970-71 to 272,625 (47.3 per cent) in 1980-81 and 368,840 (64.0 per cent) in 1984-85. By January 1986, 380,091 (66 per cent) villages in the country had been electrified. The population covered by these villages is considerably more than 70 per cent of the rural population. (The latest figures for the union territories are not available.) India expects to have 486,941 villages electrified by the end of the Seventh Plan (84.5 per cent of all villages) and total village electrification by 1994-95. Three states (Punjab, Haryana and Kerala) now have all their villages electrified. Five other states that are expected to have complete electrification by 1990 are Andhra Pradesh, Karnataka, Gujarat, Himachal Pradesh and Tamil Nadu. It is also expected that by 1990 nearly all the villages in Assam (97.6 per cent), Jammu & Kashmir (99.6 per cent), Maharashtra (99.5 per cent) and Nagaland (97.1 per cent) will be electrified. The Seventh Plan outlay on rural electrification is Rs 20,920 million. Rural electrification should, however, be understood in the context of its definition: a village is considered as having been electrified even if a single connection is supplied to it. On the other hand, even to achieve this, the capital expenditure has to be quite heavy.

With the accelerated pace of electrification in the seventies and eighties, the number of new electricity connections to houses has increased at the rate of one million or so rural households a year. But the number of new households has been increasing at a rate of a little more than 2 million households a year.

There is one aspect of rural electrification that dims its success: the painfully frequent breakdown of power supply. This requires a major drive to improve transmission and distribution.

Sources: 1. *Current Energy Scene in India, July 1986* and *Basic Statistics Relating to the Indian Economy, Vol. 2: States, September 1986,* Centre for Monitoring Indian Economy.

2. *Seventh Five Year Plan 1985-90, Vol. II,* Government of India Planning Commission.

Table 63

RURAL ELECTRIFICATION

	Villages electrified by January end 1986	Villages electrified[1] — 1977	Total number of households in electrified villages — 1977 (in million)	% of total households using power in electrified villages — 1977	% of power used for domestic purposes by these households — 1977
States					
Andhra Pradesh	23,261	18,658	6.06	17	50
Assam	12,819	1,941	0.51	7	95
Bihar	34,879	18,235	3.85	16	32
Gujarat	16,856	7,271	2.20	41	62
Haryana	6,731	6,739	1.40	29	39
Himachal Pradesh	15,065	8,426	0.44	52	98
Jammu & Kashmir	5,743	2,854	0.36	46	97
Karnataka	23,272	14,821	3.53	22	65
Kerala	1,268	1,177	2.98	20	74
Madhya Pradesh	42,538	14,165	2.46	15	43
Maharashtra	33,629	21,617	5.16	24	40
Manipur	655	240	0.06	10	69
Meghalaya	1,288	N.A.	N.A.	N.A.	N.A.
Nagaland	721	N.A.	N.A.	N.A.	N.A.
Orissa	24,228	N.A.	1.74	7	86
Punjab	12,126	12,280	2.05	46	37
Rajasthan	20,421	8,576	1.49	28	39
Sikkim	207	N.A.	N.A.	N.A.	N.A.
Tamil Nadu	15,707	15,477	6.90	24	47
Tripura	1,978	N.A.	0.05	14	90
Uttar Pradesh	64,840	32,644	7.84	12	29
West Bengal	19,559	11,033	1.93	13	47
Union Territories	2,300				
Andaman & Nicobar	—	N.A.	N.A.	N.A.	N.A.
Arunachal Pradesh	—	N.A.	N.A.	N.A.	N.A.
Chandigarh	—	N.A.	N.A.	N.A.	N.A.
Dadra & Nagar Haveli	—	N.A.	N.A.	N.A.	N.A.
Delhi	—	N.A.	0.11	35	67
Goa, Daman & Diu	—	N.A.	0.12	37	91
Lakshadweep	—	N.A.	N.A.	N.A.	N.A.
Mizoram	—	N.A.	N.A.	N.A.	N.A.
Pondicherry	—	N.A.	0.06	30	71
INDIA	380,091	203,252[2]	51.30	21	69

Note: [1] Estimated.

[2] Including other states and union territories for which break-up figures are not available.

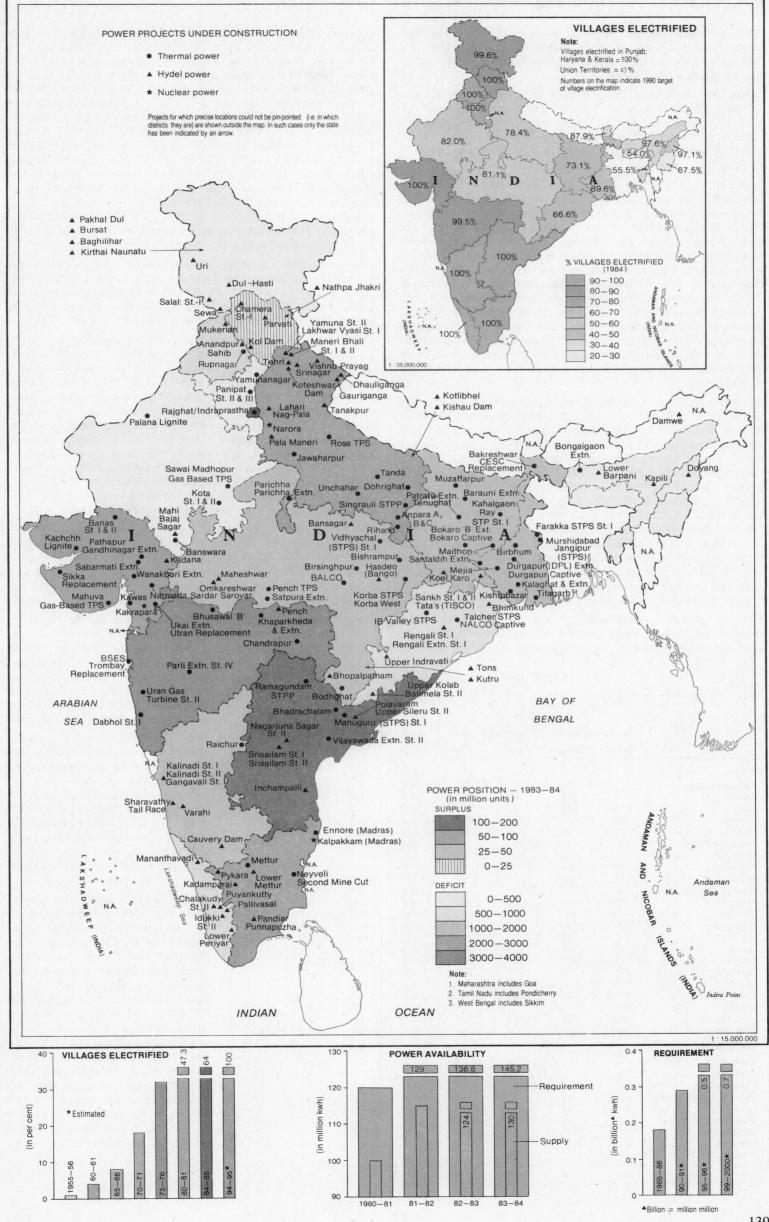

POWER PROJECTS UNDER CONSTRUCTION

● Thermal power

▲ Hydel power

★ Nuclear power

Projects for which precise locations could not be pin-pointed (i.e. in which districts they are) are shown outside the map. In such cases only the state has been indicated by an arrow.

▲ Pakhal Dul
▲ Bursat
▲ Baghilihar
▲ Kirthai Naunatu

VILLAGES ELECTRIFIED

Note:
Villages electrified in Punjab, Haryana & Kerala = 100%
Union Territories = 43%
Numbers on the map indicate 1990 target of village electrification.

99.6%
100%
100%
100%
N.A.
78.4%
87.9%
97.6%
N.A.
82.0%
54.0%
97.1%
81.1%
73.1%
55.5%
67.5%
100%
I N D I A
89.6%
99.5%
66.6%
N.A.
100%
N.A.
100%
100%

% VILLAGES ELECTRIFIED (1984)

90—100
80—90
70—80
60—70
50—60
40—50
30—40
20—30

1 : 35,000,000

LAKSHADWEEP (INDIA)

ANDAMAN AND NICOBAR ISLANDS (INDIA)

Map labels (power projects)

Uri
Dul-Hasti
Nathpa Jhakri
Salal St-I
Sewa
Chamera St-I
Mukerian
Parvati
Anandpur Sahib
Kol Dam
Yamuna St. II
Lakhwar Vyasi St. I
Maneri Bhali St. I & II
Rupnagar
Tehri
Yamunanagar
Srinagar
Vishnu Prayag
Panipat St. II & III
Koteshwar Dam
Dhauliganga
Gauriganga
Rajghat/Indraprastha
Lahari Nag-Pala
Tanakpur
Kotlibhel
Kishau Dam
Palana Lignite
Damwe
Pala Maneri
Rosa TPS
Bakreshwar CESC Replacement
Bongaigaon Extn.
Jawaharpur
Lower Barpani
N.A.
Kapili
Doyang
Sawai Madhopur Gas Based TPS
Tanda
Muzaffarpur
Kota St. I & II
Parichha
Parichha Extn.
Unchahar
Dohrighat
Patratu Extn.
Barauni Extn.
Kahalgaon
Mahi Bajaj Sagar
Singrauli STPP
Tenughat
Ray
STP St. I
Banas St. I & II
Bansagar
Anpara A, B&C
Bokaro 'B' Ext.
Farakka STPS St. I
Kachchh Lignite
Pathapur
Gandhinagar Extn.
Vidhyachal (STPS) St. I
Rihand
Bokaro Captive
Murshidabad Jangipur (STPS)
Sabarmati Extn.
Kadana
Banswara
Bishrampur
Santaldih Extn.
Maithon
Birbhum
Sikka Replacement
Wanakbori Extn.
Maheshwar
Birsinghpur
Hasdeo (Bango)
Mejia
Koel Karo
Durgapur (DPL) Extn.
Durgapur Captive
Mahuva Gas-Based TPS
Omkareshwar
Sardar Sarovar
Pench TPS
BALCO
Kalaghat & Extn.
Kawas
Narmada
Satpura Extn.
Korba STPS
Korba West
Sankh St. I & II
Tata's (TISCO)
Kishtobazar
Titagarh
Kakrapara
Pench
Khaparkheda & Extn.
Bhimkund
Bhusawal 'B'
IB Valley STPS
Talcher STPS
NALCO Captive
Ukai Extn.
Utran Replacement
Chandrapur
Rengali St. I
Rengali Extn. St. I
BSES Trombay Replacement
Parli Extn. St. IV
Upper Indravati
Tons
Kutru
Uran Gas Turbine St. II
Bhopalpatnam
Upper Kolab
Balimela St. II
Ramagundam STPP
Bodhghat
Polavaram
Upper Sileru St. II (STPS) St. I
Dabhol St. I
Bhadrachalam
Manuguru
Nagarjuna Sagar St. II
Vijayawada Extn. St. II
Raichur
Srisailam St. I
Srisailam St. II
Kalinadi St. I
Kalinadi St. II
Gangavali St. I
Inchampalli
Sharavathy Tail Race
Varahi
Ennore (Madras)
Kalpakkam (Madras)
Cauvery Dam
Mananthavadi
Mettur
Pykara
Lower Mettur
Neyveli Second Mine Cut
Kadamparai
Puyankutty
Chalakudy St. II
Pallivasal
Idukki St. II
Pandiar Punnapuzha
Lower Periyar

ARABIAN SEA

BAY OF BENGAL

LAKSHADWEEP (INDIA)

ANDAMAN AND NICOBAR ISLANDS (INDIA)

Andaman Sea

N.A.

Indira Point

INDIAN OCEAN

POWER POSITION — 1983—84
(in million units)

SURPLUS

100—200
50—100
25—50
0—25

DEFICIT

0—500
500—1000
1000—2000
2000—3000
3000—4000

Note:
1. Maharashtra includes Goa
2. Tamil Nadu includes Pondicherry
3. West Bengal includes Sikkim

1 : 15,000,000

VILLAGES ELECTRIFIED

(in per cent)

*Estimated

47.3
64
100

1955—56
60—61
65—66
70—71
75—76
80—81
84—85
94—95*

POWER AVAILABILITY

(in million kwh)

129
136.8
145.2
124
130

Requirement

Supply

1980—81
81—82
82—83
83—84

REQUIREMENT

(in billion▲ kwh)

0.5
0.7

1985—86
90—91*
95—96*
99—2000*

▲Billion = million million

139

BANKING AND THE PEOPLE

From the earliest years of formalized, western-style banking in India, banks were urban-based. But since nationalization in July 1969 they have moved into rural areas in very large numbers, following changes in government's financial policy.

In June 1969, there were 8262 scheduled commercial banks (inclusive of branch offices) in the country, each serving an average of 65,000 persons (1.6 banks to 100,000). In the next 17 years they registered a growth of 542 per cent, reaching a total of 53,085 in March 1986, each bank serving an average of 13,000 persons (7 banks to 100,000). Much of this growth has been in rural areas.

According to 1981 figures, there were 38,047 banks for 120 million households. The states and union territories doing much worse than the national average of 3147 households per bank were Mizoram (5425), Manipur (4532), Assam (4305) and Tripura (4268). The developed states such as Maharashtra, Tamil Nadu and Gujarat were only slightly better than the national average. The smaller states and union territories generally did best on this count: Goa (819), Chandigarh (1026), Delhi (1245), Lakshadweep (1327), Punjab (1497), Himachal Pradesh (1606), etc. Despite massive branch expansion, West Bengal, Bihar, Uttar Pradesh, Madhya Pradesh, Orissa and Andhra Pradesh still face poor population coverage. In this connection it is heartening to note that the Reserve Bank of India has shifted the emphasis from population coverage to need-based criteria while laying down guidelines for the Branch Expansion Policy in 1985-89.

Deposits in 1970 totalled Rs 60,280 million (Rs 100 per capita). This rose to Rs 773,820 million in 1985 (Rs 1038 per capita). The share of rural bank offices in total deposits was 3.1 per cent and 13.6 per cent respectively in June 1969 and June 1985. The highest per capita deposit among the major states in 1970 was Rs 288 (Maharashtra); in 1985 it was Rs 2270 (Punjab). The lowest was Rs 21 in 1970 (Orissa) and Rs 175 in 1985 (Manipur). Advances, which in 1970 amounted to Rs 45,790 million (76 per cent of all deposits), rose to Rs 531,220 million in 1985 (68.6 per cent of all deposits). The per capita advances for the same years were Rs 84 and Rs 712 respectively, with the highest among the major states amounting to Rs 262 (in Maharashtra) in 1970 and Rs 1897 (in Maharashtra again) in 1985. Savings deposits claimed the highest share (66.7 per cent) in the total number of accounts, but term deposits had the largest share (56 per cent) in the total amount of deposits.

With the upsurge in rural banking, there has been a significant diversification of bank credit to previously neglected sectors, such as agriculture, small industries, transport, professionals and self-employed persons, retail trade, export and food procurement. There are now 25 major categories of banks and credit finance institutions, including the Reserve Bank of India, Industrial Development Bank of India, Export and Import Bank of India and the National Bank for Agricultural and Rural Development. Their lending operations in 1984 totalled Rs 949,140 million, of which Rs 338,040 million or 35.6 per cent was for industry, Rs 191,250 m or 20 per cent for agriculture, Rs 108,270 m or 11.4 per cent for trade, Rs 39,670 m or 4.2 per cent for power, Rs 29,060 m or 3 per cent for transport, Rs 19,820 m or 2.1 per cent for housing, and Rs 223,040 m or 23.5 per cent for other sectors of the economy. Scheduled commercial banks in 1984 disbursed the largest amount of these loans, Rs 478,060 million or 50.4 per cent, followed by Cooperative Credit Societies (Rs 111,220 million or 11.7 per cent) and the IDBI (Rs 63,440 million or 6.7 per cent). The lowest lending has been by the National Film Development Corporation (Rs 100 million or 0.01 per cent).

The statewise per capita deposits are, generally, larger than the per capita advances in most states and union territories. The highest per capita deposit is in Delhi (Rs 10,115 in 1985) and the lowest in Manipur (Rs 175). Chandigarh has the highest per capita advances drawn (Rs 26,183), while the lowest is in Manipur (Rs 125).

The regional imbalances in banking facilities are glaring in the states as well as the union territories. The problem persists essentially because the banks are, to a great extent, still urban biased, with higher concentrations in metropolitan areas. In June 1985, Bombay alone accounted for 13.1 per cent of the total deposits, and, taken together, three metropolitan centres — Bombay, Delhi and Calcutta — accounted for 29.2 per cent of the total deposits of the country. The top ten metropolitan centres — Bombay, Delhi, Calcutta, Madras, Bangalore, Hyderabad, Ahmadabad, Pune, Lucknow and Kanpur — accounted for 39.3 per cent of the deposits and the top 50 urban centres for 53.5 per cent. However, there has been a discernible change during recent years: the total deposits of the top 50 urban centres has declined significantly, from 74 per cent in 1962 to 62.6 per cent in 1974 and 53.5 per cent in 1985.

In terms of districts, 50 districts, with 23 per cent of the country's population, deposited 63.4 per cent of the total deposits in 1985, but many of the 50 major urban centres are in these districts. The total deposits from the top 50 urban centres amounted to Rs 414,320 million, whereas it was Rs 490,970 million in the top 50 districts.

The operations of scheduled commercial banks during 1985-86 showed a distinct slowing down in the rates of growth of both deposits and advances. The resources position of the banks was, however, comfortable — their recourse to borrowings from the Reserve Bank of India was very modest while they augmented their liquidity substantially with a view to maintaining their cash reserve ratio.

During 1985-86, the aggregate deposits of scheduled banks rose by 17.3 per cent. This rate was lower than that recorded during 1984-85 (19.2 per cent) and 1983-84 (18 per cent). There has also been a distinct shift from demand to time deposits. The increase in demand deposits was only 7.1 per cent, while time deposits increased by 19.7 per cent as against 17.9 per cent in 1984-85.

Borrowings from the Reserve Bank of India registered a steep decline of Rs 6040 million in contrast to an increase of Rs 2220 million in the previous year. The scheduled banks strengthened their cash position considerably. The absolute increases in cash and balances with the RBI was Rs 42,160 million (53.2 per cent) as of March 1986. Consequently, the cash-deposit ratio rose sharply to 14.3 from 11 in the previous year. Investment by scheduled commercial banks in government and other approved securities rose by only Rs 24,340 million as against Rs 68,920 million in 1985. The growth rate of investment decelerated sharply to 8.7 per cent.

Source: 1. *Report on Currency and Finance 1985-86, Vol. I & II*, Reserve Bank of India.

BANK AVAILABILITY

Table 64

	Number of Bank offices				Number of Bank offices per 100,000 population			
	September 1968		September 1985		September 1968		September 1985	
	Rural	Urban	Rural	Urban	Rural	Urban	Rural	Urban
States								
Andhra Pradesh	128	371	2,508	1,566	0.4	4.6	5.8	10.6
Assam	20	44	583	322	0.2	3.4	3.0	13.3
Bihar	43	200	3,133	946	0.1	3.6	4.8	8.9
Gujarat	180	473	1,560	1,486	1.0	6.7	6.2	12.2
Haryana	35	99	648	448	0.5	5.8	5.9	12.6
Himachal Pradesh	N.A.	N.A.	457	67	N.A.	N.A.	10.8	18.1
Jammu & Kashmir	1	16	492	192	N.A.	2.3	9.8	13.0
Karnataka	176	484	2,216	1,668	0.8	7.1	8.0	13.1
Kerala	150	268	1,194	1,508	0.9	8.1	5.5	27.4
Madhya Pradesh	63	239	2,613	1,047	0.2	3.8	5.8	8.1
Maharashtra	122	868	2,198	2,639	0.4	5.8	5.1	10.4
Manipur	N.A.	N.A.	43	17	N.A.	N.A.	3.9	3.8
Meghalaya	N.A.	N.A.	85	37	N.A.	N.A.	7.2	12.3
Nagaland	N.A.	N.A.	45	21	N.A.	N.A.	6.0	12.3
Orissa	25	69	1,276	415	0.1	4.1	5.2	10.3
Punjab	73	213	1,122	868	0.7	6.5	8.8	15.9
Rajasthan	118	207	1,788	840	0.6	4.7	6.7	7.3
Sikkim	N.A.	N.A.	14	5	N.A.	N.A.	4.6	7.1
Tamil Nadu	159	824	1,761	2,142	0.6	7.1	5.2	12.1
Tripura	N.A.	N.A.	72	32	N.A.	N.A.	3.6	11.4
Uttar Pradesh	146	503	4,875	2,243	0.2	4.1	5.0	9.1
West Bengal	44	398	1,572	1,667	0.1	3.8	3.6	10.2
Union Territories								
Andaman & Nicobar	N.A.	N.A.	9	4	N.A.	N.A.	5.6	6.6
Arunachal Pradesh	N.A.	N.A.	47	2	N.A.	N.A.	7.2	3.3
Chandigarh	N.A.	N.A.	19	85	N.A.	N.A.	63.3	16.0
Dadra & Nagar Haveli	N.A.	N.A.	5	1	N.A.	N.A.	4.5	12.5
Delhi	N.A.	N.A.	249	78	959	N.A.	7.3	19.5	13.8
Goa, Daman & Diu	N.A.	N.A.	178	81	N.A.	N.A.	23.1	19.7
Lakshadweep	N.A.	N.A.	5	—	N.A.	N.A.	21.7	—
Mizoram	N.A.	N.A.	40	7	N.A.	N.A.	10.0	3.5
Pondicherry	N.A.	N.A.	28	34	N.A.	N.A.	7.0	8.5
INDIA	1,561	5,570	30,664	21,349	0.4	5.3	5.6	11.0

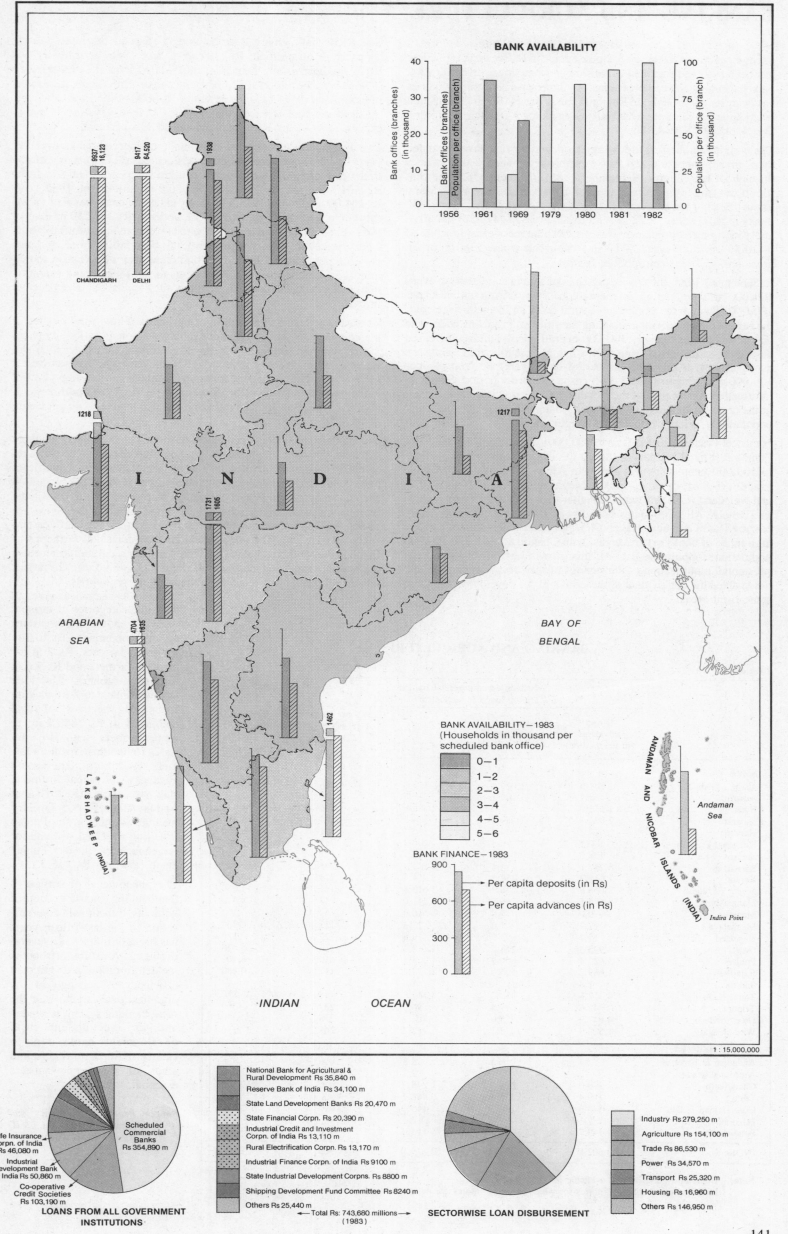

BANK AVAILABILITY

Bank offices (branches)
(in thousand)

Population per office (branch)
(in thousand)

1956 1961 1969 1979 1980 1981 1982

CHANDIGARH   DELHI

I N D I A

ARABIAN
SEA

BAY OF
BENGAL

LAKSHADWEEP (INDIA)

ANDAMAN AND NICOBAR ISLANDS (INDIA)

Andaman
Sea

Indira Point

BANK AVAILABILITY—1983
(Households in thousand per
scheduled bank office)

0—1
1—2
2—3
3—4
4—5
5—6

BANK FINANCE—1983

900

600

300

0

→ Per capita deposits (in Rs)

→ Per capita advances (in Rs)

INDIAN   OCEAN

1 : 15,000,000

**LOANS FROM ALL GOVERNMENT
INSTITUTIONS**

Scheduled
Commercial
Banks
Rs 354,890 m

Life Insurance
Corpn. of India
Rs 46,080 m

Industrial
Development Bank
of India Rs 50,860 m

Co-operative
Credit Societies
Rs 103,190 m

National Bank for Agricultural &
Rural Development Rs 35,840 m

Reserve Bank of India Rs 34,100 m

State Land Development Banks Rs 20,470 m

State Financial Corpn. Rs 20,390 m

Industrial Credit and Investment
Corpn. of India Rs 13,110 m

Rural Electrification Corpn. Rs 13,170 m

Industrial Finance Corpn. of India Rs 9100 m

State Industrial Development Corpns. Rs 8800 m

Shipping Development Fund Committee Rs 8240 m

Others Rs 25,440 m ←— Total Rs: 743,680 millions →
(1983)

SECTORWISE LOAN DISBURSEMENT

Industry Rs 279,250 m

Agriculture Rs 154,100 m

Trade Rs 86,530 m

Power Rs 34,570 m

Transport Rs 25,320 m

Housing Rs 16,960 m

Others Rs 146,950 m

141

BANKING AND AGRICULTURE

Since the nationalization of the major commercial banks in 1969, a phenomenal transformation has taken place in Indian banking. There has been both geographical expansion and functional diversification into challenging and complex areas. An impressive array of institutions has been set up through the Regional Rural Banks (RRBs), sponsored by commercial banks, and the National Bank for Agriculture and Rural Development (NABARD).

The commercial banks, following this development, have expanded their branch network and have established over 52,000 offices throughout the country (September 1985). Of these, 58.9 per cent (or 30,664) are in rural areas, most of them opened since 1969. Many urban branches are also close to rural areas. In 1969, there were 1832 banks in rural areas, 3322 in semi-urban areas, 1447 in urban areas and 1661 in metropolitan centres. In 1982 the number had increased to 20,398 in rural areas, 8763 in semi-urban areas, 5360 in urban areas and 4659 in metropolitan centres.

In September 1985, there were 5.6 bank offices in rural areas to every 100,000 rural population, an increase in 17 years from 0.4 banks per 100,000. Though the expansion in rural areas has been thirteen-fold, the rural-urban disparity in banking facilities still remains wide. The number of offices to every 100,000 overall population has increased during the same period from 1.4 to 7.0, almost four-fold, while the urban offices in this 17 year period have increased from 11 per 100,000 people against the 5.3 offices in the earlier period. Statewise information is indicative of the fact that rural-urban disparity is highest in the north-eastern states of Assam (3 bank offices to 100,000 rural population), Tripura (3.6) and Manipur (3.9).

Himachal Pradesh (10.8 offices to 100,000), Jammu & Kashmir (9.8), Punjab (8.8) and Karnataka (8) top the list in terms of bank offices to 100,000 people in rural areas. In Andhra Pradesh and Madhya Pradesh (5.8 offices to 100,000), Meghalaya (7.2), Rajasthan (6.7) and Nagaland (6.0) this index is higher than that of the country (5.6) as a whole. All other states have lower ratios. On the other hand, barring Dadra & Nagar Haveli (4.5), all other union territories have high ratios of banks in rural areas, with Chandigarh (63.3 to 100,000 population) topping the list. All this growth has resulted in a very substantial mobilization of the savings of the rural communities and has enabled the deployment of these savings for development in these same rural areas.

The RRBs have emerged as an integral part of the multi-agency approach to rural credit. By the end of June 1986, as many as 193 RRBs had been established, covering 341 districts. The total number of RRB branches stood at 12,606 at the end of 1985. These succeeded in mobilizing deposits to the extent of Rs 9600 million and deployed credit to the tune of Rs 10,810 million. At these levels, the credit-deposit ratio of the RRBs worked out to 113 per cent.

The distribution of credit by scheduled commercial banks over the years has shown differential increases/decreases in terms of sectors. In the case of agriculture, bank credit grew very substantially, from Rs 5010 million in 1972, to Rs 56,390 million in 1982, and Rs 154,000 million in 1983. However, industrial credit has also shown substantial growth, right from 1950, and was Rs 279,250 million in 1983. But in percentage terms of the total bank credit, industrial credit, which peaked in 1972 (6.4 per cent), has been falling since, and was only 3.8 per cent in 1983. Trade and other sectors have also shown increasing credits over the years, though not to the same extent as industrial credit. But unlike agriculture, they too have shown a decline in percentage.

The average credit to a hectare of agricultural land for 17 states in the country, for which data are available, amounts to Rs 245 (June 1981), with the largest sum being advanced in Kerala (Rs 682) and the smallest in Madhya Pradesh (Rs 91), the ratio of largest to smallest being 7.5:1. The credit per capita, on the other hand, averages Rs 68, being highest in Punjab (Rs 198) and lowest in Bihar and Jammu & Kashmir (Rs 25 each). The ratio of highest to lowest credit per capita is 7.9:1.

Regionwise data on loans and advances disbursed to farmers with three different sizes of landholdings: 2 ha (small farmers), 2-4 ha (medium farmers), and more than 4 ha (large farmers) indicate significant variations. The southern region has drawn a total of Rs 2737.8 million for all categories together, the northern region Rs 1364.8 m, the western region Rs 999 m, the central region Rs 903.3 m, the eastern region Rs 378 m, the union territories Rs 41.4 m, and the north-eastern region Rs 17.7 m. The largest total advance for small farmers was made in the southern region (Rs 1766.5 m); for medium farmers also the southern region advance was the largest (Rs 437.6 m). The largest advance for large farmers was in the northern region (Rs 771.1 m).

The smallest advances given to the three categories of farmers were all in the north-eastern region, amounting to Rs 10.1 m for small farmers, Rs 2 m for medium farmers and Rs 5.6 m for large farmers. The total advances made to farmers with 2 or less hectares of land amounted to Rs 2840.2 m; to large farmers with 4+ ha, Rs 2416 m; and to medium farmers, Rs 1186 m. The largest advances for small farmers, statewise, were made in Tamil Nadu (totalling Rs 605.8 m), for medium farmers in the Punjab (totalling Rs 223.1 m), and for large farmers in the Punjab again (totalling Rs 466.5 m).

The phenomenal expansion of the banking industry coupled with its functional diversification in the last fifteen years has brought in its wake several problems. A natural offshoot is the deterioration in operational efficiency. A change in the organizational structure of the rural banking system is needed and an independent rural development bank, with a clearly defined rural role and areas of operation, appears essential.

BANKING AND AGRICULTURE

Table 65

	Scheduled commercial banks: outstanding loans to agriculture — 1982				Public Sector Banks
	Direct advances (in Rs. million)	Indirect advances (in Rs. million)	Total (in Rs. million)	Rs. per hectare of gross cropped area	Aggregate advances — 1984 (in Rs. million)
States					
Andhra Pradesh	5,841.0	1,026.1	6,867.1	526	30,530
Assam	133.7	105.1	238.8	70	3,810
Bihar	1,615.9	971.0	2,586.9	243	11,990
Gujarat	2,085.3	1,089.2	3,174.5	291	25,520
Haryana	1,919.3	241.0	2,160.3	371	8,860
Himachal Pradesh	190.2	90.5	280.7	295	1,860
Jammu & Kashmir	158.0	30.1	188.1	192	1,340
Karnataka	3,387.3	714.4	4,101.7	365	27,760
Kerala	1,915.3	159.1	2,074.4	714	14,300
Madhya Pradesh	2,056.6	578.9	2,635.5	121	14,050
Maharashtra	3,674.5	1,566.5	5,241.0	257	108,320
Manipur	9.0	7.0	16.0	67	160
Meghalaya	22.2	1.0	23.2	113	260
Nagaland	15.7	3.9	19.6	122	220
Orissa	935.3	253.6	1,188.9	136	6,420
Punjab	3,627.4	600.9	4,228.3	610	18,160
Rajasthan	1,669.2	505.6	2,174.8	117	10,880
Sikkim	1.8	—	1.8	—	—
Tamil Nadu	4,317.6	852.2	5,169.8	748	37,850
Tripura	42.6	47.4	90.0	237	390
Uttar Pradesh	4,647.9	1,755.0	6,402.9	258	30,330
West Bengal	1,508.9	473.3	1,982.2	268	37,170
Union Territories				3,388[1]	
Andaman & Nicobar	2.3	6.5	8.8	N.A.	N.A.
Arunachal Pradesh	—	2.1	2.1	N.A.	N.A.
Chandigarh	390.0	858.3	1,248.3	N.A.	10,110
Delhi	140.4	328.3	468.7	N.A.	36,990
Goa, Daman & Diu	143.3	2.3	145.6	N.A.	2,110
Mizoram	2.1	—	2.1	N.A.	N.A.
Pondicherry	149.7	3.3	153.0	N.A.	540
INDIA	40,602.5	12,272.6	52,875.1	300	439,900[2]

Note: [1] Break-up figures are not available for union territories.
[2] Including other union territories for which break-up figures are not available.

Source: *Report on Currency and Finance 1985-86, Vol. I & II,* Reserve Bank of India.

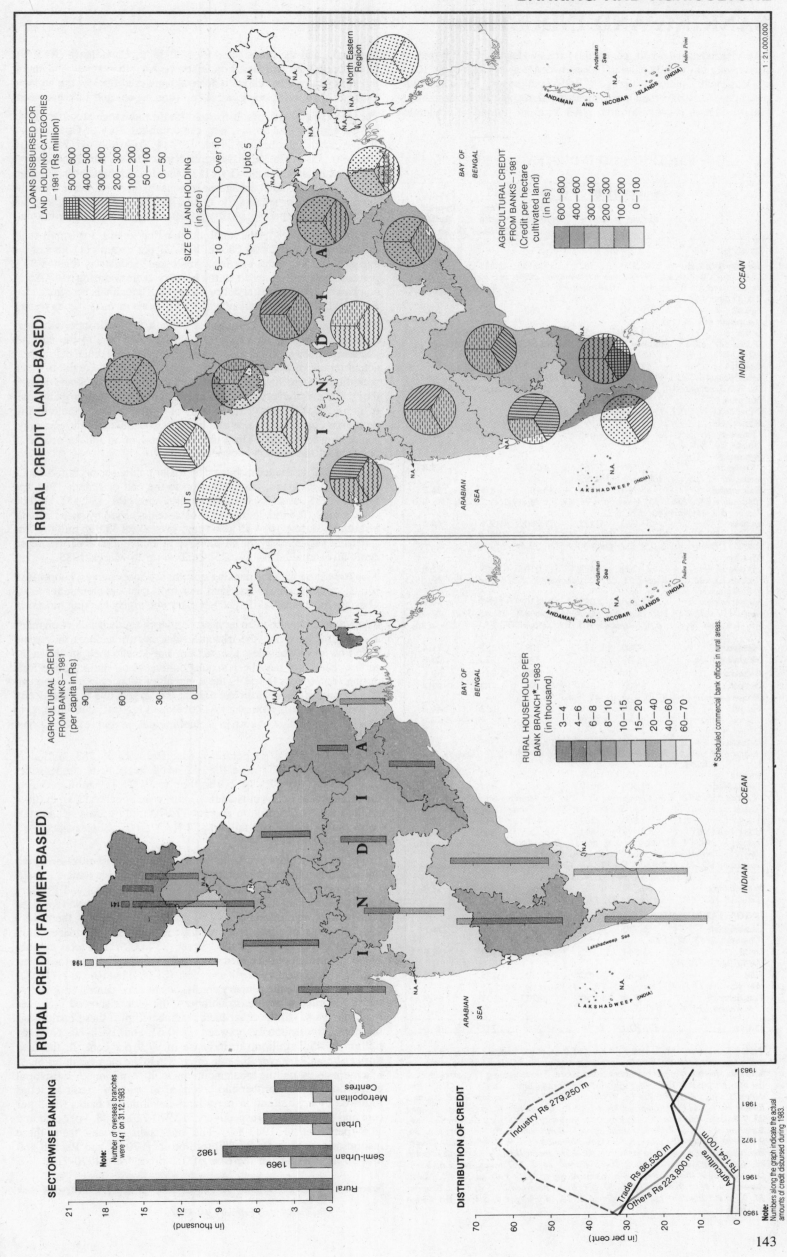

1 : 21,000,000

RURAL CREDIT (LAND-BASED)

LOANS DISBURSED FOR
LAND HOLDING CATEGORIES
– 1981 (Rs million)

500–600
400–500
300–400
200–300
100–200
50–100
0–50

SIZE OF LAND HOLDING
(in acre)

Over 10
5–10
Upto 5

AGRICULTURAL CREDIT
FROM BANKS –1981
(Credit per hectare
cultivated land)
(in Rs)

600–800
400–600
300–400
200–300
100–200
0–100

North Eastern
Region

BAY OF BENGAL

ANDAMAN AND NICOBAR ISLANDS (INDIA)

Andaman Sea

Indira Point

INDIAN OCEAN

ARABIAN SEA

LAKSHADWEEP (INDIA)

UTs

RURAL CREDIT (FARMER-BASED)

AGRICULTURAL CREDIT
FROM BANKS –1981
(per capita in Rs)

90
60
30
0

RURAL HOUSEHOLDS PER
BANK BRANCH* – 1983
(in thousand)

3–4
4–6
6–8
8–10
10–15
15–20
20–40
40–60
60–70

* Scheduled commercial bank offices in rural areas.

141

198

BAY OF BENGAL

Andaman Sea

Indira Point

ANDAMAN AND NICOBAR ISLANDS (INDIA)

INDIAN OCEAN

Lakshadweep Sea

ARABIAN SEA

LAKSHADWEEP (INDIA)

SECTORWISE BANKING

Note:
Number of overseas branches
were 141 on 31.12.1983

(in thousand)

21
18
15
12
9
6
3
0

Rural
Semi-Urban
Urban
Metropolitan
Centres

1969
1982

DISTRIBUTION OF CREDIT

Industry Rs 279,250 m

Trade Rs 86,530 m

Others Rs 223,800 m

Agriculture Rs 154,100 m

(in per cent)

70
60
50
40
30
20
10
0

1950
1961
1972
1981
1983

Note:
Numbers along the graph indicate the actual
amounts of credit disbursed during 1983.

BANKING AND INDUSTRY

Statewise industrial credit, per capita, varies widely. In 1981, it ranged from a very low of fifty paise in Lakshadweep to a very high of Rs 1058 in Delhi, with a national average of Rs 168 per person. The states with per capita industrial credit above the national average are Maharashtra, Gujarat, Tamil Nadu, Haryana, Uttar Pradesh, Punjab, Kerala and

BANKING AND INDUSTRY — 1985

Table 66

	DEPOSITS		ADVANCES		Credit-Deposit Ratio (in per cent)
	Rs. million	Per capita (in Rs.)	Rs. million	Per capita (in Rs.)	
STATES					
Andhra Pradesh	45,210	782	35,390	612	78.3
(December 1984 details: Hyderabad 360 reporting offices in the city, Rs. 117 m. deposits and Rs. 93 m. credit, Vishakhapatnam 64, 28, 19; Vijayawada 81, 15, 9.6; Guntur 54, 12, 15; Kakinada 29, 7.1, 3.8)					
Assam	9,110	418	4,850	222	53.2
(Guwahati 63, 26, 14)					
Bihar	34,620	454	14,430	189	41.7
(Patna 117, 49, 26; Dhanbad 27, 21, 5.8; Ranchi 51, 21, 7.2; Jamshedpur 47, 18, 5.5)					
Gujarat	49,750	1,345	27,210	735	54.7
(Ahmadabad 372, 98, 76; Vadodara 128, 66, 35; Surat 100, 31, 13; Rajkot 67, 18, 7.8; Jamnagar 41, 12, 4.1; Anand 45, 11.6, 5.6; Bhavnagar 41, 10.1, 6.1;)					
Haryana	14,310	987	9,680	668	67.6
(Faridabad 34, 10.7, 13.3; Ambala 46, 11.5, 4.5)					
Himachal Pradesh	4,810	1,046	2,120	461	44.0
(Shimla 24, 8.5, 4.1)					
Jammu & Kashmir	7,550	1,161	3,230	497	42.8
(Srinagar 82, 25, 17; Jammu 57, 16, 5.5)					
Karnataka	38,190	943	33,100	817	86.7
(Bangalore 486, 138, 136; Mangalore 91, 19, 13; Mysore 88, 15, 11; Dharwar 76, 14, 6.3; Belgaum 60, 109, 5.2)					
Kerala	31,020	1,132	21,260	776	68.5
(Cochin 153, 31, 35; Trivandrum 109, 29, 21; Trichur 41, 10.8, 6; Kozhikode 67, 9.8, 7.1; Kottayam 27, 8.9, 5.1; Quilon 39, 7.6, 12.8)					
Madhya Pradesh	27,200	478	17,030	299	62.6
(Indore 100, 28, 19; Bhopal 92, 25, 13; Jabalpur 71, 19, 9.7; Gwalior 49, 15, 8.4; Raipur 36, 9.4, 5.7)					
Maharashtra	142,860	2,089	129,730	1,897	90.8
(Bombay 1,025, 885, 981; Pune 208, 72, 52; Nagpur 119, 35, 18; Thane 49, 18, 7.3; Kolhapur 54, 9.1, 6.6; Nashik 43, 8.5, 5.1; Solapur 47, 8.5, 4.1; Pimpri 26, 11.3, 7.1)					
Manipur	280	175	200	125	71.4
Meghalaya	1,240	824	330	220	26.6
Nagaland	710	791	280	311	39.4
Orissa	8,820	311	8,190	288	92.8
(Bhubaneshwar 54, 19, 15; Cuttack 45, 9.4, 9.1)					
Punjab	41,310	2,270	18,410	1,012	44.6
(Jalandar 100, 40, 17; Ludhiana 103, 39, 26; Amritsar 111, 36, 14; Patiala 55, 18, 9.7;)					
Rajasthan	19,760	517	13,930	365	70.5
(Jaipur 129, 38, 30; Jodhpur 53, 12, 5.8; Ajmer 38, 8.3, 3.9; Udaipur 36, 8.1, 5.5; Kota 44, 7.7, 7.8)					
Sikkim	420	1,051	60	150	14.3
Tamil Nadu	49,500	957	48,820	944	98.6
(Madras 555, 177, 210; Coimbatore 113, 26, 35; Madurai 96, 16, 14; Tiruchchirappalli 71, 12, 6.9; Salem 55, 8.7, 6.6; Erode 37, 6.9, 5.6)					
Tripura	850	369	620	943	72.9
Uttar Pradesh	77,490	643	38,110	316	49.2
(Lucknow 149, 74, 32; Kanpur 220, 66, 36; Dehra Dun 45, 35, 3.9; Varanasi 102, 28, 7.9; Allahabad 91, 27, 10; Agra 87, 23, 11; Meerut 69, 20, 7.3; Gorakhpur 40, 13, 4; Bareilly 51, 12, 3.8; Ghaziabad 48, 12, 8.2; Moradabad 48, 11.1, 5.7; Muzaffarnagar 30, 7.2, 4.1)					
West Bengal	80,770	1,367	41,950	710	51.9
(Calcutta 763, 485, 362; Haora 55, 19, 4.4)					
UNION TERRITORIES					
Chandigarh	5,730	9,550	15,710	26,183	274.2
(Chandigarh 88, 51, 100)					
Delhi	73,840	10,115	43,560	5,967	58.9
(Delhi 934, 681, 410)					
Goa	6,480	5,400	2,130	1,775	32.9
(Margao 21, 11.1, 4.7; Panaji 25, 10, 5.4; Vasco da Gama 19, 7.2, 4.7)					
Pondicherry	1,070	1,789	620	1,033	57.9
(Pondicherry 24, 6.8, 3.7)					
INDIA	773,820	1,038	531,220	712	68.6

Note: The above table is based on a list of the top 100 banking centres receiving deposits and the top 100 banking centres giving credits. Only 83 of the centres on both lists are common and have been listed above. The remaining 17 centres on each list are: **Deposits: Bih.:** *Muzaffarpur 38 reporting offices in the city; Rs. 9.8m. deposits;* **Guj.:** *Gandhidham 12, 7, Navasari 28, 11.2; Porbandar 28, 8.2;* **Har.:** *Rohtak 36, 7.9;* **H.P.:** *Hoshiarpur 21, 8.8;* **M.P.:** *Bhilainagar 28, 8.4;* **Meg.:** *Shillong 25, 7.6;* **Ori.:** *Raurkela 31, 7.2;* **Pun.:** *Phagwara 19, 10.1;* **U.P.:** *Aligarh 38, 10.4; Saharanpur 38,8.6;* **W.B.:** *Asansol 26, 8.6; Barddhaman 25, 7.4; Durgapur 36, 9.7; South Suburbs (Behala) 28, 8.5;* **Goa.:** *Mapuca 14, 8.9.* **Credit: Bih.:** *Bokaro Steel City 16 reporting offices in the city, Rs. 4.5m. credit; Hatia 2, 5.3;* **Guj.:** *Bharuch 21, 6.6; Gandhinagar 15, 6.4;* **Har.:** *Karnal 24, 5.2; Panipat 24, 4.6;* **Kar.:** *Doorvaninagar (Bangalore) 1, 3.8;* **Ker.:** *Alleppey 33, 5; Palghat 27, 4.1;* **Mah.:** *Aurangabad 41, 5.9;* **M.P.:** *Ujjain 37, 5.1;* **Pun.:** *Bhatinda 28, 5.3;* **T.N.:** *Tiruppur 27, 5;* **U.P.:** *Bhadohi 11, 3.9; Bidhuna 2, 10; Modinagar 9, 5.4;* **W.B.:** *Khana 11, 16.2.*

Karnataka and the union territories of Delhi, Chandigarh (Rs 930), Goa, Daman & Diu and Pondicherry. All other states and union territories have a per capita industrial bank credit below the national average, the lowest being in Sikkim (one rupee) and Lakshadweep.

The per capita bank credit to small-scale industries stood at Rs 45 for the country as a whole, with the union territory of Goa, Daman & Diu having the highest level, Rs 414, followed by Chandigarh (Rs 345), Delhi (Rs302), Dadra & Nagar Haveli and Pondicherry. Among the states, Punjab (Rs 111), Haryana, Maharashtra, Gujarat, Kerala, Tamil Nadu and Karnataka have received more than the average credit for small-scale industries. Sikkim (seventy paise per person) and Lakshadweep (fifty paise) are again at the very bottom.

As of March 1985, the total outstanding credit for small-scale industries stood at Rs 66,080 million, 21 per cent higher than at the end of 1983-84. Bank credit to small-scale industries formed 35.9 per cent of the total credit to the priority sectors during 1984-85 as compared to 36.6 per cent during 1983-84. Though this percentage has decreased, the total amount of aid to small-scale industry has increased.

The stepping up of the annual bank credit to small-scale industries is quite in keeping with their growing contribution to the national economy. Even so, the level of assistance remained much below their actual requirements. As several of these units operate under difficult conditions, there has been increased incidence of sickness among them: one out of every 13 units. This amounts to a very large number of units, 98,000 out of 1,275,000, in 1984-85. Nevertheless, this sector has emerged as a new and dynamic constituent of the economy, accounting for half of India's industrial production and for nearly 25 per cent of the country's exports.

The financial assistance disbursed by major public-sector, non-banking financial institutions, cumulative up to the end of March 1981 and March 1985 was Rs 79,299 million and Rs 184,531 million respectively. The distribution of assistance disbursed by such financial institutions during 1984-85 alone amounted to Rs 33,995 million. The assistance disbursed by the all-India, term-lending financial institutions cumulative totalled Rs 141,757 million up to March 1985.

Also coexisting with the banking systems are cooperatives, cooperative banks, credit societies and land and industrial development banks. They are functionally different, yet carry out largely banking functions.

Cooperatives have been under heavy criticism for their failure on many a score, for example, the irksome rules and procedures they seem to follow for any business transaction. But despite their apathy in the earlier years, they have been making commendable and serious efforts during recent years to rectify the regional imbalances in development. If wide disparities nevertheless persist, it is largely due to a real dearth of sound proposals from the less developed regions, due, in part, to lack of development of entrepreneurship, and, in part, to institutional deficiencies.

Deposits in central cooperatives rose from Rs 19,783 million in 1979-80 to Rs 31,837 m in 1982-83 while in the state cooperatives they went up from Rs 14,239 million to Rs 21,169 million during the same years. Loans outstanding in the case of central cooperatives totalled Rs 26,175 million and Rs 47,070 million, and in the state cooperatives Rs 15,102 million and Rs 29,160 million respectively in 1979-80 and 1983-84.

From the particulars available for five categories of rural-oriented, non-scheduled financial institutions (State Cooperative Banks, Central Cooperative Banks, Primary Credit Societies, Central and Primary Land Development Banks, and Industrial Cooperative Banks) for three time periods, the following may be gathered: in 1960-61, there were 234,892 units of these five institutions, of which primary credit societies numbered by far the largest (234,000). The total number of the five institutions declined in 1970-71 to 182,277 (again the largest number was of primary credit societies: 181,000) but rose to 1,118,000 in 1983-84. While the primary credit societies and central cooperative banks recorded a decline in numbers, the others showed increases over the years. On the other hand, membership of these institutions increased tremendously between 1960-61 and 1970-71, from 24 million to 45.3 million, an increase of 87.9 per cent. In the next thirteen years, however, the rate of growth in the membership slowed to a mere doubling, the total amounting to 90.7 million. A substantial proportion of the membership belonged to primary credit societies followed by central and primary land development banks (88.3 per cent and 11.3 per cent respectively in 1970-71 and 88.3 per cent and 11.4 per cent in 1983-84). The outstanding loans from these institutions stood at Rs 35,780 million in 1970-71 and Rs 148,460 million in 1983-84, an increase of 3.1 times from 1970-71.

Sources: 1. *Report on Currency and Finance 1985-86, Vol. I & II.* Reserve Bank of India.
2. *Business India, February 9-2-1987.*

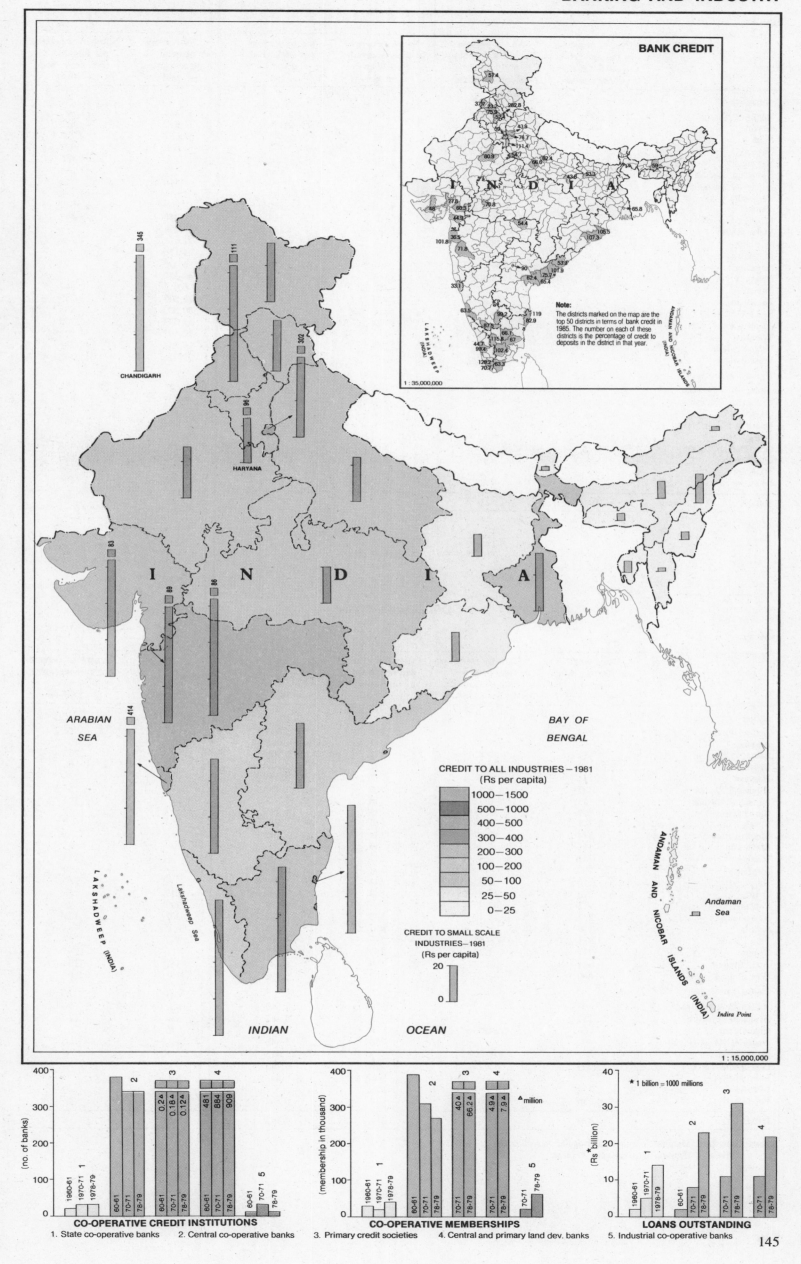

BANK CREDIT

Note:
The districts marked on the map are the top 50 districts in terms of bank credit in 1985. The number on each of these districts is the percentage of credit to deposits in the district in that year.

1 : 35,000,000

ARABIAN
SEA

CHANDIGARH

HARYANA

I N D I A

LAKSHADWEEP (INDIA)

Lakshadweep Sea

BAY OF
BENGAL

CREDIT TO ALL INDUSTRIES – 1981
(Rs per capita)

1000 – 1500
500 – 1000
400 – 500
300 – 400
200 – 300
100 – 200
50 – 100
25 – 50
0 – 25

CREDIT TO SMALL SCALE
INDUSTRIES – 1981
(Rs per capita)

20

ANDAMAN AND NICOBAR ISLANDS (INDIA)

Andaman
Sea

Indira Point

INDIAN   OCEAN

1 : 15,000,000

CO-OPERATIVE CREDIT INSTITUTIONS
(no. of banks)

CO-OPERATIVE MEMBERSHIPS
(membership in thousand)

LOANS OUTSTANDING
(Rs *billion)
* 1 billion = 1000 millions

1. State co-operative banks   2. Central co-operative banks
3. Primary credit societies   4. Central and primary land dev. banks   5. Industrial co-operative banks

145

Table VI

EDUCATION & HEALTH

	No. of persons in the age group of 5-9 per primary school	No. of persons in the age group of 10-14 per secondary school	No. of persons in the age group of 15-24 per college	No. of persons per registered doctor — 1984	No. of persons per hospital & dispensary — 1985	Number of persons per hospital & dispensary bed — 1985
States						
Andhra Pradesh	190	802	11,370	2,142	41,344	1,597
Assam	N.A.	N.A.	N.A.	2,479	37,847	1,812
Bihar	211	632	18,523	3,445	62,052	2,770
Gujarat	412	260	18,180	2,103	13,009	923
Haryana	362	837	11,539	N.A.	45,171	1,766
Himachal Pradesh	N.A.	N.A.	N.A.	N.A.	15,231	1,045
Jammu & Kashmir	N.A.	N.A.	N.A.	2,114	9,516	1,574
Karnataka	230	330	9,508	1,691	22,625	1,189
Kerala	417	710	18,704	1,880	18,169	612
Madhya Pradesh	138	576	18,657	8,604	61,248	3,277
Maharashtra	245	376	11,787	1,845	9,052	771
Manipur	N.A.	N.A.	N.A.	N.A.	21,917	1,284
Meghalaya	N.A.	N.A.	N.A.	N.A.	21,126	709
Nagaland	N.A.	N.A.	N.A.	N.A.	8,035	797
Orissa	118	379	18,068	3,148	46,329	2,223
Punjab	168	583	11,982	765	8,912	847
Rajasthan	241	625	22,731	3,695	23,507	1,757
Sikkim	N.A.	N.A.	N.A.	N.A.	80,000	838
Tamil Nadu	208	640	23,824	1,425	48,453	1,200
Tripura	N.A.	N.A.	N.A.	N.A.	10,502	1,821
Uttar Pradesh	241	790	9,788	4,460	49,224	2,281
West Bengal	167	893	16,197	1,617	71,376	1,143
Union Territories						
Andaman & Nicobar	N.A.	N.A.	N.A.	N.A.	4,081	274
Arunachal Pradesh	N.A.	N.A.	N.A.	N.A.	6,862	623
Chandigarh	N.A.	N.A.	N.A.	N.A.	21,428	469
Dadra & Nagar Haveli	N.A.	N.A.	N.A.	N.A.	25,000	3,571
Delhi	N.A.	N.A.	N.A.	N.A.	11,889	498
Goa, Daman & Diu	N.A.	N.A.	N.A.	N.A.	10,344	423
Lakshadweep	N.A.	N.A.	N.A.	N.A.	10,000	800
Mizoram	N.A.	N.A.	N.A.	N.A.	16,216	534
Pondicherry	N.A.	N.A.	N.A.	N.A.	13,333	241
INDIA	194	522	13,591	2,457	25,503	1,353

COMMUNICATION

Table VII

	Road length per 1000 sq. kms. (in km.) — 1981-82	Railway route length per 1000 sq. kms. (in km.) — 1984-85	No. of persons per registered motor vehicle — 1982-83	No. of persons per post office — 1983-84	No. of persons per telephone — 1984-85	Number of persons per cinema house (in thousand) — 1985-86
States						
Andhra Pradesh	468	18.03	138	3,395	231	24
Assam	754	29.59	252	6,293	887[1]	109
Bihar	481	30.82	333	6,706	931	207
Gujarat	309	28.74	67	4,149	120	68
Haryana	544	34.14	85	5,546	—	138
Himachal Pradesh	367	4.57	716	1,850	—	164
Jammu & Kashmir	53	0.35	136	4,362	320	191
Karnataka	587	15.75	75	4,056	177	32
Kerala	2,744	23.49	116	5,576	160	20
Madhya Pradesh	244	13.02	134	5,195	480	105
Maharashtra	585	17.59	59	5,580	77	51
Manipur	242	—	142	2,702	—	133
Meghalaya	237	—	N.A.	3,181	—	150
Nagaland	370	0.53	272	3,162	—	180
Orissa	770	12.71	349	3,622	617	156
Punjab	923	42.78	39	4,639	197[2]	98
Rajasthan	212	16.42	136	3,771	302	151
Sikkim	159	—	N.A.	3,174	N.A.	133
Tamil Nadu	1,021	30.05	120	4,187	147	24
Tripura	797	1.20	235	3,577	—	287
Uttar Pradesh	521	30.23	189	6,386	557	134
West Bengal	640	42.25	132	6,998	171	87
Union Territories						
Andaman & Nicobar	85	—	70	2,500	N.A.	50
Arunachal Pradesh	152	—	1,195	3,167	N.A.	117
Chandigarh	1,250	84.00	6	11,363	20	63
Dadra & Nagar Haveli	492	—	196	3,448	N.A.	80
Delhi	7,895	110.00	9	11,413	18	100
Goa, Daman & Diu	2,034	19.75	26	4,743	N.A.	40
Lakshadweep	—	—	N.A.	2,857	N.A.	—
Mizoram	119	—	N.A.	1,872	N.A.	120
Pondicherry	4,286	54.00	16	6,382	N.A.	13
INDIA	470	18.82	104	4,942	196	58

Note: [1] Includes Manipur, Meghalaya, Nagaland and Tripura.
[2] Includes Haryana and Himachal Pradesh.

The Produce

This section comprises 78 maps & 132 charts. It examines what man and nature, resources and infrastructure have been able to produce for India's people and assesses how far production meets the country's requirements.

> **Note:**
> Data used for the states in this section tends to be quite variable, ranging from 1971-72 to 1985-86. This has been due to non-availability of data, partly because such data was not collected during later periods. For instance, the last Livestock Census was in 1977 and no census has been held since then. In the case of the Industries section, we have tried to give production figures, value added by manufacture, number of factories and number of workers in each state for each product. But in every map at least one of these items is missing in certain states due to nonavailability of information. Tamil Nadu, for instance, one of the leading manufacturers of handlooms in the country is unable to provide information about value added in manufacture. Where districtwise information is provided, attention should be paid again to the note on page 15.

LIVESTOCK

Animal husbandry is a vital part of the national, especially rural, economy. It accounts for a tenth of the national income. The contribution made to this economy by cattle and buffalo in particular is considerable, for they are the main source of draught power in agricultural operations and rural transportation. They provide essential food — milk and meat — as also large quantities of animal by-products, such as hides, bones and manure. Meat and meat products alone earn for India about Rs 400 million in foreign exchange.

The twelfth quinquennial livestock census was conducted in 1977, but a few states had to postpone it for a variety of reasons; hence data in consolidated form are available only for 1972, the year of the earlier census. However, reliable statewise estimates are available in the statistical abstracts published from time to time and these have been used here.

India has the largest and most varied animal resources in the world. There were in 1977, 180.1 million cattle, 62 million buffalo, 75.6 million goats and 40.9 million sheep in the country. India has about one-sixth of the cattle, half the buffalo, and one-fifth of the goat population of the world. The poultry population was 160.9 million.

The density of cattle population per 100 hectares of gross cropped area in India is 12.8. The largest number of cattle are in Madhya Pradesh (26.3 m or 14.6 per cent of all cattle in India). The shares of the other states are: Uttar Pradesh 14.3 per cent, Maharashtra 8.4 per cent, Bihar 8.4 per cent, Rajasthan 7.2 per cent, Orissa 6.7 per cent, Andhra Pradesh 6.7 per cent, West Bengal 6.6 per cent, Tamil Nadu 6 per cent, Karnataka 5.7 per cent, Assam 3.7 per cent and Gujarat 3.3 per cent. Of the 62 million buffalo in the country, 22.6 per cent are in Uttar Pradesh, 11.5 per cent in Andhra Pradesh and 9.4 per cent in Madhya Pradesh.

Livestock density is greatest in Maharashtra (11 head of livestock per hectare of arable land), closely followed by Haryana (10 head/ha). Livestock density in other parts of the country is: Karnataka and Rajasthan 5 each, Orissa 4, Tripura, West Bengal, Uttar Pradesh, Bihar, Assam and Tamil Nadu 3 each, Andhra Pradesh, Himachal Pradesh, Madhya Pradesh and Sikkim 2 each, and in the rest less than two to a hectare. Of the union territories, Pondicherry has 4, and Delhi, Lakshadweep and Dadra & Nagar Haveli two each per hectare of arable land.

A Special Livestock Production Programme was initiated by the Ministry of Agriculture in 1975-76 and is in operation in 183 districts in 21 states and 4 union territories. In 1982-83, a provision of Rs 75 million was made against which a sum of Rs 54.8 million was spent, covering 54,190 beneficiaries. Improved breeding facilities are made available to owners of breedable cattle through a network of over 15,000 artificial insemination centres and 60 frozen semen banks and bull stations. There has been a steady increase in the number of livestock, from 293 million in 1951 to 369.5 million in 1977.

As the quality of Indian livestock is poor, the desired result of a well-developed livestock industry has not yet been achieved. The average annual yield of a cow in milk is 157 kgs, which is one of the lowest in the world (compare it to 4154 kgs in USA, 3959 kgs in UK and 3902 kgs in Denmark).

The dairy industry has, however, received a great boost in recent years through Operation Flood, first initiated by the cooperatives of Gujarat. There are now a large number of milk-supply schemes in the public and cooperative sectors throughout the country. There are 233 dairy plants in the public and cooperative sectors in addition to 141 in the private sector. The installed capacity of the plants is 13.6 million litres a day.

Milk production contributes no less than Rs 6200 million to the national dividend. This is obtained from the milk produced by 67 million cows and buffalo maintained in nearly 40 million small, fragmented holdings throughout the country.

Milk is obtained in India from cows, buffalo, goat and, in Rajasthan, even from camel. The respective shares of the three main milk producers are 63.8 per cent from buffalo, 33.5 per cent from cows and 2.7 per cent from goat. Buffalo milk predominates in Gujarat, Andhra Pradesh, Uttar Pradesh, Maharashtra, Haryana and Punjab. Cow's milk predominates in West Bengal, Orissa and Kerala.

India's poultry population has grown substantially over recent decades, from 73 million in 1951 to 160.9 million in 1977. But there are only 26 poultry birds for every 100 persons in India as against 375 in Canada and 540 in Denmark. The annual laying capacity of an Indian hen is 60 small eggs in a year as against 180 large eggs of the white leghorn. Nevertheless, poultry production has made rapid strides in the last decade. The production of broilers had reached 35 million in 1981-82 compared to 4 million in 1971, and production of eggs increased from 2500 million in 1971 to about 13,000 million in 1980-81. However, the per capita availability is very low compared to world standards. The availability of eggs in India is estimated at 12 eggs per person per year as against 295 in the USA, 282 in Canada and 249 in Germany, but this has to be seen in the context of the large number of vegetarians in India for whom even an egg is taboo. The availability of poultry meat is also low, being estimated at 131 gms per person a year.

In 1977, the 12th livestock census year, the state with the largest number of poultry birds was Andhra Pradesh (13.4 per cent of all poultry birds in India — 21.6 m). The share of the other states was: Maharashtra 11.7 per cent, West Bengal 9.6 per cent, Tamil Nadu 8.9 per cent, Bihar 8.7 per cent, Kerala 8.3 per cent, Assam 6.5 per cent, Karnataka 6.0 per cent, and Orissa 5.9 per cent. All other states and territories have less than 8 million birds each, the lowest number being in Lakshadweep (26,000).

Annual egg production for the end of the Seventh Plan has been targeted at 19,900 million, Andhra Pradesh setting itself the highest target — 3600 million eggs or 18.1 per cent of what the country expects to produce. High targets have also been set by West Bengal (11.2 per cent), Tamil Nadu (10 per cent), Maharashtra (9.8 per cent), Kerala (9.3 per cent), Bihar (7.4 per cent) and Punjab (7.3 per cent).

Source: 1. *Directory & Yearbook 1983,* Times of India.
2. *Statistical Abstract India 1984,* Ministry of Planning. Quoting from Directorate of Economics and Statistics, Ministry of Agriculture.
3. *Seventh Five Year Plan 1985, Vol. II,* Government of India Planning Commission.

LIVESTOCK AND POULTRY — 1977

Table 67 (in thousand numbers)

States	Cattle	Buffalo	Sheep	Goat	Other Livestock	Poultry
Andhra Pradesh	12,041	7,163	7,064	4,364	840	21,609
Assam	6,603	730	59	1,657	553	10,450
Bihar	15,162	4,463	1,150	9,925	1,057	14,076
Gujarat	6,006	3,473	1,592	3,084	251	3,426
Haryana	2,442	2,940	542	519	461	1,403
Himachal Pradesh	2,106	560	1,055	1,335	39	330
Jammu & Kashmir	2,138	500	1,216	692	113	2,040
Karnataka	10,222	3,278	4,536	3,388	376	9,696
Kerala	3,006	454	3	1,683	173	13,389
Madhya Pradesh	26,253	5,845	968	6,725	538	7,156
Maharashtra	15,218	3,899	2,636	7,563	327	18,751
Manipur	294	52	2	16	143	938
Meghalaya	477	40	20	119	157	1,073
Nagaland	93	8	*	24	250	715
Orissa	12,120	1,359	1,431	3,417	299	9,490
Punjab	3,312	4,110	498	722	355	5,540
Rajasthan	12,896	5,072	9,938	12,307	1,146	1,590
Sikkim	158	5	16	89	24	221
Tamil Nadu	10,801	3,077	5,289	4,202	776	14,347
Tripura	592	14	3	199	45	665
Uttar Pradesh	25,773	13,965	2,059	8,462	2,085	5,497
West Bengal	11,878	824	793	5,211	379	15,492
Union Territories						
Andaman & Nicobar	27	10	*	18	21	184
Arunachal Pradesh	168	12	20	76	307	764
Chandigarh	3	12	1	2	1	51
Dadra & Nagar Haveli	38	3	*	13	1	45
Delhi	49	109	9	20	19	217
Goa, Daman & Diu	122	40	1	21	78	413
Lakshadweep	1	*	*	5	—	26
Mizoram	49	2	1	23	48	1,128
Pondicherry	92	10	5	39	2	148
INDIA	180,140	62,029	40,907	75,620	10,844	160,870

Note: * Less than 500

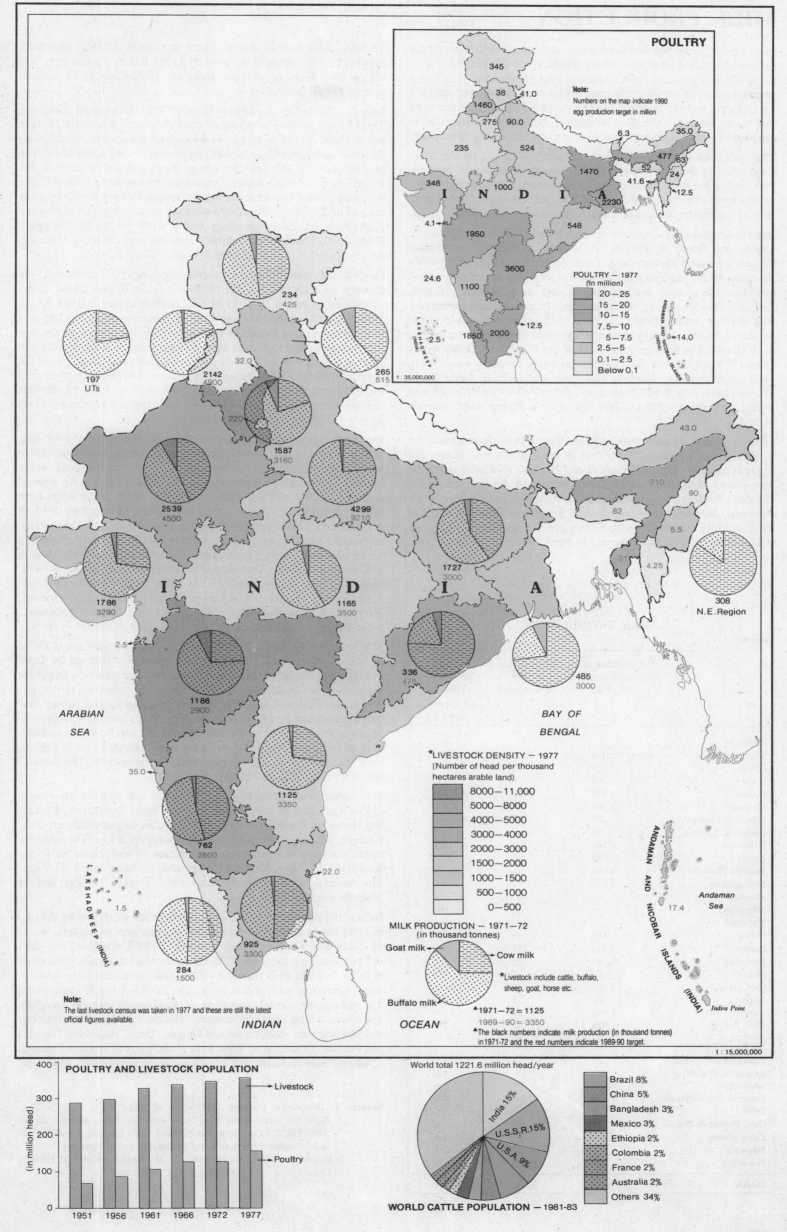

POULTRY

Note:
Numbers on the map indicate 1990
egg production target in million

345
38 41.0
1460
275 90.0
235 524
6.3 35.0
477 63
348 I N D I 1470 52 24
1000 A 41.6 12.5
4.1 548
1950
24.6 3600
1100
2.5 1850 2000 12.5
14.0

POULTRY — 1977
(in million)
20 — 25
15 — 20
10 — 15
7.5 — 10
5 — 7.5
2.5 — 5
0.1 — 2.5
Below 0.1

1 : 35,000,000

234
425

197
UTs

2142
4900

32.0

265
515

220

1587
3160

27 43.0

710
90

2539
4500

4299
9210

82

5.5

1786
3290

I N D I A

1165
3500

1727
3000

4.25

31

2.5

308
N.E. Region

35.0

1186
2900

ARABIAN
SEA

336
475

485
3000

BAY OF
BENGAL

*LIVESTOCK DENSITY — 1977
(Number of head per thousand
hectares arable land)
8000 — 11,000
5000 — 8000
4000 — 5000
3000 — 4000
2000 — 3000
1500 — 2000
1000 — 1500
500 — 1000
0 — 500

1125
3350

22.0

762
2600

ANDAMAN AND NICOBAR ISLANDS (INDIA)

Andaman
Sea

17.4

1.5

284
1500

925
3300

MILK PRODUCTION — 1971–72
(in thousand tonnes)
Goat milk → ← Cow milk

*Livestock include cattle, buffalo,
sheep, goat, horse etc.

Buffalo milk

▲ 1971–72 = 1125
1989–90 = 3350
▲The black numbers indicate milk production (in thousand tonnes)
in 1971-72 and the red numbers indicate 1989-90 target.

Indira Point

Note:
The last livestock census was taken in 1977 and these are still the latest
official figures available.

INDIAN OCEAN

1 : 15,000,000

POULTRY AND LIVESTOCK POPULATION

400
300
200
100
0
(in million head)

1951 1956 1961 1966 1972 1977

→ Livestock
→ Poultry

World total 1221.6 million head/year

India 15%
U.S.S.R.15%
U.S.A. 9%

Brazil 8%
China 5%
Bangladesh 3%
Mexico 3%
Ethiopia 2%
Colombia 2%
France 2%
Australia 2%
Others 34%

WORLD CATTLE POPULATION — 1981-83

MILK PRODUCTION

India has had in the past few years a very ambitious dairy development programme called Operation Flood. Phase III of this programme — the world's largest and most ambitious milk programme — is being negotiated amidst considerable controversy relating to commodity aid for dairying and the fact that the Operation Flood programme does not centre on cross-breeding strategies. But whatever may be its drawbacks, this programme has very greatly helped the rural poor.

The manifesto — 'EEC milk out of India' — issued by the India Committee on the Netherlands (ICN), reiterates in its fourth demand, that, during the third phase of Operation Flood, a stop should be put to the export of animal feed from India to EEC countries. The present state of undernourishment of Indian cattle makes this a crucial issue.

India in fact has a good number of milch breeds. The best Indian cattle breeds are found in Punjab, Rajasthan, Saurashtra (peninsular Gujarat), parts of Maharashtra, Tamil Nadu, Karnataka and Andhra Pradesh. In the eastern parts of the country and the coastal areas, however, the animals are nondescript and are poor milk yielders. *Sahiwal, Red Singhi* and *Deoni* are some of the outstanding breeds. The *Sahiwal* yields between 2725 and 3175 kgs of milk in a lactation period of 300 to 325 days. The average yield of *Gir,* another breed, is 1675 kgs. The average milk yield of the *Singhi* and *Deoni* varies from 1135 to 1725 kgs. Several general utility breeds are also fairly good milk yielders. *Hariana* (Haryana and Delhi), *Kankrej* and *Tharparker* (Gujarat and Rajasthan), *Ongole* (Andhra Pradesh) and others give about 1500 to 2000 kgs of milk during their lactation periods.

Draught breeds are poor milk yielders, but the bullocks are excellent draught animals. About 42 per cent of the cattle in the country are draught animals. The *Amritmahals* and *Hallikar* (Karnataka) are the best-known draught breeds. Other important breeds are *Malvi* (Madhya Pradesh), *Nagori* (Rajasthan), *Kenketha* and *Ponwar* (Uttar Pradesh), *Bahaur* (Bihar), *Bargur* and *Kangayam* (Tamil Nadu) and *Khillari* (Maharashtra). These breeds are as important as the great milk yielders, for farmers depend on them for their field operations.

MILK PRODUCTION

Table 68

	Number of milk coop. — 1985	Membership in 1985 (in thousand)	Milk production (1984-85) (in thousand tonnes)
States			
Andhra Pradesh	2,518	215	2,500
Assam	124	6	550
Bihar	769	16	2,330
Gujarat	7,973	1,150	3,100
Haryana	2,383	146	2,400
Himachal Pradesh	78	9	400
Jammu & Kashmir	86	4	340
Karnataka	2,289	374	1,900
Kerala	468	68	1,220
Madhya Pradesh	1,685	77	2,780
Maharashtra	1,925	383	2,360
Manipur	—	—	70
Meghalaya	—	—	60
Nagaland	—	—	N.A.
Orissa	285	11	330
Punjab	3,818	184	3,810
Rajasthan	2,060	123	3,500
Sikkim	107	5	20
Tamil Nadu	4,249	572	2,850
Tripura	61	3	20
Uttar Pradesh	2,503	120	7,100
West Bengal	978	49	2,210
Union Territories			
Andaman & Nicobar	20	1	N.A.
Arunachal Pradesh	—	—	N.A.
Chandigarh	—	—	N.A.
Dadra & Nagar Haveli	—	—	N.A.
Delhi	—	—	N.A.
Goa, Daman & Diu	66	6	N.A.
Lakshadweep	—	—	N.A.
Mizoram	—	—	N.A.
Pondicherry	53	9	N.A.
INDIA	34,494	3,531	39,860

To look after animal health, there are some 15,000 veterinary hospitals/dispensaries in the country (1981-82) as against only 904 during the 1930s. In addition, there are 15,500 stockmen centres/veterinary aid centres.

Intensive Cattle Development Projects (ICDP), formulated during the Third Five Year Plan to increase the milk production potential of cows and buffalo, operate in 90 places throughout the country. In addition, there are 625 key village blocks functioning on the same lines. The two projects (ICDP and the village blocks) together account for 15 million breedable bovines. Development work at the Central Cattle Breeding Farms has achieved breakthroughs in milk production. At the end of the Fourth Plan there were six such farms, and during the Fifth Plan six cattle breeding farms were set up at Suratgarh (Rajasthan), Dhamroad (Gujarat), Koraput and Chiplima (Orissa), Avadi (Tamil Nadu) and Andesh Nagar (Uttar Pradesh).

Despite concerted efforts at breeding and dairy development, the country took about 32 years to double its milk production, from 17 million tonnes in 1950-51 to 34.7 million tonnes in 1982-83. In 1984-85, milk production was 39.9 million tonnes. This is targeted to reach 51 million tonnes in 1989-90. While the growth in the first fifteen years after 1951 was almost negligible, milk production has spurted since 1966. In 1950-51 per capita milk production was of the order of 47.1 kgs. This, owing to population increase, has risen only marginally to 48.5 kgs in 1982-83. However, in 1984-85, the per capita production was 54.6 kgs, which is expected to increase to 63 kgs in 1989-90, despite population growth.

During the planning era, in keeping with the rising demand for milk and dairy development programmes, plan outlays and expenditure have been steadily increasing, although expenditure has never exceeded plan outlays. The outlay for the Sixth Plan was Rs 4603 m and it is Rs 7518 m for the Seventh Plan. Almost all the states have shown improvements in milk production in 1984-85 over their 1983-84 performance. The largest producer was Uttar Pradesh (7.1 m tonnes), accounting for 17.8 per cent of total production. The only other state which produced nearly a tenth of the total production was Punjab with 3.8 million tonnes. The output of other major producers was: Rajasthan 3.5 m tonnes, Tamil Nadu 2.9 m tonnes, Madhya Pradesh 2.8 m tonnes, Gujarat 2.6 m tonnes, Andhra Pradesh 2.5 m tonnes, Haryana 2.4 m tonnes, Maharashtra 2.4 m tonnes, Bihar 2.3 m tonnes, West Bengal 2.2 m tonnes and Karnataka 1.9 m tonnes.

The Seventh Plan targets for milk production have been set at a total of 51 million tonnes. The highest state target is that set by Uttar Pradesh, 9.2 million tonnes. The average milk production target for the country is 1.6 million tonnes for each of the states and territories. The states, besides Uttar Pradesh, that have set targets higher than this average are Punjab (4.9 m tonnes), Rajasthan (4.5 m tonnes), Madhya Pradesh (3.5 m tonnes), Andhra Pradesh (3.4 m tonnes), Gujarat and Tamil Nadu (3.3 m tonnes each), Haryana (3.1 m tonnes), Bihar and West Bengal (3 m tonnes each), Maharashtra (2.9 m tonnes) and Karnataka (2.6 m tonnes).

Per capita milk production in 1984-85 was highest in Punjab (213.4 kgs), followed by Haryana (172.7 kgs), Rajasthan (93.6 kgs) and Himachal Pradesh (87 kgs). However, the largest producer, Uttar Pradesh, has a per capita production of only 59.8 kgs. The southern states too have low per capita production: Tamil Nadu 56.1 kgs, Karnataka 47.7 kgs, Kerala 45.4 kgs and Andhra Pradesh 43.9 kgs. The lowest per capita milk production is in Tripura (0.7 kgs) and the all-India average is 54.4 kgs.

India is the third largest producer of milk in the world, its production in 1985 (41 m tonnes) accounting for 8.1 per cent of the total world production of 506.8 m tonnes. The USSR (98.2 m tonnes) and USA (65 m tonnes) are the leaders. But despite substantial milk production in India, availability is poor. However, with increasing dairy development, a national milk grid is emerging, though it is still incomplete. The metropolitan centres now form the focal points with medium cities and towns acting as feeder centres. The milk grid enables Bombay and Maharashtra, Ahmadabad and Gujarat, Delhi, Rajasthan, Haryana and Punjab, Tamil Nadu, West Bengal and parts of Andhra Pradesh, Madhya Pradesh, Karnataka and Orissa to be well-served with milk.

Sources: 1. *Directory & Yearbook 1983*, Times of India.
2. *Basic Statistics Relating to the Indian Economy, Vol.1: All India, August 1986, Vol. 2: States, September 1986* and *World Economy & India's Place in it, October 1986*, Centre for Monitoring Indian Economy.
3. *Seventh Five Year Plan 1985-90, Vol. II*, Government of India Planning Commission.

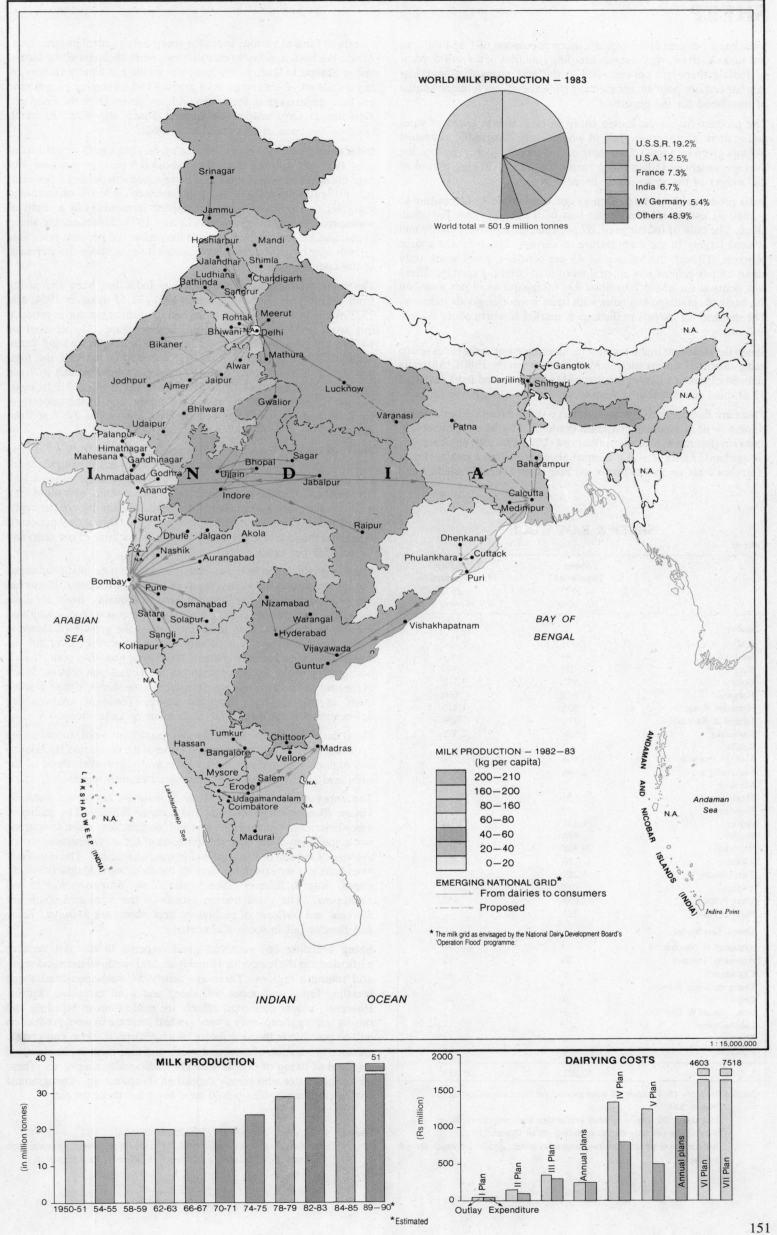

WORLD MILK PRODUCTION — 1983

U.S.S.R. 19.2%
U.S.A. 12.5%
France 7.3%
India 6.7%
W. Germany 5.4%
Others 48.9%

World total = 501.9 million tonnes

MILK PRODUCTION — 1982–83
(kg per capita)

200—210
160—200
80—160
60—80
40—60
20—40
0—20

EMERGING NATIONAL GRID*

→ From dairies to consumers
--→ Proposed

* The milk grid as envisaged by the National Dairy Development Board's 'Operation Flood' programme.

1 : 15,000,000

ARABIAN SEA

BAY OF BENGAL

INDIAN OCEAN

ANDAMAN AND NICOBAR ISLANDS (INDIA)

Andaman Sea

Indira Point

LAKSHADWEEP (INDIA)

Lakshadweep Sea

MILK PRODUCTION

40

30

20

10

0

(in million tonnes)

1950-51 54-55 58-59 62-63 66-67 70-71 74-75 78-79 82-83 84-85 89-90*

51

*Estimated

DAIRYING COSTS

2000

1500

1000

500

0

(Rs million)

I Plan II Plan III Plan Annual plans IV Plan V Plan Annual plans VI Plan VII Plan

4603 7518

Outlay Expenditure

SHEEP

India has 3 per cent of the world's sheep population of 1,134 million, and ranks sixth among the sheep-breeding countries of the world. Most of India's sheep are concentrated in the arid zones that have low rainfall and are poor in agriculture. Here they form a major source of livelihood for the people.

The productivity of the Indian sheep stock is low in terms of wool and mutton. The average yield of wool is only 700 grams, as against 5-7 kgs given by some of the more exotic breeds. As for mutton, the average weight of Indian sheep varies from 25 to 30 kgs, a third of the weight of the more exotic breeds.

India produces about 36 million kgs of wool annually (according to a 1980-81 estimate), the greater part of it coming from Rajasthan alone. The bulk of Indian wool, 57 per cent, is, however, coarse and is used largely in the manufacture of carpets, druggets and coarse blankets. Though the balance of 43 per cent is apparel wool, only about 15 per cent of this apparel wool is of combing quality. There is a demand for about 70 million kgs of apparel wool per year, on the basis of spindlage available with India's woollen goods industry. The demand for carpets in the export market is worth about Rs 800 million.

Sheep development in the country includes breeding indigenous sheep using imported strains such as Merino. In the early 1980s, Australia allowed export of a limited number of Merino rams and India imported 93 of them and distributed them to various breeding farms in India.

There are five centrally sponsored, large sheep-breeding farms in the country — at Mamidipally (Andhra Pradesh), Challekeri (Karnataka), Daksum (Jammu & Kashmir), Bhaisora (Uttar Pradesh) and Fatehpur (Rajasthan). There is also a Central Sheep Breeding Farm at Hissar (Haryana), set up in 1969 to breed and supply rare Corriedale and crossbred rams to various states for sheep development programmes. Action has been initiated to establish two more sheep breeding farms, one at Shivpuri (Madhya Pradesh) and another at Chatra (Bihar). A large-scale sheep shearing, wool grading and marketing programme has been undertaken in Rajasthan with assistance from the Food and Agricultural Organization. A Central Sheep and Wool Research Institute has been in existence since 1962.

India's sheep population has shown a modest but steady upward trend over 1951-66, growing at a rate of about 0.5 per cent per year, but over the next few years (1966-72), it registered a decline of 5 per cent. It is difficult to explain such a decline except in terms of increased slaughter of sheep and a general decline in numbers as a result of widespread drought. Between 1972 and 1977, however, the sheep population has grown once again at the earlier 0.5 per cent rate. This growth may be attributed to increasing efforts at sheep development in the country.

The total production of raw wool in India has been increasing from year to year: 27 million kgs in 1951, 32.55 m kgs in 1961, and 35.2 m kgs in 1972. Sixteen states and four union territories produce this wool — Rajasthan, with the largest share (41 per cent or 14.4 m kgs), Andhra Pradesh (8.2 per cent or 2.88 m kgs) and Tamil Nadu (7.7 per cent or 2.7 m kgs) accounting for half the total production. The other substantial producers are Karnataka (6.7 per cent or 2.4 m kgs) and Uttar Pradesh (6 per cent or 2.1 m kgs). Nearly 72 per cent of the raw wool production in the country is from adult sheep, 11.5 per cent is from lambs and the rest from pulled wool.

The distribution of wool production would seem to truly reflect the distribution of sheep in the country. Rajasthan has the largest stock (24 per cent of India's total), followed by Andhra Pradesh (17 per cent) and Tamil Nadu (13 per cent). Other states with more than a million sheep are Karnataka (11 per cent), Maharashtra (6 per cent), Uttar Pradesh (5 per cent), Bihar (2.8 per cent), Gujarat (3.9 per cent), Himachal Pradesh (2.6 per cent), Jammu & Kashmir (3 per cent) and Orissa (3.5 per cent).

The sheep population in India consists mainly of poor quality animals. But the temperate, hilly areas (such as Jammu & Kashmir, Himachal Pradesh and western Uttar Pradesh) have good quality sheep, for these areas have a favourable climate for raising sheep as well as year-round grazing facilities. In such temperate areas, the growth of fleece is rapid, being a necessity as it serves as a protective cover for the animal. In arid and semi-arid areas, such fleece growth is poor. India's best quality sheep are, thus, found in the Himalayan region. Sheep in the arid and barren hills are considered a menace, for their grazing leads to degradation of vegetal cover. Thus, in areas where deforestation is rampant, it is inadvisable to raise sheep.

From the point of view of quality and quantity of wool, Indian sheep may be classified into four groups: those of the temperate Himalayan and high elevation areas; those of the arid north-west; those of the semi-arid south; and those of the humid eastern region.

The important sheep breeds of the Himalayan region are *Kashmir Valley, Bhadarwah, Bhakarwal* and *Rampur-Bushair*. The quality of sheep deteriorates towards the east and, despite steps taken to upgrade stock, the sheep of the Uttarkhand region of Uttar Pradesh are inferior to those of Jammu & Kashmir and Himachal Pradesh. The important wool breeds of the arid north-west are the *Jaisalmeri, Malpuri, Sonadi, Pugal, Magra, Bikaneri, Shekhawati, Lohi, Marwari, Kutchi* and *Kathiawari*. The good mutton breeds of the semi-arid south are *Deccani* and *Nellore*; of secondary importance are *Mandya, Yalag* and *Bandur* (all in south Karnataka).

Sheep breeding has received great impetus in the last decade, particularly in the temperate Himalayan, arid north-western and semi-arid southern regions. There are nearly 50 state-sponsored sheep breeding farms and about 600 sheep and wool extension centres. However, unless concerted efforts are made both in breeding and raising quality sheep, only a very gradual increase in wool production will be possible in the near future. Such efforts, if taken, can greatly enhance production of meat and wool, and thus raise the income and standard of living of people who are traditionally shepherds. There are certain castes who greatly depend on sheep rearing. A programme with a community bias would be a boon for these people.

SHEEP & RAW WOOL

Table 69

States	Sheep Population — 1977 (in '000)	Total Raw Wool Production — 1972 (Tonne)
Andhra Pradesh	7,064	2,884
Assam	59	—
Bihar	1,150	547
Gujarat	1,592	3,718[1]
Haryana	542	666
Himachal Pradesh	1,055	1,413
Jammu & Kashmir	1,216	954
Karnataka	4,536	2,375
Kerala	3	2
Madhya Pradesh	968	70
Maharashtra	2,636	1,439
Manipur	2	—
Meghalaya	20	—
Nagaland	*	—
Orissa	1,431	4
Punjab	498	729
Rajasthan	9,938	14,418[2]
Sikkim	16	2,696
Tamil Nadu	5,289	—
Tripura	3	2,101
Uttar Pradesh	2,059	452
West Bengal	793	—
Union Territories		
Andaman & Nicobar	*	—
Arunachal Pradesh	20	12
Chandigarh	1	—
Dadra & Nagar Haveli	*	—
Delhi	9	48
Goa, Daman & Diu	1	2
Lakshadweep	*	—
Mizoram	1	—
Pondicherry	5	2
INDIA[3]	40,907	35,169[4]

Note: [1] Includes 1814 tonnes of wool production from migratory sheep.
* Below 500.
[2] Includes 1500 tonnes of wool production from migratory sheep.
[3] Totals may not tally due to rounding off of figures.
[4] Includes wool production from migratory sheep, details for which are not available.

Sources: 1. *Directory & Yearbook 1983*, Times of India.
2. *Statistical Abstract India 1984*, Ministry of Planning. Quoting from Directorate of Marketing & Inspection, Ministry of Agriculture.

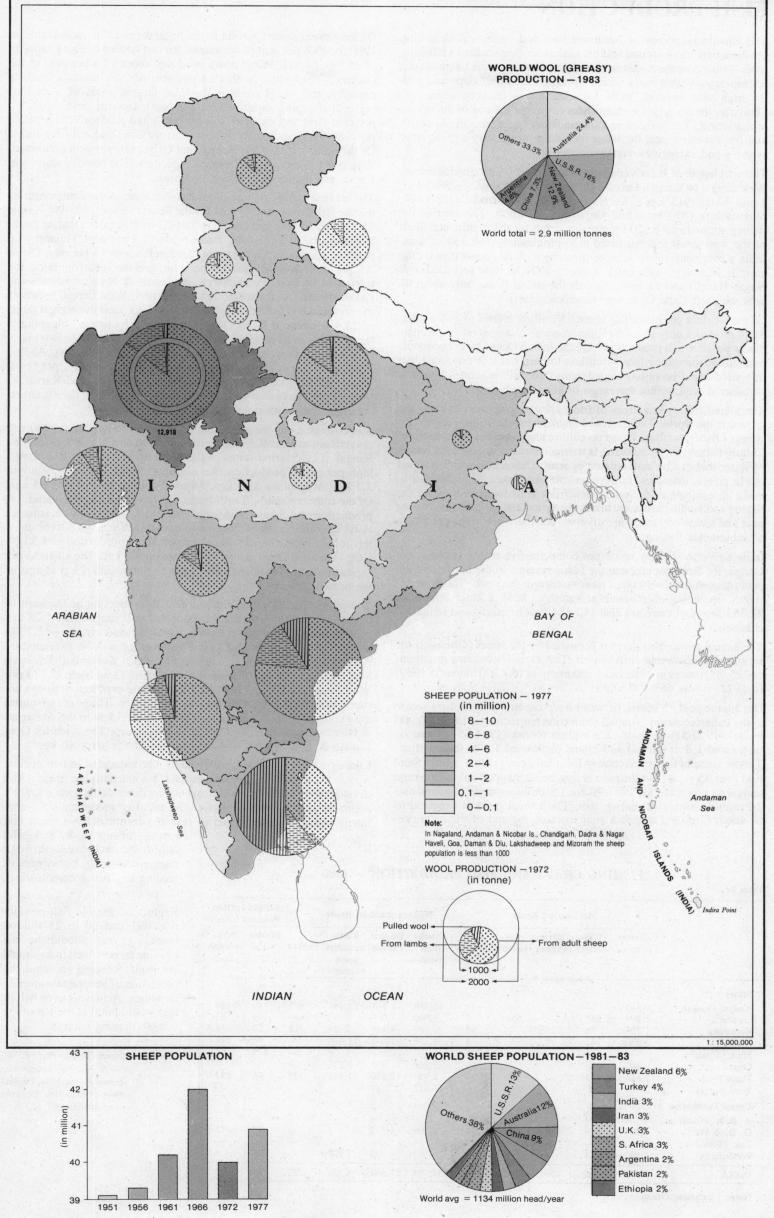

WORLD WOOL (GREASY) PRODUCTION — 1983

Others 33.3%
Australia 24.4%
U.S.S.R. 16%
New Zealand 12.9%
China 1.3%
Argentina 4.8%

World total = 2.9 million tonnes

ARABIAN
SEA

BAY OF
BENGAL

I N D I A

12,918

ANDAMAN AND NICOBAR ISLANDS (INDIA)

Andaman
Sea

Indira Point

LAKSHADWEEP (INDIA)

Lakshadweep
Sea

SHEEP POPULATION — 1977
(in million)

	8 — 10
	6 — 8
	4 — 6
	2 — 4
	1 — 2
	0.1 — 1
	0 — 0.1

Note:
In Nagaland, Andaman & Nicobar Is., Chandigarh, Dadra & Nagar Haveli, Goa, Daman & Diu, Lakshadweep and Mizoram the sheep population is less than 1000

WOOL PRODUCTION —1972
(in tonne)

Pulled wool
From lambs
From adult sheep

1000
2000

INDIAN   OCEAN

1 : 15,000,000

SHEEP POPULATION

(in million)

43
42
41
40
39

1951 1956 1961 1966 1972 1977

WORLD SHEEP POPULATION—1981—83

Others 38%
U.S.S.R.13%
Australia 12%
China 9%

New Zealand 6%
Turkey 4%
India 3%
Iran 3%
U.K. 3%
S. Africa 3%
Argentina 2%
Pakistan 2%
Ethiopia 2%

World avg = 1134 million head/year

FISH PRODUCTION

The aquatic resources of India are vast and varied. With a long coastline extending around half the mainland, besides the rich fishing areas surrounding the Andaman & Nicobar Islands and the Lakshadweep Archipelago, with a fairly wide continental shelf and slope and with the high seas beyond, India has rich marine fishery resources. Similarly, the country's inland fishery resources are one of the richest in the world. The inshore and inland waters comprise the brackish and fresh water types, the latter including the country's great river systems and extensive irrigation network.

The total length of India's coastline is about 5,600 kms divided among West Bengal (400 kms), Orissa (720 kms), Andhra Pradesh (960 kms), Tamil Nadu (912 kms), Kerala (480 kms), Karnataka (288 kms), Maharashtra (840 kms) and Gujarat (1000 kms). The area of the fishing grounds up to 200 metres in depth on the continental shelf of the west coast are estimated at approximately 168,350 sq kms. India's continental shelf is more prominent on this coast than on the east. It is widest (193 kms) at about 20°N latitude (off Dadra & Nagar Haveli) and narrows towards the south, being only about 48 kms wide off Cape Comorin (Kanniyakumari).

Of the world's production of about 75 million tonnes of fish a year, India produces only about 3 million tonnes. A conservative estimate places possible fish production from the Indian Ocean by conventional methods at something like 14 million tonnes, but it is estimated that this yield could be raised considerably: 20 million tonnes a year is considered well within the range of accomplishment.

The inland fishery resources of India are also potentially among the richest in the world. The fresh and brackish inland waters sustain two types of fisheries: those based on culture and those based on capture. Culture fishery, or pisciculture, is normally practised in smaller bodies of water that can be manipulated by man. Capture fishery is carried out in rivers, estuaries, large reservoirs and lakes. It is difficult to make an estimate of the total catch from inland waters because the fishing and landing areas are highly dispersed and isolated, the fishing gear and tackle employed are diverse, and there is a high percentage of subsistence fishing.

India's riverine fishery resources comprise five major systems: the Ganga, the Brahmaputra and the Indus systems in the north, and the peninsular east and west coast river systems in the south. The principal rivers, including their main tributaries, have a total length of 27,359 kms and there are also 112,654 kms of canals and irrigation channels.

The annual marketable surplus furnished by the states is the basis for estimating India's inland fish catch. This varied between a minimum of 215,700 tonnes in 1952 and a maximum of 693,200 tonnes in 1969. In 1972, it was 665,800 tonnes.

During the past 35 years, fisheries have become a significant sector in the Indian economy. Annual production has increased from 750,000 tonnes (1950-51) to nearly 2.9 million tonnes (1984-85) of marine (estimated 1.8 m tonnes) and inland (estimated 1.1 m tonnes) fish. This increase in fish production in India has been gradual, nearly four-fold over 35 years, but the pace is now being stepped up: the average annual targets fixed for 1985-90 are 1.8 million tonnes of inland and 2.2 million tonnes of marine fish. The annual yield is expected to be about 6 million tonnes a year towards the end of the century.

Of the littoral states, Kerala is the biggest producer, accounting in 1980 for 29.6 per cent of the marine harvest (with a freezing capacity of 35.9 per cent). Maharashtra produces about 19.1 per cent of the country's production (with 19.4 per cent of the country's freezing capacity) and Tamil Nadu is the third largest producer with 15.6 per cent (freezing capacity 12.1 per cent). Gujarat producing 13.2 per cent (freezing capacity 6.2 per cent) and Andhra Pradesh 10.2 per cent (freezing capacity 5.8 per cent) are other major fishing states. Of the other states, West Bengal and Orissa (10 per cent combined) have comparatively high freezing capacities. The last two states and Kerala are traditionally fish-eating areas.

The leading marine fishing states have set themselves commensurate marine fish production targets for the Seventh Plan, 1985-90: Kerala 25.7 per cent of the country's target of 2.2 million tonnes marine catch, Maharashtra 20.5 per cent, Tamil Nadu 14.1 per cent, Gujarat 13.6 per cent, Karnataka 11.1 per cent, Andhra Pradesh 9.1 per cent, Orissa 4.5 per cent, West Bengal 3.6 per cent, and the rest from the union territories of Goa, Pondicherry, Andaman & Nicobar Islands and Lakshadweep. As for inland fish production, West Bengal — where the preference is for fresh-water fish — has set itself the highest target of 28.9 per cent of India's target of 1.8 million tonnes, planning to take maximum advantage of its location in an area where two major river systems flow into the sea. Tamil Nadu (13.9 per cent), Andhra Pradesh (11.1 per cent), Bihar (8.6 per cent), Orissa (6.1 per cent), Uttar Pradesh (5.6 per cent), Assam (5.3 per cent) and Karnataka (4.2 per cent) have also set themselves substantial inland fishing targets commensurate with their inland water resources.

It is estimated that the per capita inland fish production in 1990 will range from around 0.33 kgs in Punjab to as much as 8 kgs in West Bengal. The riverine states of the north and north-east will have a high per capita production, for example: Bihar 1.83 kgs, Manipur 7.1 kgs, Meghalaya 3.53 kgs, Assam 3.88 kgs and Sikkim 2.5 kgs. Of the southern states, Tamil Nadu will have a per capita inland fish production of 4.5 kgs and Andhra Pradesh 3.16 kgs while Karnataka (1.68 kgs) and Kerala (1.5 kgs) will have less than 2 kgs. Of the union territories, Pondicherry will have a per capita production of 4.3 kgs, Goa, Daman & Diu 3.6 kgs, and Mizoram 5.7 kgs. The all-India per capita production of inland fish will be an estimated 2.2 kgs as against the present 1.4 kgs.

The corresponding per capita marine fish production on the basis of targets set for 1990 by the littoral states will range from 1.23 kgs in West Bengal to 19 kgs in Kerala, with a national average of 2.7 kgs as against the present 2.1 kgs. This average will be exceeded by Andhra Pradesh (3.2 kgs), Gujarat (7.4 kgs), Karnataka (5.5 kgs), Maharashtra (6.06 kgs), Orissa (3.24 kgs) and Tamil Nadu (5.57 kgs). West Bengal (1.23 kgs) is about the only state expected to have a per capita marine catch less than the national average. The island territories are expected to have a per capita catch of 83.3 kgs in the Andaman & Nicobar Islands and 180 kgs in Lakshadweep. The catch for Goa, Daman & Diu will be 46.4 kgs and for Pondicherry 40 kgs.

Like agriculture, fish culture could also be expanded in two distinct ways: by extending operations and by intensifying them. The development of inland fisheries and culture depends upon the country's ability to raise seed fish output. The trend is undoubtedly towards intensive fish culture. Fisheries research institutes have found that catches of up to 8000 kgs/ha could be achieved through intensive stocking, higher rates of feeding and better combinations of species.

Experts on marine fish production feel that 20 to 25 million tonnes a year should be the ultimate target which India should set itself, keeping in mind the protection of marine resources in its waters. Achieving even half of that would put it at the top of the world's fishing nations.

Sources: 1. *Seventh Five Year Plan 1985-90*, Vol. II, Planning Commission.
2. *Annual Report 1984*, Department of Fisheries, Government of India.

FISHING CRAFT AND FISH PRODUCTION — 1980

Table 70

	Mechanized Boats					Non-mechanized Boats				Total production (in thousand tonnes)	
	Traw-lers	Gill-netters	Purse seiners	Dal-netters	Others	Plank-built boats	Dugout canoes	Kattu-marams (cata-marans)	Others	Mecha-nized	Non-mecha-nized
States											
Andhra Pradesh	580	—	—	—	—	11,359	1,781	22,198	675	19	110
Gujarat	1,209	1,547	—	650	7	3,040	1,080	—	—	104	63
Karnataka	1,553	28	325	—	98	1,747	4,454	23	718	29	58.5
Kerala	2,630	362	37	—	9	4,376	10,415	11,480	—	96	279
Mah. (Total)	—	—	—	—	4,718	—	—	—	—	179	63
Orissa	350	119	—	—	—	3,262	186	6,276	4	—	—
Tamil Nadu	2,614	143	—	—	—	8,957	2,210	31,851	325	46	152
West Bengal	—	740	—	—	—	3,972	89	—	—	6[1]	27[1]
Union Territories											
A. & N. (Total)	—	—	—	—	10	—	—	—	—	—	—
G. D. & D.	494	274	66	—	74	1,108	1,397	8	—	17	12
Lak. (Total)	—	—	—	—	213	—	—	—	—	—	—
Pondicherry	160	3	—	—	—	83	72	1,595	—	—	8
INDIA	9,590	3,016	428	650	5,129	37,904	21,684	73,431	1,722	496	772.5

Note: [1] Including Orissa.

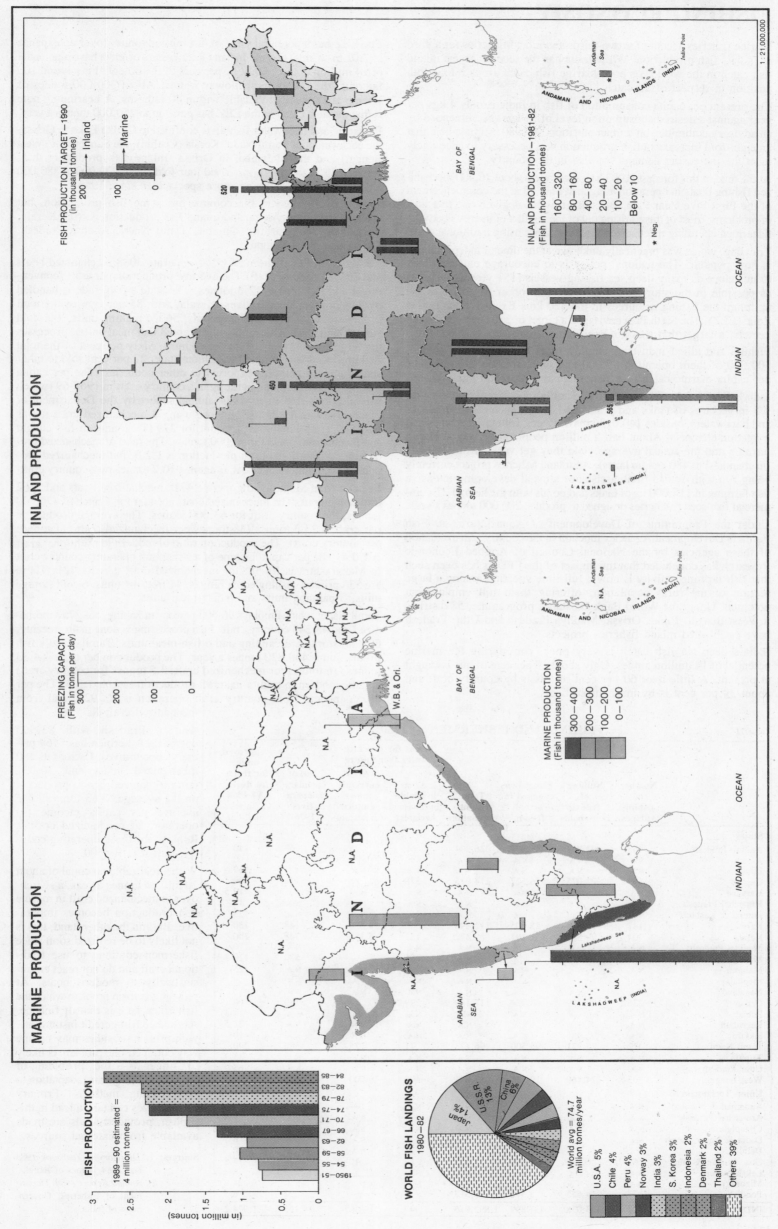

FISHING ECONOMY

Marine fisheries account for two-thirds and inland fisheries for a third of India's fish production. While India has the second largest inland fish catch in the world, its poor marine fish catch drops it to eighth position in respect of all catch.

The present per capita consumption of fish in India is only 4 kgs per year against a desired consumption level of 31 kgs recommended by an advisory committee on human nutrition. A 1980 estimate held that an eight-fold increase in fish production was necessary to adequately meet the increasing demand for fish in the country.

At the time of this forecast (1980-81), there were about 16,400 mechanized fishing boats in operation as against only 13 at the commencement of the First Five Year Plan. But this growth has given rise to a host of problems, most of them arising out of the conflict of interests between fishermen operating mechanized craft and those using traditional craft.

Trawling, which was practically unknown at the time of independence, is now popular. The national policy is to encourage entrepreneurs to introduce as many deep sea fishing vessels (151 vessels in 1981) as possible in the shortest possible time, in order to utilize to the optimum the fishing resources in the 320 kms Exclusive Economic Zone (EEZ). The Sixth Plan envisaged a target of 350 deep-sea fishing vessels, a target that was not achieved.

Fishing and allied industries employ more than a million people, 500,000 of them operating 85,000 and more traditional craft. The per capita marine catch of a fisherman is 2000-2500 kgs a year.

India's inland fisheries extend over 3 million hectares of reservoirs, 1.5 m hectares of tanks and ponds and 1.4 m hectares of estuaries and backwaters, besides 140,013 kms of rivers, tributaries, canals and irrigation channels. About half a million people are engaged in fish farming and the annual average yield they get varies from 140 kgs (in channels) to 480 kgs (in lakes). An inland fisheries project currently being run with World Bank assistance aims at developing intensive fish farming in 117,000 ha of tanks and ponds with the help of 27 commercial fish-seed hatcheries designed to produce 200,000 tonnes a year.

Under the Programme of Development of Aquaculture, 50 Fish Farmers Development Agencies function in the country. An evaluation of these agencies by the National Council of Applied Economic Research has concluded that the impact of the FFDAs has been such that fish farming activity is now a full-time vocation among a large section of the rural population, offering them full employment potential. Under the World Bank assistance programme, 58 districts in West Bengal, Bihar, Orissa, Madhya Pradesh and Uttar Pradesh have established inland fisheries projects.

India's deep-sea fish catch is very poor, considering the marine potential of 14 million tonnes. Only about 10 per cent of this is caught at present. A little over 60 per cent of this is by country craft and about 39 per cent is by mechanized boats.

The EEZ has a potential yield of 4.5 million tonnes (over an expanse of 2.02 m sq kms) and the Indian Ocean waters of India have a potential of 14 million tonnes. Only 11 per cent of the potential is beyond 100 fathoms, the rest is in shallower waters. About 300,000 tonnes of shrimp are annually available within 40 fathoms. A heartening note is that the tuna catch in the EEZ has gone up to 250,000 tonnes a year.

Estuarine and backwater fishing is confined to Chilka Lake in Orissa, the backwaters of Tamil Nadu, Kerala (Cochin), the Sunderbans (West Bengal) and the Mahanadi in Orissa. Inland fish production has, between 1951 and 1980, increased four-fold, from 220,000 to 880,000 tonnes, the increase being more spectacular since 1966.

Almost every state in the country has some fish production, but 90 per cent of the marine and inland fish production is from Kerala, Karnataka, Maharashtra, Gujarat, Tamil Nadu, Andhra Pradesh, Orissa and West Bengal.

Kerala's 131,101 fishermen (1980) operate 3038 mechanized boats and 26,271 country craft. The average production per craft (conventional and modern) is 12.8 tonnes. The state's fish trade is handled by 748 primary cooperatives. Kerala, with 85 per cent of the total processing units (117 freezing units, 56 ice making units, 141 cold storage units, 42 fish canning units, and 3 fish meal units), processes the largest amount of fish in the country. Sixty per cent of the total fish produced is consumed within the state, 22 per cent sold to other states, and 18 per cent exported to other countries. The per capita income of a fisherman operating in a country craft in 1968-69 (which was also reported in its 1984 annual report by the Department of Fisheries) was Rs 216. For those using a mechanized craft it was Rs 75/-. Traditional craft account for 279,000 tonnes of fish caught and mechanized boats for 96,000 tonnes. The ratio of mechanized production to traditional craft production is 1:2.9, but mechanized production is 32 tonnes per boat as against 10.2 tonnes by a country craft.

Fishermen in Maharashtra operate 4718 mechanized boats and 5662 traditional craft. The mechanized boats account for a catch of 170,000 tonnes, the country craft for 63,000 tonnes. The average production per craft is 23.3 tonnes (38 tonnes per mechanized craft, 11 tonnes per country craft). The production ratio of modern to traditional craft is 1:0.4. The per capita income of a fisherman operating country craft in Maharashtra is Rs 58; for mechanized craft it is Rs 121. There are 530 primary fishing cooperatives, 41 freezing units, 6 cold storage units, 3 canning units and 6 fish-meal units.

Tamil Nadu, which employs 96,500 persons in fishing, has 2757 mechanized and 43,343 country craft. Fish processing is done in 46 freezing, 60 cold storage, 3 canning and 6 fish-meal units. Tamil Nadu's fish production is 400,820 tonnes a year. The production per craft is 4.33 tonnes (16 tonnes for mechanized craft, 3.5 tonnes for country craft) and the modern to traditional craft production ratio is 1:3.3. The per capita income from country craft operation is Rs 924, and from mechanized craft Rs 558.

Andhra Pradesh with 83,903 operating fishermen has 1169 primary cooperatives. Each of its 580 mechanized boats and 36,031 non-mechanized boats, produces, on the average, 3.53 tonnes. The average per capita income for operators of mechanized craft is Rs 479; for those operating country craft it is Rs 600.

With a realizable potential of about 20 million tonnes a year, a switch-over to mechanized craft in marine fish production becomes imperative. But, on the other hand, this is not likely to be realized soon since fishermen continue to use traditional craft and do not react enthusiastically to modern boats. At present, the main problem which the fishermen face is that of finance. Assistance from credit institutions, with proper subsidies may be one way out. Yet, one problem is likely to linger — the problem of attitudinal change and adaptation to new fishing methods. Primary cooperatives can take a lead in this project, provided funds are made available from external sources.

Sources: 1. *Directory & Yearbook 1983 and 1984*, Times of India.
2. *Annual Report 1984*, Department of Fisheries, Government of India.

Table 71

FISHING AND FISHERMEN — 1980

| | Number of fishing villages | Number of fishing households | Fishermen engaged in actual fishing | Total fishing population | Areas for brackish-water fish culture | | No. of primary cooperatives —1976-77 | Approved outlay for fisheries during VI Plan (in Rs. million) |
					Area available (in thousand hectares)	Area currently under culture (in hectare)		
States								
Andhra Pradesh	453	72,862	83,903	326,304	64	25	1,169	140
Assam	N.A.	N.A.	N.A.	N.A.	N.A.	N.A.	N.A.	60
Bihar	N.A.	N.A.	N.A.	N.A.	N.A.	N.A.	N.A.	69.6
Gujarat	179	23,075	36,527	114,015	378	88	138	200
Haryana	N.A.	N.A.	N.A.	N.A.	N.A.	N.A.	N.A.	19.5
Himachal Pradesh	N.A.	N.A.	N.A.	N.A.	N.A.	N.A.	N.A.	18
Jammu & Kashmir	N.A.	N.A.	N.A.	N.A.	N.A.	N.A.	N.A.	23
Karnataka	147	15,638	25,005	112,893	8	4,800	148	130.3
Kerala	304	99,894	131,101	639,872	122	5,117	748	200
Madhya Pradesh	N.A.	N.A.	N.A.	N.A.	N.A.	N.A.	N.A.	70
Maharashtra	N.A.	N.A.	N.A.	N.A.	81	—	530	120.7
Manipur	N.A.	N.A.	N.A.	N.A.	N.A.	N.A.	N.A.	25
Meghalaya	N.A.	N.A.	N.A.	N.A.	N.A.	N.A.	N.A.	9
Nagaland	N.A.	N.A.	N.A.	N.A.	N.A.	N.A.	N.A.	7
Orissa	236	20,329	30,724	117,144	299	—	199	100
Punjab	N.A.	N.A.	N.A.	N.A.	N.A.	N.A.	N.A.	17.5
Rajasthan	N.A.	N.A.	N.A.	N.A.	N.A.	N.A.	N.A.	22.5
Sikkim	N.A.	N.A.	N.A.	N.A.	N.A.	N.A.	N.A.	10
Tamil Nadu	422	75,721	96,500	96,173	71	25	500	240
Tripura	N.A.	N.A.	N.A.	N.A.	N.A.	N.A.	N.A.	33.3
Uttar Pradesh	N.A.	N.A.	N.A.	N.A.	N.A.	N.A.	N.A.	65.9
West Bengal	303	14,169	19,756	83,561	405	9,000	672	270
Union Territories								
Andaman & Nicobar	N.A.	N.A.	N.A.	N.A.	N.A.	—	10	20
Arunachal Pradesh	N.A.	N.A.	N.A.	N.A.	N.A.	N.A.	N.A.	7.8
Chandigarh	N.A.	N.A.	N.A.	N.A.	N.A.	N.A.	N.A.	1
Dadra & Nagar Haveli	N.A.	N.A.	N.A.	N.A.	N.A.	N.A.	N.A.	6
Delhi	N.A.	N.A.	N.A.	N.A.	N.A.	N.A.	N.A.	40
Goa, Daman & Diu	61	6,725	8,871	39.912	19	10	14	17.5
Lakshadweep	N.A.	N.A.	N.A.	N.A.	N.A.	—	2	6
Mizoram	N.A.	N.A.	N.A.	N.A.	N.A.	N.A.	N.A.	24.1
Pondicherry	27	4,625	5,512	25,312	N.A.	—	32	0.5
INDIA	2,132	333,038	437,899	1,892,916	1,447	19,065	4,153	1,974.2

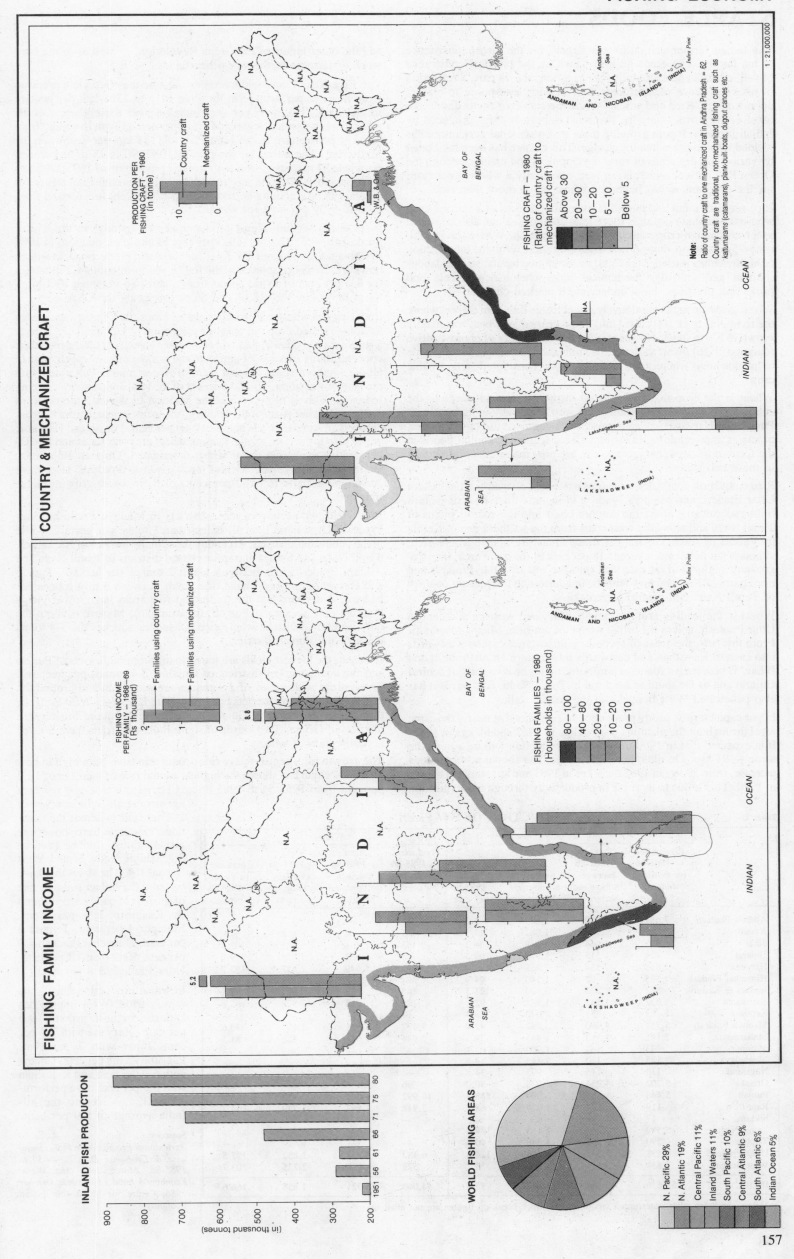

COUNTRY & MECHANIZED CRAFT

PRODUCTION PER
FISHING CRAFT – 1980
(in tonne)

Country craft
Mechanized craft

FISHING CRAFT – 1980
(Ratio of country craft to
mechanized craft)

Above 30
20–30
10–20
5–10
Below 5

Note:

Ratio of country craft to one mechanized craft in Andhra Pradesh = 62.
Country craft are tradional, non-mechanized fishing craft such as
kattumarans (catamarans), plank-built boats, dugout canoes etc.

BAY OF BENGAL

ANDAMAN AND NICOBAR ISLANDS (INDIA)
Andaman Sea

ARABIAN SEA

Lakshadweep Sea

LAKSHADWEEP (INDIA)

INDIAN OCEAN

I N D I A

1 : 21,000,000

FISHING FAMILY INCOME

FISHING INCOME
PER FAMILY – 1968–69
(Rs thousand)

Families using country craft
Families using mechanized craft

FISHING FAMILIES – 1980
(Households in thousand)

80–100
40–80
20–40
10–20
0–10

BAY OF BENGAL

ANDAMAN AND NICOBAR ISLANDS (INDIA)
Andaman Sea

ARABIAN SEA

Lakshadweep Sea

LAKSHADWEEP (INDIA)

INDIAN OCEAN

I N D I A

INLAND FISH PRODUCTION

(in thousand tonnes)

900
800
700
600
500
400
300
200

1951 56 61 66 71 75 80

WORLD FISHING AREAS

N. Pacific 29%
N. Atlantic 19%
Central Pacific 11%
Inland Waters 11%
South Pacific 10%
Central Atlantic 9%
South Atlantic 6%
Indian Ocean 5%

157

STAPLE FOODS

The Indian farmer still cultivates largely for the direct subsistence of his family. The crops he grows, within the physical constraints of soil, climate and availability of irrigation, are, in part, a reflection of his subsistence requirements and, in part, a response to market opportunities. Rice and wheat — one the staple of South India, the other of the North — are the two most important crops produced by Indian farmers. It is in growing these that traditional agriculture has yielded most to pressures for change. This change has been facilitated by various programmes of rural development and mainly through the Green Revolution, which is, in fact, essentially a wheat revolution, for the programme has benefited wheat much more than rice.

Rice requires a hot and humid climate, and thrives as a rainfed crop where the annual rainfall is more than 200 cms. It can also be raised as a transplanted crop with the help of irrigation. Where the rainfall is 65-75 cms, short-duration varieties are raised. With the introduction of high-yielding varieties (HYVs), the economic inputs, such as labour (human and animal), mechanical aids, chemical nutrients and protection practices, have undergone a marked change.

Wheat grows in areas of relatively dry climate. It seldom thrives where the rainfall is over 100 cms. Until a few years ago, wheat was mainly grown in North India. However, with the availability of photosensitive (duration-fixed) dwarf varieties, its cultivation has spread far and wide to include areas where, until recently, it was considered an unsuitable crop.

Wheat is the crop that has ushered in the greatest advances in the country's foodgrain production. This crop, especially in its heartland, Punjab and Haryana, has seen greater changes in the use of techniques, practices and technologies, and, as a consequence, capital goods in the form of tractors, harvesters, etc, are now major economic inputs in these two states.

Rice cultivation dominates in regions extending eastwards from eastern Uttar Pradesh and Madhya Pradesh to include the whole of eastern India, southwards along the Andhra coast into the coastal plains of Tamil Nadu and along the west coast from the southern end of Kerala to Valsad in south Gujarat. If those areas are added where rice, although not dominant, exceeds 15 per cent of the sown area, the rice-growing region can be extended into the rest of Uttar Pradesh and even parts of Haryana and Punjab, in addition to the plateau country of Tamil Nadu and central Karnataka.

Wheat is the leading crop in Punjab, Haryana, western and central Uttar Pradesh, and in the broad wedge of northern Madhya Pradesh. From this belt, the areas of secondary importance extend westwards into eastern Rajasthan and eastwards into eastern Uttar Pradesh and Bihar. Wheat is also found as a minor crop in the lava soils of central Gujarat and as far south as Madurai in Tamil Nadu. Eastwards it has also penetrated West Bengal and the Assam valley.

In per capita terms, paddy production has remained at nearly the same level throughout the planning era, 93 kgs in 1952 and 98 kgs in 1983. It is expected that in 1990 the level of production will be almost the same — 91 kgs. On the other hand, wheat has shown a tremendous increase, from 20 kgs in 1952 to 59 kgs in 1983 and an estimated 69 kgs in 1990. This is due to increase in productivity through modernization and the other inputs of the Green Revolution, as well as to the hard work of farmers in the wheat heartland.

The Five-Year Plan yield averages indicate that rice yield went up from 817 kgs per hectare in the First Plan to 1361 kgs per hectare in the Sixth Plan, a 65 per cent increase over these years. On the other hand, wheat has registered a 140 per cent growth in yields, from 738 kgs per hectare in the First Plan to 1771 kgs per hectare in the Sixth Plan. The target set for rice for 1989-90 is 75 m tonnes and for wheat 57 m tonnes, with a per hectare production of 1695-1705 kgs for rice and 2000-2035 kgs for wheat. The Seventh Plan target for increase in productivity is from 13.8 to 17 per cent for rice and from 8.05 to 9.99 per cent for wheat.

India is the second-largest producer of rice (paddy) in the world, producing 90.5 m tonnes (19.5 per cent of world production) in 1985 to China's 170 m tonnes (36.7 per cent of world production). In wheat, however, India occupies only the fourth position, with 44.2 m tonnes (or 8.6 per cent of world production) against 86 m tonnes by China, 83 m tonnes by the USSR and 66 m tonnes by the USA.

Both rice and wheat are mainly irrigated crops, though they are grown as non-irrigated crops in parts of the country. Distinctively, rice is a summer crop (kharif) and wheat a winter crop (rabi) and hence their concentrations in areas of summer and winter rains respectively. In all the southern states (except Kerala) rice is mainly cultivated with the help of irrigation, 60-100 per cent of the total area under rice being irrigated. This is more or less the position in the extreme north as well. In all other parts of the country, the gross irrigated area to gross cropped area under rice is between 0-60 per cent. Punjab and Haryana, which also grow rice, are the leading wheat growing states and 80-100 per cent of their area under wheat is irrigated. Gujarat, Rajasthan, Uttar Pradesh, Bihar, Orissa and Andhra Pradesh also have considerable parts (60-100 per cent) of their wheat-growing areas irrigated.

Rice yield per hectare in most districts in Madhya Pradesh, Uttar Pradesh, Orissa and Bengal, is less than 1 tonne per hectare, while in the peninsular districts, excepting those in the interior of Karnataka, Tamil Nadu and Kerala and some coastal districts in Tamil Nadu and Andhra Pradesh, it is between 1 and 2 tonnes per hectare. Punjab and Haryana, however, have rice yields between 3 to 4 tonnes per hectare. It appears that the traditional rice areas have lower yields than the new acceptors (Punjab, for example). Mysore is about the only traditional district where rice production touches the 3-4 tonne mark of the Punjab districts.

In wheat, the highest yields are found in the western districts of Punjab and the southernmost districts of Gujarat (3-4 tonnes per hectare). Those areas with yields of 2-3 tonnes are mostly in and around the heartland and in the districts bordering Bangladesh (in West Bengal), the northern districts of Gujarat, and Mayurbhanj and Sundargarh districts of Orissa. Two-thirds of south India's districts have little or no cultivation of wheat.

The growth rates registered in rice production have been high in new areas — Punjab has shown the highest annual rate of increase of 12.2 per cent from 1953-54 to 1985-86 and Haryana a rate of 9 per cent against the all-India average of 2.8 per cent, whereas the traditional rice areas have shown low growth rates: Andhra Pradesh 3.1; Tamil Nadu 3; and West Bengal 1.4. The states that have bettered the Indian average rate of growth in paddy are: Jammu & Kashmir 3.9 per cent, Manipur 4 per cent, Tripura 3 per cent, Himachal Pradesh 4.6 per cent, Rajasthan 4.2 per cent, Uttar Pradesh 3.8 per cent.

In wheat production during the same period, West Bengal has had the highest growth, 9.3 per cent. Haryana with 7.4 per cent, Bihar with 7 per cent, Punjab with 6.8 per cent, Gujarat with 6.9 per cent and Rajasthan with 6.5 per cent have performed well, considering the all-India average of 5.9 per cent.

Sources:
Agricultural Production in India, State-wise & Crop-wise Data: 1949-50 to 1985-86, March 1987 and *World Economy & India's Place in it*, October 1986, Centre for Monitoring Indian Economy.

Table 72

PRODUCTION OF STAPLES

	Rice				Wheat			
	Production — 1985-86 (in thousand tonnes)	Area — 1981-86 average (in thousand hectares)	Yield — 1981-86 average (in kilogram/hectare)	Increase in yield — 1953-86 (in per cent)	Production — 1985-86 (in thousand tonnes)	Area — 1981-86 average (in thousand hectares)	Yield — 1981-86 average (in kilogram/hectare)	Increase in yield — 1953-86 (in per cent)
Andhra Pradesh	7,659	3,710	1,680	64.2	—	—	—	—
Assam	2,847	2,332	1,082	12.5	155	N.A.	N.A.	N.A.
Bihar	6,075	5,064	938	56.3	3,143	1,791	1,535	192.4
Gujarat	454	515	1,270	108.2	783	632	2,055	329.9
Haryana	1,636	534	2,565	171.4	5,257	1,649	2,617	193.7
Himachal Pradesh	125	92	1,140	94.9	304	364	948	96.3
Jammu & Kashmir	587	271	2,120	121.3	168	213	887	7.1
Karnataka	1,872	1,132	1,943	92	123	318	572	163.6
Kerala	1,163	751	1,670	79.4	—	—	—	—
Madhya Pradesh	5,759	4,900	882	42.5	4,127	3,531	1,107	130.6
Maharashtra	2,182	1,517	1,445	79.1	644	1,041	852	81.3
Manipur	333	164	1,700	63.1	—	—	—	—
Meghalaya	125	110	1,145	14.5	—	—	—	—
Nagaland	130	115	992	34.8	—	—	—	—
Orissa	5,202	4,234	1,008	80.0	96	N.A.	N.A.	N.A.
Punjab	5,448	1,483	3,090	288.7	10,992	3,061	3,155	196.5
Rajasthan	119	144	1,076	68.7	3,918	1,897	1,779	144.7
Sikkim	17	N.A.	N.A.	N.A.	—	—	—	—
Tamil Nadu	5,599	2,267	2,165	108.6	—	—	—	—
Tripura	390	285	1,340	40.8	—	—	—	—
Uttar Pradesh	8,198	5,343	1,260	124.6	6,482	8,237	1,852	127.5
West Bengal	7,834	5,136	1,350	33.7	578	290	2,235	203.7
INDIA[1]	58,636	40,363	1,402	83.5	44,069	23,321	1,858	166.6

Note: [1] Includes other states/union territories for which break-up figures are not available.

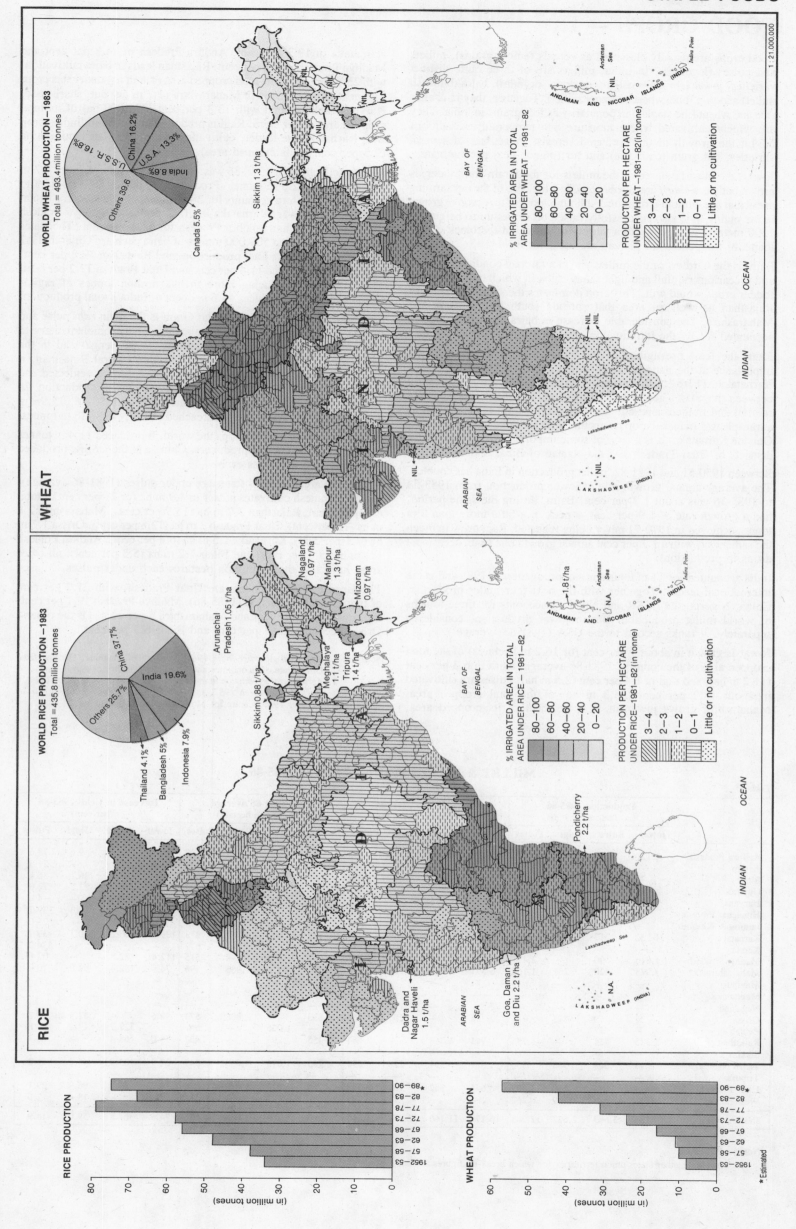

WHEAT

RICE

WORLD WHEAT PRODUCTION — 1983
Total = 493.4 million tonnes

China 16.2%
U.S.A. 13.3%
India 8.6%
Canada 5.5%
U.S.S.R. 16.8%
Others 39.6

WORLD RICE PRODUCTION — 1983
Total = 435.8 million tonnes

China 37.7%
India 19.6%
Indonesia 7.9%
Bangladesh 5%
Thailand 4.1%
Others 25.7%

% IRRIGATED AREA IN TOTAL
AREA UNDER WHEAT — 1981–82
80–100
60–80
40–60
20–40
0–20

PRODUCTION PER HECTARE
UNDER WHEAT –1981–82 (in tonne)
3–4
2–3
1–2
0–1
Little or no cultivation

% IRRIGATED AREA IN TOTAL
AREA UNDER RICE – 1981–82
80–100
60–80
40–60
20–40
0–20

PRODUCTION PER HECTARE
UNDER RICE – 1981–82 (in tonne)
3–4
2–3
1–2
0–1
Little or no cultivation

Sikkim 1.3 t/ha

Arunachal Pradesh 1.05 t/ha
Nagaland 0.97 t/ha
Manipur 1.3 t/ha
Mizoram 0.97 t/ha
Meghalaya 1.1 t/ha
Tripura 1.4 t/ha
Sikkim 0.88 t/ha

Pondicherry 2.2 t/ha

Dadra and Nagar Haveli 1.5 t/ha
Goa, Daman and Diu 2.2 t/ha

RICE PRODUCTION
WHEAT PRODUCTION
(in million tonnes)

1952-53
57-58
62-63
67-68
72-73
77-78
82-83
89-90 *

* Estimated

1 : 21,000,000

ANDAMAN AND NICOBAR ISLANDS (INDIA)
Andaman Sea
Indira Point
NIL

BAY OF BENGAL
ARABIAN SEA
INDIAN OCEAN
LAKSHADWEEP (INDIA)
Lakshadweep Sea
N.A.
1.8 t/ha

159

FOOD CROPS — I

Food crops are usually classified as cereals (edible grains), millets and pulses (legumes). In India, millets are of three quite distinct varieties, jowar (*sorghum vulgare* or grain sorghum), bajra (bullrush millet) and ragi (buckwheat or finger millet). Together, they rank after rice and wheat (the staples) in popularity as foodgrains in India. They are widely cultivated where moisture and temperature conditions inhibit the growth of the preferred cereals. Pulses are of several varieties, but gram (chickpea) and tur (pigeon-pea) predominate.

Jowar, the most favoured of the millets for foodgrain and fodder, is grown both as a *kharif* and a *rabi* crop. It is the crop of the dry farming areas that are without irrigation. Although most of this crop is grown in the plains, it is also successfully raised on slopes up to heights of 1200 metres. It is grown in a variety of soils, but the black clayey loams of the peninsular plateau are ideal for it.

Bajra is the hardiest of the millets and can tolerate conditions of low and precarious rainfall and light sandy soils, in which other food and fodder crops would wilt. This crop dominates the semi-arid areas of Rajasthan as well an area that spreads south into Gujarat and Maharashtra. The considerable increase in bajra area since 1969 is accounted for by the introduction of HYVs.

Ragi, the least prestigious of the major foodgrains and of little importance at the national level, is the dominant crop in southern Karnataka. There, in the extensive thin-soiled interfluve (areas in between rivers) of a gently undulating country, it thrives on moderate rainfall and limited tank and canal irrigation. Ragi is also found as a transplanted irrigated crop competing with *rabi* rice in certain areas. Outside Karnataka, it is a crop of some importance only in the hilly districts of Uttar Pradesh and the plateau of Bihar and Orissa.

Between 1950-51 and 1981-82, millet production in India has doubled. The average annual growth rate of jowar production from 1953-54 to 1985-86 was about 1.7 per cent. Bajra, during the same period, had a growth rate of 1.9 per cent, a peak of a little more than 8 m tonnes achieved in 1970-71 never being repeated. Ragi grew in much the same way, with a 1.7 per cent annual growth rate (with occasional drops in production).

India accounted for 11 m tonnes millet production in 1985 and is the international leader. Together with the next four major producers, China, Nigeria, the USSR and Niger, it accounts for three-fourths of world millet production. But in jowar production, considered separately, it ranks second to the USA (1981-83 average.).

Jowar is grown in about 9 per cent (or 16.2 m hectares) of the total cropped area of the country, 1981-86 average; bajra in 6.3 per cent (11.2 m ha); and ragi in 1.4 per cent (2.5 m ha). Pulses are cultivated in about 13.2 per cent (23.3 m ha) of the total cropped area. Maharashtra grows jowar in 32.3 per cent of its cropped area,

Karnataka in 19.3 per cent, Andhra Pradesh in 14.8 per cent and Madhya Pradesh in 9.6 per cent. Rajasthan leads in bajra cultivation, with 25.6 per cent of its total cropped area (4.8 m ha) under this crop. The next largest area is in Maharashtra (1.7 m ha) but, share-wise, Haryana ranks second with 13.3 per cent (775,000 ha) of its total cropped area under bajra. Ragi is predominantly a South Indian crop with Karnataka the leader, cultivating it in more than 1 million ha (9.5 per cent of its cropped area).

Millet production for 1985-86 was: jowar 10.1 m tonnes, bajra 3.7 m tonnes and ragi 2.5 m tonnes. Production of pulses was about 13 m tonnes. Maharashtra accounts for 38.8 per cent of the country's total production of jowar, Karnataka for 13.2 per cent, Madhya Pradesh for 17.9 per cent and Andhra Pradesh for 11.1 per cent. The states producing more than 500,000 tonnes of bajra each are Gujarat, 17.2 per cent of country's total production, and Rajasthan 19.8 per cent. Maharashtra produces 11.4 per cent and Uttar Pradesh 17.2 per cent. The only state producing more than a million tonnes of ragi is Karnataka, which produces 44.6 per cent of India's total production.

Pulses are seasonal crops in India. Gram is the main *rabi* pulse and tur the main *kharif* pulse. *Rabi* pulses are often complementary to rice cultivation in Uttar Pradesh, Bihar, West Bengal and in the Mahanadi delta in Orissa. In a belt through central Rajasthan, it alternates with *kharif* bajra. *Kharif* pulses tend to be neglected and are low-yielding. Pulses show great fluctuations in production. In 1950-51, production was 8.4 m tonnes and in 1985-86, 12.96 m tonnes, but peak production was reached in 1975-76 — 13 m tonnes.

In pulses production, India tops the world. It produced 11.4 m tonnes in 1982. The next major producers, China and the USSR, produced only about 6.5 m tonnes each.

Substantial areas in most states are under pulses (1981-86 average). Madhya Pradesh cultivates pulses in 4.9 m ha (or 21 per cent of total cropped area), Rajasthan 3.7 m ha (15.7 per cent), Maharashtra 2.8 m ha (12 per cent), Uttar Pradesh 3 m ha (12.8 per cent), Orissa 1.7 m ha (7.1 per cent), Karnataka 1.5 m ha (6.4 per cent), Andhra Pradesh 1.3 m ha (5.6 per cent) and Bihar 1.2 m ha (5.3 per cent); all other areas have less than a million hectares each under pulses.

The largest pulse producers are Uttar Pradesh, with 21.7 per cent of India's total production (1985-86), Madhya Pradesh 19.1 per cent, Rajasthan 13.5 per cent and Maharashtra 9 per cent. Of the others, Orissa accounts for 7 per cent and Bihar 6.3 per cent.

Sources: 1. *Statistical Abstract India 1984*, Ministry of Planning. Quoting from Directorate of Economics and Statistics, Ministry of Agriculture.
2. *Agricultural Production in India State-wise & Crop-wise Data: 1949-50 to 1985-86, March 1987*, and *World Economy & India's Place in it, October 1986*, Centre for Monitoring Indian Economy.

MILLET & PULSES PRODUCTION — 1985-86

Table 73

	Production 1985-86 (in thousand tonnes)				Area 1981-86 average (in thousand hectares)				Yield 1981-86 average (in kilogram/hectare)				Increase in yield[1] 1953-86 (in per cent)			
	Jowar	Bajra	Ragi	Pulses	Jowar	Bajra	Ragi	Pulses	Jowar	Bajra	Ragi	Pulses	Jowar	Bajra	Ragi	Pulses
Andhra Pradesh	1,121	219	226	607	1,937	457	240	1,363	652	650	1,020	400	28.9	50.5	9.6	78.6
Assam	—	—	—	67	—	—	—	128	—	—	—	438	—	—	—	21.3
Bihar	—	7	95	822	—	12	143	1,230	—	583	644	630	—	23.3	36.1	36.4
Gujarat	354	635	33	344	928	1,390	46	757	535	925	957	602	173.0	148.7	22.2	72.0
Haryana	28	317	—	669	131	775	—	777	230	606	—	525	-20.0	53.8	—	-4.0
Himachal Pradesh	—	—	5	11	—	—	8	46	—	—	750	217	—	—	27.3	-39.7
Jammu & Kashmir	—	6	—	28	—	18	—	48	—	388	—	625	—	-23.3	—	25.0
Karnataka	1,332	217	1,125	433	2,165	527	1,068	1,488	763	470	1,160	373	109.0	106.2	110.1	39.7
Kerala	—	—	—	20	—	171	—	31	—	—	—	677	—	—	—	334.0
Madhya Pradesh	1,816	90	—	2,477	2,087	—	—	4,887	820	615	—	515	112.4	82.5	—	33.8
Maharashtra	3,923	420	261	1,165	6,591	1,682	226	2,796	696	370	1,066	396	74.9	182.4	62.0	10.0
Manipur	—	—	—	—	—	—	—	—	—	—	—	—	—	—	—	—
Meghalaya	—	—	—	—	—	—	—	—	—	—	—	—	—	—	—	—
Nagaland	—	—	—	—	—	—	—	—	—	—	—	—	—	—	—	—
Orissa	35	10	212	911	35	—	286	1,654	830	—	808	573	98.6	—	52.5	3.2
Punjab	—	27	—	202	—	47	—	231	—	1,065	—	667	—	327.7	—	2.8
Rajasthan	375	731	—	1,756	945	4,768	—	3,664	455	293	—	432	75.0	68.4	—	76.3
Sikkim	—	—	—	—	—	—	—	—	—	—	—	—	—	—	—	—
Tamil Nadu	693	368	358	356	689	331	228	694	795	1,015	1,368	347	21.9	103.0	45.2	76.2
Tripura	—	—	—	—	—	—	—	—	—	—	—	—	—	—	—	—
Uttar Pradesh	434	633	170	2,808	645	962	159	2,973	755	810	1,030	860	19.7	39.7	96.9	29.1
West Bengal	—	—	9	262	—	—	15	398	—	—	600	585	—	—	13.4	-2.0
INDIA[2]	10,123	3,683	2,522	12,963	16,174	11,160	2,454	23,266	695	505	1,065	525	74.6	67.8	59.0	16.9

Note: [1] Estimated.
[2] Including other states/union territories for which break-up figures are not available.

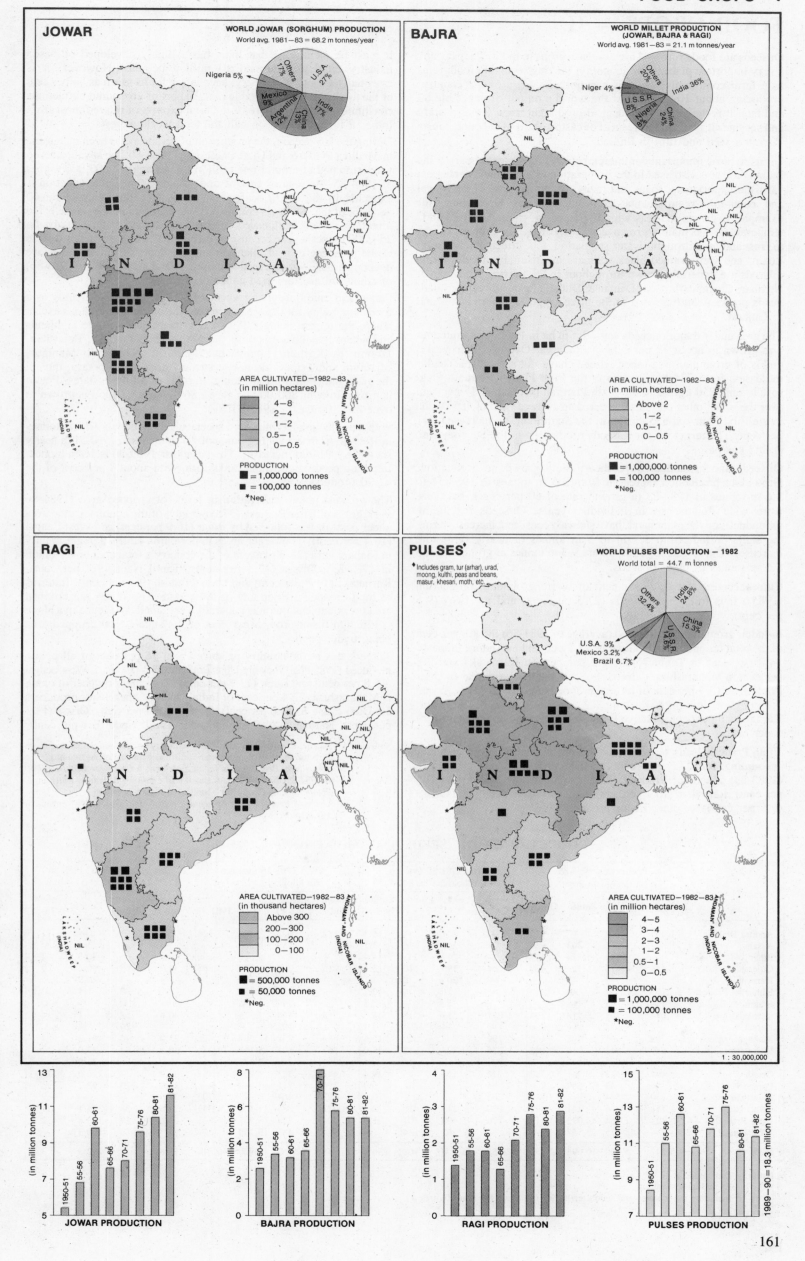

JOWAR

WORLD JOWAR (SORGHUM) PRODUCTION
World avg. 1981—83 = 68.2 m tonnes/year

Nigeria 5%
Others 17%
U.S.A. 27%
Mexico 9%
Argentina 12%
China 13%
India 17%

I N D I A

NIL

LAKSHADWEEP (INDIA)
NIL

ANDAMAN AND NICOBAR ISLANDS (INDIA)
NIL

AREA CULTIVATED—1982—83
(in million hectares)
4—8
2—4
1—2
0.5—1
0—0.5

PRODUCTION
■ = 1,000,000 tonnes
▪ = 100,000 tonnes
*Neg.

BAJRA

**WORLD MILLET PRODUCTION
(JOWAR, BAJRA & RAGI)**
World avg. 1981—83 = 21.1 m tonnes/year

Niger 4%
Others 20%
India 36%
U.S.S.R. 8%
Nigeria 8%
China 24%

I N D I A

NIL

LAKSHADWEEP (INDIA)
NIL

ANDAMAN AND NICOBAR ISLANDS (INDIA)
NIL

AREA CULTIVATED—1982—83
(in million hectares)
Above 2
1—2
0.5—1
0—0.5

PRODUCTION
■ = 1,000,000 tonnes
▪ = 100,000 tonnes
*Neg.

RAGI

NIL
NIL
NIL
NIL

I N D I A

LAKSHADWEEP (INDIA)
NIL

ANDAMAN AND NICOBAR ISLANDS (INDIA)
NIL

AREA CULTIVATED—1982—83
(in thousand hectares)
Above 300
200—300
100—200
0—100

PRODUCTION
■ = 500,000 tonnes
▪ = 50,000 tonnes
*Neg.

PULSES*

* Includes gram, tur (arhar), urad, moong, kulthi, peas and beans, masur, khesari, moth, etc.

WORLD PULSES PRODUCTION — 1982
World total = 44.7 m tonnes

Others 32.4%
India 24.8%
China 15.3%
U.S.S.R. 14.6%
U.S.A. 3%
Mexico 3.2%
Brazil 6.7%

I N D I A

LAKSHADWEEP (INDIA)
NIL

ANDAMAN AND NICOBAR ISLANDS (INDIA)
*

AREA CULTIVATED—1982—83
(in million hectares)
4—5
3—4
2—3
1—2
0.5—1
0—0.5

PRODUCTION
■ = 1,000,000 tonnes
▪ = 100,000 tonnes
*Neg.

1 : 30,000,000

JOWAR PRODUCTION
(in million tonnes)
13 / 11 / 9 / 7 / 5
1950-51, 55-56, 60-61, 65-66, 70-71, 75-76, 80-81, 81-82

BAJRA PRODUCTION
(in million tonnes)
8 / 6 / 4 / 2
1950-51, 55-56, 60-61, 65-66, 70-71, 75-76, 80-81, 81-82

RAGI PRODUCTION
(in million tonnes)
4 / 3 / 2 / 1 / 0
1950-51, 55-56, 60-61, 65-66, 70-71, 75-76, 80-81, 81-82

PULSES PRODUCTION
(in million tonnes)
15 / 13 / 11 / 9 / 7
1950-51, 55-56, 60-61, 65-66, 70-71, 75-76, 80-81, 81-82
1989—90 = 18.3 million tonnes

FOOD CROPS — II

Oilseeds are food crops and are also cash crops in India. They occupy a very important place in the economy, ranking second to foodgrains as a farm commodity. India is a major oilseeds-growing country, producing about 10 per cent of the world's output of vegetable oil and fats. It is the largest producer of groundnut, rape seed, mustard and sesame. It is second in respect of castor, third in coconut, fourth in cotton seed and fifth in linseed.

Oil from these important commercial crops is an essential part of the Indian diet. It is also used in the preparation of medicines, perfumes, varnishes, lubricants, soaps and many other products. The physical and economic requirements for oilseeds vary. Groundnut, for example, is grown in areas with 50-75 cms of rainfall and 20-30°C temperature. It is mainly grown as a *kharif* crop. The crop succumbs to frost and long drought; water stagnation, too, affects it adversely. Loams and well-drained black soils are most suitable for it. While oilseeds are generally grown throughout India, barring Andaman & Nicobar, Chandigarh, Goa, Daman & Diu and Lakshadweep, groundnut is predominantly grown in the three bigger southern states and in Gujarat.

The area under major oilseeds was 19.1 m ha in 1982-83. Groundnut was grown in about 38 per cent of this area. Over the Plan years, yields of groundnut and other oilseeds have not shown spectacular increases: from 724 kgs/hectare in the First Plan, groundnut yield rose to only 850 kgs/ha in the Sixth Plan, an increase of 17 per cent over the years; other oilseeds registered an average yield of 465 kgs/ha in the First Plan and 655 kgs/ha in the Sixth Plan, an increase of 41 per cent. The target set for oilseeds production in the Seventh Plan is 887 kgs/ha.

Oilseeds and groundnut have shown slow growth in production. Groundnut production increased from 3.5 m tonnes in 1950-51 to 5.6 m tonnes in 1982-83 (a growth rate of 60 per cent), but there were wide fluctuations in individual years. Oilseeds (excluding groundnut), on the other hand, have shown a consistent increase during the same period, from 1.7 m to 5 m tonnes (a nearly three-fold increase). The targets for 1989-90 are 9.3 m tonnes of groundnut and 8.6 m tonnes of other oilseeds.

China accounts for 27.7 per cent of world groundnut production (1985). It is followed by India (27.3 per cent) and the USA (9.5 per cent).

In India, groundnut is cultivated (average of 1981-86) in 18.8 per cent of the total cropped area in Gujarat, 12.2 per cent in Andhra Pradesh, 14.8 per cent in Tamil Nadu, 7.2 per cent in Karnataka and 3.8 per cent in Maharashtra. Oilseeds are cultivated in vast areas of Uttar Pradesh — 15.7 per cent of its total cropped area (1982-83) — Assam 7.1 per cent, Jammu & Kashmir 5.3 per cent, Orissa 6.4 per cent, Karnataka 5 per cent, Madhya Pradesh 8.5 per cent and Maharashtra 6 per cent.

Uttar Pradesh leads the states in the production of other oilseeds, accounting for 33.3 per cent of the country's total production (1982-83). Madhya Pradesh (15.8 per cent), Maharashtra (10.5 per cent), Rajasthan (10.1 per cent), Gujarat (9.5 per cent) and Orissa (6.5 per cent) are other major producers.

It must be mentioned that India has a highly developed oil-based industry employing around 10 million persons. However, it is essentially a food-oil industry, accounting for as much as 83 per cent of the total supply of vegetable oils. Export of groundnut extractions constitutes the major share of the total exports of oilseed meal from India; it forms more than half the total world exports.

During the last decade, the production of soyabean has been increasing in Madhya Pradesh and Uttar Pradesh. In 1985-86, the soyabean crop was estimated at around one million tonnes against 800,000 tonnes of 1984-85. As regards sesame and sunflower, a slight reduction is expected in the *kharif* season. At present (1986), only about 60 per cent of the country's cottonseed is crushed and the balance is fed direct to cattle. Cottonseed oil production has stagnated around 250,000 tonnes a year because of difficulties in exporting this oil due to the presence in it of aflatoxin and the fact that Indian prices are not comparable with international ones. About 1.1 m tonnes (1985-86) of edible oil are imported annually at present.

Potato and onion are major vegetable crops, cultivated mainly as a food commodity for the urban market. The onion is ubiquitous except where the terrain is high. Potato is grown mainly in the highlands and where the climate is mild. Kerala, Andhra Pradesh, Rajasthan, Jammu & Kashmir, Arunachal Pradesh, Nagaland, Manipur, Mizoram and Tripura do not report any area under potato, though the requisite conditions are found in parts of all these states. Potato is grown in over 800,000 ha (1981-86 average). Onion is grown in over 236,100 ha (in 1982-83).

Potato production has increased spectacularly, touching 10.7 m tonnes in 1985-86 from a small beginning of 1.7 m tonnes in 1950-51 (a nearly six and a half fold increase). The target for 1989-90 is 16 m tonnes. In potato production, India ranks fifth, with about 4 per cent of the world production.

The greatest potato cultivation in terms of cropped area (1981-86 average) is in Uttar Pradesh (37.6 per cent of the country's total area under potato). It is followed by Bihar (16.9 per cent) and West Bengal (16.7 per cent). The larger areas under onion cultivation are mainly in Maharashtra (23.3 per cent of the country's total area under onion in 1982-83), Orissa (12.5 per cent), Tamil Nadu (9.4 per cent), Karnataka (14.8 per cent) and Uttar Pradesh (9.3 per cent). In terms of total cropped area in any of these states, however, the cropped area for potato and onion is generally very small. In Meghalaya alone is it a significant crop, 8 per cent of the state's total cropped area being under potato.

Three states in the country produce about 76 per cent of all potato produced (1985-86). They are Uttar Pradesh (38.1 per cent), West Bengal (25.8 per cent) and Bihar (12.3 per cent). The chief producer of onion is Maharashtra (22.4 per cent of India's total). Other major producers are Uttar Pradesh (14.6 per cent), Gujarat (16.7 per cent), Orissa (13.3 per cent), Karnataka (6.9 per cent) and Tamil Nadu (8.1 per cent).

Sources: 1. *Statistical Abstract India 1984*, Ministry of Planning. Quoting from Directorate of Economics and Statistics, Ministry of Agriculture.
2. *Agricultural Production in India, State-wise & Crop-wise Data: 1949-50 to 1984-85, February 1986* and *World Economy & India's Place in it, October 1986*, Centre for Monitoring Indian Economy.
3. *Area and Production of Principal Crops in India 1981-84*, Ministry of Agriculture.

Table 74

PRODUCTION OF OIL SEEDS AND FOOD CROPS — 1985-86

	Production 1985-86 (in thousand tonnes)				Area 1981-86 average (in thousand hectares)				Yield 1981-86 average (in kilogram/hectare)				Increase in yield[1] 1953-86 (in per cent)			
	Ground-nut	Oilseeds[2] 1982-83	Potato	Onion	Ground-nut	Oilseeds[2] 1982-83	Potato	Onion 1982-83	Ground-nut	Oilseeds[2] 1982-83	Potato	Onion 1982-83	Ground-nut	Oilseeds[2] 1970-83	Potato	Onion 1978-83
Andhra Pradesh	1,325	112	—	126	1,592	521	—	16.3	862	602	—	8,926	−2.5	1	—	−11
Assam	—	112	303	N.A.	—	244	46	2.1	—	59	6,500	N.A.	—	N.A.	23.4	N.A.
Bihar	—	106	1,314	127	—	219	135	14.4	—	487	9,520	6,806	—	6	−10.9	2
Gujarat	448	473	153	478	2,054	195	10	13.1	715	700	26,200	25,908	34.9	369	6.2	20
Haryana	—	142	143	N.A.	—	205	10	1.1	—	695	15,500	N.A.	—	3	40.9	N.A.
Himachal Pradesh	—	4	46	N.A.	—	16	14	0.7	—	214	3,285	N.A.	—	−8	1.1	N.A.
Jammu & Kashmir	—	67	—	N.A.	—	52	—	0.5	—	1,295	—	N.A.	—	N.A.	—	N.A.
Karnataka	1,032	208	118	197	811	557	16	35.0	953	544	7,375	6,240	40.4	58	94.1	4
Kerala	—	4	—	N.A.	—	14	—	0.4	—	502	—	N.A.	—	−24	—	N.A.
Madhya Pradesh	168	681	298	143	299	1,849	29	12.6	645	388	11,205	12,476	25.0	76	91.3	7
Maharashtra	579	541	60	643	775	1,225	12	55.0	867	603	5,167	13,356	50.5	84	−23.0	17
Meghalaya	—	—	149	—	—	—	18	—	—	—	7,823	—	—	—	33.0	—
Orissa	494	323	341	382	278	559	16	29.4	1,442	793	8,000	6,847	102.8	17	220.0	15
Punjab	44	43	669	N.A.	64	86	32	0.9	890	745	18,938	N.A.	31.9	−10	93.2	N.A.
Rajasthan	150	514	—	48	205	1,179	—	10.1	693	464	—	3,703	38.0	−7	—	−9
Tamil Nadu	1,151	74	141	232	1,020	219	10	22.3	1,022	834	11,100	8,341	−0.6	−35	74.1	−21
Uttar Pradesh	121	1,464	4,073	419	226	3,900	301	21.9	704	393	15,777	16,521	−36.5	−17	113.8	32
West Bengal	—	175	2,757	—	—	362	134	—	—	485	19,790	—	—	N.A.	100.3	—
INDIA[3]	5,547	5,000	10,696	2,870	7,358	11,709	801	236.1	858	552	13,800	10,714	15.6	12	92.3	3

Notes: [1] Estimated.
[2] Excluding groundnut.
[3] Including other states and union territories for which break-up figures are not available.

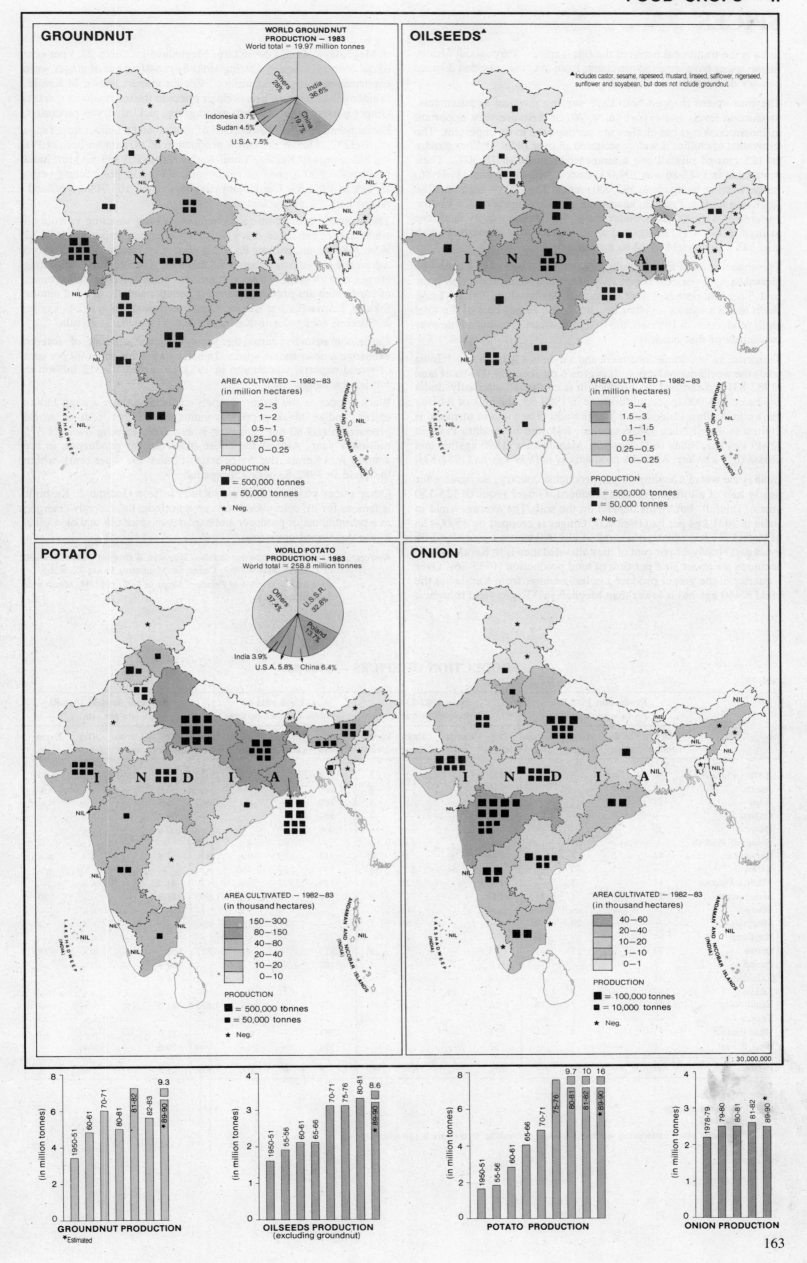

GROUNDNUT

WORLD GROUNDNUT PRODUCTION — 1983
World total = 19.97 million tonnes

- India 36.6%
- China 19.7%
- U.S.A. 7.5%
- Sudan 4.5%
- Indonesia 3.7%
- Others 28%

AREA CULTIVATED — 1982–83
(in million hectares)
- 2–3
- 1–2
- 0.5–1
- 0.25–0.5
- 0–0.25

PRODUCTION
- ■ = 500,000 tonnes
- ▪ = 50,000 tonnes
- ★ = Neg.

OILSEEDS▲

▲ Includes castor, sesame, rapeseed, mustard, linseed, safflower, nigerseed, sunflower and soyabean, but does not include groundnut.

AREA CULTIVATED — 1982–83
(in million hectares)
- 3–4
- 1.5–3
- 1–1.5
- 0.5–1
- 0.25–0.5
- 0–0.25

PRODUCTION
- ■ = 500,000 tonnes
- ▪ = 50,000 tonnes
- ★ = Neg.

POTATO

WORLD POTATO PRODUCTION — 1983
World total = 258.8 million tonnes

- U.S.S.R. 32.8%
- Poland 13.7%
- China 6.4%
- U.S.A. 5.8%
- India 3.9%
- Others 37.4%

AREA CULTIVATED — 1982–83
(in thousand hectares)
- 150–300
- 80–150
- 40–80
- 20–40
- 10–20
- 0–10

PRODUCTION
- ■ = 500,000 tonnes
- ▪ = 50,000 tonnes
- ★ = Neg.

ONION

AREA CULTIVATED — 1982–83
(in thousand hectares)
- 40–60
- 20–40
- 10–20
- 1–10
- 0–1

PRODUCTION
- ■ = 100,000 tonnes
- ▪ = 10,000 tonnes
- ★ = Neg.

1 : 30,000,000

GROUNDNUT PRODUCTION *Estimated — (in million tonnes); 1950-51, 60-61, 70-71, 80-81, 81-82, 82-83, *89-90 9.3

OILSEEDS PRODUCTION (excluding groundnut) — (in million tonnes); 1950-51, 55-56, 60-61, 65-66, 70-71, 75-76, 80-81, *89-90 8.6

POTATO PRODUCTION — (in million tonnes); 1950-51, 55-56, 60-61, 65-66, 70-71, 75-76 9.7, 80-81 10, 81-82, *89-90 16

ONION PRODUCTION — (in million tonnes); 1978-79, 79-80, 80-81, 81-82, *89-90

163

SPICES

India is the traditional home of the finest spices. Throughout history Indian spices have attracted merchants from over the seas; that demand has not decreased.

The four spices mapped here have varying physical requirements, production levels and export value. All of them are very important in Indian cooking, but chillies are perhaps the most important. The cultivation of chillies is widely scattered all over India. Chillies require 60-125 cms of rainfall and a temperature range of 10-30°C. Their production in 1985-86 was 709,000 tonnes, with the current (1981-86) annual average being about 592,000 tonnes. The yield is highest, 1280 kgs/ha, in Andhra Pradesh, against the national average of 725 kgs/ha. An area of 816,000 ha is under chilli. Chilli productivity is almost as high in Bihar (1180 kgs/ha) as it is in Andhra Pradesh. Punjab also has a high yield of 1140 kgs/ha.

The annual growth rate of chilli production between 1953 and 1986 increased by 2 per cent per year, while the area of production grew by 1.3 per cent per year. Andhra·Pradesh, Maharashtra, Orissa, Tamil Nadu and Karnataka together accounted for 79 per cent of the total chilli production in 1985-86, the first three states growing a little over two-thirds of that quantity.

Turmeric, an important condiment and a dye, is a spice in which India leads the world in production. It is grown on about 86,000 ha of land (1982-83) and most of the production is consumed internally. India produced 244,000 tonnes of turmeric in 1984-85, the bulk of it being grown in Andhra Pradesh and Tamil Nadu. The yield in turmeric is highest in Tamil Nadu (3626 kgs/ha), followed by Andhra Pradesh (2949 kgs/ha), Bihar 1667 (kgs/ha), Maharashtra (1590 kgs/ha) and Orissa (877 kgs/ha). All-India productivity is 1939 kgs/ha (1982-83).

India is the world's leading ginger-producing country, accounting for nearly half of all the dry ginger produced. Ginger requires 125-130 cms of rainfall, but yields depend on the soil. The average yield in India is 2031 kgs per ha (1982-83). Ginger is cropped on 45,000 ha of land (1982-83) which yields around 113,000 tonnes a year (1981-86 average). Nearly 27 per cent of the cultivated area is in Kerala, which accounts for about 30.8 per cent of total production (1985-86). Over a quarter of the ginger produced in India comes from Kerala but the yield (2460 kgs/ha) is lower than Meghalaya. Ginger yield is highest in Meghalaya, 4138 kgs/hectare. Meghalaya produces 22.3 per cent of the country's ginger. During 1980-81, 6840 tonnes of ginger were exported, earning Rs 37 million. Cochin ginger, grown in Kerala, is among the finest and gets premier prices in the international market. Ginger is exported in different ways: green, pickled, dry or processed.

Cardamom requires 150-600 cms of rainfall and a temperature range of 10-32°C. Almost the entire production of cardamom is raised in the hill ranges of Kerala, Tamil Nadu, Karnataka and Sikkim. India produces 8,800 tonnes of cardamom, the yield per hectare being 85 kgs in 1982-83. Cardamom is cultivated on 103,300 ha of land. About half the production is exported.

The aromatic cardamom seed contains a strong smelling volatile oil whose rate of extraction is 2-8 per cent. Cardamom exports in 1981-82 were 2484 tonnes, earning Rs 312 million. Cardamom export has seen a dramatic growth since then, with the West Asian countries importing 80 per cent of India's total cardamom exports. About 8800 tonnes of cardamom are produced annually, almost equally divided among Sikkim, Karnataka and Kerala. Cardamom production is 241 kgs/ha in Sikkim, 60 kgs/ha in Karnataka and 57 kgs/ha in Kerala.

Cardamom usually earns the second highest amount of foreign exchange among Indian spices. Indian exports are about 90 per cent of world exports. Cardamom in 1981-82 earned Rs 312 million in foreign exchange.

Black pepper is the largest foreign exchange earner among Indian spices. Indian black pepper accounts for 20 per cent of world production and 80 per cent of it is exported, earning over Rs 375 million a year. Almost the entire black pepper production in the country is in Kerala, the 'Spice State of India' (97.2 per cent), where the yield in 1982-83 was 258 kgs/ha.

Other spices produced in India include saffron (Jammu & Kashmir is famous for it), celery seed and cummin. India has recently emerged as a potential major producer and exporter of spice oils and oleoresins to the developed countries.

Sources: 1. *Agricultural Production in India, State-wise & Crop-wise Data: 1949-50 to 1985-86, March 1987*, Centre for Monitoring Indian Economy.
2. *Area and Production of Principal Crops in India 1981-84*, Ministry of Agriculture.

PRODUCTION OF SPICES — 1982-83

Table 75

	Production 1982-83 (in thousand tonnes)				Area 1982-83 (in thousand hectares)				Yield 1982-83 (in kilogram/hectare)				Increase in yield 1970-83 (in per cent)			
	Dry Chillies[1]	Cardamom	Dry Ginger[1]	Turmeric	Dry Chillies[2]	Cardamom	Dry Ginger[2]	Turmeric	Dry Chillies[2]	Cardamom	Dry Ginger	Turmeric	Dry Chillies[3]	Cardamom	Dry Ginger	Turmeric
Andhra Pradesh	309	—	6	80	171	—	2	22	1,280	—	N.A.	2,949	96.9	—	N.A.	24
Assam	6	—	—	5	10	—	—	8	600	—	N.A.	—	54.6	—	—	N.A.
Bihar	11	—	1	7	11	—	1	4	1,180	—	N.A.	1,667	75.6	—	N.A.	N.A.
Gujarat	9	—	—	—	14	—	—	—	857	—	—	—	59.3	—	—	—
Haryana	12	—	—	—	9	—	—	—	1,000	—	—	—	200.0	—	—	—
Himachal Pradesh	—	—	1	—	—	—	N.A.	—	—	—	N.A.	—	—	—	N.A.	—
Karnataka	41	2	4	9	146	30	2	2	315	60	N.A.	N.A.	-7.4	N.A.	N.A.	N.A.
Kerala	—	3	40	6	—	54	12	3	—	57	2,460	N.A.	—	159	52	N.A.
Madhya Pradesh	11	—	3	—	52	—	2	—	289	—	N.A.	—	-41.3	—	N.A.	—
Maharashtra	72	—	1	13	143	—	1	8	504	—	N.A.	1,590	18.9	—	N.A.	20
Manipur	—	—	N.A.	—	—	—	1	—	—	—	N.A.	—	—	—	N.A.	—
Meghalaya	—	—	29	2	—	—	6	1	—	—	4,138	N.A.	—	—	N.A.	N.A.
Nagaland	—	—	—	—	—	—	1	—	—	—	N.A.	—	—	—	N.A.	—
Orissa	77	—	15	41	75	—	5	25	825	—	N.A.	877	106.8	—	N.A.	-28
Punjab	8	—	—	—	7	—	—	—	1,140	—	—	—	62.9	—	—	—
Rajasthan	31	—	—	—	39	—	—	—	667	—	—	—	14.6	—	—	—
Sikkim	—	4	9	—	—	15	3	—	—	241	N.A.	—	—	N.A.	N.A.	—
Tamil Nadu	58	N.A.	2	70	68	5	N.A.	10	720	N.A.	N.A.	3,626	-41.5	N.A.	N.A.	-4
Tripura	—	—	1	—	—	—	1	—	—	—	N.A.	—	—	—	N.A.	—
Uttar Pradesh	23	—	4	—	24	—	1	—	792	—	N.A.	—	32.0	—	N.A.	—
West Bengal	31	—	7	—	31	—	3	—	774	—	N.A.	—	-50.6	—	N.A.	—
INDIA[4]	709	9	130	244	816	103	45	86	725	85	2,031	1,939	23.9	107	48	4

Note: [1] 1985-86
[2] 1981-86 average
[3] 1953-86
[4] Includes other states/union territories for which break-up figures are not available.

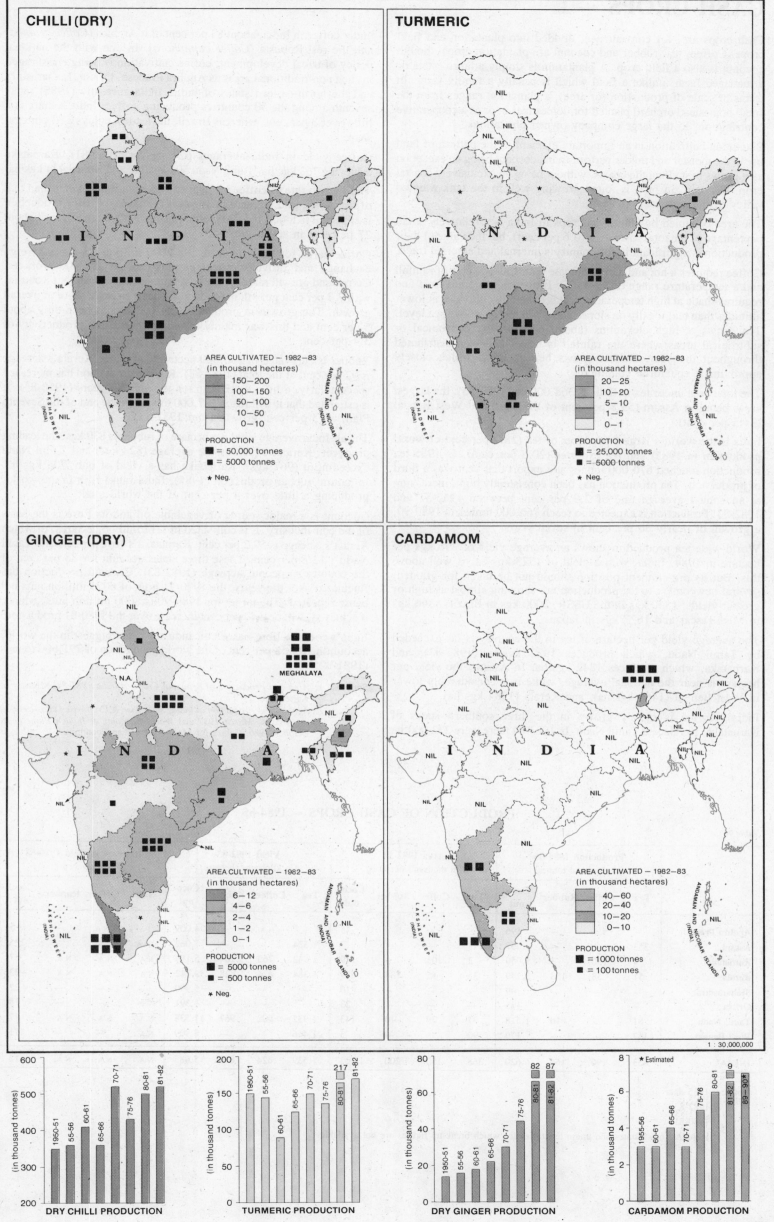

CHILLI (DRY)

I N D I A

AREA CULTIVATED – 1982–83
(in thousand hectares)
150—200
100—150
50—100
10—50
0—10

PRODUCTION
■ = 50,000 tonnes
▪ = 5000 tonnes
★ Neg.

NIL

LAKSHADWEEP
(INDIA)

ANDAMAN AND NICOBAR ISLANDS
(INDIA)
NIL

TURMERIC

I N D I A

NIL
NIL
NIL
NIL
NIL
NIL
NIL
NIL

AREA CULTIVATED – 1982–83
(in thousand hectares)
20—25
10—20
5—10
1—5
0—1

PRODUCTION
■ = 25,000 tonnes
▪ = 5000 tonnes
★ Neg.

LAKSHADWEEP
(INDIA)
NIL

ANDAMAN AND NICOBAR ISLANDS
(INDIA)
NIL

GINGER (DRY)

I N D I A

NIL
NIL
NIL
N.A.
NIL
NIL

MEGHALAYA

NIL

AREA CULTIVATED – 1982–83
(in thousand hectares)
6—12
4—6
2—4
1—2
0—1

PRODUCTION
■ = 5000 tonnes
▪ = 500 tonnes
★ Neg.

LAKSHADWEEP
(INDIA)
NIL

ANDAMAN AND NICOBAR ISLANDS
(INDIA)
NIL

CARDAMOM

I N D I A

NIL
NIL
NIL
NIL
NIL
NIL
NIL
NIL
NIL
NIL
NIL
NIL

AREA CULTIVATED – 1982–83
(in thousand hectares)
40—60
20—40
10—20
0—10

PRODUCTION
■ = 1000 tonnes
▪ = 100 tonnes

LAKSHADWEEP
(INDIA)

ANDAMAN AND NICOBAR ISLANDS
(INDIA)
NIL

1 : 30,000,000

DRY CHILLI PRODUCTION
(in thousand tonnes)
1950-51, 55-56, 60-61, 65-66, 70-71, 75-76, 80-81, 81-82

TURMERIC PRODUCTION
(in thousand tonnes)
1950-51, 55-56, 60-61, 65-66, 70-71, 75-76, 80-81 (217), 81-82

DRY GINGER PRODUCTION
(in thousand tonnes)
1950-51, 55-56, 60-61, 65-66, 70-71, 75-76, 80-81 (82), 81-82 (87)

CARDAMOM PRODUCTION
(in thousand tonnes)
★ Estimated
1955-56, 60-61, 65-66, 70-71, 75-76, 80-81, 81-82 (9), 89—90 ★

165

CASH CROPS — I

Cash crops are, for convenience, divided into plantation and field crops. Coffee, tea, rubber and coconut are plantation crops, though coconut is also a field crop. A plantation is simply a large estate or a managed farm, unlike a field which is usually a family farm. In terms of scale of production (or area), a plantation ranges from the small homestead orchard planted for domestic use, or as a cooperative contribution, to the large company-owned plantation.

Plantation cultivation is an important element in the pattern of land use in the moister and milder parts of India, coconut being an exception as it grows best along the coasts, without being an exclusively coastal crop. Coconut, in fact, is found growing well in the tank-watered valley of the Karnataka plateau.

The area occupied by plantation crops in India is a relatively small percentage of the total sown area (0.6 per cent), but the value of their production is significant in the country's internal and external trade.

Coffee requires a hot and humid climate with 150-200 cms of rainfall and a temperature range of 15-30° C. It does not tolerate frost and requires shade at high temperatures. It is generally cultivated at lower altitudes than tea, ideally on slopes 300-1800 metres above sea level. Tea grows at high elevations (above 600 metres) in tropical or subtropical areas where the rainfall is heavy and well-distributed throughout the year. Rubber grows best in areas which closely approximate equatorial conditions.

The total area under tea in India is 368,000 ha (1982-83), the largest areas being in Assam (53.8 per cent of the total) and West Bengal (24.5 per cent).

India is the world's largest producer of tea (28.7 per cent of world production in 1985) and tea exporter (20.5 per cent). In 1985 tea production touched 670,000 tonnes and export that year was a third of production. Tea production has been consistently high, increasing at an annual average rate of 2.9 per cent between 1952-53 and 1982-83. Production is expected to reach 766,000 tonnes in 1989-90, a growth of nearly 36 per cent in seven years.

World-wide tea production shows an average yield of 796 kgs per hectare in 1984. India, with a yield of 1573 kgs/ha, is well above this. But its pre-eminent position should not be taken for granted; several newcomers to tea production are showing almost as high or greater yields: 1890 kgs in the USSR, 2000 kgs in Brazil, 1846 kgs in Madagascar and 1677 kgs in Japan.

The average yield per hectare of tea in the country is far exceeded by Tamil Nadu, which produces 1932 kgs/ha (1982-83), and Karnataka, which produces 1842 kgs/ha. In Assam, the yield per hectare is near the national average, while it is considerably lower in West Bengal (1466 kgs/ha), and Kerala (1384 kgs/ha).

India's coffee is mainly grown in the three southern states of Karnataka, Kerala and Tamil Nadu. In 1983-84 there were 232,000 ha under coffee in India, about 83 per cent of it Arabica (*Coffee arabica*) and the rest Robusta (*Coffee canophora*). In line with the national policy of tribal development, coffee cultivation is being encouraged in such non-traditional areas as Andhra Pradesh, Orissa, Maharashtra and the north-eastern states of India. India currently (1985) ranks seventh among the 50 countries producing coffee; India's share is a little over 3 per cent, whereas Brazil, the leader, enjoys a 31 per cent share.

Coffee yield in India averages 624 kgs/ha (1980-81), Karnataka achieving 763 kgs/ha, Tamil Nadu 502 kgs/ha, and Kerala 424 kgs/ha.

The growth rate of coffee production has been much better than that of tea, a five-fold increase in 30 years (1955-85). Coffee production is expected to touch 165,000 tonnes in 1989-90, an increase of about 27 per cent in seven years.

Coffee earns for India a steady Rs 2000 million a year in foreign exchange and provides direct employment to 270,000 workers. Compound growth rates for the past four decades reveal that Robusta, with 7.3 per cent growth in production, has a sizeable share of overall growth. The growth in production of Arabica has been only about 5 per cent and this was mainly due to an increase in productivity of 3.4 per cent.

Rubber is cultivated in 200,000 hectares, of which Kerala's acreage is 92 per cent of the total (1982-83). Rubber production has increased spectacularly, a little more than ten times in 33 years (1950-83). It is estimated that it will reach 187,000 tonnes by the end of the Seventh Plan, a 13 per cent growth from 1982-83.

The all-India average yield per hectare of rubber is 830 kgs, the leading producer, Kerala, equalling the average (828 kgs) and Tamil Nadu exceeding it (967 kgs). Karnataka has a yield of only 704 kgs/ha. In natural rubber production (1985), India ranks fifth in the world, producing a little over 4 per cent of the world total.

Coconut is a major source of vegetable oil and its fibre is the basis of the coir industry. It is cultivated in 1.11 million hectares, of which Kerala's acreage is 59.2 per cent, Karnataka's 16.1 per cent and Tamil Nadu's 12.8 per cent. These three states account for 88 per cent of the country's coconut acreage (1982-83). Coconut production has fluctuated over the years, the 1970-71 high of 6075 million nuts not being repeated. The target for 1989-90 is 8000 million nuts, which, if achieved, will be a 40 per cent increase over the 1980-81 production.

India's coconut-fibre-based coir industry is the biggest in the world, accounting for 58 per cent of the world production of 291,600 tonnes (1981-82).

Sources: 1. *Area and Production of Principal Crops in India 1981-84,* Ministry of Agriculture.
2. *Agricultural Production in India, State-wise & Crop-wise Data: 1949-50 to 1985-86, March 1987* and *World Economy & India's Place in it, October 1986,* Centre for Monitoring Indian Economy.

PRODUCTION OF CASH CROPS — 1984-85

Table 76

	Production 1985-86 (in thousand tonnes)				Area 1982-83 (in thousand hectares)				Yield 1982-83 (in kilogram / hectare)				Increase in yield 1970-83 (in per cent)			
	Tea[1]	Coffee	Rubber[2]	Coconut	Tea	Coffee	Rubber	Coconut	Tea	Coffee[3]	Rubber	Coconut	Tea	Coffee	Rubber	Coconut
Andhra Pradesh	—	N.A.	—	196	—	N.A.	—	45	—	N.A.	—	4,009	—	N.A.	—	−5
Assam	353	—	—	57	198	—	—	7	1,534	—	—	7,062	N.A.	—	—	135
Karnataka	4	130	3	1,050	2	105	4	179	1,842	763	704	5,199	N.A.	N.A.	N.A.	−7
Kerala	51	38	153	3,149	35	55	184	659	1,384	424	828	3,712	N.A.	N.A.	N.A.	−33
Maharashtra	—	—	—	99	—	—	—	10	—	—	—	5,990	—	—	—	22
Orissa	—	—	—	135	—	—	—	23	—	—	—	4,391	—	—	—	18
Tamil Nadu	81	22	10	1,518	37	29	10	143	1,932	502	967	11,538	N.A.	N.A.	N.A.	17
West Bengal	157	—	—	170	90	—	—	3	1,466	—	—	8,909	N.A.	—	—	171
INDIA[4]	653	190	166	6,620	368	190	200	1,113	1,539	624	830	5,088	N.A.	N.A.	N.A.	−12

Notes: [1] 1985 data
[2] 1982-83
[3] 1980-81
[4] Including other states and union territories for which break-up figures are not available.

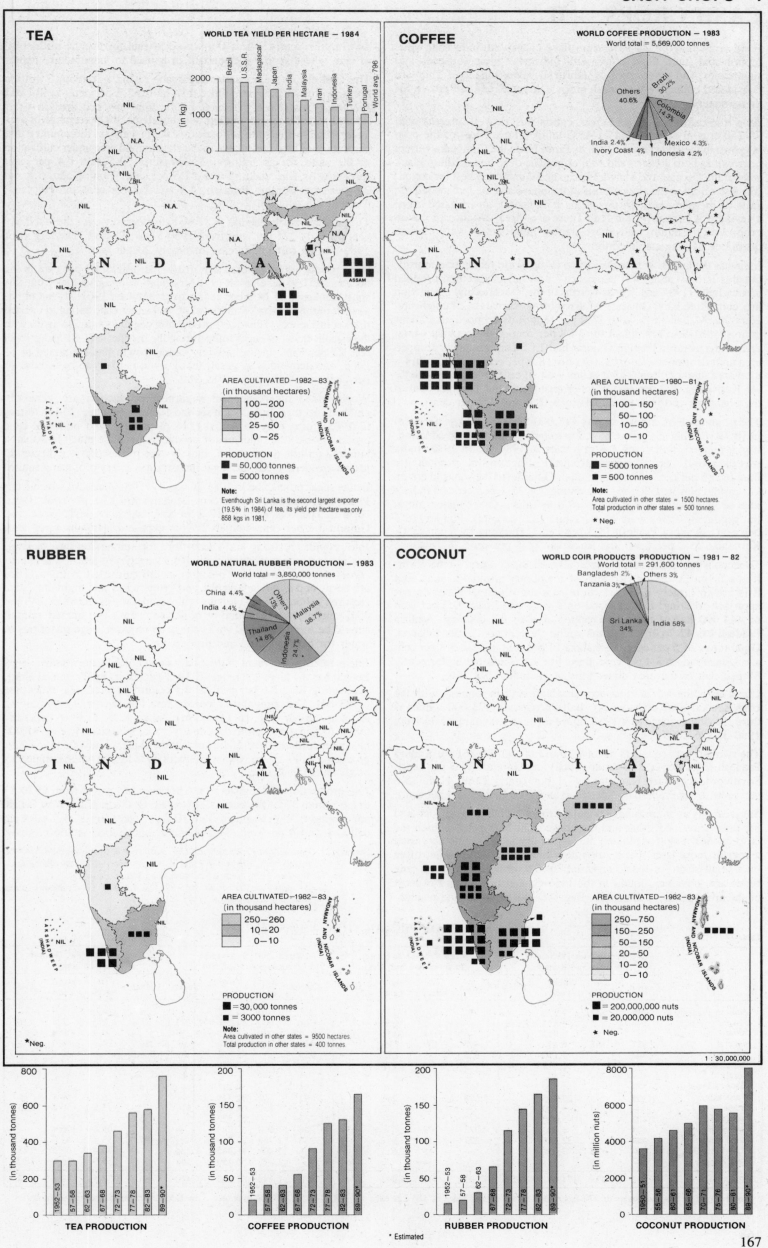

TEA

WORLD TEA YIELD PER HECTARE – 1984

(in kg)

Brazil, U.S.S.R., Madagascar, Japan, India, Malaysia, Iran, Indonesia, Turkey, Portugal

World avg. 796

AREA CULTIVATED–1982–83
(in thousand hectares)
- 100—200
- 50—100
- 25—50
- 0—25

PRODUCTION
- ■ = 50,000 tonnes
- ▪ = 5000 tonnes

Note:
Eventhough Sri Lanka is the second largest exporter (19.5% in 1984) of tea, its yield per hectare was only 858 kgs in 1981.

ASSAM

COFFEE

WORLD COFFEE PRODUCTION – 1983
World total = 5,569,000 tonnes

Brazil 30.2%, Colombia 14.3%, Others 40.6%, India 2.4%, Ivory Coast 4%, Mexico 4.3%, Indonesia 4.2%

AREA CULTIVATED–1980–81
(in thousand hectares)
- 100—150
- 50—100
- 10—50
- 0—10

PRODUCTION
- ■ = 5000 tonnes
- ▪ = 500 tonnes

Note:
Area cultivated in other states = 1500 hectares.
Total production in other states = 500 tonnes.
★ Neg.

RUBBER

WORLD NATURAL RUBBER PRODUCTION – 1983
World total = 3,850,000 tonnes

China 4.4%, India 4.4%, Others 13%, Malaysia 38.7%, Thailand 14.8%, Indonesia 24.7%

AREA CULTIVATED–1982–83
(in thousand hectares)
- 250—260
- 10—20
- 0—10

PRODUCTION
- ■ = 30,000 tonnes
- ▪ = 3000 tonnes

Note:
Area cultivated in other states = 9500 hectares.
Total production in other states = 400 tonnes.

*Neg.

COCONUT

WORLD COIR PRODUCTS PRODUCTION – 1981 – 82
World total = 291,600 tonnes

Bangladesh 2%, Others 3%, Tanzania 3%, Sri Lanka 34%, India 58%

AREA CULTIVATED–1982–83
(in thousand hectares)
- 250—750
- 150—250
- 50—150
- 20—50
- 10—20
- 0—10

PRODUCTION
- ■ = 200,000,000 nuts
- ▪ = 20,000,000 nuts

★ Neg.

1 : 30,000,000

TEA PRODUCTION
(in thousand tonnes)
1952–53, 57–58, 62–63, 67–68, 72–73, 77–78, 82–83, 89–90*

COFFEE PRODUCTION
(in thousand tonnes)
1952–53, 57–58, 62–63, 67–68, 72–73, 77–78, 82–83, 89–90*

RUBBER PRODUCTION
(in thousand tonnes)
1952–53, 57–58, 62–63, 67–68, 72–73, 77–78, 82–83, 89–90*

COCONUT PRODUCTION
(in million nuts)
1950–51, 55–56, 60–61, 65–66, 70–71, 75–76, 80–81, 89–90*

* Estimated

CASH CROPS — II

Field crops may be classified into fibre crops with industrial value (cotton and jute), food crops with industrial value (oilseeds and sugarcane), and other crops of industrial value (tobacco). They are considered collectively as 'cash crops', because they are grown for their monetary value.

As a tropical or subtropical crop, cotton thrives well in areas with 75-250 cms of rainfall and 21-45°C temperature. Most of the crop is grown in the *kharif* season, but in Tamil Nadu the Cambodia variety is grown as both a *kharif* and *rabi* crop. The deep and medium black soils of the Deccan, the Malwa Plateau and Gujarat are best for cotton. It is also raised in the alluvial soils of the plains and the red and laterite soils of the peninsula. In the plains it is mainly an irrigated crop, elsewhere it is mostly rain-fed. There are three varieties of cotton crop: long, medium and short staple. India grows very limited quantities of long-staple cotton.

India has the largest area under cotton in the world (about one-fourth), but this is only about 4.4 per cent (7.8 m ha) of the total cropped area in India (1984-86). Three-fourths of the total area and production are contributed by Gujarat, Maharashtra, Karnataka, Punjab and Haryana. Maharashtra, with 13.2 per cent of its cropped area under cotton (1981-85), accounted for 34.8 per cent of the country's total area under cotton. Though Gujarat had 13.2 per cent of its cropped area under cotton it accounted for only 18.6 per cent of the country's total area under cotton. Other states with significant cotton acreages are Karnataka (8.2 per cent of its cropped area and 11.9 per cent of the country's cotton area) and Punjab (8.9 and 7.9 per cent).

The production of cotton lint was 518,000 tonnes (170 kgs per bale) in 1950-51. This has grown to 1.5 m tonnes in 1985-86. Production, it is estimated, will reach 1.6 m tonnes in 1989-90. In 1985, India was the world's fourth-largest producer of cotton lint, producing a little over 8 per cent of the world total. Gujarat led the states in cotton lint production in 1985-86, accounting for 23.1 per cent of India's total production of 1.5 m tonnes (8.6 m bales). Other major cotton lint producing states were Maharashtra (22 per cent), Punjab (16.3 per cent), Andhra Pradesh (8.3 per cent), Haryana (8.7 per cent), Karnataka (5.8 per cent) and Rajasthan (6.4 per cent).

Tobacco is grown, to some extent, all over the country. In the north, it is a favourite on the rich silts of temporary alluvial islands. But it is grown largely in areas where *rabi* night temperatures do not approach freezing. It is cropped on a very small part (0.3 per cent or 444,000 ha) of the total cropped area of the country. Andhra Pradesh has 42.7 per cent of the country's total area under tobacco, Gujarat has 24.5 per cent, Karnataka 11 per cent, Orissa 3.8 per cent and Uttar Pradesh 3.4 per cent, these five states accounting for nearly 85 per cent of the area under tobacco in India (1981-86).

China leads the world in tobacco production (2 m tonnes), with the USA second (702,000 tonnes). India producing 473,000 tonnes is third, having reached this 1985 figure through gradual growth from 1950-51 (261,000 tonnes).

The two largest producers of tobacco in India (1985-86) are Andhra Pradesh (32.1 per cent of India's total) and Gujarat (38.3 per cent). Small producers are Karnataka (7.1 per cent), Tamil Nadu (3.6 per cent), Uttar Pradesh (6.4 per cent) and West Bengal (3.4 per cent).

Sugarcane is an increasingly popular crop with the growing demand for factory refined sugar and the consequent high prices which the crop fetches. Sugar is planted during the dry season to be harvested a year or more later. Cane cultivation is widespread in India, often on a small-scale for local consumption. The concentrations of cane fields are, however, found in the Indo-Gangetic Plain and in most of the river deltas of the south. Bagasse, the crushed cane residue,

was, until recently, burnt and served as fuel for the the mills, but, of late, where supply is abundant, it is used to manufacture paper.

The total cropped area under sugarcane in India is about 3 million hectares, of which a little more than half (52.8 per cent) is in Uttar Pradesh. This is 6.6 per cent, of the total cropped area in Uttar Pradesh. In other major cane-producing states, the percentage of area under cane in the state to total area under cane in the country and percentage of land under cane in the state to total area under cultivation in the state are as follows: Maharashtra, 9.5 and 1.4 per cent respectively; Karnataka, 5.6 and 1.5 per cent; Tamil Nadu, 5.8 and 2.6 per cent; Andhra Pradesh, 4.9 and 1.2 per cent and Haryana, 4.3 and 2.3 per cent.

The production of cane has increased considerably over the last thirty years, nearly 225 per cent from 1950-51 to 1981-82. The target for 1989-90 is a further increase of nearly 20 per cent.

Cane production registered a dramatic increase of 93 per cent in the decade 1951-61 as a result of diversification in agriculture, but this spurt slackened to 14.9 per cent growth between 1961-71, mainly as a result of the farmers' withdrawal of land under cane owing to internal market influences. However, production began looking up again with the establishment of sugar mills during the decade 1971-81 and growth was 22 per cent. In the last few years growth has continued to be good, and demand is so great, that growth is likely to be around 41 per cent over the period 1981-90.

World-wide, sugarcane and sugarbeet production are generally clubbed together. India is the second-largest producer (174 m tonnes in 1985), next only to Brazil (246 m tonnes), and both are cane producers. All the other major producers produce much less than a hundred million tonnes each. India's cane production is expected to grow considerably, as the sugar industry has a very promising future.

Indian sugar production in 1985-86 was 171.7 m tonnes, of which Uttar Pradesh produced 42.6 per cent, Maharashtra 13.8 per cent, Tamil Nadu 12.9 per cent, Karnataka 7.3 per cent, Andhra Pradesh 5.6 per cent, Gujarat 3.8 per cent, Punjab 2.9 per cent and Haryana 3 per cent.

Jute provides a strong and cheap fibre for conversion into sacking, hessian and carpet-packing cloth. Two varieties of jute — *Corchorus capsularis* and *Corchorus olitorius* — are cultivated in India. They are hardy and highly adaptable, and grow in low and highlands. Jute requires a hot, humid climate with temperatures between 24° and 35°C and a relative humidity of 90 per cent. The annual rainfall should be more than 150 cm as the crop requires large quantities of water for both growing and processing.

Jute is mainly cultivated in the eastern and north-eastern states. West Bengal has the largest acreage (8.3 per cent of its total cropped area), accounting for 62.6 per cent of the country's total area under jute (1983-84). The comparative percentages for other major producing states are: Bihar (1.1; 16.1 per cent), Assam (2.9; 13.6 per cent) and Orissa (0.5; 5.4 per cent). The country's total area under jute is 741,000 ha. In jute productivity, India holds third place worldwide (1981-83), producing 1140 kgs/ha. China produces 4488 kgs/ha while Bangladesh produces 1556 kgs/ha.

Jute production has registered a 630 per cent growth from 1950-51 to 1985, from 270,000 tonnes to 2 m tonnes. Of this tonnage, West Bengal contributes 67.5 per cent, Assam 10.8 per cent, Bihar 15.7 per cent and Orissa 4.4 per cent. Meghalaya's contribution is 8640 tonnes.

Sources: 1. *Agricultural Production in India, State-wise & Crop-wise data: 1949-50 to 1985-86, March 1987* and *World Economy & India's Place in it, October 1986,* Centre for Monitoring Indian Economy.
2. *Seventh Five Year Plan 1985-90, Vol. I* and *II,* Government of India Planning Commission.

Table 77

CASH CROP PRODUCTION — 1985-86

	Production 1985-86 (in thousand tonnes)				Area 1981-86 average (in thousand hectares)				Yield 1981-86 average (in kilogram/hectare)				Increase in yield 1953-86 (in per cent)			
	Cotton[1]	To-bacco	Sugar-cane	Jute[2]	Cotton	To-bacco	Sugar-cane	Jute[3]	Cotton	To-bacco	Sugar-cane	Jute[3]	Cotton	To-bacco	Sugar-cane	Jute[4]
Andhra Pradesh	719	141	9,575	—	509	190	152	—	268	1,010	74,572	—	378.6	29.0	11.3	—
Assam	—	3	1,971	1,178	—	5	50	101	—	800	44,260	1,527	—	2.0	71.4	16
Bihar	—	15	3,937	1,714	—	14	121	119	—	1,000	33,388	1,116	—	30.9	90.6	14
Gujarat	1,987	168	6,490	—	1,443	109	95	—	215	1,715	70,600	—	80.7	24.4	53.0	—
Haryana	745	—	5,150	—	355	—	132	—	330	—	41,833	—	1,000.0	—	30.9	—
Karnataka	503	31	12,574	—	923	49	172	—	116	653	79,058	—	78.5	36.9	48.4	—
Kerala	—	—	422	—	—	—	N.A.	—	—	—	N.A.	—	—	—	N.A.	—
Madhya Pradesh	277	—	1,519	—	563	—	46	—	84	—	30,217	—	10.5	—	17.9	—
Maharashtra	1,895	5	23,705	—	2,698	10	295	—	92	600	92,719	—	41.5	−16.0	38.5	—
Orissa	—	10	3,700	484	—	17	52	40	—	530	64,038	1,519	—	−27.8	50.4	13
Punjab	1,402	—	5,050	—	617	—	90	—	322	—	62,133	—	6.6	—	151.2	—
Rajasthan	474	3	1,010	—	372	3	33	—	226	1,000	40,788	—	48.7	105.8	106.4	—
Tamil Nadu	549	16	22,165	—	227	12	181	—	288	1,750	97,718	—	97.3	53.4	45.7	—
Uttar Pradesh	—	28	73,058	N.A.	—	15	1,635	9	—	1,400	46,485	1,724	—	80.4	67.3	23
West Bengal	—	15	812	7,390	—	14	20	464	—	1,072	56,150	1,544	—	54.9	36.6	30
INDIA[5]	8,612	439	171,681	10,952	7,763	444	3,097	741	170	1,140	57,600	1,470	86.8	37.5	79.6	24

Note: [1] Lint in thousand bales of 170 kgs each. [2] in thousand bales of 180 kgs each. [3] 1983-84. [4] 1970-84 average. [5] Includes other states and union territories.

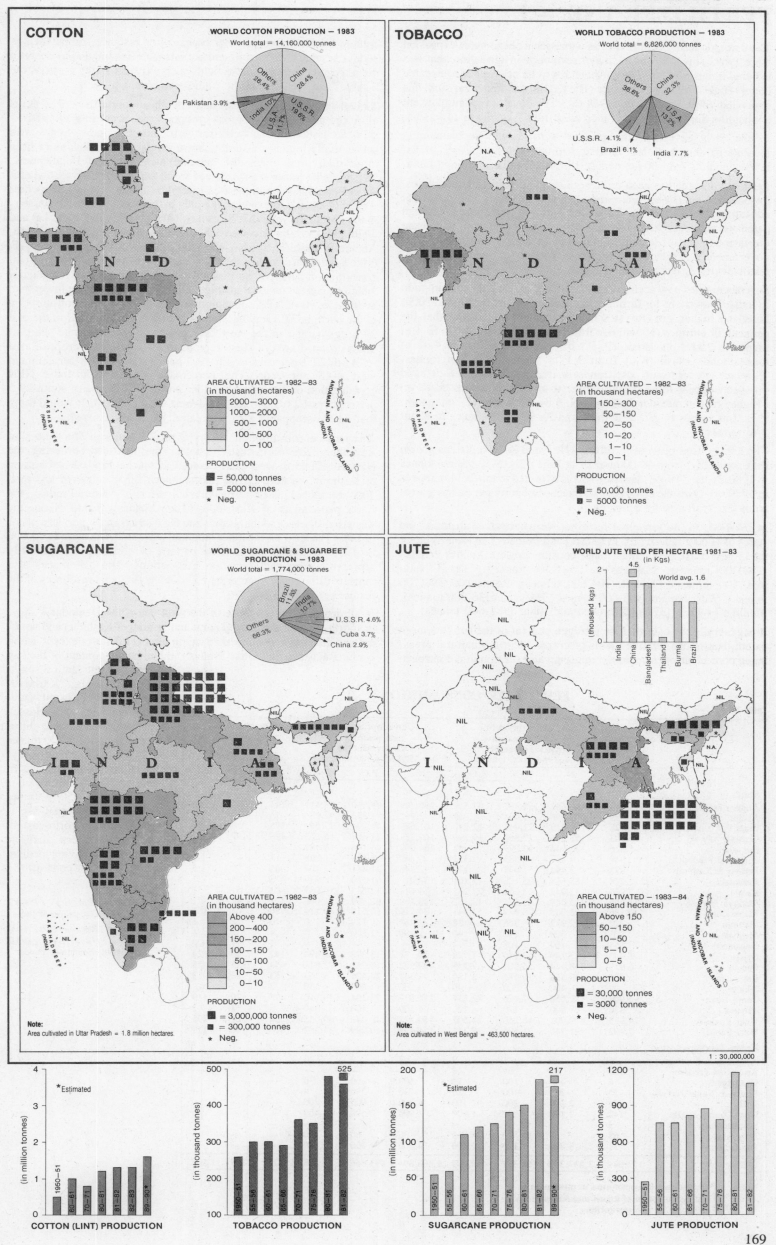

COTTON

WORLD COTTON PRODUCTION — 1983
World total = 14,160,000 tonnes

Others 26.4% | China 28.4% | Pakistan 3.9% | India 10% | U.S.A 11.7% | U.S.S.R 19.6%

AREA CULTIVATED — 1982–83
(in thousand hectares)
- 2000–3000
- 1000–2000
- 500–1000
- 100–500
- 0–100

PRODUCTION
- = 50,000 tonnes
- = 5000 tonnes
- ★ Neg.

TOBACCO

WORLD TOBACCO PRODUCTION — 1983
World total = 6,826,000 tonnes

Others 38.6% | China 32.3% | U.S.A 13.2% | U.S.S.R. 4.1% | Brazil 6.1% | India 7.7%

AREA CULTIVATED — 1982–83
(in thousand hectares)
- 150–300
- 50–150
- 20–50
- 10–20
- 1–10
- 0–1

PRODUCTION
- = 50,000 tonnes
- = 5000 tonnes
- ★ Neg.

SUGARCANE

WORLD SUGARCANE & SUGARBEET PRODUCTION — 1983
World total = 1,774,000 tonnes

Brazil 11.8% | India 10.7% | Others 66.3% | U.S.S.R. 4.6% | Cuba 3.7% | China 2.9%

AREA CULTIVATED — 1982–83
(in thousand hectares)
- Above 400
- 200–400
- 150–200
- 100–150
- 50–100
- 10–50
- 0–10

PRODUCTION
- = 3,000,000 tonnes
- = 300,000 tonnes
- ★ Neg.

Note:
Area cultivated in Uttar Pradesh = 1.8 million hectares.

JUTE

WORLD JUTE YIELD PER HECTARE 1981–83
(in Kgs)

World avg. 1.6

India | China 4.5 | Bangladesh | Thailand | Burma | Brazil

AREA CULTIVATED — 1983–84
(in thousand hectares)
- Above 150
- 50–150
- 10–50
- 5–10
- 0–5

PRODUCTION
- = 30,000 tonnes
- = 3000 tonnes
- ★ Neg.

Note:
Area cultivated in West Bengal = 463,500 hectares.

1 : 30,000,000

COTTON (LINT) PRODUCTION
(in million tonnes)
*Estimated
1950–51, 60–61, 70–71, 80–81, 81–82, 82–83, 89–90*

TOBACCO PRODUCTION
(in thousand tonnes)
525
1950–51, 55–56, 60–61, 65–66, 70–71, 75–76, 80–81, 81–82

SUGARCANE PRODUCTION
(in million tonnes)
*Estimated
217
1950–51, 55–56, 60–61, 65–66, 70–71, 75–76, 80–81, 81–82, 89–90*

JUTE PRODUCTION
(in thousand tonnes)
1950–51, 55–56, 60–61, 65–66, 70–71, 75–76, 80–81, 81–82

FERTILIZER USE — I

The domestic fertilizer industry has witnessed a phenomenal expansion since 1969. This followed the major revolution in agriculture that was ushered in during 1966-1969. About 1.9 m ha of cropped area came under High Yielding Varieties (HYVs) in the first year and this expanded to 9.3 m ha by 1968-69. The level of plant nutrient application during this period increased from 7 kgs/ha to 11 kgs/ha.

In the early seventies, the private sector began supplementing institutional distribution of fertilizers to farmers. By the late 1970s, it was handling over 60 per cent of all fertilizer distribution in India.

Fertilizer demand has been influenced by greater irrigation facilities, introduction of HYVs, increased cropping intensity and efforts at educating the farmer in advanced farming techniques through extension services. Where manufacturers have taken pains to educate the farmers, fertilizer consumption has increased substantially.But government's policy of subsidizing production and controlling fertilizer prices has also helped greatly.

Seventeen states now account for 97 per cent of fertilizer consumption in India. Projections for fertilizer demand for the period 1984 to 2000, based on higher growth rates in irrigated areas, indicate that the demand for nitrogen (N) will rise from 6.2 m tonnes in 1984 to 13.4 m tonnes in 2000, phosphate (P_2O_5) from 2.03 m tonnes to 4.42 m tonnes, and potash (K_2O) from 1.1 m tonnes to 2.3 m tonnes. However, in terms of past trends in irrigated areas, the demand prospects appear to be slightly lower than the projections based on a higher growth rate in irrigation — N: 5.8 m tonnes to 11.7 m tonnes; P: 1.9 m tonnes to 3.8 m tonnes; and K: 990,000 tonnes to about 2 m tonnes.

The total consumption of fertilizers (N, P and K) in India has grown astronomically, from 66,000 tonnes in 1951-52 to 5.52 million tonnes in 1980-81, an 84-fold increase! This is expected to touch 14 m tonnes in 1989-90. Over the same period, the consumption per ha has grown from 0.6 kgs/ha to 48.2 kgs/ha.

In 1984-85, 8.2 m tonnes of fertilizer were consumed in India, the states of Uttar Pradesh (1.6 m tonnes) and Punjab (1 m tonnes) being the largest consumers. Other major consumers were Andhra Pradesh (980,300 tonnes), Maharashtra (581,300 tonnes), Tamil Nadu (690,500 tonnes), Gujarat (504,600 tonnes), Karnataka (590,700 tonnes), West Bengal (405,000 tonnes), Haryana (336,000 tonnes), Madhya Pradesh (372,600 tonnes) and Bihar (381,600 tonnes).

It is generally the case that the nitrogen content of the soil to a large extent determines relative yields of crops. Most agricultural areas, being poor in nitrogen content, require greater substitution of nitrogen content by fertilizers or other means. For instance, crops such as paddy, sesame etc. do not facilitate nitrogen fixation and to enhance the nitrogen content in these fields such nitrogen-fixing crops as the legumes are included in the traditional cropping cycle.

The all-India average of total nutrient consumption (N + P + K) is 46.3 kgs per hectare of gross cropped area, according to 1984-85 statistics issued by the Fertiliser Association of India. States with substantially higher consumption are Punjab, 151.2 kgs/ha (a little more than three times the national average), and Pondicherry, 281.5 kgs/ha (more than six times the national average). Elsewhere, West Bengal has the highest consumption in the east (54.8 kgs/ha), Assam (4 kgs/ha), the lowest, not only in this region but in the country; in the north, Haryana (57.7 kgs/ha), Uttar Pradesh (65.1 kgs/ha) and Delhi (93.4 kgs/ha) are above the national average; and in the south, Andhra Pradesh (75 kgs/ha), Karnataka (52.6 kgs/ha) and Tamil Nadu (100 kgs/ha) are higher than the national average and show good consumption patterns. Surprisingly, in the west, except for Gujarat which equals the national average, all the other states have lower than average levels of NPK consumption. The average NPK consumption in the north is 77.3 kgs/ha, in the south it is 70.4 kgs/ha, in the east 29.6 kgs/ha and in the west 23.2 kgs/ha. The percentage share of gross cropped area of these regions to all-India gross cropped area is: north 22.3 per cent, south 19.3 per cent, east 17.8 per cent and west 40.5 per cent. The percentage share of all-India NPK consumption on the other hand is: north 37.2 per cent, south 29.3 per cent, east 11.3 per cent, and west 20.4 per cent. It would seem that where croppable area is greater, less NPK is used.

Districtwise consumption shows a mixed pattern. The whole of Rajasthan, the north-eastern states, the districts bordering the Himalaya in north-west Uttar Pradesh, Himachal Pradesh and Jammu & Kashmir, western Orissa and southern Bihar consume less than 5000 tonnes per district. The deltaic districts of Andhra Pradesh, the flood plain areas of Karnataka (Tungabhadra), Tamil Nadu, the western districts of Maharashtra, and the districts in the northern plains of Uttar Pradesh, on the other hand, all seem to be consuming 25,000 to 175,000 tonnes each. This pattern of consumption generally represents crop and irrigation associations. The differences also indicate differences in enterprise and in the financial resources of the farmers.

In all districts, except those of the west coast, the Damodar Valley, the north-western districts (Gujarat and Jammu & Kashmir) and north-eastern districts and a few districts in almost all states except Tamil Nadu, Rajasthan and Uttar Pradesh, the nitrogen content of the soil is low. In the entire south, barring Kerala, Kanniya-kumari district in Tamil Nadu, Uttar Kannad and Hassan districts of Karnataka, nitrogen use is above 5,000 tonnes per district, while in the hilly, plateau and desert districts of north and central India it is between 1000 - 5000 tonnes per district. Nitrogen use is low only in the north-eastern and north-western districts (these districts having medium to high nitrogen content).

Sources:

1. *Fertiliser Statistics 1984-85,* Fertiliser Assosiation of India.

2. *Fertiliser Requirements: Long Term Perspective, 1986,* Srivatsava, Economic Times.

FERTILIZER CONSUMPTION

Table 78

	Gross culti-vated area 1981-82 ('000 ha)	Consumption in kharif season (April-September) — 1984 ('000 tonnes)			Consumption in rabi season (October-March) — 1984-85 ('000 tonnes)			Targets of consumption 1985-90 ('000 tonnes)		
		N	P_2O_5	K_2O	N	P_2O_5	K_2O	N	P_2O_5	K_2O
States										
Andhra Pradesh	13,047	364,662	155,402	47,569	280,299	98,950	33,411	1,143	481	158
Assam	3,439	3,586	1,004	1,469	3,952	1,833	1,954	76	40	24
Bihar	10,628	119,519	21,063	12,270	161,689	43,394	23,663	595	250	82
Gujarat	10,903	152,526	77,369	19,399	167,789	71,407	16,073	498	197	51
Haryana	5,826	106,643	10,099	1,942	166,002	46,146	5,687	405	112	28
Himachal Pradesh	949	9,487	583	322	6,840	2,620	1,898	18	3	3
Jammu & Kashmir	978	3,494	478	120	2,148	1,441	194	45	9	5
Karnataka	11,228	183,938	108,106	72,101	127,017	61,180	38,339	475	184	137
Kerala	2,905	33,043	19,009	22,265	24,614	13,633	15,081	102	51	51
Madhya Pradesh	21,756	113,349	55,875	13,252	110,269	68,809	11,040	521	241	77
Maharashtra	20,386	253,005	78,649	52,441	124,210	50,616	40,514	980	307	182
Manipur	240	2,932	365	38	264	139	20	10	2	1
Meghalaya	203[1]	500	400	40	1,100	1,200	10	4	2	1
Nagaland[2]	164	182	108	—	—	—	—	1	1	Neg.
Orissa	8,743[1]	37,249	13,984	7,709	33,178	13,398	8,494	142	40	24
Punjab	6,929	338,144	56,261	9,549	420,582	210,246	12,823	953	345	28
Rajasthan	18,596	50,868	16,186	2,187	99,541	34,426	3,363	353	83	13
Sikkim	92[1]	400	200	40	200	200	50	2	1	Neg.
Tamil Nadu	6,909	182,558	67,670	82,317	211,317	70,204	76,450	476	120	119
Tripura	380	800	170	100	1,300	400	340	5	2	2
Uttar Pradesh	24,773	479,901	48,280	24,505	760,122	240,033	60,065	2,006	562	395
West Bengal	7,402	91,745	31,594	28,310	154,499	60,299	39,282	476	160	119
Union Territories								19[3]	6[3]	5[3]
Andaman & Nicobar	36	6	1	9	2	2	20	—	—	—
Arunachal Pradesh	152	83	61	48	—	—	—	—	—	—
Chandigarh	4	400	100	10	500	200	30	—	—	—
Dadra & Nagar Haveli	25[1]	200	100	20	30	30	4	—	—	—
Delhi	84	2,200	200	100	4,200	1,000	200	—	—	—
Goa, Daman & Diu	142	2,100	900	1,900	900	500	400	—	—	—
Lakshadweep	3	—	—	—	—	—	—	—	—	—
Mizoram	68	50	20	30	30	60	20	—	—	—
Pondicherry	51	3,800	1,200	1,900	4,300	1,300	1,800	—	—	—
INDIA	177,041	2,537,370	765,337	402,002	2,866,894	1,093,666	391,225	9,305	3,197	1,505

Note: [1] Forecast data has been used in estimation.
[2] Total consumption of kharif and rabi seasons.
[3] Total for all union territories.

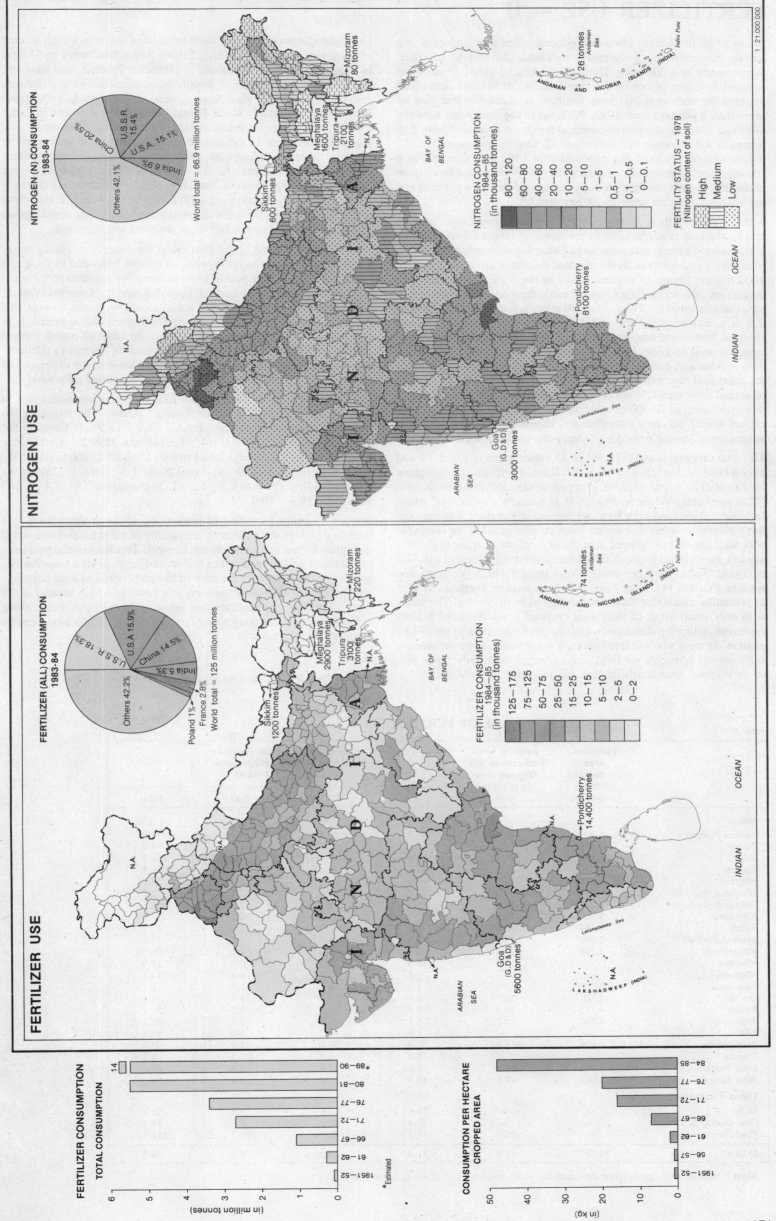

FERTILIZER USE

NITROGEN USE

NITROGEN (N) CONSUMPTION
1983-84

- U.S.S.R. 15.4%
- U.S.A. 15.1%
- India 6.9%
- China 20.5%
- Others 42.1%

World total = 66.9 million tonnes

NITROGEN CONSUMPTION
1984-85
(in thousand tonnes)

80-120 | 60-80 | 40-60 | 20-40 | 10-20 | 5-10 | 1-5 | 0.5-1 | 0.1-0.5 | 0-0.1

FERTILITY STATUS — 1979
(Nitrogen content of soil)

High | Medium | Low

Mizoram 80 tonnes
Meghalaya 1600 tonnes
Tripura 2100 tonnes
Sikkim 600 tonnes
Pondicherry 8100 tonnes
Goa (G,D&D) 3000 tonnes
Andaman 26 tonnes

FERTILIZER (ALL) CONSUMPTION
1983-84

- U.S.A 15.9%
- China 14.5%
- India 5.3%
- France 2.8%
- Poland 1%
- U.S.S.R 18.3%
- Others 42.2%

World total = 125 million tonnes

FERTILIZER CONSUMPTION
1984-85
(in thousand tonnes)

125-175 | 75-125 | 50-75 | 25-50 | 15-25 | 10-15 | 5-10 | 2-5 | 0-2

Mizoram 220 tonnes
Meghalaya 2900 tonnes
Tripura 3100 tonnes
Sikkim 1200 tonnes
Pondicherry 14,400 tonnes
Goa (G,D&D) 5600 tonnes
Andaman 74 tonnes

FERTILIZER CONSUMPTION

TOTAL CONSUMPTION

(in million tonnes)

14

89-90*
80-81
76-77
71-72
66-67
61-62
1951-52

*Estimated

CONSUMPTION PER HECTARE
CROPPED AREA

(in kg)

84-85
76-77
71-72
66-67
61-62
56-57
1951-52

171

FERTILIZER USE — II

Prices of all fertilizers, whether produced in India or imported, are strictly controlled by government to enable farmers to get their requirements at a fair price. During July-August 1984, government reduced fertilizer prices by 7.5 per cent. A further discount of 10 per cent the same year increased fertilizer consumption that year by more than 8 per cent over 1983. Whereas in the four years between 1980 and 1983, production of chemical fertilizers increased from 3 m tonnes to 4.4 m tonnes and utilization of capacity improved from 57 per cent to 66 per cent, the production in 1983-84 increased, as a consequence of government's policy, by 4.5 m tonnes for foodgrains alone. But capacity utilization that year came down by 4 per cent and the share of imports grew to 34 per cent.

A major share of these increases in consumption, capacity utilization and production is attributable to the needs of larger farmers. Small and medium farmers, and some large farmers, use substantial amounts of manures and compost. It is estimated that there are annually about 1000 million tonnes of organic waste in the form of crop residue; in addition, 300-400 million tonnes of cattle dung and animal droppings are available annually. These materials provide 6 m tonnes of nitrogen, 2.5 m tonnes of phosphate and 4.5 m tonnes of potassium. Rural compost from waste amounts to about 50 m tonnes, while urban waste compost used in agriculture is about 15 m tonnes. But, before the crop residue and dung are made into manure, they meet at least half the rural fuel requirements. However, the entire 300-400 m tonnes of animal dung available are now being considered for biogas production (equivalent of 70,000 m cubic metres of methane or 160 m tonnes of fuel wood), an energy supplement. Biogas plants have the added advantage of leaving a residue of slurry that could be used as manure.

The total cropped area (1975-76) in 15 major states is 171 m ha and only a third of this (56.3 m ha) is fertilized. Fertilizer use is highest in Punjab (76.3 per cent of cropped area), followed by Kerala (72.6 per cent). While in Punjab it is modern methods of wheat farming that demand high fertilizer inputs, in Kerala it is mainly the large plantation areas that utilize them on a large scale. In Haryana, with 48.7 per cent of cropped area under fertilizer, it is, as in Punjab, modern agricultural methods that determine the high level of use. But in Tamil Nadu (55.4 per cent), West Bengal (49.9 per cent) and Andhra Pradesh (41.1 per cent) it is both modern methods as well as plantation needs that make for substantial fertilizer use. The states with only small areas of their total cropped areas fertilized mainly represent field crop dominance and the predominance of small and medium farmers who use fertilizer on a small scale and prefer manure and compost nutrients for field crops. Assam, with only 5 per cent of its cropped area fertilized, has the least use of fertilizer.

Only nine districts in the whole of India have soil with a high content of phosphate: Kheda in Gujarat, Sirohi, Sawai Madhopur and Jaipur in Rajasthan, Kinnaur and Kullu in Himachal Pradesh, and Kamrup, Lakhimpur and Dibrugarh in Assam. Most districts with low nitrogen content have also low phosphate content, the exceptions being some districts in Rajasthan, some Himalayan foothill districts in Uttar Pradesh, the western districts in Madhya Pradesh and western Tamil Nadu (all medium). In the case of certain areas, those with medium nitrogen have low phosphate: districts in Orissa and the western districts in Maharashtra. As for phosphate use, the pattern is very mixed, though high-content districts generally consume less and low-content districts (generally) use 3000 tonnes and upwards. In the whole of India, there are quite a few districts where phosphate consumption is less than 1000 tonnes, but these districts are scattered.

As for potash, a third of the districts in the country — contiguously in three patches: in the north-west, in central India and in the south — have soils with high potash content. Very few districts in the south and several districts in the north (Jammu & Kashmir, Himachal Pradesh, Uttar Pradesh and north-eastern India) have low potash content. All others have medium potash content. The highest use of potash is in the south (15-20 thousand tonnes), in medium and high potash content districts in Tamil Nadu, and one district each in Karnataka (Bellary) and Andhra Pradesh (Guntur). Most medium-content districts and high-content districts use low to medium quantities of potash.

The three nutrients are used in the following consumption ratios in India — 5.8 nitrogen : 2.2 phosphate : 1 potash. The ratios for the major agricultural areas are : Punjab 33.9 : 11.9 : 1, Orissa 4.3 : 1.7 :1, West Bengal 3.9 : 1.4 : 1, Haryana 35.8 : 7.4 : 1, Uttar Pradesh 14.7 : 3.4 : 1, Andhra Pradesh 8 : 3.1 : 1, Karnataka 2.8 : 1.5 : 1, Kerala 1.5 : 0.9 : 1, Tamil Nadu 2.5 : 0.9 : 1, Gujarat 9 : 4.2 :1, Madhya Pradesh 9.2 : 5.1 : 1, Maharashtra 3.8 : 1.4 : 1 and Rajasthan 31.4 : 10.7 : 1.

It must be pointed out that in some parts of the country there is a tendency to over-use fertilizers, despite the fact that the recommended quantities for use are known to the farmers. This has created problems of chemical pollution of water bodies and land. Such a situation has arisen, on the one hand, because of the constraints of area coverage by agricultural extension workers, as a result of which several of the plant protection practices are not known. On the other hand, there is a misconception among many farmers that fertilizers alone can do wonders in crop returns.

Source: *Fertiliser Statistics 1984-85*, Fertiliser Association of India.

USE OF FERTILIZERS

Table 79

States	Fertilized area (in thousand hectares) — 1975-76	Ratio of ferti-lized area to total cropped area (in per cent) — 1975-76	Consumption of Plant Nutrients per unit of gross cropped area (in kg./ha.) — 1984-85			
			N	P₂O₅	K₂O	Total
States						
Andhra Pradesh	5,403	41.7	49.3	19.5	6.2	75.0
Assam	156	4.9	2.2	0.8	1.0	4.0
Bihar	3,984	35.3	26.4	6.1	3.4	35.9
Gujarat	4,395	43.1	29.4	13.6	3.3	46.3
Haryana	2,655	48.7	46.8	9.6	1.3	57.7
Himachal Pradesh	255	27.6	17.2	3.4	2.3	22.9
Jammu & Kashmir	263	28.0	22.4	5.9	1.3	29.6
Karnataka	3,727	33.4	27.7	15.1	9.1	52.0
Kerala	2,164	72.6	19.8	11.2	12.9	43.9
Madhya Pradesh	2,306	10.8	10.3	5.7	1.1	17.1
Maharashtra	5,368	27.3	19.6	6.3	4.6	28.5
Manipur	—	—	13.3	2.1	0.2	15.6
Meghalaya	—	—	7.8	5.7	0.8	14.3
Nagaland	—	—	1.1	0.6	—	1.7
Orissa	1,601	20.7	8.0	3.1	1.8	12.9
Punjab	4,773	76.3	109.5	38.5	3.2	151.2
Rajasthan	3,446	20.1	8.1	2.8	0.3	11.2
Sikkim	—	—	7.6	4.9	1.0	13.5
Tamil Nadu	4,008	55.4	57.0	20.0	23.0	100.0
Tripura	—	—	5.4	1.5	1.3	8.2
Uttar Pradesh	7,414	32.1	50.1	11.6	3.4	65.1
West Bengal	3,963	49.8	33.3	12.4	9.1	54.8
Union Territories						
Delhi	—	—	75.4	14.8	3.2	93.4
Goa, Daman & Diu	—	—	20.8	10.3	8.6	39.7
Pondicherry	—	—	159.1	49.7	72.7	281.5
INDIA¹	56,257	32.9	31.0	10.6	4.7	46.3

Note: ¹ Includes other states/union territories for which break-up figures are not available.

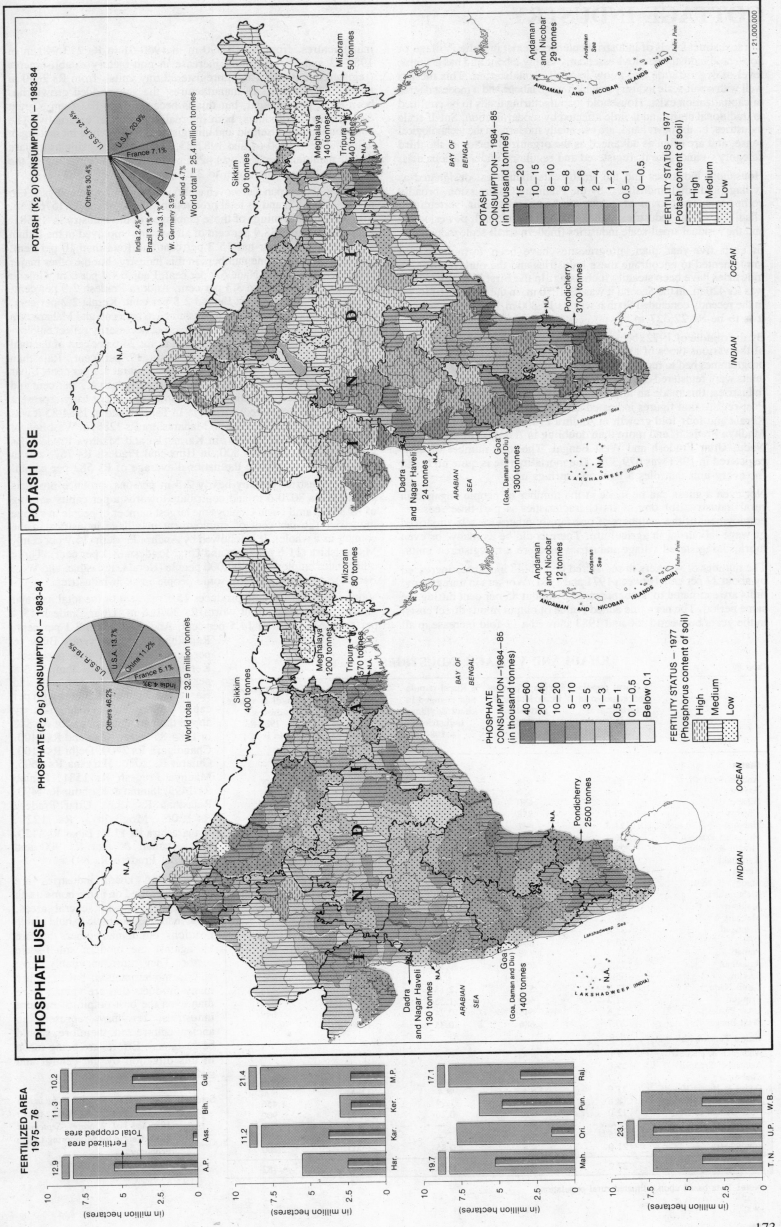

POTASH USE

POTASH (K₂O) CONSUMPTION — 1983-84

World total = 25.4 million tonnes

- U.S.A. 20.9%
- U.S.S.R. 24.4%
- France 7.1%
- Poland 4.7%
- W. Germany 3.9%
- China 3.1%
- Brazil 3.1%
- India 2.4%
- Others 30.4%

POTASH CONSUMPTION — 1984–85 (in thousand tonnes)
15–20, 10–15, 8–10, 6–8, 4–6, 2–4, 1–2, 0.5–1, 0–0.5

FERTILITY STATUS — 1977 (Potash content of soil)
High, Medium, Low

Sikkim 90 tonnes
Meghalaya 140 tonnes
Mizoram 50 tonnes
Tripura 140 tonnes
Andaman and Nicobar 29 tonnes
Pondicherry 3700 tonnes
Dadra and Nagar Haveli 24 tonnes
Goa (Goa, Daman and Diu) 1300 tonnes

PHOSPHATE USE

PHOSPHATE (P₂O₅) CONSUMPTION — 1983-84

World total = 32.9 million tonnes

- U.S.A. 13.7%
- China 11.2%
- France 5.1%
- India 4.3%
- U.S.S.R. 19.5%
- Others 46.2%

PHOSPHATE CONSUMPTION—1984–85 (in thousand tonnes)
40–60, 20–40, 10–20, 5–10, 3–5, 1–3, 0.5–1, 0.1–0.5, Below 0.1

FERTILITY STATUS — 1977 (Phosphorus content of soil)
High, Medium, Low

Sikkim 400 tonnes
Meghalaya 1200 tonnes
Mizoram 80 tonnes
Tripura 570 tonnes
Andaman and Nicobar 12 tonnes
Pondicherry 2500 tonnes
Dadra and Nagar Haveli 130 tonnes
Goa (Goa, Daman and Diu) 1400 tonnes

FERTILIZED AREA 1975–76

Fertilized area / Total cropped area (in million hectares)

A.P. 12.9, Ass., Bih. 11.3, Guj. 10.2
Har., Kar., Ker. 11.2, M.P. 21.4
Mah. 19.7, Ori., Pun., Raj. 17.1
T.N., U.P. 23.1, W.B.

COTTAGE INDUSTRY

Three distinct levels of industrial enterprise exist in India. Village or household industry employs essentially family labour and has a simple level of organization and a low level of capitalization. This merges well with small-scale industry in which wage labour and a modest degree of capitalization exist. Household manufacturing tends to be confined to traditional crafts and is little affected by modernization. Small-scale industries, on the other hand, are essentially modern, in the technological sense, and are often as advanced as the organized sector — the third category — and, like it, registered and regulated under various acts.

Household industries and the small-scale industries, designated as 'village and small industries' by the Planning Commission, broadly comprise seven groups of industries: handicrafts, coir, sericulture, khadi and village industries, handlooms (all traditional), powerlooms, and the residual small-scale industries (modern small-scale industries).

In each five-year plan, programmes have been formulated and implemented to encourage these industries and the outlays on these industries have been steadily increasing. In the First Plan, the outlay was Rs 420 m, in the Second it was Rs 1870 m, in the Fifth Rs 6110 m, in the recently concluded Sixth it was Rs 17,800 m, and for the Seventh it is to be Rs 27,527 m.

By the middle of 1972, because all those units which wished to avail of the various types of assistance extended through small industries programmes had to register with the directorates of industries, 250,000 units were registered country-wide. Together with the organized small industries, this made an impressive total of 500,000 units that year. Post-registration figures indicated abnormally high rises in Bihar and Kerala and four-fold growth in Andhra Pradesh, about three-fold in Madhya Pradesh, and more than doubling in Assam, Gujarat, Tamil Nadu, Uttar Pradesh and West Bengal. The total number of units registered in 1983 was 679,396. The mortality rate is quite high, but for every unit that dies a new one springs up.

Not even a guess can be made at the number of people engaged in rural industry, for one of its characteristics is part-time seasonal employment of the members of one family otherwise self-employed or wage-labouring in agriculture. There could be as many, or even more, unregistered village industries as there are registered units.

The number of workers in registered household industries increased by about 27 per cent between 1971 and 1981. Workers in unregistered units are estimated to have increased by about 85 per cent during the same period. The net value added (value of output minus direct costs) in the years between 1960 and 1983 showed a 12-fold increase in all manufactures, from Rs 18,590 m in 1960-61 to Rs 213,940 m in 1982-83 and about 10-fold increase in non-factory establishments (registered household and unregistered tiny units), from Rs 7850 m to Rs 75,500 m. In 'all manufactures' the value added growth has been quite phenomenal, but this is because there has been a great demand for its products, both internally as well as externally. The contribution of household and unregistered units, while increasing in value between 1960-61 and 1982-83, has been decreasing as a share of 'all manufactures'. It was 42.2 per cent at the beginning of the period and had come down to 35.2 per cent by 1982-83.

The khadi (homespun and woven cloth) industry in India employs 1.3 m people (1984-85) and its total production value is about Rs 1576.2 m. The per capita earnings of those employed in the industry is Rs 582. Uttar Pradesh has 38.9 per cent of all persons employed in the Indian khadi industry. Bihar has 15.3 per cent and Rajasthan 10 per cent. All states employ some numbers in this industry, but the other major employers are: Tamil Nadu 6.7 per cent, Punjab 4.4 per cent, Gujarat 3.7 per cent, Karnataka 3.3 per cent, Andhra Pradesh 2.9 per cent, Assam 2.9 per cent, West Bengal 2.8 per cent, Kerala 2.2 per cent, Haryana 1.6 per cent, Jammu & Kashmir 1.6 per cent, and Maharashtra 1.1 per cent. However, production does not necessarily reflect employment. Value-wise, Uttar Pradesh accounts for 24.6 per cent of the total national khadi production, Tamil Nadu 17 per cent, Rajasthan 10.7 per cent, West Bengal 7.3 per cent, Gujarat 6.1 per cent, Bihar 6 per cent, Karnataka 5.5 per cent, Andhra Pradesh 4.7 per cent and Punjab 4.6 per cent. Again, production levels do not truly represent per capita (per employee) earnings. In Tamil Nadu in 1984-85 it was Rs 1616, in Gujarat Rs 1329, in Maharashtra Rs 1286, in West Bengal Rs 1178, in Kerala Rs 983, in Karnataka and Madhya Pradesh Rs 971 each, in Tripura Rs 800, in Himachal Pradesh Rs 767, and in Delhi Rs 650, all above the national average of Rs 582 per capita.

Village industries employ nearly 2.5 m persons, produce products valued at Rs 8070.6 m and contribute towards a per capita earning of Rs 888. Tamil Nadu employs the largest number of people in village industries (26 per cent of employment in village industries in the country as a whole). It is followed by Andhra Pradesh (15.9 per cent), Maharashtra (11.6 per cent) and Uttar Pradesh (8.1 per cent). Three other states employ 100-200,000 people (Kerala, Rajasthan and West Bengal). All states employ some people in such industries.

Value-wise, Tamil Nadu produces 15.5 per cent of the total national village industries' output worth Rs 8070.6 m, Uttar Pradesh 15.4 per cent, Maharashtra 14.5 per cent, Andhra Pradesh 8.1 per cent, Rajasthan 7.2 per cent, Bihar 6 per cent, Karnataka 5.9 per cent, Kerala 4.3 per cent, Punjab 3.8 per cent, Gujarat 3.4 per cent and Haryana 3.2 per cent. States with per capita earnings from village industry above the national average of Rs 888 in 1984-85 were: Nagaland Rs 2400, Chandigarh Rs 2400, Delhi Rs 2100, Gujarat Rs 2036, Haryana Rs 1605, Madhya Pradesh Rs 1581, Punjab Rs 1493, Jammu & Kashmir Rs 1490, Rajasthan Rs 1339, Uttar Pradesh Rs 1306, Meghalaya Rs 1220, Maharashtra Rs 1136, Bihar Rs 1120, Goa, Daman & Diu Rs 900 and Himachal Pradesh Rs 891.

Village and khadi industries are generally carried on in the home using traditional tools. In several areas, these industries continue to hold viable positions, but in many others they are up against the modern small-scale sector. They must, inevitably, give way to modernization, but there are many products that are valued more than even the best output of modern industries, for they represent an ancient culture and, therefore, should be sustained. Efforts are being made in this direction by the central and state governments.

Sources: *Basic Statistics Relating to the Indian Economy*, Vol. 1: *All India*, August 1986 and Vol. 2: *States*, September 1986, Centre for Monitoring Indian Economy.

Table 80

KHADI AND VILLAGE INDUSTRIES

	% Rural population engaged in agricultural activities — 1982	Earnings/person in agricultural activities — 1982 (in Rs.)	% Rural population engaged in khadi & village industries — 1984	Production/person in khadi & village industries — 1984-85 (in Rs.)	Earnings/person in khadi & village industries — 1984-85 (in Rs.)
States					
Andhra Pradesh	36.9	836	1.00	1,698	690
Assam	N.A.	656	N.A.	867	257
Bihar	26.2	450	0.43	2,046	516
Gujarat	27.0	858	0.29	5,095	1,571
Haryana	21.3	1,463	0.56	4,737	1,245
Himachal Pradesh	26.1	495	1.26	2,942	877
Jammu & Kashmir	22.4	801	0.99	3,676	998
Karnataka	31.7	837	0.49	4,115	853
Kerala	12.9	631	0.81	2,172	891
Madhya Pradesh	13.7	605	0.09	4,544	1,373
Maharashtra	35.4	922	0.70	4,001	1,143
Manipur	33.3	745	1.66	3,756	561
Meghalaya	38.1	591	0.42	4,780	1,220
Nagaland	34.4	388	0.41	5,700	1,700
Orissa	23.9	684	0.30	1,441	473
Punjab	22.3	1,756	0.91	3,278	1,001
Rajasthan	25.7	771	0.82	3,149	863
Sikkim	35.1	N.A.	N.A.	N.A.	N.A.
Tamil Nadu	33.9	522	2.18	2,073	774
Tripura	22.2	608	1.35	2,441	656
Uttar Pradesh	25.6	648	0.74	2,298	649
West Bengal	20.7	686	0.35	1,680	590
Union Territories					
Andaman & Nicobar	9.0	N.A.	—	—	—
Arunachal Pradesh	39.0	N.A.	0.16	100	100
Chandigarh	6.7	N.A.	3.23	20,200	2,400
Dadra & Nagar Haveli	31.0	N.A.	N.A.	N.A.	N.A.
Delhi	8.6	N.A.	1.94	7,767	1,456
Goa, Daman & Diu	12.0	N.A.	0.26	4,100	900
Lakshadweep	N.A.	N.A.	—	—	—
Mizoram	37.5	N.A.	—	—	—
Pondicherry	21.0	N.A.	0.35	3,900	2,200
INDIA	27.2	710	0.73	2,546	782

Note: Data based upon estimated rural population.

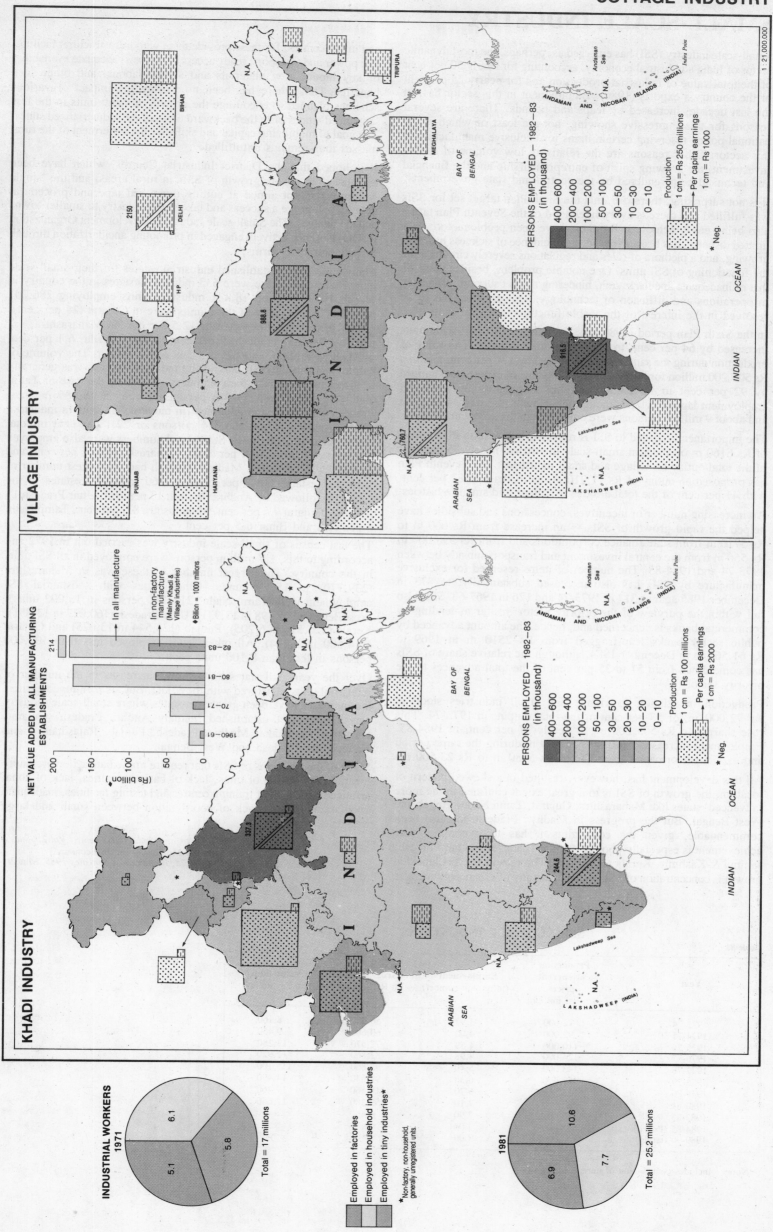

VILLAGE INDUSTRY

PERSONS EMPLOYED — 1982–83
(in thousand)

400–600
200–400
100–200
50–100
30–50
10–30
0–10

Production
1 cm = Rs 250 millions

Per capita earnings
1 cm = Rs 1000

* Neg.

KHADI INDUSTRY

NET VALUE ADDED IN ALL MANUFACTURING
ESTABLISHMENTS

In all manufacture

In non-factory
manufacture
(Mainly Khadi &
Village industries)

◆Billion = 1000 millions

200
150
100
50
0

(Rs billion◆)

82–83   214
80–81
70–71
1960–61

PERSONS EMPLOYED — 1982–83
(in thousand)

400–600
200–400
100–200
50–100
20–30
10–20
0–10

Production
1 cm = Rs 100 millions

Per capita earnings
1 cm = Rs 2000

* Neg.

INDUSTRIAL WORKERS
1971

6.1
5.1
5.8

Total = 17 millions

1981

10.6
6.9
7.7

Total = 25.2 millions

Employed in factories

Employed in household industries

Employed in tiny industries*

*Non-factory, non-household,
generally unregistered units.

SMALL-SCALE INDUSTRY

Small-scale industry (SSI) has emerged as, perhaps, the most dynamic sector of India's industrial economy, accounting for nearly 50 per cent of the total value of industrial production and for nearly 25 per cent of the country's exports (1986). Employment in this sector has, in the last decade, increased by leaps and bounds. There are several reasons for this impressive showing, not the least of which is the national policy reserving certain items for exclusive manufacture in this sector. Other reasons are the relatively low volume of capital investment, the growing spirit of entrepreneurship and the financial and technical help rendered by the central and state governments.

It is not surprising, therefore, that the Sixth Plan target set for SSIs was fulfilled and there is every likelihood of the Seventh Plan target also being easily achieved. But there have been problems too: only limited dispersal has been achieved, the incidence of sickness has been growing, and a plethora of rules and regulations severely circumscribe the functioning of SSI units. One notable problem, besides these, is that of inadequate ancillarization, hindering market support, efficiency in operations and diffusion of technology. This problem has to be resolved in the interest of the viable functioning of SSI units.

In the Sixth Plan period (1980-85), the number of small-scale units increased by 64 per cent, to almost 1.3 million units. The value of production during the same period increased by 134 per cent, totalling Rs 505,200 million for the five years. The value of exports also jumped by 92 per cent to Rs 23,500 million. Comparatively speaking, employment lagged behind, though the growth rate was 34 per cent and about 9 million persons were employed in the sector in 1984-85.

The importance accorded to SSI is reflected in the Sixth Plan outlay of Rs 6160 m on modern small-scale industries, nearly 35 per cent of the total outlay on village and small industries. The Seventh Plan has proposed an outlay of Rs 11,210 m, an increase of 82 per cent, with 41 per cent of the total allotted to village and small industries.

An increasing number of incentives, concessions and subsidies have helped the rapid growth of SSIs — an increase from Rs 660 m to Rs 5710 m from state finance corporations and from Rs 590,000 to Rs 550 m from the central investment and transport subsidy between 1973-74 and 1984-85. The number of items reserved for exclusive manufacture by SSIs has also increased substantially: to 873 in December 1985 against 313 in 1973-74 and 126 in 1967-68. SSIs also fall within the purview of the priority sector insofar as lending by commercial banks is concerned and, hence, the amount advanced by public sector banks leap-frogged from Rs 2510 m in 1969 to Rs 61,560 m in December 1984, although the relative share of SSIs has come down from 57 to 35 per cent of the total advances to the priority sector.

Production in relation to village and small industries stood at Rs 72,000 m, or 53 per cent of the total output, in 1973-74. This rose sharply to Rs 505,200 m, or nearly 77 per cent, in 1984-85. Employment increased from 3.97 m to 9 m during the same period and export performance soared from Rs 5380 m to Rs 23,500 m.

All this development has, however, resulted in a skewed pattern of location; the growth of SSI is to a great extent confined to the more 'advanced' states like Maharashtra, Gujarat, Tamil Nadu, Punjab and West Bengal. But the progress in Madhya Pradesh has also been commendable, given the constraints it has to overcome. The achievement is especially poor in Assam, Bihar, Himachal Pradesh, Jammu & Kashmir, Kerala and Orissa. The reason for this lopsided growth is concentration of SSIs in the proximity of urban agglomerates and urban centres, for these provide them with infrastructural facilities like power and transport, ready access to markets, adequate availability of key inputs, raw materials and skilled labour, and other such benefits. Thus, there has been no worthwhile impact of various schemes planned to encourage the growth of small units in the less 'advanced' states or in the backward areas of the industrialized states. The goal of mobilizing capital and skills for the betterment of the rural masses thus remains unfulfilled.

It is hoped that the District Industrial Centres, which have been established to ensure growth of SSIs in rural areas, and the 'single window' concept aimed at cutting down red tape and procedural delays, will prove a success and take small industry to smaller towns and the villages. The Small-scale Industries Development Organization (SIDO) has been actively engaged in promoting ancillarization through its subcontracting firms.

Many states have established industrial estates for their small-scale industrial units. There were 633 industrial estates in the country in March 1979, housing 18,421 industrial units employing 286,201 persons. The majority of these units were in Gujarat (24 per cent), 15 per cent in Andhra Pradesh, 12.5 per cent in Maharashtra, 9.2 per cent in Uttar Pradesh, 9 per cent in Tamil Nadu, 6.1 per cent in Madhya Pradesh and 6.3 per cent in Karnataka. The volume of employment in the SSI units in the industrial estates was generally in direct proportion to their numbers, though in the case of Tamil Nadu they employed 53,446 persons as against 13,886 in Uttar Pradesh. The largest employment in modern SSI units in industrial estates was in Gujarat (63,234 porsons or 22.1 per cent of total employment in such units). Substantial numbers were also employed in Andhra Pradesh (16.1 per cent), Maharashtra (12.1 per cent) and Karnataka (6 per cent). Madhya Pradesh had the largest number of industrial estates (13.7 per cent of total industrial estates in the country), followed by Andhra Pradesh 13 per cent, Uttar Pradesh 11 per cent, Gujarat 9.5 per cent, Maharashtra 8.5 per cent, Tamil Nadu 8 per cent and Bihar 6.5 per cent.

The last census of small-scale industry was carried out in 1972 and according to this, 1.7 million persons were employed in all SSI units in the country. The largest number of these was in Maharashtra (239,770) in 15,358 units. Other states with substantial SSI employment were Tamil Nadu (215,182 persons in 16,002 units); West Bengal (176,198 in 13,931); Uttar Pradesh (160,027 in 12,851); Kerala (126,514 in 6,205); Punjab (123,544 in 13,675) and Gujarat (114,500 in 9,904). All other states employed less than 100,000 persons in less than 8,100 units each.

Over the years, all states have shown increases in the number of industrial units registered with the SIDO. But, as mentioned earlier, the increases have been greater in states where small-scale industry was already well-established, namely Andhra Pradesh, Gujarat, Haryana, Karnataka, Madhya Pradesh, Punjab, Rajasthan, Tamil Nadu, Uttar Pradesh and West Bengal.

Some of the problems of SSIs that require immediate action are faulty location and design of units, lack of basic amenities, lack of zonal arrangements, lack of training centres and testing facilities, inadequate ancillarization, and lack of coordination between small and large industries.

Sources: 1. *Basic Statistics Relating to the Indian Economy, Vol.2: States, September 1986,* Centre for Monitoring Indian Economy.
2. *Small Scale Industries in India, Handbook of Statistics 1985,* Ministry of Industry.

SMALL-SCALE INDUSTRIES

Table 81

Year	Production [1] at current prices (in Rs. million)	Employment [1] (in million number)	Gross output (in Rs. million)	Value added (in Rs. million)
1973-74	72,000	3.97	50,550	8,160
1974-75	92,000	4.00	60,680	9,450
1975-76	110,000	4.59	77,570	13,280
1976-77	124,000	4.98	82,860	12,920
1977-78	143,000	5.40	95,860	15,100
1978-79	157,900	6.38	99,630	14,490
1979-80	216,350	6.70	116,660	17,460
1980-81	280,600	7.10	159,280	22,650
1982-83	350,000	7.90	N.A.	N.A.
1984-85 (Provisional)	505,200	9.00	N.A.	N.A.
1989-90 (target)	802,000	9.00	N.A.	N.A.

Note: [1] Includes contribution of unregistered units.

SMALL SCALE INDUSTRY

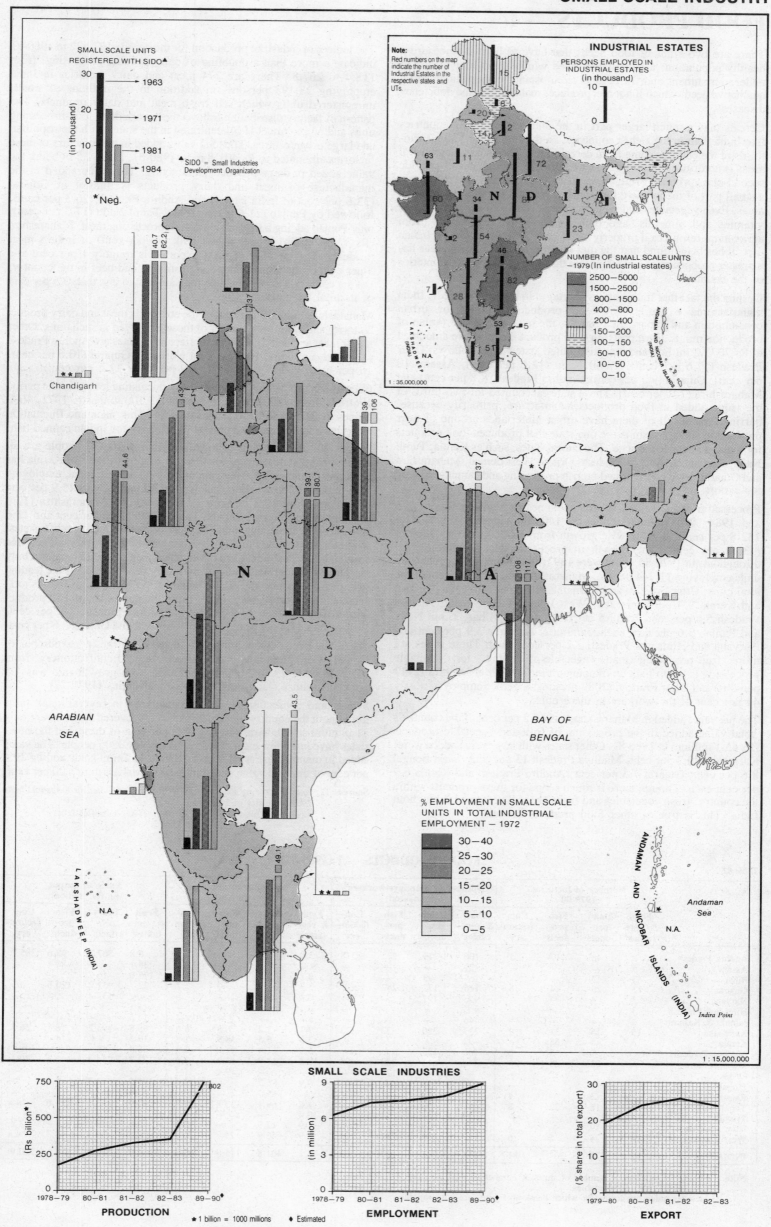

SMALL SCALE UNITS
REGISTERED WITH SIDO▲

30
20
10
0
(in thousand)

← 1963
← 1971
← 1981
← 1984

▲ SIDO = Small Industries
Development Organization

* Neg.

Chandigarh

INDUSTRIAL ESTATES

Note:
Red numbers on the map
indicate the number of
Industrial Estates in the
respective states and
UTs.

PERSONS EMPLOYED IN
INDUSTRIAL ESTATES
(in thousand)

10
0

NUMBER OF SMALL SCALE UNITS
—1979 (In industrial estates)

2500–5000
1500–2500
800–1500
400–800
200–400
150–200
100–150
50–100
10–50
0–10

ANDAMAN AND
NICOBAR ISLANDS
(INDIA)

LAKSHADWEEP
(INDIA)

N.A.

1 : 35,000,000

40.7 62.2

37

39 106

43

44.6

39.7 80.7

37

108 117

43.5

49.1

I N D I A

ARABIAN
SEA

BAY OF
BENGAL

LAKSHADWEEP
(INDIA)

N.A.

% EMPLOYMENT IN SMALL SCALE
UNITS IN TOTAL INDUSTRIAL
EMPLOYMENT – 1972

30–40
25–30
20–25
15–20
10–15
5–10
0–5

ANDAMAN AND NICOBAR ISLANDS (INDIA)

Andaman
Sea

N.A.

Indira Point

1 : 15,000,000

SMALL SCALE INDUSTRIES

750
500
250
0
(Rs billion*)

802

1978–79 80–81 81–82 82–83 89–90♦

PRODUCTION

★ 1 billion = 1000 millions   ♦ Estimated

9
6
3
0
(in million)

1978–79 80–81 81–82 82–83 89–90♦

EMPLOYMENT

30
20
10
0
(% share in total export)

1979–80 80–81 81–82 82–83

EXPORT

FOOD PRODUCTS — I

There are two kinds of food needs that have to be met to ensure a healthy population: a quantitative need which if not met results in undernourishment and, in certain circumstances, starvation; and a qualitative need, which if not met produces malnutrition and deficiency diseases.

Cereals play a much larger part in the diet of developing countries like India than in the richer countries. In the Indian diet, about two-thirds of the calories come from cereal foods. The Indian diet in rural areas is notoriously lacking in animal protein, though pulses provide second-class protein. Fruit, meat, dairy products and fish form only a small part of the average Indian diet, and that too mainly in urban areas. Even vegetables and milk, both important sources of protective vitamins and minerals, and both relatively easy to supply from agricultural resources if properly developed, are deficient in the Indian diet. It has been found that the diet of people in the littoral areas has a greater proportion of animal protein than that of people in the interior of the country.

Despite the fact that India suffers shortages in basic food needs, India manufactures a variety of food products, mainly for urban consumption and export. In 1980-81, the net value added (value of production minus direct costs) in food product manufacture amounted to Rs 7072.2 million and was distributed statewise as follows: Uttar Pradesh (17.6 per cent), Tamil Nadu (12.6 per cent), Assam (10 per cent), Gujarat (9.1 per cent), Andhra Pradesh (8.8 per cent) and Maharashtra (7.9 per cent). These states accounted for two-thirds of the value added in food products manufacture, primarily because, barring Assam, all of them have urban hinterlands around or near production sites. Assam is the one state that produces food products for a market outside the state. The littoral states, such as Kerala, Tamil Nadu, Karnataka, Andhra Pradesh, Orissa, West Bengal, Maharashtra and Gujarat, have concentrated on fish processing and canning, mainly for export.

Processed fruit and vegetable production nearly trebled between 1975 and 1984, from 49,500 tonnes to 140,000 tonnes (growth rate: 182.8 per cent), a value-wise growth from Rs 260 m to Rs 1380 m (430.8 per cent). The growth in processing has been consistent throughout. In 1979-80, there were 139 fruit and vegetable processing units employing 11,744 people. Maharashtra had 18.7 per cent of the total units, Uttar Pradesh and Karnataka 10.6 per cent each, Gujarat and Kerala 9.4 per cent each, West Bengal 8.6 per cent, Andhra Pradesh 5.8 per cent each, Delhi 5 per cent, Assam, Bihar, Tamil Nadu and Punjab 3.6 per cent each, Jammu & Kashmir 2.9 per cent and Haryana and Himachal Pradesh 2.2 per cent each. These states are more fruit-producing than vegetable-producing, for vegetable processing is yet to pick up momentum in India. Maharashtra (2719 persons) and Uttar Pradesh (2089 persons) together employed nearly 41 per cent of the workers in these units.

The net value added in Maharashtra was 41.2 per cent of the country's total value added in the processing of fruit and vegetable products (Rs 45.1 million) in 1980-81. Other states with large value added were: Karnataka 20.2 per cent, Madhya Pradesh 13 per cent, West Bengal 8.6 per cent, Gujarat 7.3 per cent, Andhra Pradesh and Kerala 6.7 per cent each. Though there is much scope for these products within the country, fresh vegetables and fruits are still preferred throughout India. This is true of other food products as well.

The indices of industrial production for the decade 1971-72 to 1980-81 indicate a more than a doubling of dairy product processing, from 115.4 to 267.8. There are 274 meat and dairy factories in India employing 38,193 persons, in addition to the millions of small, unregistered units which sell fresh meat and dairy products. The pattern of factory distribution follows the pattern of all food-processing units and employment is concentrated in the states with metropolitan and large urban centres. The total value added in manufacture in these factories amounted to Rs 401.8 m in 1980-81, Rs 354.5 m being the value added in dairy product manufacture. The value added in the manufacture of meat and dairy products is highest in Gujarat (33.6 per cent of India's total) and Andhra Pradesh (26.3 per cent), followed by Punjab (21.6 per cent) and Tamil Nadu (17.2 per cent), only Punjab adding a negligible value in processing meat. Maharashtra (Rs 37.5 million) processes the bulk (79 per cent) of India's meat products. The quantity of meat exported is roughly 1 per cent and 5 per cent of the total production of mutton and beef in the country, but the country's exports of livestock amount to less than 0.5 per cent of its total livestock holding.

Maharashtra has 21.2 per cent of the country's meat and dairy product factories and employs a quarter of those employed in such units. Other major meat and dairy product manufacturing states are Andhra Pradesh (17.5 per cent units; 10.4 per cent workers), Gujarat (10.6 per cent; 16 per cent) and West Bengal (8.9 per cent; 11.5 per cent).

Canned fish production has shown very fluctuating levels for the period 1966-80, from 7800 tonnes in 1966 to 13,100 tonnes in 1971, 4600 tonnes in 1975, and 5200 tonnes in 1980. This slack and fluctuation indicates a falling demand outside the country for Indian canned fish.

India's 207 fish canning factories employing 10,478 people are all in the littoral states except for three in Madhya Pradesh. Kerala has the largest number, 32.9 per cent of the total units in India, employing 21.3 per cent of the workers, followed by Karnataka, with 15.9 per cent units employing 10.2 per cent of the workers, Maharashtra (12.6 per cent and 17.6 per cent) and Tamil Nadu (10.6 per cent and 12.6 per cent). The remaining units (28.5 per cent of the total) are in 6 states/union territories and employ 38.3 per cent of the workers. The three units in Madhya Pradesh are based largely on inland fishing and only marginally on marine fishing from nearby states. Kerala is the leading state in fish produce and adds a value of 33.6 per cent of India's total value added in processing. Other leading value adders in fish production are Tamil Nadu 17.9 per cent, Andhra Pradesh 12.3 per cent, Maharashtra 11.4 per cent, Orissa 7.9 per cent and Gujarat 7.8 per cent.

There has been a slow but steady increase in the index number of industrial production of cocoa, chocolate, and confectionery (from 111.5 in 1971-72 to 116.5 in 1980-81). The growth rate was 4.5 per cent, with a slack only in the late seventies (1976-77).

Sweets and confectionery are manufactured in several small units throughout the country, but the larger and registered factories are only 81 in number. Maharashtra has 28.3 per cent of these and the other states have just a few each. These units employ 3205 people. The value added in manufacture is Rs 46.6 m, 8 units in Tamil Nadu adding 29.4 per cent of the total value added in India to Maharashtra's 20 per cent.

Sources: 1. *Annual Survey of Industries 1979-80 & 1980-81, Statistical Abstract India 1984,* Ministry of Planning.
2. *Economic Times 12-7-1985 & 24-10-1985,* Times of India

Table 82

FOOD PRODUCTS — I (1980-81)

	Number of factories 1979-80				Number of mandays-workers [1] (in thousand)				Net value added (in Rs. million)				Value of output (in Rs. million)			
	Fruit & vegetables[2]	Dairy products	Fish products	Confectionery	Fruit & vegetables[2]	Dairy products	Fish products	Confectionery	Fruit & vegetables[2]	Dairy products	Fish products	Confectionery	Fruit & vegetables[2]	Dairy products	Fish products	Confectionery
Andhra Pradesh	8	48	17	3	106	1,219	275	378	3.0	105.7	14.2	7.3	8.8	567.8	98.6	385.7
Assam	5	—	—	—	229	—	—	—	N.A.	—	—	—	38.1	—	—	—
Bihar	5	13	—	9	167	93	—	—	0.2	0.8	—	2.8	175.6	16.2	—	—
Gujarat	13	29	14	8	115	1,172	242	N.A.	3.3	134.9	9.0	2.7	22	1,952.2	120.6	N.A.
Haryana	3	4	—	—	N.A.	197	—	—	0.4	20.2	—	—	N.A.	135.4	—	—
Himachal Pradesh	3	3	—	—	N.A.	N.A.	—	—	-0.1	N.A.	—	—	N.A.	N.A.	—	—
Jammu & Kashmir	4	—	—	—	46	—	—	—	1.4	—	—	—	14.5	—	—	—
Karnataka	15	15	33	9	237	280	32	123	9.1	-2.2	6.3	6.6	68.8	194.2	16.3	36.5
Kerala	13	5	68	—	92	N.A.	122	—	3.0	8.5	38.9	—	19.5	N.A.	214	—
Madhya Pradesh	N.A.	7	3	6	166	123	N.A.	N.A.	5.9	11.4	N.A.	0.9	68.4	68	N.A.	N.A.
Maharashtra	26	58	26	23	340	611	302	94	18.6	-11.9	13.2	9.5	123.3	234.4	142.1	30.1
Orissa	—	—	12	—	—	—	127	—	—	—	9.1	—	—	—	82.9	—
Punjab	5	11	—	—	290	394	—	—	0.9	86.7	—	—	191.5	653.7	—	—
Rajasthan	—	13	—	4	—	163	—	N.A.	—	5.7	—	N.A.	—	163.4	—	—
Tamil Nadu	5	18	22	8	261	206	103	N.A.	1.8	69.1	20.8	13.7	281.9	283.1	127.6	N.A.
Tripura	N.A.	—	—	—	9	—	—	—	0.4	—	—	—	1.1	—	—	—
Uttar Pradesh	15	19	—	5	248	294	—	12	-1.0	17.9	—	-0.1	49.6	249.9	—	2.3
West Bengal	12	24	5	6	152	1,139	282	N.A.	3.9	-46.9	21	2.7	21.8	44.3	48.7	N.A.
INDIA[3]	139	274	207	81	2,458	6,641	1,227	501	45.1	401.8	115.9	46.6	1,084.9	5,570.7	814	220.9

Note: [1] Number of workers × number of shifts × number of working days.
[2] Canning of fruits & vegetables.
[3] Includes union territories for which break-up figures are not available.

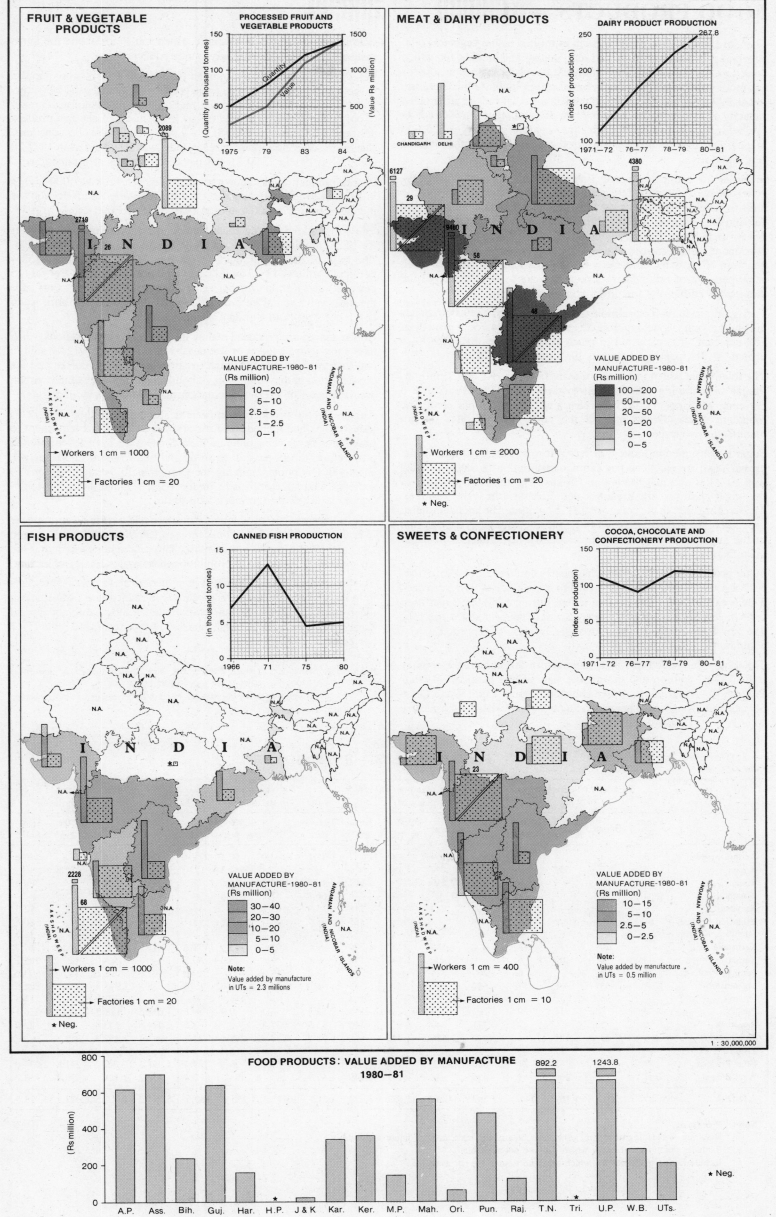

FRUIT & VEGETABLE PRODUCTS

PROCESSED FRUIT AND VEGETABLE PRODUCTS

(Quantity in thousand tonnes) / (Value Rs million)

Quantity
Value

1975 79 83 84

2089

N.A.

2749

26

I N D I A

N.A.

N.A.

LAKSHADWEEP (INDIA)

ANDAMAN AND NICOBAR ISLANDS (INDIA)

VALUE ADDED BY MANUFACTURE-1980-81
(Rs million)
- 10—20
- 5—10
- 2.5—5
- 1—2.5
- 0—1

→ Workers 1 cm = 1000

→ Factories 1 cm = 20

MEAT & DAIRY PRODUCTS

DAIRY PRODUCT PRODUCTION

(index of production)

267.8

250
200
150
100
1971-72 76-77 78-79 80-81

N.A.

CHANDIGARH DELHI

6127

29

9480

58

48

I N D I A

N.A.

4380

N.A.

N.A.

LAKSHADWEEP (INDIA)

ANDAMAN AND NICOBAR ISLANDS (INDIA)

VALUE ADDED BY MANUFACTURE-1980-81
(Rs million)
- 100—200
- 50—100
- 20—50
- 10—20
- 5—10
- 0—5

→ Workers 1 cm = 2000

→ Factories 1 cm = 20

★ Neg.

FISH PRODUCTS

CANNED FISH PRODUCTION

(in thousand tonnes)

15
10
5
0
1966 71 75 80

N.A.

N.A.

★

I N D I A

N.A.

2228

68

N.A.

N.A.

LAKSHADWEEP (INDIA)

ANDAMAN AND NICOBAR ISLANDS (INDIA)

VALUE ADDED BY MANUFACTURE-1980-81
(Rs million)
- 30—40
- 20—30
- 10—20
- 5—10
- 0—5

Note:
Value added by manufacture in UTs = 2.3 millions

→ Workers 1 cm = 1000

→ Factories 1 cm = 20

★ Neg.

SWEETS & CONFECTIONERY

COCOA, CHOCOLATE AND CONFECTIONERY PRODUCTION

(index of production)

150
100
50
0
1971-72 76-77 78-79 80-81

N.A.

N.A.

I N D I A

23

5

N.A.

N.A.

LAKSHADWEEP (INDIA)

ANDAMAN AND NICOBAR ISLANDS (INDIA)

VALUE ADDED BY MANUFACTURE-1980-81
(Rs million)
- 10—15
- 5—10
- 2.5—5
- 0—2.5

Note:
Value added by manufacture in UTs = 0.5 million

→ Workers 1 cm = 400

→ Factories 1 cm = 10

1 : 30,000,000

FOOD PRODUCTS : VALUE ADDED BY MANUFACTURE 1980—81

(Rs million)

892.2 1243.8

800
600
400
200
0

A.P. Ass. Bih. Guj. Har. H.P. J & K Kar. Ker. M.P. Mah. Ori. Pun. Raj. T.N. Tri. U.P. W.B. UTs.

★ ★

★ Neg.

FOOD PRODUCTS — II

There are certain food products such as refined edible vegetable oils, beverages (processed coffee and tea), sugar, and alcohol for which there is a very good market within the country. Vanaspati, one of the more popular edible oils, is based on groundnut oil. There are (by a 1979-80 count) 65 vanaspati units in the country with an installed capacity of 1.35 m tonnes. Taking into account other types of edible oils being used at present, the *effective* vanaspati capacity is 823,000 tonnes. Part of this capacity is utilized to manufacture other refined vegetable oils.

In order to reduce regional imbalances in vanaspati capacity, government has permitted the setting up of new, economically viable units in states which do not yet have any units but have a reasonable demand for vanaspati, such as Himachal Pradesh, Jammu & Kashmir, Assam and Orissa. Encouragement is also being given to ensure that uneconomic units are raised to economically viable capacities. The quantum of imported edible oils for use in vanaspati has been fixed at 60 per cent of the industry's requirements so as to conserve the indigenous edible oils for direct consumption.

Vanaspati production has almost trebled since 1961 (339,000 tonnes during the year), touching 909,000 tonnes in 1984-85, a growth of 168 per cent. It is expected to reach 1.21 m tonnes by 1989-90 (a growth of 257 per cent between 1961 and 1990).

The total value added in vanaspati manufacture is Rs 664.7 million, Punjab contributing 8.8 per cent towards this, Uttar Pradesh 8.6 per cent, Maharashtra 14.5 per cent and Gujarat 10 per cent. The mountain states, Orissa and all the north-eastern states do not manufacture vanaspati.

Sugar is consumed in India in different forms: as refined mill sugar, brown sugar or gur. The per capita consumption of sugar in 1979 was 10.2 kgs and in 1980 it was 7.5 kgs. The main by-products of the sugar industry are molasses and bagasse (the cane residue). Molasses are used in the country's 131 distilleries. The annual installed capacity of these distilleries is 744,513 litres. Bagasse is used as fuel in the sugar mills and as a supplementary raw material in paper manufacture.

Sugar production in the country has nearly doubled since 1961, from 3.03 m tonnes to 5.9 m tonnes in 1984 (a growth rate of 94.7 per cent). This is expected to reach 10.2 m tonnes during 1989-90. Sugar used to be in very short supply till a few years ago, but near self-sufficiency has now been reached, although sugar is also an important export commodity.

The total value added in sugar manufacture throughout the country is Rs 1573.2 million and the greater part of this is added in Uttar Pradesh (44.9 per cent). Other contributors are Tamil Nadu (13.6 per cent) and Bihar (11.5 per cent). Sugar is processed in almost every

state in India; Jammu & Kashmir, Himachal Pradesh and the north eastern territories are the only non-producers.

Beverages are a daily need in most families. Coffee is more popular in south Indian households and tea in north Indian households. However, both beverages have an all-India market. The internal consumption of coffee in 1977-78 was 44.1 m kgs (35 per cent of national production) and 47.7 m kgs in 1980-81 (39.7 per cent of national production). The estimated consumption of tea in India was 300.2 m kgs in 1977-78 and 358 m kgs in 1980-81 (63 per cent of national production).

Coffee production has shown a steady increase from 19,000 tonnes in 1950-51 to 190,000 tonnes in 1984-85. Similarly tea production has increased from 278,000 tonnes in 1950-51 to 653,000 tonnes in 1985.

The value added during manufacture in tea production in India is Rs 1225.7 million (1980-81) and Rs 50.2 million in coffee production. The bulk of the value added in tea production is in Assam (52.4 per cent), followed by Tamil Nadu (16.0 per cent), West Bengal (15 per cent) and Kerala (7.7 per cent). Karnataka (62.9 per cent) adds the most value in coffee production and is followed by Tamil Nadu (24.1 per cent) and Kerala (7.8 per cent).

Coffee and tea are produced mainly in rural areas (in plantations which have their own factories or use nearby ones), but are largely sold in and mainly distributed through urban centres. Total coffee and tea consumption in the villages is, of course, much higher than in urban areas because of greater population numbers in the countryside.

Annual alcohol production in India is about 535 million litres. This industry has had a chequered history due to India's commitment to prohibition never being quite fulfilled. Some states have never imposed prohibition, others have done so on and off. Country liquor (arrack and toddy) is popular among the economically weaker sections of society, while factory-made liquor (called Indian-made 'foreign' liquor) is drunk by the more well-to-do.

It is in beer and ale production that there has been a steep increase over the years, from 5.4 m bulk litres to 138.9 m bulk litres between 1957 and 1980-81, a 26-fold increase. This production picked up in the seventies with the lifting of prohibition in several parts of the country.

The total value added in production of alcohol and alcoholic products in India is Rs 661 million. The greatest value added is in Uttar Pradesh (25.6 per cent of the Indian total). Other major value adders are Karnataka (19.9 per cent), Maharashtra (12.6 per cent) and Bihar (11.7 per cent).

Sources: 1. *Statistical Abstract India 1984*, and *Annual Survey of Industries 1979-80, & 1980-81*, Ministry of Planning.
2. *Seventh Five Year Plan 1985-90, Vol. I & II*, Government of India Planning Commission.

FOOD PRODUCTS — II (1980-81)

Table 83

	Number of Factories — 1979-80				Number of mandays-workers (in thousand)				Net value added (in Rs. million)				Value of output (in Rs. million)			
	Vanas-pati	Sugar	Coffee & Tea[1]	Alco-hol[2]	Vanas-pati	Sugar	Coffee & Tea[1]	Alco-hol[2]	Vanas-pati	Sugar	Coffee & Tea[1]	Alco-hol[2]	Vanas-pati	Sugar	Coffee & Tea[1]	Alco-hol[2]
Andhra Pradesh	5	21	4	16	455	2,167	N.A.	247	45.8	104.0	30.1	29.4	429.9	636.7	N.A.	115.8
Assam	—	—	533	N.A.	—	—	13,188	N.A.	—	—	642.3	2	—	—	3,337.8	N.A.
Bihar	N.A.	292	—	9	N.A.	2,516	—	601	14.8	181.0	—	77.3	N.A.	569.5	—	360.6
Gujarat	9	12	4	—	632	1,090	641	—	66.3	57.8	0.8	—	971.9	669.5	425.6	—
Haryana	3	7	—	5	N.A.	823	—	218	3.9	83.2	—	24.9	N.A.	305.1	—	92.3
Himachal Pradesh	—	—	8	4	—	—	N.A.	155	—	—	0.2	15.7	—	—	N.A.	80.5
Jammu & Kashmir	—	—	—	N.A.	—	—	—	57	—	—	—	2.5	—	—	—	12.5
Karnataka	4	20	30	14	N.A.	2,560	1,337	1,373	N.A.	103.3	43.9	131.3	N.A.	878.3	155.0	67.7
Kerala	—	4	105	6	—	N.A.	1,190	114	—	7.6	98.0	40.8	—	N.A.	627.2	105.2
Madhya Pradesh	4	5	—	6	209	374	—	89	13.2	19.5	—	19.0	307.9	84.2	—	21.6
Maharashtra	8	63	4	17	602	7,238	322	513	96.9	24.1	N.A.	83.3	1,358.8	3,975.4	354.7	322.5
Orissa	—	N.A.	—	9	—	50	—	104	—	N.A.	—	N.A.	—	7.1	—	18.3
Punjab	8	6	14	9	302	749	N.A.	818	58.2	36.7	0.9	59.8	1,098.6	188.0	N.A.	336.1
Rajasthan	6	3	—	7	171	340	—	195	35.9	7.3	—	3.5	494.6	43.4	—	22.0
Tamil Nadu	4	23	171	4	N.A.	2,686	1,748	N.A.	38.8	213.4	208.5	21.4	N.A.	1,192.3	919.8	N.A.
Tripura	—	—	27	—	—	—	99	—	—	—	5.4	—	—	—	10.4	—
Uttar Pradesh	7	86	3	21	530	13,470	141	1,644	145.1	707.1	26.5	169.4	1,391.3	3,353	271.3	2,075.0
West Bengal	7	3	266	9	496	N.A.	7,193	1,149	57.3	N.A.	183.3	33.4	732.2	N.A.	1,970.3	343.2
INDIA[3]	65	551[4]	1,169	142[4]	3,397	34,063	25,859	7,277	664.7[4]	1,573.2[4]	1,275.9[4]	660.9[4]	6,785.2	11,902.5	8,072.1	3,973.3

Note: [1] Processing.
[2] Distilling & blending of spirits, wine, malt liquors and malt, country liquor and toddy.
[3] Excludes union territories for which data are not available.
[4] Includes union territories for which break-up figures are not available.

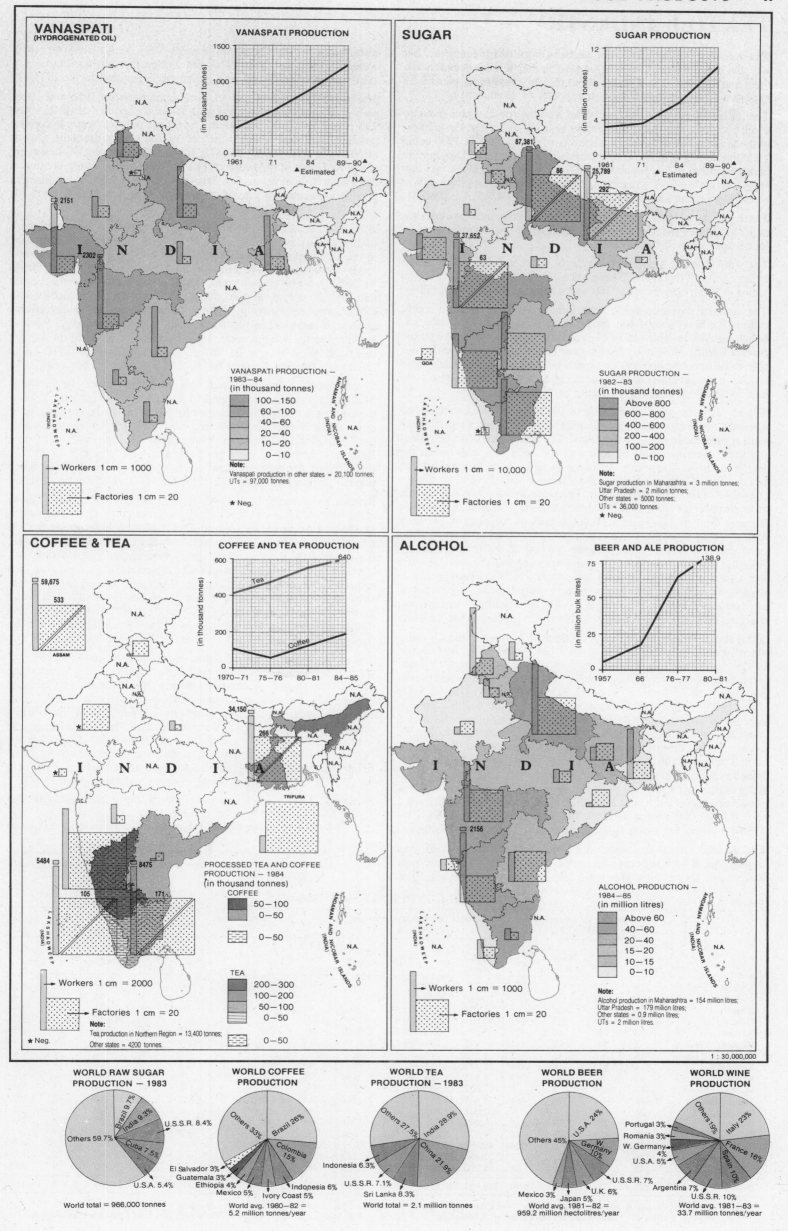

VANASPATI (HYDROGENATED OIL)

VANASPATI PRODUCTION
(in thousand tonnes)
1500 | 1000 | 500 | 0
1961 71 84 89—90 ▲ Estimated

N.A. N.A. N.A.
N.A. N.A. N.A.
2151
I N D I A
2302
N.A.
N.A.
N.A.

Workers 1 cm = 1000
Factories 1 cm = 20
★ Neg.

VANASPATI PRODUCTION — 1983—84
(in thousand tonnes)
100—150
60—100
40—60
20—40
10—20
0—10

Note:
Vanaspati production in other states = 20,100 tonnes;
UTs = 97,000 tonnes.

SUGAR

SUGAR PRODUCTION
(in million tonnes)
12 | 8 | 4 | 0
1961 71 84 89—90 ▲ Estimated

N.A.
87,381
86
25,789
292
37,652
63
I N D I A
N.A. N.A.
N.A.
GOA
N.A.
★

Workers 1 cm = 10,000
Factories 1 cm = 20

SUGAR PRODUCTION — 1982—83
(in thousand tonnes)
Above 800
600—800
400—600
200—400
100—200
0—100

Note:
Sugar production in Maharashtra = 3 million tonnes;
Uttar Pradesh = 2 million tonnes;
Other states = 5000 tonnes;
UTs = 36,000 tonnes.
★ Neg.

COFFEE & TEA

COFFEE AND TEA PRODUCTION
(in thousand tonnes)
600 | 400 | 200 | 0
640
Tea
Coffee
1970—71 75—76 80—81 84—85

59,675
533
ASSAM
N.A.
N.A.
N.A.
N.A.
★
N.A.
34,150
266
I N D I A
N.A.
N.A.
N.A.
N.A.
TRIPURA
5484
8475
105
171
★

Workers 1 cm = 2000
Factories 1 cm = 20
★ Neg.

PROCESSED TEA AND COFFEE
PRODUCTION — 1984
(in thousand tonnes)
COFFEE
50—100
0—50
0—50
TEA
200—300
100—200
50—100
0—50
0—50

Note:
Tea production in Northern Region = 13,400 tonnes;
Other states = 4200 tonnes.

ALCOHOL

BEER AND ALE PRODUCTION
(in million bulk litres)
75 | 50 | 25 | 0
138.9
1957 66 76—77 80—81

N.A.
N.A.
I N D I A
N.A.
N.A.
N.A.
2156

Workers 1 cm = 1000
Factories 1 cm = 20

ALCOHOL PRODUCTION — 1984—85
(in million litres)
Above 60
40—60
20—40
15—20
10—15
0—10

Note:
Alcohol production in Maharashtra = 154 million litres;
Uttar Pradesh = 179 million litres;
Other states = 0.9 million litres;
UTs = 2 million litres.

1 : 30,000,000

WORLD RAW SUGAR PRODUCTION — 1983
Brazil 9.7%
India 9.3%
U.S.S.R. 8.4%
Cuba 7.5%
U.S.A. 5.4%
Others 59.7%
World total = 966,000 tonnes

WORLD COFFEE PRODUCTION
Brazil 26%
Colombia 15%
Indonesia 6%
Ivory Coast 5%
Mexico 5%
Ethiopia 4%
Guatemala 3%
El Salvador 3%
Others 33%
World avg. 1980—82 = 5.2 million tonnes/year

WORLD TEA PRODUCTION — 1983
India 28.9%
China 21.9%
Sri Lanka 8.3%
U.S.S.R. 7.1%
Indonesia 6.3%
Others 27.5%
World total = 2.1 million tonnes

WORLD BEER PRODUCTION
U.S.A. 24%
W. Germany 10%
U.S.S.R. 7%
U.K. 6%
Japan 5%
Mexico 3%
Others 45%
World avg. 1981—82 = 959.2 million hectolitres/year

WORLD WINE PRODUCTION
Italy 23%
France 16%
Spain 10%
U.S.S.R. 10%
Argentina 7%
U.S.A. 5%
W. Germany 4%
Romania 3%
Portugal 3%
Others 19%
World avg. 1981—83 = 33.7 million tonnes/year

TEXTILE INDUSTRY — I

The excellence of Indian textiles has been extolled for centuries. But with the industrial revolution in the west, Indian textiles began to lose their market overseas. It was at about this time that India became an exporter of cotton rather than an exporter of cloth.

The Indian industrial revolution began with the cotton textile industry about 130 years ago. India now enjoys near self-sufficiency in textiles, with the estimated per capita availability of cotton cloth in 1980 being 11.25 metres. In addition, cotton fabrics, yarn, apparel and hosiery are being exported in large quantities. Of particular note is the fact that India's exports of readymade garments have grown considerably in recent years, from Rs 140 million in 1971-72 to Rs 4295 m in 1980-81.

There are, by a 1980 count, some 136,000 authorized powerlooms and 700 knitting machines besides a number of unauthorized power-looms operating in India.

The wool industry began a century ago, but there was no significant expansion until the 1950s. There were only 16 units in the organized sector in 1950; now there are 360 with a capacity of 422,990 spindles.

The synthetics or man-made fibre industry is of even more recent origin. The first factory went into production only in 1950. Synthetics, being durable, reliable and of high quality, have expanded the textile market significantly, even though their prices are 4 to 5 times higher than those of cotton textiles. The synthetic textiles industry made rapid strides in the seventies at the expense of cotton textiles, and by 1980 had achieved an annual production of 192,000 tonnes.

The textile policy announced in June 1985 augurs well for increased domestic production and an adequate supply of synthetics and filament yarn at reasonable prices in the future. The policy also plans on retaining the pre-eminent role of cotton as the main raw material in the textile industry. The policy must be viewed from the fact that growth in capacity of man-made fibre and filament plants would be almost four-fold in the next three years. This growth would be continuous from the present till the end of the decade. It is estimated that over the next three to four years, 160,000 tonnes of polyester staple fibre and an equal amount of nylon and polyester filament yarn will be consumed every year. But a recent survey of the textile industry indicates that the majority of the cotton spinning mills are endangered. For a total of 15.77 m spindles and nearly 250,000 looms, some in composite mills and the rest in the powerloom sector, a process of attrition has already set in. A fall in demand has already lowered prices, margins thinning for the better mills and disappearing for the rest.

The value of textile output has increased 11½ times in the 25 years since 1960-61, from Rs 10,930 m to Rs 124,700 m. In the first decade of this period the increase was very small, but in the next 15 years it has been phenomenal. The income generated during the same period has grown about 9 times, from Rs 3210 m in 1960-61 to Rs 5970 m in 1970-71 and Rs 28,950 m in 1984-85. Though this growth represents growth in all sectors of the textile industry, the major contribution is from synthetics rather than from cotton yarn or cloth or woollens. Cotton yarn production grew from 590 m kgs in 1951 to 860 m kgs in 1961, 880 m kgs in 1971, and 1434 m kgs in 1985. Woollen fabrics on the other hand showed a two-fold increase from 8.2 m metres in 1956 to about 19 m metres in 1978. But nylon filament production increased from a mere 200 tonnes in 1962 to 10,300 tonnes in 1971, 22,000 tonnes in 1981 and 32,000 tonnes in 1984. The 1989-90 target for nylon filament has been set at 56,000 tonnes.

The decadal growth in cotton yarn production during 1951-61 was of the order of 45.8 per cent while in the following decade, this growth was a mere 2.3 per cent. This latter decade was replete with industrial

disturbance which hit cotton textiles very hard. However, with corrective measures in the 1970s and continued production in the early 1980s, production registered a growth of 63 per cent during 1971-85.

The growth in woollen fabrics production between 1956 and 1978, owing to increasing demand and fast developing industries, was 132 per cent. In the meantime, synthetic fabrics were becoming popular and nylon filament thus registered a very dramatic growth of 5050 per cent during 1961-71. In the following decade, however, the growth was of the order of 114 per cent, representing continued demand in the internal market. Between 1981 and 1984, nylon filament production grew at a rate of 45 per cent. The estimated growth for 1981-90 is 156 per cent.

The net value added by the Indian textiles industry in the manufacture of all textiles was Rs 17.057.7 m in 1980-81. Statewise, Maharashtra in the same period added Rs 4760.9 m or 27.9 per cent of India's total), Gujarat Rs 3942.2 m (or 23.1 per cent) and Tamil Nadu Rs 2552.3 m (or 15 per cent). These three states have the largest number of cotton textile mills, in Bombay, Ahmadabad and Coimbatore respectively. Other states contributing a significant amount of value addition by textile manufacture are: Madhya Pradesh Rs 881.5 m (or 5.2 per cent), Uttar Pradesh Rs 838.3 m (or 4.9 per cent) and West Bengal Rs 781.9 m (or 4.6 per cent). The union territories together account for Rs 362.5 m (or 2.1 per cent).

Woollen textile are produced only in a few states. In 1980-81, Maharashtra had a value added of Rs 174.8 m, or 33.5 per cent of the country's total value added of Rs 521.9 m. The contributions of other states were: Punjab Rs 152.5 m (or 29.2 per cent), Gujarat Rs 75.6 m (or 14.5 per cent), Haryana Rs 41.9 m (or 8.0 per cent), Rajasthan Rs 36.5 m (or 7 per cent) and Uttar Pradesh Rs 23.5 m (or 4.5 per cent). Significantly, those states with a large sheep population play little or no role in the manufacture of woollen textiles.

Synthetic yarn and textile production in the country is restricted to eight states: Maharashtra is the largest producer with 40,262 tonnes in 1983-84, followed by Gujarat 23,526 tonnes, Uttar Pradesh 8126 tonnes, West Bengal 5913 tonnes, Madhya Pradesh 5297 tonnes, Rajasthan 4052 tonnes, Punjab 3622 tonnes and Tamil Nadu 1745 tonnes. When polyester staple fibre production is included, the sequence does not change much; the states with the larger concentrations of textile mills also have more synthetic fibre mills and hence produce higher quantities of synthetics. Tamil Nadu produces 7086 tonnes, Maharashtra 6939 tonnes, Uttar Pradesh 6813 tonnes, Rajasthan 3847 tonnes and Gujarat 3362 tonnes.

The share of textile exports to total merchandise exports in 1982 for the world was 3.6 per cent, but India achieved a 15.2 per cent share of this. This, however, compares poorly with Pakistan, whose textile exports' share to total merchandise exports' share was 41.9 per cent. In 1981, India had 211,700 looms or 7.3 per cent of the world's installed weaving capacity (2.9 million looms). The share of automatic looms in India was 21.9 per cent (or 46,362 looms). China has the highest installed weaving capacity, with about 60 per cent of its looms being automatic. In wool production, India accounts for 1.3 per cent (or 37,700 tonnes) of the world's output and ranks sixth (1983).

Sources: 1. *Annual Survey of Industries 1979-80 & 1980-81* and *Statistical Abstract India 1984*, Ministry of Planning.
2. *Seventh Five Year Plan 1985-90, Vol. I & II*, Government of India Planning Commission.
3. *Basic Statistics Relating to the Indian Economy, Vol. 1: All India, August 1986*, and *Vol. 2: States, September 1986*, Centre for Monitoring Indian Economy.

Table 84

TEXTILE INDUSTRY — I (1980-81)

	Number of Factories — 1979-80				Number of mandays-workers (in thousand)				Net value added (in Rs. million)				Value of output (in Rs. million)			
	All textiles[1]	Cotton[2]	Wool[3]	Synthetics[4]	All textiles[1]	Cotton[2]	Wool[3]	Synthetics[4]	All textiles[1]	Cotton[2]	Wool[3]	Synthetics[4]	All textiles[1]	Cotton[2]	Wool[3]	Synthetics[4]
Andhra Pradesh	359	354	—	5	8,773	8,410	—	N.A.	387	350	—	N.A.	1,556	1,529	—	N.A.
Assam	19	19	—	—	434	434	—	—	21	21	—	—	65	65	—	—
Bihar	39	39	N.A.	—	1,270	1,077	193	—	47	34	N.A.	—	141	122	19	—
Gujarat	1,564	1,129	5	430	66,981	57,356	823	6,964	3,942	3,138	76	613	16,672	11,837	237	3,841
Haryana	227	142	60	25	5,827	3,255	356	2,010	245	126	42	65	1,353	741	129	452
Himachal Pradesh	3	—	3	—	123	—	123	—	−2	—	−2	—	24	—	24	—
Jammu & Kashmir	16	—	13	3	1,578	—	168	N.A.	36	—	7	16	169	—	24	N.A.
Karnataka	614	575	—	39	11,271	9,934	—	N.A.	547	499	—	2	1,700	1,214	—	N.A.
Kerala	48	43	—	5	4,643	4,408	—	N.A.	321	250	—	41	884	803	—	N.A.
Madhya Pradesh	445	429	8	8	17,623	15,653	103	1,711	882	702	3	172	2,828	1,964	13	831
Maharashtra	1,425	931	30	464	72,645	55,693	2,378	11,648	4,761	3,145	175	1,264	16,267	10,172	692	4,807
Orissa	24	24	—	—	2,643	2,643	—	—	68	66	—	—	303	303	—	—
Punjab	929	217	287	425	9,014	4,744	3,228	361	495	249	153	59	2,348	1,169	878	134
Rajasthan	508	455	39	14	9,377	5,590	503	3,060	773	264	37	405	2,683	999	159	1,142
Tamil Nadu	1,156	932	N.A.	224	40,093	36,146	76	3,089	2,552	2,289	N.A.	226	9,313	88,263	7	903
Uttar Pradesh	226	193	21	12	21,640	19,558	1,381	443	838	785	24	8	3,285	2,986	172	59
West Bengal	243	225	N.A.	18	18,200	14,814	131	2,520	782	555	2	190	2,863	1,920	19	800
INDIA[5]	7,893	5,728	469	1,696	299,931	246,876	9,463	32,030	17,058	12,792	522	3,094	63,524	125,081	2,373	13,037

Note: [1] Cotton, wool, silk & synthetic fibre textiles.   [2] Cotton ginning and baling, spinning on charkha, weaving and finishing on powerlooms.
[3] Woollen textiles.   [4] Man-made textiles, printing & bleaching.   [5] Includes other states and union territories for which break-up figures are not available.

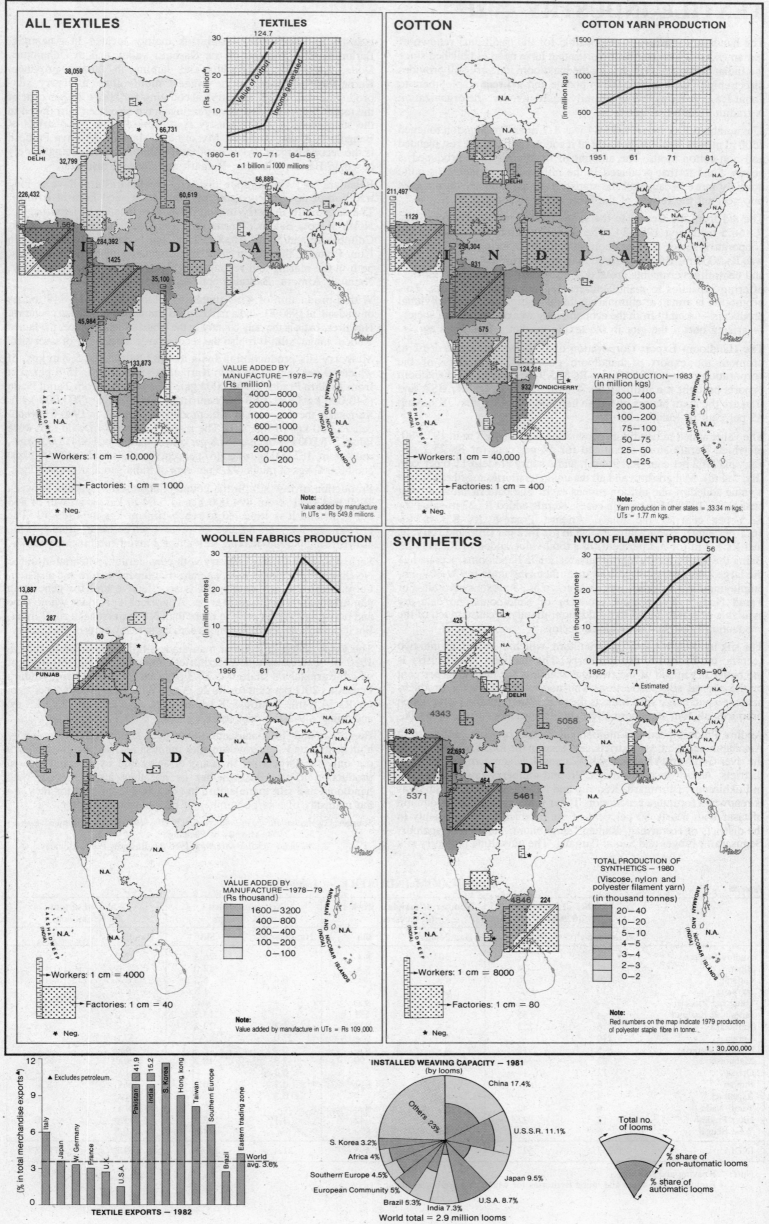

ALL TEXTILES

TEXTILES

(Rs billion)
Value of output
Income generated
124.7

1960–61 70–71 84–85
△1 billion = 1000 millions

38,059
66,731
32,799
226,432
1,564
284,392
1425
35,100
45,984
133,873
56,889
60,619

INDIA

DELHI

VALUE ADDED BY
MANUFACTURE—1978–79
(Rs million)
4000–6000
2000–4000
1000–2000
600–1000
400–600
200–400
0–200

LAKSHADWEEP (INDIA) N.A.

ANDAMAN AND NICOBAR ISLANDS (INDIA) N.A.

Workers: 1 cm = 10,000
Factories: 1 cm = 1000

★ Neg.

Note:
Value added by manufacture
in UTs = Rs 549.8 millions.

COTTON

COTTON YARN PRODUCTION

(in million kgs)
1500
1000
500

1951 61 71 81

211,497
1129
259,304
931
575
124,216
932 PONDICHERRY

INDIA

DELHI

YARN PRODUCTION—1983
(in million kgs)
300–400
200–300
100–200
75–100
50–75
25–50
0–25

LAKSHADWEEP (INDIA) N.A.

ANDAMAN AND NICOBAR ISLANDS (INDIA) N.A.

Workers: 1 cm = 40,000
Factories: 1 cm = 400

★ Neg.

Note:
Yarn production in other states = .33.34 m kgs;
UTs = 1.77 m kgs.

WOOL

WOOLLEN FABRICS PRODUCTION

(in million metres)
30
20

1956 61 71 78

13,887
287
60

PUNJAB

INDIA

VALUE ADDED BY
MANUFACTURE—1978–79
(Rs thousand)
1600–3200
400–800
200–400
100–200
0–100

LAKSHADWEEP (INDIA) N.A.

ANDAMAN AND NICOBAR ISLANDS (INDIA) N.A.

Workers: 1 cm = 4000
Factories: 1 cm = 40

★ Neg.

Note:
Value added by manufacture in UTs = Rs 109,000.

SYNTHETICS

NYLON FILAMENT PRODUCTION

(in thousand tonnes)
30
20
10
56

1962 71 81 89–90△
△ Estimated

425
4343
5058
430
22,693
464
5371
5461
4846 224

DELHI

INDIA

TOTAL PRODUCTION OF
SYNTHETICS — 1980
(Viscose, nylon and
polyester filament yarn)
(in thousand tonnes)
20–40
10–20
5–10
4–5
3–4
2–3
0–2

LAKSHADWEEP (INDIA) N.A.

ANDAMAN AND NICOBAR ISLANDS (INDIA) N.A.

Workers: 1 cm = 8000
Factories: 1 cm = 80

★ Neg.

Note:
Red numbers on the map indicate 1979 production
of polyester staple fibre in tonne.

1 : 30,000,000

TEXTILE EXPORTS — 1982

(% in total merchandise exports△)
12
9
6
3
0

▲ Excludes petroleum.

Italy
Japan
W. Germany
France
U.K.
U.S.A.
Pakistan 41.9
India 15.2
S. Korea
Hong kong
Taiwan
Southern Europe
Brazil
Eastern trading zone
World avg. 3.6%

INSTALLED WEAVING CAPACITY — 1981
(by looms)

Others 23%
China 17.4%
U.S.S.R. 11.1%
Japan 9.5%
U.S.A. 8.7%
India 7.3%
Brazil 5.3%
European Community 5%
Southern Europe 4.5%
Africa 4%
S. Korea 3.2%

World total = 2.9 million looms

Total no. of looms
% share of non-automatic looms
% share of automatic looms

183

TEXTILE INDUSTRY — II

The handloom industry is responsible for the traditional renown of Indian textiles. Despite the factories that have been established since the Indian Industrial Revolution, the handloom industry still provides direct employment to over 10 m people and indirect employment to about 1.5 m persons. Being a family-based activity, the organization is tradition-bound.

The number of looms in 1983-84 was 1.7 m and production touched 3500 m metres, about 80 per cent of it cotton cloth and the rest blended and non-cotton cloth. The amount of handloom cloth produced is nearly equal to that produced in the mill sector; in fact, it is quite possibly more because all the statistics on the industry do not take into account the results of the large unorganized sector.

The aid from the Centre to the states for the handloom sector was Rs 66.5 m during 1983-84. The flow of institutional credit to the cooperative sector for handloom manufacture during the same year was Rs 2000 m, Rs 555 m more than in the previous year. The state and central governments are active promoters of handloom fabrics, offering subsidies to manufacturers and discounts to buyers. This promotion is aimed at eliminating the hardships faced by individual producers — mainly from the economically weaker sections of society — arising out of the glut in the textile market a few years ago.

The Handloom Export Corporation of India was set up in 1962 to promote the export of handloom goods. The turnover of the corporation during 1981-82 was Rs 875.9 m; in 1982-83, handloom exports brought it a profit of Rs 4.1 m. The UK, USSR, USA and Iran are the main buyers of Indian handloom fabrics, the UK buying 35 per cent of India's exports.

The value added in handloom production was Rs 46.5 m in 1979-80, of which Kerala alone accounted for 54 per cent (or Rs 25.2 m), Haryana 20.2 per cent (Rs. 9.4 m), and Andhra Pradesh 11.6 per cent (Rs. 5.4 m). Maharashtra and all the union territories together showed a value added of one million rupees each. The pattern showed great changes in 1980-81 in certain cases. Kerala added Rs 27 m value (or 52.5 per cent of the total), Andhra Pradesh Rs 8.3 m (or 16.1 per cent), Maharashtra Rs 7.3 m (or 14.2 per cent) and Haryana just Rs 1.7 m (or 3.3 per cent) in a total value added of Rs 51.4 m. Being the state with the highest production in handlooms, Kerala had the largest number of organized manufacturing units (1979-80), 110 employing 3799 workers with an average of 35 persons per unit. For Tamil Nadu, an important producer of handlooms, no data are available on value added in production, mainly because much of the production is in the unorganized sector.

The silk industry is based on sericulture, which in India falls into two sectors: mulberry and non-mulberry. The mulberry silk industry is the better organized of the two and accounts for nearly 90 per cent of the natural silk production. The non-mulberry sector produces *tasar, eri* and *muga* silk. India is the only country in the world which produces all four commercially-exploited varieties of natural silk.

Andhra Pradesh produces mulberry and tasar raw silk: mulberry in Anantapur and Chittoor districts, and tasar in the forest regions astride the river Godavari in Karimnagar, Warangal, Khammam and Adilabad districts. Assam produces muga, eri and mulberry raw silk, mostly in Lakhimpur, Dibrugarh, Kamrup and Nowgong districts. Sibsagar is renowned for muga production. Bihar ranks first in the production of tasar with nearly 80 per cent of the production being mainly in the districts of Hazaribag, Ranchi, Singhbhum, Palamu, Bhagalpur, Begusarai, Munger and Santal Pargana. The univoltine mulberry silk industry in Jammu & Kashmir is mainly located in Anantnag, Baramulla and Srinagar in the Kashmir valley and in Udhampur, Doda, Riasi, Punch, Kathua and Jammu in Jammu province. Karnataka, however, is the biggest multivoltine mulberry silk-producing state in the country, accounting for about 70 per cent of the total production. Bivoltine rearings have also been introduced in the state recently. The industry is concentrated in the districts of Bangalore, Kolar, Tumkur, Mysore and Mandya. Madhya Pradesh is the second largest producer of tasar silk, production being largely based in Bastar, Bilaspur, Raigarh and Surguja districts.

Bhandara and Chandrapur districts in Maharashtra have a long tradition of tasar culture and there is a tasar seed farm in Armori. The sericulture industry in Tamil Nadu is located in Dharmapuri, North Arcot, Salem, Tirunelveli, Periyar, Madurai, Thanjavur, Coimbatore and South Arcot districts but is desultorily pursued. In Uttar Pradesh, the industry has been organized on cooperative lines, both in the rearing and reeling sectors, in Dehra Dun, Saharanpur, Nainital, Almora, Bahraich and Tehri-Garwal districts.

With a production of 4593 tonnes of mulberry silk and 448 tonnes of wild silk in 1980-81, India ranked third among silk-producing countries. However, India is the only country in the world which produces the famed golden 'muga' silk. It is also the second largest producer of tasar silk.

Mulberry silk production in India in 1983-84 totalled 5.7 m kgs, of which 59 per cent came from Karnataka (3.3 m kgs), 15.8 per cent from Andhra Pradesh (900,000 kgs), 11.3 per cent from Tamil Nadu (640,000 kgs) and 12.3 per cent from West Bengal (700,000 kgs). Karnataka, the only state in India producing 'dupion', in 1980-81 produced 75,000 kgs of this silk. The major producers of wild silk were Bihar, 275,000 kgs of tasar (65.8 per cent of India's total of 418,000 kgs), and Assam, 182,000 kgs of eri (67 per cent of India's total of 270,000 kgs) and 50,000 kgs of muga (92.5 per cent of India's total of 54,000 kgs).

Production of raw silk in the country increased 5½ times between 1950 and 1981, from 894,000 kgs to 5.04 m kgs, a growth of 464 per cent, and it is expected to reach 10.9 m kgs by 1989-90. The growth over the years has not, however, been commensurate with the efforts being put into making sericulture a major rural income-earner.

Sericulture is a cottage industry with considerable potential in India. Needing an agricultural base, an industrial superstructure and an essentially labour-intensive set-up, it is an effective tool for generating gainful employment in rural areas. At present it provides whole-time and part-time employment to more than 3.8 m persons in rural India. But it can absorb very many more persons in many parts of the country.

The value added in silk and man-made silk textiles manufacture in 1980-81 (excluding union territories) amounted to Rs 130.9 m, of which Karnataka's share was 22.3 per cent (or Rs 29.2 m), Andhra Pradesh's 20.1 per cent (or Rs 26. 3 m), West Bengal's 18.9 per cent (or Rs 24.8 m), Jammu & Kashmir's 10.2 per cent (or Rs 13.3 m) and Bihar's 9.2 per cent (or Rs 12 m).

Handloom and silk manufacture have been assiduously promoted by both central and state governments and it should be possible to reach not only the targets set in terms of sales and exports but also in production levels. But whether or not they are achieved, both handloom and silk manufacture have to be encouraged for they are the mainstay of a large number of people in the country.

Sources: 1. *Annual Survey of Industries 1979-80 & 1980-81* and *Statistical Abstract India 1984*, Ministry of Planning.
2. *All India Textiles Directory 1986-87*, Business Press, Bombay.

Table 85

HANDLOOM INDUSTRY — 1980-81

	Number of Factories 1979-80		Number of mandays-workers (in thousand)		Net value added (in Rs. million)		Value of output (in Rs. million)	
	Handlooms[1]	Silk	Handlooms[1]	Silk	Handlooms[1]	Silk	Handlooms[1]	Silk
Andhra Pradesh	25	62	307	N.A.	8.3	26.3	16.7	N.A.
Bihar	—	24	—	N.A.	—	12.0	—	N.A.
Gujarat	10	16	N.A.	N.A.	0.6	5.3	N.A.	N.A.
Haryana	10	N.A.	N.A.	N.A.	1.7	1.5	N.A.	N.A.
Himachal Pradesh	—	3	—	N.A.	—	N.A.	—	N.A.
Jammu & Kashmir	—	35	—	533	—	13.3	—	33.9
Karnataka	16	253	N.A.	921	1.2	29.2	N.A.	92.5
Kerala	110	—	567	—	27.0	—	37.0	—
Madhya Pradesh	5	13	N.A.	N.A.	0.9	2.3	N.A.	N.A.
Maharashtra	23	45	N.A.	99	7.3	10.2	N.A.	42.4
Orissa	12	—	N.A.	—	0.3	—	N.A.	—
Punjab	8	N.A.	N.A.	N.A.	0.7	0.3	N.A.	N.A.
Rajasthan	8	—	N.A.	—	0.3	—	N.A.	—
Tamil Nadu	6	42	N.A.	N.A.	N.A.	2.8	N.A.	N.A.
Uttar Pradesh	18	22	83	42	3.1	2.9	15.0	4.0
West Bengal	6	4	230	N.A.	N.A.	24.8	0.5	N.A.
INDIA[2]	257	522	1,187	1,595	51.4	130.9	69.2	172.8

Note: [1] Weaving and finishing
[2] Excludes other states and union territories for which data are not available.

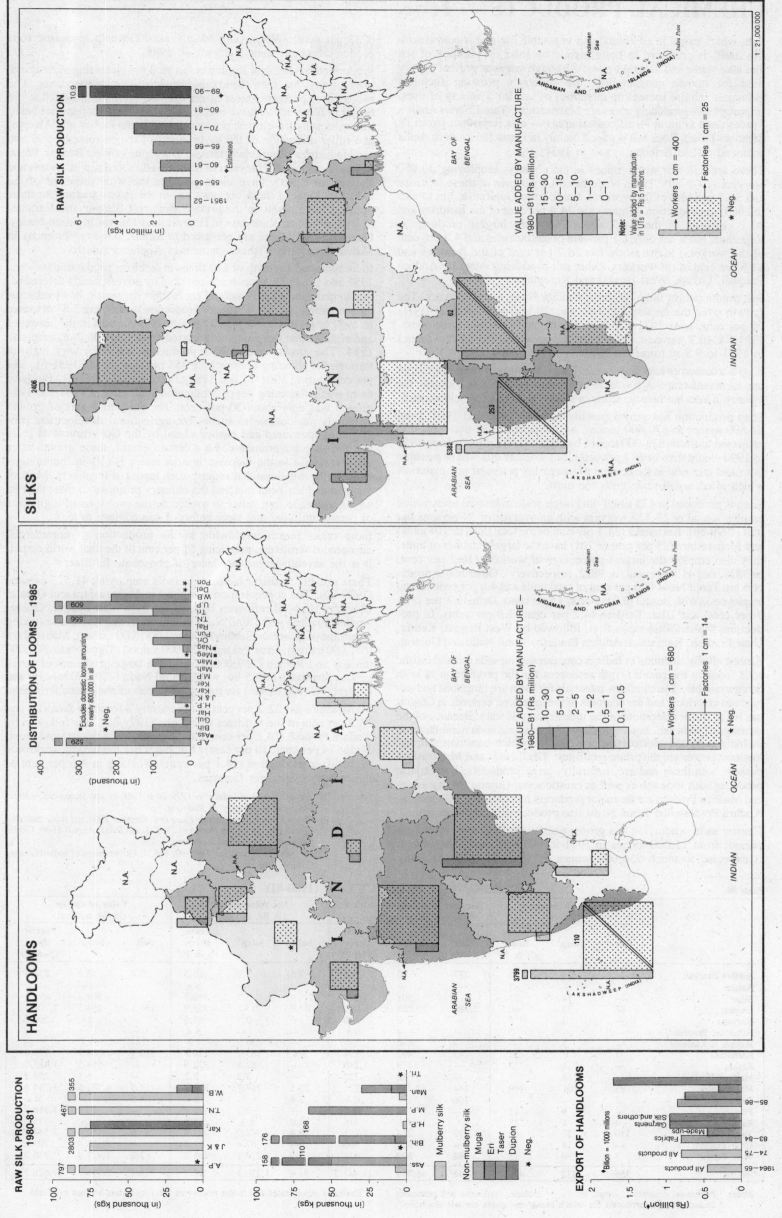

SILKS

RAW SILK PRODUCTION
(in million kgs)

1951-52
55-56
60-61
65-66
70-71
80-81
89-90
10.9
← Estimated

VALUE ADDED BY MANUFACTURE
1980-81 (Rs million)
15-30 10-15 5-10 1-5 0-1

Note:
Value added by manufacture in UTs = Rs 5 millions.

Workers 1 cm = 400
Factories 1 cm = 25
* Neg.

2406
5382
253
62

HANDLOOMS

DISTRIBUTION OF LOOMS — 1985
(in thousand)

*Excludes domestic looms amounting to nearly 800,000 in all
* Neg.

A.P. 629
Ass.*
Bih.
Guj.
Har.
H.P.*
J & K
Kar.
Ker.
M.P.
Mah.
Man.*
Meg.*
Nag.*
Ori.
Pun.
Raj.
T.N. 556
U.P. 609
W.B.
Del.*
Pon.*

VALUE ADDED BY MANUFACTURE —
1980-81 (Rs million)
10-30 5-10 2-5 1-2 0.5-1 0.1-0.5

Workers 1 cm = 680
Factories 1 cm = 14
* Neg.

3799
110

RAW SILK PRODUCTION
1980-81
(in thousand kgs)

A.P. 797
J & K 2803
Kar.
T.N. 467
W.B. 355
*

(in thousand kgs)

Ass. 158 176
Bih. 110
H.P. 168
M.P.
Man.
Tri. *

Mulberry silk
Non-mulberry silk
Muga
Eri
Taser
Dupion
* Neg.

EXPORT OF HANDLOOMS
(Rs billion*)
*Billion = 1000 millions

1964-65 All products
74-75   All products
83-84   Made-ups Fabrics Silk and others
85-86   Garments

CHEMICAL PRODUCTS — I

Salt, which exists in all animal and vegetable life and is coeval with life itself, is produced in India mainly by solar evaporation of sea and lake water and natural brines. Although salt is at present mainly used for human consumption, the country's growing chemical industries require increasing quantities of salt for a variety of uses. Important salt-producing states are Maharashtra, Tamil Nadu, Andhra Pradesh and Gujarat — all coastal areas — and Rajasthan (from its Sambhar Lake). Rock salt is mined mainly in Himachal Pradesh. India achieved self-sufficiency in salt in 1951.

There are 356 salt works/mines in the country, employing 30,850 workers (1979-80). Tamil Nadu has 30.9 per cent of these factories and 31.3 per cent of the workers, but follows Gujarat (6.3 m tonnes in 1985) in production. Gujarat has 24.4 per cent of the factories and 40.1 per cent of the workforce. The third largest producer is Rajasthan, but it has only 1.1 per cent of the factories and 4.2 per cent of the workers. Maharashtra has 26.7 per cent of the factories and 11.3 per cent of the workers. Other salt-producing states are Andhra Pradesh, Orissa, West Bengal and Karnataka.

Salt production in India has been steadily rising over the years. The growth over the decade 1951-61 was 25 per cent, over 1961-71, 56 per cent, over 1971-81, 65 per cent, and over the three years, 1981-84, 10.7 per cent. Production has increased from 2.8 m tonnes in 1951 to 9.9 m tonnes in 1984.

Soap is a consumer item in daily use. Large quantities of vegetable oils go into its manufacture. A few large private and public units dominate the industry in India but there are scores of small-scale units meeting local needs.

Soap production has grown steadily between 1951 and 1984, from 84,800 tonnes to 370,400 tonnes, an increase of 337 per cent. It is projected to touch 625,000 tonnes by 1989-90 (the growth from 1951 to 1990 would then be 637 per cent), but whether this will be possible is a moot question in the context of increasing prices of raw materials which affect small-scale units the most.

Soap is produced in 432 small- and large-scale industrial units which employ a total of 23,523 workers with an average of 54 persons per unit (1979-80). Karnataka (24.1 per cent of the total units or 104 units) and Maharashtra (23 per cent or 101) have the largest number of units; they also employ the largest percentage of workers (16.2 per cent, or 3828, and 41.8 per cent, or 9822, respectively). Other major producers are Tamil Nadu (8.6 per cent of total units and 4.8 per cent of total employees), West Bengal (6.9 per cent; 15 per cent), Delhi (5.3 per cent; 3 per cent) and Uttar Pradesh (4.2 per cent; 3.6 per cent). In production, Maharashtra ranks first, followed by West Bengal, Kerala, Uttar Pradesh, Karnataka, Andhra Pradesh, Tamil Nadu and Gujarat.

Among alkalis, attention in India is concentrated on soda ash and caustic soda (sodium hydroxide), rich resources for the production of both being available. The chief raw materials for them are limestone and salt (sodium chloride), and the nearness of the limestone deposits in Gujarat has led to the development of the industry in this state. Because of the shortage of soda ash, however, most of the caustic soda manufactured in India is by the electrolytic process, for which common salt and low-cost power are the prime requisites. Tamil Nadu and Maharashtra possess both these and are, naturally, large producers, but Gujarat produces both soda ash as well as caustic soda. Gujarat, Maharashtra and Madhya Pradesh are the major producers of caustic soda. Rajasthan, Andhra Pradesh and Tamil Nadu also produce substantial quantities.

Caustic soda production has grown nearly 47 times over a 34-year period: from 15,000 tonnes in 1951 to 698,600 tonnes in 1984. It is expected to touch 950,000 tonnes by 1989-90 (1951 to 1990:

6233 per cent). Worldwide, India ranked eleventh in caustic soda production (2 per cent of 35 mt) in 1985.

Production of chemical fertilizers, so vital for increasing agricultural productivity, has seen remarkable growth since independence. The two most important centres of nitrogenous fertilizer production are Sindri in Bihar and Alwaye in Kerala. Large fertilizer factories based on low-cost hydro-electric power (Nangal in the Punjab and Alwaye) and fuller utilization of the by-products of the petroleum, coal and steel industries (Raurkela in Orissa, Durgapur in West Bengal, Bhilai in Madhya Pradesh, Neyveli in Tamil Nadu, Namrup in Assam and Barauni in Bihar), have made possible the wide dispersal of the industry. Most of these big units are in the public sector, but there are also large plants in the private sector (at Kanpur and Varanasi in Uttar Pradesh, Vadodara in Gujarat, and Madras in Tamil Nadu). Phosphatic fertilizers are produced in Sindri, Alwaye, Trombay in Maharashtra, and Vishakhapatnam in Andhra Pradesh.

India achieved a growth of 130 times in fertilizer production between 1951 and 1984. This was made possible by government's determination to make the country self-reliant in this vital sector, however great the increasing demand. Fertilizer production was a mere 3.87 m tonnes in 1951 and grew to 93.9 m tonnes in 1971. It was in the seventies and eighties that the production soared, reaching 506.7 m tonnes in 1984. The growth of fertilizer production has been very high all through the planning era reaching 455 per cent during 1951-61, 337 per cent during 1961-71 and 315 per cent during 1971-81. But later there was a flattening out of production and between 1981 and 1984 growth was only about 30 per cent. The early high rate of growth is largely attributable to successive agricultural development programmes sponsored and pushed ahead by the Government of India and the state governments. To a certain extent, these growth rates are a response to the increase in area under HYVs in Indian agriculture, for it is these that require such inputs of fertilizers. Most of this demand has been met and the industry promises to meet the rest within a decade, but it has to overcome the problems of high costs of petroleum, coal and steel before it can achieve this.

India ranks fourth worldwide in the production of commercial nitrogenous fertilizers, producing 51 per cent of the total world output. It is the seventh-largest producer of phosphate fertilizer.

There are 492 fertilizer factories in India employing 41,513 persons (1979-80). The concentration is highest in Maharashtra and Gujarat (Bombay High oil refineries being close by), West Bengal, Orissa, Andhra Pradesh and Tamil Nadu (coal-based) and Karnataka (taking advantage of power availability). Gujarat (885,000 tonnes), Maharashtra (651,000 tonnes), Uttar Pradesh (538,000 tonnes), Tamil Nadu (504,000 tonnes) and Punjab (369,000 tonnes) are large producers of nitrogenous fertilizers (1985-86) while Tamil Nadu (220,000 tonnes) and Gujarat (389,000 tonnes) are the big producers of phosphatic fertilizers.

Maharashtra has 22.2 per cent of the country's fertilizer factories and 13.8 per cent of the fertilizer workforce (1979-80). It is followed by Andhra Pradesh (14.2 per cent units; 5.5 per cent workers) and Tamil Nadu (14 per cent; 10 per cent). But Bihar has the largest workforce in fertilizer production (14.1 per cent), working in 4.6 per cent of the country's fertilizer factories.

Sources: 1. *Annual Survey of Industries 1979-80 & 1980-81* and *Statistical Abstract India 1984*, Ministry of Planning.
2. *World Economy and India's Place in it, October 1986,* and *Basic Statistics Relating to the Indian Economy, Vol. 1: All India, August 1986*, Centre for Monitoring Indian Economy.
3. *Seventh Five Year Plan 1985-90, Vol. 1*, Government of India Planning Commission.

Table 86

CHEMICAL PRODUCTS — I (1980-81)

	Number of factories — 1979-80			Number of mandays – workers (in thousand)			Net value added (in Rs. million)			Value of output (in Rs. million)		
	Salt	Soap[1]	Ferti-lizers (N & P)[2]	Salt	Soap[1]	Ferti-lizers (N & P)[2]	Salt	Soap[1]	Ferti-lizers (N & P)[2]	Salt	Soap[1]	Ferti-lizers (N & P)[2]
Andhra Pradesh	47	20	70	277	80	766	7.6	3.9	126.3	9.4	13.3	874.4
Assam	—	—	3	—	—	N.A.	—	—	N.A.	—	—	N.A.
Bihar	—	14	23	—	501	1,879	—	5.7	−389.5	—	369.8	498.2
Gujarat	87	33	60	2,380	205	1,029	67.9	43.2	358.9	130.1	254.5	2,702.3
Haryana	—	3	8	—	N.A.	360	—	−0.3	−65.5	—	N.A.	509.5
Himachal Pradesh	—	—	3	—	—	N.A.	—	—	N.A.	—	—	N.A.
Jammu & Kashmir	—	5	—	—	N.A.	—	—	30.6	—	—	N.A.	—
Karnataka	3	104	31	145	492	194	N.A.	105.9	124.8	115.1	309.9	729.4
Kerala	—	11	24	—	479	1,248	—	82.6	322.8	—	554.5	1,483.1
Madhya Pradesh	—	18	7	—	52	322	—	3.2	−2.2	—	6.1	259.8
Maharashtra	95	101	109	244	83,333	1,771	20.3	717.7	828.1	13.5	4,232.5	3,764.5
Orissa	7	—	3	166	—	N.A.	3.4	—	N.A.	4.6	—	N.A.
Punjab	—	9	11	—	46	72.2	—	−2.2	116.8	—	38.4	1,318.9
Rajasthan	4	6	13	N.A.	N.A.	716	20.0	N.A.	67.5	N.A.	N.A.	735.0
Tamil Nadu	110	37	69	1,588	224	1,211	28.4	42.0	21.2	43.6	150.0	3,584.3
Uttar Pradesh	—	18	39	—	272	806	—	37.4	71.5	—	326.4	709.2
West Bengal	3	30	15	98	899	916	1.9	234.8	−131.3	2.8	1,429.4	316.2
INDIA[4]	**356[3]**	**432**	**492**	**4,898[3]**	**86,710**	**11,441.2[3]**	**151.6**	**1,270.5**	**2,109.2**	**319.1[3]**	**7,764.8**	**17,628.7**

Note: [1] Perfumes, cosmetics, soaps, etc. [2] Includes fertilizers and pesticides. [3] Excludes other states and union territories for which data are not available. [4] Includes union territories for which break-up figures are not available.

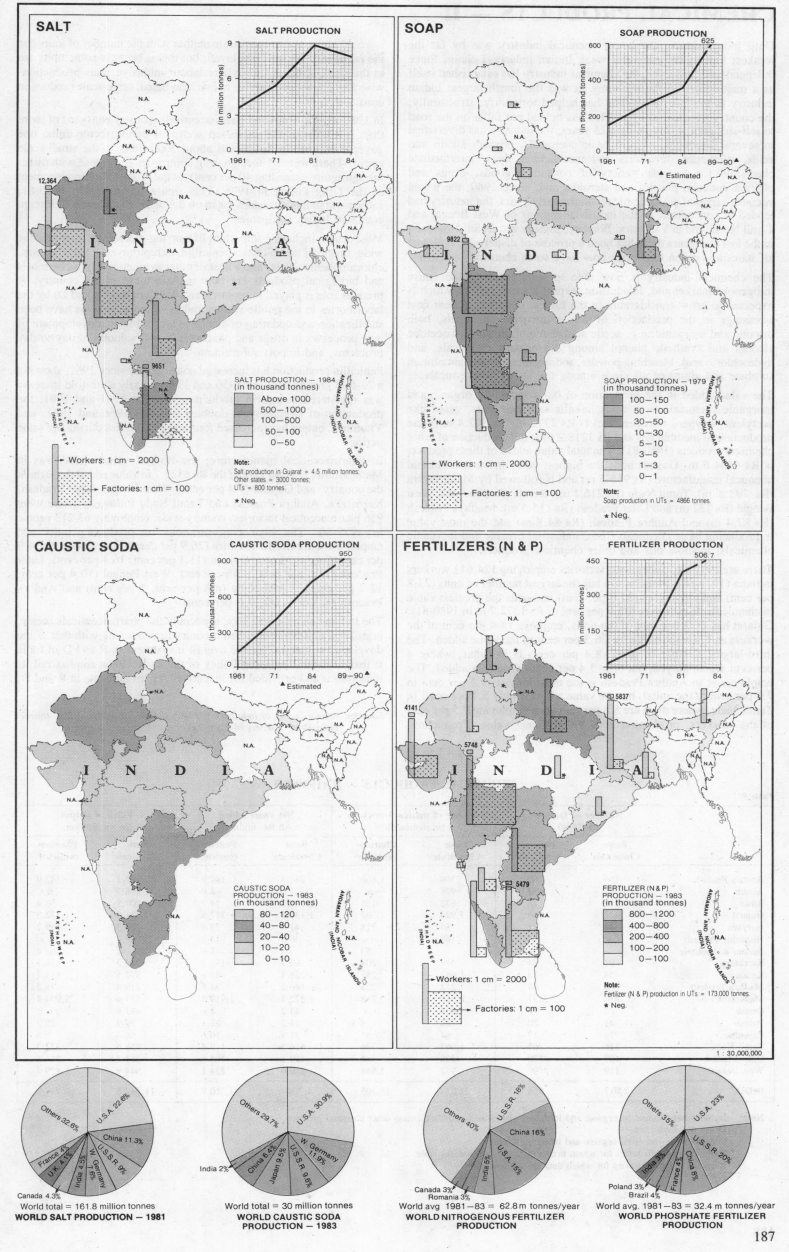

SALT

SALT PRODUCTION
(in million tonnes)

SALT PRODUCTION — 1984
(in thousand tonnes)

Above 1000
500—1000
100—500
50—100
0—50

Workers: 1 cm = 2000

Factories: 1 cm = 100

Note:
Salt production in Gujarat = 4.5 million tonnes;
Other states = 3000 tonnes;
UTs = 800 tonnes;
★ Neg.

12,364
9651

SOAP

SOAP PRODUCTION
(in thousand tonnes)

625

▲ Estimated

SOAP PRODUCTION — 1979
(in thousand tonnes)

100—150
50—100
30—50
10—30
5—10
3—5
1—3
0—1

Workers: 1 cm = 2000

Factories: 1 cm = 100

Note:
Soap production in UTs = 4866 tonnes.

★ Neg.

9822

CAUSTIC SODA

CAUSTIC SODA PRODUCTION
(in thousand tonnes)

950

▲ Estimated

CAUSTIC SODA
PRODUCTION — 1983
(in thousand tonnes)

80—120
40—80
20—40
10—20
0—10

FERTILIZERS (N & P)

FERTILIZER PRODUCTION
(in million tonnes)

506.7

FERTILIZER (N & P)
PRODUCTION — 1983
(in thousand tonnes)

800—1200
400—800
200—400
100—200
0—100

Workers: 1 cm = 2000

Factories: 1 cm = 100

Note:
Fertilizer (N & P) production in UTs = 173,000 tonnes.

★ Neg.

4141
5748
5837
5479

1 : 30,000,000

WORLD SALT PRODUCTION — 1981
World total = 161.8 million tonnes
U.S.A. 22.6%
China 11.3%
U.S.S.R. 9%
W. Germany 7%
India 4.5%
U.K. 4.5%
France 4%
Canada 4.3%
Others 32.6%

WORLD CAUSTIC SODA PRODUCTION — 1983
World total = 30 million tonnes
U.S.A. 30.9%
W. Germany 11.9%
U.S.S.R. 9.6%
Japan 9.5%
China 6.4%
India 2%
Others 29.7%

WORLD NITROGENOUS FERTILIZER PRODUCTION
World avg. 1981–83 = 62.8 m tonnes/year
U.S.S.R. 18%
China 16%
U.S.A. 15%
India 5%
Canada 3%
Romania 3%
Others 40%

WORLD PHOSPHATE FERTILIZER PRODUCTION
World avg. 1981–83 = 32.4 m tonnes/year
U.S.A. 23%
U.S.S.R. 20%
China 8%
France 4%
Brazil 4%
Poland 3%
India 3%
Others 35%

CHEMICAL PRODUCTS — II

Until independence, the heavy chemical industry was by far the weakest link in the generally weak Indian industrial chain. Since independence, however, the chemical industry has established itself as a major force in the economy (it was the fourth largest Indian industry in the late 1970s) and has helped to modify, structurally, the country's industrial base. This has helped put India on the road to self-sufficiency. In less than 25 years, the industry has diversified in several directions, such as basic chemicals, primary alkalis and acids, synthetics, fertilizers and insecticides, dyes, intermediate chemicals for a wide variety of consumer goods, drugs and pharmaceuticals, plastics and steroids, and, since 1980, the whole versatile range of petrochemicals. In the early years, the industry had tended to concentrate in the industrial regions of West Bengal and Tamil Nadu, and in Vadodara, Bombay, Mysore and Delhi. However, in the last few years there has been a process of dispersal and almost all states and union territories now have some chemical industries.

The chemical industry is now able to supply almost the entire indigenous market and, in addition, export a small surplus which is expected to grow considerably in the future. India has a clear cost advantage in the production of labour-intensive chemicals, both organic and inorganic, e.g. acetic acid, butanol, acetone, diacetone alcohol and synthetic phenol among the organic chemicals, and hydrochloric acid, bleaching powder, sodium bicarbonate, aluminium sulphate and alums of all kinds among the inorganic chemicals.

The value added in the production of basic industrial organic and inorganic chemicals (e.g. acids, alkalis and their salts, gases like acetylene, oxygen, nitrogen, etc) is Rs 2779 m, Rs 307.4 m in the production of inedible oils and Rs 1218.2 m in the production of other chemical products (1980-81). The total value added of these products is Rs 4304.6 m. Gujarat adds the highest value in basic industrial chemical manufacture (Rs 955.5 m) and is followed by Maharashtra (Rs 793.8 m), Tamil Nadu (Rs 216.1 m), Kerala (Rs 206.7 m), West Bengal (Rs 132 m) and Uttar Pradesh (Rs 115.5 m). Madhya Pradesh (Rs 87.4 m) and Andhra Pradesh (Rs 84.8 m) add the most value in producing basic chemicals (basic industrial organic and inorganic chemicals, inedible oils and other chemical products).

There are 2017 basic chemicals factories employing 124,611 workers in India (1979-80). Maharashtra has the largest number of units (23.8 per cent) and employees (22.7 per cent) and adds the greatest value in chemicals manufacture (31.9 per cent or Rs 1372.7 m in 1980-81). Gujarat has 21.3 per cent of the units, employs 18.4 per cent of the workers and is responsible for 26.3 per cent of the value added. The third-largest number of units (8.4 per cent) is in Bihar, where 4 per cent are employed, but only 3.4 per cent of value is added. The employment in Andhra Pradesh is the third largest (9.5 per cent in 7.7 per cent of the units), but the value added is only 5.7 per cent. In Tamil Nadu, where there are 5.8 per cent units employing 9.5 per cent of the workers, production adds 9.9 per cent value. Production,

it would seem, is commensurate neither with the number of units nor the number of persons employed; but this is because some units are in the small-scale sector and are labour-intensive, thus production-wise they are not comparable to machine-based, large-scale production units.

In 1980-81, the drugs and pharmaceuticals industry consisted of more than 150 units in the organized sector, 4 public sector units, one government-managed unit and about 3000 units in the small-scale sector. There were 31 foreign drug companies (i.e. those with direct foreign equity exceeding 40 per cent), at the time of the announcement of the Drug Policy in 1978. This figure fell to 23 in 1980-81, subsequent to new decisions taken on the permissible levels of foreign equity in Indian companies.

Most drugs, including important life-saving medicines, covering the wide range of antibiotics (penicillin, streptomycin, tetracyclines, chloramphenicols etc) and a host of other synthetics, phytochemicals and biological products are now manufactured in the country. A premier role is played in research and development (R and D) by the laboratories in the public sector, and the main objectives have been stabilization and updating of available technologies, development of new processes in drugs and pharmaceuticals, solution of day-to-day problems, and import substitution.

Penicillin production has increased considerably since 1961, showing a 23-fold growth between 1956 and 1981. Nearly a six-fold increase was registered in aspirin production between 1961 and 1981; the production of streptomycin doubled between 1966 and 1981; and Vitamin C output has increased four and a half times during the same period.

In pharmaceutical manufacture, the highest value added was in Maharashtra (57.8 per cent of the Rs 3320.7 m value added throughout the country) and Gujarat (9.4 per cent). West Bengal, Uttar Pradesh, Karnataka, Andhra Pradesh and Tamil Nadu followed. There were 956 pharmaceutical factories, country-wide, employing 63,213 people in 1979-80. The high concentrations of producing units and employment are in Maharashtra (26.9 per cent of the units with 34.9 per cent of the workers), Gujarat (11.1 per cent; 16.4 per cent), Uttar Pradesh (12.6 per cent; 9.4 per cent), West Bengal (10.4 per cent; 12.7 per cent), Tamil Nadu (6.8 per cent; 6 per cent) and Andhra Pradesh (6.3 per cent; 7.6 per cent).

The Indian chemical industry, especially the pharmaceuticals sector, is still at a disadvantage when it comes to competing with that of the developed countries, for the overall investment in R and D in India is insignificant. The Drug Policy of the early 1980s emphasized its importance and provided for incentives for investments in R and D.

Sources: *Annual Survey of Industries 1979-80 & 1980-81* and *Statistical Abstract India 1984*, Ministry of Planning.

CHEMICAL PRODUCTS — II (1980-81)

Table 87

	Number of factories 1979-80		Number of mandays-workers (in thousand)		Net value added (in Rs. million)		Value of output (in Rs. million)	
	Basic Chemicals[1]	Pharma-ceuticals[2]	Basic Chemicals[3]	Pharma-ceuticals[2]	Basic Chemicals[1]	Pharma-ceuticals[2]	Basic Chemicals[3]	Pharma-ceuticals[2]
Andhra Pradesh	156	60	306	1,666	244.3	149.9	152.1	742.9
Assam	14	3	926	N.A.	2.6	-4.0	208.7	N.A.
Bihar	170	18	632	154	146.8	14.7	207.7	50.4
Gujarat	421	106	3,924	1,362	1,130.1	312.9	4,431.0	1,372.3
Haryana	48	21	64	215	41.4	23.6	19.7	109.7
Himachal Pradesh	N.A.	4	169	N.A.	N.A.	14.7	43.3	N.A.
Jammu & Kashmir	—	6	—	153	—	-1.0	—	260.4
Karnataka	91	34	374	204	63.7	151.2	264.3	238.3
Kerala	34	35	1,042	281	228.1	48.6	597.5	113.4
Madhya Pradesh	62	44	212	221	180.2	34.8	219.4	73.2
Maharashtra	480	257	2,310	5,268	1,372.7	1,919.3	3,557.4	5,994.4
Orissa	26	16	634	N.A.	43.2	4.6	482.6	N.A.
Punjab	41	20	69	8.1	24.2	27.1	39.9	20.2
Rajasthan	76	17	64	61	31.1	10.6	56.4	9.6
Tamil Nadu	116	65	1,067	736	426.5	119.4	758.6	372.3
Uttar Pradesh	107	120	508	145.7	159.1	194.1	313.4	844.5
West Bengal	119	99	952	1,943	210.6	224.1	544.9	1,429.4
INDIA[4]	2,017	956	12,211	15,105	4,304.6[5]	3,320.7	11,542.6	10,944.0

Note: [1] Includes basic industrial organic and inorganic chemicals, inedible oils and other chemical products.
[2] Drugs and medicines.
[3] Includes basic industrial organic and inorganic chemicals.
[4] Includes union territories for which break-up figures are not available.
[5] Excludes union territories for which data are not available.

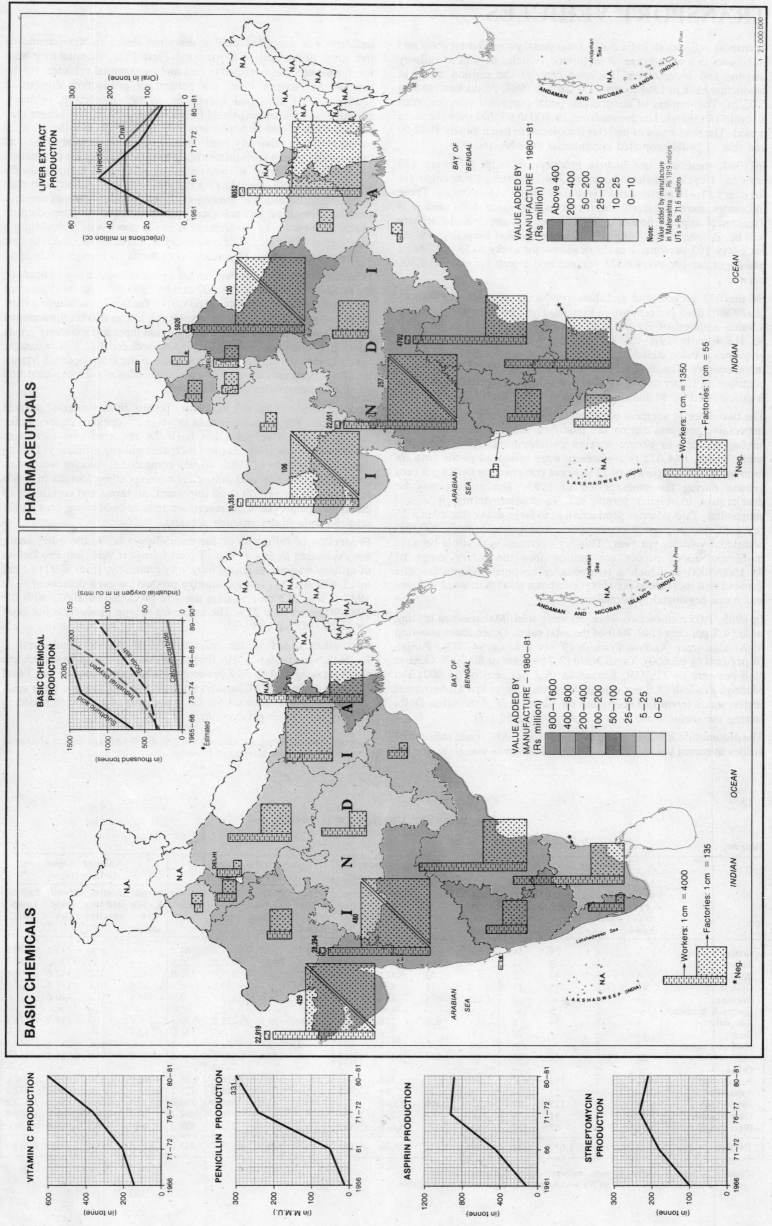

PHARMACEUTICALS

VALUE ADDED BY MANUFACTURE – 1980–81
(Rs million)

Above 400
200–400
50–200
25–50
10–25
0–10

Note:
Value added by manufacture in Maharashtra = Rs 1919 millions
UTs = Rs 71.6 millions

→ Workers: 1 cm = 1350
Factories: 1 cm = 55

* Neg.

LIVER EXTRACT PRODUCTION

(Oral in tonne)
(Injections in million cc)

1951 61 71–72 80–81

Injection
Oral

BASIC CHEMICALS

VALUE ADDED BY MANUFACTURE – 1980–81
(Rs million)

800–1600
400–800
200–400
100–200
50–100
25–50
5–25
0–5

→ Workers: 1 cm = 4000
Factories: 1 cm = 135

* Neg.

BASIC CHEMICAL PRODUCTION

(in thousand tonnes)
(Industrial oxygen in m cu mtrs)

1965–66 73–74 84–85 89–90

Sulphuric acid
Industrial oxygen
Soda ash
Calcium carbide

♦ Estimated

VITAMIN C PRODUCTION
(in tonne)

1966 71–72 76–77 80–81

PENICILLIN PRODUCTION
(in M.M.U.)

1961 71–72 80–81

ASPIRIN PRODUCTION
(in tonne)

1961 66 71–72 80–81

STREPTOMYCIN PRODUCTION
(in tonne)

1966 71–72 76–77 80–81

TRANSPORT VEHICLES

Transport vehicles in India range from pedal-propelled bicycles and rickshaws to a wide range of motorized vehicles as well as railway wagons and locomotives. Bicycles crossed the million mark in production back in 1961 (actual: 1.05 m). In 1984, production reached 5.95 m. The number of commercial pedal-propelled vehicles, such as cycle-rickshaws, has grown from 28,400 to 93,200 over the same period. The production of bicycles is expected to touch 8 m by 1989-90 and that of pedal-propelled commercial vehicles 160,000.

In 1984, there was one bicycle produced in India for every 123 persons, 1 cycle-rickshaw for every 7840 persons, 1 passenger car for every 11,410 persons and 1 jeep for every 30,440 persons. These indicators show a greatly improved picture since 1961 and it is expected to improve further by 1990, when it is anticipated that there will be, despite the growing population in India, 1 bicycle produced for every 103 persons, 1 cycle rickshaw for every 5138 persons, 1 passenger car for every 6323 persons and 1 jeep for every 18,300 persons.

Of the 672 bicycle and rickshaw producing units in the country (1979-80), 61.5 per cent are in Punjab. Their production resulted in a value addition of Rs 144.4 m (46.1 per cent of India's total of Rs 313.4 m in 1980-81, the highest in the country). Haryana, generates a value added of 21.2 per cent, the second largest return in the country. The third major cycle-producing state is Tamil Nadu, contributing 10 per cent of value added in the country in this industry. It is followed by Maharashtra, Gujarat and Uttar Pradesh.

The two-wheeler segment of the automobile industry has registered impressive progress during the past five years, with Government having decided to permit foreign collaboration. After the 1985 achievement, 644,675 two-wheelers were produced in the first six months of 1986, a growth of 21 per cent compared to the 28 per cent growth during the same period of 1985. The growth may be decelerative in relative terms, but, in absolute terms, it is still substantial. Two-wheeler production is so large today that even a 25 per cent rise in present consumption levels would mean 300,000 additional vehicles per year. This high demand is possible because more and more people are getting into the salary range of Rs 1500-3000 at which a two-wheeler becomes affordable. The demand will increase as Rs 1500 becomes a minimum wage in more and more organized industrial units.

In 1985, 1.05 million two-wheelers were sold; Maharashtra leading with 14.3 per cent (149,700) of the total sales. Other states showing high sales were Andhra Pradesh (9 per cent or 94,700), Punjab (8 per cent or 83,600), Tamil Nadu (7.7 per cent or 80,300), Gujarat (6.8 per cent or 71,700), Karnataka (6.2 per cent or 64,000) and Madhya Pradesh (5.8 per cent or 60,400). Due to low Government levies, union territories accounted for 12 per cent of the sales, Delhi having the major share with 8.2 per cent or 86,000.

The automobile industry also has, against all odds, made substantial strides in recent years. Till the 1980s, its progress was slow, for the industry was accorded only a marginal status in the scheme of industrial development. During the First Plan, manufacture with foreign tie-ups was confined to cars and commercial vehicles. By the end of the Second Plan, the pattern of production diversified somewhat. Of the total output of 69,490 motorized vehicles, commercial vehicles accounted for nearly 40 per cent, passenger cars for 28 per cent and two wheelers for 24 per cent. But the scale of investment was relatively small considering the potential as well as the development of such infrastructure as roads. Thus the production of vehicles of all types showed only a modest increase during the 30 years ending 1980. Recognizing the vital role this industry can play in the industrialization of the country through the creation of employment opportunities, technological upgradation and the strengthening of the road network, government is in the process of formulating a new automobile policy. The new policy will take into account the cost to the country of any liberalization in terms of foreign exchange.

However, if the Seventh Plan demand projections are any indication, the projected demand of 154,000 cars by 1989-90 is far less than even the current licensed capacity (216,000) of the five manufacturing units in the country. Those waiting to enter the Indian market the moment the new policy is announced may find that there are too many goods chasing too few buyers. In 1985 India produced 132,370 passenger cars and sold 102,456 of them, the share of the Indo-Japanese Maruti being 48 per cent, as against Hindustan Motor's 23 per cent and Premier Automobiles' 29 per cent.

Recessionary conditions, however, persist in commercial vehicle production. The discordant note of stagnant demand in this sector during the last five years has been the result of cost increases, government's credit squeeze and increased railway efficiency. During the first six months of 1986, 46,679 commercial vehicles were sold, compared to 50,848 sold during the corresponding months in 1985. In a sector where credit is all important, all terms and conditions of credit and the prevalent interest structure have been taking on a highly sensitive role in determining demand.

Production of rolling stock for the railways is, on the other hand, not expanding as fast as that of road transport vehicles. Production of railway wagons increased only very gradually, from 8200 in 1961 to 15,400 in 1984, and locomotive production even decreased — in 1984 it was 212 only against the 252 produced in 1961, with the 1989-90 target only 225. The target for wagon production has been set at 20,000 for 1989-90.

The value added in the manufacture of railway equipment is Rs 2443.5 m (1980-81). The four states that register the largest returns are West Bengal, with 36.8 per cent of India's total value added, Tamil Nadu 10.8 per cent, Maharashtra 8.7 per cent and Bihar 4.6 per cent. These four states account for 60.9 per cent of the value added in railway equipment production.

Sources: *Annual Survey of Industries 1979-80 & 1980-81*, and *Statistical Abstract India 1984*, Ministry of Planning.

TRANSPORT VEHICLES — 1980-81

Table 88

	Number of Factories — 1979-80				Number of mandays-workers (in thousand)				Net value added (in Rs. million)				Value of output (in Rs. million)			
	Bicycles & cycle rick-shaws	Motor-ized two wheelers	Motor vehi-cles	Railway equip-ment[1]	Bicycle & cycle rick-shaws	Motor-ized two wheelers	Motor vehi-cles	Railway equip-ment[1]	Bicycle & cycle rick-shaws	Motor-ized two wheelers	Motor vehi-cles	Railway equip-ment[1]	Bicycle & cycle rick-shaws	Motor-ized two wheelers	Motor vehi-cles	Railway equip-ment[1]
Andhra Pradesh	5	4	28	14	N.A.	N.A.	295	1,450	4.1	15.5	31.6	62.8	N.A.	N.A.	143.5	229.1
Assam	—	—	3	4	—	—	N.A.	509	—	—	0.8	20.8	—	—	N.A.	52.3
Bihar	6	3	44	7	N.A.	63	3,477	494	0.2	N.A.	645.9	112.9	N.A.	3.7	2,942.6	62.6
Gujarat	43	16	75	17	148	118	892	1,392	9.8	0.7	60.8	66.7	32.6	30.1	237.7	127.0
Haryana	5	14	N.A.	3	1,168	624	137	N.A.	66.5	88.4	118.7	8.0	346.9	368.3	440.6	N.A.
Jammu & Kashmir	—	—	3	—	—	—	N.A.	—	—	—	2.1	—	—	—	N.A.	—
Karnataka	4	19	54	8	N.A.	620	208	1,569	N.A.	39.8	279.2	79.5	N.A.	191.7	819.9	178.9
Kerala	11	—	—	20	51	—	—	N.A.	3.7	—	—	N.A.	12.4	—	—	N.A.
Madhya Pradesh	4	5	30	N.A.	N.A.	N.A.	758	469	0.2	3.0	56.6	13.3	N.A.	N.A.	232.2	539.9
Maharashtra	23	57	220	45	198	1,611	9,199	5,517	13.4	282.5	1,440.2	213.1	86.3	937.1	6,085.2	833.6
Orissa	—	—	4	—	—	—	N.A.	—	—	—	0.9	—	—	—	N.A.	—
Punjab	413	7	158	N.A.	1,827	N.A.	1,406	N.A.	144.4	8.1	126.9	26.3	1,002.0	N.A.	377.6	N.A.
Rajasthan	—	—	28	4	—	—	99	3,745	—	—	19.5	125.2	—	—	20	971.2
Tamil Nadu	19	38	125	27	338	660	5,048	7,235	30.5	40.6	886.0	264.0	173.5	214.9	3,652.6	1,323.8
Uttar Pradesh	41	36	51	8	167	785	335	5,149	9.2	37.2	56.6	492.1	29.2	238.9	188.3	1,862.8
West Bengal	32	3	67	127	128	475	3,616	15,103	3.2	N.A.	335.2	899.3	10.8	110.1	1,524.6	N.A.
INDIA[2]	672	233	1,057	312	4,348	4,842	30,030	46,563	313.4	520.9	4,139.7	2,443.5	1,810.1	2,158.4	16,880.9	6,838.3

Note: [1] Includes locomotives, parts, railway wagons and coaches.
[2] Includes union territories for which break-up figures are not available.

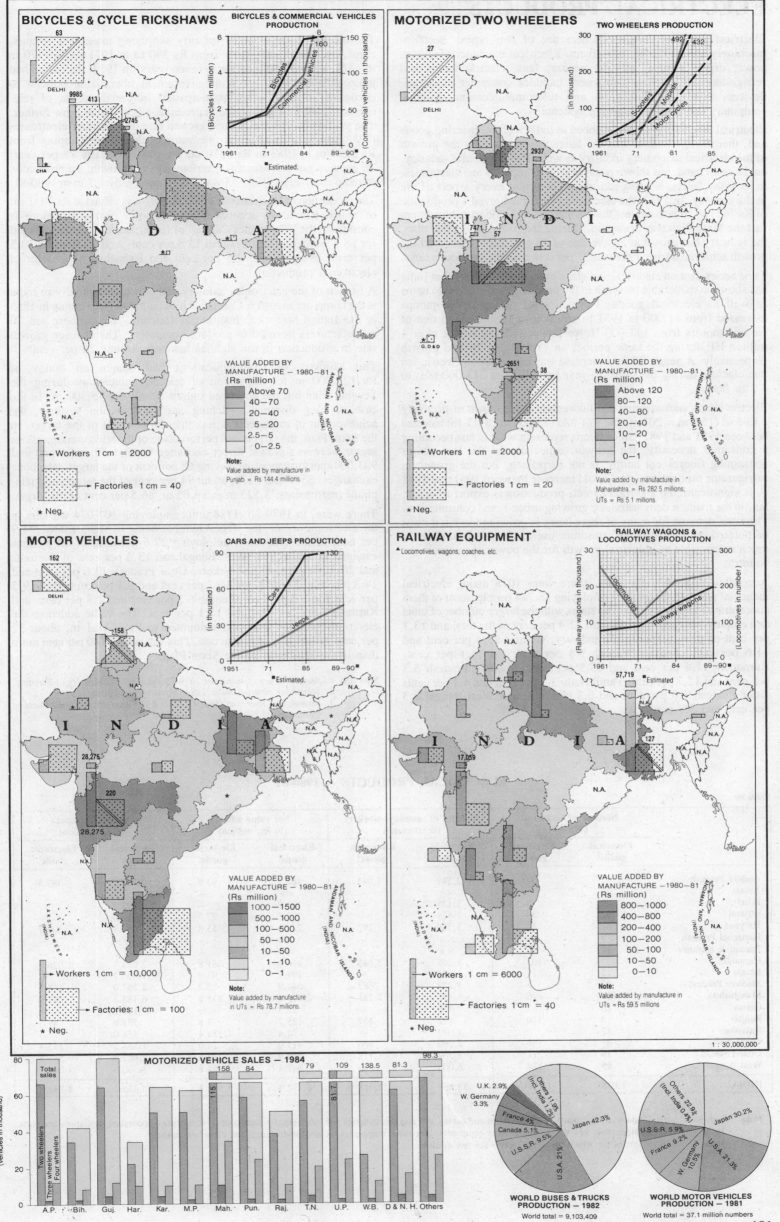

BICYCLES & CYCLE RICKSHAWS

63
DELHI

9985   413   N.A.
2745   N.A.
CHA

N.A.

I N D I A

N.A.   N.A.   N.A.

N.A.

N.A.

LAKSHADWEEP (INDIA)   N.A.

ANDAMAN AND NICOBAR ISLANDS (INDIA)   N.A.

BICYCLES & COMMERCIAL VEHICLES PRODUCTION

Bicycles
Commercial Vehicles

1961   71   84   89–90
■ Estimated.

VALUE ADDED BY MANUFACTURE – 1980–81 (Rs million)
Above 70
40–70
20–40
5–20
2.5–5
0–2.5

→ Workers 1 cm = 2000
→ Factories 1 cm = 40
★ Neg.

Note:
Value added by manufacture in Punjab = Rs 144.4 millions.

MOTORIZED TWO WHEELERS

27
DELHI

N.A.

2937
7471   57

I N D I A

N.A.

N.A.

G. D & D

2651   38

N.A.

LAKSHADWEEP (INDIA)   N.A.

ANDAMAN AND NICOBAR ISLANDS (INDIA)   N.A.

TWO WHEELERS PRODUCTION

492   432
Scooters
Mopeds
Motor cycles

1961   71   81   85

VALUE ADDED BY MANUFACTURE – 1980–81 (Rs million)
Above 120
80–120
40–80
20–40
10–20
1–10
0–1

→ Workers 1 cm = 2000
→ Factories 1 cm = 20
★ Neg.

Note:
Value added by manufacture in Maharashtra = Rs 282.5 millions; UTs = Rs 5.1 millions.

MOTOR VEHICLES

162
DELHI

158   ★
N.A.

I N D I A

28,275
220
28,275

★

N.A.

LAKSHADWEEP (INDIA)   N.A.   N.A.

ANDAMAN AND NICOBAR ISLANDS (INDIA)   N.A.

CARS AND JEEPS PRODUCTION

130
Cars
Jeeps

1951   71   85   89–90
■ Estimated.

VALUE ADDED BY MANUFACTURE –1980–81 (Rs million)
1000–1500
500–1000
100–500
50–100
10–50
1–10
0–1

→ Workers 1 cm = 10,000
→ Factories: 1 cm = 100
★ Neg.

Note:
Value added by manufacture in UTs = Rs 78.7 millions.

RAILWAY EQUIPMENT▲

▲ Locomotives, wagons, coaches, etc.

N.A.

57,719   Estimated

I N D I A

17,059

127

N.A.

LAKSHADWEEP (INDIA)   N.A.

ANDAMAN AND NICOBAR ISLANDS (INDIA)   N.A.

RAILWAY WAGONS & LOCOMOTIVES PRODUCTION

Locomotives
Railway wagons

1961   71   84   89–90 ■

VALUE ADDED BY MANUFACTURE –1980–81 (Rs million)
800–1500
400–800
200–400
100–200
50–100
10–50
0–10

→ Workers 1 cm = 6000
→ Factories 1 cm = 40
★ Neg.

Note:
Value added by manufacture in UTs = Rs 59.5 millions.

1 : 30,000,000

MOTORIZED VEHICLE SALES – 1984

Total sales
Two wheelers
Three wheelers
Four wheelers

158   84   115   79   109   138.5   81.3   98.3
81.7

A.P.   Bih.   Guj.   Har.   Kar.   M.P.   Mah.   Pun.   Raj.   T.N.   U.P.   W.B.   D & N. H. Others

WORLD BUSES & TRUCKS PRODUCTION – 1982
World total = 9,103,409

Japan 42.3%
U.S.A. 21%
U.S.S.R. 9.5%
Canada 5.1%
France 4%
W. Germany 3.3%
U.K. 2.9%
Others (incl. India 1.2%) 11.9%

WORLD MOTOR VEHICLES PRODUCTION – 1981
World total = 37.1 million numbers

Japan 30.2%
U.S.A. 21.3%
W. Germany 10.5%
France 9.2%
U.S.S.R. 5.9%
Others (incl. India 0.4%) 22.9%

191

ELECTRICAL PRODUCTS

Electrical products considered here are of two types: electrical machinery and electronic equipment. Electrical machinery includes power driven pumps, electric motors, fans, electric lamps and refrigerators. Electronic equipment includes consumer electronic devices, mass communication and telecommunication systems, computers and distance learning systems.

Electrical machinery is often clubbed in India with engineering goods and, therefore, whatever ails the latter industry affects the growth of the electrical machinery industry as well. The electronics industry, on the other hand, has shown positive growth impulses and electronics, in its many facets, is being introduced in almost every aspect of life in the country. In 1982, the electronics sector achieved a production of Rs 7431 m. It was against this background of modest achievement that the Sixth Plan was drawn up. During the five years of this plan, the industry has grown at an average of 28 per cent a year. And the growth achieved during 1984 was 37 per cent over the previous year's.

Time series data on electrical and electronic goods indicate that India has shown considerable increase in the production of electronic items as well as electrical goods. Production of power-driven pumps increased from 41,000 in 1951 to 466,000 in 1980-81, and that of electric motors from 143,000 Horse Power (HP) to a little over 4 million HP during the same period, an 11-fold and 28-fold growth respectively. A nearly 20-fold increase was registered in electric fan manufacture during the same 30-year period, from 213,000 fans to 4.2 m fans.

Electric lamps, incandescent and fluorescent, have shown respectively a 13-fold (15.5 m to 201 m) and 374-fold (75,000 to 28.03 m) increase between 1951 and 1981. Since lamps represent what is fast becoming a consumer necessity, the growth, especially of the low power consuming fluorescent lamps, is not surprising. But the growth in refrigerator production — a 93-fold increase between 1961 and 1981 — is significant. To a certain extent, production is export-oriented, but, in the main, it demonstrates a growing urban-based consumerism. The rural demand for many of these items is growing too but it must be pointed out that growing consumer use of electrical goods is as much responsible as industrial growth for the power shortage in the country.

According to a 1979-80 count, there were 1078 major electrical industrial plants in the country employing 85,764 people, most of them concentrated in urban areas. The states with the larger number of units and employment are: Maharashtra 12.4 per cent of the total and 13.1 per cent of the workers in the industry; Gujarat 23.4 per cent and 11.6 per cent; Andhra Pradesh 9.5 per cent and 14.1 per cent; Karnataka 11.8 per cent and 13.5 per cent; Madhya Pradesh 5.3 per cent and 12.7 per cent; Tamil Nadu 3.3 per cent and 4.1 per cent; Uttar Pradesh 3.6 per cent and 6.2 per cent; and West Bengal 2.3 per cent and 3.9 per cent.

The electronic components industry supplying integrated circuits, chips, transistors etc., grew from Rs 390 m in 1971 to Rs 1170 m in 1978, a three-fold increase in seven years. In 1984, the output had jumped to Rs 3000 m. With government liberalizing policies to encourage private sector participation in certain areas of telecommunications, the increase in production is bound to grow further. The year 1984 saw the highest growth rate of consumer electronics (TV, radio etc.) production, the value of the output jumping from Rs 3300 m in 1983 to Rs 5870 m, a growth of nearly 78 per cent. To mention just one item that contributed to this growth: the production of radio receivers in 1984-85 was 7 m, against only 1.7 m in 1980-81.

The production of all electronic goods in 1982 was valued at Rs 7431 m, of which Karnataka accounted for 35.4 per cent (Bangalore is considered the 'Electronics Capital' of India). Maharashtra accounted for 18.5 per cent, Uttar Pradesh 13.8 per cent, Andhra Pradesh 10.5 per cent, West Bengal 4.5 per cent and Rajasthan 4 per cent of electronics production.

A branch of the electronics industry that is growing rapidly in India is the computer industry. Computer installation in India began in 1955 at the Indian Statistical Institute at Calcutta. Today there are 70 manufacturers licensed to produce computers. The average growth rate in production in the eighties has been around 25 per cent.

The Ministry of Communications had sought an outlay of Rs 125,000 m for expansion of telecommunications during the Seventh Plan but the approved outlay is only Rs 45,300 m. In the case of long distance switching and transmission systems, the achievement of this sector is less than 50 per cent of the target. In the Sixth Plan, the targets and performance of the telecommunications sector were as follows: direct exchange lines: 1.33 m (target) and 903,000 (performance), achieving 68 per cent of the target; telephone exchanges: 3500 and 3278, about 94 per cent of the target; and telephone instruments: 3.525 m and 3.05 m, 86.5 per cent of the target.

There were, in 1979-80, 1734 units employing 107,074 workers in the electronic sector, the largest number being in Maharashtra (34.9 per cent of the total units employing 27.6 per cent of all persons engaged in the sector). West Bengal had 13.5 per cent of the units and 20.4 per cent of the workers, Uttar Pradesh 12.6 per cent and 14.3 per cent, Tamil Nadu 10.7 per cent and 9.1 per cent, Delhi 9.2 per cent and 3.3 per cent, Punjab 6.3 per cent and 2.8 per cent, and Karnataka 5.9 per cent and 15.1 per cent. The value added in the electronic industry that year amounted to Rs 1538.1 m, about 22 per cent of the value of production. This was nearly 140 per cent more than in the final year of the Sixth Plan.

Sources: 1. *Annual Survey of Industries 1979-80* and *1980-81*, Ministry of Planning.
2. *Statistical Abstract India 1984*, Ministry of Planning.
3. *Seventh Five Year Plan 1985-90*, *Vol. I & II*, Government of India Planning Commission.

ELECTRICAL PRODUCTS — 1980-81

Table 89

	Number of factories 1979-80		Number of mandays-workers (in thousand)		Net value added (in Rs. million)		Value of output (in Rs. million)	
	Electrical goods[1]	Electronic goods[2]	Electrical goods[1]	Electronic goods[2]	Electrical goods[1]	Electronic goods[2]	Electrical goods[1]	Electronic goods[2]
Andhra Pradesh	102	37	2,540	1,043	461.3	93.9	2,385.5	393.3
Assam	3	—	N.A.	—	N.A.	—	N.A.	—
Bihar	41	5	1,135	—	141.8	—	682.7	—
Gujarat	252	37	2,679	—	307.5	27.7	1,312.0	—
Haryana	62	17	1,389	292	228.3	53.8	954.1	204.7
Himachal Pradesh	7	—	N.A.	—	4.1	—	N.A.	—
Jammu & Kashmir	10	—	N.A.	—	4.7	—	N.A.	—
Karnataka	127	103	3,208	3,688	496.6	361.8	2,025.7	1,026.7
Kerala	25	11	1,394	—	198.7	97.9	1,060.7	—
Madhya Pradesh	57	5	3,229	683	621.9	23.3	2,267.0	118.6
Maharashtra	134	606	8,335	2,281	2,015.6	235.8	6,185.8	797.0
Orissa	15	3	138	—	32.9	—	160.8	—
Punjab	13	110	832	137	125.7	7.8	571.6	49.5
Rajasthan	42	4	1,047	—	74.4	134.4	458.0	—
Tamil Nadu	36	186	2,290	866	415.0	78.0	1,182.9	183.4
Uttar Pradesh	39	218	4,364	1,718	659.9	272.5	2,859.5	581.8
West Bengal	25	234	6,005	850	847.4	68.8	3,390.5	216.7
INDIA[3]	1,078	1,734	23,689	13,318	6,772.4	1,566.8	25,455.2	4,782.8

Note: [1] Includes electrical industrial machinery, insulated wires and cables, dry and wet batteries and other electric components, accessories, apparatus, appliances and parts.
[2] Includes consumer electronics and telecommunication equipment, electronic computers, control instruments and other equipment.
[3] Includes union territories for which break-up figures are not available.

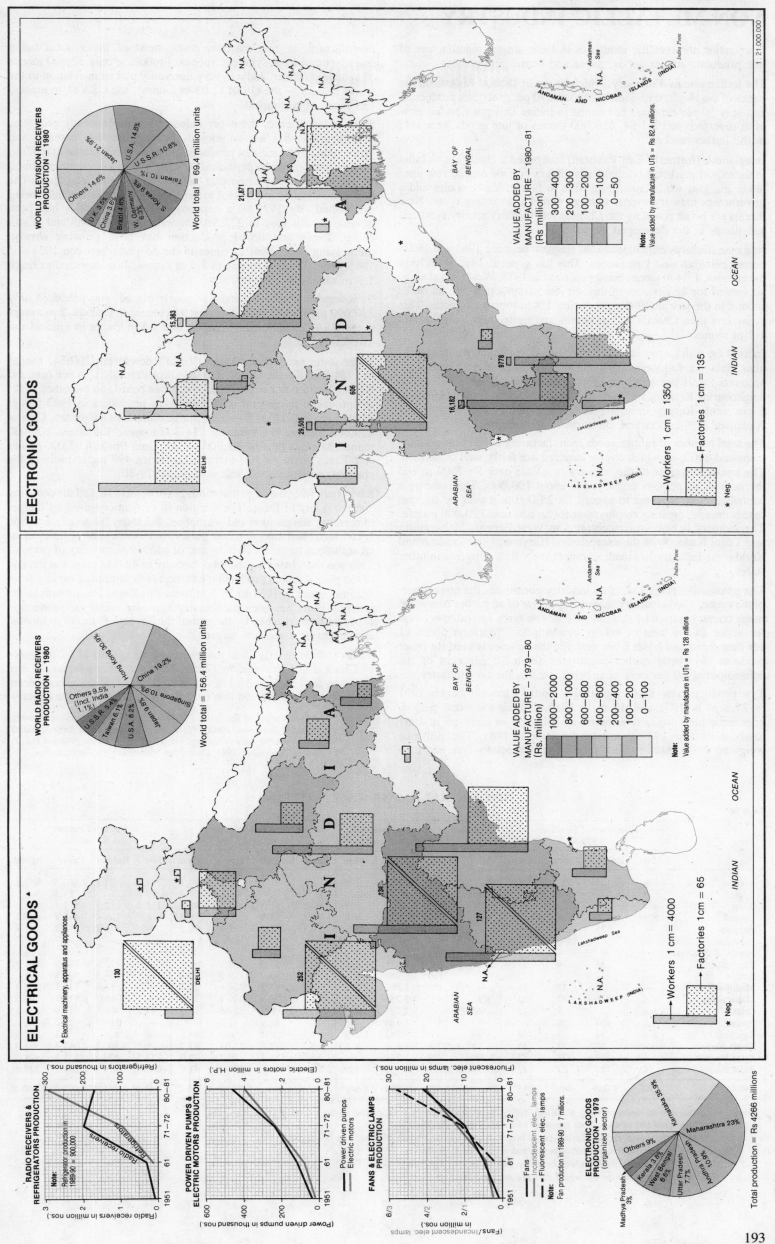

ELECTRONIC GOODS

ELECTRICAL GOODS ▲

▲ Electrical machinery, apparatus and appliances.

WORLD TELEVISION RECEIVERS PRODUCTION – 1980

U.S.A. 14.6%
Japan 21.9%
U.S.S.R. 10.6%
Taiwan 10.1%
S. Korea 9.8%
W. Germany 6.3%
Brazil 4.6%
China 3.6%
U.K. 3.9%
Others 14.6%

World total = 69.4 million units

WORLD RADIO RECEIVERS PRODUCTION – 1980

Hong Kong 30.8%
China 19.2%
Singapore 10.9%
Japan 10.9%
U.S.A. 8.2%
Taiwan 6.1%
U.S.S.R. 5.4%
Others 9.5%
(Incl. India 1.1%)

World total = 156.4 million units

VALUE ADDED BY MANUFACTURE – 1980-81 (Rs million)

300 – 400
200 – 300
100 – 200
50 – 100
0 – 50

Note:
Value added by manufacture in UTs = Rs 824 millions

VALUE ADDED BY MANUFACTURE – 1979-80 (Rs. million)

1000 – 2000
800 – 1000
600 – 800
400 – 600
200 – 400
100 – 200
0 – 100

Note:
Value added by manufacture in UTs = Rs 128 millions

Workers 1 cm = 1350
Factories 1 cm = 135
* Neg.

Workers 1 cm = 4000
Factories 1cm = 65
* Neg.

BAY OF BENGAL
ARABIAN SEA
INDIAN OCEAN
Lakshadweep Sea
ANDAMAN AND NICOBAR ISLANDS (INDIA)
LAKSHADWEEP (INDIA)
Andaman Sea
Indira Point

RADIO RECEIVERS & REFRIGERATORS PRODUCTION

(Refrigerators in thousand nos.)
(Radio receivers in million nos.)

Radio receivers
Refrigerators

Note:
Refrigerator production in 1989-90 = 900,000

1951 61 71-72 80-81

POWER DRIVEN PUMPS & ELECTRIC MOTORS PRODUCTION

(Electric motors in million H.P.)
(Power driven pumps in thousand nos.)

Power driven pumps
Electric motors

1951 61 71-72 80-81

FANS & ELECTRIC LAMPS PRODUCTION

(Fluorescent elec. lamps in million nos.)
(Fans/Incandescent elec. lamps in million nos.)

Fans
Incandescent elec. lamps
Fluorescent elec. lamps

Note:
Fan production in 1989-90 = 7 millions

1951 61 71-72 80-81

ELECTRONIC GOODS PRODUCTION – 1979 (organized sector)

Karnataka 35.9%
Maharashtra 23%
Uttar Pradesh 7.7%
West Bengal 6.6%
Andhra Pradesh 3.9%
Kerala 3.9%
Madhya Pradesh 3%
Others 9%

Total production = Rs 4266 millions

NON-METALLIC INDUSTRY

Four major non-metallic industries in India are the manufacture of jute products, rubber goods, paper and board, and cement.

The Indian jute mill industry, which began in 1855 at Rishra, on the banks of the Hugli, traditionally exported 70 per cent of its production and only 30 per cent was consumed in India. This position has now been reversed; in 1983-84, 950,000 tonnes of jute goods were sold in the Indian market.

Bangladesh (formerly East Pakistan) has posed a challenge to India in the export market, having the advantages of good quality raw jute, lower margins while converting fibre to fabric, lower wages and a government-industry export strategy aimed at lowering costs. New threats are being posed by the Chinese jute industry and by synthetic substitutes in the developed world.

Jute manufactures remained almost stagnant between 1961 and 1971, output being around 1 m tonnes. This has gone up very slightly in the eighties, 1.34 m tonnes being produced in 1983. West Bengal alone accounted for 86 per cent of this. In jute manufactures, India ranks second in the world (1985), producing 1.8 m tonnes of goods. This is far less than China's 3.2 m tonnes and more than Bangladesh's 1.2 m tonnes.

India is the fifth largest natural rubber-producing country in the world after Malaysia, Indonesia, Thailand and Sri Lanka. Kerala state alone accounts for 91 per cent of the total area under rubber in India. The neighbouring Kanniyakumari district in Tamil Nadu accounts for about 5 per cent. Rubber cropping is being experimented with in the Andaman & Nicobars and the north-eastern region.

The total number of rubber goods manufacturing units in the country is around 2000, of which over a hundred are fairly well organized. The annual turnover of the industry is a little over Rs 7500 m and the number of persons employed is about 100,000. The industry's contribution is expected to exceed Rs 25,000 m a year by the end of this decade, creating employment for an additional 20,000 people. The industry is heavily concentrated in West Bengal, Maharashtra and Tamil Nadu. With the exception of Haryana, Uttar Pradesh and Kerala, the industry has made comparatively little progress in other states.

The automotive tyres and tubes industry constitutes the largest part of the rubber goods industry: about 52 per cent of all rubber consumed in the country is used by this industry. Cycle tyres and tubes account for about 14 per cent of rubber consumption, footwear about 11 per cent, belts and hoses 6 per cent and the balance is used for other products. Synthetic rubber constitutes about 20 per cent of the consumption, 60 per cent of it being used in the tyre industry.

Tyre production has built up gradually from 4.82 m numbers in 1951 to 27.3 m in 1971 and 40 m in 1984. During the same period, pneumatic tube production increased from 5.7 m numbers in 1951 to about 19 m in 1971 and about the same in 1981. The millstone weighing down rubber goods production, especially tyre and tube manufacture, is raw material costs, most of these items being manufactured from synthetic rubber. Producing only 32,800 tonnes of synthetic rubber, India is very unequally placed in relation to the world leaders — the USSR (2.03 m tonnes), the USA (1 m tonnes) and Japan (1 m tonnes).

Since independence, the paper industry has made steady progress, but though it has almost closed the gap between demand and production, it has still not produced material of international quality. Power shortages cause periodical deficits, adding to the burden caused by the necessity of importing newsprint and quality paper. The newsprint situation is expected to improve with several new mills being established. By 1979-80 a capacity of 150,000 tonnes had been developed; this rose to 222,500 tonnes during 1984-85 and is still rising. Paper and board production has been growing slowly, multiplying only about 11 times in the 35 years between 1951 and 1985, from 130,000 tonnes to 1.5 m tonnes. It is expected to reach 1.8 m tonnes in 1989-90.

In newsprint production India is poorly placed, and produced only 220,000 tonnes in 1984, as against 9 m tonnes in Canada, 5 m tonnes in the USA, 2.6 m tonnes in Japan, and 1.5 m tonnes in Finland and Sweden.

Three states produce the bulk of India's newsprint (200,547 tonnes in 1984-85): Kerala 32 per cent, Madhya Pradesh 32.4 per cent and Karnataka 35.6 per cent. Paper and paper board, on the other hand, are produced by several states, the large producers in 1983 being: Andhra Pradesh 216,000 tonnes, Maharashtra 166,700 tonnes, Gujarat 119,800 tonnes, West Bengal 114,200 tonnes, Karnataka 100,000 tonnes, Madhya Pradesh 92,700 tonnes, Uttar Pradesh 93,000 tonnes and Tamil Nadu 86,500 tonnes. There are 499 paper mills in the country, employing 70,362 workers (1979-80).

It was only five years ago that shortages of cement rocked the economy and the polity of India. The mention of cement conjured up images of scarcity, adulteration and corruption. But today the supply situation is comfortable thanks to a new policy framework. The Sixth Plan set an ambitious target of 20 m tonnes of additional capacity of cement. This was met, bringing the total capacity to 42.5 m tonnes at the end of the plan. Cement production has increased somewhat rapidly from 3.25 m tonnes in 1951 to 29.5 m tonnes in 1984. The growth in the last few years has been particularly impressive and the production expected to be reached by the end of the Seventh Plan, 49 m tonnes, appears to be a realistic target.

India produces nearly 26 m tonnes of cement against 142.2 m tonnes by China, 135.1 m tonnes by China, 72.9 m tonnes by Japan, 70.3 m tonnes by the USA and 37.2 m tonnes by Italy (1985). India is expected to climb much higher on this ladder before the end of the century.

Sources: 1 *Annual Survey of Industries 1979-80 & 1980-81*, Ministry of Planning.
2. *World Economy and India's Place in it, October 1986 & Basic Statistics Relating to the Indian Economy, Vol. 1: All India, August 1986* and *Vol. 2: States, September 1986*, Centre for Monitoring Indian Economy.

NON-METALLIC INDUSTRY — (1980-81)

Table 90

	Number of factories 1979-80				Number of mandays-workers (in thousand)				Net value added (in Rs. million)				Value of output (in Rs. million)			
	Jute[1]	Rubber[2]	Paper[3]	Cement[4]	Jute[1]	Rubber[2]	Paper[3]	Cement[4]	Jute[1]	Rubber[2]	Paper[3]	Cement[4]	Jute[1]	Rubber[2]	Paper[3]	Cement[4]
Andhra Pradesh	9	21	32	15	5,715	196	3,153	1,289	210.6	25.1	212.1	190.7	744.3	74.3	1,005.9	647.2
Assam	3	—	3	N.A.	456	—	519	N.A.	11.4	—	1.5	32.2	27.7	—	85.5	N.A.
Bihar	N.A.	9	18	17	1,917	43	1,677	1,599	66.7	11.0	142.8	41.3	159.4	12.7	313.1	398.6
Gujarat	N.A.	76	78	32	N.A.	350	1,412	1,078	0.5	24.6	136.1	130.6	N.A.	126.4	444.4	629.2
Haryana	—	35	26	7	—	1,321	1,549	592	—	252.6	192.3	96.2	—	976.2	872.7	351.9
Himachal Pradesh	—	—	—	3	—	—	—	131	—	—	—	N.A.	—	—	—	127.3
Jammu & Kashmir	—	—	—	N.A.	—	—	—	N.A.	—	—	—	1.7	—	—	—	N.A.
Karnataka	—	35	19	21	—	385	2,275	1,417	—	27.0	132.9	105.3	—	472.6	762.9	401.7
Kerala	—	122	7	3	—	950	786	N.A.	—	168.7	109.0	10.8	—	864.8	400.8	N.A.
Madhya Pradesh	—	13	18	75	—	119	1,962	2,847	—	3.7	224.4	25.1	—	13.9	738.1	1,246.1
Maharashtra	N.A.	230	87	34	N.A.	3,451	3,261	1,015	3.9	426.3	489.8	35.8	N.A.	2,063.0	1,572.4	292.5
Orissa	N.A.	N.A.	7	12	560	N.A.	2,393	425	13.5	0.1	157.5	14.1	44.2	N.A.	638.9	246.5
Punjab	—	104	20	—	—	577	140		—	46.9	2.8	—	—	218.7	25.2	—
Rajasthan	N.A.	14	7	54	N.A.	N.A.	41	1,279	1.1	−203.1	4.9	264.6	N.A.	N.A.	5.3	666.7
Tamil Nadu	N.A.	57	37	15	54	1,275.8	1,655	1,818	2.7	272.6	223.3	75.7	1.1	8,500.8	2,067.3	901.9
Uttar Pradesh	4	69	77	12	1,686	729	1,322	N.A.	88.0	181.3	70.7	75.0	253.5	134.8	335.8	N.A.
West Bengal	74	121	57	12	66,347	2,799	3,376	485	2,576.0	322.7	257.4	41.9	6,948	1,419.6	884.5	229.6
INDIA[5]	90[6]	1,049	499	315[6]	76,630[6]	13,552	26,074	18,641[6]	3,002.9[6]	1,843.3	2,235.9	1,325.7[6]	8,148.8[6]	9,265.2	9,009.2	7,183.6[6]

Note : [1] Includes jute & mesta pressing & baling, dyeing & bleaching, hemp & other coarse jute goods and jute bags.
[2] Includes tyres and tubes, rubber footwear and other rubber products.
[3] Includes pulp, paper and newsprint, paper boards, containers and boxes of paper and board.
[4] Includes cement, lime & plaster, asbestos and other cement products.
[5] Includes union territories for which break-up figures are not available.
[6] Excludes union territories for which data are not available.

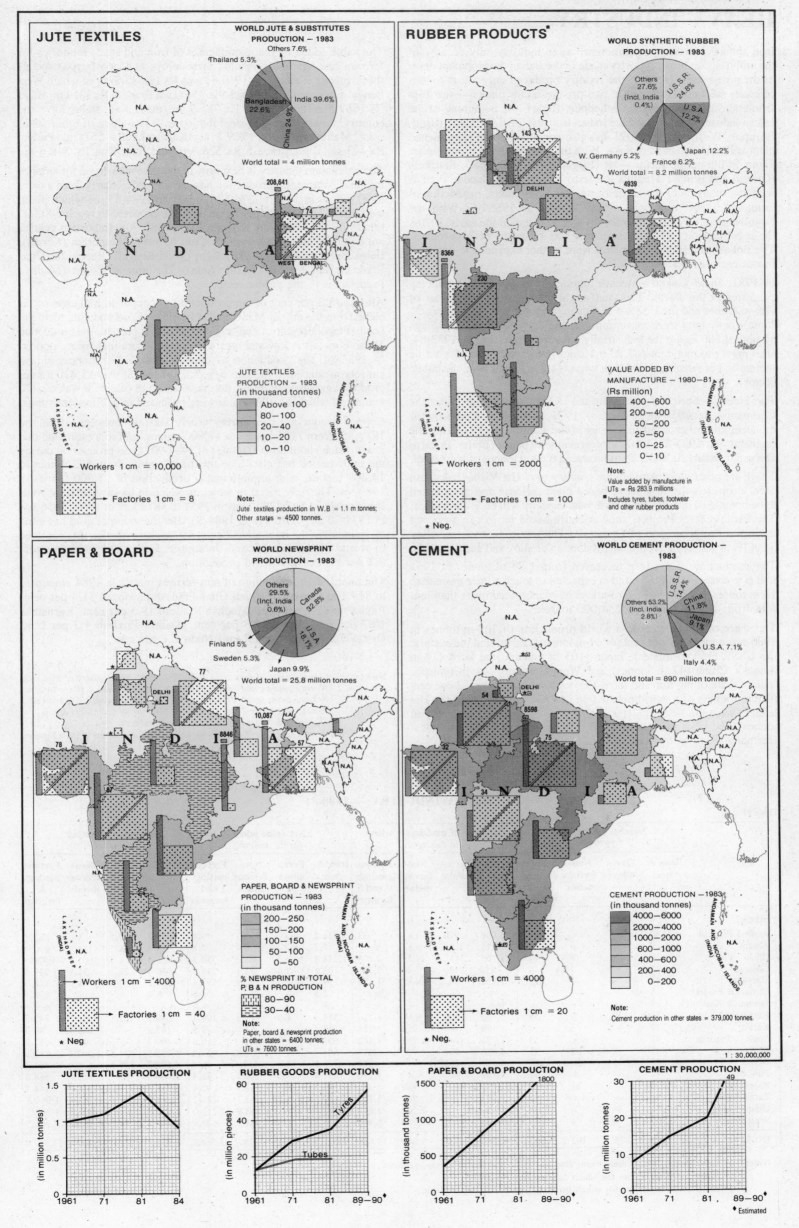

JUTE TEXTILES

WORLD JUTE & SUBSTITUTES
PRODUCTION — 1983
Others 7.6%
Thailand 5.3%
Bangladesh 22.6%
India 39.6%
China 24.9%

World total = 4 million tonnes

208,641
74
WEST BENGAL

N.A.

LAKSHADWEEP
(INDIA)
ANDAMAN AND NICOBAR ISLANDS (INDIA)

JUTE TEXTILES
PRODUCTION — 1983
(in thousand tonnes)
Above 100
80 — 100
20 — 40
10 — 20
0 — 10

Workers 1 cm = 10,000
Factories 1 cm = 8

Note:
Jute textiles production in W.B. = 1.1 m tonnes;
Other states = 4500 tonnes.

RUBBER PRODUCTS

WORLD SYNTHETIC RUBBER
PRODUCTION — 1983
Others 27.6% (Incl. India 0.4%)
U.S.S.R. 24.8%
U.S.A. 12.2%
Japan 12.2%
W. Germany 5.2%
France 6.2%

World total = 8.2 million tonnes

N.A. 143
DELHI
4939
8366
230

LAKSHADWEEP
(INDIA)
ANDAMAN AND NICOBAR ISLANDS (INDIA)

VALUE ADDED BY
MANUFACTURE — 1980–81
(Rs million)
400 — 600
200 — 400
50 — 200
25 — 50
10 — 25
0 — 10

Note:
Value added by manufacture in
UTs = Rs 283.9 millions
■ Includes tyres, tubes, footwear
and other rubber products

Workers 1 cm = 2000
Factories 1 cm = 100

★ Neg.

PAPER & BOARD

WORLD NEWSPRINT
PRODUCTION — 1983
Others 29.5% (Incl. India 0.6%)
Canada 32.8%
U.S.A. 18.1%
Finland 5%
Sweden 5.3%
Japan 9.9%

World total = 25.8 million tonnes

77
DELHI
10,087
8846
57
78
87

LAKSHADWEEP
(INDIA)
ANDAMAN AND NICOBAR ISLANDS (INDIA)

PAPER, BOARD & NEWSPRINT
PRODUCTION — 1983
(in thousand tonnes)
200 — 250
150 — 200
100 — 150
50 — 100
0 — 50

% NEWSPRINT IN TOTAL
P, B & N PRODUCTION
80 — 90
30 — 40

Workers 1 cm = 4000
Factories 1 cm = 40

★ Neg

Note:
Paper, board & newsprint production
in other states = 6400 tonnes;
UTs = 7600 tonnes.

CEMENT

WORLD CEMENT PRODUCTION —
1983
Others 53.2% (Incl. India 2.8%)
U.S.S.R. 14.4%
China 11.8%
Japan 9.1%
U.S.A. 7.1%
Italy 4.4%

World total = 890 million tonnes

DELHI
54
8598
32
75
34

LAKSHADWEEP
(INDIA)
ANDAMAN AND NICOBAR ISLANDS (INDIA)

CEMENT PRODUCTION — 1983
(in thousand tonnes)
4000 — 6000
2000 — 4000
1000 — 2000
600 — 1000
400 — 600
200 — 400
0 — 200

Workers 1 cm = 4000
Factories 1 cm = 20

★ Neg.

Note:
Cement production in other states = 379,000 tonnes.

1 : 30,000,000

JUTE TEXTILES PRODUCTION
(in million tonnes)
1.5
1
0.5
0
1961 71 81 84

RUBBER GOODS PRODUCTION
(in million pieces)
60
40
20
0
Tyres
Tubes
1961 71 81 89–90

PAPER & BOARD PRODUCTION
(in thousand tonnes)
1800
1500
1000
500
0
1961 71 81 89–90

CEMENT PRODUCTION
(in million tonnes)
49
30
20
10
0
1961 71 81 89–90
◆ Estimated

195

HEAVY INDUSTRY

Iron and steel came into prominence as an industry only because of the emphasis on heavy industry made in the Indian development plans of the post-fifties. In 1947, the country produced only 1.2 m tonnes of crude steel (steel ingots) in two private sector plants — the Tata Iron and Steel Company (Jamshedpur, Bihar), the beginning of the Indian industrial dream, and the Indian Iron and Steel Company (Kulti-Burnpur, West Bengal). Steel was also produced in the state-owned plant at Bhadravati in Karnataka. With the establishment of three other plants in the public sector — at Durgapur (West Bengal), Raurkela (Orissa) and Bhilai (Madhya Pradesh) — during the Second Plan as well as expansion of the capacities of the older plants, production of crude steel rose to about 6.5 m tonnes in 1965-66. With the establishment of the Bokaro Steel Plant in Bihar in the early 1970s, the ingot capacity of all these plants was raised to 11.4 m tonnes. The Bokaro and Bhilai steel plants were further expanded to 4 m steel ingots each in 1981-82.

In 1981, India ranked sixteenth among the major steel-producing countries of the world. The total world output was of the order of 708 m tonnes and the USSR was the top producer with 149 m tonnes. India's per capita steel consumption in the early eighties was 15 kgs as against 600 kgs in the industrially advanced countries. In 1983-84, steel ingot production was 7.93 m tonnes (6 m tonnes from the public sector) and of saleable steel 6.4 m tonnes (4.8 m tonnes from the public sector).

The Steel Authority of India Limited (SAIL) was set up under the Companies Act 1956 on 24 January, 1973, as a holding company for the steel and other input industries in the public sector. During the financial year 1981-82, SAIL's integrated steel plants (Bhilai, Bokaro, Durgapur, Raurkela and Salem) produced 6.64 m tonnes of ingot steel.

There are other steel plants in the country too. The Vishakhapatnam Steel Project is the first shore-based plant in India and is expected to be completed in 1987 when its annual capacity will be 3.4 m tonnes. The Paradip Steel Project, another shore-based project, is to be set up in the Daitari region of Orissa. At Durgapur, there is an Alloy Steel Plant which pioneered the production of alloy and special steels.

The production of pig iron increased from 1.74 m tonnes in 1951 to 8.6 m tonnes in 1980-81 and is expected to touch a little more than 13 m tonnes in 1989-90. Ferro-alloys have increased more than ten-fold from 24,000 tonnes to 254,000 tonnes.

The share of Indian steel in a world production of 705 m tonnes in 1985 was a mere 1.7 per cent (or 11 m tonnes), far behind the leaders: the USSR (156 m tonnes), Japan (105 m tonnes), the USA (79 m tonnes), China (47 m tonnes) and West Germany (41 m tonnes). India's share of pig iron and ferro-alloys was slightly better, 2 per cent (9.9 m tonnes) of a world production of 483 m tonnes in 1985. The leaders were the USSR (110.9 m tonnes — 1984 data), Japan (82 m tonnes), the USA (45.4 m tonnes), China (45 m tonnes) and West Germany (31.7 m tonnes).

The value added by the manufacture of iron and steel, foundries for ferrous castings and forgings, ferro-alloys and non-ferrous metals (basic metals & alloys) in 1980-81 was Rs 12,342.5 m, of which West Bengal accounted for Rs 2033.9 m, Madhya Pradesh Rs 2219 m, Bihar Rs 1667.5 m and Orissa Rs 781.7 m (these four states have the country's major iron and steel industries). The share of other states was: Maharashtra Rs 1985.7 m, Karnataka Rs 561 m, Punjab Rs 443 m, Uttar Pradesh Rs 526.7 m and Rajasthan Rs 348.6 m.

The aluminium industry is next only to the steel industry in importance in India. Aluminium production has grown spectacularly from a mere 3900 tonnes in 1951 to 276,500 tonnes in 1984-85, of which about a quarter was from the public sector. It is expected to reach 500,000 tonnes by the end of the Seventh Plan (1989-90). Important centres with aluminium smelters are Hirakud (Orissa), Alupuram (Kerala), Belgaum (Karnataka), Jaykaynagar (West Bengal), Renukoot (Uttar Pradesh) and Mettur (Tamil Nadu). The unit at Korba (Madhya Pradesh) is in the public sector.

Almost 33.3 per cent of the production of copper in India has so far come from the unit at Maubhandar in Bihar. Lead and zinc form yet another essential pair of non-ferrous metals. At present, the production of these metals is low and most requirements are met from imports. In 1983-84, the production of copper was 35,370 tonnes, almost entirely produced in the public sector, of lead it was 15,420 tonnes produced exclusively by the public sector, and of zinc it was 60,170 tonnes, of which 53,700 tonnes were produced in public sector units.

Copper production has grown nearly six-fold between 1951 and 1984-85, from 7196 tonnes to 41,000 tonnes, and it is estimated that it will reach about 86,000 tonnes in 1989-90. Lead production during the same period has also gone up considerably, from 873 tonnes to 14,200 tonnes, with an anticipated production of 27,000 tonnes in 1989-90. The only metal in which production has dwindled is gold. It has dropped from 7040 kgs in 1951 to 4872 kgs in 1961, 3656 kgs in 1971-72 and 2036 kgs in 1984-85. But the value of gold has gone up so much in the internal and external markets, that profits continue to accrue from its production. In copper, India produces only about 0.2 per cent of the world production.

The total Indian production of non-ferrous metals in 1984 amounted to 314,100 tonnes, of which Uttar Pradesh produced 31.4 per cent, Rajasthan 22.7 per cent, Madhya Pradesh 19.5 per cent, Karnataka 10.7 per cent, Kerala 5.8 per cent, Andhra Pradesh 4.3 per cent, Orissa 3.5 per cent and Tamil Nadu 2.1 per cent.

Sources: 1 *Annual Survey of Industries 1979-80 & 1980-81*, Ministry of Planning.
2. *World Economy and India's Place in it, October 1986 & Basic Statistics Relating to the Indian Economy, Vol. 1: All India, August 1986*, and *Vol. 2 States, September 1986*, Centre for Monitoring Indian Economy.
3. *Directory & Yearbook 1984*, Times of India.
4. *Seventh Five Year Plan 1985-90, Vol. I & II*, Government of India Planning Commission.

Table 91

HEAVY INDUSTRY — 1980-81

	Number of factories 1979-80				Number of mandays-workers (in thousand)				Net value added (in Rs. million)				Value of output (in Rs. million)			
	Iron & steel	Ferro-alloys	Non-ferrous metals[1]	Ferrous castings and forgings	Iron & steel	Ferro-alloys	Non-ferrous metals[1]	Ferrous castings and forgings	Iron & steel	Ferro alloys	Non-ferrous metals[1]	Ferrous castings and forgings	Iron & steel	Ferro alloys	Non-ferrous metals[1]	Ferrous castings & forgings
States																
Andhra Pradesh	41	33	N.A.	29	2,036	976	N.A.	931	126.2	N.A.	118.3	79.6	627.1	712.5	N.A.	283.9
Assam	9	N.A.	N.A.	N.A.	11	N.A.	N.A.	N.A.	7.2	N.A.	0.4	7.8	3.4	N.A.	N.A.	N.A.
Bihar	76	4	6	44	21,167	N.A.	1,287	3,808	1,234.7	−7.3	82.2	350.3	8,812.4	N.A.	603.1	1,420
Gujarat	75	7	7	48	950	20	47	2,528	103.1	3.4	33.1	248.3	956.4	22.1	31.6	929.8
Haryana	48	4	4	35	1,152	N.A.	N.A.	1,133	159.7	1.3	22.9	197.4	906	—	N.A.	118.6
Himachal Pradesh	5	—	—	N.A.	65	—	—	N.A.	2.5	—	—	4.6	6.4	—	—	N.A.
Jammu & Kashmir	4	—	—	N.A.	N.A.	—	—	N.A.	1.7	—	—	1.4	N.A.	—	—	N.A.
Karnataka	39	10	9	38	3,186	600	N.A.	1,193	354.9	1.6	17.1	109.4	1,659.5	725.1	N.A.	349.5
Kerala	19	—	3	7	199	—	294	114	39.5	—	57	13.6	185.8	—	258.4	86.2
Madhya Pradesh	92	—	N.A.	30	7,772	—	1,195	1,462	2,607.4	—	−499.3	110.9	6,309.3	—	505	517.2
Maharashtra	145	7	38	91	4,080	401	1,735	6,979	681.3	49.8	331.5	923.0	3,955.5	334.5	1,303.8	3,552.1
Orissa	17	5	N.A.	17	6,599	462	N.A.	810	702.5	76.2	−438	46.7	4,599.2	315.8	N.A.	187.3
Punjab	272	—	N.A.	30	2,159	—	N.A.	1,276	361.4	—	0.2	81.4	2,069	—	N.A.	368.8
Rajasthan	50	—	11	12	416	—	769	389	56.8	—	251.7	40.1	231.2	—	571.9	144.5
Tamil Nadu	84	—	N.A.	55	1,093	—	366	3,105	128.3	—	63.2	292.7	683.8	—	364.4	1,094.8
Uttar Pradesh	154	—	4	68	1,684	—	7,272	2,612	195.7	—	118.9	212.0	1,055.3	—	932.2	1,097.3
West Bengal	165	8	9	143	15,819	158	1,264	8,820	1,256.5	4.6	164.9	607.9	6,305.8	84.5	682.2	2,261.5
INDIA[3]	1,331	78[2]	103	663	66,784	1,835[2]	9,642	35,616	8,041.3	205.2[2]	718	3,366.7	38,855.3	1,220[2]	6,822.9	13,547.4

Note: [1] Includes copper, brass, aluminium, zinc and other non-ferrous metal industries.
[2] Excludes union territories for which data are not available.
[3] Includes union territories for which break-up figures are not available.

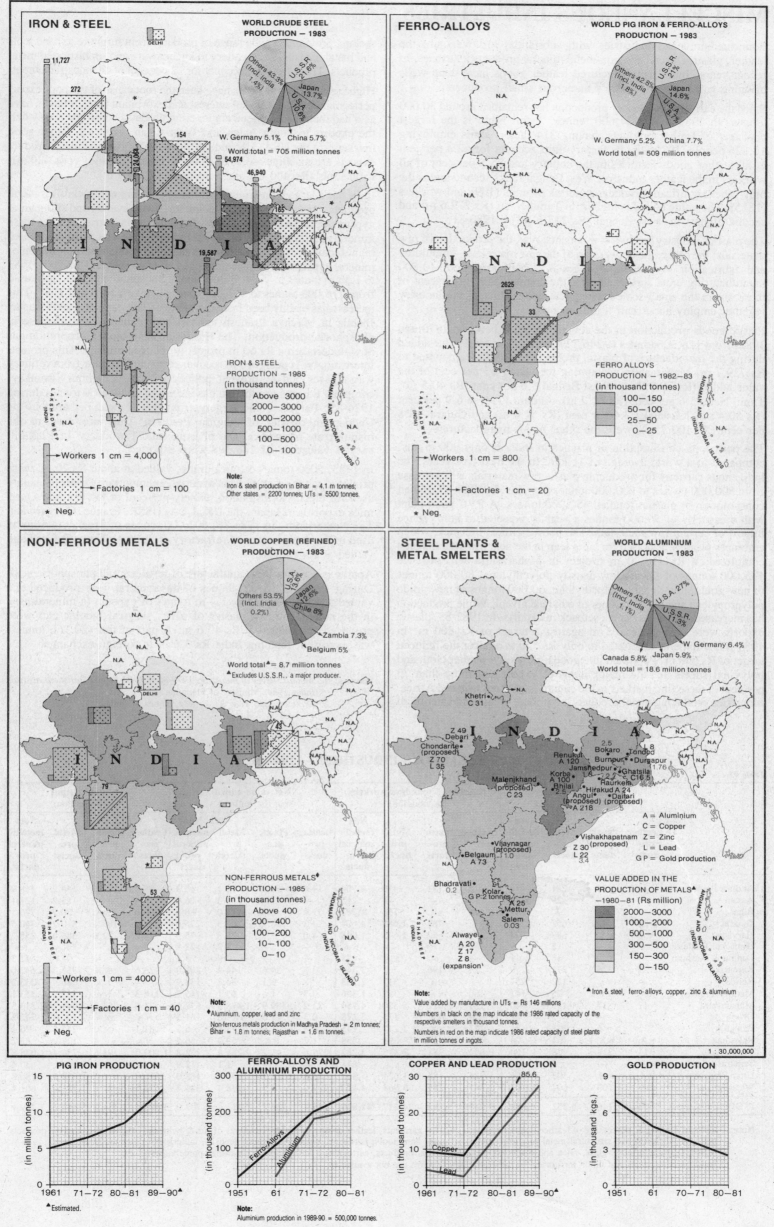

IRON & STEEL

WORLD CRUDE STEEL PRODUCTION — 1983

Others 43.3% (Incl. India 1.4%)
U.S.S.R. 21.6%
Japan 13.7%
U.S.A. 10.6%
W. Germany 5.1%
China 5.7%

World total = 705 million tonnes

11,727
272
24,064
154
54,974
46,940
165
19,587
DELHI

IRON & STEEL PRODUCTION — 1985 (in thousand tonnes)

Above 3000
2000–3000
1000–2000
500–1000
100–500
0–100

→ Workers 1 cm = 4,000
→ Factories 1 cm = 100
* Neg.

Note:
Iron & steel production in Bihar = 4.1 m tonnes;
Other states = 2200 tonnes; UTs = 5500 tonnes.

FERRO-ALLOYS

WORLD PIG IRON & FERRO-ALLOYS PRODUCTION — 1983

Others 42.5% (Incl. India 1.6%)
U.S.S.R. 21%
Japan 14.6%
U.S.A. 8.7%
W. Germany 5.2%
China 7.7%

World total = 509 million tonnes

2625
33

FERRO ALLOYS PRODUCTION — 1982-83 (in thousand tonnes)

100–150
50–100
25–50
0–25

→ Workers 1 cm = 800
→ Factories 1 cm = 20
* Neg.

NON-FERROUS METALS

WORLD COPPER (REFINED) PRODUCTION — 1983

U.S.A. 13.6%
Japan 12.6%
Chile 8%
Others 53.5% (Incl. India 0.2%)
Zambia 7.3%
Belgium 5%

World total ▲ = 8.7 million tonnes
▲ Excludes U.S.S.R., a major producer.

DELHI
79
53

NON-FERROUS METALS ♦ PRODUCTION — 1985 (in thousand tonnes)

Above 400
200–400
100–200
10–100
0–10

→ Workers 1 cm = 4000
→ Factories 1 cm = 40
* Neg.

Note:
♦ Aluminium, copper, lead and zinc
Non-ferrous metals production in Madhya Pradesh = 2 tonnes;
Bihar = 1.8 m tonnes; Rajasthan = 1.6 m tonnes.

STEEL PLANTS & METAL SMELTERS

WORLD ALUMINIUM PRODUCTION — 1983

U.S.A. 27%
U.S.S.R. 11.3%
W. Germany 6.4%
Japan 5.9%
Canada 5.8%
Others 43.6% (Incl. India 1.1%)

World total = 18.6 million tonnes

Khetri C 31
Z 49 Debari
Chondarite (proposed) Z 70 L 35
Renukut A 120
Bokaro 2.5
Tonoda 1.8
Burnpur Durgapur
Jamshedpur 1.8
Ghatsila 1.76
Malenjkhand (proposed) C 23
Korba A 100
Bhilai 2.5
C 16.5
Raurkela
Hirakud A 24
Daitari (proposed)
Angul (proposed) A 218
Vishakhapatnam (proposed)
Z 30 L 22 3.4
Vijaynagar (proposed)
Belgaum A 73
11.0
Bhadravati 0.2
Kolar G P.2 tonnes
A 25
Mettur
Salem 0.03
Alwaye A 20 Z 17 Z 8 (expansion)

A = Aluminium
C = Copper
Z = Zinc
L = Lead
G P = Gold production

VALUE ADDED IN THE PRODUCTION OF METALS ▲ —1980–81 (Rs million)

2000–3000
1000–2000
500–1000
300–500
150–300
0–150

▲ Iron & steel, ferro-alloys, copper, zinc & aluminium

Note:
Value added by manufacture in UTs = Rs 146 millions
Numbers in black on the map indicate the 1986 rated capacity of the respective smelters in thousand tonnes.
Numbers in red on the map indicate 1986 rated capacity of steel plants in million tonnes of ingots.

1 : 30,000,000

PIG IRON PRODUCTION

(in million tonnes)
15
10
5
0
1961 71–72 80–81 89–90 ▲
▲ Estimated.

FERRO-ALLOYS AND ALUMINIUM PRODUCTION

(in thousand tonnes)
300
200
100
0
1951 61 71–72 80–81
Ferro-Alloys
Aluminium

Note:
Aluminium production in 1989-90 = 500,000 tonnes.

COPPER AND LEAD PRODUCTION

(in thousand tonnes)
30
20
10
0
1961 71–72 80–81 89–90 ▲
85.6
Copper
Lead

GOLD PRODUCTION

(in thousand kgs.)
9
6
3
0
1951 61 70–71 80–81

197

MISCELLANEOUS INDUSTRIES

Four medium-sized industries with substantial turnovers are the leather, plastics, metal and non-metal mineral products industries. In recent years, India's manufactured leather goods have done well, bringing gains to the country's numerous small producers.

Since the early 1970s, leather production has remained around 30,000 tonnes. In 1982 it was 27,600 tonnes. Tamil Nadu is the largest producer of leather products, having 314 factory units employing 14,826 people (1979-80). Its factory units account for 62.4 per cent of all leather goods factories in the country and 75.5 per cent of all workers in the leather industry. West Bengal ranks second and is the only other significant producer in terms of units (109) and workers (1815). The value added is highest in Tamil Nadu (Rs 279.6 m) and second highest in West Bengal (Rs 217.6 m) in 1980-81.

The plastics industry has two components: one, the production of raw materials; the other, the conversion of the raw materials — moulding and fabrication — into various goods. There are about 10,000 manufacturing units spread all over the country and 90 per cent of these are in the small-scale sector. The two sectors of the industry together employ about half a million people.

Plastic goods production in the country doubled between 1970 and 1982, from 126,306 tonnes to 246,117 tonnes. The total value added during the manufacture of plastic products in 1980-81 amounted to Rs 729.7 m, Maharashtra accounting for nearly 53 per cent of the value added (Rs. 389.9 m). West Bengal 5.6 per cent (Rs 40.5 m), Tamil Nadu 5.6 per cent (Rs 41.2 m), Andhra Pradesh 6.9 per cent (Rs 50.4 m), Karnataka 7.4 per cent (Rs 54.6 m) and Gujarat 13.8 per cent (Rs 100.7 m), were the other major manufacturers.

The per capita consumption of plastics in India is only 300 grams, compared to a world average of 11 kgs. In the next five years, the indigenous capacity for producing plastic raw materials will increase from 300,000 tonnes to 600,000 tonnes per year. In 1984-85, Indian consumption of plastics totalled 450,000 tonnes. A PVC resin plant with a capacity of 55,000 tonnes a year is expected to help reduce import of certain raw materials. The petrochemical industry, which embraces plastics, is poised for a big leap in the next few years. When completed, a Rs 11,700 m project in Maharashtra will produce 135,000 tonnes of linear, low-density polyethylene, 80,000 tonnes of new grade, low-density polyethylene, 60,000 tonnes of new grade polypropylene and 55,000 tonnes of ethylene glycol. While production has increased, there has been a setback in exports. In 1982-83, plastic exports were only Rs 626.5 m against a target of Rs 800 m. In 1983-84, shipments amounted to only Rs 520 m against the reduced target of Rs 650 m. Polylined jute goods, polyvinyl leather cloth and polyvinyl sheets are the major items of export and more than 75 per cent of these find markets in the countries of West Asia and Africa. Industries producing metal products have come of age and cater to all sectors, providing a wide range of products from furniture to hand tools and metal utensils, from cutlery to kitchenware and sophisticated metal products. These factories account for 32 per cent of all industrial output.

High costs of raw materials have been the root cause of the poor export performance. High rates of interest and exorbitant foreign tariffs have also had their effect. Metal engineering products constitute nearly half the exports to South Asian and West Asian countries. Other good markets are the USA, UK and USSR. India's principal metal product exports are machine tools (Rs 300 m), sanitary castings (Rs 420 m), hand tools (Rs 400 m) and auto parts (Rs 400 m).

India has a large industrial sector manufacturing non-metallic mineral products, using limestone and dolomite, mica, kyanite and sillimanite, gypsum, apatites and phosphates. Some of the industries that use these minerals are cement, fertilizer, refractory, and electrical equipment manufacture. With the increase in cement production, production of mineral gypsum has also been stepped up from 990,000 tonnes in 1983 to 1.38 m tonnes in 1984. Production of phosphorite has also increased from 776,000 tonnes to 957,000 tonnes between 1983 and 1984. This increase has mainly been from the mines of Thamarakotra and Kanpur. Jhabua in Madhya Pradesh too has contributed to the increase in phosphorite production. The Hindustan Petroleum Corporation has now undertaken a Rs 60 m project to recover sulphur. This project, interestingly, is designed to recover elemental sulphur from refinery waste gases and to prevent pollution in the Chembur (Bombay) complex. Barytes production has increased to 446,000 tonnes during 1984-85. Barytes come largely from the Mangampet area in the Cuddappah district of Andhra Pradesh. They are used in the manufacture of medical and pharmaceutical products. In 1982-83, barytes valued at Rs 148,398 were exported.

In 1985, 5000 tonnes of mica crude, valued at about Rs 20 m, were produced in the country. This material, an important export, is mainly used for electrical insulation. Ninety per cent of the country's total mica export is to Japan, the UK, USA, USSR, France, Hungary and The Netherlands. In 1984-85, 602,000 tonnes of fireclay (a mineral used in the manufacture of refractory arc lamps) were produced and valued at Rs 18.6 m.

Apatite is used for the manufacture of fertilizer and phosphoric acid. During 1984-85, 16,000 tonnes of this mineral were produced and valued at Rs 4.3 m. Also, 1.4 m tonnes of gypsum (a mineral used in the manufacture of plaster of paris, surgical moulds etc) were produced and valued at Rs 40.6 m. A part of this (20,766 tonnes) was exported, fetching India Rs 3.6 m in foreign exchange.

Sources: 1. *Annual Survey of Industries 1979-80 & 1980-81,* and *Statistical Abstract India 1984*, Ministry of Planning.
2. *Directory and Year Book 1984*, Times of India.

MISCELLANEOUS INDUSTRIES — 1980-81

Table 92

	Number of factories 1979-80				Number of mandays-workers (in thousand)				Net value added (in Rs. million)				Value of output (in Rs. million)			
	Leather products[1]	Plastic products	Metal products[2]	Non-metallic mineral products[3]	Leather products[1]	Plastic products	Metal products[2]	Non-metallic mineral products[3]	Leather products[1]	Plastic products	Metal products[2]	Non-metallic mineral products[3]	Leather products[1]	Plastic products	Metal products[2]	Non-metallic mineral products[3]
Andhra Pradesh	24	24	281	476	216	132	470	2,704	14.6	50.4	50.1	65.3	58.4	184.6	154.2	458.9
Assam	—	—	29	16	—	—	74	215	—	—	8.3	−26.9	—	—	21.2	123.1
Bihar	N.A.	N.A.	270	849	475	N.A.	136	6,381	37.9	0.4	17.0	88.3	200.6	N.A.	20.7	776.1
Gujarat	22	22	662	795	N.A.	488	1,685	4,604	3.3	100.7	207.3	275.6	N.A.	189.4	510.7	740.5
Haryana	N.A.	N.A.	687	74	31	50	1,675	1,048	4.0	5.7	225.8	75.5	24.2	22.6	586.6	258.2
Himachal Pradesh	—	—	8	9	—	—	N.A.	N.A.	—	—	1.2	21.8	—	—	N.A.	N.A.
Jammu & Kashmir	—	—	18	11	—	—	N.A.	175	—	—	2.9	3.6	—	—	N.A.	12.3
Karnataka	5	5	271	279	59	196	702	3,544	3.8	54.6	144.2	208.0	23.6	N.A.	355.3	642.0
Kerala	—	N.A.	122	257	—	N.A.	312	2,363	—	1.1	35.2	138.9	—	N.A.	86.4	233.2
Madhya Pradesh	N.A.	N.A.	75	141	180	N.A.	392	1,396	0.7	1.4	65.1	145.2	118.3	N.A.	153.2	353.1
Maharashtra	17	17	1,195	840	187	2,017	8,330	4,750	22.2	389.9	1,478.6	532.1	71.6	1,054.7	3,949.3	7,313.5
Orissa	4	4	72	81	N.A.	N.A.	133	3,278	0.3	3.5	11.8	181.1	N.A.	N.A.	245.5	742.2
Punjab	5	5	528	31	227	N.A.	1,217	145	8.8	1.9	114.3	6.2	81.1	N.A.	245.5	20.1
Rajasthan	N.A.	N.A.	101	294	N.A.	27	393	858	1.7	N.A.	50.9	56.5	N.A.	2.4	127.9	83.2
Tamil Nadu	314	314	428	190	3,651	282	2,475	3,282	279.6	41.2	285.5	322.8	1,809.0	91.6	754.4	1,025.8
Tripura	—	—	3	N.A.	—	—	N.A.	841	—	—	1.1	12.9	—	—	N.A.	30.4
Uttar Pradesh	N.A.	N.A.	503	494	1,925	183	1,621	5,368	82.5	23.9	137.8	168.0	556.0	N.A.	339.1	780.8
West Bengal	109	109	602	182	2,845	255	3,967	4,420	217.6	40.5	378.4	240.1	632.6	121.8	1,181.4	601.4
INDIA[4]	503	146	6,232	5,080	9,945	3,748	24,217	41,806	685.3	729.7	3,312.4	2,707.0	3,608.6	1,864.2	8,651.2	7,269.1

Note: [1] Includes tanning & processing of leather, leather footwear, leather garments, leather consumer goods, garments of fur & pelts and other leather and fur products.
[2] Includes fabricated and structural metal products, metal furniture, hand tools, hardware, enamelling, metal utensils, cutlery and other metal products.
[3] Includes structural clay, glass, mica and other non-metallic mineral products, earthenware and pottery, chinaware, porcelainware, etc.
[4] Includes other states and union territories for which break-up figures are not available.

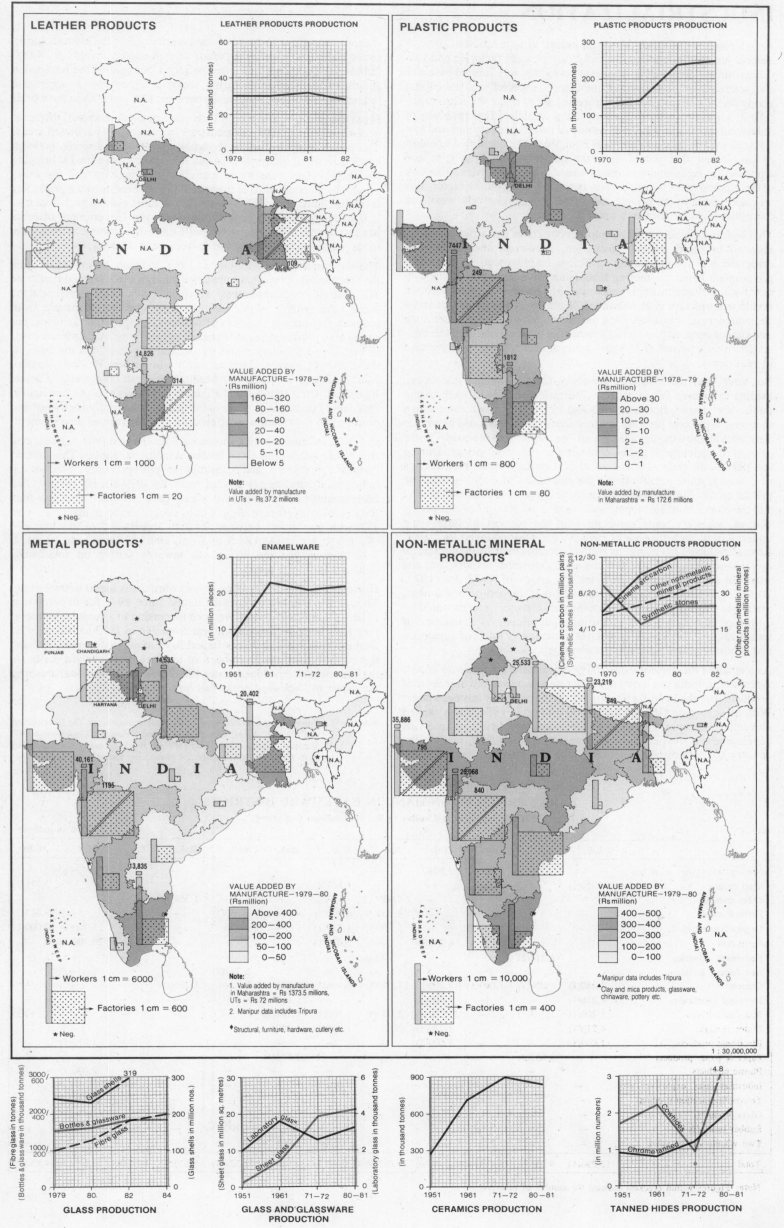

LEATHER PRODUCTS

LEATHER PRODUCTS PRODUCTION

(in thousand tonnes)

VALUE ADDED BY MANUFACTURE—1978–79 (Rs million)
- 160–320
- 80–160
- 40–80
- 20–40
- 10–20
- 5–10
- Below 5

→ Workers 1 cm = 1000
→ Factories 1 cm = 20

★ Neg.

Note:
Value added by manufacture in UTs = Rs 37.2 millions

PLASTIC PRODUCTS

PLASTIC PRODUCTS PRODUCTION

(in thousand tonnes)

VALUE ADDED BY MANUFACTURE—1978–79 (Rs million)
- Above 30
- 20–30
- 10–20
- 5–10
- 2–5
- 1–2
- 0–1

→ Workers 1 cm = 800
→ Factories 1 cm = 80

★ Neg.

Note:
Value added by manufacture in Maharashtra = Rs 172.6 millions

METAL PRODUCTS♦

ENAMELWARE

(in million pieces)

VALUE ADDED BY MANUFACTURE—1979–80 (Rs million)
- Above 400
- 200–400
- 100–200
- 50–100
- 0–50

→ Workers 1 cm = 6000
→ Factories 1 cm = 600

★ Neg.

Note:
1. Value added by manufacture in Maharashtra = Rs 1373.5 millions, UTs = Rs 72 millions
2. Manipur data includes Tripura

♦ Structural, furniture, hardware, cutlery etc.

NON-METALLIC MINERAL PRODUCTS▲

NON-METALLIC PRODUCTS PRODUCTION

(Cinema arc carbon in million pairs) (Synthetic stones in thousand kgs)

(Other non-metallic mineral products in million tonnes)

Cinema arc carbon
Other non-metallic mineral products
Synthetic stones

VALUE ADDED BY MANUFACTURE—1979–80 (Rs million)
- 400–500
- 300–400
- 200–300
- 100–200
- 0–100

→ Workers 1 cm = 10,000
→ Factories 1 cm = 400

★ Neg.

△ Manipur data includes Tripura
▲ Clay and mica products, glassware, chinaware, pottery etc.

1 : 30,000,000

GLASS PRODUCTION

(Fibre glass in tonnes) (Bottles & glassware in thousand tonnes)

(Glass shells in million nos.)

Glass shells 319
Bottles & glassware
Fibre glass

GLASS AND GLASSWARE PRODUCTION

(Sheet glass in million sq. metres)

(Laboratory glass in thousand tonnes)

Laboratory glass
Sheet glass

CERAMICS PRODUCTION

(in thousand tonnes)

TANNED HIDES PRODUCTION

(in million numbers)

Cowhides
Chrome tanned
4.8

INDUSTRIALIZATION

At independence, India did have an industrial foundation of sorts — several jute and cotton textile mills, some sugar factories and two major iron and steel plants. However, even this very limited base was characterized by disparity and imbalance, especially in the capital goods sector. Government therefore sought to make the nation self-reliant, using economic planning as a vehicle. The problem of industrialization was perceived as one of insufficient savings and low investment. The Second Five Year Plan, therefore, identified a pattern of investment that would optimize output in the long run, and emphasized the capital and intermediate goods industries. The task of developing these industries was entrusted to the public sector; this was acceptable to the private sector, for it was relatively weak and the policy did not encroach upon its domain.

This policy, complemented by import restrictions and strict licensing, resulted in rapid import-substitution. However, the policy was criticized for not generating sufficient employment. In response, government began promoting labour-intensive strategies along with new fiscal and administrative measures (the handloom sector in cotton textile manufacture is an example). With the establishment of state-sponsored research institutions and financial support through state investment corporations, India's industrial output grew until the mid-sixties, over a period of 15 years, at an annual rate of 7 to 8 per cent in real terms.

But after this, the industrial economy faced a demand recession, leading to a sharp fall in capacity utilization and severe cutbacks in public investment. Between 1965 and 1975, the growth came down to around 4 per cent per year and this deceleration was aided by slower agricultural growth, devaluation of the rupee and relatively faster growth of agricultural prices compared to industrial prices. During this period, the policy for industrial development underwent a series of changes, most notable being the relaxation of controls over the private sector. Simultaneously, conscious efforts were made to protect and develop a 'modern' small-scale sector. As a result of this policy change, ancillary units came up with the support of large-scale undertakings and the banks which were nationalized in 1969. Liberalization of licensing, as well as controls over foreign trade, in the mid-seventies, reoriented the policy towards export, but the situation became critical with the oil crises of 1973 and 1979. However, the Sixth Plan saw a fresh effort at stepped-up investment, mainly in the energy industries of the public sector, accompanied by a further liberalization of licensing and pricing in a number of industries in 1980: several industries immediately initiated expansion and modernization, upgrading technology.

As a result of the present industrial policy and the emphasis on import substitution, India can today manufacture a wide variety of industrial products. But the overall effect of this on the economy has been poor. With an aggregate GNP of $ 197,000 million in 1984, India ranked eleventh in the world. However, the more appropriate indicator of development is the per capita income which was $ 260, ranking the country 161, almost at the bottom of the list. This poor performance comes into sharper focus when the contribution of the manufacturing sector to the GDP is considered: this amounted to only $ 24,000 million in 1984, accounting for merely 1 per cent of the total value generated in the manufacturing sector in the world. In aggregate industrial production, India ranks between 15th and 20th in the world.

From the mid-60s, constant efforts were made at industrial dispersal under a scheme to develop industries in industrially backward areas. This policy was based on the recognition of the wide disparity between states in terms of industrial progress. The policy also aimed at bringing the employment benefits of manufacturing activity and the other growth benefits obtained in modern industrial centres to the grassroots level and more especially to areas which most needed such benefits. It is in relation to this backwardness that the Government of India has introduced a District Industrial Centre Programme with a view to developing industrialization in backward districts.

The Industrial Policy Resolution, 1980, indicated as one of its aims the 'correction of regional imbalance through a preferential treatment of industrially backward areas' which had been identified and selected using such criteria as per capita income, per capita income from industry and mining, number of workers in registered factories, per capita annual consumption of electricity, and length of surfaced roads in relation to population and the area of the state. On the basis of these criteria, several districts, mainly in Andhra Pradesh, Assam, Bihar, Jammu & Kashmir, Madhya Pradesh, Nagaland, Orissa, Rajasthan, Uttar Pradesh and all union territories except Chandigarh, Delhi and Pondicherry, were identified as industrially backward and preferential treatment was offered for their industrial development.

In 1980 a locational policy was mooted for the first time — a concrete step towards bringing about balanced industrialization. This offered further concessions and subsidies to entrepreneurs who set up industries in identified backward areas. By 1979-80, 10,064 units had been established in backward areas, each of which had less than Rs 1 million investment. Bigger units established up to that time were: 177 with investment between Rs 1 m and Rs 1.5 m in each; 276 (Rs 1.5 to 5 m), 148 (Rs 5 to 10 m) and 143 (above Rs 10 m). It was obvious that the thrust was towards setting up small-scale industries in these areas.

However, interest in investing in backward areas seems to be waning. Though the number of licenses issued since 1975 for undertakings to be set up in such areas have been increasing in a cumulative sense, they have been declining each year from 216 in 1975 to 102 in 1979. In these five years, licenses issued have amounted to only 712 and the majority of them have been in industrially backward states. In the industrially developed states like Tamil Nadu and Maharashtra, the move to backward areas has been slow.

Sources: 1. *Report on Industrial Dispersals*, National Committee on The Development of Backward Areas, Government of India Planning Commission.
2. *Shape of Things to Come, July 1986* and *World Economy & India's Place in it, October 1986*, Centre for Monitoring Indian Economy.

LARGE PROJECTS ON HAND IN BACKWARD DISTRICTS
(Each with a capital outlay of Rs. 1000 million and above)

Table 93

(Rs. in million)

	A.P.	Ass.	Bih.	Guj.	Kar.	M.P.	Mah.	Ori.	Raj.	T.N.	U.P.	W.B.
Mining, quarrying, oil & gas	—	—	—	1,004(1)	—	—	—	—	—	—	—	—
Coal mining	1,320(1)	—	2,173(1)	—	—	21,869(8)	—	—	—	3,348(1)	—	3,775(2)
Other mining	—	—	—	—	1,035(1)	—	—	—	1,358(1)	—	—	—
Ferrous metals	—	—	—	—	1,800(1)	6,900(1)	4,822(2)	3,060(1)	—	—	—	19,625(3)
Organic chemicals	—	—	5,260(2)	—	5,130(4)	1,720(1)	—	—	—	—	5,700(2)	10,000(1)
Fertilizers	—	—	—	6,250(2)	—	5,871(1)	—	—	—	—	20,700(3)	—
Petroleum refineries	—	4,514(2)	—	—	9,000(1)	—	—	—	—	—	—	—
Non-ferrous metals	—	—	—	—	—	—	—	25,367(2)	—	—	—	—
Cement	1,350(1)	1,000(1)	1,000(1)	2,200(2)	1,050(1)	6,106(5)	1,300(1)	—	1,255(1)	—	—	—
Inorganic chemicals	4,216(1)	—	—	1,600(1)	—	—	—	—	—	—	—	—
Man-made fibres	1,200(1)	1,393(1)	—	1,050(1)	1,000(1)	1,500(1)	1,190(1)	—	—	—	—	1,250(1)
Motor vehicles	4,210(1)	—	—	1,240(1)	—	1,010(1)	1,100(1)	—	1,020(1)	—	—	—
Electronic products	1,000(1)	—	—	3,000(2)	—	—	—	—	—	—	2,010(1)	—
Paper & paper products	—	3,052(1)	—	—	—	—	1,553(1)	—	—	—	—	—
Plastic products	—	—	—	—	—	—	—	—	—	—	2,000(1)	—
Industrial gases, explosives, etc.	—	—	—	—	5,000(1)	—	—	—	—	—	—	—
Telecommunication products	—	—	—	—	—	—	—	—	—	—	1,770(1)	—
Glass products	—	—	—	—	—	—	—	—	—	—	1,440(1)	—
Rubber products	—	—	—	—	—	—	—	1,100(1)	—	—	—	—
Two wheelers & bicycles	—	—	—	—	—	1,005(1)	—	—	—	—	—	—
Total	13,296(6)	9,955(5)	8,433(4)	16,344(10)	19,015(9)	49,976(19)	10,970(7)	29,527(4)	3,633(3)	3,348(1)	33,620(9)	34,650(7)

Note: Figures within brackets indicate the number of projects.

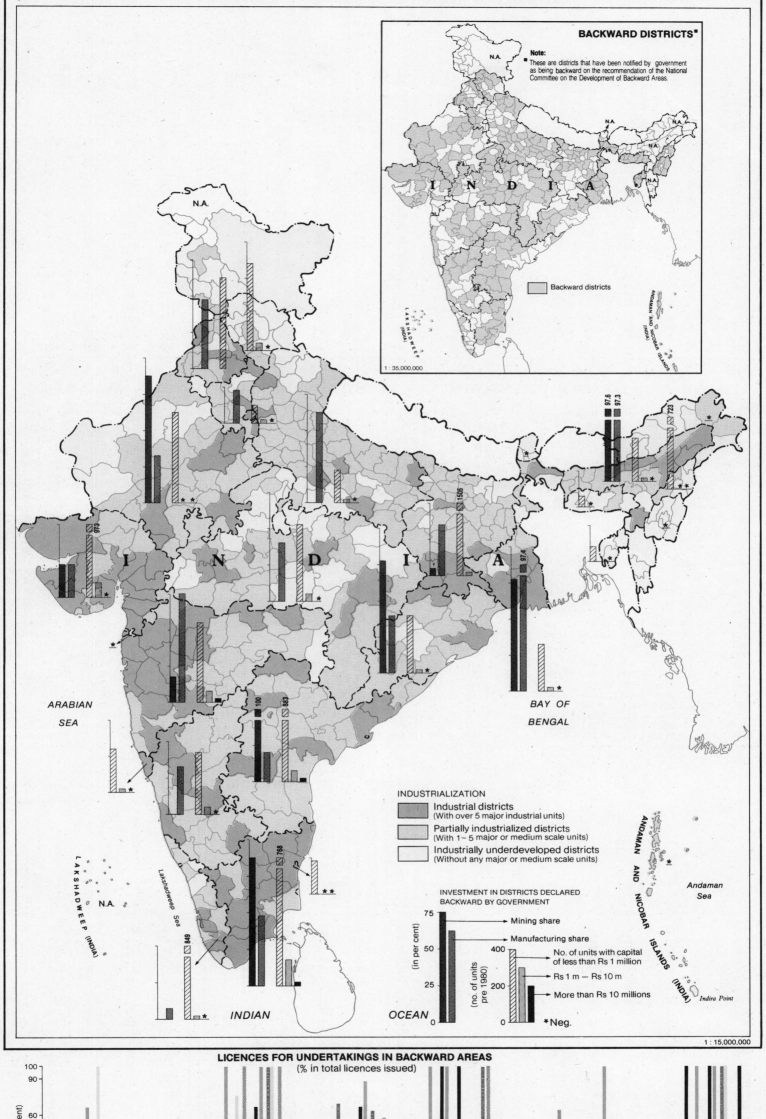

BACKWARD DISTRICTS■

Note:
■ These are districts that have been notified by government as being backward on the recommendation of the National Committee on the Development of Backward Areas.

☐ Backward districts

1 : 35,000,000

N.A.

ARABIAN SEA

BAY OF BENGAL

LAKSHADWEEP (INDIA)

Lakshadweep Sea

N.A.

INDIAN

OCEAN

INDUSTRIALIZATION

Industrial districts
(With over 5 major industrial units)

Partially industrialized districts
(With 1–5 major or medium scale units)

Industrially underdeveloped districts
(Without any major or medium scale units)

INVESTMENT IN DISTRICTS DECLARED
BACKWARD BY GOVERNMENT

75

50

25

0

(in per cent)

→ Mining share

→ Manufacturing share

400

200

0

(no. of units
pre 1980)

→ No. of units with capital
of less than Rs 1 million

→ Rs 1 m – Rs 10 m

→ More than Rs 10 millions

ANDAMAN AND NICOBAR ISLANDS (INDIA)

Andaman Sea

Indira Point

*Neg.

1 : 15,000,000

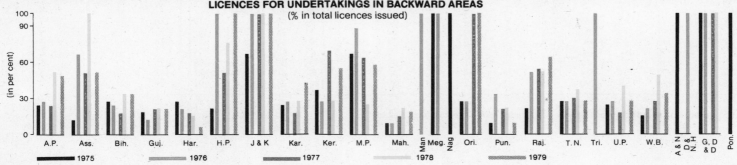

LICENCES FOR UNDERTAKINGS IN BACKWARD AREAS
(% in total licences issued)

(in per cent)

100

30

0

A.P. | Ass. | Bih. | Guj. | Har. | H.P. | J & K | Kar. | Ker. | M.P. | Mah. | Man. | Meg. | Nag. | Ori. | Pun. | Raj. | T.N. | Tri. | U.P. | W.B. | A & N / D & N | D & / N / G, D & D | Pon.

■ 1975   ▨ 1976   ▦ 1977   ▨ 1978   ▨ 1979

REGISTERED WORKING FACTORIES — 30th JUNE 1983

Table VIII

STATES/UTs	Food products	Beverages, tobacco & tobacco products	Cotton textiles	Wool, silk & synthetic fibre textiles	Jute textiles	Textile products including clothing	Wood & wood products, furniture & fixtures	Paper & paper products, printing, publishing & allied industries	Leather (tanned hide), leather & fur products (except repairs)	Rubber, plastic, petroleum & coal products	Chemicals & chemical products	Non-metallic mineral products	Basic metals & alloys	Metal products	Non-electrical machinery	Electrical machinery, apparatus and appliances	Transport equipment	Other industries & services
A.P.	8,520	601	512	68	23	34	1,682	1,651	21	142	357	696	331	392	2,077	148	107	834
Ass.	883	3	16	1	3	2	323	24	N.A.	15	16	30	36	26	67	2	8	73
Bih.	22,719	68	325	29	238	89	1,930	1,280	33	620	517	1,308	557	1,570	1,909	225	45	2,536
Guj.	1,131	282	1,393	555	5	227	759	456	41	564	1,337	1,307	722	1,076	1,260	401	130	804
Har.	450	16	225	126	2	45	625	93	13	116	202	370	371	618	369	131	87	169
H.P.	17	5	N.A.	6	N.A.	3	345	7	1	3	17	12	13	1	13	5	2	77
J. & K.	82	8	2	34	N.A.	17	58	16	2	3	26	66	38	32	26	25	7	76
Kar.	1,677	173	2,239	981	2	240	17	1,480	22	201	381	417	399	321	487	213	75	531
Ker.	1,643	60	797	5	N.A.	422	2,378	881	39	928	433	437	265	618	470	142	52	822
M.P.	1,992	22	456	33	2	33	2,223	121	11	66	214	375	226	183	90	82	45	355
Mah.	1,791	63	2,164	882	5	586	2,393	1,081	72	873	987	921	955	1,514	1,532	836	265	1,345
Man.	97	N.A.	1	N.A.	N.A.	N.A.	31	6	N.A.	1	1	5	1	2	N.A.	N.A.	N.A.	8
Meg.	3	1	1	N.A.	1	N.A.	15	7	N.A.	N.A.	1	2	1	N.A.	N.A.	N.A.	N.A.	18
Ori.	291	16	43	N.A.	5	3	472	39	6	20	86	126	60	100	55	20	10	66
Pun.	1,048	17	246	955	N.A.	342	1,535	155	16	192	159	203	483	777	944	157	795	303
Raj.	1,251	20	1,814	135	1	93	1,274	966	5	48	183	487	265	124	101	72	38	279
T.N.	2,482	92	1,487	199	5	496	253	740	410	324	872	403	567	545	751	302	236	768
Tri.	43	N.A.	1	N.A.	1	1	30	3	1	1	1	2	72	3	7	N.A.	N.A.	36
U.P.	1,572	64	169	67	7	79	38	291	119	158	307	341	529	406	360	232	107	585
W.B.	931	33	259	26	136	151	308	451	154	572	416	229	840	831	609	300	210	695
UTs	432	49	101	40	5	553	145	368	14	388	200	140	324	560	293	461	401	523

ESTIMATED AVERAGE DAILY EMPLOYMENT IN REGISTERED WORKING FACTORIES — 30th JUNE 1983

Table IX

STATES/UTs	Food products	Beverages, tobacco & tobacco products	Cotton textiles	Wool, silk & synthetic fibre textiles	Jute textiles	Textile products including clothing	Wood & wood products, furniture & fixtures	Paper & paper products, printing, publishing & allied industries	Leather (tanned hide), leather & fur products (except repairs)	Rubber, plastic, petroleum & coal products	Chemicals & chemical products	Non-metallic mineral products	Basic metals & alloys	Metal products	Non-electrical machinery	Electrical machinery, apparatus and appliances	Transport equipment	Other industries & services
A.P.	121,850	155,637	38,358	3,413	14,081	2,063	11,414	23,641	899	5,016	27,826	21,565	14,683	6,702	29,113	21,489	27,397	42,006
Ass.	39,043	119	1,699	315	1,554	36	7,565	2,655	N.A.	2,921	3,253	1,601	896	222	1,835	69	4,453	2,226
Bih.	82,941	2,457	6,695	1,018	7,679	587	9,751	14,288	2,339	17,104	15,991	69,505	82,089	18,725	29,342	5,932	35,800	28,536
Guj.	67,821	9,398	260,512	39,079	66	9,268	9,091	16,675	857	17,430	63,697	44,583	24,314	26,175	43,623	17,368	9,065	33,661
Har.	13,958	2,150	23,619	5,565	651	3,171	4,443	10,937	506	9,772	7,404	20,579	18,646	14,848	28,745	20,278	12,527	16,070
H.P	331	337	N.A.	407	N.A.	47	1,228	855	13	148	634	1,028	624	452	821	145	2	4,860
J. & K.	1,777	415	2,105	3,013	N.A.	715	1,523	690	72	175	1,531	2,240	578	393	501	407	137	6,006
Kar.	65,110	15,462	71,983	16,931	23	18,801	2,546	56,876	1,414	8,939	20,209	24,183	37,314	16,092	49,963	67,813	35,247	38,187
Ker.	125,511	854	25,977	2,591	N.A.	7,920	24,816	12,436	337	11,582	16,804	15,615	5,086	5,570	6,818	8,585	3,744	15,387
M.P.	48,884	1,975	88,957	6,404	1,343	2,156	27,529	15,631	1,519	3,452	36,148	38,325	41,222	9,055	45,413	24,271	30,938	27,913
Mah.	95,591	9,074	226,552	43,136	68	21,641	15,863	52,339	2,503	45,280	136,805	41,189	72,170	64,347	114,868	68,130	82,315	104,475
Man.	442	N.A.	2	N.A.	N.A.	N.A.	191	228	N.A.	14	22	57	106	44	N.A.	N.A.	N.A.	844
Meg.	29	35	62	N.A.	130	N.A.	561	384	N.A.	N.A.	21	709	24	N.A.	N.A.	N.A.	N.A.	670
Ori.	9,918	605	13,866	N.A.	2,216	158	5,133	9,172	220	366	6,134	15,305	22,830	2,694	5,493	1,023	2,593	4,364
Pun.	26,647	3,249	29,913	33,299	N.A.	9,065	6,291	5,273	1,581	5,590	8,375	3,887	21,422	16,736	23,098	6,173	30,523	16,468
Raj.	16,656	799	44,119	17,557	20	3,911	5,321	7,204	243	2,117	6,931	19,460	13,762	3,151	6,282	2,499	3,875	28,769
T.N.	123,069	4,413	172,076	3,084	132	22,230	6,372	30,239	18,226	29,578	82,284	25,006	30,528	23,096	60,641	20,581	72,679	66,659
Tri.	1,630	N.A.	4	N.A.	750	4	433	234	147	11	66	6,781	128	64	N.A.	N.A.	N.A.	1,374
U.P.	124,996	8,139	63,791	8,963	5,645	13,971	2,430	17,778	14,590	9,643	21,258	32,888	33,199	13,385	28,117	30,146	31,117	74,343
W.B.	50,659	4,726	55,283	12,855	236,972	9,660	7,072	33,271	13,990	24,589	35,394	22,943	126,262	40,971	54,351	34,404	66,486	75,162
UTs	10,995	2,820	26,231	2,254	137	27,149	4,993	16,060	685	9,796	8,916	3,949	8,886	11,741	11,966	19,532	13,368	38,704

VALUE ADDED BY MANUFACTURE (FACTORY SECTOR) — 1981-82

Table X

(in Rs. million)

STATES/UTs	Food products	Beverages, tobacco & tobacco products	Cotton textiles	Wool, silk & synthetic fibre textiles	Jute textiles	Textile products including clothing	Wood & wood products, furniture & fixtures	Paper & paper products, printing, publishing & allied industries	Leather (tanned hide), leather & fur products (except repairs)	Rubber, plastic, petroleum & coal products	Chemicals & chemical products	Non-metallic mineral products	Basic metals & alloys	Metal products	Non-electrical machinery	Electrical machinery, apparatus and appliances	Transport equipment	Other industries & services
A.P.	722	706	310	18	180	8	21	574	7	90	551	352	318	59	656	872	244	1,270
Ass.	674	3	22	N.A.	3	1	166	11	N.A.	74	165	22	12	18	9	8	25	188
Bih.	255	125	21	8	47	2	13	331	32	773	-43	321	4,547	34	428	147	1,809	376
Guj.	709	108	2,831	1,022	Neg.	73	30	255	2	534	3,120	438	491	223	921	432	154	1,250
Har.	154	33	292	139	6	29	10	240	9	298	272	259	316	227	923	274	343	5,656
H.P.	Neg.	19	N.A.	3	N.A.	Neg.	3	8	N.A.	N.A.	33	51	9	1	44	5	N.A.	737
J. & K.	9	2	N.A.	16	N.A.	17	8	4	N.A.	1	82	4	5	3	6	6	2	74
Kar.	402	173	516	37	N.A.	52	57	278	5	133	597	369	528	156	524	542	541	1,877
Ker.	416	108	213	42	N.A.	153	123	193	Neg.	490	839	189	128	55	98	237	164	948
M.P.	209	159	549	258	26	10	40	225	30	10	431	386	3,844	59	65	729	84	7,953
Mah.	1,715	418	3,469	1,790	4	343	59	1,257	22	1,274	6,790	749	2,059	1,570	3,078	2,351	3,216	33,857
Man.	N.A.	N.A.	N.A.	N.A.	N.A.	N.A.	2	1	N.A.	N.A.	N.A.	N.A.	N.A.	2[1]	N.A.	N.A.	N.A.	30
Meg.	Neg.	1	Neg.	N.A.	N.A.	N.A.	7	4	N.A.	N.A.	N.A.	N.A.	N.A.	N.A.	N.A.	N.A.	N.A.	74
Ori.	68	6	22	—	N.A.	7	4	30	215	2	6	112	260	771	17	83	44	4	2,310
Pun.	593	83	385	374	N.A.	172	8	17	9	74	457	11	370	166	357	112	377	4,296
Raj.	147	14	262	591	36[2]	N.A.	1	32	5	1	282	155	333	53	210	238	199	3,702
T.N.	1,072	94	1,906	176	3	120	33	638	282	639	1,905	610	612	318	1,620	538	1,903	14,301
Tri.	6	Neg.	Neg.	N.A.	N.A.	N.A.	1	2	N.A.	N.A.	N.A.	18	1	1	N.A.	N.A.	N.A.	52
U.P.	1,538	280	619	164	69	141	15	237	112	733	695	246	635	126	212	1,189	43	14,849
W.B	492	94	534	241	1,834	94	42	547	216	708	810	349	2,501	468	1,067	978	1,775	14,820
UTs	184	98	190	46	-5	212	58	289	7	188	257	37	91	96	186	288	150	2,814

Note: [1] Includes non-electrical machinery   [2] Includes textile products and clothing.

The Tourist Vista

This section comprises 4 maps and 2 charts. It identifies the potential of yet another aspect of development, an area now considered the fastest growing industry in the world.

Note:

In dealing with tourist arrivals in India we have grouped the various countries into 15 international regions. Some of these regions are countries by themselves, namely the British Isles, Canada, Japan, South Africa, U.S.A. and U.S.S.R.

The other regions and the countries they include are Central & South America — Argentina, Brazil, Mexico and Trinidad.

Rest of Western Europe — Austria, Belgium, Denmark, Finland, France, Ireland, Italy, Netherlands, Norway, Portugal, Spain, Sweden, Switzerland, Greece and West Germany.

Rest of Eastern Europe — Czechoslovakia, East Germany, Hungary, Poland and Yugoslavia.

Rest of Africa — Ethiopia, Kenya, Mauritius, Nigeria, Somalia, Sudan, Egypt, Tanzania and Zambia.

West Asia — Bahrain, U.A.E,, Iraq, Jordan, Kuwait, Lebanon, Oman, Qatar, Saudi Arabia, Turkey, Syria and Yemen.

South Asia — Afghanistan, Sri Lanka, Iran and Nepal.

South-East Asia — Burma, Indonesia, Malaysia, Philippines, Singapore and Thailand.

East Asia — China, Hong Kong and Korea.

Australasia — Australia, Fiji and New Zealand.

TOURIST ARRIVALS

Travel abroad for some provides entertainment; for others it serves as a holiday; and for yet others it is a means of understanding other peoples' ways of life, culture and traditions. All such travel is called tourism. For India, tourism is the second largest foreign exchange earner. India's rich cultural heritage, historical splendour, religious life and natural beauty all provide a diversity that attracts tourists from all over the world. But tourist arrival figures in India tend to be deceptive. Many among them are Indians settled abroad returning home for a visit. There are no separate statistics covering such visitors, but, watching the scene at any entry point, it is obvious that India receives far fewer real tourists than many other developing countries. There is still much to be done to attract larger numbers of real tourists. This includes improving and expanding the infrastructure, appreciating the needs of tourists, and developing a tourist-oriented hospitality.

India experiences its peak tourist traffic during the winter months when the climate is pleasant throughout the country, but there is also appreciable traffic in summer to Jammu & Kashmir, Himachal Pradesh and other Himalayan regions. In fact, there is some tourist traffic all the year round.

Over the years, tourism has emerged, albeit very slowly, as an important element of the economy, comparable in returns to some of the major merchandise exports. Foreign exchange earnings from tourism, which amounted to about Rs 320 m in 1971-72, have been increasing, though as erratically as the tourist flow, and in 1985 touched Rs 13,000 m. Because imports needed for tourism are limited, the earnings constitute a relatively large value-added component. If it could be tapped efficiently, the potential for tourist earnings in India is very high.

International tourism has grown substantially in the last three decades, but India has not benefited as much as it should have done from this. Though the number of foreign tourist arrivals has been increasing from 16,829 in 1951 (excluding those from Pakistan and Bangladesh, from where there is an annual stream of nearly 360,000, mainly to visit relatives) to 1,259,384 in 1985 (including visitors from Pakistan and Bangladesh), the annual growth rate has been erratic. But, on an average, the rate of growth between 1961 and 1970 was 8.4 per cent per year. Though the growth continued, the overall growth rate declined considerably between 1971 and 1980. This was essentially because of a steep rise in the cost of air travel, which accounts for 90 per cent of the tourist traffic. Additionally, there was a recessionary trend in the major tourist-generating markets of Western Europe and North America, which account for nearly half the arrivals

in India. For the Seventh Plan, the targeted rate of growth is 7 per cent per year. It would be possible to achieve this, given appropriate place-marketing strategies, necessary improvements in the infrastructure, measures to remove the main constraints and major irritants to tourists, and a certain degree of relaxation of entry policies and procedures.

Of the 1,304,976 tourists who visited India in 1983, the largest number was from South Asia (40.9 per cent), Bangladesh accounting for 16.3 per cent (40 per cent of all visitors from South Asia) and Pakistan 15.8 per cent (38.6 per cent of the total from South Asia). Western Europe contributed 26.2 per cent, the UK accounted for 10.5 per cent, France 3.8, the Federal Republic of Germany 3.9 and Italy 2.1. The USA and Canada together accounted for 9.6 per cent of all arrivals, and West Asia for 7.1 per cent. The flow from Saudi Arabia was 1.9 per cent. Arrivals from Africa were 3.8 per cent (22.2 per cent of them from South Africa, mainly people of Indian descent) and from Australasia 2.3 per cent, Australia contributing nearly 79.2 per cent of them. East European visitors to India made up only 2 per cent of all arrivals, two-thirds of them from the USSR.

The average daily flow from Bangladesh was 586, Pakistan 566, UK 375, the USA 263, Sri Lanka 224, West Germany 140 and France 137. The least daily arrivals were from Trinidad, Czechoslovakia, East Germany, Ethiopia, Sudan, Turkey, Syria, Fiji and Argentina. Trinidad and Fiji, both with large numbers of persons of Indian descent, had surprisingly few of their citizens visiting India.

Tourists to India on the average spend 27 days a year in the country (1982), but there are wide variations in the duration of stay of people from different parts of the world. For instance, a North American stays between 23 and 37 days, a Central and South American between 9 and 33 days, a West European 16 to 36 days, an East European 18 to 28 days, an African 17 to 62 days, a West Asian 14 to 35 days, a South Asian 20 to 60 days, a South-east Asian 16 to 53 days, an East Asian 14 to 20 days and an Australasian 27 to 46 days. The average number of days spent is important both in respect of foreign exchange received as well as the acquisition of familiarity with the country and its people, culture and heritage.

Tourists in the age group 25-34 were predominant (27.1 per cent) in 1982, followed by those in the age group 35-44, (23.3 per cent) and the 45-54 group (16.2 per cent). A further analysis reveals that male tourists constituted the large majority (70 per cent) of total arrivals in India. While the proportion of males was 77.2 per cent for tourists from Japan, it was far less for Australia, whose women tourists represented over 40 per cent of the total from that country.

Air is the most popular mode of travel, accounting for nearly 90 per cent of arrivals since 1982. Sea traffic, which had declined in 1983, picked up in 1984 to touch nearly 6 per cent of the total arrivals.

A look at the graphics (bars) of tourist arrivals over the years indicates that the flow to India has been increasing from all continents and countries, but more slowly in the case of some than in others. The USA, the West European countries, West Asia, South-east Asia and the South Asian countries have increased their tourist flows to India much more than others.

One particular feature of planned tours of India is the emphasis on historical sites. India has not fully exploited vacation and recreation tourism. This calls for considerable improvement of the infrastructure in less accessible parts of the country and along its coast.

Sources: 1. *Statistical Abstract India 1984,* Ministry of Planning. Quoting from Ministry of Tourism & Civil Aviation.
2. *Statistical Outline of India 1986,* Tata Services Limited.

TOURIST ARRIVALS FROM ABROAD

Table 94

Nation	Number of tourists 1984 — (in thousand)	Average number of days stayed — 1982
Afghanistan	6.7	41.0
Australia	24.5	27.5
Austria	7.0	26.2
Bangladesh	247.5	N.A.
Canada	25.1	35.8
France	47.1	25.7
Germany (W)	47.9	25.0
Iran	15.3	31.3
Italy	23.6	20.7
Japan	29.6	22.0
Kenya	7.7	34.7
Malaysia	23.0	38.4
Pakistan	110.7	N.A.
Persian Gulf	41.6	N.A.
Saudi Arabia	22.4	23.8
Singapore	19.2	31.7
Spain	7.5	20.5
Sri Lanka	92.4	26.6
Sweden	7.7	26.3
Switzerland	14.9	24.7
Thailand	7.1	38.3
The Netherlands	12.1	25.0
U.K.	124.2	31.5
U.S.A.	95.7	28.9
U.S.S.R.	14.8	28.5
Others	135.4	N.A.
Total	1,210.7	28.4

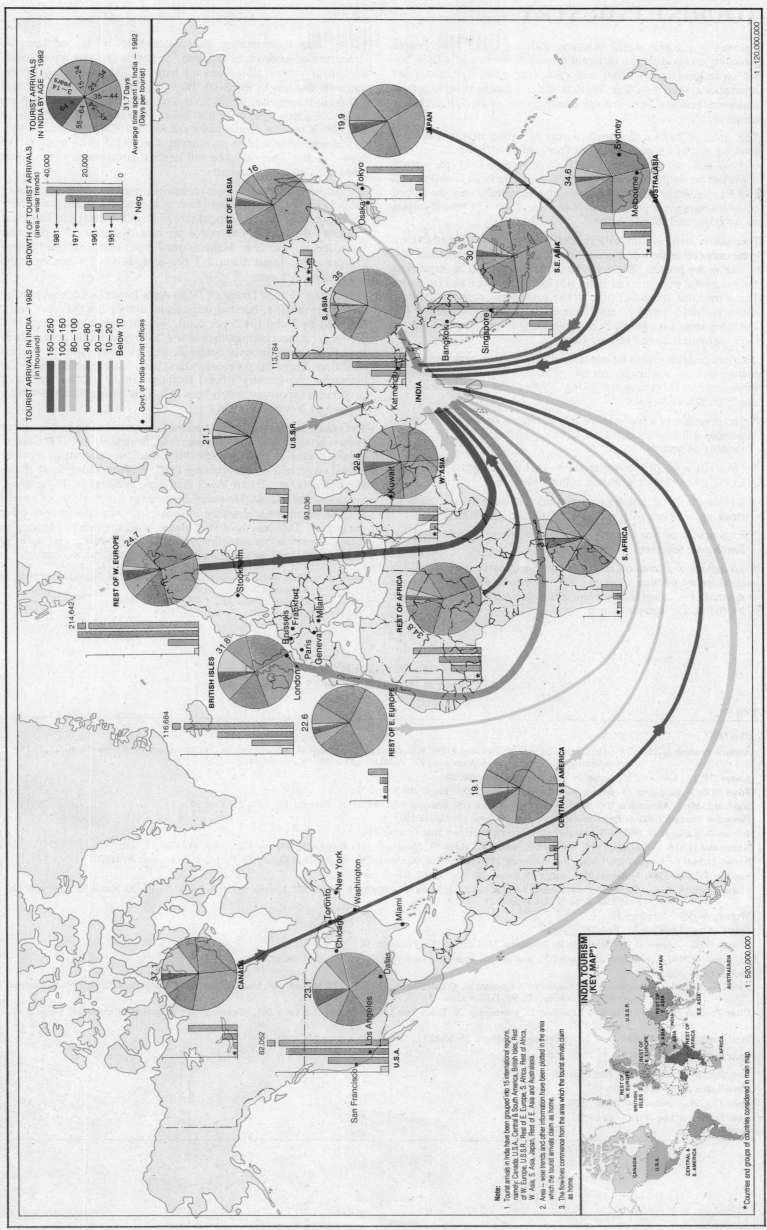

TOURIST ARRIVALS IN INDIA BY AGE — 1982

3-14 years
15-24
25-34
35-44
45-54
55-64
64+

31.7 Days
Average time spent in India — 1982
(Days per tourist)

GROWTH OF TOURIST ARRIVALS
(area – wise trends)

40,000   20,000   0

1981
1971
1961
1951
* Neg.

TOURIST ARRIVALS IN INDIA — 1982
(in thousand)

150—250
100—150
80—100
40—80
20—40
10—20
Below 10

● Govt. of India tourist offices

JAPAN 19.9
Tokyo
Osaka

REST OF E. ASIA 16

AUSTRALASIA 34.6
Sydney
Melbourne

S.E. ASIA 30
Bangkok
Singapore

S. ASIA 35
113.784
Katmandu

INDIA

U.S.S.R. 21.1

W. ASIA 22.6
Kuwait
93,036

REST OF W. EUROPE 24.7
Stockholm
214,642

Brussels
Frankfurt
Milan
Paris
Geneva

REST OF AFRICA 34.8

S. AFRICA

BRITISH ISLES 31.8
London
116,684

REST OF E. EUROPE 22.6

CENTRAL & S. AMERICA 19.1

New York
Washington
Toronto
Chicago
Miami
Dallas

CANADA 37.1

U.S.A. 23.1
82,052
San Francisco
Los Angeles

Note:

1. Tourist arrivals in India have been grouped into 15 international regions, namely Canada, U.S.A., Central & South America, British Isles, Rest of W. Europe, U.S.S.R., Rest of E. Europe, S. Africa, Rest of Africa, W. Asia, S. Asia, Japan, Rest of E. Asia and Australasia.

2. Area – wise trends and other information have been plotted in the area which the tourist arrivals claim as home.

3. The flow-lines commence from the area which the tourist arrivals claim as home.

INDIA TOURISM
(KEY MAP*)

1 : 520,000,000

U.S.S.R.
REST OF E. ASIA
JAPAN
REST OF E. EUROPE
S. ASIA
INDIA
REST OF W. EUROPE
BRITISH ISLES
REST OF AFRICA
S.E. ASIA
W. ASIA
AUSTRALASIA
S. AFRICA
U.S.A.
CANADA
CENTRAL & S. AMERICA

*Countries and groups of countries considered in main map.

1 : 120,000,000

205

TOURIST INDUSTRY

Tourism is a major world industry, with more and more people travelling on vacation as a matter of routine. Unfortunately, India has not capitalized on its assets and earned its rightful share of the international tourist market. There is still much to be done to get the maximum benefits from this important source of foreign exchange earnings.

The share of India's tourism in terms of world arrivals has been growing at a snail's pace, from 0.23 per cent in 1975 to 0.28 per cent in 1980, 0.44 per cent in 1981 and 0.46 per cent in 1984 (the growth is worked out on the basis of all arrivals, including those from Pakistan and Bangladesh). Whereas smaller and essentially less-developed countries have made great strides in tourism, India lags behind despite its diversity and tourist riches.

Tourism in India is primarily cultural tourism, since Indian culture is the only one in the world where several elements of the past continue to live in the present. But amenities are as important as attractions for successful tourism, and India still lacks amenities and facilities. The attractions India can offer relate to history, culture, tradition nature (wildlife, scenery), entertainment, recreation and leisure. The last three areas have hardly been exploited and very few facilities for their enjoyment have been provided.

Tourism priorities must be considered if India is to make tourism a major foreign exchange earner. The Seventh Plan document fortunately includes such rethinking. Some of the steps envisaged in this Plan are as follows:

* Development of a few selected tourist circuits or centres which are popular with tourists instead of spreading resources thin over a large number of resorts.

* Diversification of tourism from the traditional sight-seeing tours to the more rapidly growing holiday tourism.

* Development of non-traditional pursuits for tourists, such as trekking, winter sports wildlife safaris and beach holidays.

* Exploitation of new tourist-generating countries, such as the West, South and South-east Asian countries.

* Launching of a 'national image-building and marketing plan' in key markets by pooling resources of the various public and private agencies instead of independent and disjointed efforts undertaken by the public agencies alone.

Taking into consideration the potential to be tapped and the infrastructure needed to be installed, the central and state governments have made various allocations for tourism. The total outlay for the states on tourism in the Sixth Plan was Rs 688.6 m. The central provision for tourism in the Sixth Plan amounted to Rs 1874.6 m. In the Seventh Plan this has been enhanced to Rs 3261.6 m (Rs 1386.8 m in the central sector and Rs 1874.8 m in the state sector). In addition, quite a large investment is expected to be made by the private sector. This would be still larger if government were to treat tourism as an export industry.

The statewise plan expenditure on tourism, made by the Centre during 1980-85 and expected to bear fruit in the coming years was: Delhi 64.8 per cent of the total expenditure of Rs 688.6 m on tourism, Uttar Pradesh 4.9 per cent, Orissa 3.4 per cent, Karnataka 3.1 per cent, West Bengal 3 per cent, Maharashtra 2.6 per cent, Madhya Pradesh 2.7 per cent, Tamil Nadu 2.5 per cent, Bihar 2.5 per cent and Rajasthan 2 per cent.

Though the 'Golden Triangle', Delhi-Agra-Jaipur, is the most popular tourist trip in India, Bombay is the major entry point (37.6 per cent), followed by Delhi (34.5 per cent), with Madras a poor third (8.2 per cent). In these metropolises as well as in most tourist centres there are several western-style hotels as well as other types of accommodation. There has, in recent years, been considerable progress in increasing the country's hotel facilities. In 1971, the number of government-approved hotels was 152. This had gone up to 495 by the end of November 1986.

The number of government-approved rooms available in November 1986 was 31,954, a three-fold increase, from 9091 in 1971. Current planning aims at providing the infrastructure required to meet an annual target of 2.5 m tourists by 1990. A comparative study shows that Indian hotel rates are lower than those in Bangkok, Hong Kong, Singapore and Manila. However, the average length of stay of tourists in those countries is less than in India, for India has more to offer the tourist. This length of stay makes it all the more necessary to increase hotel accommodation and keep rates reasonable, so as to induce the tourist to make an even longer stay in the country.

Sources: 1. *Hotel and Restaurant Guide India '87*, The Federation of Hotel & Restaurant Associations of India.
2. *Economic Times 12.2.1986*, Times of India.
3. *Business India, February 24, 1986*.

ROOMS IN GOVERNMENT
APPROVED HOTELS (as on 15.7.1986)

Table 95

STATES

Andhra Pradesh (2,554) — Guntur 22; Hyderabad-Secunderabad 1,497; Kakinada 124; Khammam 47; Kottagudem 20; Machilipatnam 38; Nellore 73; Nizamabad 64; Rajahmundry 30; Tirupati 90; Vijayawada 197; Vishakhapatnam 298; Warrangal 54.

Assam (128) — Guwahati 56; Jorhat 28; Kaziranga 24; Tinsukia 20.

Bihar (433) — Bhagalpur 23; Bokaro 44; Dhanbad 34; Patna 228; Rajgir 26; Ranchi 78.

Gujarat (526) — Ahmadabad 206; Ankleshwar 30; Vadodara 148; Bhavnagar 20; Rajkot 41; Sasangir 24; Surat 27; Vapi 30.

Himachal Pradesh (202) — Dalhousie 16; Kullu 24; Parwanu 12; Shimla 160.

Jammu & Kashmir (1,209) — Gulmarg 80; Jammu 189; Pahalgam 104; Srinagar 836.

Karnataka (1,915) — Bangalore 923; Harihar 20; Hassan 47; Hubli 58; Mangalore 311; Manipal 50; Mysore 435; Udupi 44; Vilal 27.

Kerala (1,164) — Alleppey 28; Cochin 346; Kottayam 33; Kovalam 162; Kozhikode 57; Palghat 28; Quilon 100; Trichur N.A.; Trivandrum 410.

Madhya Pradesh (389) — Bhopal 45; Gwalior 46; Indore 26; Jabalpur 68; Khajuraho 204.

Maharashtra (6,153) — Aurangabad 294; Bombay 4,471; Chowdi 27; Ellora 16; Karad 19; Kolhapur 202; Lonavla 53; Matheran 64; Nagpur 20; Nashik 32; Panchgani 64; Pune 850; Roha 14; Ulhasnagar 27.

Meghalaya (59) — Shillong 59.

Orissa (43) — Gopalpur 21; Puri 22.

Punjab (376) — Amritsar 207; Bhatinda 36; Jallandhar 74; Ludhiana 25; Pathankot 34.

Rajasthan (1,305) — Bharatpur 18; Jaipur 526; Jodhpur 54; Kota 19; Mount Abu 65; Sariska 15; Udaipur 232.

Sikkim (55) — Gangtok 55.

Tamil Nadu (3,318) — Coimbatore 229; Coonoor 57; Kovalam 80; Kodaikanal 26; Madras 2,270; Madurai 174; Mahabalipuram 168; Udagamandalam 187; Pudukottai 12; Salem 40; Tiruchchirappalli 59; Tuticorin 16.

Uttar Pradesh (1,696) — Agra 647; Allahabad 14; Balarampur 20; Dehra Dun 105; Kanpur 84; Lucknow 265; Mussoorie 130; Nainital 69; Rampur 13; Ranikhet 42; Varanasi 307.

West Bengal (1,467) — Calcutta 1,246; Darjiling 127; Digha 35; Kulti 11; Shiliguri 48.

UNION TERRITORIES

Andaman & Nicobar (82) — Port Blair 82.

Chandigarh (97) — Chandigarh 97.

Delhi (7,120) — Delhi-New Delhi 7,120.

Goa, Daman & Diu (951) — Goa 951.

INDIA (31,242)

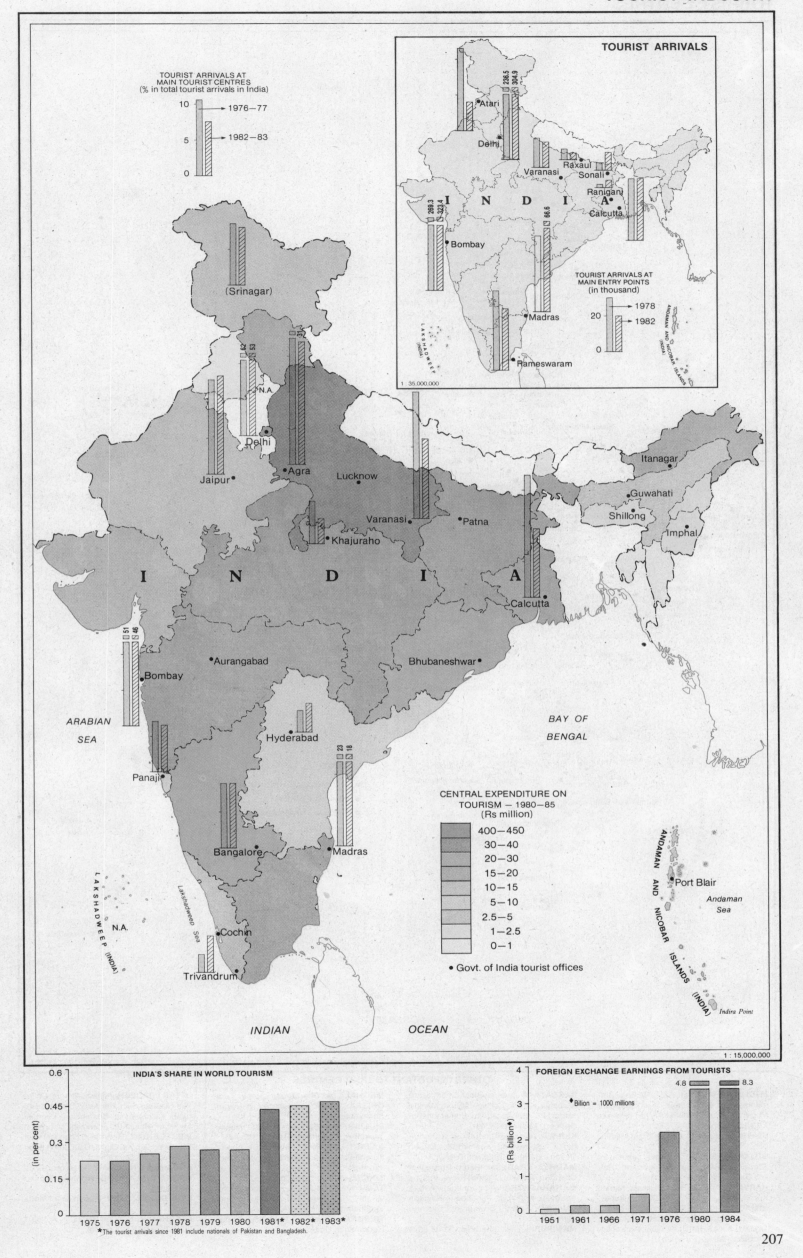

TOURIST ARRIVALS

TOURIST ARRIVALS AT
MAIN TOURIST CENTRES
(% in total tourist arrivals in India)

10
5
0
→ 1976—77
→ 1982—83

TOURIST ARRIVALS AT
MAIN ENTRY POINTS
(in thousand)

20
0
→ 1978
→ 1982

I N D I A

Atari
Delhi
Varanasi
Raxaul
Sonali
Ranigani
Calcutta
Bombay 269.3 — 323.4
Madras 66.6
Rameswaram
238.5 — 304.9

1 : 35,000,000

(Srinagar)

N.A.
Delhi
Jaipur
Agra
Lucknow
Varanasi
Patna
Khajuraho
Itanagar
Guwahati
Shillong
Imphal

I   N   D   I   A

62 — 53
31

Calcutta

Bombay 51 — 46
Aurangabad
Bhubaneshwar

ARABIAN
SEA

Hyderabad

Panaji

Bangalore
Madras 23 — 18

BAY OF
BENGAL

CENTRAL EXPENDITURE ON
TOURISM — 1980—85
(Rs million)

400—450
30—40
20—30
15—20
10—15
5—10
2.5—5
1—2.5
0—1

● Govt. of India tourist offices

ANDAMAN AND NICOBAR ISLANDS (INDIA)

Port Blair
Andaman
Sea

LAKSHADWEEP (INDIA)
N.A.
Lakshadweep Sea

Cochin

Trivandrum

INDIAN   OCEAN

Indira Point

1 : 15,000,000

INDIA'S SHARE IN WORLD TOURISM

0.6
0.45
0.3
0.15
0

(in per cent)

1975 1976 1977 1978 1979 1980 1981* 1982* 1983*

*The tourist arrivals since 1981 include nationals of Pakistan and Bangladesh.

FOREIGN EXCHANGE EARNINGS FROM TOURISTS

4
3
2
1
0

(Rs billion●)

● Billion = 1000 millions

4.8   8.3

1951 1961 1966 1971 1976 1980 1984

207

TOURIST CENTRES

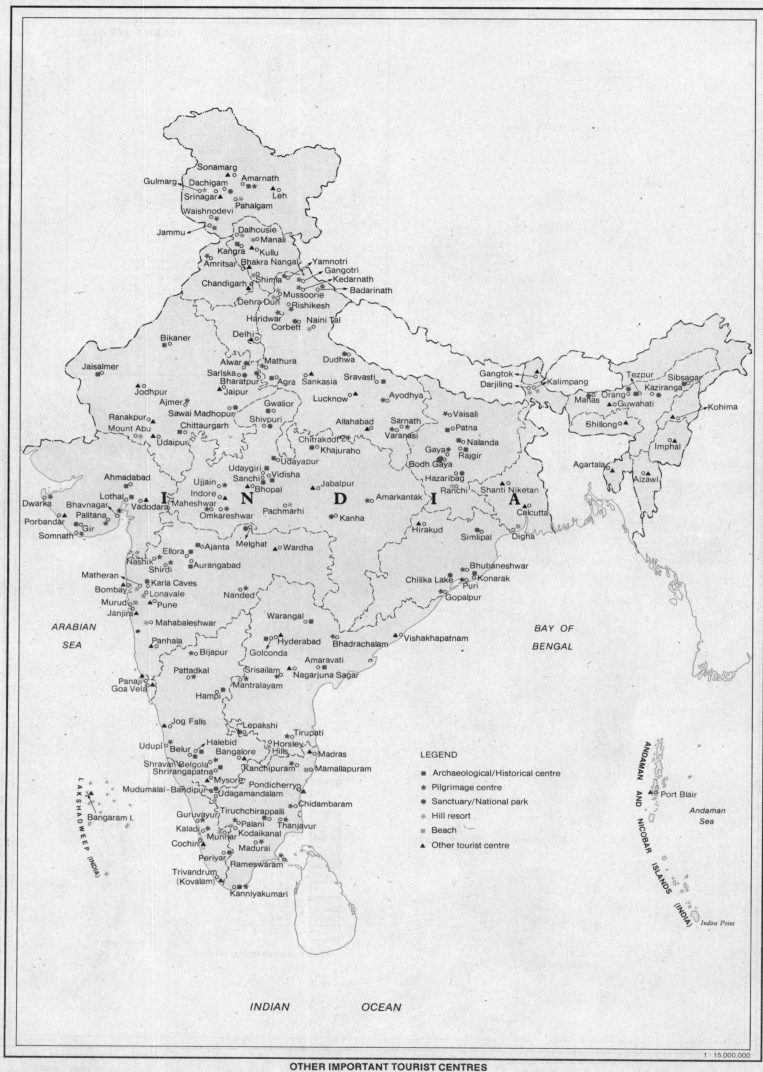

LEGEND

- ■ Archaeological/Historical centre
- ★ Pilgrimage centre
- ● Sanctuary/National park
- ✛ Hill resort
- ▪ Beach
- ▲ Other tourist centre

1 : 15,000,000

OTHER IMPORTANT TOURIST CENTRES

ANDHRA PRADESH: Araku Valley, Bhimunipatnam, Ethipothala, Machilipatnam, Medak, Palampet, Rajahmundry, Sri Kalahasti.
ASSAM: Barpeta, Hajo, Silchar.
BIHAR: Bihar Shariff, Jamshedpur, Madhubani, Palamau.
GUJARAT: Bhuj, Chorwad, Jamnagar, Girnar, Modhera, Patan, Pavagadh, Rajkot, Saputara, Surat, Veraval.
HARYANA: Ambala, Bhiwani, Hisar, Karnal, Kurukshetra, Pinjaur, Rohtak, Surajkund.
HIMACHAL PRADESH: Chail, Dharmsala, Kasauli, Kotgarh, Mandi, Palampur, Renuka.
JAMMU & KASHMIR: Achabal, Anantnag, Martand, Patnitop, Vernag.

KARNATAKA: Badami, Bidar, Dharmastala, Gulbarga, Karwar, Kollur, Madikeri, Manipal, Nandi Hills, Ranganathittu, Somanathapur, Sringeri, Talakaveri, Ullal.
KERALA: Alleppey, Cheruthuruthy, Kodungallur, Kottayam, Kozhikode, Quilon, Sultan's Battery, Thekkadi, Trichur.
MADHYA PRADESH: Bagh, Bheraghat, Chanderi, Mandu, Orchha, Rewa, Vidisha.
MAHARASHTRA: Ahmadnagar, Alibag, Amravati, Kolhapur, Panchgani, Raigadh, Ratnagiri, Satara, Sewagram, Shivneri, Vajreshwari.
MANIPUR: Loktak, Moirang, Ukhrul.

MEGHALAYA: Cherrapunji, Jowai, Tura.
ORISSA: Cuttack, Jeypore, Ratnagiri, Sambalpur.
PUNJAB: Anandpur, Dera Baba Nanak, Jalandhar, Kapurthala, Patiala, Sirhind.
RAJASTHAN: Amer, Churu, Deshnoke, Kumbhalgarh, Nathdwara, Osiyan, Pushkar, Ranthambor.
SIKKIM: Pemayangtse, Yuksom.
TAMIL NADU: Auroville, Coonoor, Pichavaram, Tiruchchendur, Tiruvannamalai, Vedantangal, Velanganni, Yercaud.
TRIPURA: Dumbur, Radhakishorepur, Rudranagar, Udaipur.

UTTAR PRADESH: Aligarh, Almora, Chitrakut, Fatehpur Sikri, Hemkund, Jhansi, Joshimath, Karnaprayag, Kushinagar, Lumbini, Mirzapur, Ranikhet, Uttarkashi, Vrindavan.
WEST BENGAL: Bishnupur, Chandannagar, Diamond Harbour, Digha, Gaur, Koch Bihar, Murshidabad, Shrirampur.
ANDAMAN & NICOBAR: Car Nicobar, Ross
ARUNACHAL PRADESH: Along, Parasuram Kund.
GOA, DAMAN & DIU: Daman, Diu, Ponda, Priol.
LAKSHADWEEP: Pitti Island.
PONDICHERRY: Karaikal.

The National Economy

This section comprises 23 maps and 24 charts. It evaluates the practical results of the present level of development as they affect the people of India and points to the potential for the future.

Note:

All monetary figures used in this section relate to the current prices, *i.e.* prices for the respective years that are given. Any exceptions are indicated separately. Amongst the abbreviations used in this section are:

GNP = Gross National Product
GDP = Gross Domestic Product
NDP = Net Domestic Product

INDUSTRIAL DEVELOPMENT

The first decade of industrial development, which was undoubtedly 1955-65, marked a departure from the previous decades of painfully slow growth. But this progress did not make a lasting impact and a slide downhill set in at the end of the Third Plan. The annual compound rate of growth of industrial output between 1947 and 1951, was 4.8 per cent, during 1951-56 it was 7.4 per cent, during 1956-61 6.8 per cent, and during 1961-65 8.9 per cent. But during 1965-70 it fell by 3.3 per cent and during 1970-74 by 2.8 per cent.

One of the most disappointing aspects of the performance of the Indian economy since the mid-1960s relates to the deceleration in the growth of output in organized industry, accompanied by sluggish investment, vast underutilization of capacity and very insignificant increase in employment in the organized (industrial) sector. The most important point to be noted is that the growth rate of manufacturing experienced a serious setback in the 1970s. One reason for believing that the slide continued beyond the mid-1970s is the relative shares and growth rates of industry in the annual domestic products. While the relative share of manufacturing increased from 11.2 per cent in 1956-57 to 14.9 per cent in 1981-82, the growth rates of the domestic product declined from 6 per cent, during 1956-66, to 4.3 per cent in the years between 1966 and 1981.

The index of industrial production in manufacturing recorded a growth rate of 4.6 per cent per year between 1959-60 and 1979-80. On the other hand, the value added and value of output at constant prices increased at a rate of 5.2 per cent and 6.4 per cent respectively. At the macro-level, however, the retrogression of the Indian economy observed between 1965 and 1975 does not seem to have persisted in later years. The consensus is that the Indian economy has been growing after 1975-76, though perhaps slower than in the years before 1965 and marginally faster than in the years since.

An analysis of the share of the states in major industrial production in 1981-82 reveals Maharashtra to be the leader in industrial production, leading in the production of at least 13 major products. Tamil Nadu and West Bengal follow some way behind. Other states with good industrial production records are Uttar Pradesh and Gujarat. Maharashtra, Tamil Nadu and Gujarat are the top three states in levels of urbanization and West Bengal is sixth. Though Uttar Pradesh ranks only 14th in urbanization, all its industries are concentrated in its cities. This would appear to indicate that in India industrialization and urbanization are closely linked.

Indices of per capita value added in the factory sector in the years between 1960-61 and 1982-83 indicate that regional inequalities have tended to decline significantly. The indices for the developed states, such as Maharashtra, Gujarat and West Bengal, that have per capita values added higher than the all-India average (100 units for 1960-61, 1970-71 and 1982-83) show a declining trend over the period. The only exceptions are Tamil Nadu and Haryana which have improved their positions in relation to the all-India average. In contrast, the indices for most of the underdeveloped states, such as Andhra Pradesh, Madhya Pradesh and Orissa, have shown significant improvements. It is obvious that much effort is going into industrializing the backward states. It is also obvious that the per capita value added in the factory sector since 1960-61 has been steadily increasing, the increase between 1960-61 and 1982-83 for India amounting to 943 per cent, from Rs 23 in 1960-61 to Rs 58 in 1970-71, Rs 178 in 1980-81 and Rs 240 in 1982-83.

The levels of industrial development based on per capita value added by manufacture in 1982-83 might indicate that Maharashtra (Rs 561), Gujarat (Rs 437), Haryana (Rs 394), Tamil Nadu (Rs 344), Himachal Pradesh (Rs 309), West Bengal (Rs 293) and Punjab (Rs 266) pay as much attention to industrial development as to population control. These seven states have a per capita level of value added by manufacture higher than the national average of Rs 240. On the other hand, the states of Bihar (Rs 157), Orissa (Rs 134), Uttar Pradesh (Rs 132), Rajasthan (Rs 112), Assam (Rs 75), Jammu & Kashmir (Rs 58), Tripura and Manipur (Rs 19 each) are hardly industrialized or, as in the case of Uttar Pradesh and Bihar, have the benefits of their moderate industrial development eroded by a burgeoning population. It must however be noted that even in the major industrial states, industrial development is entirely due to the concentrated industrialization of a few districts in each of them. In Himachal Pradesh alone has industrialization not been confined to particular districts.

In the years since 1980, industrial development in the country has been marked by some significant achievements. In the late seventies, the Indian economy overcame the energy crisis, righted the adverse balance of payments and reduced the effects of the erosion of domestic savings and investment. The rate of gross capital formation exceeded by 20 per cent that of 1966-80. Output in the registered sector grew by 7.6 per cent per year as compared to 4.6 per cent per year in the mid-1960s. This compared favourably with the targets of growth in the Fifth and Sixth Plans. This growth performance was not very different from that of other industrializing countries such as Brazil (7.8 per cent), Mexico (6.9 per cent) and even Japan (6.6 per cent), although it was lower than that of some other Asian countries: South Korea (14.5 per cent), Malaysia (10.6 per cent) and Singapore (9.3 per cent).

Overall, the GNP aggregate registered an overwhelming increase of 923 per cent between 1965 and 1986. Correspondingly, the GNP per capita grew by 558 per cent during the same period. The Gross Domestic Product during this period grew 468 per cent and is expected to touch Rs 2,468,810 m by the end of the Seventh Plan. GDP per capita increased 249 per cent and is expected to reach Rs 3027 in 1989-90.

All these facts, then, point to one conclusion: that somewhere in the late 1970s there was an upturn in the industrial economy which has, since then, been sustained, both at the aggregate and industrial group levels. However, the recovery in recent years has been only modest, the growth rate being still lower than what was achieved during the Second and Third Plans and much lower than the Plan targets.

Sources: 1. *Basic Statistics Relating to the Indian Economy, Vol. I; All India, August 1986, Vol. 2; States, September 1986*, and *Shape of Things to Come*, Centre for Monitoring Indian Economy.
2. World Bank's *World Development Report 1986*, Oxford University Press.
3. *Economic and Political Weekly Annual Number, February 1976*, Raj.

LARGE PROJECTS ON HAND[1]

Table 96   (Each with an investment of Rs. 1000 million and above)   (Rs. in million)

	A.P.	Ass.	Bih.	Guj.	Har.	Kar.	Ker.	M.P.	Mah.	Ori.	Pun.	Raj.	T.N.	U.P.	W.B.
Mining & quarrying															
Crude oil & natural gas	—	—	—	7,009(1)	—	—	—	—	—	1,220(1)	—	—	—	—	—
Coal mining	—	—	6,613(4)	1,004(1)	—	—	—	5,406(2)	—	—	—	—	3,348(1)	—	—
Manufacturing															
Ferrous metals	59,670(1)	—	33,150(3)	—	4,000(1)	5,256(2)	—	27,630(2)	1,000(1)	10,299(2)	—	—	—	—	—
Organic petro-chemicals	2,118(2)	—	3,010(1)	18,472(5)	—	—	2,602(1)	—	26,950(4)	—	—	—	7,420(2)	10,000(2)	—
Fertilizers	7,073(2)	2,485(1)	—	2,800(2)	—	—	5,634(2)	5,871(1)	—	4,160(1)	—	7,640(1)	3,800(1)	14,160(2)	—
Petroleum refineries	—	2,514(1)	—	12,551(3)	15,000(1)	—	—	—	—	—	—	—	—	—	—
Non-ferrous metals	—	—	—	—	—	—	—	—	—	—	—	1,958(1)	—	1,400(1)	—
Cement	1,241(1)	1,000(1)	—	—	—	—	—	1,000(1)	—	—	—	1,000(1)	—	—	—
Inorganic chemicals	1,140(1)	—	—	5,230(2)	—	—	—	—	—	—	—	—	—	—	—
Man-made fibres	—	1,394(1)	1,000(1)	1,050(1)	—	—	—	1,500(1)	1,200(1)	—	—	—	—	1,250(1)	1,250(1)
Motor vehicles	—	—	—	—	—	—	—	1,010(1)	1,700(1)	—	—	—	—	—	—
Electronic products	1,000(1)	—	—	—	—	—	—	—	—	—	—	—	—	1,050(1)	—
Paper & paper products	—	—	—	—	—	—	—	—	—	—	—	—	—	2,800(1)	—
Plastic products	—	—	—	—	—	—	—	—	—	—	—	—	1,681(1)	1,900(1)	—
Rail & road equipment	—	—	—	—	—	—	—	—	—	—	3,135(2)	—	—	—	—
Telecommunication products	—	—	—	—	—	1,360(1)	—	—	—	—	—	—	—	—	—
Glass products	—	—	—	—	—	—	—	—	—	—	—	—	—	1,440(1)	—
Rubber products	—	—	—	—	—	—	—	—	—	1,100(1)	—	—	—	—	—
INDIA	72,241(8)	7,393(4)	43,773(9)	48,116(15)	19,000(2)	6,616(3)	8,237(3)	42,416(8)	30,850(7)	16,779(5)	3,135(2)	10,598(3)	16,249(5)	34,000(10)	1,250(1)

Note: Figures within brackets indicate the number of projects (excluding projects in backward areas).
[1] Excludes multi-state and unallocated projects.

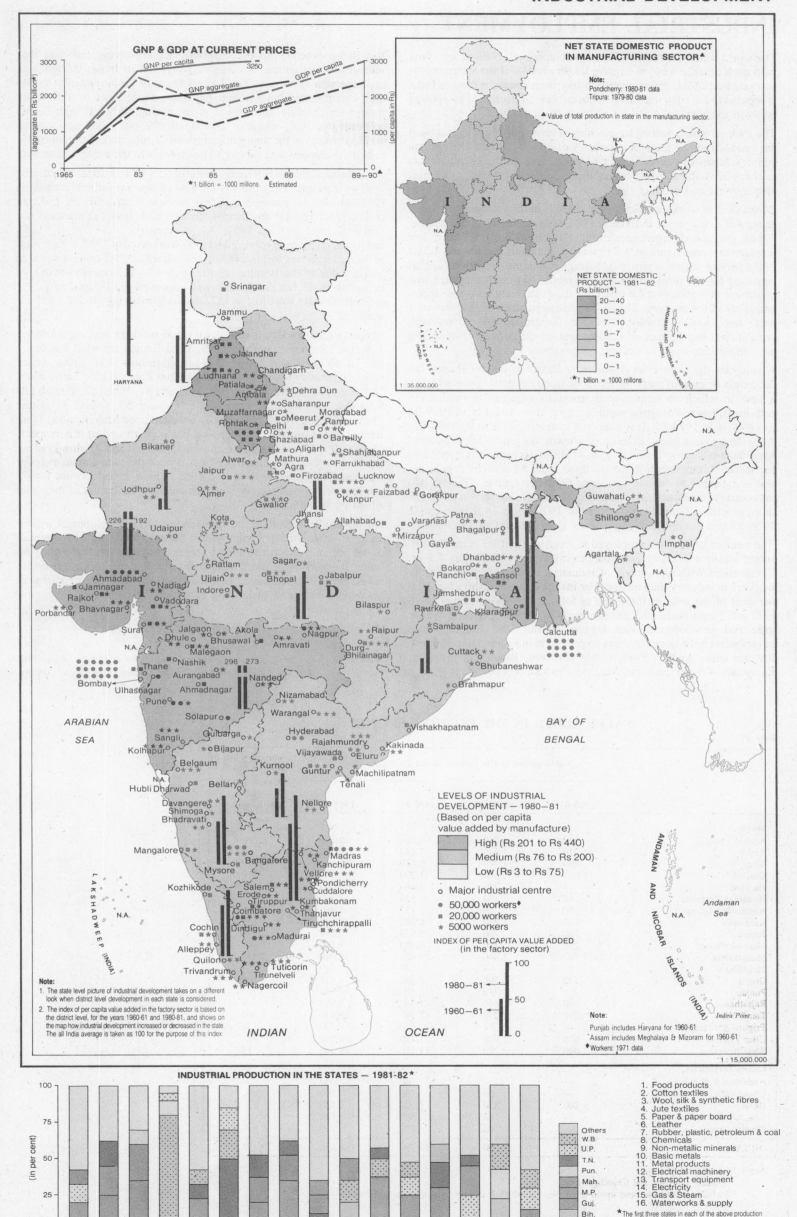

GNP & GDP AT CURRENT PRICES

GNP per capita 3250
GNP aggregate
GDP per capita
GDP aggregate

*1 billion = 1000 millions   Estimated

NET STATE DOMESTIC PRODUCT IN MANUFACTURING SECTOR▲

Note:
Pondicherry: 1980-81 data
Tripura: 1979-80 data

▲ Value of total production in state in the manufacturing sector.

I N D I A

NET STATE DOMESTIC
PRODUCT — 1981—82
(Rs billion*)

20—40
10—20
7—10
5—7
3—5
1—3
0—1

*1 billion = 1000 millons

1 : 35,000,000

LAKSHADWEEP (INDIA)
ANDAMAN AND NICOBAR ISLANDS (INDIA)

Map labels

Srinagar
Jammu
Amritsar
Jalandhar
Chandigarh
Ludhiana
Patiala
HARYANA
Ambala
Dehra Dun
Muzaffarnagar
Saharanpur
Rohtak
Meerut
Moradabad
Rampur
Delhi
Bikaner
Ghaziabad
Bareilly
Aligarh
Shahjahanpur
Alwar
Mathura
Jaipur
Agra
Firozabad
Lucknow
Jodhpur
Ajmer
Faizabad
Gorakpur
Kanpur
Gwalior
Jhansi
Allahabad
Varanasi
Patna
Bhagalpur
Guwahati
Shillong
Kota
Udaipur 226 192
Ratlam
Sagar
Mirzapur
Gaya
Dhanbad
Imphal
Agartala
Ujjain
Bhopal
Jabalpur
Bokaro
Ranchi
Asansol
Ahmadabad
Nadiad
Indore
Bilaspur
Jamshedpur
Jamnagar
Vadodara
Raurkela
Kharagpur
Rajkot
Bhavnagar
Surat
Raipur
Sambalpur
Calcutta
Porbandar
Jalgaon
Akola
Nagpur
Dhule
Bhusawal
Amravati
Durg-Bhilainagar
Cuttack
Malegaon
Nashik 296 273
Bhubaneshwar
Thane
Aurangabad
Nanded
Bombay
Ahmadnagar
Nizamabad
Brahmapur
Ulhasnagar
Pune
Warangal
Solapur
Sangli
Gulbarga
Hyderabad
Vishakhapatnam
Kolhapur
Bijapur
Rajahmundry
Belgaum
Vijayawada
Eluru
Kakinada
Kurnool
Guntur
Machilipatnam
Hubli Dharwad
Bellary
Tenali
Davangere
Nellore
Shimoga
Bhadravati
Mangalore
Madras
Bangalore
Kanchipuram
Mysore
Vellore
Pondicherry
Kozhikode
Salem
Cuddalore
Erode
Tiruppur
Kumbakonam
Coimbatore
Thanjavur
Cochin
Dindigul
Tiruchchirappalli
Alleppey
Madurai
Quilon
Trivandrum
Tuticorin
Tirunelveli
Nagercoil

ARABIAN SEA
BAY OF BENGAL
LAKSHADWEEP (INDIA)
ANDAMAN AND NICOBAR ISLANDS (INDIA)
Andaman Sea
INDIAN OCEAN
Indira Point

LEVELS OF INDUSTRIAL DEVELOPMENT — 1980-81
(Based on per capita value added by manufacture)

High (Rs 201 to Rs 440)
Medium (Rs 76 to Rs 200)
Low (Rs 3 to Rs 75)

o Major industrial centre
● 50,000 workers♦
■ 20,000 workers
★ 5000 workers

INDEX OF PER CAPITA VALUE ADDED
(in the factory sector)

100

1980—81 ←
1960—61 ←

50

0

Note:
Punjab includes Haryana for 1960-61
Assam includes Meghalaya & Mizoram for 1960-61
♦ Workers: 1971 data

Note:
1. The state level picture of industrial development takes on a different look when district level development in each state is considered.
2. The index of per capita value added in the factory sector is based on the district level, for the years 1960-61 and 1980-81, and shows on the map how industrial development increased or decreased in the state. The all India average is taken as 100 for the purpose of this index.

1 : 15,000,000

INDUSTRIAL PRODUCTION IN THE STATES — 1981-82*

(in per cent)

Others
W.B.
U.P.
T.N.
Pun.
Mah.
M.P.
Guj.
Bih.
A.P.

1. Food products
2. Cotton textiles
3. Wool, silk & synthetic fibres
4. Jute textiles
5. Paper & paper board
6. Leather
7. Rubber, plastic, petroleum & coal
8. Chemicals
9. Non-metallic minerals
10. Basic metals
11. Metal products
12. Electrical machinery
13. Transport equipment
14. Electricity
15. Gas & Steam
16. Waterworks & supply

*The first three states in each of the above production categories are individually indicated on the respective bars.

INDUSTRIAL EMPLOYMENT

The world is in the throes of an economic upheaval. Nations are vying with each other in a bid to capture larger chunks of the international market. For India, the competition has been tough, although India is reputed to have the third largest human capital (technical manpower pool) in the world.

The rock on which nations have achieved economic miracles has been the productivity of their blue-collar workers. This factor alone has enabled Japan to double its output every five years, keeping costs unchanged. India has just begun experiencing increasing industrial employment and productivity.

Of the 24.2 m workers officially on the rolls in the various established sectors of employment in India in 1984, manufacturing accounted for 6.2 m, the second largest sector after services (9.3 m). Of the 6.2 m workers in manufacturing, 1.7 m were in the public sector and 4.5 m in the private sector. Transport, storage and communication (2.9 m) provided for the third largest employment. Agriculture figures low on the list of workers on the official rolls, even though it is the largest employer in the country. This is because agriculture is, for the most part, not in the established sector (controlled by government regulations) and figures of employment on the official rolls reflect only those employed in estates, cooperatives etc.

A study by the Centre for Monitoring Indian Economy shows that the levels of employment, in terms of standard person years (SPY: a year in which an employee is *normally* engaged in production activities), have undergone rapid changes: from 151.1 m in 1979-80 to 186.7 m in 1984-85. It is expected to reach 227.1 m by 1989-90. According to this study, in terms of growth, the employment patterns have been changing differently in different sectors. In agriculture, for instance, employment has declined from 53.1 to 51.5 per cent and is expected to decline further to 50.2 per cent in 1989-90. On the contrary, manufacturing, including electricity, has shown an upturn in employment from 14.6 per cent in 1979-80 to 14.9 per cent in 1984-85 and is expected to go up to 15.4 per cent in 1989-90.

The number of workers employed in industry (that is, in the factory sector and in electricity and other utilities) in 1981-82 was 7.9 m. Those states with high levels of industrial development have more workers in industry than others. For instance, Maharashtra has 16.9 per cent of the total in India and is followed by West Bengal 11.8 per cent, Tamil Nadu 10.7 per cent and Gujarat 8.7 per cent. Where industrial development has lagged behind, the bigger states nevertheless have large numbers employed in industry: Uttar Pradesh 9.9 per cent, Andhra Pradesh 9.4 per cent, Bihar 4.7 per cent, Karnataka 4.6 per cent and Madhya Pradesh 4.3 per cent.

It will be noted from the diagrams accompanying the map that industrial employment is comparatively low in India, though it is growing at a slow rate. The largest share of employment is in the agricultural sector, but this share has been coming down over the years though in absolute numbers it is increasing. What is significant, however, is that, whereas almost all (99 per cent) agricultural employment is in the unorganized sector, a substantial amount of the industrial employment (about 30 per cent) is in the organized sector.

As industrialization increases, disturbances of industrial peace also tend to increase. The result is the loss of several million mandays. (It must also be remembered that the social structure of India is conducive to regular absenteeism though such losses in mandays are not accounted for). In 1951, 3.8 m mandays were lost in 1071 industrial disputes involving 691,000 workers. By 1982 it had risen to 74.6 m mandays lost in 2483 disputes involving 1.5 million workers (mainly due to the Bombay textile strike — 41.4 million mandays being lost in it). But the situation had improved in 1985 and only 29.2 million mandays were lost in 1522 disputes involving around a million workers.

The average mandays lost for every 100 workers was an estimated 178 (for 15 major states) in 1985. West Bengal reported a manday loss of 489 and was followed by Maharashtra, 431. Other states reported manday losses as follows: Andhra Pradesh 155, Tamil Nadu 143, Kerala 143, Karnataka 108, Haryana 87, Bihar 83, Punjab 68, Rajasthan 64, Madhya Pradesh 58, Gujarat 50, Orissa 31, Uttar Pradesh 26, and Assam 2. Heavily industrialized Maharashtra, West Bengal and Tamil Nadu indeed have a poor record in this respect. Maharashtra, West Bengal, Tamil Nadu, Kerala and Andhra Pradesh together accounted for nearly 81 per cent of the total annual average mandays lost between 1981 and 1985. These states account for 43 per cent of the total factory employment and 50 per cent of the value added by manufacture.

During the Seventh Plan, the focus in industrial development will be on small-scale industry employment because of the limited potential of the organized sector to absorb increases in a labour force that, the larger it grows in the more industrialized states, the more aggressive it becomes. It is envisaged that the rural non-farm sector will play an important role in generating additional employment.

Sources: *Basic Statistics Relating to the Indian Economy, Vol 1: All India, August 1986* and *Vol 2; States, September 1986*, Centre for Monitoring Indian Economy.

VALUE ADDED IN THE FACTORY SECTOR

Table 97

	Value added in the factory sector (in Rs. million)				Mandays lost through indus- trial disputes — 1985 (in thousand)
	1960-61	1970-71	1980-81	1982-83	
States					
Andhra Pradesh	320	1,250	5,840	10,160	2,286
Assam	310[1]	440[1]	1,160	1,530	14
Bihar	660	1,740	5,010	11,230	899
Gujarat	1,070	2,880	11,390	15,240	750
Haryana	—	700	3,460	5,200	92
Himachal Pradesh	10	130	620	1,360	—
Jammu & Kashmir	10	30	170	360	—
Karnataka	330	1,810	6,030	8,140	1,195
Kerala	280	900	3,910	4,840	—
Madhya Pradesh	250	1,120	6,020	9,560	801
Maharashtra	2,730	8,440	29,860	36,010	3,623
Meghalaya	—	—	140	90	—
Orissa	100	590	1,980	3,610	109
Punjab	300[2]	710	3,860	4,580	100
Rajasthan	100	660	3,340	3,950	948
Tamil Nadu	810	3,090	12,290	16,910	3,360
Tripura & Manipur	—	—	40	70	—
Uttar Pradesh	650	2,080	7,490	14,950	286
West Bengal	2,090	4,290	13,750	16,340	13,375
Union Territories					
Andaman & Nicobar	—	—	40	30	N.A.
Chandigarh	—	60	160	190	N.A.
Delhi	200	520	1,900	2,720	N.A.
Goa, Daman & Diu	—	10	610	740	N.A.
Pondicherry	—	40	220	220	N.A.
INDIA[3]	10,220	31,490	119,290	168,030	29,186

Note: [1] Includes Meghalaya.
  [2] Includes Haryana and Chandigarh.
  [3] Includes other states and union territories for which break-up figures are not available.

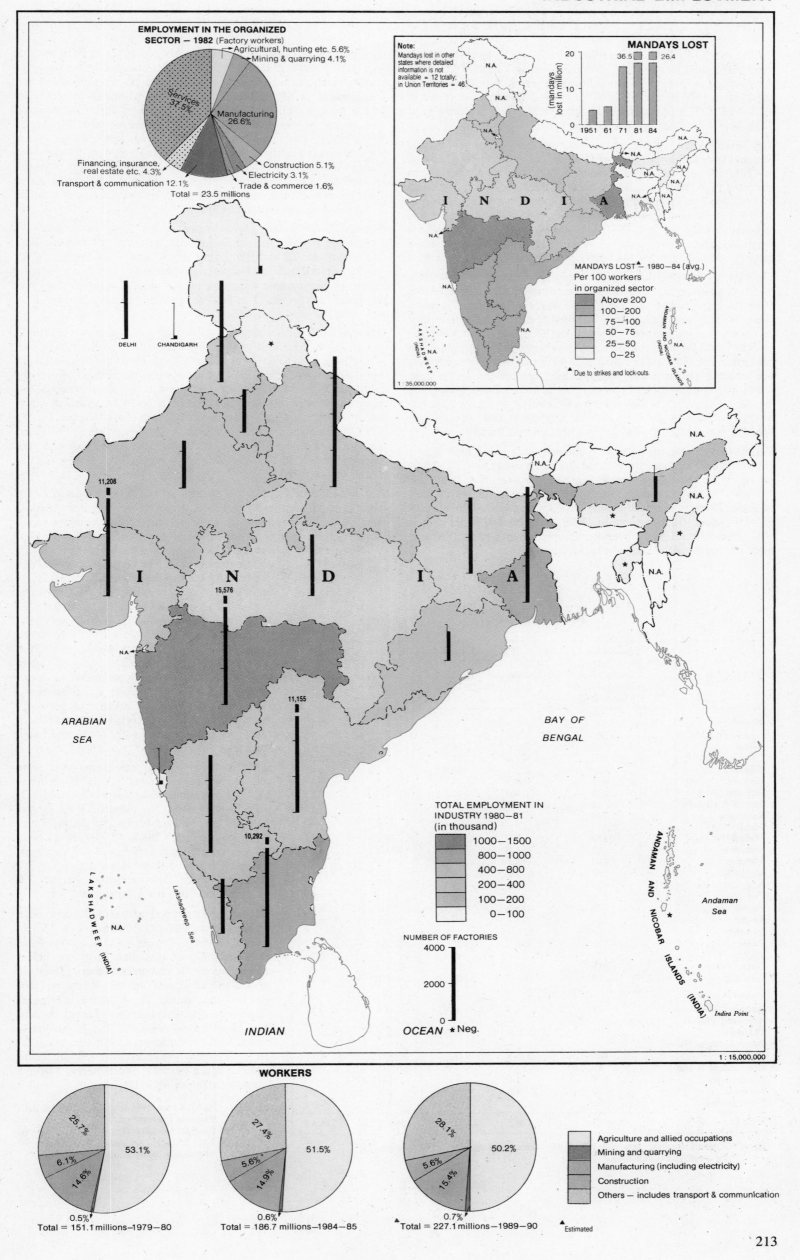

EMPLOYMENT IN THE ORGANIZED SECTOR – 1982 (Factory workers)

- Agricultural, hunting etc. 5.6%
- Mining & quarrying 4.1%
- Services 37.5%
- Manufacturing 26.6%
- Financing, insurance, real estate etc. 4.3%
- Transport & communication 12.1%
- Construction 5.1%
- Electricity 3.1%
- Trade & commerce 1.6%

Total = 23.5 millions

Note:
Mandays lost in other states where detailed information is not available = 12 totally; in Union Territories = 46

MANDAYS LOST
36.5   26.4
(mandays lost in million)
20
10
0
1951 61 71 81 84

MANDAYS LOST ▲ 1980–84 (avg.)
Per 100 workers in organized sector
- Above 200
- 100–200
- 75–100
- 50–75
- 25–50
- 0–25
▲ Due to strikes and lock-outs.

1 : 35,000,000

DELHI   CHANDIGARH

11,208
15,576
11,155
10,292

I N D I A

ARABIAN SEA

BAY OF BENGAL

LAKSHADWEEP (INDIA)
N.A.

Lakshadweep Sea

TOTAL EMPLOYMENT IN INDUSTRY 1980–81 (in thousand)
- 1000–1500
- 800–1000
- 400–800
- 200–400
- 100–200
- 0–100

NUMBER OF FACTORIES
4000
2000
0
★ Neg.

ANDAMAN AND NICOBAR ISLANDS (INDIA)

Andaman Sea

Indira Point

INDIAN OCEAN

1 : 15,000,000

WORKERS

- 25.7%
- 6.1%
- 14.6%
- 53.1%
- 0.5%

Total = 151.1 millions – 1979–80

- 27.4%
- 5.6%
- 14.9%
- 51.5%
- 0.6%

Total = 186.7 millions – 1984–85

- 28.1%
- 5.6%
- 15.4%
- 50.2%
- 0.7%

Total = 227.1 millions – 1989–90   ▲ Estimated

- Agriculture and allied occupations
- Mining and quarrying
- Manufacturing (including electricity)
- Construction
- Others – includes transport & communication

NON-WORKING POPULATION

During the Sixth Plan period, the overall employment had been expected to grow from 151.1 m Standard Person Years (SPY: a year in which an employee is *normally* engaged in production-based activities) in 1979-80 to 186.7 m SPY in 1984-85, an increase of nearly 35.6 m SPY. The growth rate of employment generation during the Sixth Plan was about 4.3 per cent per year.

The Seventh Plan, on the other hand, was, according to the 32nd National Sample Survey (NSS) of 1977-78, due to start with a backlog of 13.9 million unemployed persons, a large number of them in the unskilled pool. However, on the basis of the more recent 38th NSS of 1983-84, the Seventh Plan was to commence with a backlog of only 9.2 million unemployed as of March 1985. One reason for this reduced backlog was that 1983-84 was one of the best agricultural years and, hence, these activities generated more employment in the country, providing work for nearly 5 million additional persons.

The rate of unemployment was 3.04 per cent according to the 38th NSS. As against this, the projected rate of unemployment in March 1985, on the basis of the 32nd NSS, was 4.54 per cent. It must, however, be remembered that the national economy is still bedevilled by under-employment and low wage levels. It would, therefore, be safer to accept the levels of unemployment suggested by the 32nd NSS rather than the 38th. It was estimated in 1983 that rural unemployment was around 2.2 per cent and urban unemployment around 6.4 per cent. It was as a consequence of the very low levels of unemployment in the states studied in the 38th NSS that the overall unemployment rate worked out to 3.04 per cent. The unemployment rate among males was 3.2 per cent as against 2.7 per cent among females.

Yet another finding of the survey was that there was a high degree of unemployment among the new entrants in the age group 15-29 (6.54 per cent). This was as high as 13.4 per cent among urban youth and 4.5 per cent among rural youth. In all other age groups, the rates of unemployment recorded by the NSS are nominal (0.86 per cent in 30-34 years; 0.53 per cent in 45-49 years; and 0.58 per cent in 60+ years). Therefore the overall magnitude of employment requirement for the Seventh Plan works out to 47.58 m.

Working on the envisaged growth rate of 5 per cent in GDP and the impact of poverty eradication programmes aimed at providing self-employment and wage-employment for the poorer sections of the community, the Seventh Plan mentions that additional employment of the order of 40.36 m SPY would be generated during the Seventh Plan with an implied growth rate of 3.99 per cent per year. The special employment programmes — The National Rural Employment Programme (NREP) and the Rural Landless Employment Guarantee Programme (RLEGP) — would generate 2.26 m SPY of employment in 1989-90. The employment generation from the Integrated Rural Development Programme (IRDP) has been estimated at 3 m SPY, mainly concentrated in agriculture.

Sectorwise projected growth rates of employment reveal that agriculture is to be the single dominant generator of employment, accounting for 18 m SPY during the Seventh Plan, or nearly 45 per cent of total additional employment generation. Another significant source of employment is manufacturing, mining and quarrying; this is expected to provide 7 m SPY of additional employment (or 17.4 per cent).

The stock of educated manpower (matriculates, graduates and post-graduates) is estimated to increase during the Seventh Plan, from 47.72 m in 1985 to 64.39 m in 1990. Since all educated persons are not economically active and quite a large number of them, especially matriculates, pursue higher studies, the number of economically active educated persons in 1985 was estimated to be 30.84 m, out of which about 76 per cent were matriculates and 23 per cent graduates. During 1985-90, the addition to the economically active population of educated persons is expected to be nearly 10.6 m. The estimate of educated unemployment, 16.88 m in 1985, is expected to increase by 6.07 m at the end of 1990.

The total labour force in the country in March 1981 was 222.6 m and the population in the generally accepted working age group of 15-59 was estimated at 390 m. Those not working (or the non-working population) were thus an estimated 167.4 m. The statistics for non-workers in each state reveals that Kerala had the highest percentage of non-workers (53.5 per cent) in its 15-59 age group. Considering the emphasis on literacy in Kerala, many of its non-workers are likely to be students. This may also be true of states like West Bengal (49.4 per cent), Punjab (46.9 per cent), Haryana (45.1 per cent), Gujarat (41.7 per cent), Tamil Nadu (33 per cent) and Maharashtra (29.9 per cent). But the percentages are more likely to reflect genuine unemployment in Uttar Pradesh (43.1 per cent), Bihar (42.5 per cent), Orissa (39.4 per cent), Rajasthan (37.3 per cent), Karnataka (31.6 per cent), Madhya Pradesh (26.6 per cent) and the other states and union territories (together 19.1 per cent).

In terms of non-working population — a possible indicator of unemployment — Uttar Pradesh has 14.7 per cent of all the non-workers in India. It is followed by Bihar 9.2 per cent, West Bengal 9.0 per cent, Maharashtra 6.2 per cent, Tamil Nadu 5.6 per cent, Gujarat 4.7 per cent, Kerala 4.7 per cent, Madhya Pradesh 4.3 per cent, Andhra Pradesh 4.0 per cent, Rajasthan 3.7 per cent, Orissa 3.4 per cent, Punjab 2.6 per cent, and Haryana 1.8 per cent.

The number of job-seekers (educated) on the live registers of employment exchanges is 23.9 m (1985). Of the job-seekers in the country, those with schooling of less than 12 years (below matriculation) far outnumber others. They are more than half the total number of educated job-seekers in Jammu & Kashmir, Tripura, Meghalaya, Nagaland (the very highest), Assam, West Bengal, Goa, Andaman, Lakshadweep and Orissa, and constitute only a little less than 50 per cent of the job-seekers in Kerala, Tamil Nadu, Andhra Pradesh, Pondicherry, Madhya Pradesh, Manipur and Haryana.

JOB - SEEKERS[1]

Table 98

(in thousand)

	Job-seekers on live registers of employment exchanges as on March 1985	Educated job-seekers (December 1984)			
		Matriculates	Higher secondary	Graduates	Post graduates
States					
Andhra Pradesh	2,203	589.6	316.8	165.7	24.7
Assam	510	153.4	44.8	32.4	1.1
Bihar	2,662	1,022.4	265.7	212.5	6.5
Gujarat	642	299.1	46.3	41.1	3.6
Haryana	473	160.7	29.7	40.4	4.7
Himachal Pradesh	272	116.0	22.5	12.8	3.4
Jammu & Kashmir	69	11.4	7.8	7.4	1.1
Karnataka	788	323.4	40.1	67.5	8.6
Kerala	2,457	1,060	126.5	90.2	7.7
Madhya Pradesh	1,159	39.9	407.2	97.9	20.1
Maharashtra	2,163	736.0	95.7	133.2	12.7
Manipur	190	63.8	15.9	13.9	1.1
Meghalaya	13	3.2	0.8	0.7	0.1
Nagaland	14	1.1	0.2	0.1	—
Orissa	574	157.6	33.2	64.2	3.6
Punjab	540	169.3	47.9	58.7	8.5
Rajasthan	534	323.4	40.1	67.5	8.6
Sikkim[2]	—	—	—	—	—
Tamil Nadu	1,755	494.0	191.2	128.2	25.0
Tripura	95	19.9	16.9	5.2	0.2
Uttar Pradesh	1,959	403.4	451.6	254.0	51.0
West Bengal	4,118	953	665.4	349.2	10.7
Union Territories					
Andaman & Nicobar	11	1.2	0.8	0.6	0.1
Arunachal Pradesh[2]	—	—	—	—	—
Chandigarh	94	22.9	11.5	12.3	1.2
Dadra & Nagar Haveli[2]	—	—	—	—	—
Delhi	485	176.5	113.9	54.3	9.9
Goa, Daman & Diu	45	17.2	3.3	2.7	0.2
Lakshadweep	6	1.6	N.A.	0.1	—
Mizoram	19	4.9	1.0	0.8	0.1
Pondicherry	66	21.7	6.1	3.8	0.6
INDIA	23,915	7,126.5	3,081.5	1,907.0	216.4

Note: [1]Generally speaking, this table reflects the number seeking jobs in urban India.
[2] No employment exchange.

Sources: 1. *Seventh Plan 1985-90*, Government of India Planning Commission.
2. *Basic Statistics Relating to the Indian Economy, Vol. 2: States, September 1986*, Centre for Monitoring Indian Economy.

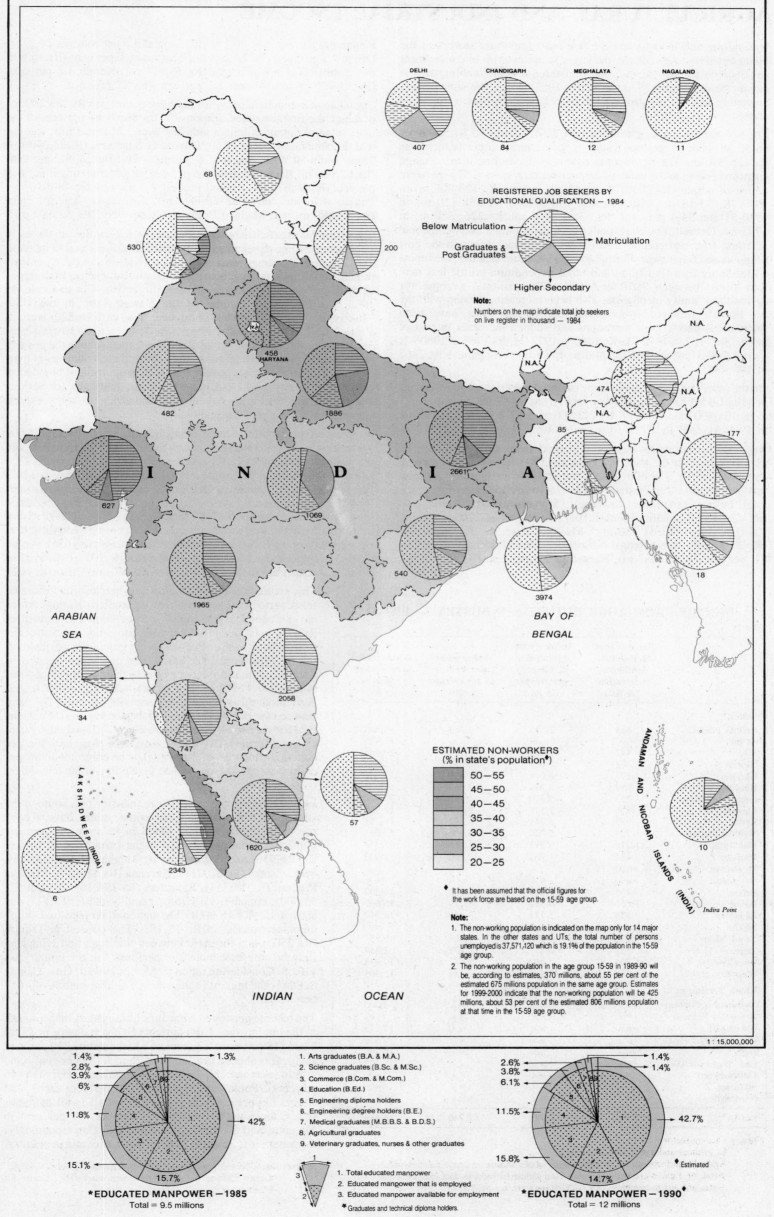

DELHI
407

CHANDIGARH
84

MEGHALAYA
12

NAGALAND
11

REGISTERED JOB SEEKERS BY
EDUCATIONAL QUALIFICATION — 1984

Below Matriculation ←

→ Matriculation

Graduates &
Post Graduates ←

→ Higher Secondary

Note:
Numbers on the map indicate total job seekers
on live register in thousand — 1984

INDIA

ARABIAN
SEA

BAY OF
BENGAL

LAKSHADWEEP (INDIA)

ANDAMAN AND NICOBAR ISLANDS (INDIA)

Indira Point

ESTIMATED NON-WORKERS
(% in state's population♦)

	50 — 55
	45 — 50
	40 — 45
	35 — 40
	30 — 35
	25 — 30
	20 — 25

♦ It has been assumed that the official figures for
the work force are based on the 15-59 age group.

Note:
1. The non-working population is indicated on the map only for 14 major
states. In the other states and UTs, the total number of persons
unemployed is 37,571,120 which is 19.1% of the population in the 15-59
age group.
2. The non-working population in the age group 15-59 in 1989-90 will
be, according to estimates, 370 millions, about 55 per cent of the
estimated 675 millions population in the same age group. Estimates
for 1999-2000 indicate that the non-working population will be 425
millions, about 53 per cent of the estimated 806 millions population
at that time in the 15-59 age group.

INDIAN OCEAN

1 : 15,000,000

1.4% ← → 1.3%
2.8% ←
3.9% ←
6% ←
11.8% ←
→ 42%
15.1% ←
15.7%

1. Arts graduates (B.A. & M.A.)
2. Science graduates (B.Sc. & M.Sc.)
3. Commerce (B.Com. & M.Com.)
4. Education (B.Ed.)
5. Engineering diploma holders
6. Engineering degree holders (B.E.)
7. Medical graduates (M.B.B.S. & B.D.S.)
8. Agricultural graduates
9. Veterinary graduates, nurses & other graduates

1. Total educated manpower
2. Educated manpower that is employed
3. Educated manpower available for employment

*EDUCATED MANPOWER — 1985
Total = 9.5 millions

* Graduates and technical diploma holders.

2.6% ← → 1.4%
3.8% ← → 1.4%
6.1% ←
11.5% ←
→ 42.7%
15.8% ←
14.7%

♦ Estimated

*EDUCATED MANPOWER — 1990♦
Total = 12 millions

AGRICULTURAL AND INDUSTRIAL INCOME

Agriculture and industry are the two most important sectors of the Indian economy. Agriculture, the larger sector in terms of employment and income for the people, has registered a near doubling of net income between April 1970 and March 1983. During the same period, however, industry has grown four and a half times in terms of value added by manufacture.

The net income accruing to farmers in 1970-71 was Rs 81,560 m or 46.52 per cent of the total value of agricultural output, which was Rs 175,310 m. The net income continued to decline in percentage terms in relation to the value of output over the years: to 37.4 per cent (value of output: Rs 331,070 m and net income Rs 123,800 m) in 1977-78, 35.13 per cent (Rs 452,130 m and Rs 158,820 m) in 1980-81 and 35.4 per cent (Rs 638,990 m and Rs 226, 380 m) in 1983-84. Overall, however, both the value of output and net income accruing from agriculture have shown increases, although the cost of inputs and other expenditure have also gone up steadily, the inputs a little more than four times and other expenditure a little less than four times between 1970 and 1984. The national average for agricultural family income has also been increasing in step with the net income received from agriculture though there have been fluctuations due to poor monsoons and sudden increases in cost of inputs: Rs 1157 in 1970-71, Rs 1480 in 1977-78, Rs 1769 in 1980-81, a drop to Rs 1588 in 1982-83 and an upward spurt again to Rs 2367 in 1983-84.

On the other hand, during the same period, industrial income in terms of value added has consistently grown, from Rs 46,190 m in 1970-71 to Rs 104,300 m in 1976-77, Rs 132,900 m in 1978-79, Rs 169,650 m in 1980-81, and Rs 277,930 m in 1984-85.

Regional variations in income generated from agriculture can be appreciated from the data available for 1981-82. The total income amounted to Rs 418,400 m and the states with high agricultural incomes were: Uttar Pradesh (Rs 63,760 m or 15.2 per cent of the total Indian agricultural income), Maharashtra 10.1 per cent (Rs 42,416 m), Andhra Pradesh 10.1 per cent (Rs 42,205 m), Bihar 7.6 per cent (Rs 31,750 m), Madhya Pradesh 7.1 per cent (Rs 29,871 m), West Bengal 6.5 per cent (Rs 27,356 m), Rajasthan 6.1 per cent (Rs 25,683 m), Karnataka 6.1 per cent (Rs 25,575 m),

Punjab 5.6 per cent (Rs 20,150 m), Gujarat 5.8 per cent (Rs 24,275), Orissa 4.8 per cent (Rs 20,233 m), Haryana 4.0 per cent (Rs 16,698 m), Tamil Nadu 4.4 per cent (Rs 18,516 m), Kerala 3.2 per cent (Rs 13,305 m) and Assam 3.1 per cent (Rs 12,869 m).

The value of output in industry for the same period was Rs 736,700 m, of which the first ten states, responsible for nearly 85 per cent of the total value of output in Indian industry, were: Maharashtra, leading with 23 per cent (Rs 169,610 m), Gujarat 11.3 per cent (Rs 83,040 m), Tamil Nadu 10.8 per cent (Rs 79,810 m), West Bengal 9.2 per cent (Rs 67,640 m), Uttar Pradesh 7.6 per cent (Rs 55,660 m), Bihar 6.1 per cent (Rs 44,670 m), Andhra Pradesh 4.7 per cent (Rs 34,910 m), Punjab 4.2 per cent (Rs 20,880 m), Karnataka 4.1 per cent (Rs 30,180 m) and Madhya Pradesh 4.1 per cent (Rs 30,380 m).

The historical experience in several countries reveals that, in the early stages of economic development, regional inequalities tend to increase and that, once a country reaches a certain high level of development, these inequalities tend to get narrowed down. But even at high stages of development, some degree of inequality exists. For example, in 1983, the per capita income of the poorest state in the USA (Mississippi $ 8155) and the richest state (Alaska $ 16,820) were in a ratio of 1:2. This inequality is not a cause for serious worry since, in such countries, the states generally operate much above the poverty line. The point that differentiates India from these countries is that, in India, the poorest states have a very high proportion of population below the poverty line. And it is this factor that calls for serious analysis of the existence, continuance and even, to some extent, aggravation of regional inequality.

In terms of per capita income from agriculture in 1981-82 (total agricultural income to total population in the state), the ratio between the richest (Haryana Rs 1212) and the poorest (Tamil Nadu Rs 295) is 4.1:1. The ratio in per capita value added from industry is 138:1 (the richest is Goa, Daman & Diu Rs 553 and the poorest is Manipur Rs 4). The two states that have the highest per capita income generation in agriculture are Punjab (Rs 1199) and Haryana. The states that have more than half the per capita agricultural income of Haryana (i.e. over Rs 600 per capita income) are Himachal Pradesh (Rs 656), Assam (Rs 653), Jammu and Kashmir (Rs 631), Manipur (Rs 631), Uttar Pradesh (Rs 621), Andhra Pradesh (Rs 618), Rajasthan (Rs 610) and Orissa (Rs 607).

INCOME FROM AGRICULTURE & INDUSTRY[1] — 1981-82

Table 99

	Income from agriculture & industry per capita (in Rs.)	Income from agriculture & industry per worker (in Rs.)	Value added per worker in agriculture[2] (in Rs.)	Value added per worker in industry (in Rs.)
States				
Andhra Pradesh	885	2,123	2,683	9,380
Assam	N.A.	N.A.	N.A.	11,475
Bihar	612	2,035	1,935	24,785
Gujarat	1,082	2,704	3,676	18,380
Haryana	1,327	4,018	7,498	21,961
Himachal Pradesh	959	3,204	3,069	10,000
Jammu & Kashmir	753	3,370	3,888	37,917
Karnataka	870	2,576	2,881	18,365
Kerala	695	4,744	4,744	14,781
Madhya Pradesh	724	2,020	1,956	23,061
Maharashtra	1,215	2,691	2,826	25,382
Manipur	624	1,957	2,180	5,000
Meghalaya	N.A.	N.A.	N.A.	N.A.
Nagaland	N.A.	N.A.	N.A.	N.A.
Orissa	855	2,871	3,137	16,500
Punjab	1,656	4,992	8,219	15,867
Rajasthan	858	3,188	3,569	18,317
Sikkim	N.A.	N.A.	N.A.	N.A.
Tamil Nadu	678	1,635	1,597	16,883
Tripura	593	2,877	2,877	2,857
Uttar Pradesh	709	2,462	2,642	19,038
West Bengal	765	2,349	3,225	15,505
Union Territories				
Andaman & Nicobar	N.A.	N.A.	N.A.	8,000
Arunachal Pradesh	N.A.	N.A.	N.A.	N.A.
Chandigarh	N.A.	N.A.	N.A.	17,273
Dadra & Nagar Haveli	N.A.	N.A.	N.A.	N.A.
Delhi	N.A.	N.A.	N.A.	14,786
Goa, Daman & Diu	N.A.	N.A.	N.A.	21,333
Lakshadweep	N.A.	N.A.	N.A.	N.A.
Mizoram	N.A.	N.A.	N.A.	N.A.
Pondicherry	N.A.	N.A.	N.A.	N.A.
INDIA[3]	834	2,740	2,740	18,424

Note: [1] Factory sector.
[2] Cultivators and agricultural workers.
[3] Agricultural income data include other states and union territories for which break-up figures are not available. But industrial income data exclude other states and union territories for which data are not available.

This picture changes dramatically when income generated from agriculture per cultivator is studied. Kerala, which has a statewide per capita income from agriculture of Rs 514, gets Rs 14,734 per cultivator. The poorest of the major states in terms of statewide per capita agricultural income (Tamil Nadu) has a per cultivator income of Rs 2572. The ratio between the highest (Kerala) and this state is 5.7:1. Punjab is the state with the second largest per cultivator agricultural income (Rs 11,449). The ratio between the richest (Kerala) and the poorest state (Nagaland Rs 1160) in per cultivator income is 12.7:1. All other states except Haryana (Rs 9714) earn less than half the per cultivator income of Kerala. The average all-India per cultivator agricultural income (excluding union territories) is Rs 4226.

Income generated per worker in industry puts some of the states in a better light. Even poorer states have a good record in terms of value added in industry per worker. Himachal Pradesh has the highest return per worker (Rs 26,956) and is followed by Meghalaya (Rs 23,333), Maharashtra (Rs 22,021), Haryana (Rs 18,404), Madhya Pradesh (Rs 18,354), Rajasthan (Rs 17,216), Gujarat (Rs 15,930), Punjab (Rs 15,755), Tamil Nadu (Rs 15,173) and Karnataka (Rs 15,000). The national average (excluding union territories) is Rs 15,188. The poorest is Tripura (Rs 2500) and the ratio between this state and Himachal Pradesh, the best industrial performer in the country, is 1:10.8. Considering union territories as well, Goa, Daman & Diu is the highest in the whole of India, with Rs 40,667 per worker.

The income generated from agriculture is less than a fourth of the total income from agriculture and industry in Tamil Nadu (18 per cent), Gujarat and Maharashtra (20 per cent each). It is about a third in West Bengal (33 per cent), and a little less than half in Bihar (49 per cent), Haryana (46 per cent), Punjab and Karnataka (45 per cent each), and Kerala (39 per cent). It is more than half in all the other states, touching 90 per cent in Tripura, 99 per cent in Manipur, and 100 per cent in Nagaland. Correspondingly, the pattern is reversed in terms of value of output in industry.

Sources: *Basic Statistics Relating to the Indian Economy, Vol.1: All India, August 1986* and *Vol.2: States, September 1986*, Centre for Monitoring Indian Economy.

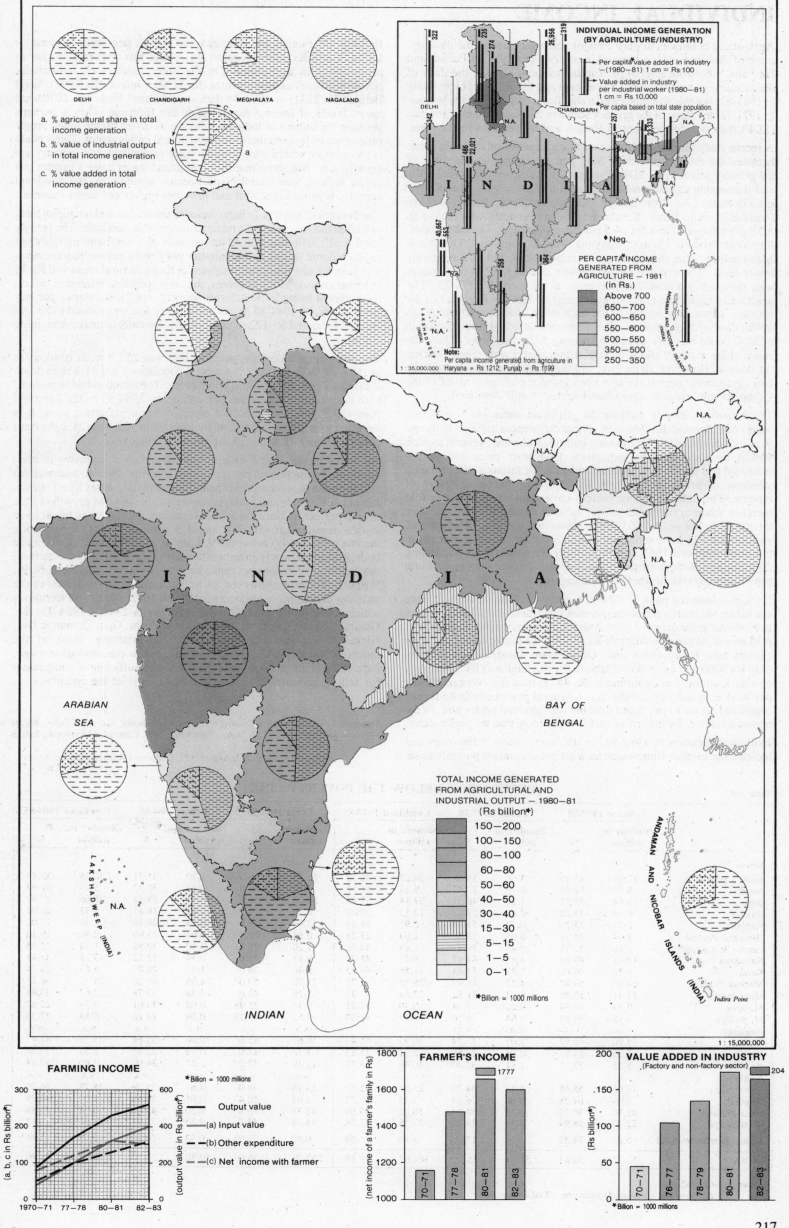

a. % agricultural share in total income generation

b. % value of industrial output in total income generation

c. % value added in total income generation

DELHI

CHANDIGARH

MEGHALAYA

NAGALAND

INDIVIDUAL INCOME GENERATION (BY AGRICULTURE/INDUSTRY)

Per capita value added in industry —(1980–81) 1 cm = Rs 100

Value added in industry per industrial worker (1980–81) 1 cm = Rs 10,000

Per capita based on total state population.

I N D I A

DELHI

CHANDIGARH

N.A.

322   235   275   319   26,956

342   274

486   22,021

40,667   553

258   257   23,333

364

◆ Neg.

PER CAPITA INCOME GENERATED FROM AGRICULTURE — 1981 (in Rs.)

Above 700
650–700
600–650
550–600
500–550
350–500
250–350

LAKSHADWEEP (INDIA)

N.A.

ANDAMAN AND NICOBAR ISLANDS (INDIA)

Note:
Per capita income generated from agriculture in Haryana = Rs 1212; Punjab = Rs 1199

1 : 35,000,000

I N D I A

ARABIAN SEA

BAY OF BENGAL

N.A.

N.A.

N.A.

N.A.

LAKSHADWEEP (INDIA)

N.A.

TOTAL INCOME GENERATED FROM AGRICULTURAL AND INDUSTRIAL OUTPUT — 1980–81 (Rs billion*)

150–200
100–150
80–100
60–80
50–60
40–50
30–40
15–30
5–15
1–5
0–1

*Billion = 1000 millions

ANDAMAN AND NICOBAR ISLANDS (INDIA)

Indira Point

INDIAN   OCEAN

1 : 15,000,000

FARMING INCOME

*Billion = 1000 millions

(a, b, c in Rs billion*)

(output value in Rs billion*)

— Output value
— (a) Input value
--- (b) Other expenditure
--- (c) Net income with farmer

1970–71   77–78   80–81   82–83

FARMER'S INCOME

(net income of a farmer's family in Rs)

1777

70–71   77–78   80–81   82–83

VALUE ADDED IN INDUSTRY
(Factory and non-factory sector)

204

(Rs billion*)

70–71   76–77   78–79   80–81   82–83

*Billion = 1000 millions

INDIVIDUAL INCOME

Agriculture continues to play a major role in the economic development of the states of the Indian Union. For example, in Punjab and Haryana, where the level of per capita income is highest, the share of the primary sector was about 44 per cent and 50 per cent respectively in 1981-82. But this share indicated a significant change from the position in 1971-72; the share of the primary sector in the State Domestic Product (SDP) then was 58 per cent in Punjab and 64 per cent in Haryana.

A recent study on SDP has indicated that a negative relationship exists between the rate of growth of SDP and the change in the share of the primary sector in the SDP. This is due to increased industrialization and decreasing agricultural contributions. This is very true of low-growth states like West Bengal, Karnataka, Tamil Nadu, Madhya Pradesh, Assam, Bihar, Kerala and Andhra Pradesh, which have an SDP growth rate of less than 4.5 per cent per annum. The other states, like Uttar Pradesh, Gujarat, Haryana, Punjab, Rajasthan, Orissa and Maharashtra, also show a negative relationship, but at a relatively lower level. The average annual growth rate of SDP in these states was between 4.5 and 7.6 per cent in the period 1971-82. The practically negligible effect of the change of SDP on the share of the primary sector may be due to the high stage of industrialization, as in the case of Maharashtra. This implies that a fast rate of growth of SDP could be more due to the non-agricultural sector. Thus, the share of the primary sector in SDP shows nearly zero change during the decade. However, in the case of Punjab and Haryana, though the non-agricultural sector has also been given a push as a result of the Green Revolution, the agricultural sector is still dominant.

These findings merely indicate that different states are at different stages of economic development. These differences are also indicated in individual incomes. To evaluate individual income, several aspects should be considered as indicators. Consumer price indices for industrial and agricultural workers indicate, for instance, the changing structure of prices and, hence, the indirectly changing nature of income. The industrial price indices have shown nearly a four-fold increase while agricultural indices have increased only a little over three-fold in the years 1965-1984. However, while agricultural and industrial prices have remained almost equal (in terms of indices) in the years between 1965 and 1971, they have later been increasing differentially, with industrial prices increasing faster than agricultural prices and overtaking them by the eighties.

Looked at from the point of view of workers in different sectors, there are substantial differentials in incomes (and these should be reflected in consumer price indices): private sector industrial employment in 1981 paid an annual average salary of Rs 11,289; the public sector paid only slightly less, Rs 10,643 a year. On the other hand, cultivation paid only Rs 3000 and agricultural labour (by the landless) Rs 1703. Any non-agricultural work paid more (Rs 4871) than the average income per worker in India (Rs 3948). It is a general practice in India for the organized sector to pay more than the unorganized sector and, in the organized sector, for the private sector to pay more than the public sector.

Individual incomes in 1982-83, in the states, show differences that indicate the sources from which such differences might possibly arise.

In 1984-85, Punjab and Haryana had a high per capita income — Rs 3835 and Rs 3296 respectively — and this was mainly due to productivity in agriculture. On the other hand, such industrially developed states as Maharashtra (Rs 3232), Gujarat (Rs 2997), West Bengal (Rs 2231) — 1983-84 data — and Tamil Nadu (Rs 2070) had lower levels of income than Punjab and Haryana. This is perhaps because the impact of industrial income is offset by the much larger proportion of less remunerative agricultural employment. Tamil Nadu is an example where contributions from agriculture and industry are equally low. But some mainly agricultural states have a greater per capita income than Tamil Nadu because agriculture, the dominant activity, is more successful and provides higher per capita incomes.

The concept of poverty in India includes destitution and unemployment as well as incomes below the minimum acceptable standards. The poverty level is officially determined on the basis of a minimum monthly per capita income linked to a minimum per capita calorie requirement. The poverty line was thus computed at Rs 65 in rural areas and Rs 75 in urban areas at 1977-78 prices, the corresponding minimum calorie requirement being 2400 calories per day and 2100 calories per day respectively. At 1984-85 prices, the poverty line was revised to Rs 107 in rural areas and Rs 122 in urban areas, the calorie intake remaining constant.

The population below the poverty line was 252.8 m in rural areas in 1977-78 (50.8 per cent of the rural population) and 51.8 m in urban areas (38.2 per cent of urban population). The situation had improved, both in absolute numbers and percentage, by 1984-85 to 222.2 m rural poor (39.9 per cent) and 50.5 m (27.7 per cent) urban poor. It is expected to improve still further by 1989-90 to 168.6 m (28.2 per cent) rural poor and 42.2 m (19.3 per cent) urban poor.

All states in the Union except two have shown increases in Net-Domestic-Product-based per capita state incomes; Punjab, Maharashtra and Haryana more than the others. Some smaller states like Tripura, Jammu & Kashmir and Manipur have also shown good growth (a little over 2 per cent). Karnataka and Rajasthan, on the other hand, have shown negative growth (0.8 and 1.3 per cent drops in per capita income respectively) between 1971 and 1981. This is perhaps because both performed poorly in agriculture and industry during this period, despite showing greater per capita income than a state like Tamil Nadu. Without exception, however, all states have shown increases in the state per capita income between 1961 and 1982. States and territories which have exceeded the all-India average (Rs 2344 in 1984-85) are: Gujarat, Punjab, Maharashtra, Haryana, Delhi, Goa, Daman & Diu, Himachal Pradesh and Pondicherry. Surprisingly, most of the industrially developed states still remain below the national average. Industrialization, it would seem, has not been sufficient to compensate for falling agricultural incomes in these parts of the country.

Sources: 1. *Basic Statistics Relating to the Indian Economy, Vol. 1: All India, August 1986* and *Vol. 2: States, September 1986,* Centre for Monitoring Indian Economy.
2. *Commerce, Vol: 143, August 1981,* Commerce Research Bureau, Bombay.

Table 100

POPULATION BELOW THE POVERTY LINE[1]

	Rural 1977-78		Urban 1977-78		Combined 1977-78		Rural 1984-85		Urban 1984-85		Combined 1984-85	
	Number in million	%	Number in million	%	Number in million	%	Number in million	%	Number in million	%	Number in million	%
States												
Andhra Pradesh	17.04	43.89	3.64	35.68	20.68	42.18	16.56	39.0	3.79	31.71	20.35	35.80
Assam	8.83	52.65	0.71	37.37	9.54	51.10	7.56	43.0	0.57	30.52	8.13	37.99
Bihar	33.84	58.91	3.29	46.07	37.14	57.49	32.25	51.0	3.41	39.88	35.66	47.87
Gujarat	9.48	43.20	2.65	29.02	12.13	39.04	6.70	28.0	1.62	18.81	8.32	22.92
Haryana	2.21	23.25	0.70	31.74	2.91	24.84	2.06	15.0	0.34	20.48	2.40	17.02
Himachal Pradesh	1.04	28.12	0.05	16.56	1.09	27.23	0.82	22.1	0.04	13.00	0.86	19.10
Jammu & Kashmir	1.46	32.75	0.44	39.33	1.89	34.06	1.20	25.71	0.30	30.88	1.50	23.43
Karnataka	12.41	49.88	3.85	43.97	16.27	48.34	9.95	37.0	3.34	32.62	13.29	33.48
Kerala	9.34	46.00	2.22	51.44	11.56	46.95	5.46	26.0	1.17	29.07	6.63	24.74
Madhya Pradesh	24.46	59.82	4.27	48.09	28.73	57.73	21.70	50.0	4.05	40.20	25.75	46.23
Maharashtra	21.41	55.85	6.13	31.62	27.54	47.71	17.29	42.0	4.38	23.78	22.67	33.83
Manipur	0.34	30.54	0.06	25.48	0.40	29.71	0.26	23.98	0.08	20.01	0.34	22.67
Meghalaya	0.55	53.87	0.04	18.16	0.59	48.03	0.52	42.29	0.04	14.26	0.56	37.33
Nagaland	—	—	0.03	4.11	—	—	N.A.	N.A.	N.A.	N.A.	N.A.	N.A.
Orissa	15.90	68.97	1.03	42.19	16.93	66.40	10.63	45.0	0.90	33.64	11.53	41.47
Punjab	1.35	11.87	0.96	24.66	2.31	15.13	1.42	11.0	0.93	22.85	2.35	13.20
Rajasthan	8.58	33.75	1.91	33.80	10.49	33.76	10.39	37.0	2.21	34.90	12.60	33.87
Sikkim	—	—	—	—	—	—	N.A.	N.A.	N.A.	N.A.	N.A.	N.A.
Tamil Nadu	17.05	55.68	6.66	44.79	23.71	52.12	14.18	44.0	4.61	35.39	18.79	36.99
Tripura	1.09	64.28	0.06	26.34	1.15	59.73	1.04	50.47	0.05	20.68	1.09	47.39
Uttar Pradesh	42.99	50.23	7.23	49.24	50.22	59.09	42.38	46.0	8.30	45.09	50.68	42.88
West Bengal	22.77	58.94	4.81	34.71	27.58	52.54	18.03	44.0	9.83	25.91	27.86	48.1
Union Territories total	0.63	34.32	1.12	17.96	1.76	21.69	0.82	26.95	0.93	14.10	1.75	15.85
INDIA	252.77	50.82	51.84	38.19	304.61	48.13	222.20	39.90	50.50	27.70	272.70	36.90

Note: [1] Estimated.
Totals may not tally due to rounding off of figures.

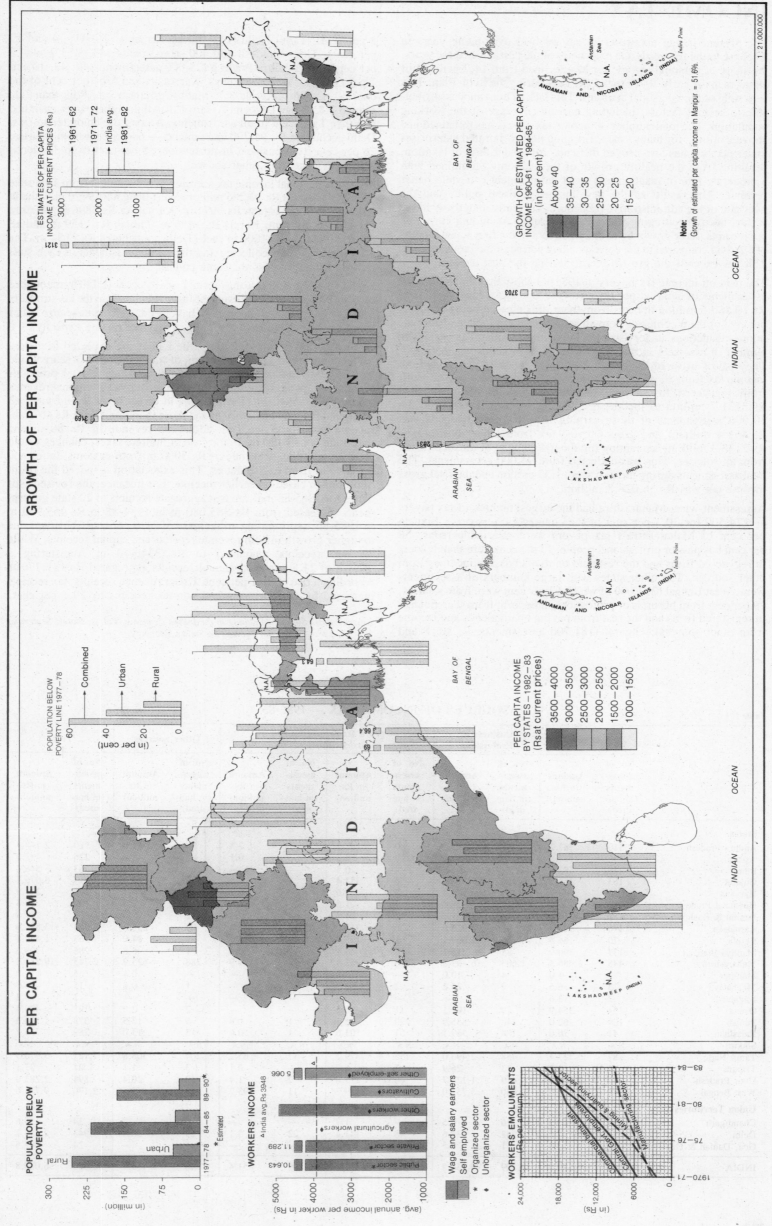

GROWTH OF PER CAPITA INCOME

ESTIMATES OF PER CAPITA INCOME AT CURRENT PRICES (Rs)
1961—62
1971—72 India avg.
1981—82

GROWTH OF ESTIMATED PER CAPITA INCOME 1960-61 — 1984-85
(in per cent)
Above 40
35—40
30—35
25—30
20—25
15—20

Note:
Growth of estimated per capita income in Manipur is 51.6%.

PER CAPITA INCOME

POPULATION BELOW POVERTY LINE 1977—78
Combined
Urban
Rural
(in per cent)

PER CAPITA INCOME BY STATES – 1982–83
(Rs at current prices)
3500—4000
3000—3500
2500—3000
2000—2500
1500—2000
1000—1500

POPULATION BELOW POVERTY LINE
Rural
Urban
(in million)
1977—78   84—85   89—90*
* Estimated

WORKERS' INCOME
▲India avg. Rs 3948
(avg. annual income per worker in Rs)
Public sector* Private sector* Agricultural workers Other workers Cultivators Other self-employed
Wage and salary earners
Self employed
* Organized sector
♦ Unorganized sector

WORKERS' EMOLUMENTS
(Rs per annum)
(in Rs)
1970—71 75—76 80—81 83—84

219

INCOME TAX

The present Indian tax structure has evolved over many years in response to the felt needs and exigencies of the past. Over these years, the share of personal income tax has declined from 14.2 per cent of all taxes in the early 1970s to 9.1 per cent during the Sixth Plan. This is in spite of high personal tax rates. This lack of buoyancy is attributable to several factors, including narrow coverage of the working population. The total number of income tax payers has remained just above a million for many years. Bringing unincorporated enterprises under taxation has also proved difficult. A broader base of taxation, resulting from the healthy growth of the economy, combined with a moderate rate of taxation and strict enforcement can yield better revenue. The results of the new long-term fiscal policy and the overhaul of the administrative procedures currently under way have, so far, been encouraging. Income tax receipts, including corporate tax receipts, have grown by over 25 per cent in the first seven months of 1985-86 compared to the corresponding period of the previous year. This far exceeds the growth record during the past decade.

The data on income tax payers in 1982-83 show that the number of assessments of salaried people was 471,900. The country obtained Rs 10,580.5 million in taxes from them, this amount being about 19 per cent of the income tax from all sources,both in India and abroad, minus deductions under Chapter IV of the Income Tax Act. The number of assessees in business or in the professions was 650,900. They paid a gross of Rs 39,731.8 million, about 71 per cent of the income tax from all sources. Those assessed for their share of profit from registered firms were 124,600 in number and they paid Rs 3131.7 million as income tax. The gross income tax from the 10,900 assessments of those earning money from dividends was Rs 451.4 million; and gross income tax from other sources was Rs 2133.4 million, accounting for about 4 per cent of the income tax from all sources. This was collected from 94,000 assessments. The total assessments during the year were 1.35 million and the total gross income tax was Rs 56,028.8 million.

Assessment-wise, Maharashtra had the largest number of tax payers — 274,500 (or 20.3 per cent of the country's tax payers). Sixteen per cent of Maharashtra's tax payers were salaried persons, 58 per cent business or professional people, 13 per cent were shareholders in registered firms and the rest paid on dividends and incomes from other sources. The other states with large numbers of assessments were: West Bengal (217,000 of which 63 per cent were from salaries, 29 per cent from business or professions, 3 per cent from shareholders in registered firms and the rest from paying on dividends and income from other sources), Gujarat (181,700 assessments: 9, 80, 8 and

3 per cent), Tamil Nadu (112,100 assessments: 26, 41, 14 and 19 per cent), Uttar Pradesh (89,400 assessments: 43, 47, 7 and 3 per cent), and Andhra Pradesh (78,000 assessments: 40, 43, 10 and 7 per cent). Of the other states, Karnataka had 5.64 per cent of the total assessments, Kerala 2.7, Madhya Pradesh 2.1, Rajasthan 2.1, Punjab 2.8, Bihar 1.94 and Assam 1.8. All the other states each had less than 1 per cent of all assessments. Among the union territories, Delhi accounted for 94,300 assessments — 66 per cent from salaries, 26 per cent from business or professions, 5 per cent from dividends and 3 per cent from other sources.

The gross personal income tax accruing to government from each state was: Maharashtra Rs 19,286 million, West Bengal Rs 9776 m, Gujarat Rs 6077 m, Tamil Nadu Rs 3971 m, Karnataka Rs 3896 m, Andhra Pradesh Rs 2148 m, Kerala Rs 1456 m, Punjab Rs 1159 m and all the rest less than Rs 1000 m each. Delhi contributed Rs 2154 m. The gross income taxes collected and the total assessments in each state followed more or less the same pattern.

Under Government's Long-Term Fiscal Policy (LTFP) announced early in 1983, several improvements have been made in the tax structure. The rates of personal income tax have been lowered to secure better compliance and improve the built-in revenue raising capacity.

The budget proposals, following the LTFP, are expected to yield a net additional revenue of Rs 4880 m, of which the centre's share would be Rs 4450 m and the states' share Rs 430 m. Along with tax proposals of the railways aggregating Rs 760 m, the total net additional revenue would be Rs 5640 m. The changes in direct taxes would yield an additional revenue of Rs 210 m and those in indirect taxes Rs 4670 m. Income tax would fetch an additional revenue of Rs 80 m and corporate tax Rs 130 m. In the case of indirect taxes, changes would yield an additional revenue of Rs 5000 m from customs duties and Rs 2540 m from excise duties. This expectation is indeed the result of optimism based on earlier receipts. For instance, the revised estimates for 1985-86 indicate that aggregate receipts of 22 state governments increased from Rs 383,080 m in 1984-85 to Rs 465,940 m in 1985-86, representing an increase of 21.6 per cent. This was due to higher growth in both revenue receipts and capital receipts. While revenue receipts amounted to Rs 334,170 m, registering a growth of 21.8 per cent, capital receipts aggregated Rs 131,770 m, recording a rise of 21.1 per cent. Under revenue receipts, tax receipts increased by 18.1 per cent and non-tax receipts by 29.1 per cent.

Source: *Basic Statistics Relating to the Indian Economy, Vol. 2: States, September 1986,* Centre for Monitoring Indian Economy.

SOURCES OF INCOME TAX — 1982-83

Table 101

	Salaries		Business & Professions		Share of profit from registered firms		Dividends		Other Sources		Total	
	No. of assessments (in hundred)	Amount (in Rs. million)	No. of assessments (in hundred)	Amount (in Rs. million)	No. of assessments (in hundred)	Amount (in Rs. million)	No. of assessments (in hundred)	Amount (in Rs. million)	No. of assessments (in hundred)	Amount (in Rs. million)	No. of assessments (in hundred)	Amount (in Rs. million)
States												
Andhra Pradesh	310	787.2	338	1,111.6	80	202.5	2	2.3	50	44.9	780	2,148.4
Assam	98	124.6	99	320.6	20	38.3	—	0.1	8	8.4	226	492.0
Bihar	248	482.9	14	62.8	—	0.5	—	—	—	0.3	263	546.4
Gujarat	163	389.4	1,460	5,277.8	150	353.4	5	5.5	39	50.6	1,817	6,076.7
Haryana	20	49.9	156	171.7	7	15.6	—	0.2	3	6.4	186	243.8
Himachal Pradesh	13	30.5	24	56.4	1	1.6	—	—	2	2.0	39	90.4
Jammu & Kashmir	8	15.9	41	205.1	12	34.9	—	—	2	1.6	63	257.5
Karnataka	146	373.5	381	2,926.6	126	315.2	12	22.4	99	258.3	764	3,896.0
Kerala	70	166.5	160	1,009.1	61	193.2	4	3.6	69	84.0	363	1,456.4
Madhya Pradesh	177	390.9	85	276.1	16	34.1	—	0.4	7	6.8	286	708.2
Maharashtra	448	1,085.8	1,603	16,469.8	356	952.4	52	151.9	288	625.9	2,745	19,285.8
Manipur	1	1.0	4	10.0	—	—	—	—	—	0.1	5	11.0
Meghalaya	3	6.5	9	25.2	2	3.5	—	—	1	0.8	15	36.0
Nagaland	2	3.6	—	—	—	0.2	—	—	—	—	2	3.8
Orissa	154	354.9	29	101.1	11	20.5	—	—	2	2.0	197	478.4
Punjab	109	262.0	243	835.9	20	48.1	1	0.8	5	12.5	378	1,159.3
Rajasthan	41	70.0	184	569.1	51	103.6	1	0.6	13	13.0	289	756.1
Sikkim	N.A.	N.A.	N.A.	N.A.	N.A.	N.A.	N.A.	N.A.	N.A.	N.A.	N.A.	N.A.
Tamil Nadu	287	727.4	458	2,490.0	155	379.2	17	65.2	204	308.8	1,121	3,970.5
Tripura	1	2.2	13	26.9	3	6.9	—	0.1	3	1.1	21	37.1
Uttar Pradesh	385	1,006.9	419	1,128.7	60	125.6	2	1.4	28	20.4	894	2,283.0
West Bengal	1,371	3,021.9	630	5,751.0	70	159.7	11	193.7	88	649.4	2,170	9,775.6
Union Territories												
Chandigarh	18	47.5	16	70.5	—	1.7	—	—	—	—	34	119:6
Delhi	622	1,137.6	242	835.6	48	141.4	22	3.1	29	36.4	943	2,154.1
Goa, Daman & Diu	24	42.0	1	0.5	—	—	—	—	—	—	24	42.5
INDIA	4,719	10,580.5	6,509	39,731.8	1,246	3,131.7	109	451.4	940	2,133.4	13,523	56,028.8

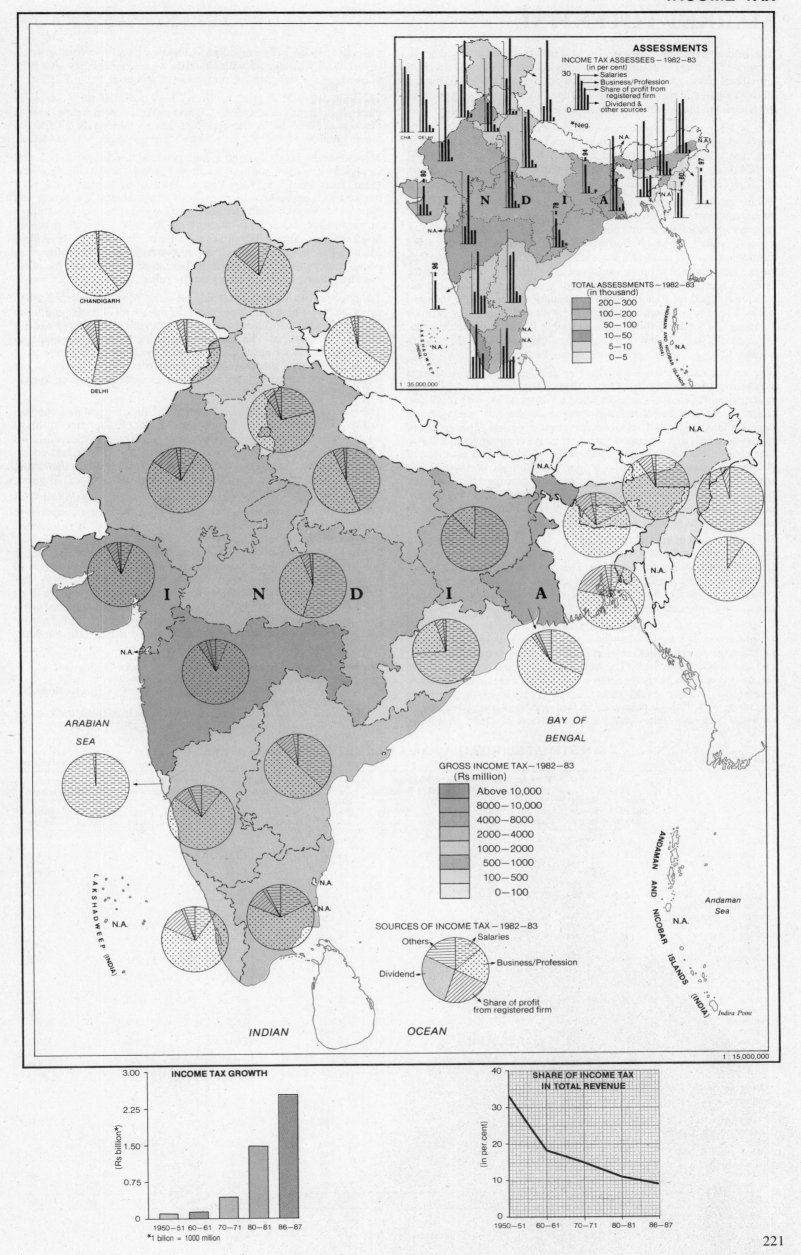

ASSESSMENTS

INCOME TAX ASSESSEES — 1982—83
(in per cent)

- Salaries
- Business/Profession
- Share of profit from registered firm
- Dividend & other sources

*Neg.

TOTAL ASSESSMENTS — 1982—83
(in thousand)

- 200 — 300
- 100 — 200
- 50 — 100
- 10 — 50
- 5 — 10
- 0 — 5

1 : 35,000,000

CHANDIGARH

DELHI

ARABIAN
SEA

I N D I A

BAY OF
BENGAL

ANDAMAN
AND
NICOBAR
ISLANDS
(INDIA)

Andaman
Sea

N.A.

LAKSHADWEEP (INDIA)

GROSS INCOME TAX — 1982—83
(Rs million)

- Above 10,000
- 8000 — 10,000
- 4000 — 8000
- 2000 — 4000
- 1000 — 2000
- 500 — 1000
- 100 — 500
- 0 — 100

SOURCES OF INCOME TAX — 1982—83

- Others
- Salaries
- Business/Profession
- Dividend
- Share of profit from registered firm

Indira Point

INDIAN   OCEAN

1 : 15,000,000

INCOME TAX GROWTH

3.00

2.25

1.50

0.75

(Rs billion*)

1950—51   60—61   70—71   80—81   86—87

*1 billion = 1000 million

SHARE OF INCOME TAX IN TOTAL REVENUE

40

30

20

10

(in per cent)

1950—51   60—61   70—71   80—81   86—87

MARKET POTENTIAL

The Indian economy is broadly divided between the rural and urban sectors. The rural economy is basically agrarian, whereas much of the urban economy is non-agricultural and industrial-based, though in most small and medium towns the economy is based on supportive activities for the surrounding rural base.

The NDP at current prices shows that the share of the primary sector (agriculture, forestry, fishing, mining and quarrying) was nearly 50 per cent in 1970-71. This gradually fell to 40 per cent in 1980-81 and 34.4 per cent in 1984-85. In monetary terms, this was Rs 173,070 m, Rs 420,220 m and Rs 603,682 m in the respective years.

With a growing awareness of the necessity for rural uplift, government has, in the past several years, been adopting several welfare measures to improve the lot of the rural people. These measures cover population control, health, education, housing, banking, financial assistance and promotion of rural and small-scale industries. The growing prosperity of the rural masses through improved agriculture, coupled with the benefits accruing from the government's welfare programmes, has metamorphosed the socio-economic and cultural conditions in the rural areas. There is yet much wanting, but the rural market today in much of India is a totally different place to what it was two decades ago. Against this background, it is not difficult to appreciate the need to recognise its growing importance to marketing men.

The Indian masses have traditionally devised appropriate market mechanisms for themselves. One of them is the *periodic market*. This system was devised in the days when the population was small and the rural markets had thresholds (population served) inadequate to permit the operation of full-time, permanent outlets. One effect of this system was that peripatetic services could congregate on a single weekday in each of a sequence of large villages. The problem of finding adequate thresholds no longer prevails, yet the system has survived mainly because it still offers opportunities for a large number of villagers to sell their produce in these markets. This 'walkabout' service functions so well that it has now been borrowed to provide such services as health, population control and postal facilities.

With the growth of substantial thresholds, several villages have grown into small and medium-sized towns. In all of them there is a demand for the wealth of Indian products. This, in the rural areas, has not merely meant fuel oils, HYVs and chemical fertilizers, but also consumer goods like transistor radios, readymade garments, clocks, electrical goods and cosmetics.

In assessing rural market potential, 12 economic indicators were weighted and aggregated district-wise to identify districts with very high market potential, high market potential, medium market potential and low market potential. (The indicators are listed on the map and are based on projections for 1986 using 1981 Census data and other available data.) The results show 27 districts with very high market potential, 47 with high potential, 65 with medium potential and 289 with low potential. The wholly urban districts of Delhi, Bombay,

Calcutta, Madras, Hyderabad and Chandigarh and five rural districts in Jammu & Kashmir for which no data were available have not been considered for the map on rural market potential.

The bulk of the 27 districts with very high potential are clustered in three areas, Tamil Nadu, around Calcutta in West Bengal and the tri-state junction of Haryana, Punjab and Rajasthan. A third of the 27 districts are in Tamil Nadu, perhaps the country's best consumer product market.

Most of the country's medium to high potential districts are to be found in peninsular India, south of Madhya Pradesh, and in the Gangetic Plain, from Punjab to West Bengal. But just as UP has some poor markets to mar the broad strip of prosperity, coastal Karnataka and parts of Maharashtra and Orissa reduce the impact of market strength in peninsular India. The majority of the low potential districts, however, are almost contiguous in parts of Gujarat, Rajasthan, Madhya Pradesh, Bihar and UP on the one hand, and in the hilly areas of the country (Himachal Pradesh, Jammu & Kashmir and the north eastern region, except Assam) on the other.

To assess urban market potential, all cities (266 excluding 3 cities in Assam for which data were not available) with 100,000 and more population were considered and classified in the same four categories as for rural markets. The indicators used are listed on the map and are based on 1981 Census data and other available data. While most indicators for rural market potential assessment were agricultural, the indicators for the urban areas have been, besides population, mainly industrial variables.

The highest market potential is found in the three biggest metropolitan centres: Calcutta, Greater Bombay and Delhi. Eight other 'million-cities' have the next highest potential. Of medium potential are 48 cities and towns, while 207 have low potential. Taking into consideration medium markets as well, the best states for urban marketing are: UP, Maharashtra, Tamil Nadu, Madhya Pradesh and Gujarat.

In rural areas, cooperative marketing has been catching up fast. The Seventh Plan aims to strengthen the primary marketing cooperative societies and make their activities more broad-based. It is intended to make them the main institutional agencies for procurement and selling operations on behalf of the central and state governments and public sector commodity corporations. Cooperatives in urban areas mainly distribute rural produce (food) and household essentials. But market potential is now being looked at in both sectors, mainly from the point of view of what it is for consumer products manufactured by the private sector. The consumer product market in India is now estimated to be the second biggest in the world in absolute numbers, though China is fast catching up.

Sources: 1. *Primary Census Abstract, Part II B(i)*, Census of India 1981.
2. *Fertiliser Statistics 1984-85*, Fertiliser Association of India.
3. *Profiles of Districts Part I* and *Part II, July 1985*, Centre for Monitoring Indian Economy.
4. *Thompson Rural Market Index '72* and *Urban Market Index '73*, Hindustan Thompson Associates Limited.

ESTIMATED RURAL-URBAN MARKET POTENTIAL — 1986

Table 102

(Population in thousand)

States	Total no. of districts	Rural Population	Very high & high market potential No. of districts	Population	Medium market potential No. of districts	Population	Urban Population	Class I Towns No. of towns	Population	Very high & high market potential No. of towns	Population	Medium market potential No. of towns	Population
States													
Andhra Pradesh	23	44,864	6	15,290	4	7,590	15,485	28	9,234	1	3,155	2	1,421
Assam	17	23,205	2	N.A.	5	N.A.	2,599	6	930	—	—	—	—
Bihar	38	67,517	4	3,818	2	6,006	11,116	17	6,148	—	—	4	3,527
Gujarat	19	26,102	2	2,987	6	9,758	12,775	14	7,469	2	4,170	3	1,814
Haryana	12	11,220	4	4,369	3	2,429	3,660	12	2,194	—	—	1	427
Himachal Pradesh	12	4,407	—	—	—	—	381	—	—	—	—	—	—
Jammu & Kashmir	14	5,336	—	—	—	—	1,556	2	1,023	—	—	1	748
Karnataka	19	28,922	3	6,195	4	7,164	13,412	18	7,947	1	3,651	3	1,640
Kerala	14	22,301	9	12,420	4	8,875	5,653	9	3,124	—	—	3	2,074
Madhya Pradesh	45	45,601	1	2,795	1	2,789	13,550	19	6,963	—	—	5	4,226
Maharashtra	30	44,723	4	8,175	6	8,742	26,392	27	20,038	2	11,915	5	3,938
Manipur	8	1,108	—	—	—	—	682	1	283	—	—	—	—
Meghalaya	5	1,239	—	—	—	—	318	1	231	—	—	—	—
Nagaland	7	788	—	—	—	—	201	—	—	—	—	—	—
Orissa	13	25,088	3	8,785	3	7,528	4,167	6	1,732	—	—	2	724
Punjab	12	13,202	9	88,509	1	1,156	5,670	9	2,828	—	—	3	1,962
Rajasthan	27	30,765	1	1,832	4	6,034	9,302	13	4,522	1	1,309	3	1,598
Sikkim	4	265	—	—	—	—	102	—	—	—	—	—	—
Tamil Nadu	18	34,559	7	20,603	5	7,639	18,185	27	12,064	1	4,889	5	3,681
Tripura	3	2,111	—	—	—	—	270	1	157	—	—	—	—
Uttar Pradesh	57	99,950	12	30,903	15	29,120	25,868	38	14,293	1	2,130	7	5,814
West Bengal	16	44,219	7	28,784	2	4,914	16,758	17	13,392	1	10,665	—	—
Union Territories													
Andaman & Nicobar	2	178	—	—	—	—	73	—	—	—	—	—	—
Arunachal Pradesh	10	682	—	—	—	—	67	—	—	—	—	—	—
Chandigarh	1	30	—	—	—	—	594	1	594	—	—	1	594
Dadra & Nagar Haveli	1	112	—	—	—	—	11	—	—	—	—	—	—
Delhi	1	460	—	—	—	—	7,440	1	7,440	1	7,440	—	—
Goa, Daman & Diu	3	795	—	—	—	—	448	—	—	—	—	—	—
Lakshadweep	1	23	—	—	—	—	29	—	—	—	—	—	—
Mizoram	3	375	—	—	—	—	164	1	125	—	—	—	—
Pondicherry	4	295	—	—	—	—	409	1	325	—	—	—	—
INDIA	439	580,442	74	235,465	65	109,744	197,337	269	123,056	11	49,324	48	34,188

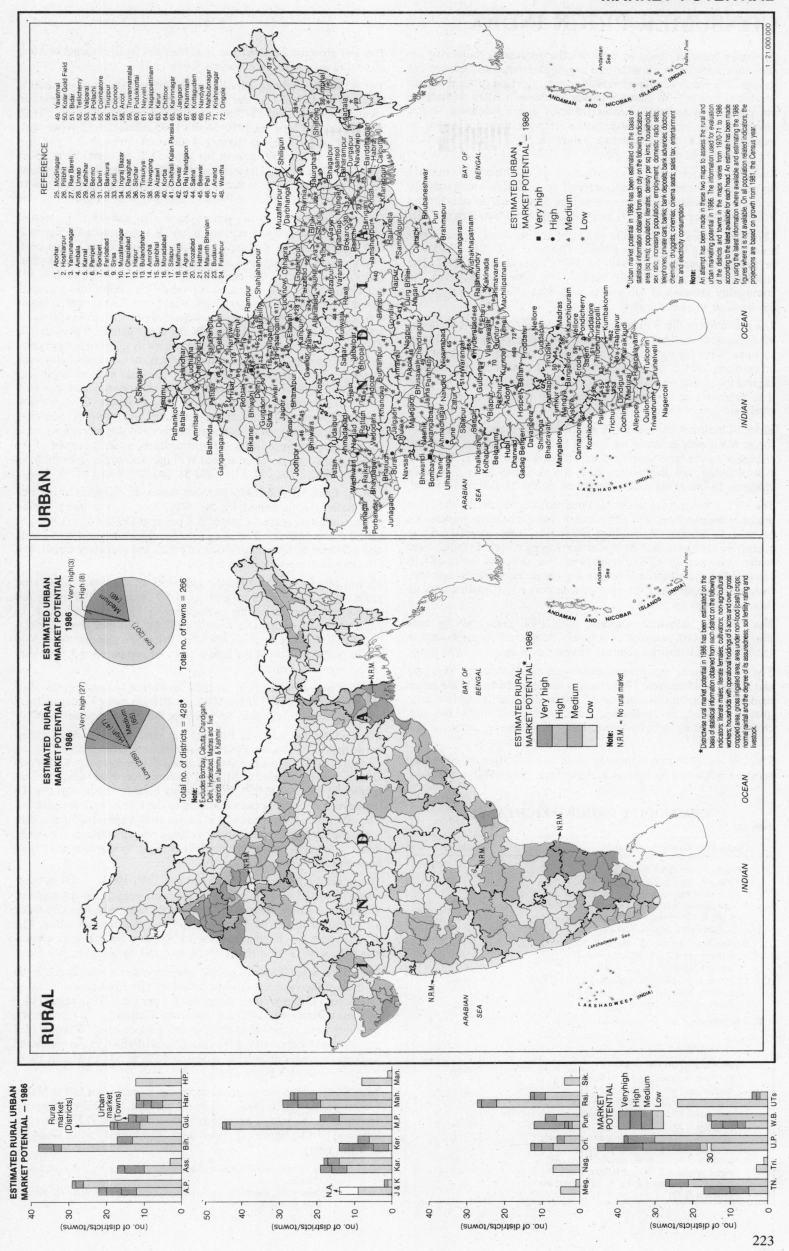

CONSUMER PRICE INDEX

An Index Number, used to measure the success or failure of an economic activity, is a specialized average designed to measure the change in a group of related variables over a period of time. By its variation, it shows the changes in magnitudes which are themselves not capable of accurate measurement or direct valuation. Index Numbers may be classified in terms of what they are meant to measure, such as price, quantity, value, etc. Price Index Numbers are dealt with here.

Stability in price levels is a reflection of a country's economic development; hence, the significance of price indices which indicate price level movements. Price indices commonly used in India are the Wholesale Price Index (WPI) and the Consumer Price Index (CPI).

The WPI, based on wholesale prices of commodities, is published weekly by the Government of India. The base year of the WPI was revised by Government in 1949, 1952-53, 1961-62 and 1970-71. The current indices are 1970-71 based. The table below gives the commodity groups and sub-groups with their respective weights on the basis of marketed value.

The limitation of the WPI is that it fails to show how different classes of consumer households are affected by price changes. Different sections of people indeed consume different commodities in varying proportions. Since, by definition, the WPI cannot be the basis to explain the impact of changes in commodity prices on family budgets, it is necessary to construct an index with particular reference to consumer prices alone.

The CPI indicates the cost of living for various sections of the work force. But it only indicates the cost of living in the current year with reference to the base year; it does not show or measure the actual cost of living, nor fluctuations in it due to causes other than price changes. In India, the CPI is published by the Ministry of Labour.

The CPI, a weighted average, is obtained by considering the retail prices of commodities. The CPI's utility is that it helps in negotiations on wages, fixation of dearness allowance, rent control, and in measuring change in real income or purchasing power. In India, 1960 is the base year for CPI computations. The different series of CPI compiled at the all-India level are for industrial workers, for non-manual employees and for agricultural workers. (Detailed data on these are given on p.234).

The CPI for industrial workers, with 1960 prices as the base (i.e. 1960 = 100), is based on a family budget survey conducted in 1958-59 in 50 industrial centres. This series is based on 100 items.

The CPI for urban non-manual employees has 1960 as the base year and relates to 45 centres. These indices measure changes in the prices paid by the middle class and cover 180 items classified in five main groups — food, beverages and tobacco; fuel and light; housing; clothing, bedding and footwear; and miscellaneous items — and 23 sub-groups.

The CPI for agricultural labour is based on retail prices of goods and services consumed by agricultural workers, with 1960 as the base.

COMMODITY GROUP WEIGHTAGES

Table 103

Commodity Group/Subgroup	Weightage
I. Primary Articles (A+B+C)	41.67
A. Food Articles (a. Food grains 12.92, b. Fruits and vegetables 6.14, c. Milk and milk products 6.15, d. Eggs, fish and meat 1.90, e. Condiments and spices 1.09, f. Other food articles 1.60)	29.80
B. Non-Food Articles (a. Fibres 3.17, b. Oilseeds 4.21, c. Other non-food articles 3.24)	10.62
C. Minerals (a. Metallic minerals 0.23, b. Other minerals 0.42, c. Petroleum crude and natural gas 0.60)	1.25
II. Fuel, Power, Light and Lubricants (a to c)	8.46
(a. Coal mining 1.15, b. Mineral oils 4.91, c. Electricity 2.40)	
III. Manufactured Products (a to k)	49.87
(a. Food products 13.32, b. Beverages, tobacco and tobacco products 2.71, c. Textiles 11.02, d. Paper and paper products 0.85, e. Leather and leather products 0.39, f. Rubber and rubber products 1.21, g. Chemicals and chemical products 5.55, h. Non-metallic mineral products 1.42, i. Basic metals, alloys and metal products 5.97, j. Machinery and transport equipment 6.71, k. Miscellaneous products 0.72)	

The four groups considered are: food; fuel and light; clothing, bedding and footwear; and miscellaneous items.

The WPI for different groups has increased over the years at variable rates. With 1970-71 as the base, the price of primary articles in 1950-51 was 56.8. In 1980-81 it was 237.5 and in 1985-86, 331.0. The annual rise between 1950-51 and 1985-86 is, thus, 5.6 per cent. The WPI of fuel, power, light and lubricants grew at 7.5 per cent in these 36 years, with indices for 1950-51 at 46, 1980-81 at 354.3 and 1985-86 at 579.9. The increase in WPI for manufactured products for this period is closer to that of primary articles (5.8 per cent), with 47.7 in 1950-51, 257.3 in 1980-81 and 342.6 in 1985-86.

Sub-groups in each of the groups have also had varying rates of increase. Of the primary articles, minerals registered a WPI growth of 9.2 per cent, eggs, fish and meat 8 per cent and pulses 7.1 per cent. The increase in foodgrains was 5.6 per cent, cereals 5.1 per cent, fruits and vegetables 6.9 per cent, and milk and milk products 6.1 per cent. On the other hand, non-food articles registered increases closer to the rate of primary articles, with fibres 5.4 per cent, oilseeds 4.6 per cent and other non-food articles 5.8 per cent. Coal registered a high of 8.2 per cent, mineral oils the equivalent (7.5 per cent) of fuel, power, light and lubricants, and electricity a low of 6.9 per cent. Of the sub-groups of manufactured products, basic metals, alloys and metal products attained a high of 8.1 per cent and miscellaneous products a low of 4.3 per cent. Sugar, khandasari and gur (6.9 per cent), beverages, tobacco and tobacco products (6.1 per cent), non-metallic mineral products and machinery and transport equipment (5.9 per cent each) show increases higher than the average increase for manufactured products. All commodities, as an aggregate of the three groups, registered an annual WPI increase of 5.9 per cent from 47.5 in 1950-51 to 257.3 in 1980-81 and 357.8 in 1985-86.

During financial year 1985-86 there was a deceleration in the rate of inflation as measured by the WPI; that is, the index for all commodities registered a rise of 3.8 per cent (1970-71 = 100), just half of that in 1984-85. This price increase, the lowest in several years, was achieved by close monitoring of prices and monetary trends, and effective demand-supply management.

Movements in WPI in 1985-86 and 1986-87 show that while food grain prices rose by 8.6 per cent as of 14 December 1985, they fell by 1.4 per cent during the period ending 13 December 1986. In contrast, non-food articles rose in 1987 by 12.8 per cent against a decline of 9.8 per cent in 1985-86. Prices of manufactured food products also rose by 11 per cent in this year compared to 5.1 per cent in the previous year.

The fall in non-food prices in 1986 was a major factor in the slow rate of increase in the WPI in 1985-86. Effective supply management of edible oils during 1985-86 has kept the increase under check.

The CPI for the states in 1983-84 for agricultural workers varied between 436 for Andhra Pradesh and 601 for Orissa, with 523 as the all-India index. Only a few states had less than the average CPI. Food CPIs however present a different picture, for these seem to bear relation to the general CPI, being mainly higher. This means that the increases have largely been confined to the CPI of primary articles. The range here is between 674 in Orissa and 461 in Andhra Pradesh. Since the food CPI represents the relative price features of the states, it is a safe guess that where the CPI is high, food prices have been high and vice versa.

The general CPI for industrial workers, with 1960 = 100 as base, was 576 in 1984, with Jammu & Kashmir registering the highest, 627, and Assam the least, 503. Kerala, 659, had the highest CPI index for food (for industrial workers); 604 was the national index. The national index for clothing, bedding and footwear for industrial workers was 578, with Gujarat (645) the highest and Assam (511) the least.

The centre-wise CPI for industrial workers indicates that during 1986 the rate of increase was highest in Indore (16.1 per cent), followed by Bhopal (15.5 per cent), Kanpur (15.3 per cent), Bhavnagar (14.5 per cent), Raniganj (14.1 per cent) and Darjiling (13.1 per cent). Declines were recorded in Rangapara (2.6 per cent), Mariani (2.2 per cent) and Labac (0.6 per cent). The all-India CPI for industrial workers increased from 586 in March 1985 to 638 in March 1986, a growth rate of 8.9 per cent.

The CPI for urban non-manual employees moved from 540 in March 1985 to 584 in March 1986, i.e., by 8.1 per cent. All centres witnessed increases in the index, but it was highest in Indore (12.9 per cent) followed by Kharagpur (12.3 per cent). The lowest was in Gulbarga (3.4 per cent).

Sources: 1. Dr. C. Selvaraj, Professor, Department of Economics, Madras Christian College.
2. Report on Currency and Finance 1985-86, Vol. I & II, Reserve Bank of India.

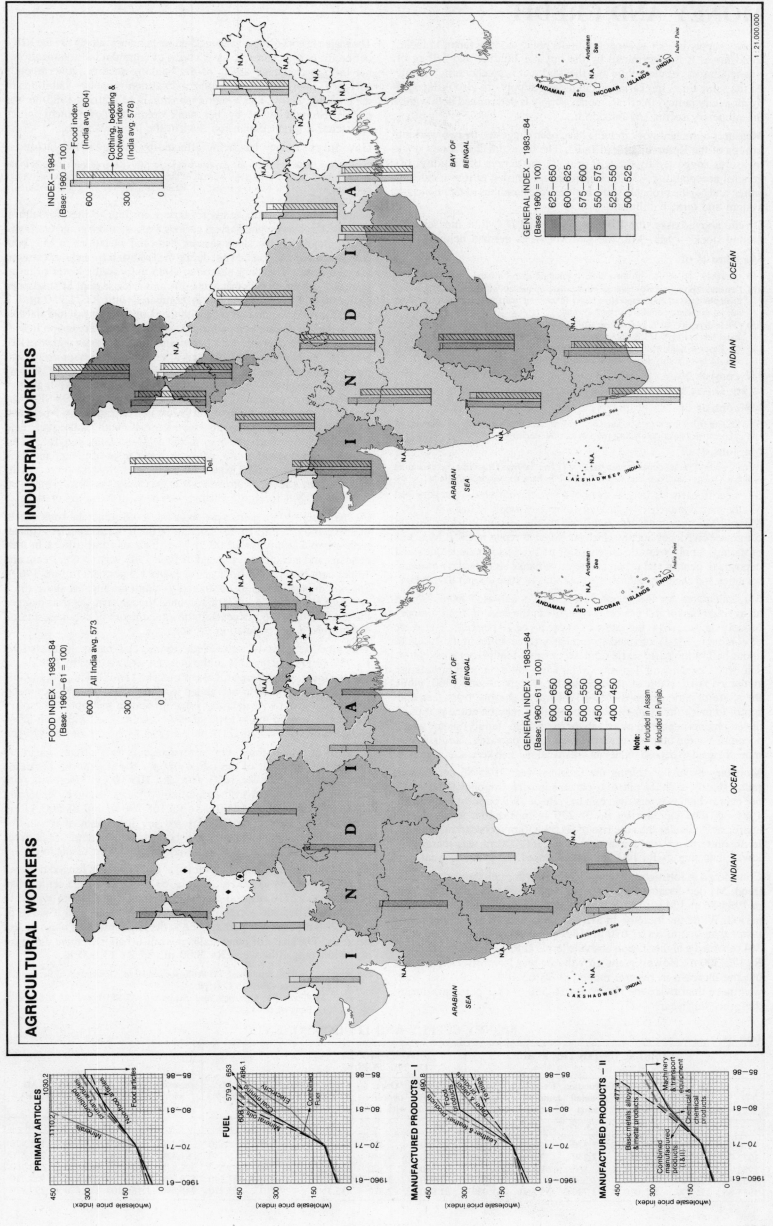

INDUSTRIAL WORKERS

INDEX — 1984
(Base: 1960 = 100)

Food index
(India avg. 604)
Clothing, bedding &
footwear index
(India avg. 578)

GENERAL INDEX — 1983–84
(Base: 1960 = 100)

625—650
600—625
575—600
550—575
525—550
500—525

AGRICULTURAL WORKERS

FOOD INDEX — 1983–84
(Base: 1960–61 = 100)

All India avg. 573

GENERAL INDEX — 1983–84
(Base: 1960–61 = 100)

600—650
550—600
500—550
450—500
400—450

Note:
★ Included in Assam
◆ Included in Punjab

PRIMARY ARTICLES
FUEL
MANUFACTURED PRODUCTS — I
MANUFACTURED PRODUCTS — II

MONEY AND CREDIT

Money supply is both an economic and a policy-control factor in India. In so far as it is determined by the buying, holding and selling of financial assets of the public and the banks, it is an economic factor. At the same time, the variations in money supply are also influenced by monetary policy. In effect, money supply is determined both within the monetary sector and outside it.

Measures of money stock in India have been suggested by two working groups of the Reserve Bank of India. The bases for the money stock measures suggested by the RBI are: (1) the degree of liquidity; (2) general acceptability; (3) the importance accorded to it as a store of value; and (4) the probable shifts between the liabilities of the monetary system and term liabilities.

On the above bases, the RBI suggested in 1977, four measures of money stock — M1, M2, M3 and M4 — as defined below:

M1 consists of:

(i) Currency and coins with the public (excluding cash on hand at all banks);
(ii) Demand deposits, such as bank accounts where immediate withdrawals are possible, at all commercial and cooperative banks (excluding inter-bank deposits, *i.e.* those due to domestic commercial banks); and
(iii) Other deposits, such as fixed or term deposits, held with the Reserve Bank of India (excluding balances in Account No. 1 of the International Monetary Fund, the Reserve Bank of India Employees' Pension, Provident and Guarantee Funds and *ad hoc* liability items which arise from time to time).

M2 consists of:

(i) M1; and (ii) Savings deposits with Post Office Savings Banks.

M3 consists of:

(i) M2; and (ii) Time deposits, such as fixed or term deposits, with all commercial and cooperative banks (excluding inter-bank time deposits).

M4 consists of:

(i) M3; and (ii) Total deposits with the Post Office Savings Organization (excluding National Savings Certificates and Post Office Savings Bank deposits included in M2 − ii).

It is clear from the definitions that M1 forms the most basic category and M4 the most comprehensive category of money supply in India. M1, M2, M3 and M4 were measures redefined by the second Working Group committee on Money Supply which published its report in 1977. Measures of money supply with the public prior to 1977 had excluded demand deposits of central and urban cooperative banks and of salary earners' societies and included inter-bank liabilities of state cooperative banks.

M1, calculated for the period before 1977, increased from 1951 at an average rate of 8.5 per cent, whereas in the case of money supply broadly defined (M4) the rate was in excess of 10 per cent. The share of currency (M1-i) declined from nearly three-fifths in the fifties to about half in the early sixties and about three-tenths in the seventies. The share of demand deposits (M1-ii) in total money in circulation remained fairly stable at about a quarter, except around 1960, while the share of time deposits (M3-ii) increased sharply from less than a fifth to more than two-fifths of the total. It may be noted that these developments happened mostly in the household (family) sector which accounts for the bulk of currency; from 1951 onwards, families were more inclined to deposit money in organizations than keep it as currency.

Monetary expansion during the financial year 1985-86 decelerated rather sharply with the rate of expansion in narrow money (M1) at 9.7 per cent, which was less than half the rate of 19.9 per cent recorded in 1984-85. The increase by Rs 38,270 m in absolute terms is also appreciably smaller than the rise of Rs 65,830 m recorded in 1984-85. In absolute terms, the expansion of Rs 160,030 m was marginally lower than that of Rs 160,580 m witnessed the previous year.

Considered in fortnightly averages, the growth rates showed a similar trend: M1 decelerating from 18.2 per cent in 1984-85 to 14.1 per cent in 1985-86 and M3 from 18.0 per cent to 17.0 per cent for the same period. While the increase in M2 was at Rs 40,410 m (9.5 per cent), lower than that of Rs 67,670 m (18.9 per cent) in the previous year, M4 recorded a higher rise in absolute terms at Rs 175,560 m as against Rs 172,300 m. However, the growth rate was lower by 2.5 per cent, but the increase in reserve money at Rs 63,160 m (20.1 per cent) was more than twice the rise of Rs 26,530 m (9.2 per cent) during the preceding year.

The major factors that bring about changes in money supply are the RBI's net credit to government, bank credit to commercial organizations, net foreign exchange assets of the banking sectors, government's currency, liabilities of the public and net non-monetary liabilities of the banking sector. The higher expansion in M3 during 1985-86 was mainly brought about by net bank credit to government which registered a larger expansion than in the previous year.

Government money has being affected by the following factors:

(a) Reserve Bank's credit to the government sector; (b) Reserve Bank's credit to the private sector; (c) Reserve Bank's credit to the banks; (d) Net change in the foreign exchange assets of the Reserve Bank; (e) Variation in the net monetary liabilities of the Reserve Bank.

Money supply in India tends to vary according to the agricultural seasons. Pre-harvest and harvest periods are busy months and end-April to end-October is the slack season. Seasonal variations in M3, both componentwise and sourcewise, during the last two busy and slack seasons are as follows: The slack season of 1985 witnessed a lower order of monetary expansion than that in the previous slack season; M3 expanded by Rs 68,950 m (6.7 per cent) as compared with Rs 71,280 m (8.2 per cent). A decline in currency with the public in contrast to a sizeable rise in the 1984 slack season brought about a lower expansion in M3. Both aggregate deposits with banks and other deposits with the RBI recorded higher increases. Source-wise, the slack season trend was more or less the same as the financial trend mentioned earlier. During the busy season of 1985-86, the expansion in M3 was higher at Rs 99,130 m as compared with Rs 85,270 m during the 1984-85 busy season. The growth rate, however, remained unchanged at 9 per cent. Of the individual components, time deposits with banks registered a sizeably lower increase than in the 1984-85 busy season. Of the sources of expansion in M3, net bank credit to government and bank credit to the commercial sector showed larger expansion than in the previous busy season; the growth rate in the case of the former, however, was lower. Net foreign exchange assets recorded a lower rise.

By end-March 1986, there were about 53,000 commercial bank offices in the country, a little more than half of them in rural centres (places with a population upto 10,000). The share of rural offices in total deposits and advances at the end of June 1985 was 13.6 per cent and 13 per cent respectively (3.1 per cent and 1.5 per cent in June 1969). Savings deposit accounts continued to claim the highest share (66.7 per cent) in the total *number* of accounts, though term deposits claimed the largest share (56 per cent) in the total *amount* of deposits, a trend discernible in all population groups.

The total deposits of scheduled commercial banks increased by Rs 727,160 m and the bulk of the increase was claimed by five states: Maharashtra, West Bengal, Uttar Pradesh, Tamil Nadu, Gujarat and Delhi, which together accounted for slightly over 60 per cent. Population-groupwise data show that urban and metropolitan centres together accounted for 64 per cent of the total deposits with semi-urban and rural offices claiming 21.7 per cent and 14.3 per cent respectively.

In terms of growth rates, the performance of rural offices was more impressive than that of the other groups. Rural deposits increased phenomenally from Rs 1450 m to Rs 105,100 m. Deposits with semi-urban and urban-metropolitan offices moved up from Rs 10,240 m and Rs 34,960 m to Rs 168,290 m and Rs 500,430 m respectively. Uttar Pradesh claimed the maximum share in the incremental deposits in respect of both rural and semi-urban areas with accretions of Rs 16,420 m and Rs 19,000 m respectively.

The share of other states in incremental rural deposits ranged between Rs 3000 m and Rs 11,000 m. Kerala (633 rural and 1655 semi-urban offices; total 2718 offices) claimed the second place with regard to incremental deposit accretion in semi-urban areas with a share of Rs 18,240 m. This was perhaps due to the large inflow of non-resident deposits. The share of other states in semi-urban incremental deposit accretion ranged between Rs 5000 m and Rs 13,000 m.

Sources: 1. Dr. E.S. Srinivasan, PG Professor and Head, Department of Economics, Madras Christian College.
2. *Report on Currency and Finance 1985-86, Vol I & II*, Reserve Bank of India.

MONEY SUPPLY AND ITS COMPONENTS

Table 104   (in Rs. million)

| | CURRENCY WITH THE PUBLIC | | | | | DEPOSIT MONEY OF THE PUBLIC | | | | | Time deposits with banks | | Total post office deposits | |
	Notes in circulation	Circulation of rupee coins	Circulation of small coins	Cash on hand with banks	Total (1+2 +3−4)	Demand deposits with banks	Other deposits with RBI	M₁ (5+6+7)	Post Office Saving Bank deposits	M₂ (8+9)		M₃ (10+11)		M₄ (12+13)
	1	2	3	4	5	6	7	8	9	10	11	12	13	14
1970-71	41,690	2,470	1,370	1,860	43,670	29,100	440	73,210	9,900	83,110	36,370	109,580	11,840	121,420
1975-76	64,920	3,310	2,240	3,430	67,040	63,850	540	131,430	14,750	146,180	91,430	222,860	31,790	254,650
1980-81	136,890	3,330	2,860	8,440	134,640	93,360	3,170	231,170	22,380	253,550	322,410	553,580	65,120	618,700
1983-84	198,880	3,860	3,330	10,540	195,530	131,950	3,180	330,660	26,480	357,140	528,330	858,990	91,120	950,110
1985-86	255,190	5,030	3,720	12,860	251,080	81,920	2,510	435,510	29,710	465,220	745,590	1,181,100	116,870	1,297,970

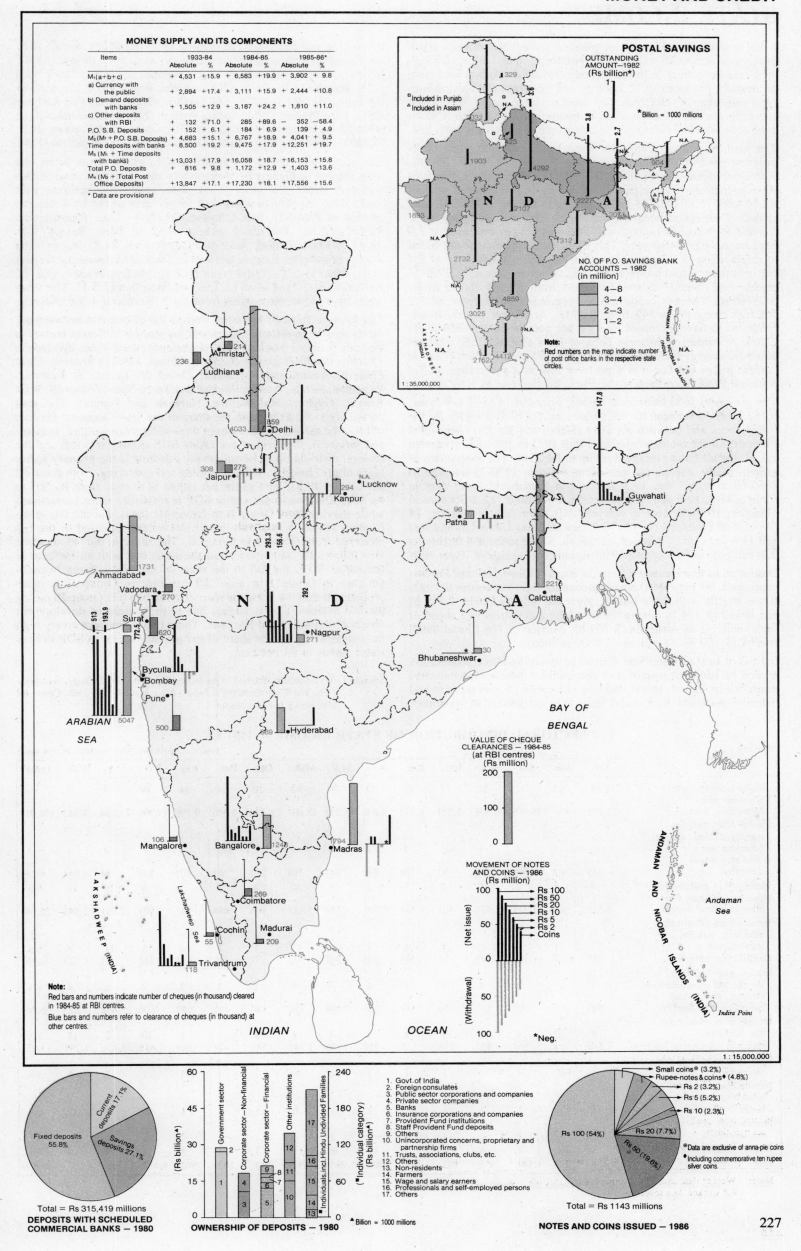

MONEY SUPPLY AND ITS COMPONENTS

Items	1933-84 Absolute	%	1984-85 Absolute	%	1985-86* Absolute	%
M₁(a+b+c)	+ 4,531	+15.9	+ 6,583	+19.9	+ 3,902	+ 9.8
a) Currency with the public	+ 2,894	+17.4	+ 3,111	+15.9	+ 2,444	+10.8
b) Demand deposits with banks	+ 1,505	+12.9	+ 3,187	+24.2	+ 1,810	+11.0
c) Other deposits with RBI	+ 132	+71.0	+ 285	+89.6	− 352	−58.4
P.O. S.B. Deposits	+ 152	+ 6.1	+ 184	+ 6.9	+ 139	+ 4.9
M₂ (M₁ + P.O. S.B. Deposits)	+ 4,683	+15.1	+ 6,767	+18.9	+ 4,041	+ 9.5
Time deposits with banks	+ 8,500	+19.2	+ 9,475	+17.9	+12,251	+19.7
M₃ (M₁ + Time deposits with banks)	+13,031	+17.9	+16,058	+18.7	+16,153	+15.8
Total P.O. Deposits	+ 816	+ 9.8	+ 1,172	+12.9	+ 1,403	+13.6
M₄ (M₃ + Total Post Office Deposits)	+13,847	+17.1	+17,230	+18.1	+17,556	+15.6

* Data are provisional

POSTAL SAVINGS
OUTSTANDING AMOUNT—1982 (Rs billion*)

* Billion = 1000 millions

☐ Included in Punjab
△ Included in Assam

NO. OF P.O. SAVINGS BANK ACCOUNTS — 1982 (in million)

- 4—8
- 3—4
- 2—3
- 1—2
- 0—1

Note:
Red numbers on the map indicate number of post office banks in the respective state circles.

1 : 35,000,000

LAKSHADWEEP (INDIA)

ANDAMAN AND NICOBAR ISLANDS

Note:
Red bars and numbers indicate number of cheques (in thousand) cleared in 1984-85 at RBI centres.

Blue bars and numbers refer to clearance of cheques (in thousand) at other centres.

VALUE OF CHEQUE CLEARANCES — 1984-85 (at RBI centres) (Rs million)

MOVEMENT OF NOTES AND COINS — 1986 (Rs million)

- Rs 100
- Rs 50
- Rs 20
- Rs 10
- Rs 5
- Rs 2
- Coins

(Net issue)
(Withdrawal)

*Neg.

ARABIAN SEA

BAY OF BENGAL

ANDAMAN Sea

LAKSHADWEEP (INDIA)

Lakshadweep Sea

INDIAN OCEAN

ANDAMAN AND NICOBAR ISLANDS (INDIA)

Indira Point

1 : 15,000,000

DEPOSITS WITH SCHEDULED COMMERCIAL BANKS — 1980

Current deposits 17.1%
Fixed deposits 55.8%
Savings deposits 27.1%

Total = Rs 315,419 millions

OWNERSHIP OF DEPOSITS — 1980

(Rs billion▲)
(*Individual category) (Rs billion▲)

1. Govt. of India
2. Foreign consulates
3. Public sector corporations and companies
4. Private sector companies
5. Banks
6. Insurance corporations and companies
7. Provident Fund institutions
8. Staff Provident Fund deposits
9. Others
10. Unincorporated concerns, proprietary and partnership firms
11. Trusts, associations, clubs, etc.
12. Others
13. Non-residents
14. Farmers
15. Wage and salary earners
16. Professionals and self-employed persons
17. Others

▲ Billion = 1000 millions

NOTES AND COINS ISSUED — 1986

- Small coins® (3.2%)
- Rupee-notes & coins◆ (4.8%)
- Rs 2 (3.2%)
- Rs 5 (5.2%)
- Rs 10 (2.3%)
- Rs 20 (7.7%)
- Rs 50 (19.6%)
- Rs 100 (54%)

Total = Rs 1143 millions

®Data are exclusive of anna-pie coins
◆Including commemorative ten rupee silver coins.

227

STATE INCOME

India over the last 35 years of planned growth has made marked progress in almost all sectors of development: in agriculture through the Green Revolution, in industry through sustained efforts at improving production, productivity, technology and modernization; and in both rural and urban development. Stil there are large gaps to be bridged and progress has yet to be made in many fields, and there is much to be done to enable the majority of the people to cross the poverty line.

The national and per capita income of a country are signs of its progress. In 1971 India's national income stood at Rs 342,350 m, with a per capita income of Rs 633. Despite the nullifying effects of an ever-increasing population, national income reached a total of Rs 504,860 m in 1982-83 and a per capita income of Rs 712. The growth of the real national income during 1983-84, according to the Central Statistical Organization (CSO) was 7.6 per cent against 1.6 per cent in the previous year. This significant rise could be attributed to the increase of 17 per cent in the foodgrains output and of 5.4 per cent in industrial production. In actual terms, that is, at 1970-71 prices, real national income increased from Rs.504,860 m to Rs 542,760 m. The per capita income registered an increase of 5.2 per cent rising to Rs 749 from Rs 712. At current prices, it was estimated to have increased by 17.8 per cent (from Rs 1868 to Rs 2201). According to Reserve Bank of India estimates, the total net domestic savings increased from Rs 258,750 m to Rs 304,770 m at current prices but this was 4 per cent less as a percentage of Net National Product, reflecting the much sharper rise in NNP.

The per capita state incomes at current prices in 1970-71 (all-India: Rs 633) varied between Rs 402 (Bihar) and Rs 1070 (Punjab) for the major states and between Rs 1199 (Delhi) and Rs 350 (Arunachal Pradesh) for the union territoties. From 1971 to 1984-85, at current prices, the all-India average went up to Rs 2344. The annual rate of growth of the per capita state income (from 1970-73 average to 1980-83 average) was 10.3 per cent in Punjab, 11.7 per cent in Maharashtra, 10.5 per cent in Jammu & Kashmir, 12.3 per cent in Manipur, 10.6 per cent in Haryana, 10.7 per cent in Gujarat, 14 per cent in Pondicherry, 11.3 per cent in Goa, Daman & and Diu and 14.6 per cent in Arunachal Pradesh. These states and territories thus had growth rates of more than the national average of 10 per cent.

Despite all its disadvantages, India, in terms of Gross National Product (aggregate, not per capita) ranked among the top 15 countries in 1982. It was twelfth with $ 186 billion, preceded only by China ($ 313 b) and Brazil ($ 284 b) among the developing countries and ahead of such countries as Australia ($ 169 b), Mexico ($ 166 b) and Saudi Arabia ($ 160 b). (1 billion = 1000 million).

Indices of state development arrived at by taking into consideration several variables for each of such development indicators as industry, transport, irrigation, power, banking, education, health services and irrigation pumpsets, have enable states to be categorized as developed,

developing and backward. According to these indices, Kerala is by far the most developed state in the country, all-around. Its index of 807 is far ahead of the others. The other developed states are: Tamil Nadu (423), Haryana (340), Punjab (306) and West Bengal (305). The developing states are: Bihar (283), Maharashtra (266), Jammu & Kashmir (256), Uttar Pradesh (250), Andhra Pradesh (246) and Karnataka (241); while the backward states are: Orissa (203), Gujarat (198), Assam (193), Himachal Pradesh (189), Madhya Pradesh (148) and Rajasthan (134).

But income-wise the pattern is different. In actual money terms, the state income is highest in Maharashtra among the 15 major states included under the analysis of state level development, with Rs 157,400 m. (Maharashtra had 15 per cent of the total national income in 1981-82). It is followed closely by Uttar Pradesh with Rs 147,560 m. The third richest state is West Bengal with Rs 87,950 m and then Andhra Pradesh with Rs 83,360 m. The income of the other states is less than half that of Maharashtra: Gujarat (1:2), Bihar (1:2.2), Tamil Nadu (1:2.4), Madhya Pradesh (1:2.4), Karnataka (1:2.7), Punjab (1:2.9) and Rajasthan (1:3.1). The other states have an income ranging from 1:4.2 (Kerala) to 1:5.6 (Assam).

The Reserve Bank of India has studied the association between per capita income in different states and the share of different sectors in the state domestic product. For convenience, the RBI has divided the 15 states into three groups *(Group I:* Orissa, Madhya Pradesh, Uttar Pradesh, Assam and Karnataka; *Group II:* Jammu & Kashmir, Rajasthan, Kerala, Andhra Pradesh and Tamil Nadu; *Group III:* West Bengal, Gujarat, Maharashtra, Haryana and Punjab). The data pertaining to the average of the share of the State Domestic Product (all income earned by the state) in three sectors (agriculture, industry and services), the level of per capita SDP at current prices and the sectoral distribution of labour reveal a decline in the primary sector (agriculture, forestry, fishing, mining and quarrying) from about 62 per cent to 49 per cent as the per capita SDP rises from Rs 505 to Rs 801. The rise in per capita SDP is relatively more pronounced while moving from Group II to Group III states, but the fall in the primary sector is relatively much less when compared to the fall observed from Group I to Group II. The share in the labour force also follows the same pattern in the sense that with an increase in per capita SDP, the fall in the share of the. labour force from 78 per cent in Group I to about 69 per cent in Group II is more pronounced than the fall in the share from Group II to Group III during the last decade. This is largely due to the pattern of development observed in Punjab and Haryana. These two states have a very high per capita SDP, but the share of the primary sector in SDP in both states stands at 63 per cent.

Sources : *Basic Statistics Relating to the Indian Economy, Vol. 2; States, September 1986,* and *World Economy & India's Place in it, October 1986,* Centre for Monitoring Indian Economy.

SECTORAL DISTRIBUTION OF STATE INCOME — 1981-82

Table 105

(Income in Rs. million, workers in thousand)

	A.P.	Ass.	Bih.	Guj.	Har.	Kar.	Ker.	M.P.	Mah.	Ori.	Pun.	Raj.	T.N.	U.P.	W.B.	India[1]
Income from agriculture	42	13	32	24	17	26	13	30	42	20	24	26	19	64	27	N.A.
Cultivators and agricultural workers	15,733	N.A.	16,409	6,602	2,227	8,877	2,806	15,272	15,007	6,450	2,860	7,196	11,596	24,134	8,483	148,018
Income from forestry, logging, fishing, mining, quarrying, etc.	3	4	7	3	Neg.	1	2	6	3	2	Neg.	1	1	3	4	N.A.
Allied agricultural activities	635	N.A.	502	280	33	594	689	541	595	273	61	407	529	184	654	6,256
Income from manufacturing	9	4	7	16	5	9	7	8	40	4	7	5	8	19	18	N.A.
Household industry and other industrial workers	2,322	N.A.	1,311	1,677	461	1,625	1,091	1,589	3,317	604	684	941	2,901	2,941	2,545	25,143
Income from construction, electricity, gas and water supply	5	1	5	5	2	6	2	5	11	1	5	4	5	11	6	N.A.
Construction workers	332	N.A.	156	164	85	245	205	283	468	79	125	191	312	345	221	3,565
Income from transport, storage and communication	4	1	4	4	1	2	3	3	9	1	2	2	4	6	6	N.A.
Transport, storage and communication workers	594	N.A.	367	415	110	337	337	329	788	140	185	257	600	638	625	6,069
Income from trade, hotels and restaurants	9	2	8	13	5	6	4	5	4	3	9	6	10	2	11	N.A.
Trade and commercial workers	1,340	N.A.	868	850	286	871	759	785	1,742	346	433	548	1,615	1,436	1,291	13,929
Income from banking and insurance	2	Neg.	1	3	1	2	1	1	10	1	1	1	2	3	4	N.A.
Income from real estate, household ownership and business service	2	Neg.	1	2	1	1	1	1	4	1	1	1	2	6	3	N.A.
Income from public administration	4	1	2	2	1	2	1	2	4	1	1	2	3	4	3	N.A.
Income from other services	4	2	4	4	2	3	4	4	10	2	4	3	3	9	6	N.A.
Other workers	1,673	N.A.	1,140	993	462	1,101	905	1,243	2,392	743	582	902	1,472	2,718	1,607	19,530

Note: [1] Worker data excludes Assam, but includes other states and union territories. All worker data relates to 1981.

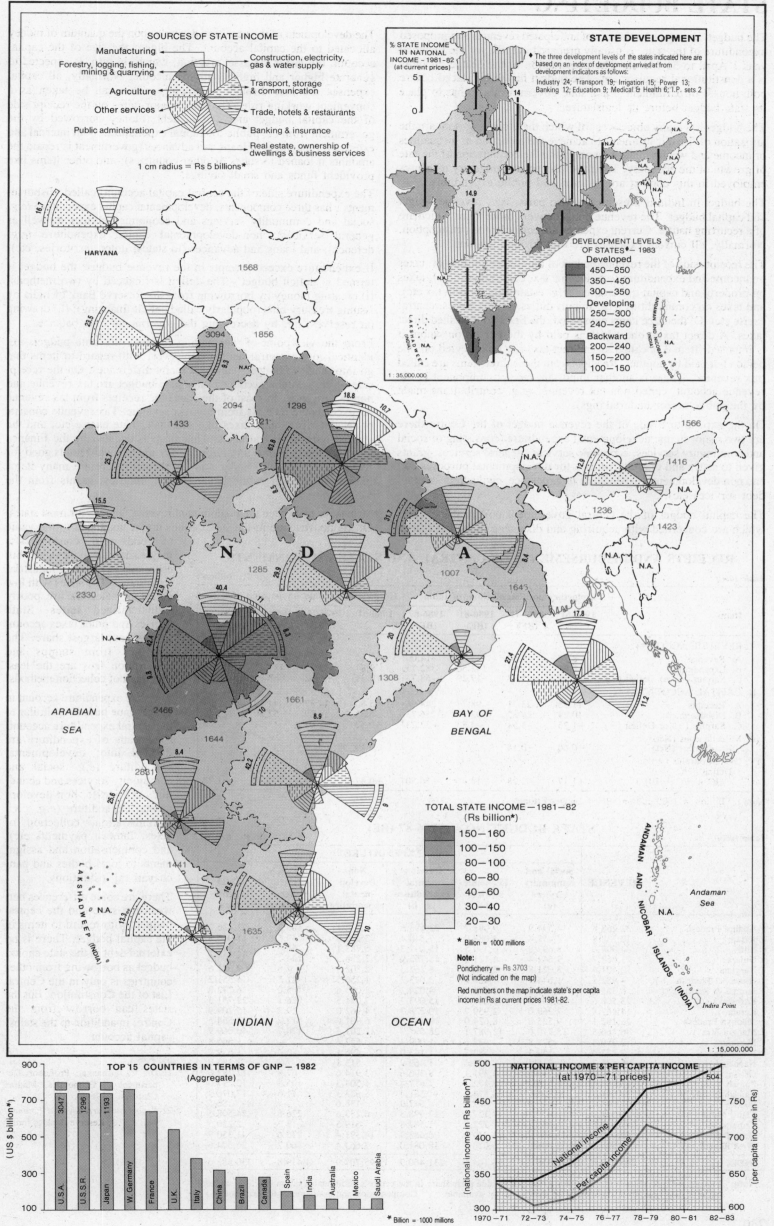

SOURCES OF STATE INCOME

Manufacturing

Forestry, logging, fishing, mining & quarrying

Agriculture

Other services

Public administration

Construction, electricity, gas & water supply

Transport, storage & communication

Trade, hotels & restaurants

Banking & insurance

Real estate, ownership of dwellings & business services

1 cm radius = Rs 5 billions*

STATE DEVELOPMENT

◆ The three development levels of the states indicated here are based on an index of development arrived at from development indicators as follows:

Industry 24; Transport 19; Irrigation 15; Power 13; Banking 12; Education 9; Medical & Health 6; I.P. sets 2

% STATE INCOME IN NATIONAL INCOME — 1981–82 (at current prices)

DEVELOPMENT LEVELS OF STATES* — 1983

Developed
450–850
350–450
300–350

Developing
250–300
240–250

Backward
200–240
150–200
100–150

1 : 35,000,000

HARYANA

ARABIAN SEA

INDIA

BAY OF BENGAL

LAKSHADWEEP (INDIA)

ANDAMAN AND NICOBAR ISLANDS (INDIA)

Andaman Sea

Indira Point

INDIAN   OCEAN

TOTAL STATE INCOME — 1981–82
(Rs billion*)

150–160
100–150
80–100
60–80
40–60
30–40
20–30

* Billion = 1000 millions

Note:
Pondicherry = Rs 3703
(Not indicated on the map)

Red numbers on the map indicate state's per capita income in Rs at current prices 1981-82.

1 : 15,000,000

TOP 15 COUNTRIES IN TERMS OF GNP — 1982
(Aggregate)

(US $ billion*)

3047 U.S.A.
1296 U.S.S.R.
1193 Japan
W. Germany
France
U.K.
Italy
China
Brazil
Canada
Spain
India
Australia
Mexico
Saudi Arabia

* Billion = 1000 millions

NATIONAL INCOME & PER CAPITA INCOME
(at 1970–71 prices)

(national income in Rs billion*)

(per capita income in Rs)

National income

Per capita income

504

1970–71   72–73   74–75   76–77   78–79   80–81   82–83

* Billion = 1000 millions

STATE BUDGETING

The budget, setting out estimates of anticipated revenues and proposed expenditure of the state, is usually prepared in India for the financial year 1 April to 31 March. The presentation of the budget in India is a constitutional requirement. The budget has to be placed before both houses of parliament. Each state government also has to place its state budget before its legislature.

The budget has many objectives, of which the most important are the utilization of fiscal instruments to achieve the reduction of inequalities of income and wealth in the economy, and the acceleration of the rate of growth of the economy. The most prominent budgetary measures employed in this context are taxation and public expenditure.

The budget in India is presented in two parts, viz., revenue budget and capital budget. The revenue budget covers those items which are of a recurring nature. Current expenses are equivalent to consumption. Normally, all current expenses are met through taxation.

The receipts side of the revenue budget of the Centre consists of: taxes on income and expenditure (e.g. income tax, corporate tax etc); taxes on property and capital transactions (e.g. estate duty, wealth tax etc); and taxes on commodities (e.g. import duties, export duties, central excise etc). Of the three items mentioned, the first two are called direct taxes. A direct tax is one which is paid by the person on whom it is imposed. Item 3 is called an indirect tax, i.e. a tax levied on one person but paid by another. Receipts from these three items are called 'tax revenue'. There is another important receipt component in the revenue account, called non-tax revenue, (e.g. contributions made by the public sector undertakings).

On the expenditure side of the revenue budget of the Centre there are two major items: developmental expenditure (consisting of social and community services, economic services, general services, grants given to states and union territories for developmental purposes etc) and non-developmental expenditure (e.g. defence, civil administration, debt services etc).

The capital budget of the central government covers those items which are concerned with acquiring and disposing of capital assets.

The development of the economy depends upon the quantum of money allocated to the capital account. The bigger the size of the capital account, the faster the asset creation, which, in turn, is expected to generate higher and higher national income. Normally, all capital expenses are met through borrowings. This can be taken as a convenient working rule. The three major items on the receipt side of the capital budget are: Public debt (money borrowed by the government from the public for capital expenditure, both internal and external); repayment of loans and advances (government is repaid the amounts it loaned to states or other countries); and other items like provident funds and small savings.

The expenditure side of the central capital account, called 'disbursements', has three components: developmental capital expenditure (e.g. social and community services and economic services as well as general services); non-developmental capital expenditure (e.g. defence); and loans and advances (to states, union territories, etc).

If expenditure exceed receipts in the revenue budget, the budget is termed a 'deficit budget'. The deficit is financed by two methods: (i) creating money by borrowing from the Reserve Bank of India by issuing treasury bills (popularly called deficit financing); (ii) drawing on reserves, i.e. by decreasing the government cash balances.

From the viewpoint of 'accounting format', the state budgets are identical to the central budget. However, with regard to items that go into the state budget, there are many differences. On the receipt side of the revenue account in the state budget are tax revenue and non-tax revenue. In most of the states the receipts from tax revenue far exceeds the receipts from non-tax revenue. Tax revenue consists of receipts from state taxes like sales tax, state excise etc, and the state's share in central taxes. This share is decided by the Finance Commission (which is a statutory body) and its award holds good for five years. Similarly, under non-tax revenue there are many items like receipts from public undertakings, lotteries, grants from the Centre etc.

State taxes have been the predominant revenue sources for most states. Of the tax revenue in the states, sales tax, a tax added to the value of goods sold, contributes a major share in most states, but especially in the more industrialized states. The states with low shares of sales tax are poorly industrialized states. State excises and other taxes account for the next largest share. The revenues from stamps and registration fees are the least important of collection methods.

The state expenditure account in the revenue budget is similar to the central expenditure account. The heads of expenditure are divided into: developmental expenditure (e.g. social and community services and economic services); non-developmental expenditure (e.g. civil administration, collection of taxes, interest payments etc); and compensation and assignments to local bodies and panchayati raj institutions.

There are some differences between the state and the central budgets with regard to items in the capital budget. There is no external debt in the state capital budget as borrowing from other countries is only in the Central List of the Constitution. But the states can borrow from the Centre, in addition to the states' capital account.

RECEIPTS AND DISBURSEMENTS OF CENTRAL AND STATE GOVERNMENTS

Table 106(i)   (in Rs. billion[1])

Items	Budgetary position of Govt. of India				Consolidated budgetary position of states			
	1960-61	1970-71 (BE)	1980-81 (BE)	1986-87 (BE)	1960-61	1970-71 (BE)	1980-81 (BE)	1986-87 (BE)
I. REVENUE ACCOUNT:								
A. Revenue	8.77	32.40	123.27	314.00	10.11	33.67	146.05	366.57
B. Expenditure	8.26	31.03	142.56	382.73	9.87	33.65	137.21	360.40
C. Surplus (+) or Deficit (−)	+0.51	+1.37	−19.29	−68.73	+0.24	+0.02	+8.84	+6.17
II. CAPITAL ACCOUNT:								
A. Receipts	11.55	23.31	98.77	248.96	5.80	14.87	44.19	117.54
B. Disbursements	10.29	26.82	93.94	216.73	6.32	16.89	57.80	125.44
C. Surplus (+) or Deficit (−)	+1.26	−3.51	+4.83	+32.23	−0.52	−2.02	−13.61	−7.90
III. Miscellaneous (Net)/ Remittances (Net)	−0.60	−0.14	—	—	−0.19	+0.03	—	—
IV. Overall Surplus (+) or Deficit (−) (IC + IIC + III)	+1.17	−2.28	−14.45	−36.50	−0.47	−1.97	−4.77	−1.73

Note : [1] Billion = 1000 million BE = Budget Estimate

STATE BUDGETING — 1986-87 (BE)

Table 106(ii)   (in Rs. million)

	REVENUE	EXPENDITURE					
		Social and community services (a)	Economic services (b)	Developmental expenditure (a+b)	Non-Develop mental expenditure	Other expenditure[1]	Total expenditure
Andhra Pradesh	32,268.8	14,594.9	9,098.7	23,693.6	8,272.1	460.5	32,426.2
Assam	10,738.8	4,452.4	3,088.0	7,540.4	3,420.8	49.6	11,010.8
Bihar	24,004.6	8,697.8	6,399.5	15,097.3	6,783.1	66.6	21,947.0
Gujarat	21,662.7	9,642.4	6,121.5	15,763.9	7,298.5	144.2	23,206.6
Haryana	10,401.4	3,021.8	3,242.5	6,264.3	2,802.5	4.8	9,071.6
Himachal Pradesh	4,982.2	1,644.4	1,428.8	3,073.2	1,256.5	21.5	4,351.2
Jammu & Kashmir	6,363.7	1,809.8	2,214.4	4,024.2	2,083.6	165.0	6,272.8
Karnataka	23,508.6	8,891.9	6,201.8	15,093.7	7,031.5	626.1	22,751.3
Kerala	14,467.8	7,269.0	2,959.7	10,228.7	4,962.0	19.2	15,209.9
Madhya Pradesh	26,182.1	9,525.8	8,624.0	18,149.8	6,901.0	513.7	25,564.5
Maharashtra	46,166.5	15,710.6	12,987.1	28,697.7	17,201.5	170.7	46,069.9
Manipur	2,359.4	571.8	437.6	1,009.4	557.4	—	1,566.8
Meghalaya	2,177.4	559.3	583.7	1,143.0	496.1	—	1,639.1
Nagaland	2,836.7	715.3	754.6	1,469.9	813.1	—	2,283.0
Orissa	12,462.0	4,915.5	3,391.1	8,306.6	3,914.6	67.8	12,283.0
Punjab	13,183.4	4,515.2	3,172.2	7,687.4	4,504.3	105.3	12,297.0
Rajasthan	16,591.0	6,806.5	4,574.9	11,381.4	5,623.3	71.6	17,076.3
Sikkim	1,044.0	240.8	407.1	647.9	138.6	—	786.5
Tamil Nadu	26,541.1	10,870.7	7,127.6	17,998.3	6,233.6	276.6	24,508.5
Tripura	2,629.6	865.1	722.7	1,587.8	619.5	13.2	2,220.5
Uttar Pradesh	40,031.6	13,464.9	13,219.0	26,683.9	14,391.4	275.6	41,350.9
West Bengal	25,540.6	12,008.8	6,072.2	18,081.0	7,663.4	760.2	26,504.6
INDIA	366,574.0	135,356.9	96,104.0	231,460.9	95,707.9	3,419.8	330,588.6

Note: Figures for individual states for 1986-87 do not include their share in the centre's additional taxation (Rs. 430 million) because data relating to the state-wise shares are not available.   [1] Compensation and assignments to local bodies.

Sources :
1. Dr. V. Jaishankar, Professor, Department of Economics, Madras Christian College.
2. *Report on Currency and Finance,* Vol. 1 & 2, Reserve Bank of India.

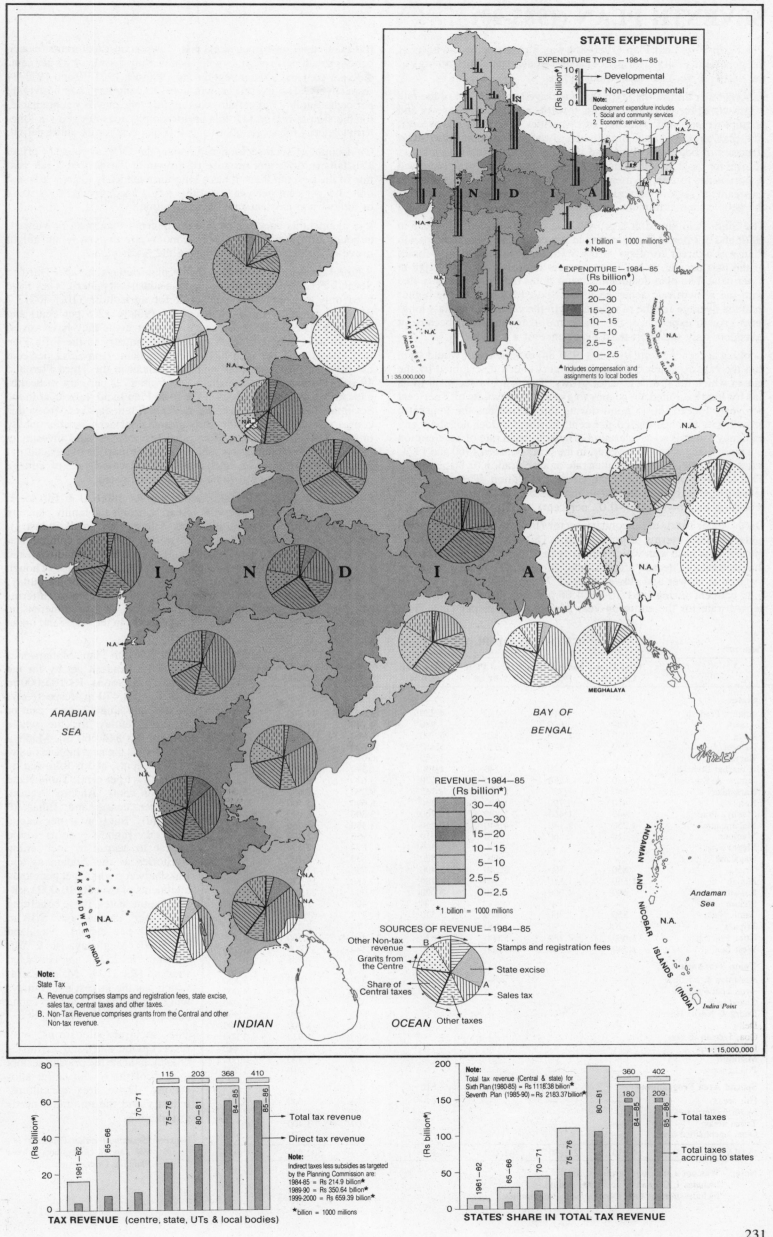

STATE EXPENDITURE

EXPENDITURE TYPES — 1984—85
(Rs billion♦)

→ Developmental
→ Non-developmental

Note:
Development expenditure includes
1. Social and community services
2. Economic services.

♦1 billion = 1000 millions
★ Neg.

▲EXPENDITURE — 1984—85
(Rs billion♦)

30—40
20—30
15—20
10—15
5—10
2.5—5
0—2.5

▲ Includes compensation and
assignments to local bodies

1 : 35,000,000

ARABIAN
SEA

BAY OF
BENGAL

MEGHALAYA

REVENUE—1984—85
(Rs billion*)

30—40
20—30
15—20
10—15
5—10
2.5—5
0—2.5

*1 billion = 1000 millions

SOURCES OF REVENUE — 1984—85

Other Non-tax
revenue
Grants from
the Centre
Share of
Central taxes
Stamps and registration fees
State excise
Sales tax
Other taxes

ANDAMAN
AND
NICOBAR
ISLANDS
(INDIA)

Andaman
Sea

N.A.

Indira Point

1 : 15,000,000

LAKSHADWEEP (INDIA)

N.A.

Note:
State Tax
A. Revenue comprises stamps and registration fees, state excise,
     sales tax, central taxes and other taxes.
B. Non-Tax Revenue comprises grants from the Central and other
     Non-tax revenue.

INDIAN   OCEAN

TAX REVENUE (centre, state, UTs & local bodies)

80
60
40
20
(Rs billion*)

1961—62   65—66   70—71   75—76   80—81   84—85   85—86
                              115   203   368   410

→ Total tax revenue
→ Direct tax revenue

Note:
Indirect taxes less subsidies as targeted
by the Planning Commission are:
1984-85 = Rs 214.9 billion*
1989-90 = Rs 350.64 billion*
1999-2000 = Rs 659.39 billion*

*billion = 1000 millions

STATES' SHARE IN TOTAL TAX REVENUE

200
150
100
50
(Rs billion*)

Note:
Total tax revenue (Central & state) for
Sixth Plan (1980-85) = Rs 1118.38 billion*
Seventh Plan (1985-90) = Rs 2183.37 billion*

1961—62   65—66   70—71   75—76   80—81   84—85   85—86
                                          360   402
                                          180   209

→ Total taxes
→ Total taxes
  accruing to states

SEVENTH PLAN (1985-90)

The Sixth Five Year Plan (1980-85) was formulated on the basis of the general overall performance, achievements and shortcomings of the earlier Five Year and Annual Plans.

Among other things, the Sixth Plan envisaged acceleration of the rate of growth of the economy, efforts towards reduction of poverty and unemployment, and reduction of disparities in income and wealth. The plan strategy aimed simultaneously at strengthening the infrastructure of both agriculture and industry, and at achieving a better climate for rapid growth of investments, output and exports. Increased opportunities for employment, especially in the rural areas and in the unorganized sector, were also aimed at with a view to securing for the people their minimum needs.

The Sixth Plan aimed at a growth rate of 5.2 per cent per year in GDP and 3.3 per cent per annum in per capita income. The Seventh Plan was a direct corollary of the Sixth Plan, in fact of the policies of the past decade. It has fixed a target rate of growth for GDP at 5 per cent. The plan document further notes that 'this rate is in line with the growth rate achieved in the Sixth Plan and a little higher than the average for the past decade.' In the context of India's long-term growth experience, this has come in for criticism because it considers only the short-term experience of a decade.

Looking at the long-term growth (1950-84) experience, it would seem that the rate at which GDP grows is very much determined by the rate at which the primary sector grows. The primary sector in India has for long exhibited a tendency to grow at the low rate of 2 per cent per year. Indeed, at no point during this period has the long-term growth rate reached even 3 per cent. Further, it does not show any tendency towards an acceleration. The long-term rate of acceleration is a poor 0.02 per cent. However, in the years between 1960 and 1971, it has experienced a fairly higher rate of acceleration (0.86 per cent). This is partly on account of the Green Revolution. Throughout the seventies and early eighties, the primary sector has experienced a mild recession in growth rate (-0.02 per cent).

On the other hand, the secondary sector (Industry) has shown much better results in terms of growth rates. This is partly on account of the very low base with which India began. But analysis also indicates that this sector, decade after decade, showed a consistent tendency to grow at slower and slower rates: 5.71 per cent during 1950-60, 5.25 per cent during 1961-70 and 4.06 per cent during 1971-84. The growth rate for the entire 34-year period was 5.06 per cent.

It is the tertiary sector (services) that shows a consistent trend towards better growth performances with each passing decade: 4.24 per cent, 4.33 per cent and 5.6 per cent during 1950-60, 1961-70 and 1971-84 respectively. The overall growth over the long term has been 4.82 per cent. But it is not certain whether this tendency is maintainable, for the simple reason that this sector simply does not show a long-term tendency for acceleration: only 0.06 per cent for the entire period.

On account of such sectoral behaviour, the GDP at constant prices also fails to offer any promise of growth at the desired 5 per cent rate of the Seventh Plan. It has a long-term tendency to grow at about only 3.5 per cent per year, and this too is accompanied by a weak or almost non-existent rate of acceleration.

It is against this background that the Seventh Plan must be viewed, although efforts to improve the economy are underway and might prove as successful as those during the Sixth Plan.

Sectoral allocations of public-sector plan outlays for the Third to Seventh Five Year Plans do not show a consistent pattern. They have been only comparatively consistent for agriculture (10.7 to 13.7 per cent), irrigation and flood control (9.4 to 12.6 per cent) and transport (15.9 to 20.6 per cent). In other areas they have varied widely: from a peak of 24.5 per cent in industry in the Fifth Plan to a low of 12.5 per cent in the Seventh Plan; from 27.2 per cent for power in the Sixth Plan to 12.6 per cent in the Third Plan and 19 per cent in the Seventh Plan; and from 17.2 per cent in health, education and social services in the Fifth Plan to 30.4 per cent in the Seventh Plan. The aim of the Seventh Plan appears, from this comparison, to be to provide, through social services, a better quality of life. Since the allocation for social services includes investment in village and small industry and since the proportion of agricultural investment is to remain unchanged, an increased rate of cottage industrialization of rural India may be anticipated.

In the Seventh Plan, the allocations are Rs 380,620 m for social services (education and scientific research, health and family planning and other social services including village and small industry), Rs 342,730 m in power, Rs 454,320 m in transport and communications, Rs 227,930 m in agriculture, Rs 224,610 m in industry and minerals and Rs 169,790 m in irrigation and flood control. Though the allocations have varied considerably in their share of the outlay, there have been consistently increasing outlays throughout in terms of rupees. Correspondingly, per capita outlays have been increasing throughout, in the states and union territories.

In the Seventh Plan, Maharashtra and Uttar Pradesh get by far the largest allocations, Rs 105,000 m and Rs 104,470 m respectively, each comprising 12.4 per cent of the total outlay for the whole country (Rs 844,660 m). Madhya Pradesh has the next highest outlay (8.3 per cent), and is followed by Gujarat (7.1 per cent), Tamil Nadu (6.8 per cent), Andhra Pradesh (6.2 per cent), and Bihar (6 per cent). But, on a per capita basis, the highest benefits would appear to accrue to such union territories as the Andamans and Lakshadweep which get per capita allotments of over Rs 10,000 each. Among the states, those benefiting most per capita are Sikkim (Rs 5750 per capita), Nagaland (Rs 4000), Meghalaya (Rs 2750), Manipur (Rs 2529), and Himachal Pradesh (Rs 2100). Maharashtra, which has the largest plan outlay, will be spending Rs 1434 per capita and Uttar Pradesh only Rs 803.

The sectoral allocations with respect to the states and territories broadly indicate the priority sectors in each of those with larger allocations. These are: agriculture, industry and the minimum needs programmes.

Sources: *Seventh Five Year Plan 1985-90, Vol. I and II*, Government of India Planning Commission.

GROWTH OF PLAN OUTLAYS

Table 107

(in Rs. million)

	First Plan 1951-56	Second Plan 1956-61	Third Plan 1961-66	Fourth Plan 1969-74	Fifth Plan 1974-79	Sixth Plan 1980-85	Seventh Plan 1985-90
States							
Andhra Pradesh	1,070	1,810	3,450	4,260	13,340	31,000	52,000
Assam	280	630	1,320	1,980	4,740	11,150	21,000
Bihar	1,020	1,770	3,320	4,790	12,960	32,250	51,000
Gujarat	990	1,470	2,380	5,450	11,670	36,800	60,000
Haryana	*	*	*	3,580	6,010	18,000	29,000
Himachal Pradesh	50	170	340	1,130	2,390	5,600	10,500
Jammu & Kashmir	130	270	610	1,620	3,620	9,000	14,000
Karnataka	940	1,390	2,510	3,740	9,980	22,650	35,000
Kerala	440	790	1,820	3,330	5,690	15,500	21,000
Madhya Pradesh	940	1,460	2,880	4,760	13,800	38,000	70,000
Maharashtra	1,250	2,140	4,340	10,050	23,480	61,750	105,000
Manipur	10	60	130	310	930	2,400	4,300
Meghalaya	*	*	*	360	890	2,350	4,400
Nagaland	*	*	110	380	830	2,100	4,000
Orissa	850	890	2,240	2,490	5,850	15,000	27,000
Punjab	1,630	1,510	2,540	4,280	10,130	19,570	32,850
Rajasthan	660	1,000	2,110	3,090	7,090	20,250	30,000
Sikkim	*	*	*	*	400	1,220	2,300
Tamil Nadu	850	1,860	3,420	5,520	11,220	31,500	57,500
Tripura	20	90	160	350	70	2,450	4,400
Uttar Pradesh	1,660	2,280	5,600	11,630	24,460	58,500	104,470
West Bengal	1,540	1,560	3,000	3,640	13,340	35,000	41,250
Union Territories				4,250			
Andaman & Nicobar	—	—	—	—	340	970	2,850
Arunachal Pradesh	—	—	—	—	300	2,120	4,000
Chandigarh	—	—	—	—	400	1,010	2,030
Dadra & Nagar Haveli	—	—	—	—	90	230	460
Delhi	—	—	—	—	3,160	8,000	20,000
Goa, Daman & Diu	—	—	—	—	850	1,920	3,600
Lakshadweep	—	—	—	—	60	200	440
Mizoram	—	—	—	—	470	1,300	2,600
Pondicherry	—	—	—	—	340	710	1,700
Special Area Programmes				7,810			
Hill areas	—	—	—	—	1,700	5,600	8,700
North-Eastern areas	—	—	—	—	900	4,700	7,560
Tribal areas	—	—	—	—	1,900	3,400	6,750 [1]
Other unclassified areas	—	—	—	—	—	300	2,580 [2]
INDIA	14,330	21,150	42,270	88,800	193,160	502,500	872,950

Note: * Was not a state during this period.
[1] Includes LIC Loan of Rs. 1000 million.
[2] Includes Border Area Development Programmes.

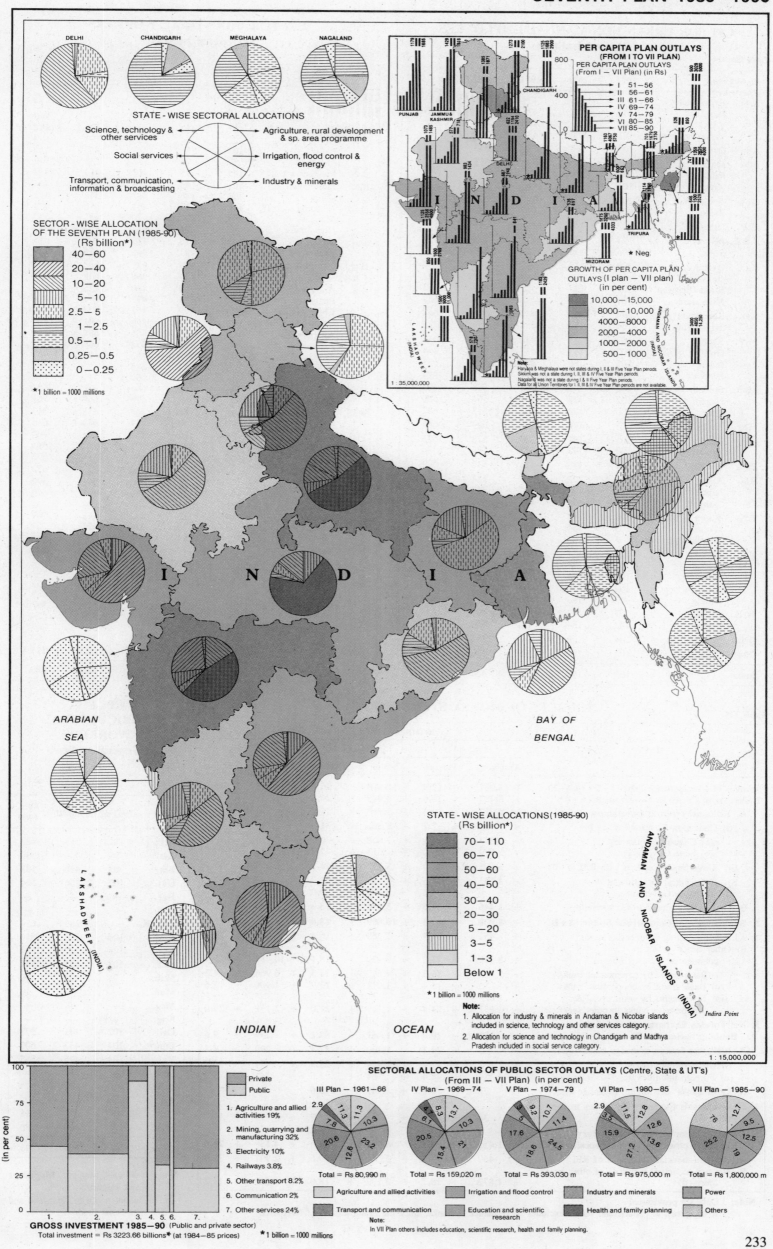

PER CAPITA PLAN OUTLAYS
(FROM I TO VII PLAN)

PER CAPITA PLAN OUTLAYS
(From I — VII Plan) (in Rs)

I	51—56
II	56—61
III	61—66
IV	69—74
V	74—79
VI	80—85
VII	85—90

STATE - WISE SECTORAL ALLOCATIONS

Science, technology & other services

Agriculture, rural development & sp. area programme

Social services

Irrigation, flood control & energy

Transport, communication, information & broadcasting

Industry & minerals

SECTOR - WISE ALLOCATION
OF THE SEVENTH PLAN (1985-90)
(Rs billion*)

40—60
20—40
10—20
5—10
2.5—5
1—2.5
0.5—1
0.25—0.5
0—0.25

*1 billion = 1000 millions

GROWTH OF PER CAPITA PLAN
OUTLAYS (I plan — VII plan)
(in per cent)

10,000—15,000
8000—10,000
4000—8000
2000—4000
1000—2000
500—1000

Note:
Haryana & Meghalaya were not states during I, II & III Five Year Plan periods.
Sikkim was not a state during I, II, III & IV Five Year Plan periods.
Nagaland was not a state during I & II Five Year Plan periods.
Data for all Union Territories for I, II, III & IV Five Year Plan periods are not available.

ARABIAN
SEA

BAY OF
BENGAL

LAKSHADWEEP (INDIA)

STATE - WISE ALLOCATIONS (1985-90)
(Rs billion*)

70—110
60—70
50—60
40—50
30—40
20—30
5—20
3—5
1—3
Below 1

*1 billion = 1000 millions

ANDAMAN AND NICOBAR ISLANDS (INDIA)

Indira Point

Note:
1. Allocation for industry & minerals in Andaman & Nicobar islands included in science, technology and other services category.
2. Allocation for science and technology in Chandigarh and Madhya Pradesh included in social service category.

INDIAN   OCEAN

1 : 15,000,000

1 : 35,000,000

GROSS INVESTMENT 1985—90 (Public and private sector)
Total investment = Rs 3223.66 billions* (at 1984—85 prices)   *1 billion = 1000 millions

SECTORAL ALLOCATIONS OF PUBLIC SECTOR OUTLAYS (Centre, State & UT's)
(From III — VII Plan) (in per cent)

Private
Public

1. Agriculture and allied activities 19%
2. Mining, quarrying and manufacturing 32%
3. Electricity 10%
4. Railways 3.8%
5. Other transport 8.2%
6. Communication 2%
7. Other services 24%

III Plan — 1961—66   Total = Rs 80,990 m
IV Plan — 1969—74   Total = Rs 159,020 m
V Plan — 1974—79   Total = Rs 393,030 m
VI Plan — 1980—85   Total = Rs 975,000 m
VII Plan — 1985—90   Total = Rs 1,800,000 m

Agriculture and allied activities
Transport and communication
Irrigation and flood control
Education and scientific research
Industry and minerals
Health and family planning
Power
Others

Note:
In VII Plan others includes education, scientific research, health and family planning.

CPI FOR URBAN NON-MANUAL EMPLOYEES
(General Indices)

Table XI
(Base 1960 = 100)

State/Centre		1961	1971	1981	1985	Feb. 1987
A.P.	Hyderabad	103	178	429	598	653
	Kurnool	102	192	421	604	665
	Vijayawada	102	181	425	620	679
	Vishakhapatnam	102	180	381	543	564
Ass.	Guwahati	105	178	380	543	591
Bih.	Muzaffarpur	101	186	428	622	672
	Patna	101	190	388	622	657
	Ranchi	104	182	416	608	654
Guj.	Ahmadabad	102	171	389	565	609
	Rajkot	104	173	415	582	609
H.P.	Shimla	103	177	376	513	554
J.&K.	Jammu	104	174	415	569	619
	Srinagar	105	190	437	605	661
Kar.	Bangalore	105	178	417	597	657
	Gulbarga	104	205	462	645	691
	Hubli-Dharwad	102	177	421	606	653
	Mangalore	104	178	423	621	659
Ker.	Kozhikode	104	185	442	593	667
	Trivandrum	104	181	458	639	725
M.P.	Bhopal	107	185	432	620	652
	Gwalior	107	193	437	611	644
	Indore	105	186	406	562	639
	Jabalpur	105	185	414	610	656
Mah.	Bombay	102	171	381	568	623
	Nagpur	102	172	391	564	590
	Pune	100	172	402	577	632
Meg.	Shillong	103	174	388	609	662
Ori.	Cuttack	105	180	387	578	625
	Sambalpur	103	181	382	567	594
Pun.	Amritsar	104	193	414	532	573
Raj.	Ajmer	104	172	393	531	561
	Jaipur	104	185	447	639	703
	Jodhpur	102	179	436	591	622
T.N.	Madras	105	184	429	619	700
	Madurai	105	164	422	630	688
	Tiruchchirappalli	105	182	422	615	689
U.P.	Agra	101	177	433	589	640
	Allahabad	101	185	425	577	637
	Kanpur	101	181	410	565	618
	Lucknow	102	172	390	556	609
	Meerut	101	186	406	529	597
W.B.	Calcutta	101	173	360	491	535
	Kharagpur	102	184	395	605	641
Del.	New Delhi	104	178	386	551	603
Cha.	Chandigarh	104	181	396	529	579
India		**103**	**178**	**403**	**574**	**624**

CPI FOR INDUSTRIAL WORKERS
(General Indices)

Table XIII
(Base 1960 = 100)

State/Centre		1961	1971	1981	1985	Feb. 1987
A.P.	Gudur	106	194	446	556	568
	Guntur	105	161	477	663	747
	Hyderabad	104	192	455	641	696
Ass.	Digboi	104	187	437	604	651
	Doom Dooma	102	204	356	493	503
	Labac	102	186	348	467	521
	Mariani	99	175	358	521	549
	Rangapara	105	179	367	483	570
Bih.	Jamshedpur	101	185	413	595	634
	Jharia	100	186	412	548	595
	Kodarma	106	211	451	616	684
	Munger	104	202	470	637	680
	Noamundi	99	196	417	557	647
Guj.	Ahmadabad	102	178	423	605	675
	Bhavnagar	102	191	445	676	742
Har.	Yamunanagar	102	199	470	622	684
J.&K.	Srinagar	104	177	460	648	743
Kar.	Ammathi	105	199	460	626	693
	Bangalore	105	193	492	694	755
	Chikmagalur	102	199	465	620	664
	Kolar Gold Field	102	186	456	641	716
Ker.	Alleppey	102	203	456	681	754
	Alwaye	104	198	452	666	764
	Mundakayam	103	197	452	603	707
M.P.	Balaghat	105	194	458	630	703
	Bhopal	108	194	459	711	796
	Gwalior	106	194	464	632	679
	Indore	106	202	475	669	761
Mah.	Bombay	103	188	447	658	739
	Nagpur	97	190	454	640	686
	Sholapur	99	195	481	663	704
Ori.	Babil	98	183	402	592	614
	Sambalpur	100	194	455	634	692
Pun.	Amritsar	102	197	456	627	672
Raj.	Ajmer	105	187	466	641	699
	Jaipur	106	188	480	662	679
T.N.	Coimbatore	101	174	466	688	732
	Coonoor	104	186	453	678	707
	Madras	103	178	435	651	704
	Madurai	105	189	460	656	708
U.P.	Kanpur	101	194	429	625	676
	Saharanpur	102	192	444	621	671
	Varanasi	102	206	490	714	791
W.B.	Asansol	99	192	444	639	657
	Calcutta	101	185	406	623	662
	Darjiling	99	171	358	529	583
	Haora	100	188	391	584	624
	Jalpaiguri	101	177	351	523	547
	Raniganj	98	186	423	584	607
UTs	Delhi	103	209	462	652	715
India		**104**	**190**	**441**	**630**	**686**

SOURCES OF MONEY STOCK (M₃)

Table XII

Items	Variations during Financial Year					
	1983-84		1984-85		1985-86¹	
	Absolute	Per cent	Absolute	Per cent	Absolute	Per cent
Sources of Money Stock (M₃) (1+2+3+4−5)	+ 13,031	+ 17.9	+ 16,058	+ 18.7	+ 16,153	+ 15.8
1. Net Bank Credit to Government (A+B)	+ 5,757	+ 16.6	+ 8,445	+ 20.8	+ 9,530	+ 19.5
A. RBI's net credit to Government (i−ii)	+ 4,311	+ 19.3	+ 3,149	+ 11.8	+ 9,127	+ 30.7
(i) Claims on Government (a+b)	+ 706	+ 2.7	+ 8,321	+ 31.0	+ 3,881	+ 11.0
(a) Central Government	− 273	− 1.1	+ 7,808	+ 30.5	+ 5,639	+ 16.9
(b) State Governments	+ 979	+ 368.0	+ 513	+ 41.2	− 1,758	+ 110.0
(ii) Government deposits with RBI (a+b)	− 3,605	− 93.4	+ 5,172	+ 2,044.3	− 5,246	− 96.7
(a) Central Government	− 3,596	− 93.5	+ 5,173	+ 2,085.9	− 5,261	− 97.0
(b) State Governments	− 9	− 64.3	− 1	− 20.0	+ 15	+ 375.0
B. Other banks' credit to Government	+ 1,446	+ 11.6	+ 5,296	+ 38.2	+ 403	+ 2.1
2. Bank Credit to Commercial Sector (A+B)	+ 8,830	+ 17.3	+ 10,809	+ 18.0	+ 10,784	+ 15.2
A. RBI's credit to commercial sector²	+ 451	+ 23.4	+ 389	+ 16.4	+ 301	+ 10.9
B. Other banks' credit to commercial sector (i+ii+iii)	+ 8,379	+ 17.0	+ 10,420	+ 18.1	+ 10,483	+ 15.4
(i) Bank credit by commercial banks	+ 5,803	+ 16.3	+ 7,660	+ 18.5	+ 6,966	+ 14.2
(ii) Bank credit by cooperative banks	+ 13	+ 13.4	+ 1,013	+ 12.7	+ 1,406	+ 15.6
(iii) Investments by commercial & cooperative banks in other securities	+ 1,633	+ 24.4	+ 1,747	+ 21.0	+ 2,111	+ 21.0
3. Net Foreign Exchange Assets of the Banking Sector (A+B)	− 182	− 10.0	+ 1,467	+ 89.1	+ 299	+ 9.6
A. RBI's net foreign exchange assets (i-ii)	− 105	− 26.1	+ 1,420	+ 87.4	+ 299	+ 9.8
(i) Gross foreign assets	+ 1,232	+ 27.4	+ 1,484	+ 25.9	+ 54	+ 0.7
(ii) Foreign liabilities	+ 1,337	+ 48.4	+ 64	+ 1.6	− 245	− 5.9
B. Other banks' net foreign exchange assets	− 77	− 77.8	+ 47	+ 213.6	—	—
4. Government's Currency Liabilities to the Public	+ 38	+ 5.6	+ 57	+ 7.9	+ 99	+ 12.7
5. Banking Sector's Net Non-monetary Liabilities other than Time Deposits (A+B)	+ 1,412	+ 9.1	+ 4,720	+ 27.8	+ 4,559	+ 21.0
A. Net non-monetary liabilities of RBI	− 460	− 8.3	+ 2,581	+ 50.6	+ 2,991	+ 38.9
B. Net non-monetary liabilities of other banks (derived)	+ 1,872	+ 18.7	+ 2,139	+ 18.0	+ 1,568	+ 11.2

Note: ¹ Data are provisional
² RBI's credit to commercial sector excludes its refinance to banks, since the establishment of NABARD.

CPI FOR AGRICULTURAL WORKERS
(General Indices)

Table XIV
(Base 1960 = 100)

	1961-62	1980-81	Feb. 1987
A.P.	101	353	483
Ass.¹	99	404	599
Bih.	103	428	583
Guj.	102	356	524
Har.	—	—	—
H.P.	—	—	—
J. & K.	N.A.	383	624
Kar.	100	398	553
Ker.	106	379	676
M.P.	103	433	582
Mah.	98	405	590
Man.	—	—	—
Meg.	—	—	—
Nag.	—	—	—
Ori.	102	441	578
Pun.²	104	437	600
Raj.	94	425	565
Sik.	—	—	—
T.N.	N.A.	391	579
Tri.	—	—	—
U.P.	N.A.	457	598
W.B.	105	400	586
INDIA	**N.A.**	**409**	**573**

Note: ¹ Includes, Manipur, Meghalaya and Tripura.
² Includes, Delhi, Haryana and Himachal Pradesh.

The International Equation

The section comprises 21 maps and 61 charts. It demonstrates the country's continued dependence on the world outside but indicates areas where this dependence is likely to be reduced.

> **Note:**
>
> This section deals with India's imports and exports and aid given and received by India. In providing information about trade and aid, only countries or regions that are major exporters or importers, donors or receivers have been considered for representation on the map.
>
> In the case of particular regions like West Asia, East Asia, Europe, North America, etc., where the trade or aid is linked with particular countries in the region, those countries have been listed in the notes given with the respective maps. But where the countries are not listed, the trade or aid connection is with most of the countries in the region.
>
> The regions and the countries they include are:
>
> South Asia — Afghanistan, Sri Lanka and Nepal.
>
> South-East Asia — Burma, Indonesia, Malaysia, Philippines, Singapore and Thailand.
>
> East Asia — China, Hong Kong, Japan and South Korea.
>
> West Asia — Bahrain, Iran, Iraq, Jordan, Kuwait, Lebanon, Oman, Qatar, Saudi Arabia, Turkey, U.A.E. and Yemen.
>
> Western Europe — Austria, Belgium, British Isles, Denmark, Finland, France, Greece, Ireland, Italy, Norway, Portugal, Spain, Sweden, Switzerland, The Netherlands and West Germany.
>
> Eastern Europe — Czechoslovakia, East Germany, Hungary, Poland, U.S.S.R. and Yugoslavia.
>
> Africa — Egypt, Ethiopia, Kenya, Mauritius, Nigeria, Somalia, South Africa, Sudan, Tanzania and Zambia.
>
> North America — Canada and U.S.A.
>
> Central & South America — Argentina, Brazil, Mexico and Trinidad.
>
> Australasia — Australia, Fiji and New Zealand.

FOREIGN TRADE

Since the attainment of independence in 1947, India's main goal has been to achieve self-reliance in the spheres of economic and technological development. From the very first Five Year Plan, this has been the prime target, necessitating modernization of agriculture and industry. As a result, import and export patterns have undergone a radical change.

To offset substantially increasing imports, government has formulated liberal policies to encourage exports. Provisions have been made for incentives to manufacturers through grants of loans, extension of technological knowhow, creation of exclusive, export-oriented, free-trade zones, liberalization of rules for the release of foreign exchange, and facilitation of foreign collaboration and joint ventures. The Export Inspection Council of India advises the government in the matter of quality control of export materials and Export Promotion Councils function for almost every material or product exported.

Despite all these facilities, India's performance on the foreign trade front has been poor right from the outset. It is not that exports have not been increasing. It is, rather, that imports have been mounting. The crunch came in the 1970s, when the cost of oil imports wiped out the benefits of an almost four-fold increase in exports (1980-81). Since then, annual imports have continued to be nearly 10 times what they were in 1970-71; but exports have also continued to improve. In 1980-81, imports grew by about 25 per cent but exports by over 50 per cent. If the trend continues and India can cut its oil bill — as is predicted might happen before the end of the century — the trade balance will be much more favourable. There can be a solution to India's balance of payments problem only if the trade deficit is brought down from nearly Rs 60,000 m at present to about Rs 20,000 m. In part, the deficit can be reduced by cutting down oil and a few other imports, but even while cutting down on these, the long-term strategy will have to be to import selected high technology items in order to bring down costs of production, improve quality and enable manufacture of goods that are not outmoded. A step in this direction has been taken with the 1986 'high tech' policy. Such a policy, however, can be sustained only if India is able to pay for these imports by enlarging exports. The goal, according to some experts, should be to raise exports to a level of 10 per cent of our GNP by 1990 and peg rationalized imports at the same level.

Government fixed the export target for 1985-86 at Rs 117,360 m, but projections indicate that it is more likely to have been around Rs 110,000 m, slightly less than exports in 1984-85. During 1984-85, India's exports recorded a rise of 20.1 per cent, to Rs 118,550 m from Rs 98,720 m in 1983-84. Yet this growth was relatively lower than in the mid-seventies, when India's exports rose by 31.9 per cent in 1974-75 and by 27.2 per cent in 1976-77.

In the Sixth Plan period, India's export performance fell far short of targets. There are as yet no signs of improvement in the Seventh Plan. Planners projected cumulative export during the Sixth Plan to be Rs 410,780 m, but realization was just about Rs 330,000 m. Exports are expected to go up by 6.8 per cent annually during 1985-90. During the first year of the Plan (1985-1986), exports have, however, shown a declining trend, falling about 12 per cent, from Rs 118,550 m in 1984-85 to an estimated Rs 104,200 m. The overall prospects for a revival in 1986-87 are uncertain. The target for 1986-87 is Rs 122,000 m.

Increased exports in the 1980s have been due to exports of crude oil since 1982-83. But with the commissioning of extra indigenous refining capacity, exports of crude oil are expected to decline, from 6.5 million tonnes to nearly 0.5 million tonnes (1985-86). This could mean that, if the present export trends continue, and less refined oil is imported, the balance of payments can dramatically improve during the Seventh Plan period.

More than half India's exports (51.5 per cent) in 1970-71 and 43.5 per cent in 1984-85 were to the USA, USSR, Japan and UK. From the largest importer of Indian goods, the USSR has now become the second largest, exchanging positions with the USA. Exports to the USA (1984-85) were 14.9 per cent, to the USSR 13.9 per cent, Japan 9 per cent and UK 5.7 per cent. Other major importers of Indian goods (1984-85) were West Germany (4 per cent), the UAE and Saudi Arabia (2.3 per cent each), France (1.8 per cent), Italy (1.7 per cent), Belgium and Singapore (1.6 per cent each).

India's imports from the USA (27.7 per cent of India's total imports) the UK (7.8 per cent), West Germany (6.6 per cent), USSR (6.5 per cent), Japan (5.1 per cent) accounted for 53.7 per cent of all imports in 1970-71. In 1984-85, the imports from these countries, in terms of shares, were USSR 10.5 per cent, USA 9.7 per cent, West Germany 7.6 per cent, Saudi Arabia 7.3 per cent, Japan 7.2 per cent, UK 5.9 per cent and Belgium 4.6 per cent. Thus the relative positions of the countries from which India imports has changed considerably.

India's major exports are (1984-85): crude oil and petroleum products 15.3 per cent (Rs 18,180 m) of total export value of Rs 118,550 m, gems and jewellery 11 per cent (Rs 13,030 m), engineering goods 10.9 per cent (Rs 13,000 m), readymade garments 8 per cent (Rs 9440 m), tea 6 per cent (Rs 7080 m), leather and manufactures 4.9 per cent (Rs 5840 m), cotton textiles 4 per cent (Rs 4720 m), basic chemicals and pharmaceuticals 3.8 per cent (Rs 4520 m), iron ore 3.8 per cent (Rs 4470 m), marine products 3.2 per cent (Rs 3840 m) and jute manufactures 2.9 per cent (Rs 3410 m). Other substantial exports from India are coffee, tobacco (unmanufactured), spices and processed foods.

Of India's imports (1984-85), the major items are petroleum and its products (31.1 per cent, or Rs 53,450 m, of total import value of Rs 171,710 m), machinery and transport equipment (15.2 per cent or Rs 26,180 m), wheat and edible products (8.2 per cent or Rs 14,160 m), unfinished stones Rs 10,280 m, iron and steel (4.5 per cent or Rs 7770 m) and non-ferrous metals (2 per cent or Rs 3450 m).

The available data show that there is something fundamentally wrong with the structure and pattern of India's exports and also the policies governing them. The extent of the ills can be gauged from the continuing deterioration of India's share of world exports which fell from 2.2 per cent in 1948-49 to 1 per cent in 1963-64 and to less than 0.5 per cent in the subsequent two decades. India's ranking among the world's trading nations slipped from 16th in 1953 to 20th in 1963 and further to 41st in 1983. Several developing countries have overtaken India in the export race.

The decline in India's share in world trade has hit all export commodities, over many of which India had a substantial hold in the early 1950s. For example, between 1970 and 1983, the share in export of tea has fallen from 33.4 per cent to 16.2 per cent. Similarly, India's share of the spices trade has gone down from 20.4 to 10.3 per cent of the world's total.

Sources: 1. *Annual Report 1985-86*, Ministry of Commerce.
2. *Report on Currency and Finance 1985-86, Vol. II*, Reserve Bank of India.
3. *Seventh Five Year Plan 1985-90*, Government of India Planning Commission.
4. *Economic Survey 1984-85*, Ministry of Finance.
5. *Foreign Trade Statistics of India, April 1984, May 1985, April 1986, April 1987 & Economic Profiles of 40 Major Countries, March 1987* and *Basic Statistics Relating to the Indian Economy, Vol. I: All India, August 1986*, Centre for Monitoring Indian Economy.

INDIA'S FOREIGN TRADE

Table 108   (in Rs. million)   (in Rs. million)

Countries	1975-76 Exports	1975-76 Imports	1980-81 Exports	1980-81 Imports	1984-85 Exports	1984-85 Imports
Argentina	39.8	102.3	52.8	180.2	10.8	622.7
Australia	482.0	1,017.0	916.0	1,701.0	1,485.0	1,924.0
Bangladesh	622.0	47.0	750.0	117.0	932.0	445.0
Belgium	454.0	865.0	1,445.0	2,959.0	1,856.0	7,930.0
Brazil	23.4	70.6	15.9	2,354.7	14.1	2,940.3
Canada	458.0	2,320.0	623.0	3,323.0	1,347.0	5,084.0
China	19.3[1]	6.3[1]	379.3	819.9	193.9	614.5
Czechoslovakia	346.0	530.1	553.0	390.1	622.3	585.2
Egypt	1,000.1	190.0	859.0	296.0	1,052.0	99.0
France	862.0	1,965.0	1,469.0	2,803.0	2,089.0	3,582.0
Germany (E)	258.7	365.9	488.9	442.7	703.0	672.9
Germany (W)	1,179.0	3,700.0	3,848.0	6,938.0	4,709.0	12,978.0
Indonesia	517.2	34.4	516.0	161.4	312.9	591.9
Iran	2,723.0	4,599.0	1,232.0	13,389.0	1,340.0	4,849.0
Iraq	639.0	2,478.0	520.0	7,525.0	487.0	6,746.0
Italy	800.2	848.3	1,515.8	2,424.8	2,031.6	2,887.4
Japan	4,328.0	3,612.0	5,078.0	7,488.0	10,610.0	12,404.0
Korea (S)	225.6	16.9	440.3	1,353.8	894.6	1,473.0
Kuwait	472.8	626.3	970.7	3,376.2	1,159.0	3,681.3
Malaysia	328.5	145.8	512.6	2,660.7	709.7	5,484.5
Mexico	13.6	144.7	46.6	407.7	11.4	109.1

Countries	1975-76 Exports	1975-76 Imports	1980-81 Exports	1980-81 Imports	1984-85 Exports	1984-85 Imports
Nepal	509.6	331.5	779.8	236.1	1,810.8	604.1
Nigeria	375.5	18.5	533.8	75.2	358.9	7.2
Pakistan	7.8	211.2	10.2	753.9	129.1	157.5
Philippines	116.9	44.6	59.5	109.2	92.2	131.0
Poland	899.0	810.4	689.7	344.1	936.6	537.9
Saudi Arabia	601.0	2,901.0	1,653.0	5,401.0	2,718.0	12,633.0
Spain	99.8	47.0	157.2	830.0	310.0	884.7
Sri Lanka	231.1	2.9	806.5	295.5	1,142.3	105.7
Sweden	133.4	684.7	283.5	849.6	360.4	1,361.3
Switzerland	585.4	554.1	1,106.2	1,206.3	1,187.4	1,628.2
Thailand	172.4	219.9	445.5	524.1	254.2	869.2
The Netherlands	822.3	638.2	1,520.4	2,145.0	1,821.2	3,640.6
U.A.E.	663.3	815.5	1,524.7	3,499.7	2,687.8	3,589.8
U.K.	4,213.0	2,840.0	3,950.0	7,310.0	6,701.0	10,188.0
U.S.A.	5,200.0	12,852.0	7,433.0	15,186.0	17,685.0	16,666.0
U.S.S.R.	4,167.0	3,098.0	12,263.0	10,137.0	16,546.0	18,034.0
Venezuela	11.0	0.9	5.1	482.6	13.3	36.0
Yugoslavia	293.4	101.7	343.6	449.1	242.5	548.2
Total[2]	40,420.0	52,650.0	67,110.0	125,490.0	118,550.0	171,710.0

Note: [1] 1977-78.
[2] Includes other countries for which break-up figures are not available.

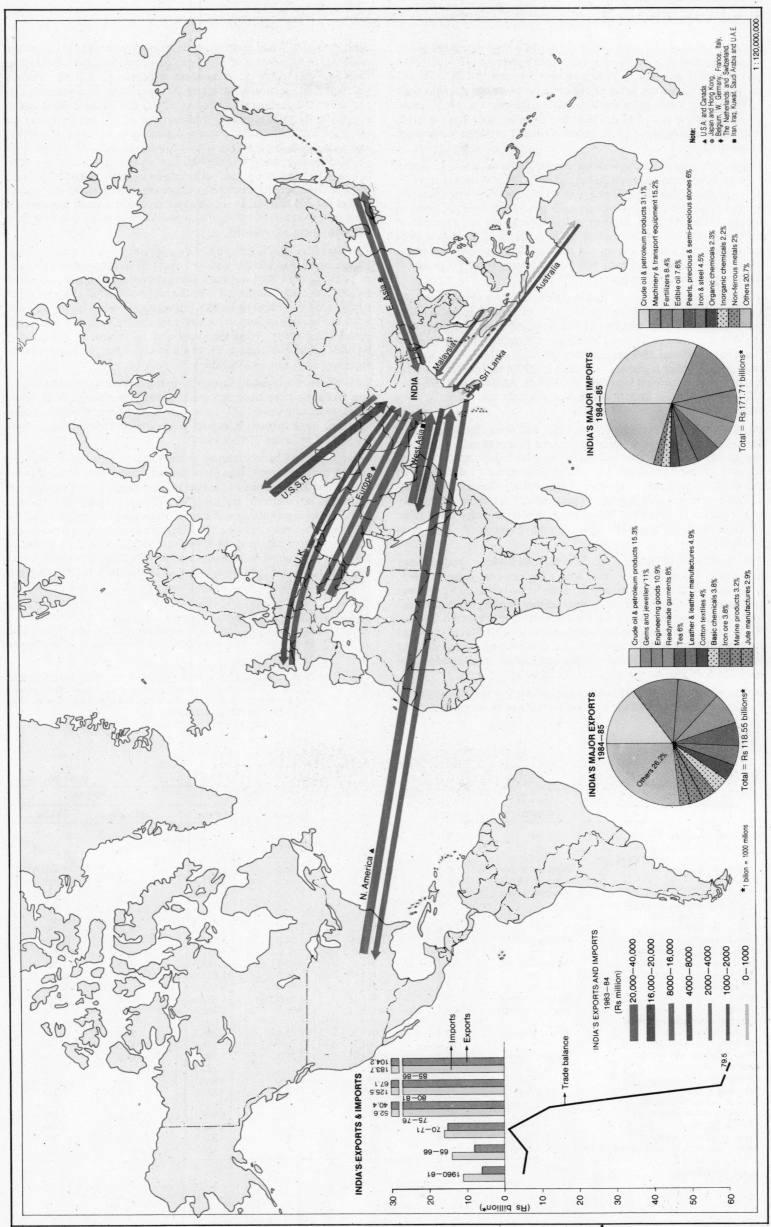

1 : 120,000,000

Note:
▲ U.S.A. and Canada.
⊕ Japan and Hong Kong.
⊖ Belgium, W. Germany, France, Italy.
◆ The Netherlands and Switzerland.
■ Iran, Iraq, Kuwait, Saudi Arabia and U.A.E.

Crude oil & petroleum products 31.1%
Machinery & transport equipment 15.2%
Fertilizers 8.4%
Edible oil 7.6%
Pearls, precious & semi-precious stones 6%
Iron & steel 4.5%
Organic chemicals 2.3%
Inorganic chemicals 2.2%
Non-ferrous metals 2%
Others 20.7%

INDIA'S MAJOR IMPORTS
1984—85

Total = Rs 171.71 billions*

Crude oil & petroleum products 15.3%
Gems and jewellery 11%
Engineering goods 10.9%
Readymade garments 8%
Tea 6%
Leather & leather manufactures 4.9%
Cotton textiles 4%
Basic chemicals 3.8%
Iron ore 3.8%
Marine products 3.2%
Jute manufactures 2.9%

INDIA'S MAJOR EXPORTS
1984—85

Others 26.2%

Total = Rs 118.55 billions*

*: 1 billion = 1000 millions

INDIA'S EXPORTS AND IMPORTS
1983—84
(Rs million)

20,000—40,000
16,000—20,000
8000—16,000
4000—8000
2000—4000
1000—2000
0—1000

INDIA'S EXPORTS & IMPORTS

Imports
Exports

Trade balance

(Rs billion*)

104.2 } 85—86
183.7
67.1 } 80—81
125.5
40.4 } 75—76
52.6

1960-61
65-66
70-71
75-76
80-81
85-86

79.5

237

EXPORTS — I

India has traditionally been one of the world's main exporters of tea, being responsible for a third of the world's exports (33.4 per cent) in 1970, 27.1 per cent in 1980 and 16.3 per cent in 1983. Though declining in share, the value of Indian tea exports has doubled since 1970. In early 1986, there was a shortfall in the tea crop, put at around 30 m kgs at the end of May 1986. But there has also been a fall in demand, which might just bring about a balance between demand and production.

Despite a host of problems with tea exports, India has not fared badly. Exports have increased from 207 m kgs in 1970 to 208 m kgs in 1983, 215 m kgs in 1984 and 220 m kgs in 1985. In the 24 years between 1960-61 and 1984-85, tea exports have grown 4.7 times in terms of money value, from Rs 1240 m to Rs 7080 m and are expected to remain at that level at the end of the Seventh Plan. In August 1986 the tea industrial units anticipated meeting that year's (1986) target of Rs 7500 m, although they had yet to deliver some 70 m kgs of tea to meet the target fixed for 1986-87.

The traditional importers of Indian (black) tea have always been the USSR and the British Isles. In 1975-76, the USSR imported 30.8 per cent of India's tea exports (Rs 729.6 m) and the UK 19.3 per cent (Rs 456.9 m).In 1982-83, the USSR and UK imported 31.6 per cent (Rs 1169.4 m) and 23.9 per cent (Rs 885.7 m) of India's tea exports respectively. Other importers — Poland Rs 227.4 m, Iran Rs 218.9 m, Egypt Rs 163.6 m, West Germany Rs 145.6 m, Afghanistan Rs 96.4 m, USA Rs 43.5 m — import less than 7 per cent each of India's tea exports (Rs 3697.5 m in 1982-83).

From a traditional exporter of raw hides and skins, India has, since the seventies, become a major competitor in world markets for the supply of finished leather and leather goods. Indian leather goods, finished leather and leather footwear accounted for 11.4 per cent of the world's total leather exports in 1970. India's share (9.3 per cent) in 1980, went down slightly to 6.3 per cent in 1983. In the last twenty years, leather exports have, however, grown in monetary value about 23 times, with rapid growth from the mid-seventies (Rs 1690 m in 1973-74).

The contribution of leather products to the total leather export earnings has almost touched 50 per cent, with a ten per cent jump in 1985. Leather product exports in 1985-86 were valued at Rs 3252.4 m and accounted for 49 per cent of the total leather exports (Rs 6625.1 m, more than even the target set for the end of the Seventh Plan — Rs 5770 m). But though there was an increase of Rs 880 m in export earnings, leather product exports failed to meet the targeted share of 53.7 per cent. This was because exports of footwear amounted to only Rs 1903.5 m against the target of Rs 2100 m.

On the whole, 1985-86 was another year of fairly good performances: the total exports exceeded the target of Rs 6050 m by nearly Rs 595 m.

While miscellaneous leather goods exports were Rs 834.2 m and represented an increase of 127.3 per cent over the target, the leather footwear and footwear-component exports, at Rs 330 m and Rs 1903.5 m, recorded 94.4 and 90.6 per cent achievements respectively. A disturbing trend, however, was that while finished leather exports at Rs 2882 m reflected an increase of 131 per cent over the target of Rs 2200 m, there was a decline both in terms of quantity and value compared to the achievements of 1984-85. Exports of finished leather totalled 1730.6 million square decimetres (sq dcm) as against 2157.4 m sq dcm, reflecting an overall drop of 20 per cent and a decline ranging from 10 to 45 per cent in individual categories, except for cow-calf skins which alone recorded a small increase (31 per cent). In terms of value, the overall decline was 7 per cent from the attainment of 1984-85.

Traditionally, the USSR, UK, Yugoslavia, East Germany, Italy, France and Japan are the leading buyers of Indian leather products. In 1975-76, the USSR was the leading importer (19.6 per cent worth Rs 393.7 m) and it continued to be the biggest buyer in 1982-83, when it bought 24.6 per cent or Rs 885.1 m worth of India's leather exports. USA too became an important buyer by 1982-83 (Rs 491.6 m worth or 13.6 per cent). West Germany also increased its share, from Rs 103.1 m (5 per cent) to Rs 414.8 m (11.5 per cent) and was the third-largest buyer in 1982-83.

India has been a regular exporter of fish and other seafoods since the late sixties and the trade has made phenomenal progress in recent years, especially since 1975. In the period 1975-76 to 1982-83, marine products have grown in export terms about 184 per cent, from Rs 1321.5 m to Rs 3751.9 m.

A large part of Indian marine product exports include frozen and canned shrimps, frogs' legs (now banned), dried fish, dried prawns, sharks'fins and fish maws. Indian marine products are exported to more than 80 countries, but the USA and Japan are the main buyers. In 1975-76, Japan accounted for 64.5 per cent (Rs 853 m) and the USA for 25.3 per cent (Rs 334.3 m) of the purchases from India. Japan's share went up to 68.9 per cent (Rs 2583.7 m) and that of the USA came down to 11.6 per cent (Rs 436.7 m) in 1982-83. Other major importers of Indian marine products are the UK (Rs 148.8 m), The Netherlands (Rs 83.7 m), Kuwait (Rs 68.2 m) and France (Rs 50.5 m).

Sources: 1. *Annual Report 1985-86*, Ministry of Commerce.
2. *Report on Currency and Finance 1985-86, Vol. II*, Reserve Bank of India.
3. *Seventh Five Year Plan 1985-90*, Government of India Planning Commission.
4. *Economic Survey 1984-85*, Ministry of Finance.
5. *Foreign Trade Statistics of India, April 1984, May 1985, April 1986, April 1987 & Economic Profiles of 40 Major Countries, March 1987* and *Basic Statistics Relating to the Indian Economy, Vol. I: All India, August 1986*, Centre for Monitoring Indian Economy.

MAJOR EXPORTS — I

Table 109

(in Rs. million)

Commodity	Units of quantity	1981-82		1983-84		1985-86[1]	
		Quantity	Value	Quantity	Value	Quantity	Value
Live animals chiefly for food	—	N.A.	52.2	N.A.	64.3	N.A.	20.7
Meat and meat preparations	—	N.A.	795.5	N.A.	683.2	N.A.	721.9
Fish, crustaceans, molluscs and other preparations	thousand tonnes	73.9	2,803.4	81.8	3,273.0	93.4	3,886.0
Rice	thousand tonnes	872.5	3,677.8	246.0	1,471.3	245.1	1,929.4
Wheat	thousand tonnes	1.4	6.0	2.5	12.9	211.6	553.0
Cereal preparations, preparations of flour, starch, fruits and vegetables	—	N.A.	89.2	N.A.	70.1	N.A.	83.7
Cashew kernels	thousand tonnes	30.7	1,815.0	39.6	1,566.2	37.5	2,153.3
Other fruits and vegetables	—	N.A.	1,060.0	N.A.	1,551.6	N.A.	1,751.6
Sugar and sugar preparations	thousand tonnes	202.1	641.8	240.0	1,398.6	19.6	109.9
Coffee and coffee substitutes	m. kgs.	73.9	1,462.9	73.5	1,832.6	84.2	2,356.4
Tea and mate	m. kgs.	213.8	3,952.0	197.5	5,013.7	213.8	6,119.1
Spices	thousand tonnes	78.1	987.6	86.0	1,092.6	84.2	2,550.0
Oil cake	thousand tonnes	824.4	1,178.5	952.3	1,462.9	665.1	1,235.4
Tobacco unmanufactured and and tobacco refuse	thousand tonnes	115.4	2,049.3	75.8	1,496.1	53.7	1,204.0
Oil seeds and oleaginous fruits	—	N.A.	364.8	N.A.	352.4	N.A.	193.8
Fixed vegetable oils and fats	thousand tonnes	15.7	±174.0	25.0	280.2	18.9	249.7
Crude animal materials	—	N.A.	286.7	N.A.	252.5	N.A.	291.2
Crude vegetable materials	—	N.A.	1,569.4	N.A.	968.1	N.A.	1,143.6
Footwear	m. pairs	12.4	361.9	7.4	232.3	4.4	332.8
Leather and leather manu-factures (except footwear)	—	N.A.	3,693.5	N.A.	3,499.3	N.A.	4,876.6
Carpets (handmade)	—	N.A.	1,812.5	N.A.	1,940.4	N.A.	2,286.1
Tobacco (manufactured)	m. kgs.	17.8	305.3	7.1	200.6	17.6	262.9
Rubber and manufactured rubber	—	N.A.	354.3	N.A.	420.8	N.A.	706.3

Note: [1] Provisional.

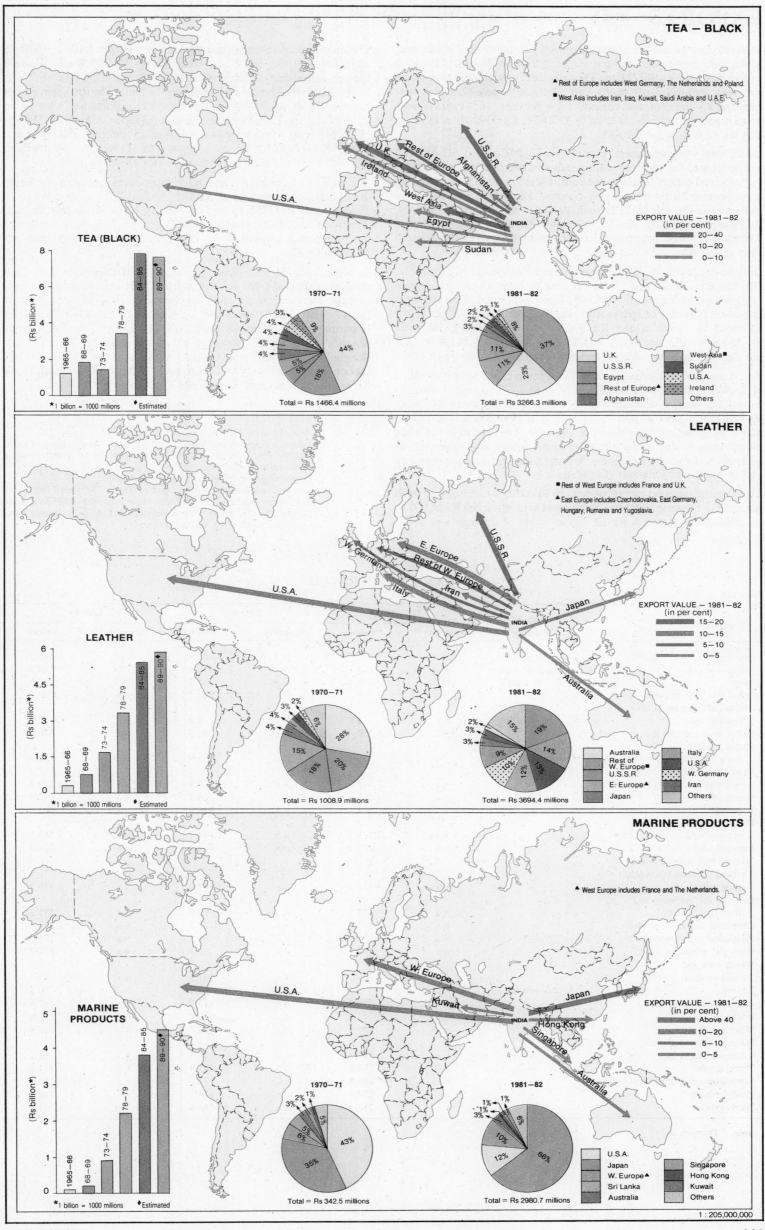

TEA — BLACK

▲ Rest of Europe includes West Germany, The Netherlands and Poland.
■ West Asia includes Iran, Iraq, Kuwait, Saudi Arabia and U.A.E.

EXPORT VALUE — 1981—82
(in per cent)
20—40
10—20
0—10

TEA (BLACK)

(Rs billion★)

8
7
6
5
4
3
2
1
0

1965—66 68—69 73—74 78—79 84—85 89—90♦

★1 billion = 1000 millions   ♦ Estimated

1970—71
3%
9%
4%
4%
4%
5%
5%
18%
44%
Total = Rs 1466.4 millions

1981—82
2% 1%
2%
3%
6%
11%
11%
23%
37%
Total = Rs 3266.3 millions

U.K.
U.S.S.R.
Egypt
Rest of Europe▲
Afghanistan

West Asia■
Sudan
U.S.A.
Ireland
Others

LEATHER

■ Rest of West Europe includes France and U.K.
▲ East Europe includes Czechoslovakia, East Germany, Hungary, Rumania and Yugoslavia.

EXPORT VALUE — 1981—82
(in per cent)
15—20
10—15
5—10
0—5

LEATHER

(Rs billion★)

6
4.5
3
1.5
0

1965—66 68—69 73—74 78—79 84—85 89—90♦

★1 billion = 1000 millions   ♦ Estimated

1970—71
3% 2%
4%
4%
6%
15%
28%
18%
20%
Total = Rs 1008.9 millions

1981—82
2%
3%
3%
15%
19%
14%
9%
10%
12%
13%
Total = Rs 3694.4 millions

Australia
Rest of
W. Europe■
U.S.S.R.
E. Europe▲
Japan

Italy
U.S.A.
W. Germany
Iran
Others

MARINE PRODUCTS

▲ West Europe includes France and The Netherlands.

EXPORT VALUE — 1981—82
(in per cent)
Above 40
10—20
5—10
0—5

MARINE PRODUCTS

(Rs billion★)

5
4
3
2
1
0

1965—66 68—69 73—74 78—79 84—85 89—90♦

★1 billion = 1000 millions   ♦ Estimated

1970—71
3% 2% 1%
5%
5%
6%
35%
43%
Total = Rs 342.5 millions

1981—82
1% 1%
3%
3%
6%
10%
12%
66%
Total = Rs 2980.7 millions

U.S.A.
Japan
W. Europe▲
Sri Lanka
Australia

Singapore
Hong Kong
Kuwait
Others

1 : 205,000,000

EXPORTS — II

Minerals and metals occupy a significant position in India's foreign trade, their export amounting to Rs 23,202 m in 1983-84 of the total export value of Rs 98,721 m. This was 23.5 per cent of all exports and was comprised of: pearls and precious stones 12.3 per cent, minerals 8.4 per cent, and chemicals 2.8 per cent. Such exports were of the order of only Rs 2475 m in 1970-71, Rs 5392 m in 1975-76 and Rs 12,681 m in 1980-81.

Of the total value of mineral exports (Rs 23,202 m) in 1983-84, iron ore accounted for 16.6 per cent, pearls and precious stones for 52.3 per cent and others for 31.1 per cent. India has been fast becoming an important iron ore exporting country. Exports of iron ore from India were valued at Rs 1173 m in 1970-71, Rs 2330 m in 1978-79, Rs 2894 m in 1980-81, Rs 3517 m in 1981-82, Rs 4470 m in 1984-85. They are expected to grow to Rs 6080 m by the end of the Seventh Plan (1980-90).

In 1970-71, Japan (Rs 391.5 m) and Eastern Europe, mainly Romania, (Rs 138.8 m) accounted for 91 per cent of India's iron ore exports (79 per cent to Japan). In 1982-83, Japan continued to be the most important importer of Indian iron ore, Rs 2615.2 m (68.7 per cent), followed by South Korea Rs 390 m (10.2 per cent), Romania Rs 354.9 m (9.3 per cent) and East Germany Rs 111.4 m (2.9 per cent). The total value of exports that year was Rs 3805 m.

Pearls and precious stones exported to foreign countries were valued at Rs 419 m in 1970-71 and Rs 1230 m in 1975-76. Since then the export has grown dramatically to reach Rs 5189 m in 1979-80. Rs 7611 m in 1981-82 and Rs 11,950 m in 1984-85. It is estimated that it will reach Rs 16,630 m in 1989-90. This increase has been almost entirely due to the import of uncut roughs and export of cut and polished diamonds. This has become a major industry in Bombay and in Surat and Palanpur (Gujarat).

This lucrative diamond export business, by a country which has very small quantities of ore-mined diamond, is estimated to have reached Rs 16,000 m in 1986. India exported Rs 13,750 m worth of diamonds in 1985.

The major countries importing finished gems from India in 1970-71 were Belgium (Rs 104.6 m), Hong Kong (Rs 100.8 m), Western Europe other than Belgium (Rs 64.8 m), the USA (Rs 46.7 m), Switzerland (Rs 31.5 m) and Japan (Rs 28.6 m). The total gem export was valued at Rs 419.4 m. By 1982-83, exports had shot up to Rs 9500 m, the USA being the largest importer, taking Rs 3178 m (33.5 per cent) worth of gems, followed by Belgium (Rs 1734 m), Hong Kong (Rs 1467 m), Japan (Rs 1196 m), Switzerland (Rs 517.3 m) and the rest of Europe (Rs 488.2 m).

The export of chemicals and related products has been moderate throughout, being valued at Rs 1555 m in 1978-79, Rs 2354 m in 1980-81 and Rs 3755 m in 1981-82. In 1984-85, India exported chemicals and chemical products totalling Rs 7230 m in value and this is expected to grow to Rs 12,240 m by the end of the Seventh Plan (1989-90).

Between 1950-51 and 1980-81, the share of chemicals and allied products exported from India rose from 1 per cent to 3.5 in its total exports. The main buyers of Indian chemicals (1982-83) are the USSR (Rs 1779 m, 51 per cent) and USA (Rs 210 m). Other countries importing Indian chemicals are the UK (Rs 71.1 m), The Netherlands (Rs 85.9 m), France (Rs 77.1 m), Saudi Arabia (Rs 57.8 m), Bangladesh (Rs 40.8 m), United Arab Emirates (Rs 33.4 m), Indonesia and Thailand (Rs 27.8 m each), Yemen (Rs 20.2 m), Hong Kong (Rs 17.6 m) and Kuwait (Rs 15.2 m).

Sources: 1. *Annual Report 1985-86*, Ministry of Commerce.
2. *Report on Currency and Finance 1985-86, Vol. II*, Reserve Bank of India.
3. *Seventh Five Year Plan 1985-90*, Government of India Planning Commission.
4. *Economic Survey 1984-85*, Ministry of Finance.
5. *Foreign Trade Statistics of India, April 1984, May 1985, April 1986, April 1987 & Economic Profiles of 40 Major Countries, March 1987* and *Basic Statistics Relating to the Indian Economy, Vol. I: All India, August 1986*, Centre for Monitoring Indian Economy.

MAJOR EXPORTS — II

Table 110

(in Rs. million)

Commodity	Units of quantity	1981-82		1983-84		1985-86[1]	
		Quantity	Value	Quantity	Value	Quantity	Value
Iron ore	m. tonnes	23.7	3,517.5	21.2	3,853.4	27.0	5,545.9
Manganese ore	thousand tonnes	552.4	148.1	574.8	179.8	379.2	168.8
Mica	m. kgs.	13.5	291.6	12.4	265.2	18.2	209.0
Other ores and minerals	—	N.A.	366.3	N.A.	368.1	N.A.	854.9
Waste and scrap metal of iron and steel	—	N.A.	7.6	N.A.	8.3	N.A.	57.5
Stone, sand and gravel	—	N.A.	258.0	N.A.	386.3	N.A.	576.6
Shellac, seedlac, gums, resins and balsams	thousand tonnes	19.0	306.3	15.6	292.9	8.9	532.6
Pearls, precious and semi-precious stones etc.	—	N.A.	7,610.7	N.A.	12,139.9	N.A.	14,299.1
Works of art	—	N.A.	1,381.4	N.A.	1,166.4	N.A.	1,233.4
Jewellery	—	N.A.	504.2	N.A.	746.6	N.A.	682.9
Metal manufactures except iron and steel	—	N.A.	2,328.1	N.A.	1,944.1	N.A.	1,563.5
Iron and steel	—	N.A.	790.5	N.A.	464.3	N.A.	481.2
Mineral fuel, lubricants and related products (excluding crude oil)	—	N.A.	286.6	N.A.	3,619.6	N.A.	5,179.2
Essential oil, perfumes and flavouring materials	—	N.A.	78.8	N.A.	90.9	N.A.	131.2
Chemicals and allied products	—	N.A.	3,641.3	N.A.	2,777.1	N.A.	2,858.9
Plastic and allied products	—	N.A.	125.6	N.A.	88.1	N.A.	155.2
Glass and glassware	—	N.A.	177.4	N.A.	101.3	N.A.	111.7
Wood, cork and allied products	—	N.A.	203.2	N.A.	107.2	N.A.	98.3
Cement	thousand tonnes	3.6	2.4	7.4	7.1	26.8	30.8
Asbestos cement articles	—	N.A.	50.4	N.A.	13.9	N.A.	12.9
Bricks etc.	—	N.A.	46.2	N.A.	22.4	N.A.	13.9
Mineral manufactures	—	N.A.	209.9	N.A.	243.7	N.A.	140.9
Sanitary, plumbing etc. fixtures and fittings	—	N.A.	43.5	N.A.	42.5	N.A.	38.4
Travel goods	—	N.A.	238.0	N.A.	243.9	N.A.	374.2
Cinematographic films (exposed)	—	N.A.	160.5	N.A.	77.6	N.A.	38.4
Printed matter	—	N.A.	159.6	N.A.	119.1	N.A.	148.7
Sports goods	—	N.A.	233.2	N.A.	144.4	N.A.	165.0

Note: [1] Provisional.

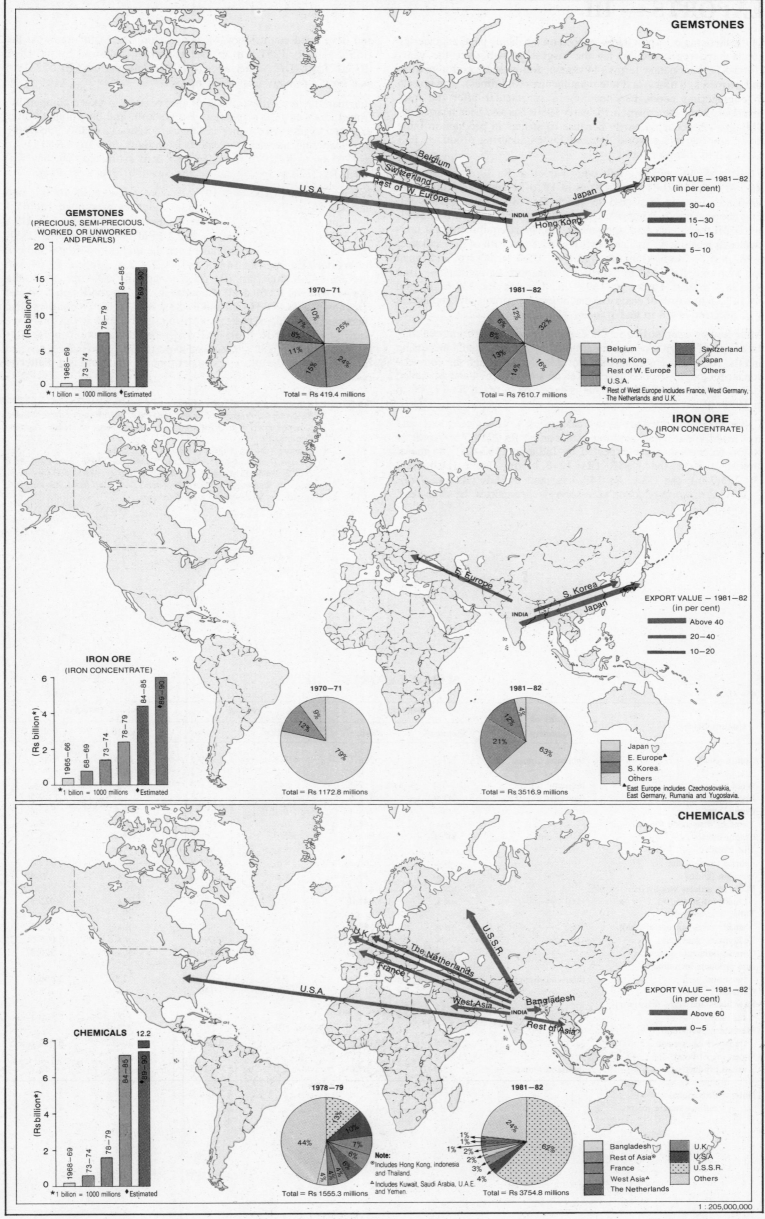

GEMSTONES

GEMSTONES
(PRECIOUS, SEMI-PRECIOUS, WORKED OR UNWORKED AND PEARLS)

EXPORT VALUE — 1981—82
(in per cent)
- 30—40
- 15—30
- 10—15
- 5—10

1970—71
- 25%
- 24%
- 15%
- 11%
- 8%
- 7%
- 10%

Total = Rs 419.4 millions

1981—82
- 32%
- 16%
- 14%
- 13%
- 8%
- 6%
- 12%

Total = Rs 7610.7 millions

Legend:
- Belgium
- Hong Kong
- Rest of W. Europe
- U.S.A.
- Switzerland
- Japan
- Others*

* Rest of West Europe includes France, West Germany, The Netherlands and U.K.

Bar chart: (Rs billion*) 20, 15, 10, 5, 0
1968—69, 73—74, 78—79, 84—85, ♦89—90
* 1 billion = 1000 millions ♦ Estimated

IRON ORE
(IRON CONCENTRATE)

IRON ORE
(IRON CONCENTRATE)

EXPORT VALUE — 1981—82
(in per cent)
- Above 40
- 20—40
- 10—20

1970—71
- 79%
- 12%
- 9%

Total = Rs 1172.8 millions

1981—82
- 63%
- 21%
- 12%
- 4%

Total = Rs 3516.9 millions

Legend:
- Japan
- E. Europe ▲
- S. Korea
- Others

▲ East Europe includes Czechoslovakia, East Germany, Rumania and Yugoslavia.

Bar chart: (Rs billion*) 6, 4, 2, 0
1965—66, 68—69, 73—74, 78—79, 84—85, ♦89—90
* 1 billion = 1000 millions ♦ Estimated

CHEMICALS

CHEMICALS

EXPORT VALUE — 1981—82
(in per cent)
- Above 60
- 0—5

1978—79
- 44%
- 13%
- 10%
- 6%
- 5%
- 4%
- 4%
- 7%

Note:
⊛ Includes Hong Kong, Indonesia and Thailand.
△ Includes Kuwait, Saudi Arabia, U.A.E. and Yemen.

Total = Rs 1555.3 millions

1981—82
- 62%
- 24%
- 4%
- 3%
- 2%
- 2%
- 1%
- 1%
- 1%

Total = Rs 3754.8 millions

Legend:
- Bangladesh
- Rest of Asia⊛
- France
- West Asia△
- The Netherlands
- U.K.
- U.S.A.
- U.S.S.R.
- Others

Bar chart: CHEMICALS 12.2 (Rs billion*) 8, 6, 4, 2, 0
1968—69, 73—74, 78—79, 84—85, ♦89—90
* 1 billion = 1000 millions ♦ Estimated

1 : 205,000,000

EXPORTS — III

Of the three major export items mapped on the facing page, engineering goods have registered by far the largest growth over the years (1965-85) in monetary terms (44 times), followed by cotton textiles and garments (8 times) and jute manufactures (0.3 times). The growth in engineering goods has been more spectacular after the mid-seventies, while the growth in the early 1980s has also been heartening. This growth occurred partly because of strides in production India made and the rising world demand for engineering products. It is expected to increase further.

Exports of engineering goods were valued at Rs 30 m in 1950-51, Rs 190 m in 1960-61, Rs 2020 m in 1970-71 and Rs 13,000 m in 1984-85, and are expected to reach Rs 18,620 m in 1989-90. Engineering goods comprise manufactured metals, iron and steel, electrical machinery and appliances, transport equipment and non-electrical machinery. India exports Rs 501.8 m worth of iron and steel (1982-83) — Saudi Arabia Rs 113.8 m, USA Rs 80.1 m and Japan Rs 39.6 m; Rs 1833 m worth of transport equipment — Iran Rs 198.1 m, Sri Lanka Rs 169.8 m, Uganda Rs 138.6 m; and Rs 1421.5 m worth of electrical machinery, apparatus and appliances — USSR Rs 714.8 m and Iraq Rs 124.7 m.

After having been stifled in the early years of the present century, the export of cotton textiles and garments has been gradually increasing and the country is emerging as a world leader, although the competition is still tough. Exports have picked up since 1973-74 and amounted to Rs 14,140 m in 1984-85. This is expected to reach Rs 17,760 m in 1989-90.

Cotton goods exported are in two categories: Cotton piecegoods, and articles of apparel and clothing accessories. Rs 2714.7 m worth of cotton piecegoods were exported by India in 1982-83. The major importers were the USSR (Rs 1218.3 m), Benin (Dahomey) (Rs 255.7 m), the USA (Rs 148.3 m) and the UK (Rs 143.9 m). Rs 6047.2 m worth of cotton accessories were exported the same year,

and developed countries were, once again, the major markets: the USSR with Rs 1281.2 m in the lead again, followed by the USA (Rs 1277 m), The Netherlands (Rs 787.7 m), the UK (Rs 552.3 m) and West Germany (Rs 337.8 m).

Jute manufacture exports have, of late, been facing keen competition from India's eastern neighbour, Bangladesh, and the competition is expected to get even tougher as China emerges as a quality jute textile producer. Jute manufacture exports from India grew from Rs 1540 m in 1965-66 to Rs 2070 m in 1984-85. It is estimated that they will touch Rs 2220 m by the end of the Seventh Five Year Plan.

Jute manufacture exports in 1978-79 were valued at Rs 1660 m and in 1982-83 at Rs 1897.4 m. Eight countries accounted for 65.4 per cent and 75.4 per cent of these exports respectively. The USA (Rs 354.1 m in 1978-79 and Rs 212.1 m in 1982-83) and the USSR (Rs 335.6 m and Rs 790 m) have been the major importers followed by Western Europe (Rs 144 m and Rs 274.2 m), Japan (Rs 64.5 m and Rs 146.8 m), Iran (Rs 58.3 m and Rs 63.4 m), Australia (Rs 47.1 m and Rs 60.5 m), Canada (Rs 37.8 m and Rs 66.1 m) and Czechoslovakia (Rs 34.6 m and Rs 88.1 m). All these countries have increased the value of their imports, but their shares in the total have shown falling trends. It must be remembered that jute manufactures have been one of India's traditional exports and the dwindling demand for them due to the entry of competitors into the field is a matter of concern.

Sources: 1. *Annual Report 1985-86*, Ministry of Commerce.
2. *Report on Currency and Finance 1985-86, Vol. II*, Reserve Bank of India.
3. *Seventh Five Year Plan 1985-90*, Government of India Planning Commission.
4. *Economic Survey 1984-85*, Ministry of Finance.
5. *Foreign Trade Statistics of India, April 1984, May 1985, April 1986, April 1987 & Economic Profiles of 40 Major Countries, March 1987* and *Basic Statistics Relating to the Indian Economy, Vol. I: All India, August 1986*, Centre for Monitoring Indian Economy.

MAJOR EXPORTS — III

Table 111

(in Rs. million)

Commodity	Units of quantity	1981-82		1983-84		1985-86[1]	
		Quantity	Value	Quantity	Value	Quantity	Value
Cotton (raw)	thousand tonnes	23.7	363.2	124.8	1,489.5	32.4	658.2
Silk (raw)	—	N.A.	0.2	N.A.	1.2	N.A.	N.A.
Wool (raw)	thousand tonnes	0.8	7.9	N.A.	0.9	N.A.	N.A.
Jute (raw)	thousand tonnes	41.6	121.0	0.7	1.5	8.0	9.5
Cotton yarn	m. kgs.	6.1	254.8	6.1	195.7	9.4	350.8
Yarns of man-made fibre	—	N.A.	146.2	N.A.	203.2	N.A.	223.2
Cotton fabrics	m. sq. mts.[2]	457.1	2,945.5	305.0	2,766.0	347.1	3,715.7
Silk fabrics	—	N.A.	323.7	N.A.	412.1	N.A.	485.2
Fabrics (man-made)	—	N.A.	360.1	N.A.	265.2	N.A.	213.3
Woollen fabrics	—	N.A.	45.7	N.A.	11.0	N.A.	34.4
Made up articles wholly or chiefly of cotton	—	N.A.	1,041.0	N.A.	763.0	N.A.	1,025.3
Tulle, lace and other small wares (except of jute)	—	N.A.	82.4	N.A.	59.0	N.A.	32.1
Readymade garments	—	N.A.	5,958.3	N.A.	6,095.9	N.A.	10,075.0
Coir manufacture	—	N.A.	170.4	N.A.	234.8	N.A.	326.3
Jute products including twist and yarn	thousand tonnes	430	2,575.1	300	1,645.2	2.5	2,696.0
Carpets	—	N.A.	2,612.0	N.A.	149.2	N.A.	106.6
Raw hides and skins (excepting fur skins)	—	N.A.	4.4	N.A.	1.2	N.A.	0.1
Machinery including transport equipment	—	N.A.	6,178.9	N.A.	4,969.8	N.A.	6,039.4
Paper, paper-board and allied products	thousand tonnes	5.4	67.0	2.0	37.3	1.9	43.7

Note: [1] Provisional.
[2] million square metres.

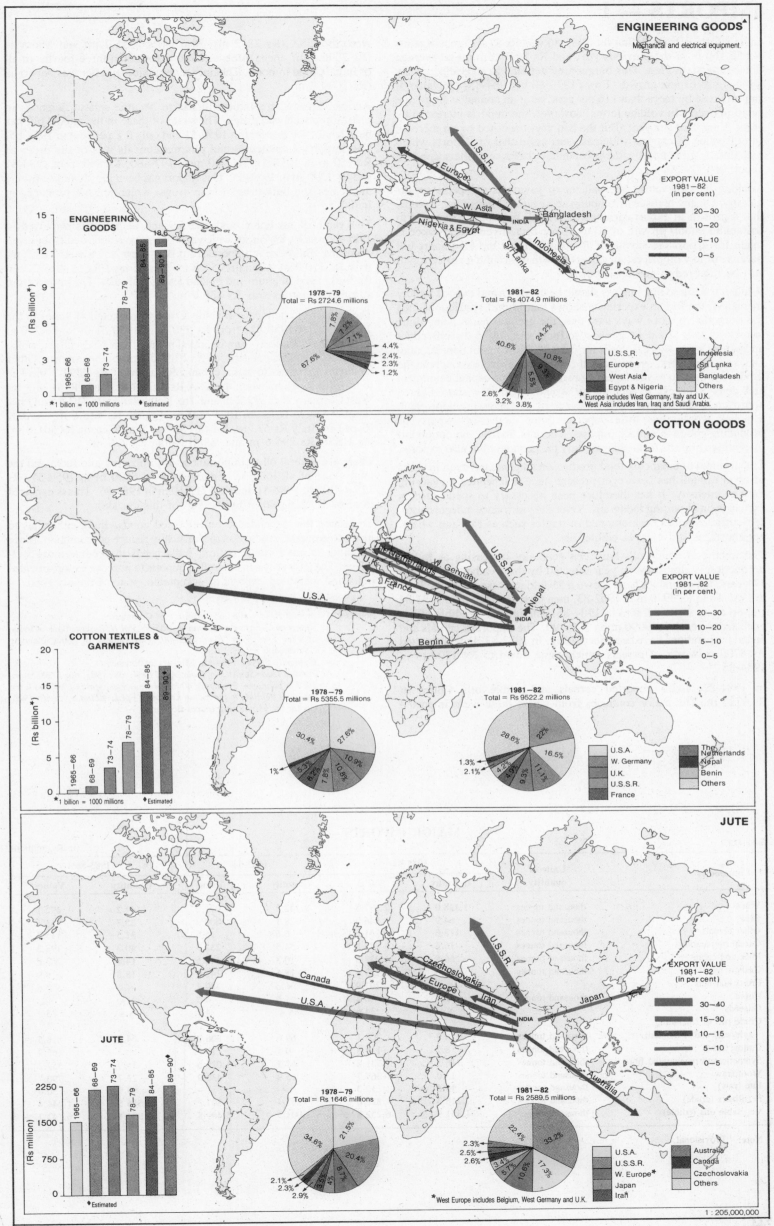

ENGINEERING GOODS
Mechanical and electrical equipment.

ENGINEERING GOODS

(Rs billion*)

15
12
9
6
3

18.6

1965—66 68—69 73—74 78—79 84—85 89—90◆

*1 billion = 1000 millions   ◆ Estimated

EXPORT VALUE
1981—82
(in per cent)

20—30
10—20
5—10
0—5

1978—79
Total = Rs 2724.6 millions

67.6%
7.8%
7.2%
7.1%
4.4%
2.4%
2.3%
1.2%

1981—82
Total = Rs 4074.9 millions

40.6%
24.2%
10.8%
9.3%
5.5%
2.6%
3.2%
3.8%

U.S.S.R.
Europe ★
West Asia ▲
Egypt & Nigeria
Indonesia
Sri Lanka
Bangladesh
Others

★ Europe includes West Germany, Italy and U.K.
▲ West Asia includes Iran, Iraq and Saudi Arabia.

COTTON GOODS

EXPORT VALUE
1981—82
(in per cent)

20—30
10—20
5—10
0—5

COTTON TEXTILES & GARMENTS

(Rs billion*)

20
15
10
5

1965—66 68—69 73—74 78—79 84—85 89—90◆

*1 billion = 1000 millions   ◆ Estimated

1978—79
Total = Rs 5355.5 millions

30.4%
27.6%
10.9%
10.8%
7.8%
6.7%
5.3%
1%

1981—82
Total = Rs 9522.2 millions

28.6%
22%
16.5%
11.1%
9.3%
4.9%
4.2%
2.1%
1.3%

U.S.A.
W. Germany
U.K.
U.S.S.R.
France
The Netherlands
Nepal
Benin
Others

JUTE

EXPORT VALUE
1981—82
(in per cent)

30—40
15—30
10—15
5—10
0—5

JUTE

(Rs million*)

2250
1500
750
0

1965—66 68—69 73—74 78—79 84—85 89—90◆

◆ Estimated

1978—79
Total = Rs 1646 millions

34.6%
21.5%
20.4%
8.7%
4%
3.5%
2.9%
2.3%
2.1%

1981—82
Total = Rs 2589.5 millions

33.2%
22.4%
17.3%
10.6%
5.7%
3.4%
2.6%
2.5%
2.3%

U.S.A.
U.S.S.R.
W. Europe ★
Japan
Iran
Australia
Canada
Czechoslovakia
Others

★ West Europe includes Belgium, West Germany and U.K.

1 : 205,000,000

243

IMPORTS — I

India's foreign trade deficit for 1985-86 was Rs 87,470 m as against Rs 53,160 m in the previous year and Rs 56,600 m on an average during the Sixth Plan. This burgeoning deficit was caused primarily by sluggish export growth. From 1979-80 to 1984-85, the value of imports rose by more than 116 per cent, or at an annual average rate of 19 per cent. In volume terms, however, the imports increased by some 3 per cent per year. But the country, benefited by an increase in unit value of exports which far surpassed that of imports which had already gone up in value substantially because of the second oil shock in 1979-80.

Looking into the future, the balance of payments prospects appear gloomy. On any realistic assumption about current invisibles, it seems that a growth in export earnings of 9 to 12 per cent is required to sustain an import growth of 10 to 11 per cent. Even if allowance is made for a possible saving of Rs 10,000 m in the oil import bill because of low oil prices, an export growth of around 6 per cent will still be required to sustain imports.

While it is realized that imports should be cut down and exports rapidly increased, the Indian experience shows that import restrictions hurt the economy in more ways than one, the greatest harm being done to exports and, as a consequence, to India's ability to be self-reliant and pay its way. There is scope for curtailing imports of non-essential and, to some extent, even essential commodities. But to condemn the productive system to shortages of raw materials and capital equipment and deny it access to technology would be a very short-sighted approach to the solution of the balance of payments problem. Such a policy would give little opportunity of increasing exports or producing goods to replace imports. There is, indeed, no option but to continue imports to sustain not only the economy but also exports.

The problem of balancing food production and demand even in years of good rainfall has been ever present since independence, varying only in intensity. It has therefore been necessary to spend foreign exchange on importing foodgrains. Over the years since independence, self-sufficiency has been attained in staples such as rice and wheat, but shortages still exist in edible oils.

The import of edible oils has been generally increasing as Indian production is almost every year affected by the weather. In the last 17 years, import values have shown a 55-fold increase, from Rs 70 m in 1955-56 to Rs 3970 m in 1982-83, then up to Rs 5400 m in 1983-84 and up again to Rs 8300 m in 1984-85. The import of edible oils is expected to reach Rs 9090 m by 1989-90. The import of cereals and cereal preparations have been decreasing. In 1982-83, India imported Rs 3700 m worth of cereals which came down to Rs 1700 m in 1984-85.

In 1981-82, India imported cereals and edible oils valued at Rs 9725 m. The major countries from which these were imported were the USA (Rs 2822 m), Brazil (Rs 2355.7 m) and Malaysia (Rs 2096.3 m), these three accounting for about three-fourths of all of India's food imports. The only other major supplier was Australia (Rs 1040.7 m).

Another area where India still has to become self-sufficient is in pharmaceuticals and drugs. The share of these in the import bill has been small, 1.9 per cent in 1950-51 and only 0.8 per cent in 1984-85. The import of medicines and pharmaceuticals was of the order of Rs 150 m in 1950-51, Rs 740 m in 1979-80, Rs 850 m in 1980-81 and Rs 1290 m in 1984-85. This import has been mostly of life-saving medicines and bulk drugs (basic drugs) which are not produced in India.

The value of medicinal and pharmaceutical products imported in 1981-82 was Rs 844 m, of which nearly half was imported from seven countries. The value of imports from West Germany was Rs 128.7 m, Italy Rs 110.2 m, China Rs 53.8 m, Switzerland Rs 53.4 m, Spain Rs 34.4 m and Belgium and Poland Rs 16.3 m each. The import from all other countries was Rs 430.9 m.

The import of chemicals in value terms skyrocketed 48 times in the 35 years after 1950-51, from Rs 530 m in 1950-51 to Rs 26,270 m in 1984-85. Chemicals were imported mainly from 12 countries, which accounted for 82.8 per cent, or Rs 10,990.7 m, of the imports. Other countries exported to India chemicals valued at Rs 2292.3 m. The four major suppliers were: the USA Rs 3135.1 m, Japan Rs 1628.2 m, West Germany Rs 1392.2 m and The Netherlands Rs 717.5 m. Other suppliers were Belgium Rs 672.9 m, France Rs 638.6 m, the UK Rs 611 m, Italy Rs 573 m, Kuwait Rs 553.8 m, Canada Rs 501.8 m, the USSR Rs 296.6 m and South Korea Rs 270 m.

There are several other important goods imported into India and the total value of all imports amounted to Rs 6500 m in 1950-51, Rs 125,490 m in 1980-81 and Rs 171,710 m in 1984-85. This is expected to cross Rs 206,940 m by the end of this decade.

To reduce this dependence on imported goods, India either has to increase domestic production or drastically reduce consumption. Since the latter is not a viable option, India needs to become more self-reliant. The answer lies in controlled imports to improve export-oriented units and increased exports of acceptable, state-of-the-art goods.

Sources: 1. *Annual Report 1985-86*, Ministry of Commerce.
2. *Report on Currency and Finance 1985-86, Vol. II*, Reserve Bank of India.
3. *Seventh Five Year Plan 1985-90*, Government of India Planning Commission.
4. *Economic Survey 1984-85*, Ministry of Finance.
5. *Foreign Trade Statistics of India, April 1984, May 1985, April 1986, April 1987 & Economic Profiles of 40 Major Countries, March 1987* and *Basic Statistics Relating to the Indian Economy, Vol. I: All India, August 1986*, Centre for Monitoring Indian Economy.

MAJOR IMPORTS — I

Table 112

(in Rs. million)

Commodity	Units of quantity	1981-82		1983-84		1985-86[1]	
		Quantity	Value	Quantity	Value	Quantity	Value
Wheat	thousand tonnes	1,328.0	2,297.5	2,142.3	5,071.2	144.2	492.1
Rice	thousand tonnes	64.9	146.9	328.1	800.0	25.7	83.1
Other cereals	thousand tonnes	113.0	251.5	6.6	15.6	31.5	115.3
Cereal preparations	thousand tonnes	16.4	75.8	54.5	232.3	40.3	198.3
Milk and cream	thousand tonnes	66.8	938.5	10.3	148.5	14.4	289.0
Cashew nuts	thousand tonnes	16.0	183.6	15.6	141.9	18.8	216.9
Fruits and nuts	—	N.A.	207.9	N.A.	389.7	N.A.	350.6
Copra	thousand tonnes	2.4	10.4	1.0	4.7	N.A.	N.A.
Oilseeds	—	N.A.	34.6	N.A.	4.7	N.A.	25.2
Crude rubber (including synthetic and reclaimed)	thousand tonnes	76.9	760.1	69.0	806.2	64.7	818.5
Cotton (raw)	thousand tonnes	8.2	118.3	0.5	12.7	6.2	65.9
Synthetic and regenerated fibre	thousand tonnes	116.5	1,731.8	67.3	1,025.6	35.9	557.2
Wool (raw)	thousand tonnes	13.0	361.7	15.9	430.9	27.7	790.7
Jute (raw)	thousand tonnes	10.4	13.3	—	—	9.9	50.0
Fertilizers (crude)	thousand tonnes	1,218.3	825.6	952.4	807.2	1,516.5	1,244.4
Vegetable oils (edible)	thousand tonnes	1,351.9	6,252.8	1,001.3	5,409.8	904.1	6,142.7

Note: [1] Provisional.

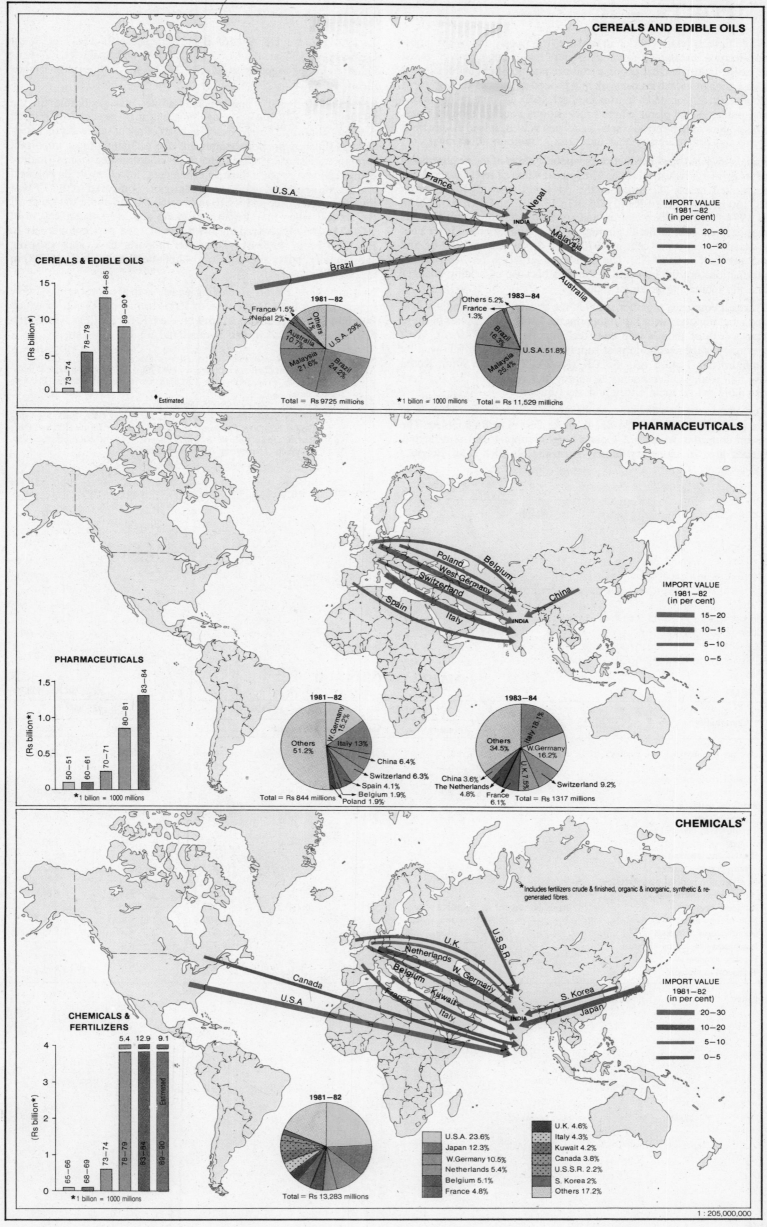

CEREALS AND EDIBLE OILS

IMPORT VALUE
1981—82
(in per cent)

20—30
10—20
0—10

CEREALS & EDIBLE OILS

(Rs billion*)

15

10

5

0

73—74 78—79 84—85 89—90♦

♦ Estimated

1981—82

France 1.5%
Nepal 2%
Others 11%
Australia 10.7%
Malaysia 21.6%
Brazil 24.2%
U.S.A. 29%

Total = Rs 9725 millions

*1 billion = 1000 millions

1983—84

Others 5.2%
France 1.3%
Brazil 16.3%
Malaysia 25.4%
U.S.A. 51.8%

Total = Rs 11,529 millions

PHARMACEUTICALS

IMPORT VALUE
1981—82
(in per cent)

15—20
10—15
5—10
0—5

PHARMACEUTICALS

(Rs billion*)

1.5

1.0

0.5

50—51 60—61 70—71 80—81 83—84

*1 billion = 1000 millions

1981—82

W.Germany 15.2%
Italy 13%
China 6.4%
Switzerland 6.3%
Spain 4.1%
Belgium 1.9%
Poland 1.9%
Others 51.2%

Total = Rs 844 millions

1983—84

Italy 18.1%
W.Germany 16.2%
Switzerland 9.2%
U.K. 7.5%
France 6.1%
The Netherlands 4.8%
China 3.6%
Others 34.5%

Total = Rs 1317 millions

CHEMICALS*

*Includes fertilizers crude & finished, organic & inorganic, synthetic & re-generated fibres.

IMPORT VALUE
1981—82
(in per cent)

20—30
10—20
5—10
0—5

CHEMICALS & FERTILIZERS

(Rs billion*)

4

3

2

1

0

65—66 68—69 73—74 78—79 83—84 89—90

5.4 12.9 9.1

Estimated

*1 billion = 1000 millions

1981—82

U.S.A. 23.6%
Japan 12.3%
W.Germany 10.5%
Netherlands 5.4%
Belgium 5.1%
France 4.8%
U.K. 4.6%
Italy 4.3%
Kuwait 4.2%
Canada 3.8%
U.S.S.R. 2.2%
S. Korea 2%
Others 17.2%

Total = Rs 13,283 millions

1 : 205,000,000

245

IMPORTS — II

While India is relatively rich in iron ore, it is poorly endowed in most non-ferrous metals and precious stones. India can, however, boast of major industries that process iron ore and non-ferrous metals and a vast cottage industries network that processes and polishes precious stones such as diamonds. In iron and steel, India has both private sector and public sector plants where processing is done on modern lines. This applies to non-ferrous metal industries as well. It is in the precious stone and pearl industry that India has a tradition all its own.

India's iron and steel production is woefully short of the requirements of its growing industrial base. The import value of iron and steel has multiplied nearly 47 times in the 34 years since 1950-51, from Rs 160 m to a peak of Rs 12,035 m in 1981-82 and then Rs 7770 m in 1984-85. The share of iron and steel imports among all imports was 2.5 per cent at the beginning of the planning era, 11 per cent a decade later, 9 per cent in 1970-71, 6.8 per cent in 1980-81 and 4.5 per cent in 1984-85. The decline is largely a result of Indian production, but the increase in value is because of pricing changes and the increasing needs of industry.

Thirteen countries together provide the greater part of India's iron and steel imports, with Japan topping the list as the supplier of more than a fifth of India's iron and steel imports (Rs 2757.9 m). West Germany is the second largest supplier (Rs 2455.8 m), followed by Belgium (Rs 1143.5 m), the UK (Rs 745.2 m) and South Korea (Rs 688.6 m). Other countries supply the remaining 35 per cent (Rs 4244 m) of India's iron and steel imports.

The import of non-ferrous metals grew a little over 13 times in the period 1950-51 to 1984-85, from Rs 260 m to Rs 3450 m. The seven countries which are India's biggest suppliers of non-ferrous metals are, in monetary terms: Australia Rs 281.7 m, Belgium

Rs 199.7 m, the UK Rs 168.8 m, Venezuela Rs 158.7 m, West Germany Rs 134 m, the USSR Rs 128.6 m and Spain Rs 127.1 m. Other countries together supply Rs 2772.4 m worth of non-ferrous metals to India.

The import of pearls and precious stones was negligible till the 1970s. Then with the growing momentum of exports of finished precious stones and pearls, Indian imports of uncut stones skyrocketed. Today, finished precious stones are one of India's most important foreign exchange earners. India imports unfinished stones (mainly diamonds) and exports them as ornaments, jewellery or as polished and cut stones; pearl exports are negligible compared to this. These imports were worth just Rs 10 m in 1950-51 and about the same till the sixties, when they began increasing and grew to Rs 250 m in 1970-71. They registered a growth rate of 1568 per cent increase to Rs 4170 m in 1980-81. Thereafter imports have kept growing substantially, but at a lesser rate (147 per cent) to reach Rs 10,280 m in 1984-85.

Non-metallic minerals (including pearls and precious stones) were imported into India to the tune of Rs 5116 m in 1981-82, of which Belgium accounted for Rs 2166.9 m, the UK for Rs 1255.5 m, South Korea for Rs 432.4 m and Switzerland for Rs 301.4 m.

Sources: 1. *Annual Report 1985-86*, Ministry of Commerce.
2. *Report on Currency and Finance 1985-86, Vol.II*, Reserve Bank of India.
3. *Seventh Five Year Plan 1985-90*, Government of India Planning Commission.
4. *Economic Survey 1984-85*, Ministry of Finance.
5. *Foreign Trade Statistics of India, April 1984, May 1985, April 1986, April 1987 & Economic Profiles of 40 Major Countries, March 1987* and *Basic Statistics Relating to the Indian Economy, Vol.I: All India, August 1986*, Centre for Monitoring Indian Economy.

MAJOR IMPORTS — II

Table 113

(in Rs. million)

Commodity	Units of quantity	1981-82		1983-84		1985-86[1]	
		Quantity	Value	Quantity	Value	Quantity	Value
Sulphur & unroasted iron pyrites	thousand tonnes	682.0	1,064.3	445.5	633.0	653.1	1,498.4
Other crude minerals	—	N.A.	487.1	N.A.	517.6	N.A.	788.3
Metalliferrous ores and metal scrap	—	N.A.	2,027.5	N.A.	1,445.7	N.A.	2,862.0
Artificial resins, plastic materials etc.	—	N.A.	1,251.7	N.A.	1,890.2	N.A.	2,914.8
Textile yarn, fabrics and made-up articles etc.	—	N.A.	967.3	N.A.	1,250.5	N.A.	1,330.4
Pearls, precious and semi-precious stones	—	N.A.	3,973.8	N.A.	10,823.8	N.A.	1,162.6
Non-metallic mineral manufactures (excluding pearls etc.)	—	N.A.	1,142.0	N.A.	1,752.2	N.A.	814.8
Iron and steel	thousand tonnes	3,165.2	12,035.4	2,195.7	9,629.0	1,996.5	12,308.5
Non-ferrous metal	—	N.A.	3,971.3	N.A.	3,690.8	N.A.	4,685.1
Manufactures of metals	—	N.A.	1,155.4	N.A.	1,477.7	N.A.	1,858.0
Machinery (excluding electric)	—	N.A.	13,843.3	N.A.	19,738.4	N.A.	24,785.7
Electric machinery	—	N.A.	2,913.2	N.A.	4,035.6	N.A.	5,671.4
Transport equipment	—	N.A.	3,050.1	N.A.	4,562.4	N.A.	4,522.2

Note: [1] Provisional.

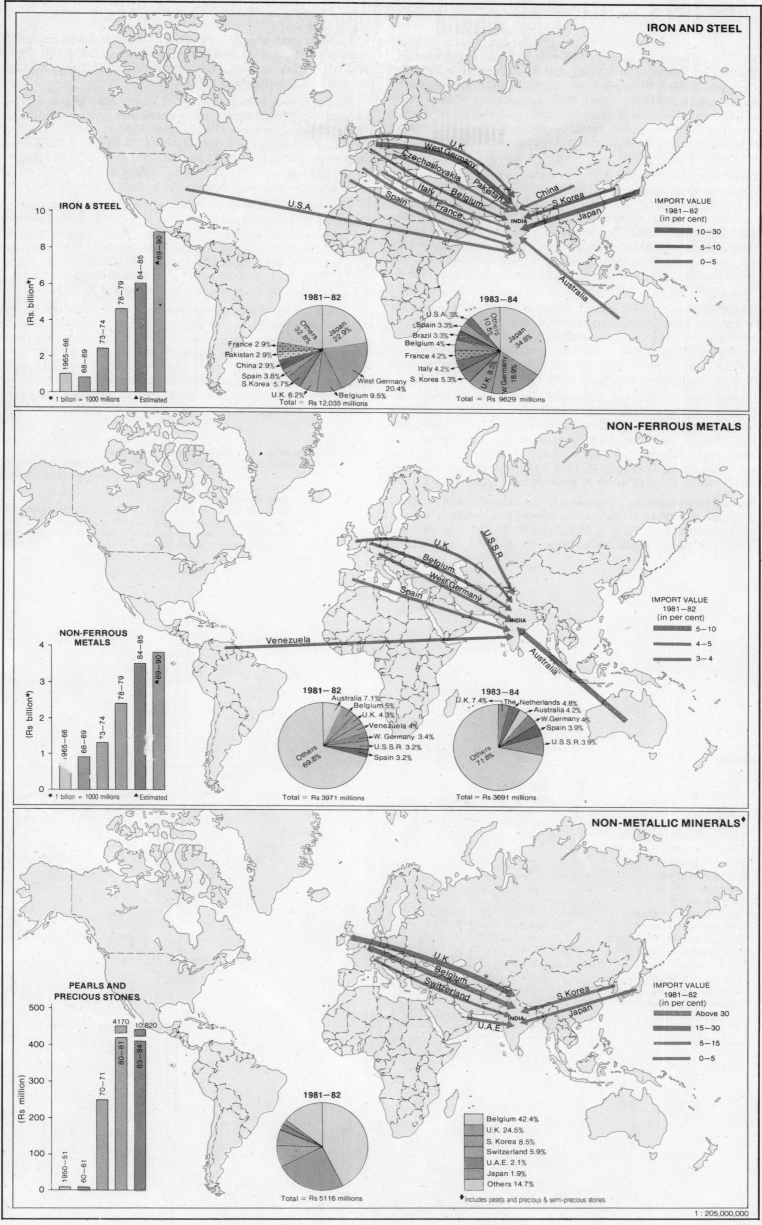

IRON AND STEEL

IMPORT VALUE
1981–82
(in per cent)
- 10–30
- 5–10
- 0–5

IRON & STEEL

(Rs. billion*)

10
8
6
4
2
0

1965–66
68–69
73–74
78–79
84–85
▲89–90

* 1 billion = 1000 millions   ▲ Estimated

1981–82

Others 32.5%
Japan 22.9%
France 2.9%
Pakistan 2.9%
China 2.9%
Spain 3.8%
S. Korea 5.7%
U.K. 6.2%
Belgium 9.5%
West Germany 20.4%

Total = Rs 12,035 millions

1983–84

U.S.A 3%
Spain 3.3%
Brazil 3.3%
Belgium 4%
France 4.2%
Italy 4.2%
S. Korea 5.3%
U.K. 8.5%
W. Germany 18.9%
Japan 34.8%
Others 10.5%

Total = Rs 9629 millions

NON-FERROUS METALS

IMPORT VALUE
1981–82
(in per cent)
- 5–10
- 4–5
- 3–4

NON-FERROUS METALS

(Rs billion*)

4
3
2
1
0

1965–66
68–69
73–74
78–79
84–85
▲89–90

* 1 billion = 1000 millions   ▲ Estimated

1981–82

Australia 7.1%
Belgium 5%
U.K. 4.3%
Venezuela 4%
W. Germany 3.4%
U.S.S.R. 3.2%
Spain 3.2%
Others 69.8%

Total = Rs 3971 millions

1983–84

U.K. 7.4%
The Netherlands 4.8%
Australia 4.2%
W. Germany 4%
Spain 3.9%
U.S.S.R. 3.9%
Others 71.8%

Total = Rs 3691 millions

NON-METALLIC MINERALS♦

IMPORT VALUE
1981–82
(in per cent)
- Above 30
- 15–30
- 5–15
- 0–5

PEARLS AND PRECIOUS STONES

(Rs million)

500
400
300
200
100
0

1950–51
60–61
70–71
80–81 4170
83–84 10,820

1981–82

Belgium 42.4%
U.K. 24.5%
S. Korea 8.5%
Switzerland 5.9%
U.A.E. 2.1%
Japan 1.9%
Others 14.7%

Total = Rs 5116 millions

♦ Includes pearls and precious & semi-precious stones.

1 : 205,000,000

247

IMPORTS — III

In the 34 years since 1950-51, the import value of petroleum products has gone up 96 times, from Rs 550 m to Rs 53,450 m in 1984-85. This import amounted to 8.5 per cent of all imports in 1950-51, 7 per cent in 1960-61 and 8.4 per cent in 1970-71. But it grew to 42 per cent in 1980-81. In 1984-85, this came down to 31 per cent as India started improving its own production, and it is likely to fall somewhat further in the near future. The value of petroleum products to be imported in 1989-90 is estimated to be about Rs 51,360 m (24.8 per cent of total imports).

Until 1973, India's import bill for crude oil and petroleum products consumed about 10 per cent of the country's total export earnings. Following the two oil shocks of 1973 and 1979 and the continued increase in domestic consumption, this ratio crossed the 60 per cent mark in the early 1980s. Since 1981-82, exports of Bombay High crude oil have provided a windfall and reduced the ratio considerably, but, with the commissioning of new refinery capacities in India, the export earnings are projected to decline to about Rs 5650 m in 1985-86 from Rs 18,180 m in 1984-85.

Recent worldwide decline in crude oil prices has provided some relief in terms of value of imports. However, domestic consumption of petroleum has been increasing at about 5 per cent per year. Hence, the import bill, even after correction for exports of some unsuitable crude and the country's growing refining capacity, will continue to be high, but never again 60 per cent.

India obtains two-thirds of its imports of petroleum products from six oil-producing countries. Iran contributes the most (24.4 per cent valued at Rs 12,651 m), followed by Saudi Arabia (15.8 per cent, Rs 8195.6 m), the USSR (8.9 per cent, Rs 4616 m), the United Arab Emirates (8.2 per cent, Rs 4233.4 m), Iraq (7.7 per cent, Rs 4007.4 m) and Venezuela (2.8 per cent, Rs 1440.3 m). The other oil producers supply Rs 16,750.4 m worth of petroleum products (1981-82).

Because of the substantial expansion in the industrial sector, capital goods and scientific instruments have increased in import value some 18 times, from Rs 1520 m in 1950-51 to Rs 29,870 m in 1984-85. Imports of capital goods amounted to Rs 22,975 m in 1981-82.

The value of goods imported from the USA that year was Rs 5611.1 m, from the UK Rs 4585.9 m, West Germany Rs 3938.5 m, Japan Rs 3169.6 m, France Rs 1424.2 m, Italy Rs 764.7 m, the USSR Rs 623 m and Switzerland Rs 514.8 m.

There has always been a shortage of paper, especially newsprint, in India. Imports of paper, paper board and paper-based articles in the 35 years between 1950-51 and 1984-85 increased nearly 18-fold, from a value of a mere Rs 100 m to Rs 120 m in 1960-61, then to Rs 251 m in 1970-71 and Rs 1870 m in 1980-81, peaking at Rs 2454 m in 1981-82 before falling to Rs 1750 m in 1984-85. This import is expected to decline slightly in the years ahead.

Imported paper cost India Rs 2454 m in 1981-82. Eight countries supplied a third of the imports, the USSR topping the list (Rs 176.7 m), followed by Sweden (Rs 156.1 m), West Germany (Rs 131.2 m), Bangladesh (Rs 72.2 m), Brazil (Rs 52 m), China (Rs 42.9 m), Yugoslavia (Rs 39.5 m) and South Korea (Rs 24.8 m). Paper supplied by other countries was valued at Rs 1758.6 m.

All these three import categories show a distinct rise in the value of imports from the 1970s, indicating that the internal demands from agriculture, industry and services have been rising faster than before and have necessitated imports to keep pace with the growth. But it must also be pointed out that these imports have done an immense amount of good to India: not only have they made India self-reliant in agriculture and some sectors of industry, but they have also made India a moderate industrial power.

Sources: 1. *Annual Report 1985-86*, Ministry of Commerce.
2. *Report on Currency and Finance 1985-86, Vol. II*, Reserve Bank of India.
3. *Seventh Five Year Plan 1985-90*, Government of India Planning Commission.
4. *Economic Survey 1984-85*, Ministry of Finance.
5. *Foreign Trade Statistics of India, April 1984, May 1985, April 1986, April 1987 & Economic Profiles of 40 Major Countries, March 1987* and *Basic Statistics Relating to the Indian Economy, Vol. I: All India, August 1986*, Centre for Monitoring Indian Economy.

MAJOR IMPORTS — III

Table 114

(in Rs. million)

Commodity	Units of quantity	1981-82		1983-84		1985-86[1]	
		Quantity	Value	Quantity	Value	Quantity	Value
Pulp and waste paper	thousand tonnes	85.6	414.4	217.1	823.0	526.1	2,359.8
Petroleum, petroleum products	m. tonnes	20.2	51,892.6	20.3	48,301.1	19.0	49,901.4
Organic chemicals	—	N.A.	2,426.8	N.A.	3,970.4	N.A.	4,373.0
Inorganic chemicals	—	N.A.	2,425.0	N.A.	2,126.1	N.A.	4,420.8
Dyeing, tanning and colouring materials	—	N.A.	248.5	N.A.	427.6	N.A.	679.7
Medicinal and pharmaceutical products	—	N.A.	844.4	N.A.	1,317.2	N.A.	1,635.7
Fertilizers (manufactured)	thousand tonnes	2,792.3	5,097.1	718.8	1,124.8	3,493.5	8,225.3
Chemical materials and products	—	N.A.	775.9	N.A.	1,228.5	N.A.	1,165.7
Paper, paper-board and allied products	thousand tonnes	412.3	2,453.8	243.5	1,725.9	208.4	1,952.4
Professional, scientific, controlling instruments, photographic, optical goods, watches and clocks	—	N.A.	2,013.2	N.A.	2,806.9	N.A.	3,479.9

Note: [1] Provisional.

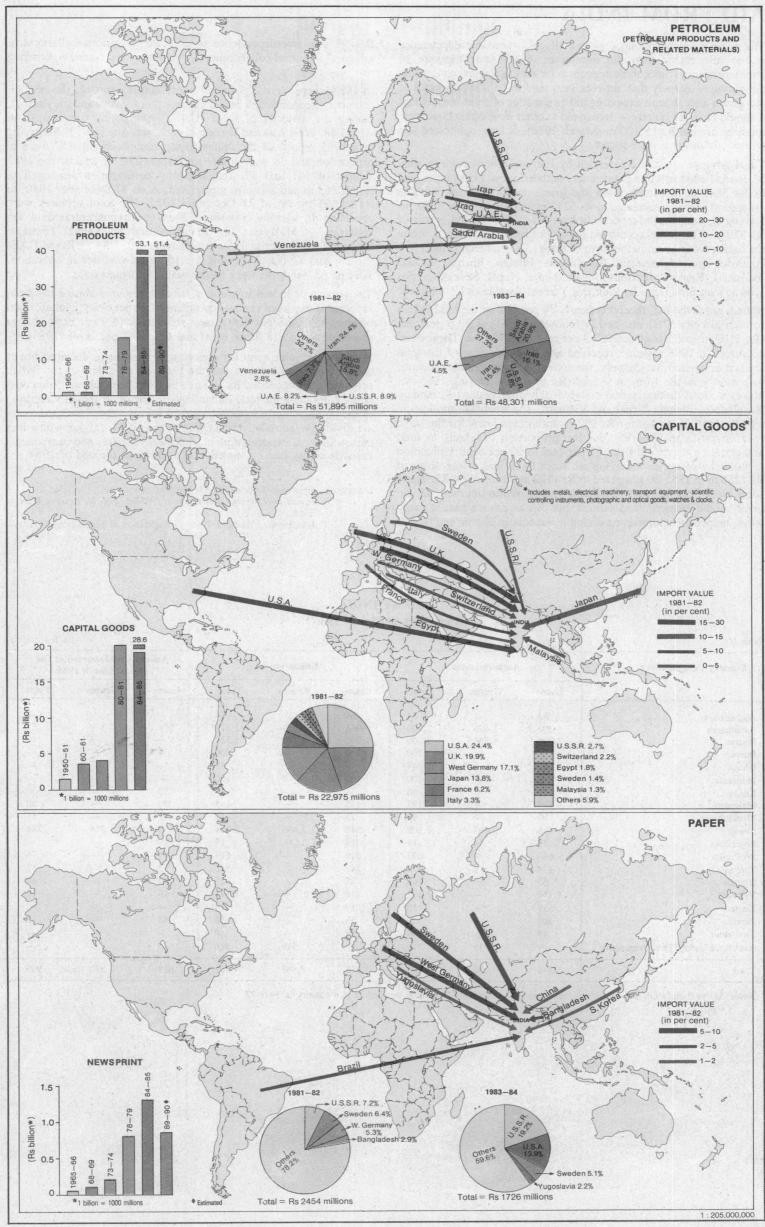

PETROLEUM (PETROLEUM PRODUCTS AND RELATED MATERIALS)

IMPORT VALUE
1981–82
(in per cent)
20–30
10–20
5–10
0–5

PETROLEUM PRODUCTS

(Rs billion*)

53.1 51.4

1965–66 | 68–69 | 73–74 | 78–79 | 84–85 | 89–90◆

*1 billion = 1000 millions ◆ Estimated

1981–82
Others 32.2%
Iran 24.4%
Saudi Arabia 15.8%
Iraq 7.7%
Venezuela 2.8%
U.A.E. 8.2%
U.S.S.R. 8.9%
Total = Rs 51,895 millions

1983–84
Others 27.3%
Saudi Arabia 20.9%
Iraq 16.1%
U.A.E. 4.5%
Iran 15.4%
U.S.S.R. 15.8%
Total = Rs 48,301 millions

U.S.S.R.
Iran
Iraq
U.A.E.
Saudi Arabia
Venezuela
INDIA

CAPITAL GOODS*

*Includes metals, electrical machinery, transport equipment, scientific controlling instruments, photographic and optical goods, watches & clocks.

IMPORT VALUE
1981–82
(in per cent)
15–30
10–15
5–10
0–5

Sweden
U.K.
W. Germany
U.S.A.
France
Italy
Switzerland
Egypt
U.S.S.R.
Japan
INDIA
Malaysia

CAPITAL GOODS

(Rs billion*)

20 28.6

1950–51 | 60–61 | 80–81 | 84–85

*1 billion = 1000 millions

1981–82
Total = Rs 22,975 millions

U.S.A. 24.4%
U.K. 19.9%
West Germany 17.1%
Japan 13.8%
France 6.2%
Italy 3.3%
U.S.S.R. 2.7%
Switzerland 2.2%
Egypt 1.8%
Sweden 1.4%
Malaysia 1.3%
Others 5.9%

PAPER

IMPORT VALUE
1981–82
(in per cent)
5–10
2–5
1–2

Sweden
U.S.S.R.
West Germany
Yugoslavia
China
Bangladesh
S. Korea
INDIA
Brazil

NEWSPRINT

(Rs billion*)

1.5
1.0
0.5

1965–66 | 68–69 | 73–74 | 78–79 | 84–85 | 89–90◆

*1 billion = 1000 millions ◆ Estimated

1981–82
Others 78.2%
U.S.S.R. 7.2%
Sweden 6.4%
W. Germany 5.3%
Bangladesh 2.9%
Total = Rs 2454 millions

1983–84
Others 59.6%
U.S.S.R. 19.2%
U.S.A. 13.9%
Sweden 5.1%
Yugoslavia 2.2%
Total = Rs 1726 millions

1 : 205,000,000

249

AID FROM INDIA

The world is fast becoming a 'global village', where interdependence is the rule rather than the exception. Aid is one measure of interdependency. Since there appears to be a universal law concerning aid-giving — namely that aid-receivers tend to be a bit resentful of aid-givers and, though expecting aid as a matter of right, expect that it should be given quietly — Indian aid to other developing countries, totalling around Rs 15,100 m (March 1986), is little publicized and treated almost as a state secret.

The aggregate assistance extended by India to other countries may be divided into loans and grants under authorization. Loans amounted to Rs 5841 m in March 1986 and grants to Rs 9259 m. But the utilization of loans amounted to Rs 5212 m and grants to Rs 8889 m, leaving an undisbursed amount in loans of Rs 629 m and grants of Rs 370 m. The grants accounted for 61 per cent of assistance, while the balance was given in the form of loans. The eighteen beneficiary countries are Afghanistan, Bangladesh, Bhutan, Burma, Ghana, Indonesia, Kenya, Mauritius, Mozambique, Nepal, Seychelles, Sri Lanka, Tanzania, Uganda, Vietnam, Yemen, Zambia and Zimbabwe.

Bhutan heads the list, receiving nearly 38 per cent of the total Indian foreign aid (Rs 5760 m), and is followed by Nepal 20.6 per cent (Rs 3109 m) and Bangladesh 17.8 per cent (Rs 2700 m). These three countries in 1986 together received about 76 per cent of the total amount authorized. Incidentally, the assistance given to these countries was mostly in the form of grants (Rs 8743 m). During 1985-86, however, fresh authorization covered only four countries, namely Bangladesh, Bhutan, Nepal and Vietnam. The financial implication was Rs 1156 m, as against Rs 977 m authorized anew for the same four countries in 1984-85. But the proportion of grants in total authorization decreased from 61 per cent to 60 per cent. Utilization of funds by countries receiving aid from India was of the order of Rs 1256 m in 1985-86 compared to Rs 1172 m during 1984-85. These sums are, of course, very small when compared to international aid given by the industrialized nations, but this assistance has, in many ways, helped the countries receiving it, especially Bhutan and Nepal.

Besides aid, interdependence is reflected in various collaborative schemes. Industrial collaboration is one of the most common, bringing two-way benefits.

In 1983, there were 235 Indian joint ventures abroad, the ESCAP countries (Economic and Social Commission for Asia and the Pacific), having the majority of them (124). Ventures elsewhere were in Africa 48, West Asia and Europe 25 each, and America 13. According to a 1985 report, of the Indian joint ventures abroad 157 were in production and 78 were under implementation as against the 1983 figures of 141 and 87. Indian equity participation amounted to Rs 662.63 m and earnings repatriated, as on 31 December 1982, to Rs 180.95 m. As of 31 December 1985, 156 joint ventures were operating in various countries. These were concentrated in ten countries — Malaysia: 25; Singapore and Sri Lanka: 16 each; Indonesia and Nigeria: 12 each; Thailand: 10; USA, UK and UAE: 9 each; and Kenya: 7. Among the 156 joint ventures in operation, 101, or 65 per cent, were in the manufacturing sector.

The pattern of Indian investment in joint ventures abroad has been in terms of export of capital equipment (63.5 per cent), capitalization of knowhow (6.8 per cent), cash remittance (8.9 per cent), bonus shares obtained (18.8 per cent) and others (2 per cent) (1983).

Some of the joint ventures operating abroad have declared bonus shares from time to time and the *pro rata* allotment to the Indian promoters has exceeded Rs 160 m which has been taken into account while computing the total Indian equity. The other benefits accruing to the country by the establishment of joint ventures over the years are dividends (cumulatively Rs 93.1 m upto 1984-85), knowhow fees etc. Additional exports of plant, machinery, spares and components towards equity had amounted to Rs 420 m at the end of 1984.

Sources: 1. *Economic Survey 1984-85*, Ministry of Finance.
2. *Report on Currency and Finance 1985-86, Vol. II*, Reserve Bank of India.
3. *Handbook of Statistics 1986*, Confederation of Engineering Industry.

AID BY INDIA
(Upto end of March 1986[1])

Table 115

(in Rs. million)

Country/Programme	Authorizations			Utilization			Amounts undisbursed at the end of March 1986		
	Loans	Grants	Total	Loans	Grants	Total	Loans	Grants	Total
Afghanistan	64	—	64	64	—	64	—	—	—
Bangladesh	1,204	1,496	2,700	966	1,347	2,313	238	149	387
Bhutan	1,206	4,554	5,760	1,206	4,551	5,757	—	3	3
Burma	200	—	200	200	—	200	—	—	—
Ghana	49	—	49	49	—	49	—	—	—
Indonesia	100	—	100	100	—	100	—	—	—
Kenya	50	—	50	—	—	—	50	—	50
Mauritius	195	—	195	145	—	145	50	—	50
Mozambique	90	—	90	73	—	73	17	—	17
Nepal	416	2,693	3,109	346	2,475	2,821	70	218	288
Seychelles	19	—	19	19	—	19	—	—	—
Sri Lanka	693	—	693	619	—	619	74	—	74
Tanzania	174	—	174	142	—	142	32	—	32
Uganda	65	—	65	25	—	25	40	—	40
Vietnam	1,157	—	1,157	1,141	—	1,141	16	—	16
Yemen	10	—	10	10	—	10	—	—	—
Zambia	100	—	100	58	—	58	42	—	42
Zimbabwe	49	—	49	49	—	49	—	—	—
Assistance under ITEC Programme	—	516	516	—	516	516	—	—	—
Total	5,841	9,259	15,100	5,212	8,889	14,101	629	370	999

Note: Aid to Bangladesh does not cover relief goods of value Rs. 170 million supplied to that country in 1971-72.

[1] Provisional

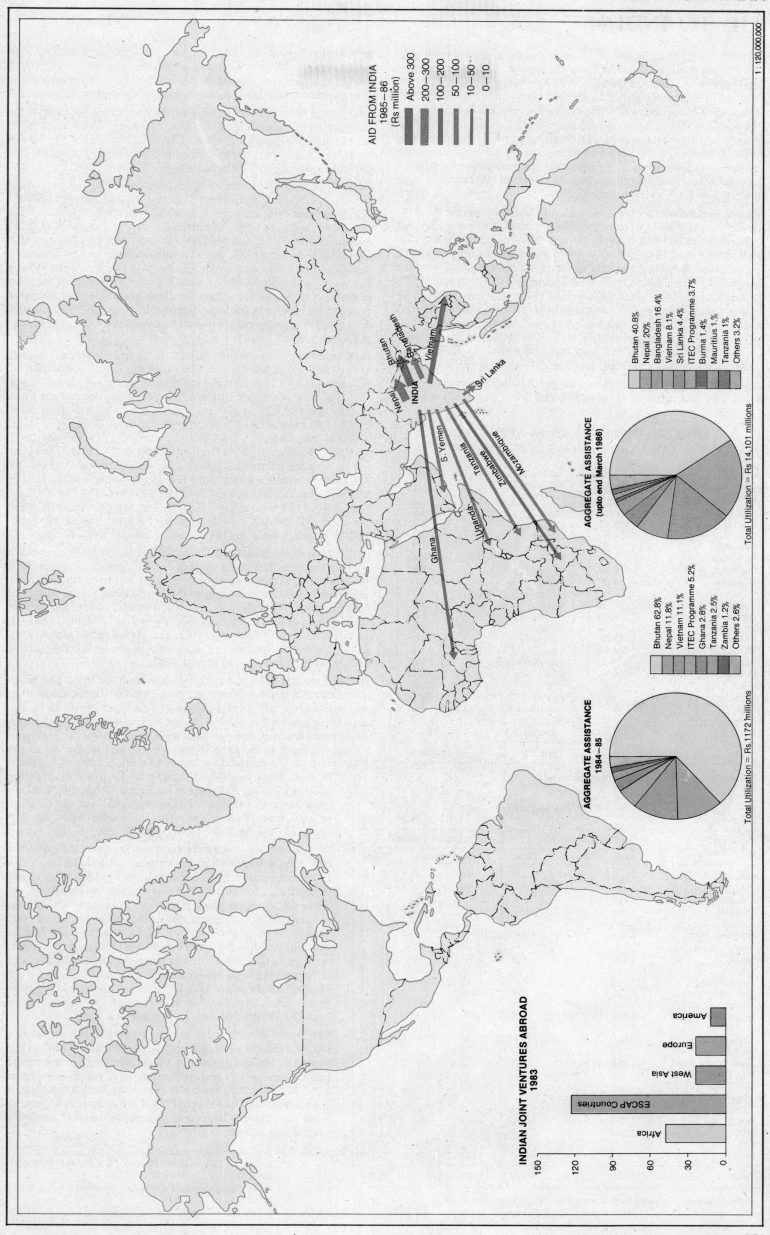

AID FROM INDIA

AID FROM INDIA
1985–86
(Rs million)

Above 300
200–300
100–200
50–100
10–50
0–10

1 : 120,000,000

Bhutan
Nepal
Bangladesh
Vietnam
INDIA
Sri Lanka
S. Yemen
Tanzania
Zimbabwe
Mozambique
Uganda
Ghana

Bhutan 40.8%
Nepal 20%
Bangladesh 16.4%
Vietnam 8.1%
Sri Lanka 4.4%
ITEC Programme 3.7%
Burma 1.4%
Mauritius 1%
Tanzania 1%
Others 3.2%

AGGREGATE ASSISTANCE
(upto end March 1986)

Total Utilization = Rs 14,101 millions

Bhutan 62.8%
Nepal 11.8%
Vietnam 11.1%
ITEC Programme 5.2%
Ghana 2.8%
Tanzania 2.5%
Zambia 1.2%
Others 2.6%

AGGREGATE ASSISTANCE
1984–85

Total Utilization = Rs 1172 millions

INDIAN JOINT VENTURES ABROAD
1983

America
Europe
West Asia
ESCAP Countries
Africa

150
120
90
60
30
0

251

AID TO INDIA

In endeavouring to attain one of its important goals, namely economic and technological self-reliance, India has needed to effect modernization in different sectors of the economy. Assistance in different forms from the developed countries has helped considerably in this effort. Such assistance includes loans, grants, credits, gifts and collaborations. But since the sixties, there has been a significant shift in the nature of such aid, India preferring non-foreign exchange expenditure with a view to stemming any depletion of the country's foreign exchange resources. The foreign exchange reserves themselves have gradually but steadily been climbing from about Rs 3190 m in 1960 to Rs 7530 m in 1970 and Rs 57,310 m during 1980.

In terms of economic performance, India can boast of a modest development, much of it achieved with the help of external assistance. Several ventures in India with foreign partners have helped to make this possible. Between 1957 and 1984, 8641 joint ventures were established in India, the UK being the partner in the largest number of them (1886) followed by the USA (1691), West Germany (1556), Japan (750), Switzerland (510), France (424), Italy (328), Sweden (170), The Netherlands (148), East Germany (144) and other countries 1034. But in terms of the Rs 618.7 m investment in joint ventures in 1983, Japan had put in the largest amount (Rs 160.77 m), followed by the USA (Rs 138.92 m), the UK (Rs 98.02 m), non-resident Indians (Rs 65.2 m), West Germany (Rs 48.42 m), and The Netherlands (Rs 26.86 m).

In the recent past, there have been changes of policy on the investment in India by multinational corporations and a few of them wound up their Indian operations. But this has not acted as a dampener to collaboration, and joint ventures are on the increase, even though foreign equity cannot exceed 40 per cent. Nearly 20 per cent of the total investment in existing collaborations is at present, foreign equity.

Joint ventures in 1984-85 accounted for around 19 per cent of the value added in Indian mining and manufacturing. These units, all in the private sector, added a value of Rs 25,290 m (11.1 per cent of the value added in the entire factory and non-factory private sector and not in mining and manufacturing).

Since November 1975, non-resident Indians (NRIs) and persons of Indian origin residing abroad have been permitted to open and maintain in India foreign currency (non-resident) accounts in designated foreign currencies, with the initial remittances received from abroad or by conversion of existing non-resident (external) accounts maintained in rupees. In February 1985, the government made an announcement clarifying that exemption under section 5(1) (xxxiii) of the Wealth Tax Act is available to the amounts lying in non-resident accounts in the case of NRIs. Since then the flow from NRIs has increased considerably. Direct investments approved during 1982-85 amounted to Rs 3159.6 m and portfolio investment totalled Rs 470 m at the end of March 1985.

Total external assistance authorized to India since independence was Rs 370,580 m at the end of March 1986. Loans constituted 82 per cent, grants 12 per cent and the balance was mostly from PL 480 funds. But the actual utilization of the funds has been of the order of only Rs 280,820 m (76 per cent). The total authorization of Rs 370,580 m originated from: the World Bank Group (IBRD and IDA) (40 per cent), the USA (18 per cent), the UK (10 per cent), the USSR (5 per cent), West Germany (5 per cent), Japan (3 per cent) and the balance from other sources. However, during 1985-86, the bulk of the assistance under loans (Rs 33,015 m) was subscribed by the IBRD (Rs 13,230 m), the USSR (Rs 8333 m), IDA (Rs 7269 m), France (Rs 1293 m), and the USA (Rs 286 m). In respect of grants (Rs 2515 m), the UK's share was Rs 1818 m, The Netherlands' Rs 273 m, the USA's Rs 188 m, and Denmark's Rs 140 m.

The volume of assistance fell from 1980-81 to 1983-84; after that it began to increase. The initial trend of decreasing assistance was partly due to the gradual slowing down of pledges and partly to substantial reduction in loan commitments by the World Bank group, Japan and West Germany. There was a steady upward trend of utilization, gradually increasing from Rs 11,386 m in 1970-80 to Rs 17,015 m in 1982-83, but it fell back in 1985-86 to Rs 16,234 m. At the end of March 1985, Rs 90,890 m was the amount of assistance available, compared to Rs 72,940 m at the end of March 1984.

There is an Aid-India Consortium, at whose meetings the donor countries make pledges. However, the aid pledged by the Consortium and even authorization for a particular year does not mean its corresponding availability to India, for after pledges are made, there are long-drawnout negotiations on signing bilateral agreements. Moreover, there is the need to use a good part of the aid to pay instalments on the outstanding loans, as well as interest payments thereon. Thus, in 1984-85, against the Consortium aid pledge of Rs 42,700 m, and gross aid utilization (including aid from other sources), of Rs 23,170 m, the actual net disbursal, or inflow of aid from 11 sources (utilization less total debt services), was only Rs 11,420 m.

One of the disquieting features of the post independence economic history of India has been her dependence on foreign economic assistance. As a result, India's outstanding external debt at the end of March 1985 was a staggering Rs 280,940 m. However, in recent years, government has been very cautious in allowing borrowings from commercial sources abroad, bearing in mind the high cost of such borrowings and the possibility of making India vulnerable to a financial crisis in the future. This procedure has been adopted despite the good credit ranking India enjoys in the international credit market. The bulk of the loans have so far been contracted by public-sector units like the Oil and Natural Gas Commission, NALCO, BHEL, Maruti-Udyog and Air India.

The World Bank in its 1986 report on the Indian economy is reported to have envisaged very large commercial borrowings abroad by India in the coming years. According to this report, India's foreign commercial borrowings, which were about $ 3.2 billion during the Sixth Plan, may have to increase to $ 17.8 b during the Seventh Plan if the country is to meet its 5 per cent growth target and finance the imports needed to sustain this economic growth.

FOREIGN INVESTMENT[1] IN APPROVED COLLABORATIONS IN INDIA

Table 116(i)

(in Rs. million)

Name of the country	1985	Name of the country	1985
Australia	N.A.	Kuwait	N.A.
Austria	10.32	Lebanon	N.A.
Belgium	26.80	Liberia	0.39
Bahamas	7.50	Malaysia	N.A.
Bulgaria	N.A.	Mexico	0.40
Bahrain	0.03	Norway	65.00
Bermuda	4.00	Portugal	8.00
Cayman Is.	N.A.	Singapore	3.70
Canada	24.70	Spain	N.A.
Denmark	2.40	Sweden	8.06
Faroe Is.	0.80	Switzerland	8.44
France	23.55	Sri Lanka	1.10
Germany (West)	118.08	Taiwan	N.A.
Finland	N.A.	The Netherlands	4.00
Greece	N.A.	U.A.E	87.14
Hong Kong	0.50	U.K.	37.06
Hungary	N.A.	U.S.A.	399.24
Italy	69.47	Yugoslavia	N.A.
Ireland	N.A.	Non-resident	
Japan	156.76	Indians	190.40
Jordan	0.50		
Korea (South)	N.A.	Total	1,258.67

Note: [1] Investment approved in that year alone

AID TO INDIA

Table 116(ii)

(in Rs. million)

Country/ Institutions	Authorized 1985-86			Utilized 1985-86		
	Loans	Grants	Total	Loans	Grants	Total
Austria	—	—	—	16	—	16
Belgium	—	—	—	19	—	19
Canada	229	—	229	309	63	372
Denmark	—	140	140	90	76	166
France	1,293	—	1,293	1,532	—	1,532
Germany (West)	1,027	61	1,088	640	7	647
Italy	49	—	49	72	—	72
Japan	824	34	858	503	62	565
Sweden	—	—	—	—	119	119
Switzerland	—	—	—	11	41	52
The Netherlands	25	273	298	241	177	418
United Kingdom	—	1,818	1,818	—	1,990	1,990
U.S.A.	286	188	474	370	64	434
U.S.S.R.	8,333	—	8,333	1,148	—	1,148
Yugoslavia	—	—	—	—	—	—
Kuwait Fund	147	—	147	272	—	272
Saudi Arabia Fund	287	—	287	1	—	1
Abu Dhabi Fund	—	—	—	—	—	—
OPEC Special Fund	—	—	—	29	—	29
International Sugar Organization	13	—	13	13	—	13
IBRD	13,230	—	13,230	2,409	—	2,409
IDA	7,269	—	7,269	5,337	—	5,337
EEC	—	—	—	—	438	438
IFAD	—	—	—	183	—	183
Total	33,015	2,515	35,530	13,197	3,037[1]	16,234

Note: [1] Including amounts carried over from the pervious year.

Sources: 1. *Economic Survey 1984-85,* Ministry of Finance.
2. *Report on Currency and Finance 1985-86, Vol. II,* Reserve Bank of India.
3. *Handbook of Statistics 1986,* Confederation of Engineering Industry.

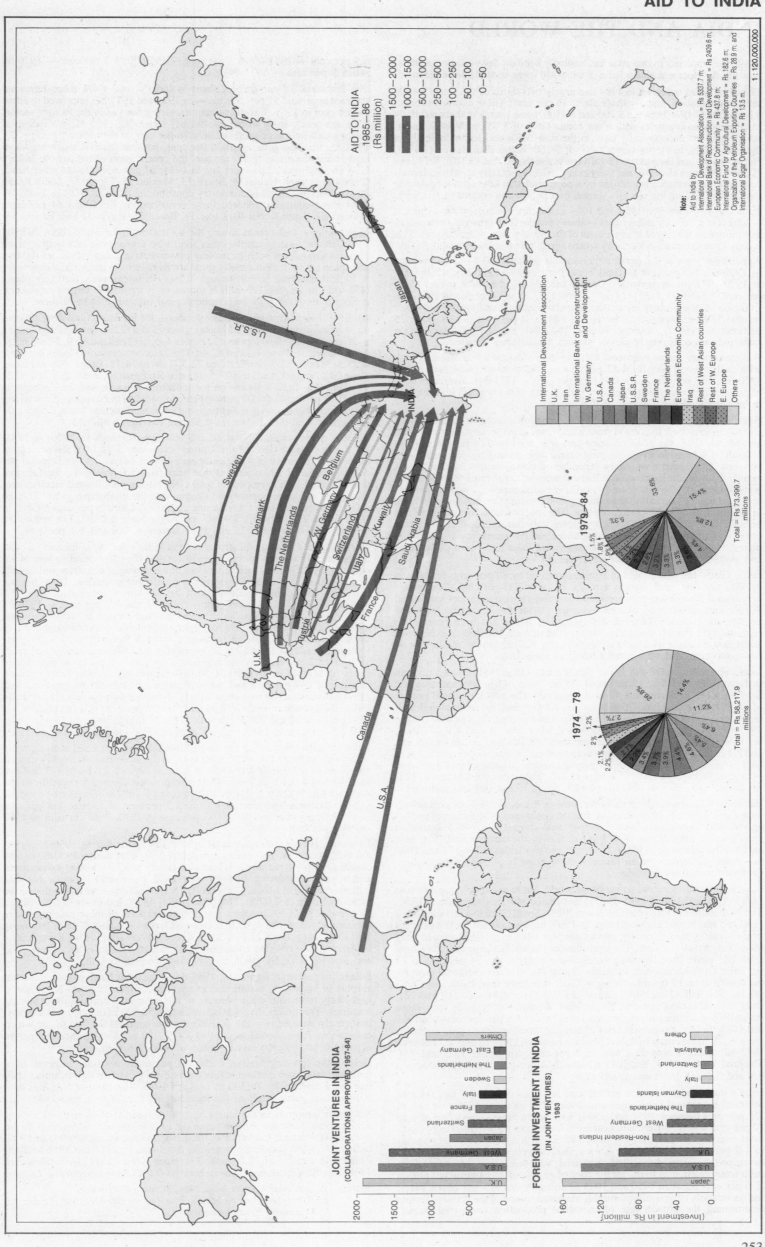

AID TO INDIA
1985—86
(Rs million)

1500—2000
1000—1500
500—1000
250—500
100—250
50—100
0—50

1 : 120,000,000

Note:
Aid to India by
International Development Association = Rs 5337.7 m;
International Bank of Reconstruction and Development = Rs 2409.6 m;
European Economic Community = Rs 437.8 m;
International Fund for Agricultural Development = Rs 182.6 m;
Organzation of the Petroleum Exporting Countries = Rs 28.9 m; and
International Sugar Organisation = Rs 13.5 m.

International Development Association
U.K.
Iran
International Bank of Reconstruction and Development
W. Germany
U.S.A.
Canada
Japan
U.S.S.R.
Sweden
France
The Netherlands
European Economic Community
Iraq
Rest of West Asian countries
Rest of W. Europe
E. Europe
Others

1979—84
Total = Rs 73,399.7 millions

33.6%
15.4%
12.8%
4.4%
3.3%
3.3%
3.2%
2.6%
1.9%
1.8%
1.5%

1974—79
Total = Rs 58,217.9 millions

26.5%
14.4%
11.2%
6.4%
5.4%
4.6%
4.4%
3.9%
3.7%
3.4%
2.3%
2.2%
2.1%
2%
1.2%
2.7%

JOINT VENTURES IN INDIA
(COLLABORATIONS APPROVED 1957-84)

Others
East Germany
The Netherlands
Sweden
Italy
France
Switzerland
West Germany
Japan
U.S.A.
U.K.

2000
1500
1000
500
0

FOREIGN INVESTMENT IN INDIA
(IN JOINT VENTURES)
1983

Others
Malaysia
Switzerland
Italy
Cayman Islands
The Netherlands
West Germany
Non-Resident Indians
U.K.
U.S.A.
Japan

160
120
80
40
0

(Investment in Rs. million)

INDIA AND THE WORLD

While the emphasis in this atlas has basically been on development in India, how this development places India in the world scene should also be recorded.

The developing countries in 1984 had nearly two-thirds of the world's 4800 million population, India's share about 16 per cent. These countries reflect a per capita GNP of between $ 100 and $ 3500, some twenty of them rightfully middle income countries with a per capita GNP of $ 751 to $ 3499. India, which took off economically only in the mid-1950s and technologically in the mid-1960s, has a per capita GNP of $ 260, better than Burma ($ 180), Nepal ($ 160) and Bangladesh ($ 130) but worse than China ($ 310), Sri Lanka ($ 360), Pakistan ($ 380), and Indonesia ($ 540). Generally speaking, India's economic performance is nullified by a population that keeps growing despite the country having the world's second best population control programme.

India's growth between 1950 and 1984 in the three sectors of development — agriculture, industry and services — shows that the GDP grows at a rate greatly determined by the rate of improvement of the primary sector (agriculture). This is only to be expected in a country where agriculture dominates economic life.

Agriculture grew at 2.2 per cent per annum during 1950-84 (2.64 per cent — 1950-60; 1.9 per cent — 1960-71; 2.3 per cent — 1971-84), while GDP at 1970-1971 prices grew at 3.56 per cent during the same period (3.53 per cent — 1950-60; 3.27 per cent — 1960-71; 3.85 per cent during 1971-84). Compared to the primary sector, the secondary sector (industry) showed better results: 5.06 per cent growth during 1950-84 (5.71 per cent — 1950-60; 5.25 per cent — 1960-71; 4.06 per cent — 1971-84). But the secondary sector rates indicate consistently slower growth decade after decade.

It is the tertiary sector (services) that has shown the most consistent growth with each passing decade: 4.82 per cent during 1950-84 (4.24 per cent — 1950-60; 4.33 per cent — 1960-71; 5.6 per cent — 1971-84). The entire hope of growing at the targeted 5 per cent growth rate of the Seventh Plan appears to be pinned on the progress of this sector.

Agriculture provides 36 per cent of India's national income; 70 per cent of all Indians earn their livelihood from it and from allied sectors. Between 1950 and March 1985, the gross cropped area increased by 32 per cent, and, as a result of a massive irrigation programme, the gross irrigated area by 16 per cent. The farmers, who once depended on moneylenders for credit, now mainly draw credit from institutional agencies at lower rates of interest. Of significance is the increased fertilizer consumption, which reflects improvements in agricultural technology — consumption has gone up from ½ kg per hectare to 47 kgs per hectare. The spectacular achievement of this period is the development of high yielding varieties of crops. HYVs today cover around 52 per cent of the area under paddy, wheat, jowar, bajra and maize. If the spectre of famine does not haunt India today as it did even as late as 1966, it is largely because of the HYVs.

In the 1960s, the wheat yield in Punjab increased by 80 per cent, production by 200 per cent, and market arrivals by over 300 per cent. The state's agricultural income and per capita income trebled at current prices and soared 70 per cent above the national average. With the exception of some neighbouring areas (Haryana, for example), it has not been possible to reproduce the Punjab model elsewhere, even in Asia. The Punjab experience is the best example of what a technologically fuelled growth model can do when superimposed on a society based on inequality.

India's agricultural GDP was 39 per cent of total GDP in 1981; it was 3 per cent in Japan and the USA, 12 in Brazil, 35 in China, 43 in Mozambique and Sri Lanka, 46 in Burma and 47 in Bangladesh. The average annual growth of agriculture during 1960-70 in India was 1.9 per cent, the same as in Brazil; in China it was 1.6 per cent, in Sri Lanka 3 and in South Korea 4.4. During 1973-84, this growth rate in India was 2.3 per cent; it was 4.9 in China, 4.1 in Sri Lanka, 4 in Brazil, 3.7 in Indonesia and Thailand, 3.1 in Bangladesh and 2.1 in South Korea. The Indian index of food production per capita, based on a percentage comparison of the average annual quantity of food produced per capita in 1982-84 and 1974-76, was 110. This index for China was 128, Sri Lanka 125, Burma 124, Brazil 115, South Korea 109 and Japan 91.

India's task for the future remains the same as in the past: a 4 per cent annual increase in agricultural production. But the compound growth rate of overall agricultural production was 2.6 per cent during 1967-86, against a demographic growth rate of 2.3 per cent over the same period. The new agricultural technology (seeds, fertilizers and pesticides) adopted since 1966 has amply rewarded Indian farmers. Now hope is pinned on technology evolved for dryland farming.

India is the tenth industrial power of the world, but in manufacturing it ranked 16th in 1960 and 18th in 1983. During that period Mexico (from 24th to 15th), The Netherlands (17th in both 1960 and 1983), Spain (23rd to 12th) and Brazil (19th to 14th) moved ahead of India. In 1960, India accounted for 0.8 per cent of the world's manufacturing production; by 1983 its share increased only marginally to 0.9 per cent. The contribution of manufacturing to India's total GDP has made little headway during these years: from 14 per cent to 15 per cent. Of the developing countries among the 30 major industrial powers, the contribution of manufacturing to GDP was more than India's in China (33 in 1960 and 42 in 1983), Brazil (26 and 27) and Mexico (19 and 22). Only Indonesia was static at 13 per cent. In terms of growth rate, these countries had significantly higher rates than India's 4.1 per cent between 1965 and 1983. In fact, the growth rates of some of the other developing countries were much higher than India's 4 per cent between 1965-73 and 4.2 between 1973-83, South Korea 21.1 and 11.8 per cent, Singapore 19.5 and 7.9, Thailand 11.4 and 8.9, Brazil 11.4 and 4.2, Algeria 10.9 and 12.6. Mexico 9.9 and 5.5 and Indonesia 9 and 12.6.

India's industrial production growth rate has been only about half the rate achieved by several developing countries. With an average growth rate of 5.8 per cent between 1951 and 1984, well below the planned rate of 8 to 10 per cent a year, different industries in India have shown differing growth rates over this period, ranging from 186 per cent a year for wrist watches to 1.0 per cent for mill-made fabrics. But, in general, India has certainly developed a fairly diversified industrial structure. It is, however, not sufficiently realized in India that the pace of growth of agricultural production (2.6 per cent between 1950 and 1985 against a desirable 4 per cent a year), and particularly of agricultural surplus marketed, is one of the decisive factors determining the growth rate of industrial production (industrial output:

6.9 per cent — 1951-60; 6.3 per cent — 1960-70; 4.2 per cent — 1970-80 and 5.5 per cent — 1974-84).

The Industrial labour force's share in India's total work force remained stagnant at around 11 per cent between 1951 and 1971 but improved to about 14 per cent in 1981. For every manufacturing worker in the factory sector, there are nearly three in the non-factory sector. In 1950-51, the industrial sector as a whole accounted for only 16 per cent of the national income. By 1984-85, this had gone up to 26 per cent, but this share is much higher in many other countries. For example, the contribution of industry to India's GDP in 1965 was 22 per cent and in 1984, 27 per cent. On the other hand, Chile 40 per cent, China 38, Brazil 33, Mexico 31, Egypt 27, South Korea 25 and Turkey 25, all had industry contributing more to GDP in 1965. By 1984 these countries bettered these contributions so: China 44 per cent, Mexico 40, South Korea 40, Chile 39, Brazil 35, Egypt 33 and Turkey 33.

In technology, India ranks among the six leaders in the world (USA, USSR, UK, Canada, France are the other five) who are capable of manufacturing complex equipment such as nuclear power stations, supersonic jet fighters and space satellites. But equally great are its failures in industrial technology. A typical example is that though it has 38 nitrogenous fertilizer plants (1984-85) producing 3.9 million tonnes (it is the fourth largest producer), it has yet to develop its own technological capability to build them.

An EMF Foundation survey, using indexes for infrastructural development (O = inadequate to 100 = adequate), places India (22.7) among the countries with low levels of development in power supply (Thailand 62.9, Brazil 65.8, South Korea 81.3 and Taiwan 82.4). The Telecommunication index for India is 23.3 (Mexico 62.7, South Korea 72.5, Taiwan 74.12, Brazil 79.6 and Hong Kong 94.5) and for roads it is 37.5 (Mexico 50, Brazil 61.3, South Korea 67.5, Thailand 80). India's allocation for industrial research and development is only 0.73 per cent of GDP, low when compared to the developed countries, but comparable to newly industrializing countries, such as South Korea 1.04 per cent, Brazil 0.6, Indonesia 0.51 and Pakistan with 0.18.

In India, the per capita GNP stood at $ 260 (approximately Rs 3200) in 1983, but the per capita GNP of the United States was $ 14,090, the per capita economic prosperity of America being 54 times India's. A tabulation of per capita income of 190 countries that year ranked India 169th, the fact that it also ranked 11th in respect of total GNP having only limited significance. Obviously the large population is eating into the productive gains. To take a few examples of other developing countries, China's per capita GNP in 1983 was $ 290, Brazil's $ 1890 and Argentina's $ 2020. The respective growth rates between 1973 and 1982 were 4.5, 2.9 and 1.1 per cent.

During the last 35 years the growth rate of India's per capita income (around 3.5 per cent) has been considerably lower than the corresponding rate of not only the capitalist developed countries, the communist countries and the oil-rich countries, but also of many other developing countries. There has not been any perceptible acceleration in the growth of national income: 3.7 per cent — 1950-61, 3.2 per cent — 1960-71 and 3.7 per cent — 1970-71 and 1985-86. Over this period, the share of agriculture in national income has declined from 51 per cent to 35 per cent. On the other hand, the shares of other sectors have shown increases: industry from 16 to 25.9 per cent, transport, trade etc from 17 to 21.4 per cent, and finance, services etc from 15.8 to 18.2 per cent. While population growth has shown an increase in index of 111 per cent (1950-51: 100), real national income has shown an increase of 247 per cent and real per capita income only 65 per cent. The growth in income has thus been largely wiped out by population growth. In India, the growth in real GNP to GDP was 3 per cent in 1985 against 3.5 per cent in 1984 and 7.8 per cent in 1983. In Brazil, however, the increase was 7.5 per cent in 1985, 4.5 per cent in 1984 and 3.2 per cent in 1983.

India's foreign trade deficit for the past five years has remained at an alarming level of about Rs 55,000 million per year. In the world context, India's exports amounted to $ 1.1 billion in 1950, $ 1.3 b in 1960, $ 2.0 b in 1970, $ 8.6 b in 1980 and $ 9.4 b in 1984. The corresponding figures for world exports were $ 61 b, $ 129 b, $ 317 b, $ 2016 b and $ 1887 b. The share of India's exports has thus been declining from 2.3 per cent in 1948 to 1.8 per cent in 1950, 1 per cent in 1960, 0.63 per cent in 1970, 0.43 per cent in 1980 and 0.4 per cent in 1985.

However, India's average annual growth in exports during 1950-85 works out to 5.7 per cent. This is an average of 1.5 per cent during 1950-60 (world annual growth 7.8 per cent), 4.3 per cent during 1960-70 (world: 9.4 per cent) and 9.3 per cent during 1970-85 (world: 12.8 per cent). Comparable data indicate that in 1955 India's exports (US $ 1.26 billion) were larger than those of South Korea ($ 0.02 b), Hong Kong ($ 0.44 b), Indonesia ($ 0.94 b) and Singapore ($ 1.1 b). In sharp contrast, the exports of these countries were substantially larger in 1985 than those of India — India $ 7.92 b, Indonesia $ 18. 6 b, Singapore $ 22.8 b, South Korea $ 29.57 b and Hong Kong $ 30.2 b. Among the exporting countries, India ranked 16th in 1950, 21st in 1960 and 31st in 1970. By 1985, its rank had dropped to 47th.

Balance of payments figures for 1983-84 on the other hand reveal a welcome surplus in invisibles, mainly due to remittances from abroad. These are at best likely to remain static because of the economic strain in oil-exporting countries. This makes the trade balance of Rs 58,700 m (exports less imports) immensely smaller, for the invisibles net around Rs 36,080 m, of which remittances from abroad amount to Rs 27,750 m. The balance on current account (trade balance less invisibles) is Rs 22,620 m.

Despite the low-key foreign trade and payments position, reserves have remained fairly buoyant and at times have even moved up significantly. They reached a peak of Rs 76,390 m in April 1986. The flow of Non-Resident Indian funds partly explains the increase (Rs 50,278.8 m, end 1985).

Not many countries in the world have improved their country-risk rating, which is an indicator of the standing of government borrowers in the international capital market. But India has. With a rating of 69.0 in 1986, India had a good standing, improving from 46th in 1984 and 1985 to 28th. China with a rating of 78.0 (rank: 22), Hong Kong 75.0 (23), South Korea 73.0 (25), Singapore and Taiwan both 68.0 (30) have comparable ratings.

Sources: 1. *Statistical Abstract India 1984*, Ministry of Planning.
2. *World Economy & India's Place in it, October 1984*, Centre for Monitoring Indian Economy.